T0280470

In *The Quantum Theory of Fields*, Nobel Laureate Steven Weinberg combines his exceptional physical insight with his gift for clear exposition to provide a self-contained, comprehensive, and up-to-date introduction to quantum field theory. Volume I introduces the foundations of the subject. The development is fresh and logical throughout, with each step carefully motivated by what has gone before, and emphasizing the reasons why such a theory should describe nature. After a brief historical outline, the book begins anew with the principles about which we are most certain, relativity and quantum mechanics, and the properties of particles that follow from these principles. Quantum field theory then emerges from this as a natural consequence. The classic calculations of quantum electrodynamics are presented in a thoroughly modern way, showing the use of path integrals and dimensional regularization. The account of renormalization theory reflects the changes in our view of quantum field theory since the advent of effective field theories. The book's scope extends beyond quantum electrodynamics to elementary particle physics and nuclear physics. It contains much original material, and is peppered with examples and insights drawn from the author's experience as a leader of elementary particle research. Problems are included at the end of each chapter. A second volume (published September, 1996) describes the modern applications of quantum field theory in today's standard model of elementary particles, and in some areas of condensed matter physics.

From reviews of Volume I

'...an impressively lucid and thorough presentation of the subject... Weinberg manages to present difficult topics with richness of meaning and marvellous clarity. Full of valuable insights, his treatise is sure to become a classic, doing for quantum field theory what Dirac's *Quantum Mechanics* did for quantum mechanics. I eagerly await the publication of the second volume.'

S. S. Schweber, *Nature*

'For over twenty years there has been no good modern textbook on the subject. For all that time, Steven Weinberg has been promising to write one. That he has finally done it is cause for celebration among those who try to teach and try to learn the subject. Weinberg's book is for serious students of field theory... it is the first textbook to treat quantum field theory the way it is used by physicists today.'

Howard Georgi, *Science*

'Steven Weinberg, who contributed to the development of quantum chromodynamics and shared the Nobel Prize in Physics for his contributions to the electroweak theory, has written a definitive text on the physical foundations of quantum field theory. His book differs significantly from the long line of previous books on quantum field theory... To summarize, *Foundations* builds the structure of quantum field theory on the sure footing of physical insight. It is beautifully produced and meticulously edited... and it is a real bargain in price. If you want to learn quantum field theory, or have already learned it and want to have a definitive reference at hand, purchase this book.'

O. W. Greenberg, *Physics Today*

'In addition to a superb treatment of all the conventional topics there are numerous sections covering areas that are not normally emphasized, such as the subject of field redefinitions, higher-rank tensor fields and an unusually clear and thorough treatment of infrared effects... this latest book reinforces his high scholarly standards. It provides a unique exposition that will prove invaluable both to new research students as well as to experienced research workers. Together with Volume II, this will become a classic text on a subject of central importance to a wide area of theoretical physics.'

M. B. Green, *CERN Courier*

'I believe that what readers will find particularly helpful in this volume is the consistency of the whole approach, and the emphasis on quantities and properties that are directly useful to particle physicists. This is particularly true for those who are interested in the more phenomenological aspects. The reader only needs limited background knowledge, and a clear line is followed throughout the book, making it easy to follow. The author presents extremely thorough but elementary discusssions of important physical questions, some of which seem to be an original way of addressing the subject.'

J. Zinn-Justin, *Physics World*

'This is a well-written book by one of the masters of the subject... it is certainly destined to become a standard text book and should find its way to the shelves of every physics library.'

J. Madore, *Classical and Quantum Gravity*

'The book starts out with an excellent historical introduction, not found anywhere else, giving citations to many by now classic papers... a valuable reference work as well as a textbook for graduate students.'

G. Roepstorff, *Zentralblatt für Mathematik*

From reviews of Volume II

'It is a majestic exposition. The two volumes are structured in a logical way. Everything is explained with incisive clarity. Weinberg always goes to the heart of any argument, and includes many things that cannot be found elsewhere in the literature. Often I find myself thinking: "Ah! Now I understand that properly."... I find it hard to imagine a better treatment of quantum field theory than Weinberg's. All serious students and researchers will want to have these volumes on their shelves.'

John C. Taylor, *Nature*

'Weinberg's *Modern Applications* goes to the boundaries of our present understanding of field theory. It is unmatched by any other book on quantum field theory for its depth, generality and definitive character, and it will be an essential reference for serious students and researchers in elementary particle physics.'

O. W. Greenberg, *Physics Today*

'... Steven Weinberg is one of our most gifted makers of theoretical tools as well as a virtuoso in their use. His new book conveys both the satisfaction of understanding nature and the feel of the atelier, for the "modern applications" of its subtitle include both the derivation of physical consequences and the development of new tools for understanding and applying field theory itself... *Modern Applications* is a splendid book, with abundant useful references to the original literature. It is a very interesting read from cover to cover, for the wholeness Weinberg's personal perspective gives to quantum field theory and particle physics.'

Chris Quigg, *Science*

'Experienced researchers and beginning graduate students will delight in the gems of wisdom to be found in these pages. This book combines exposition of technical detail with physical insight in a unique manner that confirms the promise of Volume I and I have no doubt that these two volumes will rapidly constitute the classic treatment of this important subject.'

M. B. Green, *CERN Courier*

'... a valued reference and a mine of useful information for professional field theorists.'

Tom Kibble, *New Scientist*

'... a clear presentation of the subject, explaining the underlying concepts in much depth and in an accessible style. I expect that these volumes will become the first source we turn to when trying to answer the challenging questions asked by bright postgraduates when they first encounter quantum field theory... I have no doubt that *The Quantum Theory of Fields* will soon be found on the bookshelves of most particle theorists, and that it will be one of the main sources used in the preparation of lectures on the subject for postgraduate students.'

C. T. C. Sachrajda, *The Times Higher Education Supplement*

'... Weinberg has produced a masterpiece that will be a standard reference on the field for a long time to come.'

B. E. Y. Svensson, *Elementa*

From reviews of Volume III

'... has produced a treatise that many of us had long awaited, perhaps without fully realizing it... with the publication of *The Quantum Theory of Fields*, Vol. III, has performed an analogous service for supersymmetry... Although this volume is the third in a trilogy, it is quite different from its two predecessors, and it stands on its own... May a new generation of students imbibe its content and spirit.'

Physics Today

'The third volume of *The Quantum Theory of Fields* is a self-contained introduction to the world of supersymmetry and supergravity. It will be useful both for experienced researchers in the field and for students who want to take the first steps towards learning about supersymmetry. Unlike other books in this field, it covers the wide spectrum of possible applications of supersymmetry in physics.'

Hans Peter Nilles, *Nature*

'Weinberg is of course one of the creators of modern quantum field theory, as well as of its physical culmination, the standard model of all (nongravitational) interactions. It is... very timely that this latest part of his monograph, devoted to supersymmetry and supergravity, has just appeared. As a text, it has been pretested by Weinberg for a freestanding one-year graduate course; as a clear organizing reference to this extremely vast field, it will help the experts as well... Weinberg's style of presentation is as clear and meticulous as in his previous works.'

Stanley Deser, *Journal of General Relativity and Gravitation*

'Weinberg tries to be as elementary and clear as possible and steers clear of more sophisticated mathematical tools. Together with the previous volumes, this volume will serve as an invaluable reference to researchers and a textbook for graduate students.'

G. Roepstorff, *Zentralblatt MATH*

The Quantum Theory of Fields

Volume I
Foundations

Steven Weinberg
University of Texas at Austin

CAMBRIDGE
UNIVERSITY PRESS

University Printing House, Cambridge CB2 8BS, United Kingdom

One Liberty Plaza, 20th Floor, New York, NY 10006, USA

477 Williamstown Road, Port Melbourne, VIC 3207, Australia

314-321, 3rd Floor, Plot 3, Splendor Forum, Jasola District Centre, New Delhi - 110025, India

79 Anson Road, #06-04/06, Singapore 079906

Cambridge University Press is part of the University of Cambridge.

It furthers the University's mission by disseminating knowledge in the pursuit of
education, learning and research at the highest international levels of excellence.

www.cambridge.org
Information on this title: www.cambridge.org/9780521670531

First published 1995
Eighth printing 2004
Paperback edition first published 2005
17th printing 2019

A catalogue record for this publication is available from the British Library

ISBN 978-0-521-55001-7 Volume 1 Hardback
ISBN 978-0-521-67053-1 Volume 1 Paperback
ISBN 978-0-521-55002-4 Volume 2 Hardback
ISBN 978-0-521-67054-8 Volume 2 Paperback
ISBN 978-0-521-66000-6 Volume 3 Hardback
ISBN 978-0-521-67055-5 Volume 3 Paperback
ISBN 978-0-521-67056-2 three-volume set

To Louise

Contents

Sections marked with an asterisk are somewhat out of the book's main line of development and may be omitted in a first reading.

OUTLINE OF VOLUME II

Preface To Volume I

Why another book on quantum field theory? Today the student of quantum field theory can choose from among a score of excellent books, several of them quite up-to-date. Another book will be worth while only if it offers something new in content or perspective.

As to content, although this book contains a good amount of new material, I suppose the most distinctive thing about it is its generality; I have tried throughout to discuss matters in a context that is as general as possible. This is in part because quantum field theory has found applications far removed from the scene of its old successes, quantum electrodynamics, but even more because I think that this generality will help to keep the important points from being submerged in the technicalities of specific theories. Of course, specific examples are frequently used to illustrate general points, examples that are chosen from contemporary particle physics or nuclear physics as well as from quantum electrodynamics.

It is, however, the perspective of this book, rather than its content, that provided my chief motivation in writing it. I aim to present quantum field theory in a manner that will give the reader the clearest possible idea of *why* this theory takes the form it does, and why in this form it does such a good job of describing the real world.

The traditional approach, since the first papers of Heisenberg and Pauli on general quantum field theory, has been to take the existence of fields for granted, relying for justification on our experience with electromagnetism, and 'quantize' them — that is, apply to various simple field theories the rules of canonical quantization or path integration. Some of this traditional approach will be found here in the historical introduction presented in Chapter 1. This is certainly a way of getting rapidly into the subject, but it seems to me that it leaves the reflective reader with too many unanswered questions. Why should we believe in the rules of canonical quantization or path integration? Why should we adopt the simple field equations and Lagrangians that are found in the literature? For that matter, why have fields at all? It does not seem satisfactory to me to appeal to experience; after all, our purpose in theoretical physics is not just to describe the world as we find it, but to explain — in terms of a few fundamental principles — why the world is the way it is.

The point of view of this book is that quantum field theory is the way it is because (aside from theories like string theory that have an infinite number of particle types) it is the only way to reconcile the principles of quantum mechanics (including the cluster decomposition property) with those of special relativity. This is a point of view I have held for many years, but it is also one that has become newly appropriate. We have learned in recent years to think of our successful quantum field theories, including quantum electrodynamics, as 'effective field theories,' low-energy approximations to a deeper theory that may not even be a field theory, but something different like a string theory. On this basis, the reason that quantum field theories describe physics at accessible energies is that *any* relativistic quantum theory will look at sufficiently low energy like a quantum field theory. It is therefore important to understand the rationale for quantum field theory in terms of the principles of relativity and quantum mechanics. Also, we think differently now about some of the problems of quantum field theories, such as non-renormalizability and 'triviality,' that used to bother us when we thought of these theories as truly fundamental, and the discussions here will reflect these changes. This is intended to be a book on quantum field theory for the era of effective field theories.

The most immediate and certain consequences of relativity and quantum mechanics are the properties of particle states, so here particles come first — they are introduced in Chapter 2 as ingredients in the representation of the inhomogeneous Lorentz group in the Hilbert space of quantum mechanics. Chapter 3 provides a framework for addressing the fundamental dynamical question: given a state that in the distant past looks like a certain collection of free particles, what will it look like in the future? Knowing the generator of time-translations, the Hamiltonian, we can answer this question through the perturbative expansion for the array of transition amplitudes known as the S-matrix. In Chapter 4 the principle of cluster decomposition is invoked to describe how the generator of time-translations, the Hamiltonian, is to be constructed from creation and annihilation operators. Then in Chapter 5 we return to Lorentz invariance, and show that it requires these creation and annihilation operators to be grouped together in causal quantum fields. As a spin-off, we deduce the CPT theorem and the connection between spin and statistics. The formalism is used in Chapter 6 to derive the Feynman rules for calculating the S-matrix.

It is not until Chapter 7 that we come to Lagrangians and the canonical formalism. The rationale here for introducing them is not that they have proved useful elsewhere in physics (never a very satisfying explanation) but rather that this formalism makes it easy to choose interaction Hamiltonians for which the S-matrix satisfies various assumed symmetries. In particular, the Lorentz invariance of the Lagrangian density ensures the existence of a set of ten operators that satisfy the algebra of the Poincaré group and, as we show in Chapter 3, this is the key condition that we need to prove the Lorentz invariance of the S-matrix. Quantum electrodynamics finally appears in Chapter 8. Path integration is introduced in Chapter 9, and used to justify some of the hand-waving in Chapter 8 regarding the Feynman rules for quantum electrodynamics. This is a somewhat later introduction of path integrals than is fashionable these days, but it seems to me that although path integration is by far the best way of rapidly deriving Feynman rules from a given Lagrangian, it rather obscures the quantum mechanical reasons underlying these calculations.

Volume I concludes with a series of chapters, 10–14, that provide an introduction to the calculation of radiative corrections, involving loop graphs, in general field theories. Here too the arrangement is a bit unusual; we start with a chapter on non-perturbative methods, in part because the results we obtain help us to understand the necessity for field and mass renormalization, without regard to whether the theory contains infinities or not. Chapter 11 presents the classic one-loop calculations of quantum electrodynamics, both as an opportunity to explain useful calculational techniques (Feynman parameters, Wick rotation, dimensional and Pauli–Villars regularization), and also as a concrete example of renormalization in action. The experience gained in Chapter 11 is extended to all orders and general theories in Chapter 12, which also describes the modern view of non-renormalizability that is appropriate to effective field theories. Chapter 13 is a digression on the special problems raised by massless particles of low energy or parallel momenta. The Dirac equation for an electron in an external electromagnetic field, which historically appeared almost at the very start of relativistic quantum mechanics, is not seen here until Chapter 14, on bound state problems, because this equation should not be viewed (as Dirac did) as a relativistic version of the Schrödinger equation, but rather as an approximation to a true relativistic quantum theory, the quantum field theory of photons and electrons. This chapter ends with a treatment of the Lamb shift, bringing the confrontation of theory and experiment up to date.

The reader may feel that some of the topics treated here, especially in Chapter 3, could more properly have been left to textbooks on nuclear or elementary particle physics. So they might, but in my experience these topics are usually either not covered or covered poorly, using specific dynamical models rather than the general principles of symmetry and quantum mechanics. I have met string theorists who have never heard of the relation between time-reversal invariance and final-state phase shifts, and nuclear theorists who do not understand why resonances are governed by the Breit–Wigner formula. So in the early chapters I have tried to err on the side of inclusion rather than exclusion.

Volume II will deal with the advances that have revived quantum field theory in recent years: non-Abelian gauge theories, the renormalization group, broken symmetries, anomalies, instantons, and so on.

I have tried to give citations both to the classic papers in the quantum theory of fields and to useful references on topics that are mentioned but not presented in detail in this book. I did not always know who was responsible for material presented here, and the mere absence of a citation should not be taken as a claim that the material presented here is original. But some of it is. I hope that I have improved on the original literature or standard textbook treatments in several places, as for instance in the proof that symmetry operators are either unitary or antiunitary; the discussion of superselection rules; the analysis of particle degeneracy associated with unconventional representations of inversions; the use of the cluster decomposition principle; the derivation of the reduction formula; the derivation of the external field approximation; and even the calculation of the Lamb shift.

I have also supplied problems for each chapter except the first. Some of these problems aim simply at providing exercise in the use of techniques described in the chapter; others are intended to suggest extensions of the results of the chapter to a wider class of theories.

In teaching quantum field theory, I have found that each of the two volumes of this book provides enough material for a one-year course for graduate students. I intended that this book should be accessible to students who are familiar with non-relativistic quantum mechanics and classical electrodynamics. I assume a basic knowledge of complex analysis and matrix algebra, but topics in group theory and topology are explained where they are introduced.

This is not a book for the student who wants immediately to begin calculating Feynman graphs in the standard model of weak, electromagnetic, and strong interactions. Nor is this a book for those who seek a higher level of mathematical rigor. Indeed, there are parts of this book whose lack of rigor will bring tears to the eyes of the mathematically inclined reader. Rather, I hope it will suit the physicists and physics students who want to understand *why* quantum field theory is the way it is, so that they will be ready for whatever new developments in physics may take us beyond our present understandings.

* * *

Much of the material in this book I learned from my interactions over the years with numerous other physicists, far too many to name here. But I must acknowledge my special intellectual debt to Sidney Coleman, and to my colleagues at the University of Texas: Arno Bohm, Luis Boya, Phil Candelas, Bryce DeWitt, Cecile DeWitt-Morette, Jacques Distler, Willy Fischler, Josh Feinberg, Joaquim Gomis, Vadim Kaplunovsky, Joe Polchinski, and Paul Shapiro. I owe thanks for help in the preparation of the historical introduction to Gerry Holton, Arthur Miller, and Sam Schweber. Thanks are also due to Alyce Wilson, who prepared the illustrations and typed the LaTeX input files until I learned how to do it, and to Terry Riley for finding countless books and articles. For finding various errors in the first printing of this volume, I am greatly indebted to numerous students and colleagues, especially Stephen Adler, Hideaki Aoyama, Kevin Cahill, Amir Kashani-Poor, Achim Kupferoth, Michio Masujima, Herbert Neuberger, Fabio Siringo, and San Fu Tuan. I am grateful to Maureen Storey and Alison Woollatt of Cambridge University Press for helping to ready this book for publication, and especially to my editor, Rufus Neal, for his friendly good advice.

STEVEN WEINBERG

Austin, Texas
October, 1994

Notation

Latin indices i, j, k, and so on generally run over the three spatial coordinate labels, usually taken as 1, 2, 3.

Greek indices μ, ν, etc. generally run over the four spacetime coordinate labels 1, 2, 3, 0, with x^0 the time coordinate.

Repeated indices are generally summed, unless otherwise indicated.

The spacetime metric $\eta_{\mu\nu}$ is diagonal, with elements $\eta_{11} = \eta_{22} = \eta_{33} = 1$, $\eta_{00} = -1$.

The d'Alembertian is defined as $\Box \equiv \eta^{\mu\nu}\partial^2/\partial x^\mu \partial x^\nu = \nabla^2 - \partial^2/\partial t^2$, where ∇^2 is the Laplacian $\partial^2/\partial x^i \partial x^i$.

The 'Levi–Civita tensor' $\epsilon^{\mu\nu\rho\sigma}$ is defined as the totally antisymmetric quantity with $\epsilon^{0123} = +1$.

Spatial three-vectors are indicated by letters in boldface.

A hat over any vector indicates the corresponding unit vector: Thus, $\hat{\mathbf{v}} \equiv \mathbf{v}/|\mathbf{v}|$.

A dot over any quantity denotes the time-derivative of that quantity.

Dirac matrices γ_μ are defined so that $\gamma_\mu \gamma_\nu + \gamma_\nu \gamma_\mu = 2\eta_{\mu\nu}$. Also, $\gamma_5 = i\gamma_0\gamma_1\gamma_2\gamma_3$, and $\beta = i\gamma^0$.

The step function $\theta(s)$ has the value $+1$ for $s > 0$ and 0 for $s < 0$.

The complex conjugate, transpose, and Hermitian adjoint of a matrix or vector A are denoted A^*, A^{T}, and $A^\dagger = A^{*\mathrm{T}}$, respectively. The Hermitian adjoint of an operator O is denoted $O^\dagger$, except where an asterisk is used to emphasize that a vector or matrix of operators is not transposed. +H.c. or +c.c. at the end of an equation indicates the addition of the Hermitian adjoint or complex conjugate of the foregoing terms. A bar on a Dirac spinor u is defined by $\bar{u} = u^\dagger \beta$.

Except in Chapter 1, we use units with $\hbar$ and the speed of light taken to be unity. Throughout $-e$ is the rationalized charge of the electron, so that the fine structure constant is $\alpha = e^2/4\pi \simeq 1/137$.

Numbers in parenthesis at the end of quoted numerical data give the uncertainty in the last digits of the quoted figure. Where not otherwise indicated, experimental data are taken from 'Review of Particle Properties,' *Phys. Rev.* **D50**, 1173 (1994).

1
Historical Introduction

Our immersion in the present state of physics makes it hard for us to understand the difficulties of physicists even a few years ago, or to profit from their experience. At the same time, a knowledge of our history is a mixed blessing — it can stand in the way of the logical reconstruction of physical theory that seems to be continually necessary.

I have tried in this book to present the quantum theory of fields in a logical manner, emphasizing the deductive trail that ascends from the physical principles of special relativity and quantum mechanics. This approach necessarily draws me away from the order in which the subject in fact developed. To take one example, it is historically correct that quantum field theory grew in part out of a study of relativistic wave equations, including the Maxwell, Klein–Gordon, and Dirac equations. For this reason it is natural that courses and treatises on quantum field theory introduce these wave equations early, and give them great weight. Nevertheless, it has long seemed to me that a much better starting point is Wigner's definition of particles as representations of the inhomogeneous Lorentz group, even though this work was not published until 1939 and did not have a great impact for many years after. In this book we start with particles and get to the wave equations later.

This is not to say that particles are necessarily more fundamental than fields. For many years after 1950 it was generally assumed that the laws of nature take the form of a quantum theory of fields. I start with particles in this book, not because they are more fundamental, but because what we know about particles is more *certain*, more directly derivable from the principles of quantum mechanics and relativity. If it turned out that some physical system could not be described by a quantum field theory, it would be a sensation; if it turned out that the system did not obey the rules of quantum mechanics and relativity, it would be a cataclysm.

In fact, lately there has been a reaction against looking at quantum field theory as fundamental. The underlying theory might not be a theory of fields *or* particles, but perhaps of something quite different, like strings.

From this point of view, quantum electrodynamics and the other quantum field theories of which we are so proud are mere 'effective field theories,' low-energy approximations to a more fundamental theory. The reason that our field theories work so well is not that they are fundamental truths, but that any relativistic quantum theory will look like a field theory when applied to particles at sufficiently low energy. On this basis, if we want to know why quantum field theories are the way they are, we have to start with particles.

But we do not want to pay the price of altogether forgetting our past. This chapter will therefore present the history of quantum field theory from earliest times to 1949, when it finally assumed its modern form. In the remainder of the book I will try to keep history from intruding on physics.

One problem that I found in writing this chapter is that the history of quantum field theory is from the beginning inextricably entangled with the history of quantum mechanics itself. Thus, the reader who is familiar with the history of quantum mechanics may find some material here that he or she already knows, especially in the first section, where I discuss the early attempts to put together quantum mechanics with special relativity. In this case I can only suggest that the reader should skip on to the less familiar parts.

On the other hand, readers who have no prior familiarity with quantum field theory may find parts of this chapter too brief to be altogether clear. I urge such readers not to worry. This chapter is not intended as a self-contained introduction to quantum field theory, and is not needed as a basis for the rest of the book. Some readers may even prefer to start with the next chapter, and come back to the history later. However, for many readers the history of quantum field theory should serve as a good introduction to quantum field theory itself.

I should add that this chapter is not intended as an original work of historical scholarship. I have based it on books and articles by real historians, plus some historical reminiscences and original physics articles that I have read. Most of these are listed in the bibliography given at the end of this chapter, and in the list of references. The reader who wants to go more deeply into historical matters is urged to consult these listed works.

A word about notation. In order to keep some of the flavor of past times, in this chapter I will show explicit factors of $\hbar$ and c (and even h), but in order to facilitate comparison with modern physics literature, I will use the more modern *rationalized* electrostatic units for charge, so that the fine structure constant $\alpha \simeq 1/137$ is $e^2/4\pi\hbar c$. In subsequent chapters I will mostly use the 'natural' system of units, simply setting $\hbar = c = 1$.

1.1 Relativistic Wave Mechanics

Wave mechanics started out as relativistic wave mechanics. Indeed, as we shall see, the founders of wave mechanics, Louis de Broglie and Erwin Schrödinger, took a good deal of their inspiration from special relativity. It was only later that it became generally clear that relativistic wave mechanics, in the sense of a relativistic quantum theory of a fixed number of particles, is an impossibility. Thus, despite its many successes, relativistic wave mechanics was ultimately to give way to quantum field theory. Nevertheless, relativistic wave mechanics survived as an important element in the formal apparatus of quantum field theory, and it posed a challenge to field theory, to reproduce its successes.

The possibility that material particles can like photons be described in terms of waves was first suggested[1] in 1923 by Louis de Broglie. Apart from the analogy with radiation, the chief clue was Lorentz invariance: if particles are described by a wave whose phase at position $\mathbf{x}$ and time t is of the form $2\pi(\boldsymbol{\kappa}\cdot\mathbf{x} - \nu t)$, and if this phase is to be Lorentz invariant, then the vector $\boldsymbol{\kappa}$ and the frequency ν must transform like $\mathbf{x}$ and t, and hence like $\mathbf{p}$ and E. In order for this to be possible $\boldsymbol{\kappa}$ and ν must have the same velocity dependence as $\mathbf{p}$ and E, and therefore must be proportional to them, with the same constant of proportionality. For photons, one had the Einstein relation $E = h\nu$, so it was natural to assume that, for material particles,

$$\boldsymbol{\kappa} = \mathbf{p}/h \ , \qquad \nu = E/h \ , \tag{1.1.1}$$

just as for photons. The group velocity $\partial\nu/\partial\boldsymbol{\kappa}$ of the wave then turns out to equal the particle velocity, so wave packets just keep up with the particle they represent.

By assuming that any closed orbit contains an integral number of particle wavelengths $\lambda = 1/|\boldsymbol{\kappa}|$, de Broglie was able to derive the old quantization conditions of Niels Bohr and Arnold Sommerfeld, which though quite mysterious had worked well in accounting for atomic spectra. Also, both de Broglie and Walter Elsasser[2] suggested that de Broglie's wave theory could be tested by looking for interference effects in the scattering of electrons from crystals; such effects were established a few years later by Clinton Joseph Davisson and Lester H. Germer.[3] However, it was still unclear how the de Broglie relations (1.1.1) should be modified for non-free particles, as for instance for an electron in a general Coulomb field.

Wave mechanics was by-passed in the next step in the history of quantum mechanics, the development of matrix mechanics[4] by Werner Heisenberg, Max Born, Pascual Jordan and Wolfgang Pauli in the years 1925–1926. At least part of the inspiration for matrix mechanics was the insistence that the theory should involve only observables, such as the energy levels, or emission and absorption rates. Heisenberg's 1925 paper opens with the manifesto: 'The present paper seeks to establish a basis for theoretical quantum mechanics founded exclusively upon relationships between quantities that in principle are observable.' This sort of positivism was to reemerge at various times in the history of quantum field theory, as for instance in the introduction of the S-matrix by John Wheeler and Heisenberg (see Chapter 3) and in the revival of dispersion theory in the 1950s (see Chapter 10), though modern quantum field theory is very far from this ideal. It would take us too far from our subject to describe matrix mechanics in any detail here.

As everyone knows, wave mechanics was revived by Erwin Schrödinger. In his 1926 series of papers,[5] the familiar non-relativistic wave equation is suggested first, and then used to rederive the results of matrix mechanics. Only later, in the sixth section of the fourth paper, is a relativistic wave equation offered. According to Dirac,[6] the history is actually quite different: Schrödinger first derived the relativistic equation, then became discouraged because it gave the wrong fine structure for hydrogen, and then some months later realized that the non-relativistic approximation to his relativistic equation was of value even if the relativistic equation itself was incorrect! By the time that Schrödinger came to publish his relativistic wave equation, it had already been independently rediscovered by Oskar Klein[7] and Walter Gordon,[8] and for this reason it is usually called the 'Klein–Gordon equation.'

Schrödinger's relativistic wave equation was derived by noting first that, for a 'Lorentz electron' of mass m and charge e in an external vector potential $\mathbf{A}$ and Coulomb potential ϕ, the Hamiltonian H and momentum $\mathbf{p}$ are related by*

$$0 = (H + e\phi)^2 - c^2(\mathbf{p} + e\mathbf{A}/c)^2 - m^2c^4 \,. \tag{1.1.2}$$

For a *free* particle described by a plane wave $\exp\left\{2\pi i(\boldsymbol{\kappa}\cdot\mathbf{x} - \nu t)\right\}$, the de Broglie relations (1.1.1) can be obtained by the identifications

$$\mathbf{p} = h\boldsymbol{\kappa} \to -i\hbar\nabla \;, \qquad E = h\nu \to i\hbar\frac{\partial}{\partial t} \;, \tag{1.1.3}$$

where $\hbar$ is the convenient symbol (introduced later by Dirac) for $h/2\pi$. By an admittedly formal analogy, Schrödinger guessed that an electron in the external fields $\mathbf{A}, \phi$ would be described by a wave function $\psi(\mathbf{x}, t)$ satisfying the equation obtained by making the same replacements in

* This is Lorentz invariant, because the quantities $\mathbf{A}$ and ϕ have the same Lorentz transformation property as $c\mathbf{p}$ and E. Schrödinger actually wrote H and $\mathbf{p}$ in terms of partial derivatives of an action function, but this makes no difference to our present discussion.

(1.1.2):

$$0 = \left[\left(i\hbar\frac{\partial}{\partial t} + e\phi\right)^2 - c^2\left(-i\hbar\nabla + \frac{e\mathbf{A}}{c}\right)^2 - m^2c^4\right]\psi(\mathbf{x}, t)\,. \qquad (1.1.4)$$

In particular, for the stationary states of hydrogen we have $\mathbf{A} = 0$ and $\phi = e/4\pi r$, and ψ has the time-dependence $\exp(-iEt/\hbar)$, so (1.1.4) becomes

$$0 = \left[\left(E + \frac{e^2}{4\pi r}\right)^2 - c^2\hbar^2\nabla^2 - m^2c^4\right]\psi(\mathbf{x})\,. \qquad (1.1.5)$$

Solutions satisfying reasonable boundary conditions can be found for the energy values[9]

$$E = mc^2\left[1 - \frac{\alpha^2}{2n^2} - \frac{\alpha^4}{2n^4}\left(\frac{n}{\ell + \frac{1}{2}} - \frac{3}{4}\right) + \cdots\right], \qquad (1.1.6)$$

where $\alpha \equiv e^2/4\pi\hbar c$ is the 'fine structure constant,' roughly 1/137; n is a positive-definite integer, and ℓ, the orbital angular momentum in units of $\hbar$, is an integer with $0 \leq \ell \leq n - 1$. The α^2 term gave good agreement with the gross features of the hydrogen spectrum (the Lyman, Balmer, etc. series) and, according to Dirac,[6] it was this agreement that led Schrödinger eventually to develop his non-relativistic wave equation. On the other hand, the α^4 term gave a fine structure in disagreement with existing accurate measurements of Friedrich Paschen.[10]

It is instructive here to compare Schrödinger's result with that of Arnold Sommerfeld,[10a] obtained using the rules of the old quantum theory:

$$E = mc^2\left[1 - \frac{\alpha^2}{2n^2} - \frac{\alpha^4}{2n^4}\left(\frac{n}{k} - \frac{3}{4}\right) + \cdots\right]. \qquad (1.1.7)$$

where m is the electron mass. Here k is an integer between 1 and n, which in Sommerfeld's theory is given in terms of the orbital angular momentum $\ell\hbar$ as $k = \ell + 1$. This gave a fine structure splitting in agreement with experiment: for instance, for $n = 2$ Eq. (1.1.7) gives two levels ($k = 1$ and $k = 2$), split by the observed amount $\alpha^4 mc^2/32$, or 4.53×10^{-5} eV. In contrast, Schrödinger's result (1.1.6) gives an $n = 2$ fine structure splitting $\alpha^4 mc^2/12$, considerably larger than observed.

Schrödinger correctly recognized that the source of this discrepancy was his neglect of the spin of the electron. The splitting of atomic energy levels by non-inverse-square electric fields in alkali atoms and by weak external magnetic fields (the so-called anomalous Zeeman effect) had revealed a multiplicity of states larger than could be accounted for by the Bohr–Sommerfeld theory; this led George Uhlenbeck and Samuel Goudsmit[11] in 1925 to suggest that the electron has an intrinsic angular momentum $\hbar/2$. Also, the magnitude of the Zeeman splitting[12] allowed them to estimate further that the electron has a magnetic moment

$$\mu = \frac{e\hbar}{2mc} . \tag{1.1.8}$$

It was clear that the electron's spin would be coupled to its orbital angular momentum, so that Schrödinger's relativistic equation should not be expected to give the correct fine structure splitting.

Indeed, by 1927 several authors[13] had been able to show that the spin–orbit coupling was able to account for the discrepancy between Schrödinger's result (1.1.6) and experiment. There are really two effects here: one is a direct coupling between the magnetic moment (1.1.8) and the magnetic field felt by the electron as it moves through the electrostatic field of the atom; the other is the relativistic 'Thomas precession' caused (even in the absence of a magnetic moment) by the circular motion of the spinning electron.[14] Together, these two effects were found to lift the level with total angular momentum $j = \ell + \frac{1}{2}$ to the energy (1.1.7) given by Sommerfeld for $k = \ell + 1 = j + \frac{1}{2}$, while the level with $j = \ell - \frac{1}{2}$ was lowered to the value given by Sommerfeld for $k = \ell = j + \frac{1}{2}$. Thus the energy was found to depend only on n and j, but not separately on ℓ:

$$E = mc^2 \left[1 - \frac{\alpha^2}{2n^2} - \frac{\alpha^4}{2n^4} \left(\frac{n}{j + \frac{1}{2}} - \frac{3}{4} \right) + \cdots \right] . \tag{1.1.9}$$

By accident Sommerfeld's theory had given the correct magnitude of the splitting in hydrogen ($j + \frac{1}{2}$ like k runs over integer values from 1 to n) though it was wrong as to the assignment of orbital angular momentum values ℓ to these various levels. In addition, the multiplicity of the fine structure levels in hydrogen was now predicted to be 2 for $j = \frac{1}{2}$ and $2(2j + 1)$ for $j > \frac{1}{2}$ (corresponding to ℓ values $j \pm \frac{1}{2}$), in agreement with experiment.

Despite these successes, there still was not a thorough relativistic theory which incorporated the electron's spin from the beginning. Such a theory was discovered in 1928 by Paul Dirac. However, he did not set out simply to make a relativistic theory of the spinning electron; instead, he approached the problem by posing a question that would today seem very strange. At the beginning of his 1928 paper,[15] he asks 'why Nature should have chosen this particular model for the electron, instead of being satisfied with the point charge.' To us today, this question is like asking why bacteria have only one cell: having spin $\hbar/2$ is just one of the properties that define a particle as an electron, rather than one of the many other types of particles with various spins that are known today. However, in 1928 it was possible to believe that all matter consisted of electrons, and perhaps something similar with positive charge in the atomic nucleus. Thus, in the spirit of the times in which it was asked, Dirac's question can be restated: 'Why do the fundamental constituents of matter have to have spin $\hbar/2$?'

For Dirac, the key to this question was the requirement that probabilities must be positive. It was known[16] that the probability density for the non-relativistic Schrödinger equation is $|\psi|^2$, and that this satisfies a continuity equation of the form

$$\frac{\partial}{\partial t}(|\psi|^2) - \frac{i\hbar}{2m}\nabla \cdot (\psi^*\nabla\psi - \psi\nabla\psi^*) = 0$$

so the space-integral of $|\psi|^2$ is time-independent. On the other hand, the only probability density ρ and current $\mathbf{J}$, which can be formed from solutions of the relativistic Schrödinger equation and which satisfy a conservation law,

$$\frac{\partial \rho}{\partial t} + \nabla \cdot \mathbf{J} = 0\,, \tag{1.1.10}$$

are of the form

$$\rho = N\,\mathrm{Im}\,\psi^*\left(\frac{\partial}{\partial t} - \frac{ie\phi}{\hbar}\right)\psi\,, \tag{1.1.11}$$

$$\mathbf{J} = N\,c^2\,\mathrm{Im}\,\psi^*\left(\nabla + \frac{ie\mathbf{A}}{\hbar c}\right)\psi\,, \tag{1.1.12}$$

with N an arbitrary constant. It is not possible to identify ρ as the probability density, because (with or without an external potential ϕ) ρ does not have definite sign. To quote Dirac's reminiscences[17] about this problem

> I remember once when I was in Copenhagen, that Bohr asked me what I was working on and I told him I was trying to get a satisfactory relativistic theory of the electron, and Bohr said 'But Klein and Gordon have already done that!' That answer first rather disturbed me. Bohr seemed quite satisfied by Klein's solution, but I was not because of the negative probabilities that it led to. I just kept on with it, worrying about getting a theory which would have only positive probabilities.

According to George Gamow,[18] Dirac found the answer to this problem on an evening in 1928 while staring into a fireplace at St John's College, Cambridge. He realized that the reason that the Klein–Gordon (or relativistic Schrödinger) equation can give negative probabilities is that the ρ in the conservation equation (1.1.10) involves a time-derivative of the wave function. This in turn happens because the wave function satisfies a differential equation of *second* order in the time. The problem therefore was to replace this wave equation with another one of first order in time derivatives, like the non-relativistic Schrödinger equation.

Suppose the electron wave function is a multi-component quantity $\psi_n(x)$, which satisfies a wave equation of the form,

$$i\hbar \frac{\partial \psi}{\partial t} = \mathscr{H}\psi\,, \tag{1.1.13}$$

where $\mathscr{H}$ is some matrix function of space derivatives. In order to have a chance at a Lorentz-invariant theory, we must suppose that because the equation is linear in time-derivatives, it is also linear in space-derivatives, so that $\mathscr{H}$ takes the form:

$$\mathscr{H} = -i\hbar c\boldsymbol{\alpha} \cdot \nabla + \alpha_4 mc^2\,, \tag{1.1.14}$$

where α_1, α_2, α_3, and α_4 are constant matrices. From (1.1.13) we can derive the second-order equation

$$-\hbar^2 \frac{\partial^2 \psi}{\partial t^2} = \mathscr{H}^2\psi = -\hbar^2 c^2 \alpha_i \alpha_j \frac{\partial^2 \psi}{\partial x_i \partial x_j}$$
$$-i\hbar mc^3(\alpha_i\alpha_4 + \alpha_4\alpha_i)\frac{\partial \psi}{\partial x_i} + m^2c^4\alpha_4^2\psi\,.$$

(The summation convention is in force here; i and j run over the values 1, 2, 3, or x, y, z.) But this must agree with the free-field form of the relativistic Schrödinger equation (1.1.4), which just expresses the relativistic relation between momentum and energy. Therefore, the matrices $\boldsymbol{\alpha}$ and α_4 must satisfy the relations

$$\alpha_i\alpha_j + \alpha_j\alpha_i = 2\delta_{ij}1\,, \tag{1.1.15}$$
$$\alpha_i\alpha_4 + \alpha_4\alpha_i = 0\,, \tag{1.1.16}$$
$$\alpha_4^2 = 1\,, \tag{1.1.17}$$

where δ_{ij} is the Kronecker delta (unity for $i = j$; zero for $i \neq j$) and 1 is the unit matrix. Dirac found a set of 4×4 matrices which satisfy these relations

$$\alpha_1 = \begin{bmatrix} 0 & 0 & 0 & 1 \\ 0 & 0 & 1 & 0 \\ 0 & 1 & 0 & 0 \\ 1 & 0 & 0 & 0 \end{bmatrix}, \qquad \alpha_2 = \begin{bmatrix} 0 & 0 & 0 & -i \\ 0 & 0 & i & 0 \\ 0 & -i & 0 & 0 \\ i & 0 & 0 & 0 \end{bmatrix},$$
$$\alpha_3 = \begin{bmatrix} 0 & 0 & 1 & 0 \\ 0 & 0 & 0 & -1 \\ 1 & 0 & 0 & 0 \\ 0 & -1 & 0 & 0 \end{bmatrix}, \qquad \alpha_4 = \begin{bmatrix} 1 & 0 & 0 & 0 \\ 0 & 1 & 0 & 0 \\ 0 & 0 & -1 & 0 \\ 0 & 0 & 0 & -1 \end{bmatrix}. \tag{1.1.18}$$

To show that this formalism is Lorentz-invariant, Dirac multiplied Eq. (1.1.13) on the left with α_4, so that it could be put in the form

$$\left[\hbar c\gamma^\mu \frac{\partial}{\partial x^\mu} + mc^2\right]\psi = 0\,, \tag{1.1.19}$$

where

$$\boldsymbol{\gamma} \equiv -i\alpha_4\boldsymbol{\alpha}\,, \qquad \gamma^0 \equiv -i\alpha_4\,. \tag{1.1.20}$$

(The Greek indices μ, ν, etc. will now run over the values 1, 2, 3, 0, with $x^0 = ct$. Dirac used $x_4 = ict$, and correspondingly $\gamma_4 = \alpha_4$.) The matrices γ^μ satisfy the anticommutation relations

$$\frac{1}{2}(\gamma^\mu\gamma^\nu + \gamma^\nu\gamma^\mu) = \eta^{\mu\nu} \equiv \begin{cases} +1 & \mu = \nu = 1, 2, 3 \\ -1 & \mu = \nu = 0 \\ 0 & \mu \neq \nu \end{cases}\,. \tag{1.1.21}$$

Dirac noted that these anticommutation relations are Lorentz-invariant, in the sense that they are also satisfied by the matrices $\Lambda^\mu{}_\nu\gamma^\nu$, where Λ is any Lorentz transformation. He concluded from this that $\Lambda^\mu{}_\nu\gamma^\nu$ must be related to γ^μ by a similarity transformation:

$$\Lambda^\mu{}_\nu\gamma^\nu = S^{-1}(\Lambda)\gamma^\mu S(\Lambda)\,.$$

It follows that the wave equation is invariant if, under a Lorentz transformation $x^\mu \to \Lambda^\mu{}_\nu x^\nu$, the wave function undergoes the matrix transformation $\psi \to S(\Lambda)\psi$. (These matters are discussed more fully, from a rather different point of view, in Chapter 5.)

To study the behavior of electrons in an arbitrary external electromagnetic field, Dirac followed the 'usual procedure' of making the replacements

$$i\hbar\frac{\partial}{\partial t} \to i\hbar\frac{\partial}{\partial t} + e\phi \qquad -i\hbar\nabla \to -i\hbar\nabla + \frac{e}{c}\mathbf{A} \tag{1.1.22}$$

as in Eq. (1.1.4). The wave equation (1.1.13) then takes the form

$$\left(i\hbar\frac{\partial}{\partial t} + e\phi\right)\psi = (-i\hbar c\nabla + e\mathbf{A})\cdot\boldsymbol{\alpha}\psi + mc^2\alpha_4\psi\,. \tag{1.1.23}$$

Dirac used this equation to show that in a central field, the conservation of angular momentum takes the form

$$[\mathscr{H}, -i\hbar\mathbf{r}\times\nabla + \hbar\boldsymbol{\sigma}/2] = 0\,, \tag{1.1.24}$$

where $\mathscr{H}$ is the matrix differential operator (1.1.14) and $\boldsymbol{\sigma}$ is the 4×4 version of the spin matrix introduced earlier by Pauli[19]

$$\boldsymbol{\sigma} = \begin{bmatrix} 0 & 0 & 1 & 0 \\ 0 & 0 & 0 & 1 \\ 1 & 0 & 0 & 0 \\ 0 & 1 & 0 & 0 \end{bmatrix}\boldsymbol{\alpha}\,. \tag{1.1.25}$$

Since each component of $\boldsymbol{\sigma}$ has eigenvalues ± 1, the presence of the extra term in (1.1.24) shows that the electron has intrinsic angular momentum $\hbar/2$.

Dirac also iterated Eq. (1.1.23), obtaining a second-order equation, which turned out to have just the same form as the Klein–Gordon equation (1.1.4) except for the presence on the right-hand-side of two additional terms

$$[-e\hbar c\boldsymbol{\sigma} \cdot \mathbf{B} - ie\hbar c\boldsymbol{\alpha} \cdot \mathbf{E}]\,\psi\,. \tag{1.1.26}$$

For a slowly moving electron, the first term dominates, and represents a magnetic moment in agreement with the value (1.1.8) found by Goudsmit and Uhlenbeck.[11] As Dirac recognized, this magnetic moment, together with the relativistic nature of the theory, guaranteed that this theory would give a fine structure splitting in agreement (to order $\alpha^4 mc^2$) with that found by Heisenberg, Jordan, and Charles G. Darwin.[13] A little later, an 'exact' formula for the hydrogen energy levels in Dirac's theory was derived by Darwin[20] and Gordon[21]

$$E = mc^2 \left(1 + \frac{\alpha^2}{\left\{n - j - \frac{1}{2} + \left[\left(j + \frac{1}{2}\right)^2 - \alpha^2\right]^{\frac{1}{2}}\right\}^2}\right)^{-1/2}. \tag{1.1.27}$$

The first three terms of a power series expansion in α^2 agree with the approximate result (1.1.9).

This theory achieved Dirac's primary aim: a relativistic formalism with positive probabilities. From (1.1.13) we can derive a continuity equation

$$\frac{\partial \rho}{\partial t} + \nabla \cdot \mathbf{J} = 0 \tag{1.1.28}$$

with

$$\rho = |\psi|^2\,, \qquad \mathbf{J} = c\psi^\dagger \boldsymbol{\alpha}\psi\,, \tag{1.1.29}$$

so that the positive quantity $|\psi|^2$ can be interpreted as a probability density, with constant total probability $\int |\psi|^2 d^3x$. However, there was another difficulty which Dirac was not immediately able to resolve.

For a given momentum $\mathbf{p}$, the wave equation (1.1.13) has *four* solutions of the plane wave form

$$\psi \propto \exp\left[\frac{i}{\hbar}(\mathbf{p} \cdot \mathbf{x} - Et)\right]. \tag{1.1.30}$$

Two solutions with $E = +\sqrt{\mathbf{p}^2c^2 + m^2c^4}$ correspond to the two spin states of an electron with $J_z = \pm\hbar/2$. The other two solutions have $E =$

$-\sqrt{\mathbf{p}^2c^2+m^2c^4}$, and no obvious physical interpretation. As Dirac pointed out, this problem arises also for the relativistic Schrödinger equation: for each $\mathbf{p}$, there are two solutions of the form (1.1.30), one with positive E and one with negative E.

Of course, even in classical physics, the relativistic relation $E^2 = \mathbf{p}^2c^2 + m^2c^4$ has two solutions, $E = \pm\sqrt{\mathbf{p}^2c^2+m^2c^4}$. However, in classical physics we can simply assume that the only physical particles are those with positive E. Since the positive solutions have $E > mc^2$ and the negative ones have $E < -mc^2$, there is a finite gap between them, and no continuous process can take a particle from positive to negative energy.

The problem of negative energies is much more troublesome in relativistic quantum mechanics. As Dirac pointed out in his 1928 paper,[15] the interaction of electrons with radiation can produce transitions in which a positive-energy electron falls into a negative-energy state, with the energy carried off by two or more photons. Why then is matter stable?

In 1930 Dirac offered a remarkable solution.[22] Dirac's proposal was based on the exclusion principle, so a few words about the history of this principle are in order here.

The periodic table of the elements and the systematics of X-ray spectroscopy had together by 1924 revealed a pattern in the population of atomic energy levels by electrons:[23] The maximum number N_n of electrons in a shell characterized by principal quantum number n is given by twice the number of orbital states with that n

$$N_n = 2\sum_{\ell=0}^{n-1}(2\ell+1) = 2n^2 = 2,\, 8,\, 18, \ldots . \qquad (1.1.31)$$

Wolfgang Pauli[24] in 1925 suggested that this pattern could be understood if N_n is the total number of possible states in the nth shell, and if in addition there is some mysterious 'exclusion principle' which forbids more than one electron from occupying the same state. He explained the puzzling factor 2 in (1.1.31) as due to a 'peculiar, classically non-describable duplexity' of the electron states, and as we have seen this was understood a little later as due to the spin of the electron.[11] The exclusion principle answered a question that had remained obscure in the old atomic theory of Bohr and Sommerfeld: why do not all the electrons in heavy atoms fall down into the shell of lowest energy? Subsequently Pauli's exclusion principle was formalized by a number of authors[25] as the requirement that the wave function of a multi-electron system is antisymmetric in the coordinates, orbital and spin, of all the electrons. This principle was incorporated into statistical mechanics by Enrico Fermi[26] and Dirac,[27] and for this reason particles obeying the exclusion principle are generally called 'fermions,' just as particles like photons for which the wave function is symmetric and which obey the statistics of Bose and Einstein are called 'bosons.' The exclusion principle has played a fundamental role in the theory of metals, white dwarf and neutron stars, etc., as well as in chemistry and atomic physics, but a discussion of these matters would take us too far afield here.

Dirac's proposal was that the positive energy electrons cannot fall down into negative energy states because 'all the states of negative energy are occupied except perhaps a few of small velocity.' The few vacant states, or 'holes,' in the sea of negative energy electrons behave like particles with opposite quantum numbers: positive energy and positive charge. The only particle with positive charge that was known at that time was the proton, and as Dirac later recalled,[27a] 'the whole climate of opinion at that time was against new particles' so Dirac identified his holes as protons; in fact, the title of his 1930 article[22] was 'A Theory of Electrons and Protons.'

The hole theory faced a number of immediate difficulties. One obvious problem was raised by the infinite charge density of the ubiquitous negative-energy electrons: where is their electric field? Dirac proposed to reinterpret the charge density appearing in Maxwell's equations as 'the departure from the normal state of electrification of the world.' Another problem has to do with the huge dissimilarity between the observed masses and interactions of the electrons and protons. Dirac hoped that Coulomb interactions between electrons would somehow account for these differences but Hermann Weyl[28] showed that the hole theory was in fact entirely symmetric between negative and positive charge. Finally, Dirac[22] predicted the existence of an electron–proton annihilation process in which a positive-energy electron meets a hole in the sea of negative-energy electrons and falls down into the unoccupied level, emitting a pair of gamma ray photons. By itself this would not have created difficulties for the hole theory; it was even hoped by some that this would provide an explanation, then lacking, of the energy source of the stars. However, it was soon pointed out[29] by Julius Robert Oppenheimer and Igor Tamm that electron–proton annihilation in atoms would take place at much too fast a rate to be consistent with the observed stability of ordinary matter. For these reasons, by 1931 Dirac had changed his mind, and decided that the holes would have to appear not as protons but as a new sort of positively charged particle, of the same mass as the electron.[29a]

The second and third of these problems were eliminated by the discovery of the positron by Carl D. Anderson,[30] who apparently did not know of this prediction by Dirac. On August 2, 1932, a peculiar cosmic ray track was observed in a Wilson cloud chamber subjected to a 15 kG magnetic field. The track was observed to curve in a direction that would be expected for a *positively* charged particle, and yet its range was at least ten times greater than the expected range of a proton! Both the range and the specific ionization of the track were consistent with the hypothesis that this was a new particle which differs from the electron only in the sign of its charge, as would be expected for one of Dirac's holes. (This discovery had been made earlier by P.M.S. Blackett, but not immediately published by him. Anderson quotes press reports of evidence for light positive particles in cosmic ray tracks, obtained by Blackett and Giuseppe Occhialini.) Thus it appeared that Dirac was wrong only in his original identification of the hole with the proton.

The discovery of the more-or-less predicted positron, together with the earlier successes of the Dirac equation in accounting for the magnetic moment of the electron and the fine structure of hydrogen, gave Dirac's theory a prestige that it has held for over six decades. However, although there seems little doubt that Dirac's theory will survive in some form in any future physical theory, there are serious reasons for being dissatisfied with its original rationale:

(i) Dirac's analysis of the problem of negative probabilities in Schrödinger's relativistic wave equation would seem to rule out the existence of any particle of zero spin. Yet even in the 1920s particles of zero spin were known — for instance, the hydrogen atom in its ground state, and the helium nucleus. Of course, it could be argued that hydrogen atoms and alpha particles are not elementary, and therefore do not need to be described by a relativistic wave equation, but it was not (and still is not) clear how the idea of elementarity is incorporated in the formalism of relativistic quantum mechanics. Today we know of a large number of spin zero particles — π mesons, K mesons, and so on — that are no less elementary than the proton and neutron. We also know of spin one particles — the $W^{\pm}$ and Z^0 — which seem as elementary as the electron or any other particle. Further, apart from effects of the strong interactions, we would today calculate the fine structure of 'mesonic atoms,' consisting of a spinless negative π or K meson bound to an atomic nucleus, from the stationary solutions of the relativistic Klein–Gordon–Schrödinger equation! Thus, it is difficult to agree that there is anything fundamentally wrong with the relativistic equation for zero spin that *forced* the development of the Dirac equation — the problem simply is that the electron happens to have spin $\hbar/2$, not zero.

(ii) As far as we now know, for *every* kind of particle there is an 'antiparticle' with the same mass and opposite charge. (Some purely neutral particles, such as the photon, are their own antiparticles.) But how can we interpret the antiparticles of charged *bosons*, such as the $\pi^{\pm}$ mesons or $W^{\pm}$ particles, as holes in a sea of negative energy states? For particles quantized according to the rules of Bose–Einstein statistics, there is no exclusion principle, and hence nothing to keep positive-energy particles from falling down into the negative-energy states, occupied or not. And if the hole theory does not work for bosonic antiparticles, why should we believe it for fermions? I asked Dirac in 1972 how he then felt about this point; he told me that he did not regard bosons like the pion or $W^{\pm}$ as 'important.' In a lecture[27a] a few years later, Dirac referred to the fact that for bosons 'we no longer have the picture of a vacuum with negative energy states filled up', and remarked that in this case 'the whole theory becomes more complicated.' The next section will show how the development of quantum field theory made the interpretation of antiparticles as holes unnecessary, even though unfortunately it lingers on in many textbooks. To quote Julian Schwinger,[30a] 'The picture of an infinite sea of negative energy electrons is now best regarded as a historical curiosity, and forgotten.'

(iii) One of the great successes of the Dirac theory was its correct prediction of the magnetic moment of the electron. This was particularly striking, as the magnetic moment (1.1.8) is twice as large as would be expected for the orbital motion of a charged point particle with angular momentum $\hbar/2$; this factor of 2 had remained mysterious until Dirac's theory. However, there is really nothing in Dirac's line of argument that leads unequivocally to this particular value for the magnetic moment. At the point where we brought electric and magnetic fields into the wave equation (1.1.23), we could just as well have added a 'Pauli term'[31]

$$\kappa \alpha_4 [\gamma^\mu, \gamma^\nu] \psi F_{\mu\nu} \tag{1.1.32}$$

with arbitrary coefficient κ. (Here $F_{\mu\nu}$ is the usual electromagnetic field strength tensor, with $F^{12} = B_3$, $F^{01} = E_1$, etc.) This term could be obtained by first adding a term to the free-field equations proportional to $[\gamma^\mu, \gamma^\nu](\partial^2/\partial x^\mu \partial x^\nu)\psi$, which of course equals zero, and then making the substitutions (1.1.22) as before. A more modern approach would be simply to remark that the term (1.1.32) is consistent with all accepted invariance principles, including Lorentz invariance and gauge invariance, and so there is no reason why such a term should *not* be included in the field equations. (See Section 12.3.) This term would give an additional contribution proportional to κ to the magnetic moment of the electron, so apart from the possible demand for a purely formal simplicity, there was no reason to expect any particular value for the magnetic moment of the electron in Dirac's theory.

As we shall see in this book, these problems were all eventually to be solved (or at least clarified) through the development of quantum field theory.

1.2 The Birth of Quantum Field Theory

The photon is the only particle that was known as a field before it was detected as a particle. Thus it is natural that the formalism of quantum field theory should have been developed in the first instance in connection with radiation and only later applied to other particles and fields.

In 1926, in one of the central papers on matrix mechanics, Born, Heisenberg, and Jordan[32] applied their new methods to the free radiation field. For simplicity, they ignored the polarization of electromagnetic waves and worked in one space dimension, with coordinate x running from 0 to L; the radiation field $u(x,t)$ if constrained to vanish at these endpoints thus has the same behavior as the displacement of a string with ends fixed at $x = 0$ and $x = L$. By analogy with either the case of a string or the full electromagnetic field, the Hamiltonian was taken to have the form

$$H = \frac{1}{2}\int_0^L \left\{ \left(\frac{\partial u}{\partial t}\right)^2 + c^2 \left(\frac{\partial u}{\partial x}\right)^2 \right\} dx \,. \tag{1.2.1}$$

In order to reduce this expression to a sum of squares, the field u was expressed as a sum of Fourier components with $u = 0$ at both $x = 0$ and $x = L$:

$$u(x,t) = \sum_{k=1}^{\infty} q_k(t) \sin\left(\frac{\omega_k x}{c}\right) \,, \tag{1.2.2}$$

$$\omega_k \equiv k\pi c/L \,, \tag{1.2.3}$$

so that

$$H = \frac{L}{4}\sum_{k=1}^{\infty} \left\{ \dot{q}_k^2(t) + \omega_k^2 q_k^2(t) \right\} . \tag{1.2.4}$$

Thus the string or field behaves like a sum of independent harmonic oscillators with angular frequencies ω_k, as had been anticipated 20 years earlier by Paul Ehrenfest.[32a]

In particular, the 'momentum' $p_k(t)$ canonically conjugate to $q_k(t)$ is determined, as in particle mechanics, by the condition that if H is expressed as a function of the ps and qs, then

$$\dot{q}_k(t) = \frac{\partial}{\partial p_k(t)} H(p(t), q(t)) \,.$$

This yields a 'momentum'

$$p_k(t) = \frac{L}{2}\, \dot{q}_k(t) \tag{1.2.5}$$

so the canonical commutation relations may be written

$$\left[\dot{q}_k(t), q_j(t)\right] = \frac{2}{L}\left[p_k(t), q_j(t)\right] = \frac{-2i\hbar}{L}\,\delta_{kj}\,, \tag{1.2.6}$$

$$\left[q_k(t), q_j(t)\right] = 0\,. \tag{1.2.7}$$

Also, the time-dependence of $q_k(t)$ is governed by the Hamiltonian equation of motion

$$\ddot{q}_k(t) = \frac{2}{L}\,\dot{p}_k(t) = -\frac{2}{L}\frac{\partial H}{\partial q_k(t)} = -\omega_k^2 q_k(t)\,. \tag{1.2.8}$$

The form of the matrices defined by Eqs. (1.2.6)–(1.2.8) was already known to Born, Heisenberg, and Jordan through previous work on the harmonic oscillator. The q-matrix is given by

$$q_k(t) = \sqrt{\frac{\hbar}{L\omega_k}}\left[a_k \exp(-i\omega_k t) + a_k^\dagger \exp(+i\omega_k t)\right] \tag{1.2.9}$$

with a_k a time-independent matrix and $a_k^\dagger$ its Hermitian adjoint, satisfying the commutation relations

$$\left[\,a_k\,,\,a_j^\dagger\,\right] = \delta_{kj}\,, \tag{1.2.10}$$

$$\left[\,a_k\,,\,a_j\,\right] = 0\,. \tag{1.2.11}$$

The rows and columns of these matrices are labelled with a set of positive integers n_1, n_2,..., one for each normal mode. The matrix elements are

$$(a_k)_{n'_1,n'_2,\ldots,n_1,n_2\ldots} = \sqrt{n_k}\,\delta_{n'_k,n_k-1}\prod_{j\neq k}\delta_{n'_j,n_j}\,, \tag{1.2.12}$$

$$(a_k^\dagger)_{n'_1,n'_2,\ldots,n_1,n_2\ldots} = \sqrt{n_k+1}\,\delta_{n'_k,n_k+1}\prod_{j\neq k}\delta_{n'_j,n_j}\,. \tag{1.2.13}$$

For a single normal mode, these matrices may be written explicitly as

$$a = \begin{bmatrix} 0 & \sqrt{1} & 0 & 0 & . & . & . \\ 0 & 0 & \sqrt{2} & 0 & . & . & . \\ 0 & 0 & 0 & \sqrt{3} & . & . & . \\ 0 & 0 & 0 & 0 & . & . & . \\ . & . & . & . & & & \\ . & . & . & . & & & \\ . & . & . & . & & & \end{bmatrix}, \quad a^\dagger = \begin{bmatrix} 0 & 0 & 0 & 0 & . & . & . \\ \sqrt{1} & 0 & 0 & 0 & . & . & . \\ 0 & \sqrt{2} & 0 & 0 & . & . & . \\ 0 & 0 & \sqrt{3} & 0 & . & . & . \\ . & . & . & . & & & \\ . & . & . & . & & & \\ . & . & . & . & & & \end{bmatrix}.$$

It is straightforward to check that (1.2.12) and (1.2.13) do satisfy the commutation relations (1.2.10) and (1.2.11).

The physical interpretation of a column vector with integer components $n_1, n_2, \ldots$ is that it represents a state with n_k quanta in each normal mode k. The matrix a_k or $a_k^\dagger$ acting on such a column vector will respectively lower or raise n_k by one unit, leaving all n_ℓ with $\ell \neq k$ unchanged; they may therefore be interpreted as operators which annihilate or create one quantum in the kth normal mode. In particular, the vector with all n_k equal to zero represents the vacuum; it is annihilated by any a_k.

This interpretation is further borne out by inspection of the Hamiltonian. Using (1.2.9) and (1.2.10) in (1.2.4) gives

$$H = \sum_k \hbar\omega_k \left(a_k^\dagger a_k + \tfrac{1}{2}\right) . \tag{1.2.14}$$

The Hamiltonian is then diagonal in the n-representation

$$(H)_{n'_1,n'_2,\ldots,n_1,n_2\ldots} = \sum_k \hbar\omega_k \left(n_k + \tfrac{1}{2}\right) \prod_j \delta_{n'_j\, n_j} . \tag{1.2.15}$$

We see that the energy of the state is just the sum of energies $\hbar\omega_k$ for each quantum present in the state, plus an infinite zero-point energy $E_0 = \frac{1}{2}\sum_k \hbar\omega_k$. Applied to the radiation field, this formalism justified the Bose method of counting radiation states according to the numbers n_k of quanta in each normal mode.

Born, Heisenberg, and Jordan used this formalism to derive an expression for the r.m.s. energy fluctuations in black-body radiation. (For this purpose they actually only used the commutation relations (1.2.6)–(1.2.7).) However, this approach was soon applied to a more urgent problem, the calculation of the rates for spontaneous emission of radiation.

In order to appreciate the difficulties here, it is necessary to go back in time a bit. In one of the first papers on matrix mechanics, Born and Jordan[33] had assumed in effect that an atom, in dropping from a state β to a lower state α, would emit radiation just like a classical charged oscillator with displacement

$$\mathbf{r}(t) = \mathbf{r}_{\beta\alpha} \exp(-2\pi i\nu t) + \mathbf{r}_{\beta\alpha}{}^* \exp(2\pi i\nu t) , \tag{1.2.16}$$

where

$$h\nu = E_\beta - E_\alpha \tag{1.2.17}$$

and $\mathbf{r}_{\beta\alpha}$ is the β, α element of the matrix associated with the electron position. The energy E of such an oscillator is

$$E = \frac{1}{2} m \left(\dot{\mathbf{r}}^2 + (2\pi\nu)^2 \mathbf{r}^2\right) = 8\pi^2 m\nu^2 |\mathbf{r}_{\beta\alpha}|^2 . \tag{1.2.18}$$

A straightforward classical calculation then gives the radiated power, and dividing by the energy $h\nu$ per photon gives the rate of photon emission

$$A(\beta \to \alpha) = \frac{16\pi^3 e^2 \nu^3}{3hc^3} |\mathbf{r}_{\beta\alpha}|^2 . \tag{1.2.19}$$

However, it was not at all clear why the formulas for emission of radiation by a classical dipole should be taken over in this manner in dealing with spontaneous emission.

A little later a more convincing though even less direct derivation was given by Dirac.[34] By considering the behavior of quantized atomic states in an oscillating *classical* electromagnetic field with energy density per frequency interval u at frequency (1.2.17), he was able to derive formulas for the rates $uB(\alpha \rightarrow \beta)$ and $uB(\beta \rightarrow \alpha)$ for absorption or induced emission:

$$B(\alpha \rightarrow \beta) = B(\beta \rightarrow \alpha) \simeq \frac{2\pi^2 e^2}{3h^2} |\mathbf{r}_{\beta\alpha}|^2 . \qquad (1.2.20)$$

(Note that the expression on the right is symmetric between states α and β, because $\mathbf{r}_{\alpha\beta}$ is just $\mathbf{r}_{\beta\alpha}{}^*$.) Einstein[34a] had already shown in 1917 that the possibility of thermal equilibrium between atoms and black-body radiation imposes a relation between the rate $A(\beta \rightarrow \alpha)$ of spontaneous emission and the rates uB for induced emission or absorption:

$$A(\beta \rightarrow \alpha) = \left(\frac{8\pi h\nu^3}{c^3}\right) B(\beta \rightarrow \alpha) . \qquad (1.2.21)$$

Using (1.2.20) in this relation immediately yields the Born–Jordan result (1.2.19) for the rate of spontaneous emission. Nevertheless, it still seemed unsatisfactory that thermodynamic arguments should be needed to derive formulas for processes involving a single atom.

Finally, in 1927 Dirac[35] was able to give a thoroughly quantum mechanical treatment of spontaneous emission. The vector potential $\mathbf{A}(\mathbf{x}, t)$ was expanded in normal modes, as in Eq. (1.2.2), and the coefficients were shown to satisfy commutation relations like (1.2.6). In consequence, each state of the free radiation field was specified by a set of integers n_k, one for each normal mode, and the matrix elements of the electromagnetic interaction $e\dot{\mathbf{r}} \cdot \mathbf{A}$ took the form of a sum over normal modes, with matrix coefficients proportional to the matrices a_k and $a_k^\dagger$ defined in Eqs. (1.2.10)–(1.2.13). The crucial result here is the factor $\sqrt{n_k + 1}$ in Eq. (1.2.13); the probability for a transition in which the number of photons in a normal mode k rises from n_k to $n_k + 1$ is proportional to the square of this factor, or $n_k + 1$. But in a radiation field with n_k photons in a normal mode k, the energy density u per frequency interval is

$$u(\nu_k) = \left(\frac{8\pi\nu_k^2}{c^3}\right) n_k \times h\nu_k ,$$

so the rate for emission of radiation in normal mode k is proportional to

$$n_k + 1 = \frac{c^3 u(\nu_k)}{8\pi h\nu_k^3} + 1 .$$

The first term is interpreted as the contribution of induced emission, and the second term as the contribution of spontaneous emission. Hence, without any appeal to thermodynamics, Dirac could conclude that the ratio of the rates uB for induced emission and A for spontaneous emission is given by the Einstein relation, Eq. (1.2.21). Using his earlier result (1.2.20) for B, Dirac was thus able to rederive the Born–Jordan formula[33] (1.2.19) for spontaneous emission rate A. A little later, similar methods were used by Dirac to give a quantum mechanical treatment of the scattering of radiation and the lifetime of excited atomic states,[36] and by Victor Weisskopf and Eugene Wigner to make a detailed study of spectral line shapes.[36a] Dirac in his work was separating the electromagnetic potential into a radiation field **A** and a static Coulomb potential A^0, in a manner which did not preserve the manifest Lorentz and gauge invariance of classical electrodynamics. These matters were put on a firmer foundation a little later by Enrico Fermi.[36b] Many physicists in the 1930s learned their quantum electrodynamics from Fermi's 1932 review.

The use of canonical commutation relations for q and p or a and $a^\dagger$ also raised a question as to the Lorentz invariance of the quantized theory. Jordan and Pauli[37] in 1928 were able to show that the commutators of fields at different spacetime points were in fact Lorentz-invariant. (These commutators are calculated in Chapter 5.) Somewhat later, Bohr and Leon Rosenfeld[38] used a number of ingenious thought experiments to show that these commutation relations express limitations on our ability to measure fields at spacetime points separated by time-like intervals.

It was not long after the successful quantization of the electromagnetic field that these techniques were applied to other fields. At first this was regarded as a 'second quantization'; the fields to be quantized were the wave functions used in one-particle quantum mechanics, such as the Dirac wave function of the electron. The first step in this direction seems to have been taken in 1927 by Jordan.[39] In 1928 an essential element was supplied by Jordan and Wigner.[40] They recognized that the Pauli exclusion principle prevents the occupation number n_k of electrons in any normal mode k (counting spin as well as position variables) from taking any values other than 0 or 1. The electron field therefore cannot be expanded as a superposition of operators satisfying the commutation relations (1.2.10), (1.2.11), because these relations require n_k to take all integer values from 0 to ∞. Instead, they proposed that the electron field should be expanded in a sum of operators a_k, $a_k^\dagger$ satisfying the *anticommutation* relations

$$a_k\, a_j^\dagger + a_j^\dagger a_k = \delta_{jk}\,, \tag{1.2.22}$$

$$a_k\, a_j + a_j a_k = 0 \tag{1.2.23}$$

The relations can be satisfied by matrices labelled by a set of integers

$n_1, n_2, \cdots$, one for each normal mode, each integer taking just the values zero and one:

$$(a_k)_{n'_1,n'_2,\ldots,n_1,n_2,\ldots} = \begin{cases} 1 & n'_k = 0,\ n_k = 1,\ n'_j = n_j \text{ for } j \neq k \\ 0 & \text{otherwise}, \end{cases} \tag{1.2.24}$$

$$(a^\dagger_k)_{n'_1,n'_2,\ldots,n_1,n_2,\ldots} = \begin{cases} 1 & n'_k = 1,\ n_k = 0,\ n'_j = n_j \text{ for } j \neq k \\ 0 & \text{otherwise}. \end{cases} \tag{1.2.25}$$

For instance, for a single normal mode we have just two rows and two columns, corresponding to the values unity and zero of n' and n; the a and $a^\dagger$ matrices take the form

$$a = \begin{bmatrix} 0 & 0 \\ 1 & 0 \end{bmatrix}, \qquad a^\dagger = \begin{bmatrix} 0 & 1 \\ 0 & 0 \end{bmatrix}.$$

The reader may check that (1.2.24) and (1.2.25) do satisfy the anticommutation relations (1.2.22) and (1.2.23).

The interpretation of a column vector characterized by integers $n_1, n_2, \ldots$ is that it represents a state with n_k quanta in each normal mode k, just as for bosons. The difference is, of course, that since each n_k takes only the values 0 and 1, there can be at most one quantum in each normal mode, as required by the Pauli exclusion principle. Again, a_k destroys a quantum in normal mode k if there is one there already, and otherwise gives zero; also, $a^\dagger_k$ creates a quantum in normal mode k *unless* there is one there already, in which case it gives zero. Much later it was shown by Fierz and Pauli[40a] that the choice between commutation and anticommutation relations is dictated solely by the particle's spin: commutators must be used for particles with integer spin like the photon, and anticommutators for particles with half-integer spin like the electron. (This will be shown in a different way in Chapter 5.)

The theory of general quantum fields was first laid out in 1929, in a pair of comprehensive articles by Heisenberg and Pauli.[41] The starting point of their work was the application of the canonical formalism to the fields themselves, rather than to the coefficients of the normal modes appearing in the fields. Heisenberg and Pauli took the Lagrangian L as the space-integral of a local function of fields and spacetime derivatives of fields; the field equations were then determined from the principle that the action $\int L\,dt$ should be stationary when the fields are varied; and the commutation relations were determined from the assumption that the variational derivative of the Lagrangian with respect to any field's time-derivative behaves like a 'momentum' conjugate to that field (except that commutation relations become anticommutation relations for fermion fields). They also went on to apply this general formalism to the electromagnetic and Dirac fields, and explored the various invariance and conservation laws, including the conservation of charge, momentum, and energy, and Lorentz and gauge invariance.

The Heisenberg–Pauli formalism is essentially the same as that described in our Chapter 7, and so for the present we can limit ourselves to a single example which will turn out to be useful later in this section. For a free complex scalar field $\phi(x)$ the Lagrangian is taken as

$$L = \int d^3x \left[\dot{\phi}^\dagger \dot{\phi} - c^2 (\nabla\phi)^\dagger \cdot (\nabla\phi) - \left(\frac{mc^2}{\hbar}\right)^2 \phi^\dagger \phi \right] . \tag{1.2.26}$$

If we subject $\phi(x)$ to an infinitesimal variation $\delta\phi(x)$, the Lagrangian is changed by the amount

$$\delta L = \int d^3x \left[\dot{\phi}^\dagger \delta\dot{\phi} + \dot{\phi}\delta\dot{\phi}^\dagger - c^2 \nabla\phi^\dagger \cdot \nabla\delta\phi - c^2 \nabla\phi \cdot \nabla\delta\phi^\dagger \right.$$
$$\left. - \left(\frac{mc^2}{\hbar}\right)^2 \phi^\dagger \delta\phi - \left(\frac{mc^2}{\hbar}\right)^2 \phi\delta\phi^\dagger \right] . \tag{1.2.27}$$

It is assumed in using the principle of stationary action that the variation in the fields should vanish on the boundaries of the spacetime region of integration. Thus, in computing the change in the action $\int L\,dt$, we can immediately integrate by parts, and write

$$\delta \int L\,dt = c^2 \int d^4x \left[\delta\phi^\dagger \left(\Box - \left(\frac{mc}{\hbar}\right)^2 \right) \phi + \delta\phi \left(\Box - \left(\frac{mc}{\hbar}\right)^2 \right) \phi^\dagger \right] .$$

But this must vanish for any $\delta\phi$ and $\delta\phi^\dagger$, so ϕ must satisfy the familiar relativistic wave equation

$$\left[\Box - \left(\frac{mc}{\hbar}\right)^2 \right] \phi = 0 \tag{1.2.28}$$

and its adjoint. The 'momenta' canonically conjugate to the fields ϕ and $\phi^\dagger$ are given by the variational derivatives of L with respect to $\dot{\phi}$ and $\dot{\phi}^\dagger$, which we can read off from (1.2.27) as

$$\pi \equiv \frac{\delta L}{\delta\dot{\phi}} = \dot{\phi}^\dagger , \tag{1.2.29}$$

$$\pi^\dagger \equiv \frac{\delta L}{\delta\dot{\phi}^\dagger} = \dot{\phi} . \tag{1.2.30}$$

These field variables satisfy the usual canonical commutation relations, with a delta function in place of a Kronecker delta

$$\left[\pi(\mathbf{x},t),\phi(\mathbf{y},t)\right] = \left[\pi^\dagger(\mathbf{x},t),\phi^\dagger(\mathbf{y},t)\right] = -i\hbar\delta^3(\mathbf{x}-\mathbf{y})\,, \tag{1.2.31}$$

$$\left[\pi(\mathbf{x},t),\phi^\dagger(\mathbf{y},t)\right] = \left[\pi^\dagger(\mathbf{x},t),\phi(\mathbf{y},t)\right] = 0\,, \tag{1.2.32}$$

$$\left[\pi(\mathbf{x},t),\pi(\mathbf{y},t)\right] = \left[\pi^\dagger(\mathbf{x},t),\pi^\dagger(\mathbf{y},t)\right] = \left[\pi(\mathbf{x},t),\pi^\dagger(\mathbf{y},t)\right] = 0\,, \tag{1.2.33}$$

$$\left[\phi(\mathbf{x},t),\phi(\mathbf{y},t)\right] = \left[\phi^\dagger(\mathbf{x},t),\phi^\dagger(\mathbf{y},t)\right] = \left[\phi(\mathbf{x},t),\phi^\dagger(\mathbf{y},t)\right] = 0\,. \tag{1.2.34}$$

The Hamiltonian here is given (just as in particle mechanics) by the 'sum' of all canonical momenta times the time-derivatives of the corresponding fields, minus the Lagrangian:

$$H = \int d^3x \left[\pi\dot{\phi} + \pi^\dagger\dot{\phi}^\dagger\right] - L \tag{1.2.35}$$

or, using (1.2.26), (1.2.29), and (1.2.30):

$$H = \int d^3x \left[\pi^\dagger\pi + c^2(\nabla\phi)^\dagger\cdot(\nabla\phi) + \left(\frac{m^2c^4}{\hbar^2}\right)\phi^\dagger\phi\right]\,. \tag{1.2.36}$$

After the papers by Heisenberg and Pauli one element was still missing before quantum field theory could reach its final pre-war form: a solution to the problem of the negative-energy states. We saw in the last section that in 1930, at just about the time of the Heisenberg–Pauli papers, Dirac had proposed that the negative-energy states of the electron were all filled, but with only the holes in the negative-energy sea observable, rather than the negative-energy electrons themselves. After Dirac's idea was seemingly confirmed by the discovery of the positron in 1932, his 'hole theory' was used to calculate a number of processes to the lowest order of perturbation theory, including electron–positron pair production and scattering.

At the same time, a great deal of work was put into the development of a formalism whose Lorentz invariance would be explicit. The most influential effort was the 'many-time' formalism of Dirac, Vladimir Fock, and Boris Podolsky,[42] in which the state vector was represented by a wave function depending on the spacetime and spin coordinates of all electrons, positive-energy and negative-energy. In this formalism, the total number of electrons of either positive or negative energy is conserved; for instance, production of an electron–positron pair is described as the excitation of a negative-energy electron to a positive-energy state, and the annihilation of an electron and positron is described as the corresponding deexcitation. This many-time formalism had the advantage of manifest Lorentz invariance, but it had a number of disadvantages: In particular, there was a profound difference between the treatment of the photon, described in terms of a quantized electromagnetic field, and that of the electron and positron. Not all physicists felt this to be a disadvantage;

the electron field unlike the electromagnetic field did not have a classical limit, so there were doubts about its physical significance. Also, Dirac[42a] conceived of fields as the means by which we observe particles, so that he did not expect particles and fields to be described in the same terms. Though I do not know whether it bothered anyone at the time, there was a more practical disadvantage of the many-time formalism: it would have been difficult to use it to describe a process like nuclear beta decay, in which an electron and antineutrino are created without an accompanying positron or neutrino. The successful calculation by Fermi[43] of the electron energy distribution in beta decay deserves to be counted as one of the early triumphs of quantum field theory.

The essential idea that was needed to demonstrate the equivalence of the Dirac hole theory with a quantum field theory of the electron was provided by Fock[43a] and by Wendell Furry and Oppenheimer[44] in 1933–4. To appreciate this idea from a more modern standpoint, suppose we try to construct an electron field in analogy with the electromagnetic field or the Born–Heisenberg–Jordan field (1.2.2). Since electrons carry a charge, we would not like to mix annihilation and creation operators, so we might try to write the field as

$$\psi(x) = \sum_k u_k(\mathbf{x}) e^{-i\omega_k t} a_k \,, \tag{1.2.37}$$

where $u_k(\mathbf{x})e^{-i\omega_k t}$ are a complete set of orthonormal plane-wave solutions of the Dirac equation (1.1.13) (with k now labelling the three-momentum, spin, and sign of the energy):

$$\mathscr{H} u_k = \hbar\omega_k u_k \,, \tag{1.2.38}$$

$$\mathscr{H} \equiv -i\hbar c\boldsymbol{\alpha}\cdot\nabla + \alpha_4 mc^2 \,, \tag{1.2.39}$$

$$\int u_k^\dagger u_\ell \, d^3x = \delta_{k\ell} \,, \tag{1.2.40}$$

and a_k are the corresponding annihilation operators, satisfying the Jordan–Wigner anticommutation relations (1.2.22)–(1.2.23). According to the ideas of 'second quantization' or the canonical quantization procedure of Heisenberg and Pauli,[41] the Hamiltonian is formed by calculating the 'expectation value' of $\mathscr{H}$ with a 'wave function' replaced by the quantized field (1.2.37)

$$H = \int d^3x \, \psi^\dagger \mathscr{H} \psi = \sum_k \hbar\omega_k a_k^\dagger a_k \,. \tag{1.2.41}$$

The trouble is, of course, that this is not a positive operator — half the ω_k are negative while the operators $a_k^\dagger a_k$ take only the positive eigenvalues 1 and 0. (See Eqs. (1.2.24) and (1.2.25).) In order to cure this disease, Furry and Oppenheimer picked up Dirac's idea[42] that the positron is the absence of a negative-energy electron; the anticommutation relations are symmetric between creation and annihilation operators, so they defined the positron creation and annihilation operators as the corresponding annihilation and creation operators for negative-energy electrons

$$b_k^\dagger \equiv a_k\,, \qquad b_k \equiv a_k^\dagger \qquad (\text{for } \omega_k < 0) \tag{1.2.42}$$

where the label k on b denotes a positive-energy positron mode with momenta and spin opposite to those of the electron mode k. The Dirac field (1.2.37) may then be written

$$\psi(x) = \sum_k{}^{(+)} a_k u_k(x) + \sum_k{}^{(-)} b_k^\dagger u_k(x)\,, \tag{1.2.43}$$

where $(+)$ and $(-)$ indicate sums over normal modes k with $\omega_k > 0$ and $\omega_k < 0$, respectively, and $u_k(x) \equiv u_k(\mathbf{x})e^{-i\omega_k t}$. Similarly, using the anticommutation relations for the bs, we can rewrite the energy operator (1.2.41) as

$$H = \sum_k{}^{(+)} \hbar\omega_k a_k^\dagger a_k + \sum_k{}^{(-)} \hbar|\omega_k| b_k^\dagger b_k + E_0\,, \tag{1.2.44}$$

where E_0 is the infinite c-number

$$E_0 = -\sum_k{}^{(-)} \hbar|\omega_k|\,. \tag{1.2.45}$$

In order for this redefinition to be more than a mere formality, it is necessary also to specify that the physical vacuum is a state Ψ_0 containing no positive-energy electrons or positrons:

$$a_k\Psi_0 = 0 \qquad (\omega_k > 0)\,, \tag{1.2.46}$$

$$b_k\Psi_0 = 0 \qquad (\omega_k < 0)\,. \tag{1.2.47}$$

Hence (1.2.44) gives the energy of the vacuum as just E_0. If we measure all energies relative to the vacuum energy E_0, then the physical energy operator is $H - E_0$; and Eq. (1.2.44) shows that this is a *positive* operator.

The problem of negative-energy states for a charged spin zero particle was also resolved in 1934, by Pauli and Weisskopf,[45] in a paper written in part to challenge Dirac's picture of filled negative-energy states. Here the creation and annihilation operators satisfy commutation rather than anticommutation relations, so it is not possible to interchange the roles of these operators freely, as was the case for fermions. Instead we must return to the Heisenberg–Pauli canonical formalism[41] to decide which coefficients of the various normal modes are creation or annihilation operators.

Pauli and Weisskopf expanded the free charged scalar field in plane waves in a cube of spatial volume $V \equiv L^3$:

$$\phi(\mathbf{x}, t) = \frac{1}{\sqrt{V}} \sum_{\mathbf{k}} q(\mathbf{k}, t) e^{i\mathbf{k}\cdot\mathbf{x}} \tag{1.2.48}$$

with the wave numbers restricted by the periodicity condition, that the quantities $k_j L/2\pi$ for $j = 1, 2, 3$ should be a set of three positive or negative integers. Similarly the canonically conjugate variable (1.2.29) was expanded as

$$\pi(\mathbf{x}, t) \equiv \frac{1}{\sqrt{V}} \sum_{\mathbf{k}} p(\mathbf{k}, t) e^{-i\mathbf{k}\cdot\mathbf{x}} . \tag{1.2.49}$$

The minus sign is put into the exponent here so that (1.2.29) now becomes:

$$p(\mathbf{k}, t) = \dot{q}^\dagger(\mathbf{k}, t) . \tag{1.2.50}$$

The Fourier inversion formula gives

$$q(\mathbf{k}, t) = \frac{1}{\sqrt{V}} \int d^3x \, \phi(\mathbf{x}, t) e^{-i\mathbf{k}\cdot\mathbf{x}} , \tag{1.2.51}$$

$$p(\mathbf{k}, t) = \frac{1}{\sqrt{V}} \int d^3x \, \pi(\mathbf{x}, t) e^{+i\mathbf{k}\cdot\mathbf{x}} , \tag{1.2.52}$$

and therefore the canonical commutation relations (1.2.31)–(1.2.34) yield for the qs and ps:

$$\left[p(\mathbf{k}, t), q(\mathbf{l}, t)\right] = \frac{-i\hbar}{V} \int d^3x \, e^{i\mathbf{k}\cdot\mathbf{x}} e^{-i\mathbf{l}\cdot\mathbf{x}} = -i\hbar\delta_{\mathbf{kl}} \tag{1.2.53}$$

$$\begin{aligned} \left[p(\mathbf{k}, t), q^\dagger(\mathbf{l}, t)\right] &= \left[p(\mathbf{k}, t), p(\mathbf{l}, t)\right] = \left[p(\mathbf{k}, t), p^\dagger(\mathbf{l}, t)\right] \\ &= \left[q(\mathbf{k}, t), q(\mathbf{l}, t)\right] = \left[q(\mathbf{k}, t), q^\dagger(\mathbf{l}, t)\right] = 0 \end{aligned} \tag{1.2.54}$$

together with other relations that may be derived from these by taking their Hermitian adjoints. By inserting (1.2.48) and (1.2.49) in the formula (1.2.36) for the Hamiltonian, we can also write this operator in terms of ps and qs:

$$H = \sum_{\mathbf{k}} \left[p^\dagger(\mathbf{k}, t) p(\mathbf{k}, t) + \omega_{\mathbf{k}}^2 q^\dagger(\mathbf{k}, t) q(\mathbf{k}, t)\right] , \tag{1.2.55}$$

where

$$\omega_{\mathbf{k}}^2 \equiv c^2 \mathbf{k}^2 + \left(\frac{mc^2}{\hbar}\right)^2 . \tag{1.2.56}$$

The time-derivatives of the ps are then given by the Hamiltonian equation

$$\dot{p}(\mathbf{k}, t) = -\frac{\partial H}{\partial q(\mathbf{k}, t)} = -\omega_{\mathbf{k}}^2 q^\dagger(\mathbf{k}, t) \tag{1.2.57}$$

(and its adjoint), a result which in the light of Eq. (1.2.50) is just equivalent to the Klein–Gordon–Schrödinger wave equation (1.2.28).

We see that, just as in the case of the 1926 model of Born, Heisenberg, and Jordan,[4] the free field behaves like an infinite number of coupled harmonic oscillators. Pauli and Weisskopf could construct p and q operators which satisfy the commutation relations (1.2.53)–(1.2.54) and the 'equations of motion' (1.2.50) and (1.2.57), by introducing annihilation and creation operators a, b, $a^\dagger$, $b^\dagger$ of *two* different kinds, corresponding to particles and antiparticles:

$$q(\mathbf{k},t) = i\sqrt{\frac{\hbar}{2\omega_\mathbf{k}}}\left[a(\mathbf{k})\exp(-i\omega_\mathbf{k}t) - b^\dagger(\mathbf{k})\exp(i\omega_\mathbf{k}t)\right] \tag{1.2.58}$$

$$p(\mathbf{k},t) = \sqrt{\frac{\hbar\omega_\mathbf{k}}{2}}\left[b(\mathbf{k})\exp(-i\omega_\mathbf{k}t) + a^\dagger(\mathbf{k})\exp(+i\omega_\mathbf{k}t)\right] \tag{1.2.59}$$

where

$$\left[a(\mathbf{k}), a^\dagger(\mathbf{l})\right] = \left[b(\mathbf{k}), b^\dagger(\mathbf{l})\right] = \delta_{\mathbf{kl}}\,, \tag{1.2.60}$$

$$\left[a(\mathbf{k}), a(\mathbf{l})\right] = \left[b(\mathbf{k}), b(\mathbf{l})\right] = 0\,, \tag{1.2.61}$$

$$\begin{aligned}\left[a(\mathbf{k}), b(\mathbf{l})\right] &= \left[a(\mathbf{k}), b^\dagger(\mathbf{l})\right] = \left[a^\dagger(\mathbf{k}), b(\mathbf{l})\right] \\ &= \left[a^\dagger(\mathbf{k}), b^\dagger(\mathbf{l})\right] = 0\,.\end{aligned} \tag{1.2.62}$$

It is straightforward to check that these operators do satisfy the desired relations (1.2.53), (1.2.54), (1.2.50), and (1.2.57). The field (1.2.48) may be written

$$\begin{aligned}\phi(\mathbf{x},t) = \frac{i}{\sqrt{V}}\sum_\mathbf{k}\sqrt{\frac{\hbar}{2\omega_\mathbf{k}}}\,&[a(\mathbf{k})\exp(i\mathbf{k}\cdot\mathbf{x} - i\omega_\mathbf{k}t) \\ &- b^\dagger(-\mathbf{k})\exp(-i\mathbf{k}\cdot\mathbf{x} + i\omega_\mathbf{k}t)]\end{aligned} \tag{1.2.63}$$

and the Hamiltonian (1.2.55) takes the form

$$H = \sum_\mathbf{k}\frac{1}{2}\hbar\omega_\mathbf{k}\left[b^\dagger(\mathbf{k})b(\mathbf{k}) + b(\mathbf{k})b^\dagger(\mathbf{k}) + a^\dagger(\mathbf{k})a(\mathbf{k}) + a(\mathbf{k})a^\dagger(\mathbf{k})\right]$$

or, using (1.2.60)–(1.2.62)

$$H = \sum_\mathbf{k}\hbar\omega_\mathbf{k}\left[b^\dagger(\mathbf{k})b(\mathbf{k}) + a^\dagger(\mathbf{k})a(\mathbf{k})\right] + E_0\,, \tag{1.2.64}$$

where E_0 is the infinite c-number

$$E_0 \equiv \sum_\mathbf{k}\hbar\omega_\mathbf{k}\,. \tag{1.2.65}$$

The existence of two different kinds of operators a and b, which appear in precisely the same way in the Hamiltonian, shows that this is a theory with *two* kinds of particles with the same mass. As emphasized by Pauli and Weisskopf, these two varieties can be identified as particles and the corresponding antiparticles, and if charged have opposite charges. Thus, as we stressed above, bosons of spin zero as well as fermions of spin 1/2 can have distinct antiparticles, which for bosons cannot be identified as holes in a sea of negative energy particles.

We now can tell whether a and b or $a^\dagger$ and $b^\dagger$ are the annihilation operators by taking the expectation values of commutation relations in the vacuum state Ψ_0. For instance, if $a_{\mathbf{k}}^\dagger$ were an annihilation operator it would give zero when applied to the vacuum state, so the vacuum expectation value of (1.2.60) would give

$$-||a(\mathbf{k})\Psi_0||^2 = \left(\Psi_0, \left[a(\mathbf{k}), a^\dagger(\mathbf{k})\right]\Psi_0\right) = +1 \qquad (1.2.66)$$

in conflict with the requirement that the left-hand side must be negative-definite. In this way we can conclude that it is a_k and b_k that are the annihilation operators, and therefore

$$a(\mathbf{k})\Psi_0 = b(\mathbf{k})\Psi_0 = 0\,. \qquad (1.2.67)$$

This is consistent with all commutation relations. Thus, the canonical formalism forces the coefficient of the $e^{+i\omega t}$ in the field (1.2.58) to be a creation operator, as it also is in the Furry–Oppenheimer formalism[44] for spin 1/2.

Equations (1.2.64) and (1.2.67) now tell us that E_0 is the energy of the vacuum state. If we measure all energies relative to E_0, then the physical energy operator is $H - E_0$, and (1.2.64) shows that this again is positive.

What about the problem that served Dirac as a starting point, the problem of negative probabilities? As Dirac had recognized, the only probability density ρ, which can be formed from solutions of the Klein–Gordon–Schrödinger free scalar wave equation (1.2.28), and which satisfies a conservation law of the form (1.1.10), must be proportional to the quantity

$$\rho = 2\,\mathrm{Im}\left[\phi^\dagger \frac{\partial \phi}{\partial t}\right] \qquad (1.2.68)$$

and therefore is not necessarily a positive quantity. Similarly, in the 'second-quantized' theory, where ϕ is given by Eq. (1.2.63), ρ is not a positive operator. Since $\phi^\dagger(x)$ does not commute with $\dot{\phi}(x)$ here, we can write (1.2.68) in various forms, which differ by infinite c-numbers; it proves convenient to write it as

$$\rho = \frac{i}{\hbar}\left[\frac{\partial \phi}{\partial t}\phi^\dagger - \frac{\partial \phi^\dagger}{\partial t}\phi\right]. \qquad (1.2.69)$$

The space-integral of this operator is then easily calculated to be

$$N = \int \rho\, d^3x = \sum_{\mathbf{k}} \left(a^\dagger(\mathbf{k})a(\mathbf{k}) - b^\dagger(\mathbf{k})b(\mathbf{k})\right) \qquad (1.2.70)$$

and clearly has eigenvalues of either sign.

However, in a sense this problem appears in quantum field theory for spin 1/2 as well as spin zero. The density operator $\psi^\dagger\psi$ of Dirac is indeed a positive operator, but in order to construct a physical density we ought to subtract the contribution of the filled electron states. In particular, using the plane-wave decomposition (1.2.43), we may write the total number operator as

$$N \equiv \int d^3x\, \psi^\dagger\psi = \sum_{\mathbf{k}}{}^{(+)} a^\dagger(\mathbf{k})a(\mathbf{k}) + \sum_{\mathbf{k}}{}^{(-)} b(\mathbf{k})b^\dagger(\mathbf{k})\,.$$

The anticommutation relations for the bs allow us to rewrite this as

$$N - N_0 = \sum_{\mathbf{k}}{}^{(+)} a^\dagger_{\mathbf{k}}a_{\mathbf{k}} - \sum_{\mathbf{k}}{}^{(-)} b^\dagger_{\mathbf{k}}b_{\mathbf{k}}\,, \tag{1.2.71}$$

where N_0 is the infinite constant

$$N_0 = \sum_{\mathbf{k}}{}^{(-)} 1\,. \tag{1.2.72}$$

According to Eqs. (1.2.46) and (1.2.47), N_0 is the number of particles in the vacuum, so Furry and Oppenheimer reasoned that the physical number operator is $N - N_0$, and this now has both negative and positive eigenvalues, just as for spin zero.

The solution to this problem provided by quantum field theory is that neither the ψ of Furry and Oppenheimer nor the ϕ of Pauli and Weisskopf are probability amplitudes, which would have to define conserved positive probability densities. Instead, the physical Hilbert space is spanned by states defined as containing definite numbers of particles and/or antiparticles in each mode. If Φ_n are a complete orthonormal set of such states, then a measurement of particle numbers in an arbitrary state Ψ will yield a probability for finding the system in state Φ_n, given by

$$P_n = |(\Phi_n, \Psi)|^2\,, \tag{1.2.73}$$

where (Φ_n, Ψ) is the usual Hilbert space scalar product. Hence, no question as to the possibility of negative probabilities will arise for any spin. The wave fields ϕ, ψ, etc, are not probability amplitudes at all, but *operators* which create or destroy particles in the various normal modes. It would be a good thing if the misleading expression 'second quantization' were permanently retired.

In particular, the operators N and $N - N_0$ of Eqs. (1.2.70) and (1.2.71) are not to be interpreted as total probabilities, but as number operators: specifically, the number of particles *minus* the number of antiparticles. For charged particles, the conservation of charge forces the charge operators to be proportional to these number operators, so the minus signs in (1.2.70) and (1.2.71) allow us immediately to conclude that particles and antiparticles have opposite charge. In this field-theoretic formalism, interactions contribute terms to the Hamiltonian which are of third, fourth, or higher order in field variables, and the rates of various processes are given by using these interaction operators in a time-dependent perturbation theory. The conceptual framework described in the above brief remarks will serve as the basis for much of the work in this book.

Despite its apparent advantages, quantum field theory did not immediately supplant hole theory; rather, the two points of view coexisted for a while, and various combinations of field-theoretic and hole-theoretic ideas were used in calculations of physical reaction rates. This period saw a number of calculations of cross sections to lowest order in powers of e^2 for various processes, such as $e^- + \gamma \to e^- + \gamma$ in 1929 by Klein and Nishina;[46] $e^+ + e^- \to 2\gamma$ in 1930 by Dirac;[47] $e^- + e^- \to e^- + e^-$ in 1932 by Møller;[48] $e^- + Z \to e^- + \gamma + Z$ and $\gamma + Z \to e^+ + e^- + Z$ (where Z denotes the Coulomb field of a heavy atom) in 1934 by Bethe and Heitler;[49] and $e^+ + e^- \to e^+ + e^-$ in 1936 by Bhabha.[50] (Rules for the calculation of such processes are given in Chapter 8, and worked out in detail there for the case of electron–photon scattering.) These lowest-order calculations gave finite results, in reasonable agreement with the experimental data.

Nevertheless, a general feeling of dissatisfaction with quantum field theory (whether or not in the form of hole theory) persisted throughout the 1930s. One of the reasons for this was the apparent failure of quantum electrodynamics to account for the penetrating power of the charged particles in cosmic ray showers, noted in 1936 by Oppenheimer and Franklin Carlson.[50a] Another cause of dissatisfaction that turned out to be related to the first was the steady discovery of new kinds of particles and interactions. We have already mentioned the electron, photon, positron, neutrino, and, of course, the nucleus of hydrogen, the proton. Throughout the 1920s it was generally believed that heavier nuclei are composed of protons and electrons, but it was hard to see how a light particle like the electron could be confined in the nucleus. Another severe difficulty with this picture was pointed out in 1931 by Ehrenfest and Oppenheimer:[51] the nucleus of ordinary nitrogen, N^{14}, in order to have atomic number 7 and atomic weight 14, would have to be composed of 14 protons and 7 electrons, and would therefore have to be a fermion, in conflict with the result of molecular spectroscopy[52] that N^{14} is a boson. This problem (and others) were solved in 1932 with the discovery of the neutron,[53] and by Heisenberg's subsequent suggestion[54] that nuclei are composed of protons and neutrons, not protons and electrons. It was clear that a strong non-electromagnetic force of short range would have to operate between neutrons and protons to hold nuclei together.

After the success of the Fermi theory of beta decay, several authors[54a] speculated that nuclear forces might be explained in this theory as due to the exchange of electrons and neutrinos. A few years later, in 1935,

Hideki Yukawa proposed a quite different quantum field theory of the nuclear force.[55] In an essentially classical calculation, he found that the interaction of a scalar field with nucleons (protons or neutrons) would produce a nucleon–nucleon potential, with a dependence on the nucleon separation r given by

$$V(r) \propto \frac{1}{r} \exp(-\lambda r) \tag{1.2.74}$$

instead of the $1/r$ Coulomb potential produced by electric fields. The quantity λ was introduced as a parameter in Yukawa's scalar field equation, and when this equation was quantized, Yukawa found that it described particles of mass $\hbar\lambda/c$. The observed range of the strong interactions within nuclei led Yukawa to estimate that $\hbar\lambda/c$ is of the order of 200 electron masses. In 1937 such 'mesons' were discovered in cloud chamber experiments[56] by Seth Neddermeyer and Anderson and by Jabez Curry Street and Edward Carl Stevenson, and it was generally believed that these were the hypothesized particles of Yukawa.

The discovery of mesons revealed that the charged particles in cosmic ray showers are not all electrons, and thus cleared up the problem with these showers that had bothered Oppenheimer and Carlson. At the same time, however, it created new difficulties. Lothar Nordheim[56a] pointed out in 1939 that the same strong interactions by which the mesons are copiously produced at high altitudes (and which are required in Yukawa's theory) should have led to the mesons' absorption in the atmosphere, a result contradicted by their copious appearance at lower altitudes. In 1947 it was shown in an experiment by Marcello Conversi, Ettore Pancini, and Oreste Piccioni[57] that the mesons which predominate in cosmic rays at low altitude actually interact weakly with nucleons, and therefore could not be identified with Yukawa's particle. This puzzle was cleared up by a theoretical suggestion,[58] and its subsequent experimental confirmation[59] by Cesare Lattes, Occhialini, and Cecil Powell — there are two kinds of mesons with slightly different masses: the heavier (now called the π meson or pion) has strong interactions and plays the role in nuclear force envisaged by Yukawa; the lighter (now called the μ meson, or muon) has only weak and electromagnetic interactions, and predominates in cosmic rays at sea level, being produced by the decay of π mesons. In the same year, 1947, entirely new kinds of particles (now known as K mesons and hyperons) were found in cosmic rays by George Rochester and Clifford Butler.[60] From 1947 until the present particles have continued to be discovered in a bewildering variety, but to pursue this story would take us outside the bounds of our present survey. These discoveries showed clearly that any conceptual framework which was limited to photons, electrons, and positrons would be far too narrow to be taken seriously as a fundamental theory. But an even more important obstacle was presented by a purely theoretical problem — the problem of infinities.

1.3 The Problem of Infinities

Quantum field theory deals with fields $\psi(x)$ that destroy and create particles at a spacetime point x. Earlier experience with classical electron theory provided a warning that a point electron will have infinite electromagnetic self-mass; this mass is $e^2/6\pi ac^2$ for a surface distribution of charge with radius a, and therefore blows up for $a \to 0$. Disappointingly this problem appeared with even greater severity in the early days of quantum field theory, and although greatly ameliorated by subsequent improvements in the theory, it remains with us to the present day.

The problem of infinities in quantum field theory was apparently first noted in the 1929–30 papers of Heisenberg and Pauli.[41] Soon after, the presence of infinities was confirmed in calculations of the electromagnetic self-energy of a bound electron by Oppenheimer,[61] and of a free electron by Ivar Waller.[62] They used ordinary second-order perturbation theory, with an intermediate state consisting of an electron and a photon: for instance, the shift of the energy E_n of an electron in the nth energy level of hydrogen is given by

$$\Delta E_n = \sum_{m,\lambda} \int d^3k \, \frac{| < m; \mathbf{k}, \lambda | H' | n > |^2}{E_n - E_m - |\mathbf{k}|c} \,, \tag{1.3.1}$$

where the sums and integral are over all intermediate electron states m, photon helicities λ, and photon momenta $\mathbf{k}$, and H' is the term in the Hamiltonian representing the interaction of radiation and electrons. This calculation gave a self-energy that is formally infinite; further; if this infinity is removed by discarding all intermediate states with photon wave numbers greater than $1/a$, then the self-energy behaves like $1/a^2$ as $a \to 0$. Infinities of this sort are often called ultraviolet divergences, because they arise from intermediate states containing particles of very short wavelength.

These calculations treated the electron according to the rules of the original Dirac theory, without filled negative-electron states. A few years later Weisskopf repeated the calculation of the electron self-mass in the new hole theory, with all negative-energy states full. In this case another term appears in second-order perturbation theory, which in a non-hole-theory language can be described as arising from processes in which the electron in its final state first appears out of the vacuum together with a photon and a positron which then annihilate along with the initial electron. Initially Weisskopf found a $1/a^2$ dependence on the photon wave-number cutoff $1/a$. The same calculation was being carried out (at the suggestion of Bohr) at that time by Carlson and Furry. After seeing Weisskopf's results, Furry realized that while Weisskopf had included an electrostatic term that he and Carlson had neglected, Weisskopf had made a new mistake in the calculation of the magnetic self-energy. After hearing from Furry and correcting his own error, Weisskopf found that the $1/a^2$ terms in the total mass shift cancelled! However, despite this cancellation, an infinity remained: with a wave-number cutoff $1/a$, the self-mass was found to be[63]

$$m_{em} = \frac{3\alpha}{2\pi} m \ln\left(\frac{\hbar}{mca}\right) , \qquad (1.3.2)$$

The weakening of the cut-off dependence, to $\ln a$ as compared with the classical $1/a$ or the early quantum $1/a^2$, was mildly encouraging at the time and turned out to be of great importance later, in the development of renormalization theory.

An infinity of quite a different kind was encountered in 1933, apparently first by Dirac.[64] He considered the effect of an external static nearly uniform charge density $\varepsilon(\mathbf{x})$ on the vacuum, i.e., on the negative-energy electrons in the filled energy levels of hole theory. The Coulomb interaction between $\varepsilon(\mathbf{x})$ and the charge density of the negative-energy electrons produces a 'vacuum polarization,' with induced charge density

$$\delta\varepsilon = A\varepsilon + B\left(\frac{\hbar}{mc}\right)^2 \nabla^2\varepsilon + \cdots . \qquad (1.3.3)$$

The constant B is finite, and of order α. On the other hand, A is logarithmically divergent, of order $\alpha \ln a$, where $1/a$ is the wave-number cutoff.

Infinities also seemed to occur in a related problem, the scattering of light by light. Hans Euler, Bernard Kockel, and Heisenberg[65] showed in 1935–6 that these infinities could be eliminated by using a more-or-less arbitrary prescription suggested earlier by Dirac[66] and Heisenberg[67]. They calculated an effective Lagrangian density for the non-linear electrodynamic effects produced by virtual electron–positron pairs:

$$\mathscr{L} = \frac{1}{2}\left(\mathbf{E}^2 - \mathbf{B}^2\right) + \frac{e^4\hbar}{360\pi^2 m_e^4 c^7}\left[\left(\mathbf{E}^2 - \mathbf{B}^2\right)^2 + 7\left(\mathbf{E}\cdot\mathbf{B}\right)^2\right] + \cdots , \qquad (1.3.4)$$

valid for frequencies $\nu \ll m_e c^2/h$. Soon after, Nicholas Kemmer and Weisskopf[68] presented arguments that in this case the infinities are spurious, and that Eq. (1.3.4) can be derived without any subtraction prescription.

One bright spot in the struggle with infinities was the successful treatment of *infrared* divergences, those that arise from the low-energy rather than the high-energy part of the range of integration. In 1937 it was shown by Felix Bloch and Arne Nordsieck[68a] that these infinities cancel provided one includes processes in which arbitrary numbers of low-energy photons are produced. This will be discussed in modern terms in Chapter 13.

Yet another infinity turned up in a calculation by Sidney Michael Dancoff[69] in 1939 of the radiative corrections to the scattering of electrons by the static Coulomb field of an atom. The calculation contained a mistake (one of the terms was omitted), but this was not realized until later.[69a]

Throughout the 1930s, these various infinities were seen not merely as failures of specific calculations. Rather, they seemed to indicate a gap in the understanding of relativistic quantum field theory on the most fundamental level, an opinion reinforced by the problems with cosmic rays mentioned in the previous section.

One of the symptoms of this uneasy pessimism was the continued exploration throughout the 1930s and 1940s of alternative formalisms. As Julian Schwinger[69b] later recalled, 'The preoccupation of the majority of involved physicists was not with analyzing and carefully applying the known relativistic theory of coupled electron and electromagnetic fields but with changing it.' Thus in 1938 Heisenberg[70] proposed the existence of a fundamental length L, analogous to the fundamental action h and fundamental velocity c. Field theory was supposed to work only for distances larger than L, so that all divergent integrals would effectively be cut off at distances L, or momenta h/L. Several specific proposals[70a] were made for giving field theory a non-local structure. Some theorists began to suspect that the formalism of state-vectors and quantum fields should be replaced by one based solely on observable quantities, such as the S-matrix introduced by John Archibald Wheeler[71] in 1937 and Heisenberg[72] in 1943, whose elements are the amplitudes for various scattering processes. As we shall see, the concept of the S-matrix has now become a vital part of modern quantum field theory, and for some theorists a pure S-matrix theory became an ideal, especially as a possible solution to the problems of the strong interactions.[73] In yet another direction, Wheeler and Richard Feynman[74] in 1945 attempted to eliminate the electromagnetic field, deriving electromagnetic interactions in terms of an interaction at a distance. They were able to show that a pure retarded (or pure advanced) potential could be obtained by taking into account the interaction not only between source and test charges, but also between these charges and all the other charges in the universe. Perhaps the most radical modification of quantum mechanics suggested during this period was the introduction by Dirac[75] of states of negative probability, as a means of cancelling infinities in sums over states. This idea, of an 'indefinite metric' in Hilbert space, has also flourished in quantum field theory, though not in the form originally suggested.

A more conservative idea for dealing with the infinities was also in the air during the 1930s. Perhaps these infinities could all be absorbed into a redefinition, a 'renormalization' of the parameters of the theory. For instance, it was already known that in any Lorentz-invariant classical theory the electromagnetic self-energy and self-momentum of an electron *must* take the form of corrections to the mass of the electron; hence the infinities in these quantities can be cancelled by a negative infinity in the 'bare' non-electromagnetic mass of the electron, leaving a finite measurable 'renormalized' mass. Also, Eq. (1.3.3) shows that the vacuum polarization changes the charge of the electron, from $e \equiv \int d^3x\, \varepsilon$, to

$$e_{\rm TOTAL} = \int d^3x(\varepsilon + \delta\varepsilon) = (1 + A)e\,. \tag{1.3.5}$$

Vacuum polarization gives finite results in lowest order if observables like scattering cross-sections are expressed in terms of $e_{\rm TOTAL}$ rather than e. The question was, whether all infinities in quantum field theory could be dealt with in this way. In 1936 Weisskopf[76] suggested that this is the case, and verified that known infinities could be eliminated by renormalization of physical parameters in a variety of sample calculations. However, it was impossible with the calculational techniques then available to show that infinities could always be eliminated in this way, and Dancoff's calculation[69] seemed to show that they could not.

Another effect of the appearance of infinities was a tendency to believe that any effect which turned out to be infinite in quantum field theory was actually not there at all. In particular, the 1928 Dirac theory had predicted complete degeneracy of the $2s_{1/2}$–$2p_{1/2}$ levels of hydrogen to all orders in α; any attempt at a quantum electromagnetic calculation of the splitting of these two levels ran into the problem of the infinite self-energy of a bound electron; therefore the existence of such a splitting was generally not taken seriously. Later Bethe[80] recalled that 'This shift comes out infinite in all existing theories, and has therefore always been ignored.' This attitude persisted even in the late 1930s, when spectroscopic experiments[77] began to indicate the presence of a $2s_{1/2}$–$2p_{1/2}$ splitting of order 1000 MHz. One notable exception was Edwin Albrecht Uehling,[78] who realized that the vacuum polarization effect mentioned earlier would produce a $2s_{1/2}$–$2p_{1/2}$ splitting; unfortunately, as we shall see in Chapter 14, this contribution to the splitting is much smaller than 1000 MHz, and of the wrong sign.

The gloom surrounding quantum field theory began to lift soon after World War II. On June 1–4, 1947, the Conference on the Foundations of

Quantum Mechanics at Shelter Island, NY brought theoretical physicists who had been working on the problems of quantum field theory through the 1930s together with a younger generation of theorists who had started scientific work during the war, and — of crucial importance — a few experimental physicists. The discussion leaders were Hans Kramers, Oppenheimer, and Weisskopf. One of the experimentalists (or rather theorist turned experimentalist), Willis Lamb, described a decisive measurement[79] of the $2s_{1/2}$–$2p_{1/2}$ shift in hydrogen. A beam of hydrogen atoms from an oven, many in $2s$ and $2p$ states, was aimed at a detector sensitive only to atoms in excited states. The atoms in $2p$ states can decay very rapidly to the $1s$ ground state by one-photon (Lyman α) emission, while the $2s$ states decay only very slowly by two-photon emission, so in effect the detector was measuring the number of atoms in the metastable $2s$ state. The beam was passed through a magnetic field, which added a known Zeeman splitting to any $2s_{1/2}$–$2p_{1/2}$ splitting naturally present. The beam was also exposed to a microwave-frequency electromagnetic field, with a fixed frequency $\nu \sim 10$ GHz. At a certain magnetic field strength the detector signal was observed to be quenched, indicating that the microwave field was producing resonant transitions from the metastable $2s$ state to the $2p$ state and thence by a rapid Lyman α emission to the ground state. The total (Zeeman plus intrinsic) $2s$–$2p$ splitting at this value of the magnetic field strength would have to be just $h\nu$, from which the intrinsic splitting could be inferred. A preliminary value of 1000 MHz was announced, in agreement with the earlier spectroscopic measurements.[77] The impact of this discovery can be summarized in a saying that was current in Copenhagen when I was a graduate student there in 1954: 'Just because something is infinite does not mean it is zero!'

The discovery of the Lamb shift aroused intense interest among the theorists at Shelter Island, many of whom had already been working on improved formalisms for calculation in quantum electrodynamics. Kramers described his work on mass renormalization in the classical electrodynamics of an extended electron,[79a] which showed that the difficulties associated with the divergence of the self-energy in the limit of zero radius do not appear explicitly if the theory is reexpressed so that the mass parameter in the formalism is identified with the experimental electron mass. Schwinger and Weisskopf (who had already heard rumors of Lamb's result, and discussed the matter on the trip to Shelter Island) suggested that since the inclusion of intermediate states involving positrons was known to reduce the divergence in energy level shifts from $1/a^2$ to $\ln a$, perhaps the *differences* of the shifts in atomic energy levels might turn out to be finite when these intermediate states were taken into account. (In fact, in 1946, before he learned of Lamb's experiment, Weisskopf had already assigned this problem to a graduate student, Bruce French.) Almost immediately after the conference, during a train ride to Schenectady, Hans Bethe[80] carried out a non-relativistic calculation, still without including the effects of intermediate states containing positrons, but using a simple cutoff at virtual photon momenta of order $m_e c^2$ to eliminate infinities. He obtained the encouraging approximate value of 1040 MHz. Fully relativistic calculations using the renormalization idea to eliminate infinities were soon thereafter carried out by a number of other authors,[81] with excellent agreement with experiment.

Another exciting experimental result was reported at Shelter Island by Isidor I. Rabi. Measurements in his laboratory of the hyperfine structure of hydrogen and deuterium had suggested[82] that the magnetic moment of the electron is larger than the Dirac value $e\hbar/2mc$ by a factor of about 1.0013, and subsequent measurements of the gyromagnetic ratios in sodium and gallium had given a precise value[83]

$$\mu = \frac{e\hbar}{2mc}\,[1.00118 \pm 0.00003]\,.$$

Learning of these results, Gregory Breit suggested[83a] that they arose from an order α radiative correction to the electron magnetic moment. At Shelter Island, both Breit and Schwinger described their efforts to calculate this correction. Shortly after the conference Schwinger completed a successful calculation of the anomalous magnetic moment of the electron[84]

$$\mu = \frac{e\hbar}{2mc}\left[1 + \frac{\alpha}{2\pi}\right] = \frac{e\hbar}{2mc}\,[1.001162]$$

in excellent agreement with observation. This, together with Bethe's calculation of the Lamb shift, at last convinced physicists of the reality of radiative corrections.

The mathematical methods used in this period presented a bewildering variety of concepts and formalisms. One approach developed by Schwinger[85] was based on operator methods and the action principle, and was presented by him at a conference at Pocono Manor in 1948, the successor to the Shelter Island Conference. Another Lorentz-invariant operator formalism had been developed earlier by Sin-Itiro Tomonaga[86] and his co-workers in Japan, but their work was not at first known in the West. Tomonaga had grappled with infinities in Yukawa's meson theory in the 1930s. In 1947 he and his group were still out of the loop of scientific communication; they learned about Lamb's experiment from an article in *Newsweek*.

An apparently quite different approach was invented by Feynman,[87] and described briefly by him at the Pocono Conference. Instead of introducing quantum field operators, Feynman represented the S-matrix as a functional integral of $\exp\left(iW\right)$, where W is the action integral for a set of Dirac particles interacting with a *classical* electromagnetic field, integrated over all Dirac particle trajectories satisfying certain initial and final conditions for $t \to \pm\infty$. One result of great practical importance that came out of Feynman's work was a set of graphical rules for calculating S-matrix elements to any desired order of perturbation theory. Unlike the old perturbation theory of the 1920s and 1930s, these Feynman rules automatically lumped together particle creation and antiparticle annihilation processes, and thereby gave results that were Lorentz-invariant at every stage. We have already seen in Weisskopf's early calculation[63] of the electron self-energy, that it is only in such calculations, including particles and antiparticles on the same footing, that the nature of the infinities becomes transparent.

Finally, in a pair of papers in 1949, Freeman Dyson[88] showed that the operator formalisms of Schwinger and Tomonaga would yield the same graphical rules that had been found by Feynman. Dyson also carried out an analysis of the infinities in general Feynman diagrams, and outlined a proof that these are always precisely the sort which could be removed by renormalization. One of the most striking results that could be inferred from Dyson's analysis was a criterion for deciding which quantum field theories are 'renormalizable', in the sense that all infinities can be absorbed into a redefinition of a *finite* number of coupling constants and masses. In particular, an interaction like the Pauli term (1.1.32), which would have changed the predicted magnetic moment of the electron, would spoil the renormalizability of quantum electrodynamics. With the publication of Dyson's papers, there was at last a general and systematic formalism that physicists could easily learn to use, and that would provide a common language for the subsequent applications of quantum field theory to the problems of physics.

I cannot leave the infinities without taking up a puzzling aspect of this story. Oppenheimer[61] in 1930 had already noticed that most of the ultraviolet divergence in the self-energy of a bound electron cancels when one takes the difference between the shifts of two atomic energy levels, and Weisskopf[63] in 1934 had found that most of the divergence in the self-energy of a free electron cancels when one includes intermediate states containing positrons. It would have been natural even in 1934 to guess that including positron intermediate states *and* subtracting the energy shifts of pairs of atomic states would eliminate the ultraviolet divergence in their relative energy shift.* There was even experimental evidence[77] for

* In fact, this guess would have been wrong. As discussed in Section 14.3, radiative corrections to the electron mass affect atomic energy levels not only through a shift in the electron rest energy, which is the same in all atomic energy levels, but also through a change in the electron kinetic energy, that varies from one level to another.

a $2s_{1/2}$–$2p_{1/2}$ energy difference of order 1000 MHz. So why did no one before 1947 attempt an *numerical* estimate of this energy difference?

Strictly speaking, there was one such attempt[88a] in 1939, but it focused on the wrong part of the problem, the charge radius of the *proton*, which has only a tiny effect on hydrogen energy levels. The calculation gave a result in rough agreement with the early experiments.[77] This was a mistake, as shown in 1939 by Lamb.[88b]

A fully relativistic calculation of the Lamb shift including positrons in intermediate states could have been attempted during the 1930s, using the old non-relativistic perturbation theory. As long as one keeps all terms up to a given order, old-fashioned non-relativistic perturbation theory gives the same results as the manifestly relativistic formalisms of Feynman, Schwinger, and Tomonaga. In fact, after Bethe's work, the first precise calculations[81] of the Lamb shift in the USA by French and Weisskopf and Norman Kroll and Lamb were done in just this way, though Tomonoga's group [81] in Japan was already using covariant methods to solve this and other problems.

The one missing element was confidence in renormalization as a means of dealing with infinities. As we have seen, renormalization was widely discussed in the late 1930s. But it had become accepted wisdom in the 1930s, and a point of view especially urged by Oppenheimer,[89] that quantum electrodynamics could not be taken seriously at energies of more than about 100 MeV, and that the solution to its problems could be found only in really adventurous new ideas.

Several things happened at Shelter Island to change this expectation. One was news that the problems concerning cosmic rays discussed in the previous section were beginning to be resolved; Robert Marshak presented the hypothesis[58] that there were two types of 'meson' with similar masses; the muons that had actually been observed, and the pions responsible for nuclear forces. More important was the fact that now there were reliable experimental values for the Lamb shift and the anomalous magnetic moment that forced physicists to think carefully about radiative corrections. Probably equally important was the fact that the conference brought together theorists who had in their own individual ways been thinking about renormalization as a solution to the problem of infinities. When the revolution came in the late 1940s, it was made by physicists who though mostly young were playing a conservative role, turning away from the search by their predecessors for a radical solution.

Bibliography

- □ S. Aramaki, 'Development of the Renormalization Theory in Quantum Electrodynamics,' *Historia Scientiarum* **36**, 97 (1989); *ibid.* **37**, 91 (1989). [Section 1.3.]
- □ R. T. Beyer, ed., *Foundations of Nuclear Physics* (Dover Publications, Inc., New York, 1949). [Section 1.2.]
- □ L. Brown, 'Yukawa's Prediction of the Meson,' *Centauros* **25**, 71 (1981). [Section 1.2.]
- □ L. M. Brown and L. Hoddeson, eds., *The Birth of Particle Physics* (Cambridge University Press, Cambridge, 1983). [Sections 1.1, 1.2, 1.3.]
- □ T. Y. Cao and S. S. Schweber, 'The Conceptual Foundations and the Philosophical Aspects of Renormalization Theory,' *Synthèse* **97**, 33 (1993). [Section 1.3.]
- □ P. A. M. Dirac, *The Development of Quantum Theory* (Gordon and Breach Science Publishers, New York, 1971). [Section 1.1.]
- □ E. Fermi, 'Quantum Theory of Radiation,' *Rev. Mod. Phys.* **4**, 87 (1932). [Sections 1.2 and 1.3.]
- □ G. Gamow, *Thirty Years that Shook Physics* (Doubleday and Co., Garden City, New York, 1966). [Section 1.1.]
- □ M. Jammer, *The Conceptual Development of Quantum Mechanics* (McGraw-Hill Book Co., New York, 1966). [Section 1.1.]
- □ J. Mehra, 'The Golden Age of Theoretical Physics: P.A.M. Dirac's Scientific Work from 1924 to 1933,' in *Aspects of Quantum Theory*, ed. by A. Salam and E. P. Wigner, (Cambridge University Press, Cambridge, 1972). [Section 1.1.]
- □ A. I. Miller, *Early Quantum Electrodynamics – A Source Book* (Cambridge University Press, Cambridge, UK, 1994). [Sections 1.1, 1.2, 1.3.]
- □ A. Pais, *Inward Bound* (Clarendon Press, Oxford, 1986). [Sections 1.1, 1.2, 1.3.]
- □ S. S. Schweber, 'Feynman and the Visualization of Space-Time Processes,' *Rev. Mod. Phys.* **58**, 449 (1986). [Section 1.3.]
- □ S. S. Schweber, 'Some Chapters for a History of Quantum Field Theory: 1938–1952,' in *Relativity, Groups, and Topology II*, ed. by B. S. DeWitt and R. Stora (North-Holland, Amsterdam, 1984). [Sections 1.1, 1.2, 1.3.]

□ S. S. Schweber, 'A Short History of Shelter Island I,' in *Shelter Island II*, ed. by R. Jackiw, S. Weinberg, and E. Witten (MIT Press, Cambridge, MA, 1985). [Section 1.3.]

□ S. S. Schweber, *QED and the Men Who Made It: Dyson, Feynman, Schwinger, and Tomonaga* (Princeton University Press, Princeton, 1994). [Section 1.1, 1.2, 1.3.]

□ J. Schwinger, ed., *Selected Papers in Quantum Electrodynamics* (Dover Publications Inc., New York, 1958). [Sections 1.2 and 1.3.]

□ S.-I. Tomonaga, in *The Physicist's Conception of Nature* (Reidel, Dordrecht, 1973). [Sections 1.2 and 1.3.]

□ S. Weinberg, 'The Search for Unity: Notes for a History of Quantum Field Theory,' *Daedalus*, Fall 1977. [Sections 1.1, 1.2, 1.3.]

□ V. F. Weisskopf, 'Growing Up with Field Theory: The Development of Quantum Electrodynamics in Half a Century,' 1979 Bernard Gregory Lecture at CERN, published in L. Brown and L. Hoddeson, *op. cit.*. [Sections 1.1, 1.2, 1.3.]

□ G. Wentzel, 'Quantum Theory of Fields (Until 1947)' in *Theoretical Physics in the Twentieth Century*, ed. by M. Fierz and V. F. Weisskopf (Interscience Publishers Inc., New York, 1960). [Sections 1.2 and 1.3.]

□ E. Whittaker, *A History of the Theories of Aether and Electricity* (Humanities Press, New York, 1973). [Section 1.1.]

References

1. L. de Broglie, *Comptes Rendus* **177**, 507, 548, 630 (1923); *Nature* **112**, 540 (1923); Thèse de doctorat (Masson et Cie, Paris, 1924); *Annales de Physique* **3**, 22 (1925) [reprinted in English in *Wave Mechanics*, ed. by G. Ludwig, (Pergamon Press, New York, 1968)]; *Phil. Mag.* **47**, 446 (1924).

2. W. Elsasser, *Naturwiss.* **13**, 711 (1925).

3. C. J. Davisson and L. H. Germer, *Phys. Rev.* **30**, 705 (1927).

4. W. Heisenberg, *A. Phys.* **33**, 879 (1925); M. Born and P. Jordan, *Z. f. Phys.* **34**, 858 (1925); P. A. M. Dirac, *Proc. Roy. Soc.* **A109**, 642 (1925); M. Born, W. Heisenberg, and P. Jordan, *Z. f. Phys.* **35**, 557 (1926); W. Pauli, *Z. f. Phys.* **36**, 336 (1926). These papers are reprinted in *Sources of Quantum Mechanics*, ed. by B. L. van der Waerden (Dover Publications, Inc., New York, 1968).

5. E. Schrödinger, *Ann. Phys.* **79**, 361, 489; **80**, 437; **81**, 109 (1926). These papers are reprinted in English, unfortunately in a somewhat abridged form, in *Wave Mechanics*, Ref. 1. Also see *Collected Papers on Wave Mechanics*, trans. by J. F. Schearer and W. M. Deans (Blackie and Son, London, 1928).

6. See, e.g., P. A. M. Dirac, *The Development of Quantum Theory* (Gordon and Breach, New York, 1971). Also see Dirac's obituary of Schrödinger, *Nature* **189**, 355 (1961), and his article in *Scientific American* **208**, 45 (1963).

7. O. Klein, *Z. f. Phys.* **37**, 895 (1926). Also see V. Fock, *Z. f. Phys.* **38**, 242 (1926); *ibid*, **39**, 226 (1926).

8. W. Gordon, *Z. f. Phys.* **40**, 117 (1926).

9. For the details of the calculation, see, e.g., L. I. Schiff, *Quantum Mechanics*, 3rd edn, (McGraw-Hill, Inc. New York, 1968): Section 51.

10. F. Paschen, *Ann. Phys.* **50**, 901 (1916). These experiments were actually carried out using He^+ because its fine structure splitting is 16 times larger than for hydrogen. The fine structure of spectral lines was first discovered interferometrically by A. A. Michelson, *Phil. Mag.* **31**, 338 (1891); *ibid.*, **34**, 280 (1892).

10*a*. A. Sommerfeld, *Münchner Berichte* 1915, pp. 425, 429; *Ann. Phys.* **51**, 1, 125 (1916). Also see W. Wilson, *Phil. Mag.* **29**, 795 (1915).

11. G. E. Uhlenbeck and S. Goudsmit, *Naturwiss.* **13**, 953 (1925); *Nature* **117**, 264 (1926). The electron spin had been earlier suggested for other reasons by A. H. Compton, *J. Frank. Inst.* **192**, 145 (1921).

12. The general formula for Zeeman splitting in one-electron atoms had been discovered empirically by A. Landé, *Z. f. Phys.* **5**, 231 (1921); *ibid.*, **7**, 398 (1921); *ibid.*, **15**, 189 (1923); *ibid.*, **19**, 112 (1923). At the time, the extra non-orbital angular momentum appearing in this formula was thought to be the angular momentum of the atomic 'core;' A. Sommerfeld, *Ann. Phys.* **63**, 221 (1920); *ibid.*, **70**, 32 (1923). It was only later that the extra angular momentum was recognized, as in Ref. 11, to be due to the spin of the electron.

13. W. Heisenberg and P. Jordan, *Z. f. Phys.* **37**, 263 (1926); C. G. Darwin, *Proc. Roy. Soc.* **A116**, 227 (1927). Darwin says that several authors did this work at about the same time, while Dirac quotes only Darwin.

14. L. H. Thomas, *Nature* **117**, 514 (1926). Also see S. Weinberg, *Gravitation and Cosmology*, (Wiley, New York, 1972): Section 5.1.

15. P. A. M. Dirac, *Proc. Roy. Soc.* **A117**, 610 (1928). Also see Dirac, *ibid.*, **A118**, 351 (1928), for the application of this theory to the calculation of the Zeeman and Paschen–Back effects and the relative strengths of lines within fine-structure multiplets.

16. For the probabilistic interpretation of non-relativistic quantum mechanics, see M. Born *Z. f. Phys.* **37**, 863 (1926); *ibid*, **38**, 803 (1926) (reprinted in an abridged English version in *Wave Mechanics*, Ref. 1); G. Wentzel, *Z. f. Phys.* **40**, 590 (1926); W. Heisenberg, *Z. f. Phys.* **43**, 172 (1927). N. Bohr. *Nature* **121**, 580 (1928); Naturwissenchaften **17**, 483 (1929); *Electrons et Photons – Rapports et Discussions du Ve Conseil de Physique Solvay* (Gauthier-Villars, Paris, 1928).

17. Conversation between Dirac and J. Mehra, March 28, 1969, quoted by Mehra in *Aspects of Quantum Theory*, ed. by A. Salam and E. P. Wigner (Cambridge University Press, Cambridge, 1972).

18. G. Gamow, *Thirty Years that Shook Physics*, (Doubleday and Co., Garden City, NY, 1966): p. 125.

19. W. Pauli, *Z. f. Phys.* **37**, 263 (1926); **43**, 601 (1927).

20. C. G. Darwin, *Proc. Roy. Soc.* **A118**, 654 (1928); *ibid.*, **A120**, 621 (1928).

21. W. Gordon, *Z. f. Phys.* **48**, 11 (1928).

22. P. A. M. Dirac, *Proc. Roy. Soc.* **A126**, 360 (1930); also see Ref. 47.

23. E. C. Stoner, *Phil. Mag.* **48**, 719 (1924).

24. W. Pauli, *Z. f. Phys.* **31**, 765 (1925).

25. W. Heisenberg, *Z. f. Phys.* **38**, 411 (1926); *ibid.*, **39**, 499 (1926); P. A. M. Dirac, *Proc. Roy. Soc.* **A112**, 661 (1926); W. Pauli, *Z. f. Phys.* **41**, 81 (1927); J. C. Slater, *Phys. Rev.* **34**, 1293 (1929).

26. E. Fermi, *Z. f. Phys.* **36**, 902 (1926); *Rend. Accad. Lincei* **3**, 145 (1926).

27. P. A. M. Dirac, Ref. 25.

27*a*. P. A. M. Dirac, First W. R. Crane Lecture at the University of Michigan, April 17, 1978, unpublished.

28. H. Weyl, *The Theory of Groups and Quantum Mechanics*, translated from the second (1931) German edition by H. P. Robertson (Dover Publications, Inc., New York): Chapter IV, Section 12. Also see P. A. M. Dirac, *Proc. Roy. Soc.* **A133**, 61 (1931).

29. J. R. Oppenheimer, *Phys. Rev.* **35**, 562 (1930); I. Tamm, *Z. f. Phys.* **62**, 545 (1930).

29*a*. P. A. M. Dirac, *Proc. Roy. Soc.* **133**, 60 (1931).

30. C. D. Anderson, *Science* **76**, 238 (1932); *Phys. Rev.* **43**, 491 (1933). The latter paper is reprinted in *Foundations of Nuclear Physics*, ed. by R. T. Beyer (Dover Publications, Inc., New York, 1949).

30*a*. J. Schwinger, 'A Report on Quantum Electrodynamics,' in *The Physicist's Conception of Nature* (Reidel, Dordrecht, 1973): p.415.

31. W. Pauli, *Handbuch der Physik* (Julius Springer, Berlin, 1932–1933); *Rev. Mod. Phys.* **13**, 203 (1941).

32. Born, Heisenberg, and Jordan, Ref. 4, Section 3.

32*a*. P. Ehrenfest, *Phys. Z.* **7**, 528 (1906).

33. Born and Jordan, Ref. 4. Unfortunately the relevant parts of this paper are not included in the reprint collection *Sources of Quantum Mechanics*, cited in Ref. 4.

34. P. A. M. Dirac, *Proc. Roy. Soc.* **A112**, 661 (1926): Section 5. For a more accessible derivation, see L. I. Schiff, *Quantum Mechanics*, 3rd edn. (McGraw-Hill Book Company, New York, 1968): Section 44.

34*a*. A. Einstein, *Phys. Z.* **18**, 121 (1917); reprinted in English in van der Waerden, Ref. 4.

35. P. A. M. Dirac, *Proc. Roy. Soc.* **A114**, 243 (1927); reprinted in *Quantum Electrodynamics*, ed. by J. Schwinger (Dover Publications, Inc., New York, 1958).

36. P. A. M. Dirac, *Proc. Roy. Soc.* **A114**, 710 (1927).

36*a*. V. F. Weisskopf and E. Wigner, *Z. f. Phys.* **63**, 54 (1930)

36*b*. E. Fermi, *Lincei Rend.* **9**, 881 (1929); **12**, 431 (1930); *Rev. Mod. Phys.* **4**, 87 (1932)

37. P. Jordan and W. Pauli, *Z. f. Phys.* **47**, 151 (1928).

38. N. Bohr and L. Rosenfeld, *Kon. dansk. vid. Selsk., Mat.-Fys. Medd.* **XII**, No. 8 (1933) (translation in *Selected Papers of Leon Rosenfeld*, ed. by R. S. Cohen and J. Stachel (Reidel, Dordrecht, 1979)); *Phys. Rev.* **78**, 794 (1950).

39. P. Jordan, *Z. f. Phys.* **44**, 473 (1927). Also see P. Jordan and O. Klein, *Z. f. Phys.* **45**, 751 (1929); P. Jordan, *Phys. Zeit.* **30**, 700 (1929).

40. P. Jordan and E. Wigner, *Z. f. Phys.* **47**, 631 (1928). This article is reprinted in *Quantum Electrodynamics*, Ref. 35.

40*a*. M. Fierz, *Helv. Phys. Acta* **12**, (1939); W. Pauli, *Phys. Rev.* **58**, 716 (1940); W. Pauli and F. J. Belinfante, *Physica* **7**, 177 (1940).

41. W. Heisenberg and W. Pauli, *Z. f. Phys.* **56**, 1 (1929); *ibid.*, **59**, 168 (1930).

42. P. A. M. Dirac, *Proc. Roy. Soc.* **A136**, 453 (1932); P. A. M. Dirac, V. A. Fock, and B. Podolsky, *Phys. Zeit. der Sowjetunion* **2**, 468 (1932); P. A. M. Dirac, Phys. *Zeit. der Sowjetunion* **3**, 64 (1933). The latter two articles are reprinted in *Quantum Electrodynamics*, Ref. 35, pp. 29 and 312. Also see L. Rosenfeld, *Z. f. Phys.* **76**, 729 (1932).

42*a*. P. A. M. Dirac, *Proc. Roy. Soc.* London **A136**, 453 (1932).

43. E. Fermi, *Z. f. Phys.* **88**, 161 (1934). Fermi quotes unpublished work of Pauli for the proposition that an unobserved neutral particle is emitted along with the electron in beta decay. This particle was called the neutrino to distinguish it from the recently discovered neutron.

43*a*. V. Fock, *C. R. Leningrad* 1933, p. 267.

44. W. H. Furry and J. R. Oppenheimer, *Phys. Rev.* **45**, 245 (1934). This paper uses a density matrix formalism developed by P. A. M. Dirac, Proc. Camb. *Phil. Soc.* **30**, 150 (1934). Also see R. E. Peierls, *Proc. Roy. Soc.* **146**, 420 (1934); W. Heisenberg, *Z. f. Phys.* **90**, 209 (1934); L. Rosenfeld, *Z. f. Phys.* **76**, 729 (1932).

45. W. Pauli and V. Weisskopf, *Helv. Phys. Acta* **7**, 709 (1934), reprinted in English translation in A. I. Miller, *Early Quantum Electrodynamics* (Cambridge University Press, Cambridge, 1994). Also see W. Pauli, *Ann. Inst. Henri Poincaré* **6**, 137 (1936).

46. O. Klein and Y. Nishina, *Z. f. Phys.* **52**, 853 (1929); Y. Nishina, *ibid.*, 869 (1929); also see I. Tamm, *Z. f. Phys.* **62**, 545 (1930).

47. P. A. M. Dirac, *Proc. Camb. Phil. Soc.* **26**, 361 (1930).

48. C. Møller, *Ann. d. Phys.* **14**, 531, 568 (1932).

49. H. Bethe and W. Heitler, *Proc. Roy. Soc.* **A146**, 83 (1934); also see G. Racah, *Nuovo Cimento* **11**, No. 7 (1934); *ibid.*, **13**, 69 (1936).

50. H. J. Bhabha, *Proc. Roy. Soc.* **A154**, 195 (1936).

50*a*. J. F. Carlson and J. R. Oppenheimer, *Phys. Rev.* **51**, 220 (1937).

51. P. Ehrenfest and J. R. Oppenheimer, *Phys. Rev.* **37**, 333 (1931).

52. W. Heitler and G. Herzberg, *Naturwiss.* **17**, 673 (1929); F. Rasetti, *Z. f. Phys.* **61**, 598 (1930).

53. J. Chadwick, *Proc. Roy. Soc.* **A136**, 692 (1932). This article is reprinted in *The Foundations of Nuclear Physics*, Ref. 30.

54. W. Heisenberg, *Z. f. Phys.* **77**, 1 (1932); also see I. Curie-Joliot and F. Joliot, *Compt. Rend.* **194**, 273 (1932).

54*a*. For references, see L. M. Brown and H. Rechenberg, *Hist. Stud. in Phys. and Bio. Science*, **25**, 1 (1994).

55. H. Yukawa, *Proc. Phys.-Math. Soc. (Japan)* (3) **17**, 48 (1935). This article is reprinted in *The Foundations of Nuclear Physics*, Ref. 30.

56. S. H. Neddermeyer and C. D. Anderson, *Phys. Rev.* **51**, 884 (1937); J. C. Street and E. C. Stevenson, *Phys. Rev.* **52**, 1003 (1937).

56*a*. L. Nordheim and N. Webb, *Phys. Rev.* **56**, 494 (1939).

57. M. Conversi, E. Pancini, and O. Piccioni, *Phys. Rev.* **71**, 209L (1947).

58. S. Sakata and T. Inoue, *Prog. Theor. Phys.* **1**, 143 (1946); R. E. Marshak and H. A. Bethe, *Phys. Rev.* **77**, 506 (1947).

59. C. M. G. Lattes, G. P. S. Occhialini, and C. F. Powell, *Nature* **160**, 453, 486 (1947).

60. G. D. Rochester and C. C. Butler, *Nature* **160**, 855 (1947).

61. J. R. Oppenheimer, *Phys. Rev.* **35**, 461 (1930).

62. I. Waller, *Z. f. Phys.* **59**, 168 (1930); *ibid.*, **61**, 721, 837 (1930); *ibid.*, **62**, 673 (1930).

63. V. F. Weisskopf, *Z. f. Phys.* **89**, 27 (1934), reprinted in English translation in *Early Quantum Electrodynamics*, Ref. 45; *ibid.*, **90**, 817 (1934). The electromagnetic self energy is calculated only to lowest order in α in these references; the proof that the divergence

is only logarithmic in all orders of perturbation theory was given by Weisskopf; *Phys. Rev.* **56**, 72 (1939). (This last article is reprinted in *Quantum Electrodynamics*, Ref. 35).

64. P. A. M. Dirac, XVII Conseil Solvay de Physique, p. 203 (1933), reprinted in *Early Quantum Electrodynamics*, Ref. 45. For subsequent calculations based on less restrictive assumptions, see W. Heisenberg, *Z. f. Phys.* **90**, 209 (1934); *Sachs. Akad. Wiss.* **86**, 317 (1934); R. Serber, *Phys. Rev.* **43**, 49 (1935); E. A. Uehling. *Phys. Rev.* **48**, 55 (1935); W. Pauli and M. Rose, *Phys. Rev.* **49**, 462 (1936). Also see Furry and Oppenheimer, Ref. 44; Peierls. Ref. 44; Weisskopf, Ref. 63.

65. H. Euler and B. Kockel, *Naturwiss.* **23**, 246 (1935); W. Heisenberg and H. Euler, *Z. f. Phys.* **98**, 714 (1936).

66. P. A. M. Dirac, *Proc. Camb. Phil. Soc.* **30**, 150 (1934).

67. W. Heisenberg, *Z. f. Phys.* **90**, 209 (1934).

68. N. Kemmer and V. F. Weisskopf. *Nature* **137**, 659 (1936).

68*a*. F. Bloch and A. Nordsieck, *Phys. Rev.* **52**, 54 (1937). Also see W. Pauli and M. Fierz, *Nuovo Cimento* **15**, 167 (1938), reprinted in English translation in *Early Quantum Electrodynamics*, Ref. 45.

69. S. M. Dancoff, *Phys. Rev.* **55**, 959 (1939).

69*a*. H. W. Lewis, *Phys. Rev.* **73**, 173 (1948); S. Epstein, *Phys. Rev.* **73**, 177 (1948). Also see J. Schwinger, Ref. 84; Z. Koba and S. Tomonaga, *Prog. Theor. Phys.* **3**/3, 290 (1948).

69*b*. J. Schwinger, in *The Birth of Particle Physics*, ed. by L. Brown and L. Hoddeson (Cambridge University Press, Cambridge, 1983): p. 336.

70. W. Heisenberg, *Ann. d. Phys.* **32**, 20 (1938), reprinted in English translation in *Early Quantum Electrodynamics*, Ref. 45.

70*a*. G. Wentzel, *Z. f. Phys.* **86**, 479, 635 (1933); *Z. f. Phys.* **87**, 726 (1034); M. Born and L. Infeld, *Proc. Roy. Soc.* **A150**. 141 (1935); W. Pauli, *Ann. Inst. Henri Poincaré* **6**, 137 (1936).

71. J. A. Wheeler, *Phys. Rev.* **52**, 1107 (1937).

72. W. Heisenberg, *Z. f. Phys.* **120**, 513, 673 (1943); *Z. Naturforsch.* **1**, 608 (1946). Also see C. Møller, *Kon. Dansk. Vid. Sels. Mat.-Fys. Medd.* **23**, No. 1 (1945); *ibid.* **23**, No. 19, (1946).

73. See, e.g., G. Chew, *The S-Matrix Theory of Strong Interactions* (W. A. Benjamin, Inc. New York, 1961).

74. J. A. Wheeler and R. P. Feynman, *Rev. Mod. Phys.* **17**, 157 (1945), *ibid.*, **21**, 425 (1949). For further references and a discussion of the application of action-at-a-distance theories in cosmology, see S. Weinberg *Gravitation and Cosmology*, (Wiley, 1972): Section 16.3.

75. P. A. M. Dirac, *Proc. Roy. Soc.* **A180**, 1 (1942). For a criticism, see W. Pauli, *Rev. Mod. Phys.* **15**, 175 (1943). For a review of classical theories of this type, and of yet other attempts to solve the problem of infinities, see R. E. Peierls in *Rapports du 8^{m}e Conseil de Physique Solvay 1948* (R. Stoops, Brussels, 1950): p. 241.

76. V. F. Weisskopf, *Kon. Dan. Vid. Sel., Mat.-fys. Medd.* **XIV**, No. 6 (1936), especially p. 34 and pp. 5–6. This article is reprinted in *Quantum Electrodynamics*, Ref. 35, and in English translation in *Early Quantum Electrodynamics*, Ref. 45. Also see W. Pauli and M. Fierz, Ref. 68*a*; H. A. Kramers, Ref. 79*a*.

77. S. Pasternack, *Phys. Rev.* **54**, 1113 (1938). This suggestion was based on experiments of W.V. Houston, *Phys. Rev.* **51**, 446 (1937); R. C. Williams, *Phys. Rev.* **54**, 558 (1938). For a report of contrary data, see J. W. Drinkwater, O. Richardson, and W. E. Williams, *Proc. Roy. Soc.* **174**, 164 (1940).

78. E. A. Uehling, Ref. 64.

79. W. E. Lamb, Jr and R. C. Retherford, *Phys. Rev.* **72**, 241 (1947). This article is reprinted in *Quantum Electrodynamics*, Ref. 35.

79*a*. H. A. Kramers, *Nuovo Cimento* **15**, 108 (1938), reprinted in English translation in *Early Quantum Electrodynamics*, Ref. 45; *Ned. T. Natwink.* **11**, 134 (1944); *Rapports du 8^{m}e Conseil de Physique Solvay 1948* (R. Stoops, Brussels, 1950).

80. H. A. Bethe, *Phys. Rev.* **72**, 339 (1947). This article is reprinted in *Quantum Electrodynamics*, Ref. 35.

81. J. B. French and V. F. Weisskopf; *Phys. Rev.* **75**, 1240 (1949); N. M. Kroll and W. E. Lamb, *ibid.*, **75**, 388 (1949); J. Schwinger, *Phys. Rev.* **75**, 898 (1949); R. P. Feynman, *Rev. Mod. Phys.* **20**, 367 (1948); *Phys. Rev.*, **74**, 939, 1430 (1948); **76**, 749, 769 (1949); **80**, 440 (1950); H. Fukuda, Y. Miyamoto, and S. Tomonaga, *Prog. Theor. Phys. Rev. Mod. Phys.* **4**, 47, 121 (1948). The article by Kroll and Lamb is reprinted in *Quantum Electrodynamics*, Ref. 35.

82. J. E. Nafe, E. B. Nelson, and I. I. Rabi, *Phys. Rev.* **71**, 914 (1947); D. E. Nagel, R. S. Julian, and J. R. Zacharias, *Phys. Rev.* **72**, 973 (1947).

83. P. Kusch and H. M. Foley, *Phys. Rev.* **72**, 1256 (1947).

83*a*. G. Breit, *Phys. Rev.* **71**, 984 (1947). Schwinger in Ref. 84 includes a corrected version of Breit's results.

84. J. Schwinger, *Phys. Rev.* **73**, 416 (1948). This article is reprinted in *Quantum Electrodynamics*, Ref. 35.

85. J. Schwinger, *Phys. Rev.* **74**, 1439 (1948); *ibid.*, **75**, 651 (1949); *ibid.*, **76**, 790 (1949); *ibid.*, **82**, 664, 914 (1951); *ibid.*, **91**, 713 (1953); *Proc. Nat. Acad. Sci.* **37**, 452 (1951). All but the first two of these articles are reprinted in *Quantum Electrodynamics*, Ref. 35.

86. S. Tomonaga, *Prog. Theor. Phys. Rev. Mod. Phys.* **1**, 27 (1946); Z. Koba, T. Tati, and S. Tomonaga, *ibid.*, **2**, 101 (1947); S. Kanesawa and S. Tomonaga, *ibid.*, **3**, 1, 101 (1948); S. Tomonaga, *Phys. Rev.* **74**, 224 (1948); D. Ito, Z. Koba, and S. Tomonaga, *Prog. Theor. Phys.* **3**, 276 (1948); Z. Koba and S. Tomonaga, *ibid.*, **3**, 290 (1948). The first and fourth of these articles are reprinted in *Quantum Electrodynamics*, Ref. 35.

87. R. P. Feynman, *Rev. Mod. Phys.* **20**, 367 (1948); *Phys. Rev.* **74**, 939, 1430 (1948); *ibid.*, **76**, 749, 769 (1949); *ibid* **80**, 440 (1950). All but the second and third of these articles are reprinted in *Quantum Electrodynamics*, Ref. 35.

88. F. J. Dyson, *Phys. Rev.* **75**, 486, 1736 (1949). These articles are reprinted in *Quantum Electrodynamics*, Ref. 35.

88*a*. H. Fröhlich, W. Heitler, and B. Kahn, *Proc. Roy. Soc.* **A171**, 269 (1939); *Phys. Rev.* **56**, 961 (1939).

88*b*. W. E. Lamb, Jr, *Phys. Rev.* **56**, 384 (1939); *Phys. Rev.* **57**, 458 (1940).

89. Quoted by R. Serber, in *The Birth of Particle Physics*, Ref. 69*b*, p. 270.

2

Relativistic Quantum Mechanics

The point of view of this book is that quantum field theory is the way it is because (with certain qualifications) this is the only way to reconcile quantum mechanics with special relativity. Therefore our first task is to study how symmetries like Lorentz invariance appear in a quantum setting.

2.1 Quantum Mechanics

First, some good news: quantum field theory is based on the same quantum mechanics that was invented by Schrödinger, Heisenberg, Pauli, Born, and others in 1925–26, and has been used ever since in atomic, molecular, nuclear, and condensed matter physics. The reader is assumed to be already familiar with quantum mechanics; this section provides only the briefest of summaries of quantum mechanics, in the generalized version of Dirac.[1]

(i) Physical states are represented by rays in Hilbert space. A Hilbert space is a kind of complex vector space; that is, if Φ and Ψ are vectors in the space (often called 'state-vectors') then so is $\xi\Phi + \eta\Psi$, for arbitrary complex numbers ξ, η. It has a norm*: for any pair of vectors there is a complex number (Φ, Ψ), such that

$$(\Phi, \Psi) = (\Psi, \Phi)^* , \tag{2.1.1}$$

$$(\Phi, \xi_1\Psi_1 + \xi_2\Psi_2) = \xi_1(\Phi, \Psi_1) + \xi_2(\Phi, \Psi_2) , \tag{2.1.2}$$

$$(\eta_1\Phi_1 + \eta_2\Phi_2, \Psi) = \eta_1^*(\Phi_1, \Psi) + \eta_2^*(\Phi_2, \Psi) . \tag{2.1.3}$$

The norm (Ψ, Ψ) also satisfies a positivity condition: $(\Psi, \Psi) \geq 0$, and vanishes if and only if $\Psi = 0$. (There are also certain technical assumptions that allow us to take limits of vectors within Hilbert space.) A *ray* is a

* We shall often use the Dirac bra-ket notation: instead of (Ψ_1, Ψ_2), we may write $\langle 1|2\rangle$.

set of normalized vectors (i.e., $(\Psi, \Psi) = 1$) with Ψ and Ψ' belonging to the same ray if $\Psi' = \xi\Psi$, where ξ is an arbitrary complex number with $|\xi| = 1$.

(ii) Observables are represented by Hermitian operators. These are mappings $\Psi \rightarrow A\Psi$ of Hilbert space into itself, linear in the sense that

$$A(\xi\Psi + \eta\Phi) = \xi A\Psi + \eta A\Phi \tag{2.1.4}$$

and satisfying the reality condition $A^\dagger = A$, where for any linear operator A the adjoint $A^\dagger$ is defined by

$$(\Phi, A^\dagger\Psi) \equiv (A\Phi, \Psi) = (\Psi, A\Phi)^* . \tag{2.1.5}$$

(There are also technical assumptions about the continuity of $A\Psi$ as a function of Ψ.) A state represented by a ray $\mathscr{R}$ has a definite value α for the observable represented by an operator A if vectors Ψ belonging to this ray are eigenvectors of A with eigenvalue α:

$$A\Psi = \alpha\Psi \quad \text{for} \quad \Psi \text{ in } \mathscr{R} . \tag{2.1.6}$$

An elementary theorem tells us that for A Hermitian, α is real, and eigenvectors with different αs are orthogonal.

(iii) If a system is in a state represented by a ray $\mathscr{R}$, and an experiment is done to test whether it is in any one of the different states represented by mutually orthogonal rays $\mathscr{R}_1, \mathscr{R}_2, \ldots$ (for instance, by measuring one or more observables) then the probability of finding it in the state represented by $\mathscr{R}_n$ is

$$P(\mathscr{R} \rightarrow \mathscr{R}_n) = |(\Psi, \Psi_n)|^2 , \tag{2.1.7}$$

where Ψ and Ψ_n are any vectors belonging to rays $\mathscr{R}$ and $\mathscr{R}_n$, respectively. (A pair of rays is said to be orthogonal if the state-vectors from the two rays have vanishing scalar products.) Another elementary theorem gives a total probability unity:

$$\sum_n P(\mathscr{R} \rightarrow \mathscr{R}_n) = 1 \tag{2.1.8}$$

if the state-vectors Ψ_n form a complete set.

2.2 Symmetries

A symmetry transformation is a change in our point of view that does not change the results of possible experiments. If an observer O sees a system in a state represented by a ray $\mathscr{R}$ or $\mathscr{R}_1$ or $\mathscr{R}_2 \ldots$, then an equivalent observer O' who looks at the *same* system will observe it in a different state, represented by a ray $\mathscr{R}'$ or $\mathscr{R}'_1$ or $\mathscr{R}'_2 \ldots$, respectively, but the two observers must find the same probabilities

$$P(\mathscr{R} \to \mathscr{R}_n) = P(\mathscr{R}' \to \mathscr{R}'_n) . \tag{2.2.1}$$

(This is only a necessary condition for a ray transformation to be a symmetry; further conditions are discussed in the next chapter.) An important theorem proved by Wigner[2] in the early 1930s tells us that for any such transformation $\mathscr{R} \to \mathscr{R}'$ of rays we may define an operator U on Hilbert space, such that if Ψ is in ray $\mathscr{R}$ then $U\Psi$ is in the ray $\mathscr{R}'$, with U either *unitary* and *linear*

$$(U\Phi, U\Psi) = (\Phi, \Psi) , \tag{2.2.2}$$

$$U(\xi\Phi + \eta\Psi) = \xi U\Phi + \eta U\Psi \tag{2.2.3}$$

or else *antiunitary* and *antilinear*

$$(U\Phi, U\Psi) = (\Phi, \Psi)^* , \tag{2.2.4}$$

$$U(\xi\Phi + \eta\Psi) = \xi^* U\Phi + \eta^* U\Psi . \tag{2.2.5}$$

Wigner's proof omits some steps. A more complete proof is given at the end of this chapter in Appendix A.

As already mentioned, the adjoint of a linear operator L is defined by

$$(\Phi, L^\dagger\Psi) \equiv (L\Phi, \Psi) . \tag{2.2.6}$$

This condition cannot be satisfied for an antilinear operator, because in this case the right-hand side of Eq. (2.2.6) would be linear in Φ, while the left-hand side is antilinear in Φ. Instead, the adjoint of an antilinear operator A is defined by

$$(\Phi, A^\dagger\Psi) \equiv (A\Phi, \Psi)^* = (\Psi, A\Phi) . \tag{2.2.7}$$

With this definition, the conditions for unitarity or antiunitarity both take the form

$$U^\dagger = U^{-1} . \tag{2.2.8}$$

There is always a trivial symmetry transformation $\mathscr{R} \to \mathscr{R}$, represented by the identity operator $U = 1$. This operator is, of course, unitary and linear. Continuity then demands that any symmetry (like a rotation or translation or Lorentz transformation) that can be made trivial by a continuous change of some parameters (like angles or distances or velocities) must be represented by a linear unitary operator U rather than one that is antilinear and antiunitary. (Symmetries represented by antiunitary antilinear operators are less prominent in physics; they all involve a reversal in the direction of time's flow. See Section 2.6.)

In particular, a symmetry transformation that is infinitesimally close to being trivial can be represented by a linear unitary operator that is infinitesimally close to the identity:

$$U = 1 + i\epsilon t \tag{2.2.9}$$

with ϵ a real infinitesimal. For this to be unitary and linear, t must be Hermitian and linear, so it is a candidate for an observable. Indeed, most (and perhaps all) of the observables of physics, such as angular momentum or momentum, arise in this way from symmetry transformations.

The set of symmetry transformations has certain properties that define it as a *group*. If T_1 is a transformation that takes rays $\mathscr{R}_n$ into $\mathscr{R}'_n$, and T_2 is another transformation that takes $\mathscr{R}'_n$ into $\mathscr{R}''_n$, then the result of performing both transformations is another symmetry transformation, which we write T_2T_1, that takes $\mathscr{R}_n$ into $\mathscr{R}''_n$. Also, a symmetry transformation T which takes rays $\mathscr{R}_n$ into $\mathscr{R}'_n$ has an inverse, written T^{-1}, which takes $\mathscr{R}'_n$ into $\mathscr{R}_n$, and there is an identity transformation, $T = 1$, which leaves rays unchanged.

The unitary or antiunitary operators $U(T)$ corresponding to these symmetry transformations have properties that mirror this group structure, but with a complication due to the fact that, unlike the symmetry transformations themselves, the operators $U(T)$ act on vectors in the Hilbert space, rather than on rays. If T_1 takes $\mathscr{R}_n$ into $\mathscr{R}'_n$, then acting on a vector Ψ_n in the ray $\mathscr{R}_n$, $U(T_1)$ must yield a vector $U(T_1)\Psi_n$ in the ray $\mathscr{R}'_n$, and if T_2 takes this ray into $\mathscr{R}''_n$, then acting on $U(T_1)\Psi_n$ it must yield a vector $U(T_2)U(T_1)\Psi_n$ in the ray $\mathscr{R}''_n$. But $U(T_2T_1)\Psi_n$ is also in this ray, so these vectors can differ only by a phase $\phi_n(T_2, T_1)$

$$U(T_2)U(T_1)\Psi_n = e^{i\phi_n(T_2,T_1)}U(T_2T_1)\Psi_n\,. \tag{2.2.10}$$

Furthermore, with one significant exception, the linearity (or antilinearity) of $U(T)$ tells us that these phases are independent of the state Ψ_n. Here is the proof. Consider any two different vectors Ψ_A, Ψ_B, which are not proportional to each other. Then, applying Eq. (2.2.10) to the state $\Psi_{AB} \equiv \Psi_A + \Psi_B$, we have

$$\begin{aligned} e^{i\phi_{AB}}\, U(T_2T_1)(\Psi_A + \Psi_B) &= U(T_2)U(T_1)(\Psi_A + \Psi_B) \\ &= U(T_2)U(T_1)\Psi_A + U(T_2)U(T_1)\Psi_B \\ &= e^{i\phi_A}U(T_2T_1)\Psi_A + e^{i\phi_B}U(T_2T_1)\Psi_B. \end{aligned} \tag{2.2.11}$$

Any unitary or antiunitary operator has an inverse (its adjoint) which is also unitary or antiunitary. Multiplying (2.2.11) on the left with $U^{-1}(T_2T_1)$, we have then

$$e^{\pm i\phi_{AB}}(\Psi_A + \Psi_B) = e^{\pm i\phi_A}\Psi_A + e^{\pm i\phi_B}\Psi_B\,, \tag{2.2.12}$$

the upper and lower signs referring to $U(T_2T_1)$ unitary or antiunitary, respectively. Since Ψ_A and Ψ_B are linearly independent, this is only possible if

$$e^{i\phi_{AB}} = e^{i\phi_A} = e^{i\phi_B} \,. \tag{2.2.13}$$

So as promised, the phase in Eq. (2.2.10) is independent of the state-vector Ψ_n, and therefore this can be written as an operator relation

$$U(T_2)U(T_1) = e^{i\phi(T_2,T_1)}U(T_2T_1) \,. \tag{2.2.14}$$

For $\phi = 0$, this would say that the $U(T)$ furnish a representation of the group of symmetry transformations. For general phases $\phi(T_2, T_1)$, we have what is called a projective representation, or a representation 'up to a phase'. The structure of the Lie group cannot by itself tell us whether physical state-vectors furnish an ordinary or a projective representation, but as we shall see, it can tell us whether the group has any intrinsically projective representations at all.

The exception to the argument that led to Eq. (2.2.14) is that it may not be possible to prepare the system in a state represented by $\Psi_A + \Psi_B$. For instance, it is widely believed to be impossible to prepare a system in a superposition of two states whose total angular momenta are integers and half-integers, respectively. In such cases, we say that there is a 'superselection rule' between different classes of states,[3] and the phases $\phi(T_2, T_1)$ *may* depend on which of these classes of states the operators $U(T_2)U(T_1)$ and $U(T_2, T_1)$ act upon. We will have more to say about these phases and projective representations in Section 2.7. As we shall see there, any symmetry group with projective representations can always be enlarged (without otherwise changing its physical implications) in such a way that its representations can all be defined as non-projective, with $\phi = 0$. Until Section 2.7, we will just assume that this has been done, and take $\phi = 0$ in Eq. (2.2.14).

There is a kind of group, known as a *connected Lie group*, of special importance in physics. These are groups of transformations $T(\theta)$ that are described by a finite set of real continuous parameters, say θ^a, with each element of the group connected to the identity by a path within the group. The group multiplication law then takes the form

$$T(\bar{\theta})T(\theta) = T\Big(f(\bar{\theta}, \theta)\Big) \tag{2.2.15}$$

with $f^a(\bar{\theta}, \theta)$ a function of the $\bar{\theta}$s and θs. Taking $\theta^a = 0$ as the coordinates of the identity, we must have

$$f^a(\theta, 0) = f^a(0, \theta) = \theta^a \,. \tag{2.2.16}$$

As already mentioned, the transformations of such continuous groups must be represented on the physical Hilbert space by unitary (rather than antiunitary) operators $U(T(\theta))$. For a Lie group, these operators can be represented in at least a finite neighborhood of the identity by a power series

$$U\Big(T(\theta)\Big) = 1 + i\theta^a t_a + \tfrac{1}{2}\,\theta^b\theta^c t_{bc} + \cdots , \tag{2.2.17}$$

where t_a, $t_{bc} = t_{cb}$, etc. are operators independent of the θs, with t_a Hermitian. Suppose that the $U(T(\theta))$ form an ordinary (i.e., not projective) representation of this group of transformations, i.e.,

$$U\Big(T(\bar\theta)\Big)\,U\Big(T(\theta)\Big) = U\Big(T(f(\bar\theta,\theta))\Big) . \tag{2.2.18}$$

Let us see what this condition looks like when expanded in powers of θ^a and $\bar\theta^a$. According to Eq. (2.2.16), the expansion of $f^a(\bar\theta,\theta)$ to second order must take the form

$$f^a(\bar\theta,\theta) = \theta^a + \bar\theta^a + f^a{}_{bc}\bar\theta^b\theta^c + \cdots \tag{2.2.19}$$

with real coefficients $f^a{}_{bc}$. (The presence of any terms of order θ^2 or $\bar\theta^2$ would violate Eq. (2.2.16).) Then Eq. (2.2.18) reads

$$\begin{aligned}\Big[1 + i\bar\theta^a t_a + \tfrac{1}{2}\,\bar\theta^b\bar\theta^c t_{bc} + \cdots\Big] &\times \Big[1 + i\theta^a t_a + \tfrac{1}{2}\theta^b\theta^c t_{bc} + \cdots\Big] \\ = 1 + i\Big(\theta^a + \bar\theta^a &+ f^a{}_{bc}\bar\theta^b\theta^c + \cdots\Big)\, t_a \\ + \tfrac{1}{2}(\theta^b + \bar\theta^b + \cdots)&\,(\theta^c + \bar\theta^c + \cdots)t_{bc} + \cdots \end{aligned} \tag{2.2.20}$$

The terms of order $1, \theta, \bar\theta, \theta^2$, and $\bar\theta^2$ automatically match on both sides of Eq. (2.2.20), but from the $\bar\theta\theta$ terms we obtain a non-trivial condition

$$t_{bc} = -t_b t_c - i\, f^a{}_{bc}\, t_a . \tag{2.2.21}$$

This shows that if we are given the structure of the group, i.e., the function $f(\theta,\bar\theta)$, and hence its quadratic coefficient $f^a{}_{bc}$, we can calculate the second-order terms in $U(T(\theta))$ from the generators t_a appearing in the first-order terms. However, there is a consistency condition: the operator t_{bc} must be *symmetric* in b and c (because it is the second derivative of $U(T(\theta))$ with respect to θ^b and θ^c) so Eq. (2.2.21) requires that

$$[t_b,\ t_c] = i\, C^a{}_{bc}\, t_a , \tag{2.2.22}$$

where $C^a{}_{bc}$ are a set of real constants known as *structure constants*

$$C^a{}_{bc} \equiv -f^a{}_{bc} + f^a{}_{cb} . \tag{2.2.23}$$

Such a set of commutation relations is known as a *Lie algebra.* In Section 2.7 we will prove in effect that the commutation relation (2.2.22) is the single condition needed to ensure that this process can be continued: the complete power series for $U(T(\theta))$ may be calculated from an infinite sequence of relations like Eq. (2.2.21), provided we know the first-order terms, the generators t_a. This does not necessarily mean that the operators $U(T(\theta))$ are uniquely determined for all θ^a if we know the t_a, but it does mean that the $U(T(\theta))$ are uniquely determined in at least a finite neighborhood of the coordinates $\theta^a = 0$ of the identity, in such a way that Eq. (2.2.15) is satisfied if $\theta, \bar{\theta}$, and $f(\theta, \bar{\theta})$ are in this neighborhood. The extension to all θ^a is discussed in Section 2.7.

There is a special case of some importance, that we will encounter again and again. Suppose that the function $f(\theta, \bar{\theta})$ (perhaps just for some subset of the coordinates θ^a) is simply additive

$$f^a(\theta, \bar{\theta}) = \theta^a + \bar{\theta}^a \, . \tag{2.2.24}$$

This is the case for instance for translations in spacetime, or for rotations about any one fixed axis (though *not* for both together). Then the coefficients $f^a{}_{bc}$ in Eq. (2.2.19) vanish, and so do the structure constants (2.2.23). The generators then all commute

$$[t_b, \, t_c] = 0 \, . \tag{2.2.25}$$

Such a group is called *Abelian*. In this case, it is easy to calculate $U(T(\theta))$ for all θ^a. From Eqs. (2.2.18) and (2.2.24), we have for any integer N

$$U\Big(T(\theta)\Big) = \left[U\left(T\left(\frac{\theta}{N}\right)\right)\right]^N \, .$$

Letting $N \to \infty$, and keeping only the first-order term in $U(T(\theta/N))$, we have then

$$U\Big(T(\theta)\Big) = \lim_{N\to\infty}\left[1 + \frac{i}{N}\theta^a t_a\right]^N$$

and hence

$$U\Big(T(\theta)\Big) = \exp(i t_a \theta^a) \, . \tag{2.2.26}$$

2.3 Quantum Lorentz Transformations

Einstein's principle of relativity states the equivalence of certain 'inertial' frames of reference. It is distinguished from the Galilean principle of relativity, obeyed by Newtonian mechanics, by the transformation connecting coordinate systems in different inertial frames. If x^μ are the coordinates in one inertial frame (with x^1, x^2, x^3 Cartesian space coordinates, and $x^0 = t$ a time coordinate, the speed of light being set equal to unity) then in any other inertial frame, the coordinates x'^μ must satisfy

$$\eta_{\mu\nu} dx'^\mu dx'^\nu = \eta_{\mu\nu} dx^\mu dx^\nu \tag{2.3.1}$$

or equivalently

$$\eta_{\mu\nu} \frac{\partial x'^\mu}{\partial x^\rho} \frac{\partial x'^\nu}{\partial x^\sigma} = \eta_{\rho\sigma} \, . \tag{2.3.2}$$

Here $\eta_{\mu\nu}$ is the diagonal matrix, with elements

$$\eta_{11} = \eta_{22} = \eta_{33} = +1, \quad \eta_{00} = -1 \tag{2.3.3}$$

and the summation convention is in force: we sum over any index like μ and ν in Eq. (2.3.2), which appears twice in the same term, once upstairs and once downstairs. These transformations have the special property that the speed of light is the same (in our units, equal to unity) in all inertial frames;* a light wave travelling at unit speed satisfies $|d\mathbf{x}/dt| = 1$, or in other words $\eta_{\mu\nu}dx^\mu dx^\nu = d\mathbf{x}^2 - dt^2 = 0$, from which it follows that also $\eta_{\mu\nu}dx'^\mu dx'^\nu = 0$, and hence $|d\mathbf{x}'/dt'| = 1$.

Any coordinate transformation $x^\mu \to x'^\mu$ that satisfies Eq. (2.3.2) is *linear*[3a]

$$x'^\mu = \Lambda^\mu{}_\nu x^\nu + a^\mu \tag{2.3.4}$$

with a^μ arbitrary constants, and $\Lambda^\mu{}_\nu$ a constant matrix satisfying the conditions

$$\eta_{\mu\nu}\Lambda^\mu{}_\rho\Lambda^\nu{}_\sigma = \eta_{\rho\sigma} \,. \tag{2.3.5}$$

For some purposes, it is useful to write the Lorentz transformation condition in a different way. The matrix $\eta_{\mu\nu}$ has an inverse, written $\eta^{\mu\nu}$, which happens to have the same components: it is diagonal, with $\eta^{00} = -1$, $\eta^{11} = \eta^{22} = \eta^{33} = +1$. Multiplying Eq. (2.3.5) with $\eta^{\sigma\tau}\Lambda^\kappa{}_\tau$ and inserting parentheses judiciously, we have

$$\eta_{\mu\nu}\Lambda^\mu{}_\rho(\Lambda^\nu{}_\sigma\Lambda^\kappa{}_\tau\eta^{\sigma\tau}) = \Lambda^\kappa{}_\rho = \eta_{\mu\nu}\eta^{\nu\kappa}\Lambda^\mu{}_\rho \,.$$

Multiplying with the inverse of the matrix $\eta_{\mu\nu}\Lambda^\mu{}_\rho$ then gives

$$\Lambda^\nu{}_\sigma\Lambda^\kappa{}_\tau\eta^{\sigma\tau} = \eta^{\nu\kappa} \,. \tag{2.3.6}$$

These transformations form a group. If we first perform a Lorentz transformation (2.3.4), and then a second Lorentz transformation $x'^\mu \to x''^\mu$, with

$$x''^\mu = \bar{\Lambda}^\mu{}_\rho x'^\rho + \bar{a}^\mu = \bar{\Lambda}^\mu{}_\rho(\Lambda^\rho{}_\nu x^\nu + a^\rho) + \bar{a}^\mu$$

then the effect is the same as the Lorentz transformation $x^\mu \to x''^\mu$, with

$$x''^\mu = (\bar{\Lambda}^\mu{}_\rho\Lambda^\rho{}_\nu)x^\nu + (\bar{\Lambda}^\mu{}_\rho a^\rho + \bar{a}^\mu) \,. \tag{2.3.7}$$

(Note that if $\Lambda^\mu{}_\nu$ and $\bar{\Lambda}^\mu{}_\nu$ both satisfy Eq. (2.3.5), then so does $\bar{\Lambda}^\mu{}_\rho\Lambda^\rho{}_\nu$, so this *is* a Lorentz transformation. The bar is used here just to distinguish

* There is a larger class of coordinate transformations, known as *conformal transformations*, for which $\eta_{\mu\nu}dx'^\mu dx'^\nu$ is proportional though generally not equal to $\eta_{\mu\nu}dx^\mu dx^\nu$, and which therefore also leave the speed of light invariant. Conformal invariance in two dimensions has proved enormously important in string theory and statistical mechanics, but the physical relevance of these conformal transformations in four spacetime dimensions is not yet clear.

one Lorentz transformation from the other.) The transformations $T(\Lambda, a)$ induced on physical states therefore satisfy the composition rule

$$T(\bar{\Lambda}, \bar{a})T(\Lambda, a) = T(\bar{\Lambda}\Lambda, \bar{\Lambda}a + \bar{a})\,. \tag{2.3.8}$$

Taking the determinant of Eq. (2.3.5) gives

$$(\text{Det}\Lambda)^2 = 1 \tag{2.3.9}$$

so $\Lambda^\mu{}_\nu$ has an inverse, $(\Lambda^{-1})^\nu{}_\rho$ which we see from Eq. (2.3.5) takes the form

$$(\Lambda^{-1})^\rho{}_\nu = \Lambda_\nu{}^\rho \equiv \eta_{\nu\mu}\eta^{\rho\sigma}\Lambda^\mu{}_\sigma\,. \tag{2.3.10}$$

The inverse of the transformation $T(\Lambda, a)$ is seen from Eq. (2.3.8) to be $T(\Lambda^{-1}, -\Lambda^{-1}a)$, and, of course, the identity transformation is $T(1, 0)$.

In accordance with the discussion in the previous section, the transformations $T(\Lambda, a)$ induce a unitary linear transformation on vectors in the physical Hilbert space

$$\Psi \to U(\Lambda, a)\Psi\,.$$

The operators U satisfy a composition rule

$$U(\bar{\Lambda}, \bar{a})U(\Lambda, a) = U(\bar{\Lambda}\Lambda, \bar{\Lambda}a + \bar{a})\,. \tag{2.3.11}$$

(As already mentioned, to avoid the appearance of a phase factor on the right-hand side of Eq. (2.3.11), it is, in general, necessary to enlarge the Lorentz group. The appropriate enlargement is described in Section 2.7.)

The whole group of transformations $T(\Lambda, a)$ is properly known as the *inhomogeneous Lorentz group*, or *Poincaré group*. It has a number of important subgroups. First, those transformations with $a^\mu = 0$ obviously form a subgroup, with

$$T(\bar{\Lambda}, 0)\, T(\Lambda, 0) = T(\bar{\Lambda}\Lambda, 0), \tag{2.3.12}$$

known as the *homogeneous Lorentz group*. Also, we note from Eq. (2.3.9) that either $\text{Det}\Lambda = +1$ or $\text{Det}\Lambda = -1$; those transformations with $\text{Det}\Lambda = +1$ obviously form a subgroup of either the homogeneous or the inhomogeneous Lorentz groups. Further, from the 00-components of Eqs. (2.3.5) and (2.3.6), we have

$$(\Lambda^0{}_0)^2 = 1 + \Lambda^i{}_0\Lambda^i{}_0 = 1 + \Lambda^0{}_i\Lambda^0{}_i\,. \tag{2.3.13}$$

with i summed over the values 1, 2, and 3. We see that either $\Lambda^0{}_0 \geq +1$ or $\Lambda^0{}_0 \leq -1$. Those transformations with $\Lambda^0{}_0 \geq +1$ form a subgroup. Note that if $\Lambda^\mu{}_\nu$ and $\bar{\Lambda}^\mu{}_\nu$ are two such Λs, then

$$(\bar{\Lambda}\Lambda)^0{}_0 = \bar{\Lambda}^0{}_0\Lambda^0{}_0 + \bar{\Lambda}^0{}_1\Lambda^1{}_0 + \bar{\Lambda}^0{}_2\Lambda^2{}_0 + \bar{\Lambda}^0{}_3\Lambda^3{}_0\,;.$$

But Eq. (2.3.13) shows that the three-vector $(\Lambda^1{}_0, \Lambda^2{}_0, \Lambda^3{}_0)$ has length $\sqrt{(\Lambda^0{}_0)^2 - 1}$, and similarly the three-vector $(\bar{\Lambda}^0{}_1, \bar{\Lambda}^0{}_2, \bar{\Lambda}^0{}_3)$ has length

$\sqrt{(\bar{\Lambda}^0{}_0)^2 - 1}$, so the scalar product of these two three-vectors is bounded by

$$|\bar{\Lambda}^0{}_1\Lambda^1{}_0 + \bar{\Lambda}^0{}_2\Lambda^2{}_0 + \bar{\Lambda}^0{}_3\Lambda^3{}_0| \leq \sqrt{(\Lambda^0{}_0)^2 - 1}\sqrt{(\bar{\Lambda}^0{}_0)^2 - 1}\,, \tag{2.3.14}$$

and so

$$(\bar{\Lambda}\Lambda)^0{}_0 \geq \bar{\Lambda}^0{}_0\Lambda^0{}_0 - \sqrt{(\Lambda^0{}_0)^2 - 1}\sqrt{(\bar{\Lambda}^0{}_0)^2 - 1} \;\; \geq 1\,.$$

The subgroup of Lorentz transformations with $\text{Det}\,\Lambda = +1$ and $\Lambda^0{}_0 \geq +1$ is known as the *proper orthochronous Lorentz group.* Since it is not possible by a continuous change of parameters to jump from $\text{Det}\,\Lambda = +1$ to $\text{Det}\,\Lambda = -1$, or from $\Lambda^0{}_0 \geq +1$ to $\Lambda^0{}_0 \leq -1$, any Lorentz transformation that can be obtained from the identity by a continuous change of parameters must have $\text{Det}\,\Lambda$ and $\Lambda^0{}_0$ of the same sign as for the identity, and hence must belong to the proper orthochronous Lorentz group.

Any Lorentz transformation is either proper and orthochronous, or may be written as the product of an element of the proper orthochronous Lorentz group with one of the discrete transformations $\mathscr{P}$ or $\mathscr{T}$ or $\mathscr{P}\mathscr{T}$, where $\mathscr{P}$ is the space inversion, whose non-zero elements are

$$\mathscr{P}^0{}_0 = 1,\; \mathscr{P}^1{}_1 = \mathscr{P}^2{}_2 = \mathscr{P}^3{}_3 = -1\,, \tag{2.3.15}$$

and $\mathscr{T}$ is the time-reversal matrix, whose non-zero elements are

$$\mathscr{T}^0{}_0 = -1,\; \mathscr{T}^1{}_1 = \mathscr{T}^2{}_2 = \mathscr{T}^3{}_3 = 1\,. \tag{2.3.16}$$

Thus the study of the whole Lorentz group reduces to the study of its proper orthochronous subgroup, plus space inversion and time-reversal. We will consider space inversion and time-reversal separately in Section 2.6. Until then, we will deal only with the homogeneous or inhomogeneous proper orthochronous Lorentz group.

2.4 The Poincaré Algebra

As we saw in Section 2.2, much of the information about any Lie symmetry group is contained in properties of the group elements near the identity. For the inhomogeneous Lorentz group, the identity is the transformation $\Lambda^\mu{}_\nu = \delta^\mu{}_\nu$, $a^\mu = 0$, so we want to study those transformations with

$$\Lambda^\mu{}_\nu = \delta^\mu{}_\nu + \omega^\mu{}_\nu\,, \qquad a^\mu = \epsilon^\mu\,, \tag{2.4.1}$$

both $\omega^\mu{}_\nu$ and ϵ^μ being taken infinitesimal. The Lorentz condition (2.3.5) reads here

$$\eta_{\rho\sigma} = \eta_{\mu\nu}(\delta^\mu{}_\rho + \omega^\mu{}_\rho)(\delta^\nu{}_\sigma + \omega^\nu{}_\sigma)$$
$$= \eta_{\sigma\rho} + \omega_{\sigma\rho} + \omega_{\rho\sigma} + O(\omega^2)\,.$$

We are here using the convention, to be used throughout this book, that indices may be lowered or raised by contraction with $\eta_{\mu\nu}$ or $\eta^{\mu\nu}$

$$\omega_{\sigma\rho} \equiv \eta_{\mu\sigma}\omega^\mu{}_\rho$$
$$\omega^\mu{}_\rho \equiv \eta^{\mu\sigma}\omega_{\sigma\rho}\,.$$

Keeping only the terms of first order in ω in the Lorentz condition (2.3.5), we see that this condition now reduces to the antisymmetry of $\omega_{\mu\nu}$

$$\omega_{\mu\nu} = -\omega_{\nu\mu}\,. \tag{2.4.2}$$

An antisymmetric second-rank tensor in four dimensions has $(4 \times 3)/2 = 6$ independent components, so including the four components of ϵ^μ, an inhomogeneous Lorentz transformation is described by $6 + 4 = 10$ parameters.

Since $U(1,0)$ carries any ray into itself, it must be proportional to the unit operator,* and by a choice of phase may be made equal to it. For an infinitesimal Lorentz transformation (2.4.1), $U(1+\omega,\epsilon)$ must then equal **1** plus terms linear in $\omega_{\rho\sigma}$ and ϵ_ρ. We write this as

$$U(1+\omega,\epsilon) = 1 + \tfrac{1}{2}i\,\omega_{\rho\sigma}J^{\rho\sigma} - i\epsilon_\rho P^\rho + \cdots\,. \tag{2.4.3}$$

Here $J^{\rho\sigma}$ and P^ρ are ω- and ϵ-independent operators, and the dots denote terms of higher order in ω and/or ϵ. In order for $U(1+\omega,\epsilon)$ to be unitary, the operators $J^{\rho\sigma}$ and P^ρ must be Hermitian

$$J^{\rho\sigma\dagger} = J^{\rho\sigma}\quad,\quad P^{\rho\dagger} = P^\rho\,. \tag{2.4.4}$$

Since $\omega_{\rho\sigma}$ is antisymmetric, we can take its coefficient $J^{\rho\sigma}$ to be antisymmetric also

$$J^{\rho\sigma} = -J^{\sigma\rho}\,. \tag{2.4.5}$$

As we shall see, P^1, P^2, and P^3 are the components of the momentum operators, J^{23}, J^{31}, and J^{12} are the components of the angular momentum vector, and P^0 is the energy operator, or *Hamiltonian*.**

* In the absence of superselection rules, the possibility that the proportionality factor may depend on the state on which $U(1,0)$ acts can be ruled out by the same reasoning that we used in Section 2.2 to rule out the possibility that the phases in projective representations of symmetry groups may depend on the states on which the symmetries act. Where superselection rules apply, it may be necessary to redefine $U(1,0)$ by phase factors that depend on the sector on which it acts.

** We will see that this identification of the angular-momentum generators is forced on us by the commutation relations of the $J^{\mu\nu}$. On the other hand, the commutation relations do not allow us to distinguish between P^μ and $-P^\mu$, so the sign for the $\epsilon_\rho P^\rho$ term in (2.4.3) is a matter of convention. The consistency of the choice in (2.4.3) with the usual definition of the Hamiltonian P^0 is shown in Section 3.1.

We now examine the Lorentz transformation properties of $J^{\rho\sigma}$ and P^ρ. We consider the product

$$U(\Lambda,a)\, U(1+\omega,\epsilon)U^{-1}(\Lambda,a)\,,$$

where $\Lambda^\mu{}_\nu$ and a^μ are here the parameters of a new transformation, unrelated to ω and ϵ. According to Eq. (2.3.11), the product $U(\Lambda^{-1},-\Lambda^{-1}a)\,U(\Lambda,a)$ equals $U(1,0)$, so $U(\Lambda^{-1},-\Lambda^{-1}a)$ is the inverse of $U(\Lambda,a)$. It follows then from (2.3.11) that

$$U(\Lambda,a)U(1+\omega,\epsilon)U^{-1}(\Lambda,a) = U\left(\Lambda(1+\omega)\Lambda^{-1},\Lambda\epsilon-\Lambda\omega\Lambda^{-1}a\right)\,. \tag{2.4.6}$$

To first order in ω and ϵ, we have then

$$U(\Lambda,a)\left[\tfrac{1}{2}\omega_{\rho\sigma}J^{\rho\sigma}-\epsilon_\rho P^\rho\right]U^{-1}(\Lambda,a) = \tfrac{1}{2}(\Lambda\omega\Lambda^{-1})_{\mu\nu}J^{\mu\nu} -(\Lambda\epsilon-\Lambda\omega\Lambda^{-1}a)_\mu P^\mu\,. \tag{2.4.7}$$

Equating coefficients of $\omega_{\rho\sigma}$ and ϵ_ρ on both sides of this equation (and using (2.3.10)), we find

$$U(\Lambda,a)J^{\rho\sigma}U^{-1}(\Lambda,a) = \Lambda_\mu{}^\rho\Lambda_\nu{}^\sigma(J^{\mu\nu}-a^\mu P^\nu+a^\nu P^\mu)\,, \tag{2.4.8}$$

$$U(\Lambda,a)P^\rho\, U^{-1}(\Lambda,a) = \Lambda_\mu{}^\rho P^\mu\,. \tag{2.4.9}$$

For homogeneous Lorentz transformations (with $a^\mu=0$), these transformation rules simply say that $J^{\mu\nu}$ is a tensor and P^μ is a vector. For pure translations (with $\Lambda^\mu{}_\nu=\delta^\mu{}_\nu$), they tell us that P^ρ is translation-invariant, but $J^{\rho\sigma}$ is not. In particular, the change of the space–space components of $J^{\rho\sigma}$ under a spatial translation is just the usual change of the angular momentum under a change of the origin relative to which the angular momentum is calculated.

Next, let's apply rules (2.4.8), (2.4.9) to a transformation that is itself infinitesimal, i.e., $\Lambda^\mu{}_\nu=\delta^\mu{}_\nu+\omega^\mu{}_\nu$ and $a^\mu=\epsilon^\mu$, with infinitesimals $\omega^\mu{}_\nu$ and ϵ^μ unrelated to the previous ω and ϵ. Using Eq. (2.4.3), and keeping only terms of first order in $\omega^\mu{}_\nu$ and ϵ^μ, Eqs. (2.4.8) and (2.4.9) now become

$$i\left[\tfrac{1}{2}\omega_{\mu\nu}J^{\mu\nu}-\epsilon_\mu P^\mu\,,\,J^{\rho\sigma}\right] = \omega_\mu{}^\rho J^{\mu\sigma}+\omega_\nu{}^\sigma J^{\rho\nu}-\epsilon^\rho P^\sigma+\epsilon^\sigma P^\rho\,, \tag{2.4.10}$$

$$i\left[\tfrac{1}{2}\omega_{\mu\nu}J^{\mu\nu}-\epsilon_\mu P^\mu\,,\,P^\rho\right] = \omega_\mu{}^\rho P^\mu\,. \tag{2.4.11}$$

Equating coefficients of $\omega_{\mu\nu}$ and ϵ_μ on both sides of these equations, we find the commutation rules

$$i\,[J^{\mu\nu},J^{\rho\sigma}] = \eta^{\nu\rho}J^{\mu\sigma}-\eta^{\mu\rho}J^{\nu\sigma}-\eta^{\sigma\mu}J^{\rho\nu}+\eta^{\sigma\nu}J^{\rho\mu}\,, \tag{2.4.12}$$

$$i\,[P^\mu\,,\,J^{\rho\sigma}] = \eta^{\mu\rho}P^\sigma-\eta^{\mu\sigma}P^\rho\,, \tag{2.4.13}$$

$$[P^\mu\,,\,P^\rho] = 0\,. \tag{2.4.14}$$

This is the Lie algebra of the Poincaré group.

In quantum mechanics a special role is played by those operators that are *conserved*, i.e., that commute with the energy operator $H = P^0$. Inspection of Eqs. (2.4.13) and (2.4.14) shows that these are the momentum three-vector

$$\mathbf{P} = \left\{P^1, P^2, P^3\right\} \tag{2.4.15}$$

and the angular-momentum three-vector

$$\mathbf{J} = \left\{J^{23}, J^{31}, J^{12}\right\} \tag{2.4.16}$$

and, of course, the energy P^0 itself. The remaining generators form what is called the 'boost' three-vector

$$\mathbf{K} = \left\{J^{01}, J^{02}, J^{03}\right\} . \tag{2.4.17}$$

These are *not* conserved, which is why we do not use the eigenvalues of $\mathbf{K}$ to label physical states. In a three-dimensional notation, the commutation relations (2.4.12), (2.4.13), (2.4.14) may be written

$$[J_i, J_j] = i\,\epsilon_{ijk} J_k \,, \tag{2.4.18}$$
$$[J_i, K_j] = i\,\epsilon_{ijk} K_k \,, \tag{2.4.19}$$
$$[K_i, K_j] = -i\,\epsilon_{ijk} J_k \,, \tag{2.4.20}$$
$$[J_i, P_j] = i\,\epsilon_{ijk} P_k \,, \tag{2.4.21}$$
$$[K_i, P_j] = -iH\delta_{ij} \,, \tag{2.4.22}$$
$$[J_i, H] = [P_i, H] = [H, H] = 0 \,, \tag{2.4.23}$$
$$[K_i, H] = -i\,P_i \,, \tag{2.4.24}$$

where i, j, k, etc. run over the values 1, 2, and 3, and ϵ_{ijk} is the totally antisymmetric quantity with $\epsilon_{123} = +1$. The commutation relation (2.4.18) will be recognized as that of the angular-momentum operator.

The pure translations $T(1, a)$ form a subgroup of the inhomogeneous Lorentz group with a group multiplication rule given by (2.3.8) as

$$T(1, \bar{a})T(1, a) = T(1, \bar{a} + a). \tag{2.4.25}$$

This is additive in the same sense as (2.2.24), so by using (2.4.3) and repeating the same arguments that led to (2.2.26), we find that finite translations are represented on the physical Hilbert space by

$$U(1, a) = \exp(-iP^\mu a_\mu). \tag{2.4.26}$$

In exactly the same way, we can show that a rotation $R_{\boldsymbol{\theta}}$ by an angle $|\boldsymbol{\theta}|$ around the direction of $\boldsymbol{\theta}$ is represented on the physical Hilbert space by

$$U(R_{\boldsymbol{\theta}}, 0) = \exp(i\mathbf{J} \cdot \boldsymbol{\theta}). \tag{2.4.27}$$

It is interesting to compare the Poincaré algebra with the Lie algebra of the symmetry group of Newtonian mechanics, the Galilean group. We could derive this algebra by starting with the transformation rules of the Galilean group and then following the same procedure that was used here to derive the Poincaré algebra. However, since we already have Eqs. (2.4.18)–(2.4.24), it is easier to obtain the Galilean algebra as a low-velocity limit of the Poincaré algebra, by what is known as an *Inönü–Wigner contraction*[4,5]. For a system of particles of typical mass m and typical velocity v, the momentum and the angular-momentum operators are expected to be of order $\mathbf{J} \sim 1$, $\mathbf{P} \sim mv$. On the other hand, the energy operator is $H = M + W$ with a total mass M and non-mass energy W (kinetic plus potential) of order $M \sim m$, $W \sim mv^2$. Inspection of Eqs. (2.4.18)–(2.4.24) shows that these commutation relations have a limit for $v << 1$ of the form

$$[J_i, J_j] = i\,\epsilon_{ijk}\,J_k\,, \qquad [J_i, K_j] = i\,\epsilon_{ijk}\,K_k\,, \qquad [K_i, K_j] = 0\,,$$
$$[J_i, P_j] = i\,\epsilon_{ijk}\,P_k\,, \qquad [K_i, P_j] = -i\,M\delta_{ij}\,,$$
$$[J_i, W] = [P_i, W] = 0\,, \qquad [K_i, W] = -i\,P_i\,,$$
$$[J_i, M] = [P_i, M] = [K_i, M] = [W, M] = 0\,,$$

with $\mathbf{K}$ of order $1/v$. Note that the product of a translation $\mathbf{x} \to \mathbf{x} + \mathbf{a}$ and a 'boost' $\mathbf{x} \to \mathbf{x} + \mathbf{v}t$ should be the transformation $\mathbf{x} \to \mathbf{x} + \mathbf{v}t + \mathbf{a}$, but this is not true for the action of these operators on Hilbert space:

$$\exp(-i\mathbf{K}\cdot\mathbf{v})\,\exp(-i\mathbf{P}\cdot\mathbf{a}) = \exp(iM\mathbf{a}\cdot\mathbf{v}/2)\,\exp\left(-i(\mathbf{K}\cdot\mathbf{v} + \mathbf{P}\cdot\mathbf{a})\right)\,.$$

The appearance of the phase factor $\exp(iM\mathbf{a}\cdot\mathbf{v}/2)$ shows that this is a projective representation, with a superselection rule forbidding the superposition of states of different mass. In this respect, the mathematics of the Poincaré group is simpler than that of the Galilean group. However, there is nothing to prevent us from formally enlarging the Galilean group, by adding one more generator to its Lie algebra, which commutes with all the other generators, and whose eigenvalues are the masses of the various states. In this case physical states provide an ordinary rather than a projective representation of the expanded symmetry group. The difference appears to be a mere matter of notation, except that with this reinterpretation of the Galilean group there is no need for a mass superselection rule.

2.5 One-Particle States

We now consider the classification of one-particle states according to their transformation under the inhomogeneous Lorentz group.

The components of the energy-momentum four-vector all commute with each other, so it is natural to express physical state-vectors in terms of eigenvectors of the four-momentum. Introducing a label σ to denote all other degrees of freedom, we thus consider state-vectors $\Psi_{p,\sigma}$ with

$$P^\mu \Psi_{p,\sigma} = p^\mu \Psi_{p,\sigma} \,. \tag{2.5.1}$$

For general states, consisting for instance of several unbound particles, the label σ would have to be allowed to include continuous as well as discrete labels. We take as part of the definition of a *one*-particle state, that the label σ is purely discrete, and will limit ourselves here to that case. (However, a specific bound state of two or more particles, such as the lowest state of the hydrogen atom, is to be considered as a one-particle state. It is not an *elementary* particle, but the distinction between composite and elementary particles is of no relevance here.)

Eqs. (2.5.1) and (2.4.26) tell us how the states $\Psi_{p,\sigma}$ transform under translations:

$$U(1,a)\Psi_{p,\sigma} = e^{-ip\cdot a}\Psi_{p,\sigma} \,.$$

We must now consider how these states transform under homogeneous Lorentz transformations.

Using (2.4.9), we see that the effect of operating on $\Psi_{p,\sigma}$ with a quantum homogeneous Lorentz transformation $U(\Lambda, 0) \equiv U(\Lambda)$ is to produce an eigenvector of the four-momentum with eigenvalue Λp

$$\begin{aligned} P^\mu U(\Lambda)\Psi_{p,\sigma} &= U(\Lambda)\left[U^{-1}(\Lambda)P^\mu U(\Lambda)\right]\Psi_{p,\sigma} = U(\Lambda)(\Lambda_\rho^{-1\,\mu} P^\rho)\Psi_{p,\sigma} \\ &= \Lambda^\mu{}_\rho p^\rho U(\Lambda)\Psi_{p,\sigma} \,. \end{aligned} \tag{2.5.2}$$

Hence $U(\Lambda)\Psi_{p,\sigma}$ must be a linear combination of the state-vectors $\Psi_{\Lambda p,\sigma'}$:

$$U(\Lambda)\Psi_{p,\sigma} = \sum_{\sigma'} C_{\sigma'\sigma}(\Lambda, p)\Psi_{\Lambda p,\sigma'} \,. \tag{2.5.3}$$

In general, it may be possible by using suitable linear combinations of the $\Psi_{p,\sigma}$ to choose the σ labels in such a way that the matrix $C_{\sigma'\sigma}(\Lambda, p)$ is block-diagonal; in other words, so that the $\Psi_{p,\sigma}$ with σ within any one block by themselves furnish a representation of the inhomogeneous Lorentz group. It is natural to identify the states of a specific particle type with the components of a representation of the inhomogeneous Lorentz group which is irreducible, in the sense that it cannot be further decomposed in this way.* Our task now is to work out the structure of the

---

* Of course, different particle species may correspond to representations that are isomorphic, i.e., that have matrices $C_{\sigma'\sigma}(\Lambda, p)$ that are either identical, or identical up to a similarity transformation. In some cases it may be convenient to define particle types as irreducible representations of larger groups that contain the inhomogeneous proper orthochronous Lorentz group as a subgroup; for instance, as we shall see, for massless particles whose interactions respect the symmetry of space inversion it is customary to treat all the components of an irreducible representation of the inhomogeneous Lorentz group including space inversion as a single particle type.

coefficients $C_{\sigma'\sigma}(\Lambda, p)$ in irreducible representations of the inhomogeneous Lorentz group.

For this purpose, note that the only functions of p^μ that are left invariant by all proper orthochronous Lorentz transformations $\Lambda^\mu{}_\nu$ are the invariant square $p^2 \equiv \eta_{\mu\nu}p^\mu p^\nu$, and for $p^2 \leq 0$, also the sign of p^0. Hence, for each value of p^2, and (for $p^2 \leq 0$) each sign of p^0, we can choose a 'standard' four-momentum, say k^μ, and express any p^μ of this class as

$$p^\mu = L^\mu{}_\nu(p)k^\nu \,, \tag{2.5.4}$$

where $L^\mu{}_\nu$ is some standard Lorentz transformation that depends on p^μ, and also implicitly on our choice of the standard k^μ. We can then *define* the states $\Psi_{p,\sigma}$ of momentum p by

$$\Psi_{p,\sigma} \equiv N(p)\, U(L(p))\Psi_{k,\sigma} \,, \tag{2.5.5}$$

where $N(p)$ is a numerical normalization factor, to be chosen later. Up to this point, we have said nothing about how the σ labels are related for different momenta; Eq. (2.5.5) now fills that gap.

Operating on (2.5.5) with an arbitrary homogeneous Lorentz transformation $U(\Lambda)$, we now find

$$\begin{aligned} U(\Lambda)\Psi_{p,\sigma} &= N(p)\, U(\Lambda L(p))\Psi_{k,\sigma} \\ &= N(p)\, U(L(\Lambda p))\, U(L^{-1}(\Lambda p)\Lambda L(p))\, \Psi_{k,\sigma} \,. \end{aligned} \tag{2.5.6}$$

The point of this last step is that the Lorentz transformation $L^{-1}(\Lambda p)\Lambda L(p)$ takes k to $L(p)k = p$, and then to Λp, and then back to k, so it belongs to the subgroup of the homogeneous Lorentz group consisting of Lorentz transformations $W^\mu{}_\nu$ that leave k^μ invariant:

$$W^\mu{}_\nu k^\nu = k^\mu. \tag{2.5.7}$$

This subgroup is called the *little group*.[5] For any W satisfying Eq. (2.5.7), we have

$$U(W)\Psi_{k,\sigma} = \sum_{\sigma'} D_{\sigma'\sigma}(W)\Psi_{k,\sigma'} \,. \tag{2.5.8}$$

The coefficients $D(W)$ furnish a representation of the little group; that is, for any elements $\bar{W}, W$ we have

$$\begin{aligned} \sum_{\sigma'} D_{\sigma'\sigma}(\bar{W}W)\Psi_{k,\sigma'} &= U(\bar{W}W)\Psi_{k,\sigma} = U(\bar{W})U(W)\Psi_{k,\sigma} \\ &= U(\bar{W})\sum_{\sigma''} D_{\sigma''\sigma}(W)\Psi_{k,\sigma''} = \sum_{\sigma'\sigma''} D_{\sigma''\sigma}(W)D_{\sigma'\sigma''}(\bar{W})\Psi_{k,\sigma'} \end{aligned}$$

and so

$$D_{\sigma'\sigma}(\bar{W}W) = \sum_{\sigma''} D_{\sigma'\sigma''}(\bar{W})D_{\sigma''\sigma}(W) \,. \tag{2.5.9}$$

In particular, we may apply Eq. (2.5.8) to the little-group transformation

$$W(\Lambda,p) \equiv L^{-1}(\Lambda p)\Lambda L(p) \tag{2.5.10}$$

and then Eq. (2.5.6) takes the form

$$U(\Lambda)\Psi_{p,\sigma} = N(p)\sum_{\sigma'} D_{\sigma'\sigma}(W(\Lambda,p))\,U\left(L(\Lambda p)\right)\Psi_{k,\sigma'}\,,$$

or, recalling the definition (2.5.5):

$$U(\Lambda)\Psi_{p,\sigma} = \left(\frac{N(p)}{N(\Lambda p)}\right)\sum_{\sigma'} D_{\sigma'\sigma}(W(\Lambda,p))\,\Psi_{\Lambda p,\sigma'}\,. \tag{2.5.11}$$

Apart from the question of normalization, the problem of determining the coefficients $C_{\sigma'\sigma}$ in the transformation rule (2.5.3) has been reduced to the problem of finding the representations of the little group. This approach, of deriving representations of a group like the inhomogeneous Lorentz group from the representations of a little group, is called the method of induced representations.[6]

Table 2.1 gives a convenient choice of the standard momentum k^μ and the corresponding little group for the various classes of four-momenta.

Of these six classes of four-momenta, only (a), (c), and (f) have any known interpretations in terms of physical states. Not much needs to be said here about case (f) — $p^\mu = 0$; it describes the vacuum, which is simply left invariant by $U(\Lambda)$. In what follows we will consider only cases (a) and (c), which cover particles of mass $M > 0$ and mass zero, respectively.

This is a good place to pause, and say something about the normalization of these states. By the usual orthonormalization procedure of quantum mechanics, we may choose the states with standard momentum k^μ to be orthonormal, in the sense that

$$(\Psi_{k',\sigma'},\Psi_{k,\sigma}) = \delta^3(\mathbf{k}'-\mathbf{k})\delta_{\sigma'\sigma}\,. \tag{2.5.12}$$

(The delta function appears here because $\Psi_{k,\sigma}$ and $\Psi_{k',\sigma'}$ are eigenstates of a Hermitian operator with eigenvalues $\mathbf{k}$ and $\mathbf{k}'$, respectively.) This has the immediate consequence that the representation of the little group in Eqs. (2.5.8) and (2.5.11) must be unitary**

$$D^\dagger(W) = D^{-1}(W)\,. \tag{2.5.13}$$

Now, what about scalar products for arbitrary momenta? Using the unitarity of the operator $U(\Lambda)$ in Eqs. (2.5.5) and (2.5.11), we find for the

** The little groups $SO(2,1)$ and $SO(3,1)$ for $p^2 > 0$ and $p^\mu = 0$ have no non-trivial finite-dimensional unitary representations, so if there were any states with a given momentum p^μ with $p^2 > 0$ or $p^\mu = 0$ that transform non-trivially under the little group, there would have to be an infinite number of them.

Table 2.1. Standard momenta and the corresponding little group for various classes of four-momenta. Here κ is an arbitrary positive energy, say 1 eV. The little groups are mostly pretty obvious: $SO(3)$ is the ordinary rotation group in three dimensions (excluding space inversions), because rotations are the only proper orthochronous Lorentz transformations that leave at rest a particle with zero momentum, while $SO(2,1)$ and $SO(3,1)$ are the Lorentz groups in $(2+1)$- and $(3+1)$-dimensions, respectively. The group $ISO(2)$ is the group of Euclidean geometry, consisting of rotations and translations in two dimensions. Its appearance as the little group for $p^2 = 0$ is explained below.

	Standard k^μ	Little Group
(a) $p^2 = -M^2 < 0, p^0 > 0$	$(0,0,0,M)$	$SO(3)$
(b) $p^2 = -M^2 < 0, p^0 < 0$	$(0,0,0,-M)$	$SO(3)$
(c) $p^2 = 0, p^0 > 0$	$(0,0,\kappa,\kappa)$	$ISO(2)$
(d) $p^2 = 0, p^0 < 0$	$(0,0,\kappa,-\kappa)$	$ISO(2)$
(e) $p^2 = N^2 > 0$	$(0,0,N,0)$	$SO(2,1)$
(f) $p^\mu = 0$	$(0,0,0,0)$	$SO(3,1)$

scalar product:

$$\begin{aligned}(\Psi_{p',\sigma'}, \Psi_{p,\sigma}) &= N(p)\left(U^{-1}(L(p))\Psi_{p',\sigma'},\ \Psi_{k,\sigma}\right)\\ &= N(p)N^*(p')D\left(W(L^{-1}(p),p')\right)^*_{\sigma\sigma'}\ \delta^3(\mathbf{k}'-\mathbf{k})\end{aligned}$$

where $k' \equiv L^{-1}(p)p'$. Since also $k = L^{-1}(p)p$, the delta function $\delta^3(\mathbf{k}-\mathbf{k}')$ is proportional to $\delta^3(\mathbf{p}-\mathbf{p}')$. For $p' = p$, the little-group transformation here is trivial, $W(L^{-1}(p),p) = 1$, and so the scalar product is

$$(\Psi_{p'\sigma'}, \Psi_{p,\sigma}) = |N(p)|^2 \delta_{\sigma'\sigma}\delta^3(\mathbf{k}'-\mathbf{k})\,. \tag{2.5.14}$$

It remains to work out the proportionality factor relating $\delta^3(\mathbf{k}-\mathbf{k}')$ and $\delta^3(\mathbf{p}-\mathbf{p}')$. Note that the Lorentz-invariant integral of an arbitrary scalar function $f(p)$ over four-momenta with $-p^2 = M^2 \geq 0$ and $p^0 > 0$ (i.e., cases (a) or (c)) may be written

$$\begin{aligned}
&\int d^4p\,\delta(p^2+M^2)\theta(p^0)f(p) \\
&= \int d^3\mathbf{p}\,dp^0\,\delta((p^0)^2-\mathbf{p}^2-M^2)\theta(p^0)f(\mathbf{p},p^0) \\
&= \int d^3\mathbf{p}\frac{f(\mathbf{p},\sqrt{\mathbf{p}^2+M^2})}{2\sqrt{\mathbf{p}^2+M^2}}
\end{aligned}$$

($\theta(p^0)$ is the step function: $\theta(x)=1$ for $x\geq 0$, $\theta(x)=0$ for $x<0$.) We see that when integrating on the 'mass shell' $p^2+M^2=0$, the invariant volume element is

$$d^3\mathbf{p}/\sqrt{\mathbf{p}^2+M^2}\,. \tag{2.5.15}$$

The delta function is defined by

$$\begin{aligned}
F(\mathbf{p}) &= \int F(\mathbf{p}')\delta^3(\mathbf{p}-\mathbf{p}')d^3\mathbf{p}' \\
&= \int F(\mathbf{p}')\left[\sqrt{\mathbf{p}'^2+M^2}\delta^3(\mathbf{p}'-\mathbf{p})\right]\frac{d^3\mathbf{p}'}{\sqrt{\mathbf{p}'^2+M^2}}
\end{aligned}$$

so we see that the invariant delta function is

$$\sqrt{\mathbf{p}'^2+M^2}\delta^3(\mathbf{p}'-\mathbf{p}) = p^0\delta^3(\mathbf{p}'-\mathbf{p})\,. \tag{2.5.16}$$

Since p' and p are related to k' and k respectively by a Lorentz transformation, $L(p)$, we have then

$$p^0\delta^3(\mathbf{p}'-\mathbf{p}) = k^0\delta^3(\mathbf{k}'-\mathbf{k})$$

and therefore

$$(\Psi_{p',\sigma'},\Psi_{p,\sigma}) = |N(p)|^2\delta_{\sigma'\sigma}\left(\frac{p^0}{k^0}\right)\delta^3(\mathbf{p}'-\mathbf{p})\,. \tag{2.5.17}$$

The normalization factor $N(p)$ is sometimes chosen to be just $N(p)=1$, but then we would need to keep track of the p^0/k^0 factor in scalar products. Instead, I will here adopt the more usual convention that

$$N(p) = \sqrt{k^0/p^0} \tag{2.5.18}$$

for which

$$(\Psi_{p',\sigma'},\Psi_{p,\sigma}) = \delta_{\sigma'\sigma}\delta^3(\mathbf{p}'-\mathbf{p})\,. \tag{2.5.19}$$

We now consider the two cases of physical interest: particles of mass $M>0$, and particles of zero mass.

Mass Positive-Definite

The little group here is the three-dimensional rotation group. Its unitary representations can be broken up into a direct sum of irreducible unitary representations[7] $D^{(j)}_{\sigma'\sigma}(R)$ of dimensionality $2j+1$, with $j = 0, \frac{1}{2}, 1, \cdots$. These can be built up from the standard matrices for infinitesimal rotations $R_{ik} = \delta_{ik} + \Theta_{ik}$, with $\Theta_{ik} = -\Theta_{ki}$ infinitesimal:

$$D^{(j)}_{\sigma'\sigma}(1+\Theta) = \delta_{\sigma'\sigma} + \frac{i}{2}\Theta_{ik}(J^{(j)}_{ik})_{\sigma'\sigma} \,, \tag{2.5.20}$$

$$\begin{aligned}(J^{(j)}_{23} \pm iJ^{(j)}_{31})_{\sigma'\sigma} &= (J^{(j)}_{1} \pm iJ^{(j)}_{2})_{\sigma'\sigma} \\ &= \delta_{\sigma',\sigma\pm 1}\sqrt{(j\mp\sigma)(j\pm\sigma+1)} \,,\end{aligned} \tag{2.5.21}$$

$$(J^{(j)}_{12})_{\sigma'\sigma} = (J^{(j)}_{3})_{\sigma'\sigma} = \sigma\delta_{\sigma'\sigma} \,, \tag{2.5.22}$$

with σ running over the values $j, j-1, \cdots, -j$. For a particle of mass $M > 0$ and spin j, Eq. (2.5.11) now becomes

$$U(\Lambda)\Psi_{p,\sigma} = \sqrt{\frac{(\Lambda p)^0}{p^0}} \sum_{\sigma'} D^{(j)}_{\sigma'\sigma}(W(\Lambda, p))\, \Psi_{\Lambda p,\sigma'} \,, \tag{2.5.23}$$

with the little-group element $W(\Lambda, p)$ (the Wigner rotation[5]) given by Eq. (2.5.10):

$$W(\Lambda, p) = L^{-1}(\Lambda p)\Lambda L(p).$$

To calculate this rotation, we need to choose a 'standard boost' $L(p)$ which carries the four-momentum from $k^\mu = (0, 0, 0, M)$ to p^μ. This is conveniently chosen as

$$\begin{aligned} L^i{}_k(p) &= \delta_{ik} + (\gamma - 1)\hat{p}_i\hat{p}_k \\ L^i{}_0(p) &= L^0{}_i(p) = \hat{p}_i\sqrt{\gamma^2 - 1} \,, \\ L^0{}_0(p) &= \gamma \,, \end{aligned} \tag{2.5.24}$$

where

$$\hat{p}_i \equiv p_i/|\mathbf{p}|, \qquad \gamma \equiv \sqrt{\mathbf{p}^2 + M^2}/M \,.$$

It is very important that when $\Lambda^\mu{}_\nu$ is an arbitrary three-dimensional rotation $\mathscr{R}$, the Wigner rotation $W(\Lambda, p)$ is the same as $\mathscr{R}$ for all p. To see this, note that the boost (2.5.24) may be expressed as

$$L(p) = R(\hat{\mathbf{p}})B(|\mathbf{p}|)R^{-1}(\hat{\mathbf{p}}) \,,$$

where $R(\hat{\mathbf{p}})$ is a rotation (to be defined in a standard way below, in Eq.

(2.5.47)) that takes the three-axis into the direction of $\mathbf{p}$, and

$$B(|\mathbf{p}|) = \begin{bmatrix} 1 & 0 & 0 & 0 \\ 0 & 1 & 0 & 0 \\ 0 & 0 & \gamma & \sqrt{\gamma^2 - 1} \\ 0 & 0 & \sqrt{\gamma^2 - 1} & \gamma \end{bmatrix} .$$

Then for an arbitrary rotation $\mathscr{R}$

$$W(\mathscr{R}, p) = R(\mathscr{R}\hat{\mathbf{p}})B^{-1}(|\mathbf{p}|)R^{-1}(\mathscr{R}\hat{\mathbf{p}})\mathscr{R}\, R(\hat{\mathbf{p}})B(|\mathbf{p}|)R^{-1}(\hat{\mathbf{p}}) .$$

But the rotation $R^{-1}(\mathscr{R}\hat{\mathbf{p}})\mathscr{R}\, R(\hat{\mathbf{p}})$ takes the three-axis into the direction $\hat{\mathbf{p}}$, and then into the direction $\mathscr{R}\hat{\mathbf{p}}$, and then back to the three-axis, so it must be just a rotation by some angle θ around the three-axis

$$R^{-1}(\mathscr{R}\hat{\mathbf{p}})\mathscr{R}\, R(\hat{\mathbf{p}}) = R(\theta) \equiv \begin{bmatrix} \cos\theta & \sin\theta & 0 & 0 \\ -\sin\theta & \cos\theta & 0 & 0 \\ 0 & 0 & 1 & 0 \\ 0 & 0 & 0 & 1 \end{bmatrix} .$$

Since $R(\theta)$ commutes with $B(|\mathbf{p}|)$, this now gives

$$W(\mathscr{R}, p) = R(\mathscr{R}\hat{\mathbf{p}})B^{-1}(|\mathbf{p}|)R(\theta)B(|\mathbf{p}|)R^{-1}(\hat{\mathbf{p}}) = R(\mathscr{R}\hat{\mathbf{p}})R(\theta)R^{-1}(\hat{\mathbf{p}})$$

and hence

$$W(\mathscr{R}, p) = \mathscr{R}$$

as was to be shown. Thus states of a moving massive particle (and, by extension, multi-particle states) have the same transformation under rotations as in non-relativistic quantum mechanics. This is another piece of good news — the whole apparatus of spherical harmonics, Clebsch–Gordan coefficients, etc. can be carried over wholesale from non-relativistic to relativistic quantum mechanics.

Mass Zero

First, we have to work out the structure of the little group. Consider an arbitrary little-group element $W^\mu{}_\nu$, with $W^\mu{}_\nu k^\nu = k^\mu$, where k^μ is the standard four-momentum for this case, $k^\mu = (0, 0, 1, 1)$. Acting on a time-like four-vector $t^\mu = (0, 0, 0, 1)$, such a Lorentz transformation must yield a four-vector Wt whose length and scalar product with $Wk = k$ are the same as those of t:

$$(Wt)^\mu (Wt)_\mu = t^\mu t_\mu = -1 ,$$
$$(Wt)^\mu k_\mu = t^\mu k_\mu = -1 .$$

Any four vector that satisfies the second condition may be written

$$(Wt)^\mu = (\alpha, \beta, \zeta, 1 + \zeta)$$

and the first condition then yields the relation

$$\zeta = (\alpha^2 + \beta^2)/2\,. \tag{2.5.25}$$

It follows that the effect of $W^\mu{}_\nu$ on t^ν is the same as that of the Lorentz transformation

$$S^\mu{}_\nu(\alpha,\beta) = \begin{bmatrix} 1 & 0 & -\alpha & \alpha \\ 0 & 1 & -\beta & \beta \\ \alpha & \beta & 1-\zeta & \zeta \\ \alpha & \beta & -\zeta & 1+\zeta \end{bmatrix}\,. \tag{2.5.26}$$

This does not mean that W equals $S(\alpha,\beta)$, but it does mean that $S^{-1}(\alpha,\beta)W$ is a Lorentz transformation that leaves the time-like four-vector (0,0,0,1) invariant, and is therefore a pure rotation. Also, $S^\mu{}_\nu$ like $W^\mu{}_\nu$ leaves the light-like four-vector (0,0,1,1) invariant, so $S^{-1}(\alpha,\beta)W$ must be a rotation by some angle θ around the three-axis

$$S^{-1}(\alpha,\beta)W = R(\theta)\,, \tag{2.5.27}$$

where

$$R^\mu{}_\nu(\theta) \equiv \begin{bmatrix} \cos\theta & \sin\theta & 0 & 0 \\ -\sin\theta & \cos\theta & 0 & 0 \\ 0 & 0 & 1 & 0 \\ 0 & 0 & 0 & 1 \end{bmatrix}\,.$$

The most general element of the little group is therefore of the form

$$W(\theta,\alpha,\beta) = S(\alpha,\beta)R(\theta)\,. \tag{2.5.28}$$

What group is this? We note that the transformations with $\theta = 0$ or with $\alpha = \beta = 0$ form subgroups:

$$S(\bar\alpha,\bar\beta)S(\alpha,\beta) = S(\bar\alpha+\alpha,\bar\beta+\beta) \tag{2.5.29}$$

$$R(\bar\theta)R(\theta) = R(\bar\theta+\theta)\,. \tag{2.5.30}$$

These subgroups are *Abelian* — that is, their elements all commute with each other. Furthermore, the subgroup with $\theta = 0$ is *invariant*, in the sense that its elements are transformed into other elements of the same subgroup by any member of the group

$$R(\theta)S(\alpha,\beta)R^{-1}(\theta) = S(\alpha\cos\theta+\beta\sin\theta, -\alpha\sin\theta+\beta\cos\theta)\,. \tag{2.5.31}$$

From Eqs. (2.5.29)–(2.5.31) we can work out the product of any group elements. The reader will recognize these multiplication rules as those of the group $ISO(2)$, consisting of translations (by a vector (α,β)) and rotations (by an angle θ) in two dimensions.

Groups that do *not* have invariant Abelian subgroups have certain simple properties, and for this reason are called *semi-simple*. As we have seen, the little group $ISO(2)$ like the inhomogeneous Lorentz group is *not* semi-simple, and this leads to interesting complications. First, let's take a look at the Lie algebra of $ISO(2)$. For θ, α, β infinitesimal, the general group element is

$$W(\theta, \alpha, \beta)^{\mu}{}_{\nu} = \delta^{\mu}{}_{\nu} + \omega^{\mu}{}_{\nu} ,$$

$$\omega_{\mu\nu} = \begin{bmatrix} 0 & \theta & -\alpha & \alpha \\ -\theta & 0 & -\beta & \beta \\ \alpha & \beta & 0 & 0 \\ -\alpha & -\beta & 0 & 0 \end{bmatrix} .$$

From (2.4.3), we see then that the corresponding Hilbert space operator is

$$U(W(\theta, \alpha, \beta)) = 1 + i\alpha A + i\beta B + i\theta J_3 , \tag{2.5.32}$$

where A and B are the Hermitian operators

$$A = -J^{13} + J^{10} = J_2 + K_1 , \tag{2.5.33}$$

$$B = -J^{23} + J^{20} = -J_1 + K_2 , \tag{2.5.34}$$

and, as before, $J_3 = J_{12}$. Either from (2.4.18)–(2.4.20), or directly from Eqs. (2.5.29)–(2.5.31), we see that these generators have the commutators

$$[J_3, A] = +iB , \tag{2.5.35}$$

$$[J_3, B] = -iA , \tag{2.5.36}$$

$$[A, B] = 0 . \tag{2.5.37}$$

Since A and B are commuting Hermitian operators they (like the momentum generators of the inhomogeneous Lorentz group) can be simultaneously diagonalized by states $\Psi_{k,a,b}$

$$A\Psi_{k,a,b} = a\Psi_{k,a,b} ,$$

$$B\Psi_{k,a,b} = b\Psi_{k,a,b} .$$

The problem is that if we find one such set of non-zero eigenvalues of A, B, then we find a whole continuum. From Eq. (2.5.31), we have

$$U[R(\theta)]A\, U^{-1}[R(\theta)] = A\cos\theta - B\sin\theta ,$$

$$U[R(\theta)]B\, U^{-1}[R(\theta)] = A\sin\theta + B\cos\theta ,$$

and so, for arbitrary θ,

$$A\Psi^{\theta}_{k,a,b} = (a\cos\theta - b\sin\theta)\Psi^{\theta}_{k,a,b} ,$$

$$B\Psi^{\theta}_{k,a,b} = (a\sin\theta + b\cos\theta)\Psi^{\theta}_{k,a,b} ,$$

where

$$\Psi^{\theta}_{k,a,b} \equiv U^{-1}\left(R(\theta)\right)\Psi_{k,a,b} .$$

Massless particles are not observed to have any continuous degree of freedom like θ; to avoid such a continuum of states, we must require that physical states (now called $\Psi_{k,\sigma}$) are eigenvectors of A and B with $a = b = 0$:

$$A\Psi_{k,\sigma} = B\Psi_{k,\sigma} = 0 . \tag{2.5.38}$$

These states are then distinguished by the eigenvalue of the remaining generator

$$J_3\Psi_{k,\sigma} = \sigma\Psi_{k,\sigma} . \tag{2.5.39}$$

Since the momentum $\mathbf{k}$ is in the three-direction, σ gives the component of angular momentum in the direction of motion, or *helicity*.

We are now in a position to calculate the Lorentz transformation properties of general massless particle states. First note that by use of the general arguments of Section 2.2, Eq. (2.5.32) generalizes for finite α and β to

$$U(S(\alpha, \beta)) = \exp(i\alpha A + i\beta B) \tag{2.5.40}$$

and for finite θ to

$$U(R(\theta)) = \exp(iJ_3\theta) . \tag{2.5.41}$$

An arbitrary element W of the little group can be put in the form (2.5.28), so that

$$U(W)\Psi_{k,\sigma} = \exp(i\alpha A + i\beta B)\exp(i\theta J_3)\Psi_{k,\sigma} = \exp(i\theta\sigma)\Psi_{k,\sigma}$$

and therefore Eq. (2.5.8) gives

$$D_{\sigma'\sigma}(W) = \exp(i\theta\sigma)\delta_{\sigma'\sigma} ,$$

where θ is the angle defined by expressing W as in Eq. (2.5.28). The Lorentz transformation rule for a massless particle of arbitrary helicity is now given by Eqs. (2.5.11) and (2.5.18) as

$$U(\Lambda)\Psi_{p,\sigma} = \sqrt{\frac{(\Lambda p)^0}{p^0}} \exp\left(i\sigma\theta(\Lambda, p)\right) \Psi_{\Lambda p,\sigma} \tag{2.5.42}$$

with $\theta(\Lambda, p)$ defined by

$$W(\Lambda, p) \equiv L^{-1}(\Lambda p)\Lambda L(p) \equiv S\left(\alpha(\Lambda, p), \beta(\Lambda, p)\right) R\left(\theta(\Lambda, p)\right) . \tag{2.5.43}$$

We shall see in Section 5.9 that electromagnetic gauge invariance arises from the part of the little group parameterized by α and β.

At this point we have not yet encountered any reason that would forbid the helicity σ of a massless particle from being an arbitrary real number. As we shall see in Section 2.7, there are topological considerations that restrict the allowed values of σ to integers and half-integers, just as for massive particles.

To calculate the little-group element (2.5.43) for a given Λ and p, (and also to enable us to calculate the effect of space or time inversion on these states in the next section) we need to fix a convention for the standard Lorentz transformation that takes us from $k^\mu = (0, 0, \kappa, \kappa)$ to p^μ. This may conveniently be chosen to have the form

$$L(p) = R(\hat{\mathbf{p}})B(|\mathbf{p}|/\kappa) \tag{2.5.44}$$

where $B(u)$ is a pure boost along the three-direction:

$$B(u) \equiv \begin{bmatrix} 1 & 0 & 0 & 0 \\ 0 & 1 & 0 & 0 \\ 0 & 0 & (u^2+1)/2u & (u^2-1)/2u \\ 0 & 0 & (u^2-1)/2u & (u^2+1)/2u \end{bmatrix} \tag{2.5.45}$$

and $R(\hat{\mathbf{p}})$ is a pure rotation that carries the three-axis into the direction of the unit vector $\hat{\mathbf{p}}$. For instance, suppose we take $\hat{\mathbf{p}}$ to have polar and azimuthal angles θ and ϕ:

$$\hat{\mathbf{p}} = (\sin\theta\cos\phi,\ \sin\theta\sin\phi,\ \cos\theta)\,. \tag{2.5.46}$$

Then we can take $R(\hat{\mathbf{p}})$ as a rotation by angle θ around the two-axis, which takes $(0, 0, 1)$ into $(\sin\theta,\ 0, \cos\theta)$, followed by a rotation by angle ϕ around the three-axis:

$$U(R(\hat{\mathbf{p}})) = \exp(-i\phi J_3)\exp(-i\theta J_2)\,, \tag{2.5.47}$$

where $0 \le \theta \le \pi,\quad 0 \le \phi < 2\pi$. (We give $U(R(\hat{\mathbf{p}}))$ rather than $R(\hat{\mathbf{p}})$, together with a specification of the range of ϕ and θ, because shifting θ or ϕ by 2π would give the same rotation $R(\hat{\mathbf{p}})$, but a different sign for $U(R(\hat{\mathbf{p}}))$ when acting on half-integer spin states.) Since (2.5.47) is a rotation, and does take the three-axis into the direction (2.5.46), any other choice of such an $R(\hat{\mathbf{p}})$ would differ from this one by at most an initial rotation around the three-axis, corresponding to a mere redefinition of the phase of the one-particle states.

Note that the helicity is Lorentz-invariant; a massless particle of a given helicity σ looks the same (aside from its momentum) in all inertial frames. Indeed, we would be justified in thinking of massless particles of each different helicity as different species of particles. However, as we shall see in the next section, particles of opposite helicity are related by the symmetry of space inversion. Thus, because electromagnetic and gravitational forces obey space inversion symmetry, the massless particles of helicity ± 1 associated with electromagnetic phenomena are both called *photons*, and the massless particles of helicity ± 2 that are believed to be associated with gravitation are both called *gravitons*. On the other hand, the supposedly massless particles of helicity $\pm 1/2$ that are emitted in nuclear beta decay have no interactions (apart from gravitation) that respect the symmetry of space inversion, so these particles are given different names: *neutrinos* for helicity $-1/2$, and *antineutrinos* for helicity $+1/2$.

Even though the helicity of a massless particle is Lorentz-invariant, the state itself is not. In particular, because of the helicity-dependent phase factor $\exp(i\sigma\theta)$ in Eq. (2.5.42), a state formed as a linear superposition of one-particle states with opposite helicities will be changed by a Lorentz transformation into a different superposition. For instance, a general one-photon state of four-momentum p may be written

$$\Psi_{p;\alpha} = \alpha_+ \Psi_{p,+1} + \alpha_- \Psi_{p,-1} ,$$

where

$$|\alpha_+|^2 + |\alpha_-|^2 = 1 .$$

The generic case is one of *elliptic polarization*, with $|\alpha_\pm|$ both non-zero and unequal. *Circular polarization* is the limiting case where either α_+ or α_- vanishes, and *linear polarization* is the opposite extreme, with $|\alpha_+| = |\alpha_-|$. The overall phase of α_+ and α_- has no physical significance, and for linear polarization may be adjusted so that $\alpha_- = \alpha_+^*$, but the relative phase is still important. Indeed, for linear polarizations with $\alpha_- = \alpha_+^*$, the phase of α_+ may be identified as the angle between the plane of polarization and some fixed reference direction perpendicular to $\mathbf{p}$. Eq. (2.5.42) shows that under a Lorentz transformation $\Lambda^\mu{}_\nu$, this angle rotates by an amount $\theta(\Lambda, p)$. Plane polarized gravitons can be defined in a similar way, and here Eq. (2.5.42) has the consequence that a Lorentz transformation Λ rotates the plane of polarization by an angle $2\theta(\Lambda, p)$.

2.6 Space Inversion and Time-Reversal

We saw in Section 2.3 that any homogeneous Lorentz transformation is either proper and orthochronous (i.e., $\text{Det}\Lambda = +1$ and $\Lambda^0{}_0 \geq +1$) or else equal to a proper orthochronous transformation times either $\mathscr{P}$ or $\mathscr{T}$ or $\mathscr{P}\mathscr{T}$, where $\mathscr{P}$ and $\mathscr{T}$ are the space inversion and time-reversal transformations

$$\mathscr{P}^\mu{}_\nu = \begin{bmatrix} -1 & 0 & 0 & 0 \\ 0 & -1 & 0 & 0 \\ 0 & 0 & -1 & 0 \\ 0 & 0 & 0 & 1 \end{bmatrix} , \quad \mathscr{T}^\mu{}_\nu = \begin{bmatrix} 1 & 0 & 0 & 0 \\ 0 & 1 & 0 & 0 \\ 0 & 0 & 1 & 0 \\ 0 & 0 & 0 & -1 \end{bmatrix} .$$

It used to be thought self-evident that the fundamental multiplication rule of the Poincaré group

$$U(\bar{\Lambda}, \bar{a})\, U(\Lambda, a) = U(\bar{\Lambda}\Lambda, \bar{\Lambda}a + \bar{a})$$

would be valid even if Λ and/or $\bar{\Lambda}$ involved factors of $\mathscr{P}$ or $\mathscr{T}$ or $\mathscr{P}\mathscr{T}$. In particular, it was believed that there are operators corresponding to $\mathscr{P}$ and $\mathscr{T}$ themselves:

$$\mathsf{P} \equiv U(\mathscr{P},0) \qquad \mathsf{T} \equiv U(\mathscr{T},0)$$

such that

$$\mathsf{P}\, U(\Lambda,a)\mathsf{P}^{-1} = U(\mathscr{P}\Lambda\mathscr{P}^{-1},\mathscr{P}a)\,, \tag{2.6.1}$$

$$\mathsf{T}\, U(\Lambda,a)\,\mathsf{T}^{-1} = U(\mathscr{T}\Lambda\mathscr{T}^{-1},\mathscr{T}a) \tag{2.6.2}$$

for any proper orthochronous Lorentz transformation $\Lambda^\mu{}_\nu$ and translation a^μ. These transformation rules incorporate most of what is meant when we say that P or T are 'conserved'.

In 1956–57 it became understood[8] that this is true for P only in the approximation in which one ignores the effects of weak interactions, such as those that produce nuclear beta decay. Time-reversal survived for a while, but in 1964 there appeared indirect evidence[9] that these properties of T are also only approximately satisfied. (See Section 3.3.) In what follows, we will make believe that operators P and T satisfying Eqs. (2.6.1) and (2.6.2) actually exist, but it should be kept in mind that this is only an approximation.

Let us apply Eqs. (2.6.1) and (2.6.2) in the case of an infinitesimal transformation, i.e.,

$$\Lambda^\mu{}_\nu = \delta^\mu{}_\nu + \omega^\mu{}_\nu \qquad a^\mu = \epsilon^\mu$$

with $\omega_{\mu\nu} = -\omega_{\nu\mu}$ and ϵ_μ both infinitesimal. Using (2.4.3), and equating coefficients of $\omega_{\rho\sigma}$ and ϵ_ρ in Eqs. (2.6.1) and (2.6.2), we obtain the P and T transformation properties of the Poincaré generators

$$\mathsf{P}\, iJ^{\rho\sigma}\mathsf{P}^{-1} = i\mathscr{P}_\mu{}^\rho\mathscr{P}_\nu{}^\sigma J^{\mu\nu}\,, \tag{2.6.3}$$

$$\mathsf{P}\, iP^\rho\mathsf{P}^{-1} = i\mathscr{P}_\mu{}^\rho P^\mu\,, \tag{2.6.4}$$

$$\mathsf{T}\, iJ^{\rho\sigma}\mathsf{T}^{-1} = i\mathscr{T}_\mu{}^\rho\mathscr{T}_\nu{}^\sigma J^{\mu\nu}\,, \tag{2.6.5}$$

$$\mathsf{T}\, iP^\rho\mathsf{T}^{-1} = i\mathscr{T}_\mu{}^\rho P^\mu\,. \tag{2.6.6}$$

This is much like Eqs. (2.4.8) and (2.4.9), except that we have not cancelled factors of i on both sides of these equations, because at this point we have not yet decided whether P and T are linear and unitary or antilinear and antiunitary.

The decision is an easy one. Setting $\rho = 0$ in Eq. (2.6.4) gives

$$\mathsf{P}\, iH\mathsf{P}^{-1} = iH\,,$$

where $H \equiv P^0$ is the energy operator. If P were antiunitary and antilinear then it would anticommute with i, so $\mathsf{P}H\mathsf{P}^{-1} = -H$. But then for any state Ψ of energy $E > 0$, there would have to be another state $\mathsf{P}^{-1}\Psi$ of energy $-E < 0$. There are no states of negative energy (energy less than that of the vacuum), so we are forced to choose the other alternative: P *is linear and unitary, and commutes rather than anticommutes with H.*

On the other hand, setting $\rho = 0$ in Eq. (2.6.6) yields

$$\mathsf{T}\, iH\mathsf{T}^{-1} = -iH\,.$$

If we supposed that T is linear and unitary then we could simply cancel the is, and find $\mathsf{T}H\mathsf{T}^{-1} = -H$, with the again disastrous conclusion that for any state Ψ of energy E there is another state $\mathsf{T}^{-1}\Psi$ of energy $-E$. To avoid this, we are forced here to conclude that T *is antilinear and antiunitary.*

Now that we have decided that P is linear and T is antilinear, we can conveniently rewrite Eqs. (2.6.3)–(2.6.6) in terms of the generators (2.4.15)–(2.4.17) in a three-dimensional notation

$$\mathsf{P}\mathbf{J}\mathsf{P}^{-1} = +\mathbf{J}\,, \tag{2.6.7}$$

$$\mathsf{P}\mathbf{K}\mathsf{P}^{-1} = -\mathbf{K}\,, \tag{2.6.8}$$

$$\mathsf{P}\mathbf{P}\mathsf{P}^{-1} = -\mathbf{P}\,, \tag{2.6.9}$$

$$\mathsf{T}\mathbf{J}\mathsf{T}^{-1} = -\mathbf{J}\,, \tag{2.6.10}$$

$$\mathsf{T}\mathbf{K}\mathsf{T}^{-1} = +\mathbf{K}\,, \tag{2.6.11}$$

$$\mathsf{T}\mathbf{P}\mathsf{T}^{-1} = -\mathbf{P}\,, \tag{2.6.12}$$

and, as shown before,

$$\mathsf{P}H\mathsf{P}^{-1} = \mathsf{T}H\mathsf{T}^{-1} = H\,. \tag{2.6.13}$$

It is physically sensible that P should preserve the sign of $\mathbf{J}$, because at least the orbital part is a vector product $\mathbf{r}\times\mathbf{p}$ of two vectors, both of which change sign under an inversion of the spatial coordinate system. On the other hand, T reverses $\mathbf{J}$, because after time-reversal an observer will see all bodies spinning in the opposite direction. Note by the way that Eq. (2.6.10) is consistent with the angular-momentum commutation relations $\mathbf{J}\times\mathbf{J} = i\mathbf{J}$, because T reverses not only $\mathbf{J}$, but also i. The reader can easily check that Eqs. (2.6.7)–(2.6.13) are consistent with all the commutation relations (2.4.18)–(2.4.24).

Let us now consider what P and T do to one-particle states:

$\mathsf{P}: M > 0$

The one-particle states $\Psi_{k,\sigma}$ are defined as eigenvectors of $\mathbf{P}, H$, and J_3 with eigenvalues $0, M$, and σ, respectively. From Eqs. (2.6.7), (2.6.9), and (2.6.13), we see that the same must be true of the state $\mathsf{P}\Psi_{k,\sigma}$, and therefore (barring degeneracies) these states can only differ by a phase

$$\mathsf{P}\Psi_{k,\sigma} = \eta_\sigma \Psi_{k,\sigma}$$

with a phase factor ($|\eta| = 1$) that may or may not depend on the spin σ.

To see that η_σ is σ-independent, we note from (2.5.8), (2.5.20), and (2.5.21) that

$$(J_1 \pm i J_2)\Psi_{k,\sigma} = \sqrt{(j \mp \sigma)(j \pm \sigma + 1)}\, \Psi_{k,\sigma\pm1} \,, \tag{2.6.14}$$

where j is the particle's spin. Operating on both sides with P, we find

$$\eta_\sigma = \eta_{\sigma\pm1}$$

and so η_σ is actually independent of σ. We therefore write

$$\mathsf{P}\Psi_{k,\sigma} = \eta\Psi_{k,\sigma} \tag{2.6.15}$$

with η a phase, known as the *intrinsic parity*, that depends only on the species of particle on which P acts.

To get to finite momentum states, we must apply the unitary operator $U(L(p))$ corresponding to the 'boost' (2.5.24):

$$\Psi_{p,\sigma} = \sqrt{M/p^0}\, U\left(L(p)\right) \Psi_{k,\sigma} \,.$$

We note that

$$\mathscr{P}L(p)\mathscr{P}^{-1} = L(\mathscr{P}p)$$
$$\mathscr{P}p = \left(-\mathbf{p}, \sqrt{\mathbf{p}^2 + M^2}\right)$$

so using Eqs. (2.6.1) and (2.6.15), we have

$$\mathsf{P}\Psi_{p,\sigma} = \sqrt{M/p^0}\, U(L(\mathscr{P}p))\eta\Psi_{k,\sigma}$$

or in other words

$$\mathsf{P}\Psi_{p,\sigma} = \eta\Psi_{\mathscr{P}p,\sigma} \,. \tag{2.6.16}$$

T : $M > 0$

From Eqs. (2.6.10), (2.6.12), and (2.6.13), we see that the effect of T on the zero-momentum one-particle state $\Psi_{k,\sigma}$ is to yield a state with

$$\mathbf{P}(\mathsf{T}\Psi_{k,\sigma}) = 0 \,,$$
$$H(\mathsf{T}\Psi_{k,\sigma}) = M(\mathsf{T}\Psi_{k,\sigma}) \,,$$
$$J_3(\mathsf{T}\Psi_{k,\sigma}) = -\sigma(\mathsf{T}\Psi_{k,\sigma}) \,,$$

and so

$$\mathsf{T}\Psi_{k,\sigma} = \zeta_\sigma\Psi_{k,-\sigma} \,,$$

where ζ_σ is a phase factor. Applying the operator T to (2.6.14), and recalling that T anticommutes with **J** *and i*, we find

$$(-J_1 \pm i J_2)\zeta_\sigma\Psi_{k,-\sigma} = \sqrt{(j \mp \sigma)(j \pm \sigma + 1)}\zeta_{\sigma\pm1}\, \Psi_{k,-\sigma\mp1} \,.$$

Using Eq. (2.6.14) again on the left, we see that the square-root factors cancel, and so

$$-\zeta_\sigma = \zeta_{\sigma\pm1} \,.$$

We write the solution as $\zeta_\sigma = \zeta(-)^{j-\sigma}$, with ζ another phase that depends only on the species of particle:

$$\mathsf{T}\Psi_{k,\sigma} = \zeta(-)^{j-\sigma}\Psi_{k,-\sigma} \,. \tag{2.6.17}$$

However, unlike the 'intrinsic parity' η, the time-reversal phase ζ has no physical significance. This is because we can redefine the one-particle states by a change of phase

$$\Psi_{k,\sigma} \to \Psi'_{k,\sigma} = \zeta^{1/2}\Psi_{k,\sigma} \,.$$

in such a way that the phase ζ is eliminated from the transformation rule

$$\mathsf{T}\Psi'_{k,\sigma} = \zeta^{*1/2}\mathsf{T}\Psi_{k,\sigma} = \zeta^{*1/2}\zeta(-)^{j-\sigma}\Psi_{k,-\sigma} = (-)^{j-\sigma}\Psi'_{k,-\sigma} \,.$$

In what follows we will keep the arbitrary phase ζ in Eq. (2.6.17), just to keep open our options in choosing the phase of the one-particle states, but it should be kept in mind that this phase is of no real importance.

To deal with states of finite momentum, we again apply the 'boost' (2.5.24). Note that

$$\mathscr{T} L(p) \mathscr{T}^{-1} = L(\mathscr{P}p) \,,$$

$$\mathscr{P}p = \left(-\mathbf{p}, \sqrt{\mathbf{p}^2 + M^2}\right) \,.$$

(That is, changing the sign of each element of $L^\mu{}_\nu$ with an odd number of *time*-indices is the same as changing the signs of elements with an odd number of *space*-indices.) Using Eqs. (2.6.2) and (2.5.5), we have then

$$\mathsf{T}\Psi_{p,\sigma} = \zeta(-)^{j-\sigma}\Psi_{\mathscr{P}p,-\sigma} \,. \tag{2.6.18}$$

P : *M* = 0

Acting on a state $\Psi_{k,\sigma}$, that is defined as an eigenvector of P^μ with eigenvalue $k^\mu = (0, 0, \kappa, \kappa)$ and an eigenvector of J_3 with eigenvalue σ, the parity operator P yields a state with four-momentum $(\mathscr{P}k)^\mu = (0, 0, -\kappa, \kappa)$ and J_3 equal to σ. Thus it takes a state of helicity (the component of spin along the direction of motion) σ into one of helicity $-\sigma$. As mentioned earlier, this shows that the existence of a space-inversion symmetry requires that any species of massless particle with non-zero helicity must be accompanied with another of opposite helicity. Because P does not leave the standard momentum invariant, it is convenient to consider instead the operator $U(R_2^{-1})\mathsf{P}$, where R_2 is a rotation that also takes k to $\mathscr{P}k$, conveniently chosen as a rotation by 180° around the two-axis

$$U(R_2) = \exp(i\pi J_2) \,. \tag{2.6.19}$$

Since $U(R_2^{-1})$ reverses the sign of J_3, we have

$$U(R_2^{-1})\mathsf{P}\Psi_{k,\sigma} = \eta_\sigma \Psi_{k,-\sigma} \tag{2.6.20}$$

with η_σ a phase factor. Now, $R_2^{-1}\mathscr{P}$ commutes with the Lorentz 'boost' (2.5.45), and $\mathscr{P}$ commutes with the rotation $R(\hat{p})$ which takes the three-direction into the direction of $\mathbf{p}$, so by operating on (2.5.5) with P, we find for a general four-momentum p^μ

$$\mathsf{P}\Psi_{p,\sigma} = \sqrt{\frac{\kappa}{p^0}}\, U\left(R(\hat{p})R_2 B\left(\frac{|\mathbf{p}|}{\kappa}\right)\right) U(R_2^{-1})\mathsf{P}\Psi_{k,\sigma}$$

$$= \sqrt{\frac{\kappa}{p^0}}\, \eta_\sigma\, U\left(R(\hat{p})R_2\, B\left(\frac{|\mathbf{p}|}{\kappa}\right)\right) \Psi_{k,-\sigma}\,.$$

Note that $R(\hat{\mathbf{p}})R_2$ is a rotation that takes the three-axis into the direction of $-\hat{\mathbf{p}}$, but $U(R(\hat{\mathbf{p}})R_2)$ is not quite equal to $U(R(-\hat{\mathbf{p}}))$. According to (2.5.47),

$$U(R(-\hat{\mathbf{p}})) = \exp\Big(-i(\phi \pm \pi)J_3\Big)\, \exp\Big(-i(\pi-\theta)J_2\Big)$$

with azimuthal angle chosen as $\phi + \pi$ or $\phi - \pi$ according to whether $0 \leq \phi < \pi$ or $\pi \leq \phi < 2\pi$, so that it remains in the range of 0 to 2π. Then

$$\begin{aligned} U^{-1}\Big(R(-\hat{\mathbf{p}})\Big)U\Big(R(\hat{\mathbf{p}})R_2\Big) &= \exp\Big(i(\pi-\theta)J_2\Big) \\ &\times \exp\Big(i(\phi\pm\pi)J_3\Big)\exp(-i\phi J_3)\exp(-i\theta J_2)\exp(i\pi J_2) \\ &= \exp\Big(i(\pi-\theta)J_2\Big)\exp(\pm i\pi J_3)\exp\Big(i(\pi-\theta)J_2\Big)\,. \end{aligned}$$

But a rotation of $\pm 180°$ around the three-axis reverses the sign of J_2, so

$$U\Big(R(\hat{\mathbf{p}})R_2\Big) = U\Big(R(-\hat{\mathbf{p}})\Big)\exp(\pm i\pi J_3)\,. \tag{2.6.21}$$

Also, $R(-\hat{\mathbf{p}})B(|\mathbf{p}|/k)$ is just the standard boost $L(\mathscr{P}p)$ in the direction $\mathscr{P}p = (-\mathbf{p}, p^0)$. We have then finally

$$\mathsf{P}\Psi_{p,\sigma} = \eta_\sigma \exp(\mp i\pi\sigma)\Psi_{\mathscr{P}p,-\sigma} \tag{2.6.22}$$

with the phase $-\pi\sigma$ or $+\pi\sigma$ according to whether the two-component of $\mathbf{p}$ is positive or negative, respectively. This peculiar change of sign in the operation of parity for massless particles of half-integer spin is due to the convention adopted in Eq. (2.5.47) for the rotation used to define massless particle states of arbitrary momentum. Because the rotation group is not simply connected, some discontinuity of this sort is unavoidable.

T : $M = 0$

Acting on the state $\Psi_{k,\sigma}$, which has values $k^\mu = (0,0,\kappa,\kappa)$ and σ for P^μ and J_3, the time-reversal operator T yields a state which has values $(\mathscr{P}k)^\mu = (0,0,-\kappa,\kappa)$ and $-\sigma$ for P^μ and J_3. Thus T does not change the helicity $\mathbf{J}\cdot\hat{k}$, and by itself has nothing to say about whether massless particles of one helicity σ are accompanied with others of helicity $-\sigma$. Because T like P does not leave the standard four-momentum k invariant, it is convenient to consider the generator $U(R_2^{-1})\mathsf{T}$, where R_2 is the rotation (2.6.19), which also takes k into $\mathscr{P}k$. This commutes with J_3, so

$$U(R_2^{-1})\,\mathsf{T}\,\Psi_{k,\sigma} = \zeta_\sigma \Psi_{k,\sigma} \tag{2.6.23}$$

with ζ_σ another phase. Since $R_2^{-1}\mathscr{T}$ commutes with the boost (2.5.45), and $\mathscr{T}$ commutes with the rotation $R(\hat{p})$, operating with T on the state (2.5.5) gives

$$\mathsf{T}\Psi_{p,\sigma} = \sqrt{\frac{\kappa}{p_0}}\, U\left[R(\hat{p})R_2\, B\left(\frac{|\mathbf{p}|}{\kappa}\right)\right] \zeta_\sigma \Psi_{k,\sigma} \,. \tag{2.6.24}$$

Using Eq. (2.6.21), this yields finally

$$\mathsf{T}\Psi_{p,\sigma} = \zeta_\sigma \exp(\pm i\pi\sigma)\Psi_{\mathscr{P}p,\sigma} \,. \tag{2.6.25}$$

Again, the top or bottom sign applies according to whether the two-component of $\mathbf{p}$ is positive or negative, respectively.

* * *

It is interesting that the square T^2 of the time-reversal operator has a very simple action on both massive and massless one-particle states. Using Eq. (2.6.18), and recalling that T is antiunitary, we see that for massive one-particle states:

$$\mathsf{T}^2\Psi_{p,\sigma} = \mathsf{T}\zeta(-)^{j-\sigma}\Psi_{\mathscr{P}p,-\sigma} = \zeta^*(-)^{j-\sigma}\zeta(-)^{j+\sigma}\Psi_{p,\sigma}$$

or in other words

$$\mathsf{T}^2\Psi_{p,\sigma} = (-)^{2j}\Psi_{p,\sigma} \,. \tag{2.6.26}$$

We get the same result for massless particles. If the two-component of $\mathbf{p}$ is positive then the two-component of $\mathscr{P}\mathbf{p}$ is negative, and vice-versa, so Eq. (2.6.25) gives

$$\begin{aligned}\mathsf{T}^2\Psi_{p,\sigma} &= \mathsf{T}\zeta_\sigma \exp(\pm i\pi\sigma)\Psi_{\mathscr{P}p,\sigma} = \zeta_\sigma^* \exp(\mp i\pi\sigma)\zeta_\sigma \exp(\mp i\pi\sigma)\Psi_{p,\sigma} \\ &= \exp(\mp 2i\pi\sigma)\Psi_{p,\sigma} \,.\end{aligned}$$

As long as σ is an integer or half-integer, this can be written

$$\mathsf{T}^2\Psi_{p,\sigma} = (-)^{2|\sigma|}\Psi_{p,\sigma} \,. \tag{2.6.27}$$

By the 'spin' of a massless particle, we usually mean the absolute value of the helicity, so Eq. (2.6.27) is the same as Eq. (2.6.26).

This result has an interesting consequence. When T^2 acts on any state Ψ of a system of non-interacting particles, either massive or massless, it yields a factor $(-)^{2j}$ or $(-)^{2|\sigma|}$ for each particle. Hence if the state contains an odd number of particles of half-integer spin or helicity (plus any number of particles of integer spin or helicity), we get an overall change of sign

$$\mathsf{T}^2\Psi = -\Psi \,. \tag{2.6.28}$$

If we now 'turn on' various interactions, this result will be preserved, provided these interactions respect invariance under time-reversal, even if they do not respect rotational invariance. (For instance, these arguments will apply even if our system is subjected to arbitrary static gravitational and electric fields.) Now, suppose that Ψ is an eigenstate of the Hamiltonian. Since T commutes with the Hamiltonian, $\mathsf{T}\Psi$ will also be an eigenstate of the Hamiltonian. Is it the same state? If so, then $\mathsf{T}\Psi$ can differ from Ψ only by a phase

$$\mathsf{T}\Psi = \zeta\Psi ,$$

but then

$$\mathsf{T}^2\Psi = \mathsf{T}(\zeta\Psi) = \zeta^*\mathsf{T}\Psi = |\zeta|^2\Psi = \Psi ,$$

in contradiction with Eq. (2.6.28). We see that any energy eigenstate Ψ satisfying Eq. (2.6.28) must be degenerate with another eigenstate of the same energy. This is known as a 'Kramers degeneracy.'[10] Of course, this conclusion is trivial if the system is in a rotationally invariant environment, because the total angular-momentum j of any state of this system would have to be a half-integer, and there would therefore be $2j + 1 = 2, 4, \cdots$ degenerate states. The surprising result is that at least a two-fold degeneracy persists even if rotational invariance is perturbed by external fields, such as electrostatic fields, as long as these fields are invariant under T. In particular, if any particle had an electric or gravitational dipole moment then the degeneracy among its $2j + 1$ spin states would be entirely removed in a static electric or gravitational field, so such dipole moments are forbidden by time-reversal invariance.

For the sake of completeness, it should be mentioned that P and T can have more complicated effects on multiplets of particles with the same mass. This possibility will be considered in Appendix C of this chapter. No physically relevant examples are known.

2.7 Projective Representations*

We now return to the possibility mentioned in Section 2.2, that a group of symmetries may be represented projectively on physical states; that is, the elements T, $\bar{T}$, etc. of the symmetry group may be represented on the physical Hilbert space by unitary operators $U(T), U(\bar{T})$, etc., which

* This section lies somewhat out of the book's main line of development, and may be omitted in a first reading.

satisfy the composition rule

$$U(T)U(\bar{T}) = \exp\Big(i\phi(T,\bar{T})\Big)U(T\bar{T}) \tag{2.7.1}$$

with ϕ a real phase. (A bar is used here just to distinguish one symmetry operator from another.) The basic requirement that any phase ϕ in Eq. (2.7.1) would have to satisfy is the associativity condition

$$U(T_3)(U(T_2)U(T_1)) = (U(T_3)U(T_2))U(T_1)\,,$$

which imposes on ϕ the corresponding condition

$$\phi(T_2,T_1) + \phi(T_3,T_2T_1) = \phi(T_3,T_2) + \phi(T_3T_2,T_1)\,. \tag{2.7.2}$$

Of course, any phase of the form

$$\phi(T,\bar{T}) = \alpha(T\bar{T}) - \alpha(T) - \alpha(\bar{T}) \tag{2.7.3}$$

will automatically satisfy Eq. (2.7.2), but a projective representation with such a phase can be replaced with an ordinary representation by replacing $U(T)$ with

$$\tilde{U}(T) \equiv U(T)\exp\Big(i\alpha(T)\Big)$$

for which

$$\tilde{U}(T)\tilde{U}(\bar{T}) = \tilde{U}(T\bar{T})\;\;.$$

Any set of functions $\phi(T,\bar{T})$ that satisfy Eq. (2.7.2), and that differ only by functions $\Delta\phi(T,\bar{T})$ of the form (2.7.3), is called a 'two-cocycle'. A trivial cocycle is one that contains the function $\phi = 0$, and hence consists of functions of the form (2.7.3), which can be eliminated by a redefinition of $U(T)$. We are interested here in whether a symmetry group allows any non-trivial two-cocyles; that is, whether it may have a representation on the physical Hilbert space that is *intrinsically* projective, in the sense that the phase $\phi(T,\bar{T})$ *cannot* be eliminated in this way.

In order to answer this question, it is useful first to consider the effect of a phase ϕ in Eq. (2.7.1) on the commutation relations of the generators of infinitesimal transformations. When either $\bar{T}$ or T is the identity, the phase ϕ must clearly vanish

$$\phi(T,1) = \phi(1,\bar{T}) = 0\,. \tag{2.7.4}$$

When both T and $\bar{T}$ are *near* the identity the phase must be small. Using coordinates θ^a to parameterize group elements (as in Section 2.2), with $T(0) \equiv 1$, Eq. (2.7.4) tells us that the expansion of $\phi(T(\theta),T(\bar{\theta}))$ around $\theta = \bar{\theta} = 0$ must start with terms of order $\theta\bar{\theta}$:

$$\phi\Big(T(\theta),T(\bar{\theta})\Big) = f_{ab}\theta^a\bar{\theta}^b + \cdots\,, \tag{2.7.5}$$

where f_{ab} are real numerical constants. Inserting this expansion in the power series expansion of Eq. (2.7.1), and repeating the steps that led to (2.2.22), we now have

$$[t_b, t_c] = iC^a{}_{bc}t_a + iC_{bc}1 \ , \tag{2.7.6}$$

where C_{bc} is the antisymmetric coefficient

$$C_{bc} = -f_{bc} + f_{cb} \, . \tag{2.7.7}$$

The appearance of terms on the right-hand side of the commutation relation proportional to the unit element (so-called *central charges*) is the counterpart for the Lie algebra of the presence of phases in a projective representation of a group.

The constants C_{bc} as well as $C^a{}_{bc}$ are subject to an important constraint, which follows from the Jacobi identity. Taking the commutator of (2.7.6) with t_d, and adding the same expressions with b, c, d replaced with c, d, b and d, b, c, the sum of the three double-commutators on the left-hand side vanishes identically, and so

$$C^a{}_{bc}C^e{}_{ad} + C^a{}_{cd}C^e{}_{ab} + C^a{}_{db}C^e{}_{ac} = 0 \tag{2.7.8}$$

and also

$$C^a{}_{bc}C_{ad} + C^a{}_{cd}C_{ab} + C^a{}_{db}C_{ac} = 0 \, . \tag{2.7.9}$$

Eq. (2.7.9) always has one obvious class of non-zero solutions for C_{ab}:

$$C_{ab} = C^e{}_{ab}\phi_e \, , \tag{2.7.10}$$

where ϕ_e is an arbitrary set of real constants. For these solutions, we can eliminate the central charges from Eq. (2.7.6) by a redefinition of the generators

$$t_a \to \tilde{t}_a \equiv t_a + \phi_a \ . \tag{2.7.11}$$

The new generators then satisfy the commutation relations without central charges

$$[\tilde{t}_b, \tilde{t}_c] = iC^a{}_{bc}\tilde{t}_a \ . \tag{2.7.12}$$

A given Lie algebra may or may not allow solutions of Eq. (2.7.9) other than Eq. (2.7.10).

We can now state the key theorem that governs the occurrence of intrinsically projective representations. The phase of any representation $U(T)$ of a given group can be chosen so that $\phi = 0$ in Eq. (2.7.1), if two conditions are met:

(a) The generators of the group in this representation can be redefined (as in Eq. (2.7.11)), so as to eliminate all central charges from the Lie algebra.

(b) The group is simply connected, i.e., any two group elements may be connected by a path lying within the group, and any two such paths may be continuously transformed into one another. (An equivalent statement is that any loop that starts and ends at the same group element may be shrunk continuously to a point.)

This theorem is proved in Appendix B of this chapter, which also offers comments about the case of groups that are not simply connected. It shows that there are just two (not exclusive) ways that intrinsically projective representations may arise: either algebraically, because the group is represented projectively even near the identity, or topologically, because the group is not simply connected, and hence a path from 1 to T and then from T to $\bar{T}$ may not be continuously deformable into some other path from 1 to $T\bar{T}$. In the latter case, the phase ϕ in Eq. (2.7.1) depends on the particular choice of standard paths, leading from the origin to the various group elements, that are used to define the corresponding U-operators.

Let's now consider each of these possibilities in turn for the special case of the inhomogeneous Lorentz group.

(A) Algebra

With central charges, the commutation relations of the generators of the inhomogeneous Lorentz group would read

$$i[J^{\mu\nu}, J^{\rho\sigma}] = \eta^{\nu\rho} J^{\mu\sigma} - \eta^{\mu\rho} J^{\nu\sigma} - \eta^{\sigma\mu} J^{\rho\nu} + \eta^{\sigma\nu} J^{\rho\mu} + C^{\rho\sigma,\mu\nu} , \tag{2.7.13}$$

$$i[P^{\mu}, J^{\rho\sigma}] = \eta^{\mu\rho} P^{\sigma} - \eta^{\mu\sigma} P^{\rho} + C^{\rho\sigma,\mu} , \tag{2.7.14}$$

$$i[J^{\mu\nu}, P^{\rho}] = \eta^{\nu\rho} P^{\mu} - \eta^{\mu\rho} P^{\nu} + C^{\rho,\mu\nu} , \tag{2.7.15}$$

$$i[P^{\mu}, P^{\rho}] = C^{\rho,\mu} \tag{2.7.16}$$

in place of Eqs. (2.4.12)–(2.4.14). We see that the Cs also satisfy the antisymmetry conditions

$$C^{\rho\sigma,\mu\nu} = -C^{\mu\nu,\rho\sigma} , \tag{2.7.17}$$

$$C^{\rho\sigma,\mu} = -C^{\mu,\rho\sigma} , \tag{2.7.18}$$

$$C^{\rho,\mu} = -C^{\mu,\rho} . \tag{2.7.19}$$

We will now show that all these constants have additional algebraic properties that allow them to be eliminated by shifting the definitions of $J^{\mu\nu}$ and P^{μ} by constant terms. (This corresponds to redefining the phase of the operators $U(\Lambda, a)$.) To derive these properties, we apply the Jacobi identities

$$\left[J^{\mu\nu},[P^\rho,P^\sigma]\right]+\left[P^\sigma,[J^{\mu\nu},P^\rho]\right]+\left[P^\rho,[P^\sigma,J^{\mu\nu}]\right]=0, \quad (2.7.20)$$

$$\left[J^{\lambda\eta},[J^{\mu\nu},P^\rho]\right]+\left[P^\rho,[J^{\lambda\eta},J^{\mu\nu}]\right]+\left[J^{\mu\nu},[P^\rho,J^{\lambda\eta}]\right]=0, \quad (2.7.21)$$

$$\left[J^{\lambda\eta},[J^{\mu\nu},J^{\rho\sigma}]\right]+\left[J^{\rho\sigma},[J^{\lambda\eta},J^{\mu\nu}]\right]+\left[J^{\mu\nu},[J^{\rho\sigma},J^{\lambda\eta}]\right]=0\,. \quad (2.7.22)$$

(The Jacobi identity involving three Ps is automatically satisfied, and hence yields no further information.) Using Eqs. (2.7.13)–(2.7.16) in Eqs. (2.7.20)–(2.7.22), we obtain algebraic conditions on the Cs

$$0=\eta^{\nu\rho}C^{\mu,\sigma}-\eta^{\mu\rho}C^{\nu,\sigma}-\eta^{\nu\sigma}C^{\mu,\rho}+\eta^{\mu\sigma}C^{\nu,\rho}\,, \quad (2.7.23)$$

$$0=\eta^{\nu\rho}C^{\mu,\lambda\eta}-\eta^{\mu\rho}C^{\nu,\lambda\eta}-\eta^{\mu\eta}C^{\rho,\lambda\nu}+\eta^{\lambda\mu}C^{\rho,\eta\nu}$$
$$+\eta^{\lambda\nu}C^{\rho,\mu\eta}-\eta^{\eta\nu}C^{\rho,\mu\lambda}+\eta^{\rho\lambda}C^{\eta,\mu\nu}-\eta^{\rho\eta}C^{\lambda,\mu\nu}\,, \quad (2.7.24)$$

$$0=\eta^{\nu\rho}C^{\mu\sigma,\lambda\eta}-\eta^{\mu\rho}C^{\nu\sigma,\lambda\eta}-\eta^{\sigma\mu}C^{\rho\nu,\lambda\eta}+\eta^{\sigma\nu}C^{\rho\mu,\lambda\eta}$$
$$+\eta^{\eta\mu}C^{\lambda\nu,\rho\sigma}-\eta^{\lambda\mu}C^{\eta\nu,\rho\sigma}-\eta^{\nu\lambda}C^{\mu\eta,\rho\sigma}+\eta^{\nu\eta}C^{\mu\lambda,\rho\sigma}$$
$$+\eta^{\sigma\lambda}C^{\rho\eta,\mu\nu}-\eta^{\rho\lambda}C^{\sigma\eta,\mu\nu}-\eta^{\eta\rho}C^{\lambda\sigma,\mu\nu}+\eta^{\eta\sigma}C^{\lambda\rho,\mu\nu}\,. \quad (2.7.25)$$

Contracting Eq. (2.7.23) with $\eta_{\nu\rho}$ gives

$$C^{\mu,\sigma}=0\,. \quad (2.7.26)$$

On the other hand, the constants $C^{\mu,\lambda\eta}$ and $C^{\rho\sigma,\mu\nu}$ are not necessarily zero, but their algebraic structure is simple enough so that they can be eliminated by shifting the definitions of P^μ and $J^{\mu\nu}$, respectively. Contracting Eq. (2.7.24) with $\eta_{\nu\rho}$ gives

$$C^{\mu,\lambda\eta}=\eta^{\mu\eta}C^\lambda-\eta^{\mu\lambda}C^\eta\,, \quad (2.7.27)$$

$$C^\lambda\equiv\frac{1}{3}\,\eta_{\rho\nu}C^{\rho,\lambda\nu}\,. \quad (2.7.28)$$

Also, contracting (2.7.25) with $\eta_{\nu\rho}$ gives

$$C^{\mu\sigma,\lambda\eta}=\eta^{\eta\mu}C^{\lambda\sigma}-\eta^{\lambda\mu}C^{\eta\sigma}+\eta^{\sigma\lambda}C^{\eta\mu}-\eta^{\eta\sigma}C^{\lambda\mu}\,, \quad (2.7.29)$$

$$C^{\lambda\sigma}\equiv\frac{1}{2}\,\eta_{\nu\rho}C^{\lambda\nu,\sigma\rho}\,. \quad (2.7.30)$$

(These expressions automatically satisfy Eqs. (2.7.24) and (2.7.25), so there is no further information to be gained from the Jacobi identities.) We now see that if the Cs are not zero, they can be eliminated by defining new generators

$$\tilde{P}^\mu\equiv P^\mu+C^\mu\,, \quad (2.7.31)$$

$$\tilde{J}^{\mu\sigma}\equiv J^{\mu\sigma}+C^{\mu\sigma}\,, \quad (2.7.32)$$

and the commutation relations are then what they would be for an ordinary representation

$$i[\tilde{J}^{\mu\nu}, \tilde{J}^{\rho\sigma}] = \eta^{\nu\rho}\tilde{J}^{\mu\sigma} - \eta^{\mu\rho}\tilde{J}^{\nu\sigma} - \eta^{\sigma\mu}\tilde{J}^{\rho\nu} + \eta^{\sigma\nu}\tilde{J}^{\rho\mu}\,, \tag{2.7.33}$$

$$i[\tilde{J}^{\mu\nu}, \tilde{P}^{\rho}] = \eta^{\nu\rho}\tilde{P}^{\mu} - \eta^{\mu\rho}\tilde{P}^{\nu}\,, \tag{2.7.34}$$

$$i[\tilde{P}^{\mu}, \tilde{P}^{\rho}] = 0\,. \tag{2.7.35}$$

The commutation relations will always be taken in the form Eqs. (2.7.33)–(2.7.35), but with tildas dropped.

Incidentally, the fact that there is no central charge in the algebra of the $J^{\mu\nu}$ could have been immediately inferred from the fact that this algebra is of the type known as 'semi-simple'. (Semi-simple Lie algebras are those that have no 'invariant Abelian' subalgebra, consisting of generators that commute with each other and whose commutators with any other generators also belong to the subalgebra.) There is a general theorem[11] that any central charges in semi-simple Lie algebras may always be removed by a redefinition of generators, as in Eq. (2.7.32). On the other hand, the full Poincaré algebra spanned by $J^{\mu\nu}$ and P^{μ} is not semi-simple (the P^{μ} form an invariant Abelian subalgebra), and we needed a special argument to show that its central charges can also be eliminated in this way. Indeed, the non-semi-simple Galilean algebra discussed in Section 2.4 does allow a central charge, the mass M.

We see that the inhomogeneous Lorentz group satisfies the first of the two conditions needed to rule out intrinsically projective representations. How about the second?

(B) Topology

To explore the topology of the inhomogeneous Lorentz group, it is very convenient to represent homogeneous Lorentz transformations by 2×2 complex matrices. Any real four-vector V^{μ} can be used to construct an Hermitian 2×2 matrix

$$v \equiv V^{\mu}\sigma_{\mu} = \begin{pmatrix} V^0 + V^3 & V^1 - iV^2 \\ V^1 + iV^2 & V^0 - V^3 \end{pmatrix}, \tag{2.7.36}$$

where σ_{μ} are the usual Pauli matrices with $\sigma_0 \equiv 1$. Conversely any 2×2 Hermitian matrix can be put in this form, and therefore defines a real four-vector V^{μ}.

The property of Hermiticity will be preserved under the transformation

$$v \to \lambda v \lambda^{\dagger} \tag{2.7.37}$$

with λ an arbitrary complex 2×2 matrix. Furthermore, the covariant square of the four-vector is

$$V_\mu V^\mu = (V^1)^2 + (V^2)^2 + (V^3)^2 - (V^0)^2 = -\text{Det}\, v \tag{2.7.38}$$

and this determinant is preserved by the transformation (2.7.37) provided that

$$|\text{Det}\, \lambda| = 1\,. \tag{2.7.39}$$

Each complex 2×2 matrix λ satisfying Eq. (2.7.39) thus defines a real linear transformation of V^μ that leaves Eq. (2.7.38) invariant, i.e., a homogeneous Lorentz transformation $\Lambda(\lambda)$:

$$\lambda V^\mu \sigma_\mu \lambda^\dagger = (\Lambda^\mu{}_\nu(\lambda) V^\nu)\sigma_\mu\,. \tag{2.7.40}$$

Furthermore, for two such matrices λ and $\bar{\lambda}$, we have

$$\begin{aligned}(\lambda\bar{\lambda})V^\mu\sigma_\mu(\lambda\bar{\lambda})^\dagger &= \lambda(\bar{\lambda}V^\mu\sigma_\mu\bar{\lambda}^\dagger)\lambda^\dagger\\ &= \lambda\Lambda^\mu{}_\nu(\bar{\lambda})V^\nu\sigma_\mu\lambda^\dagger = \Lambda^\mu{}_\rho(\lambda)\Lambda^\rho{}_\nu(\bar{\lambda})V^\nu\sigma_\mu\end{aligned}$$

and so

$$\Lambda(\lambda\bar{\lambda}) = \Lambda(\lambda)\Lambda(\bar{\lambda})\,. \tag{2.7.41}$$

However, two λs that differ only by an overall phase have the same effect on v in Eq. (2.7.37), and so correspond to the same Lorentz transformation. It is therefore convenient to adjust the phase of the λs so that

$$\text{Det}\, \lambda = 1, \tag{2.7.42}$$

which is consistent with Eq. (2.7.41). The 2×2 complex matrices with unit determinant form a group, known as $SL(2,C)$. (The SL stands for 'special linear', with 'special' denoting a unit determinant, while C stands for 'complex'.) The group elements depend on $4-1=3$ complex parameters, or 6 real parameters, the same number as the Lorentz group. However, $SL(2,C)$ is not the same as the Lorentz group; if λ is a matrix in $SL(2,C)$, then so is $-\lambda$, and both λ and $-\lambda$ produce the same Lorentz transformation in Eq. (2.7.37). Indeed, it is easy to see that the matrix

$$\lambda(\theta) = \begin{pmatrix} e^{i\theta/2} & 0 \\ 0 & e^{-i\theta/2}\end{pmatrix}$$

produces a Lorentz transformation $\Lambda(\lambda(\theta))$ which is just a rotation by an angle θ around the three-axis, and hence $\lambda = -1$ produces a rotation by an angle 2π. The Lorentz group is not the same as $SL(2,C)$, but rather** $SL(2,C)/Z_2$, which is the group of complex 2×2 matrices with unit determinant, and with λ identified with $-\lambda$.

Now, what is the topology of the Lorentz group? By the polar decomposition theorem[12], any complex non-singular matrix λ may be written in the form

$$\lambda = u\, e^h ,$$

where u is unitary and h is Hermitian

$$u^\dagger u = 1 , \qquad h^\dagger = h .$$

Since Det u is a phase factor, and Det $\exp h = \exp \operatorname{Tr} h$ is real and positive, the condition (2.7.42) requires both

$$\operatorname{Det} u = 1 ,$$
$$\operatorname{Tr} h = 0.$$

(The factor u simply provides the rotation subgroup of the Lorentz group; if u is unitary then $\operatorname{Tr}(uvu^\dagger) = \operatorname{Tr} v$, so $V^0 = \frac{1}{2}\operatorname{Tr} v$ is left invariant by $\Lambda(u)$.) Furthermore, this decomposition is unique, so $SL(2,C)$ is topologically just the direct product (i.e., the set of pairs of points) of the space of all us and the space of all hs. Any Hermitian traceless 2×2 matrix h can be expressed as

$$h = \begin{pmatrix} c & a - ib \\ a + ib & -c \end{pmatrix}$$

with a, b, c real but otherwise unconstrained, so the space of all hs is topologically the same as ordinary three-dimensional flat space, R_3. On the other hand, any unitary 2×2 matrix with unit determinant can be expressed as

$$u = \begin{pmatrix} d + ie & f + ig \\ -f + ig & d - ie \end{pmatrix}$$

with d, e, f, g subject to the single non-linear constraint

$$d^2 + e^2 + f^2 + g^2 = 1 ,$$

so the space $SU(2)$ of all us is topologically the same as S_3, the three-dimensional surface of a spherical ball in flat four-dimensional space. Thus $SL(2,C)$ is topologically the same as the direct product $R_3 \times S_3$. This is simply connected: any curve connecting two points of R_3 or S_3 can be deformed into any other, and the same is true of the direct product. (All

** The group Z_2 consists of just elements $+1$ and -1. In general, when we write G/H, with H an invariant subgroup of G, we mean the group G with elements g and gh identified if $g \in G$ and $h \in H$. The subgroup Z_2 is trivially invariant because its elements commute with all elements of $SL(2,C)$.

spheres S_n except the circle S_1 are simply connected.) However, we are interested in $SL(2,C)/Z_2$, not $SL(2,C)$. Identifying λ with $-\lambda$ is the same as identifying the unitary factors u and $-u$ (because e^h is always positive), so the Lorentz group has the topology of $R_3 \times S_3/Z_2$, where S_3/Z_2 is the three-dimensional spherical surface with opposite points of the sphere identified. This is *not* simply connected; for instance, a path on S_3 from u to u' cannot be continuously deformed into a path on S_3 from u to $-u'$, even though these two paths link the same points of S_3/Z_2. In fact, S_3/Z_2 is *doubly* connected; the paths between any two points fall into two classes, depending on whether or not they involve an inversion $u \to -u$, and any path in one class can be deformed into another path of that class. An equivalent statement is that a double loop, that goes twice over the same path from any element back to itself, may be continuously contracted to a point. (As discussed in Appendix B, this is summarized mathematically in the statement that the fundamental group, or first homotopy group, of S_3/Z_2 is Z_2.) Similarly, the inhomogeneous Lorentz group has the same topology as $R_4 \times R_3 \times S_3/Z_2$, and is therefore also doubly connected.

Because the Lorentz group (homogeneous or inhomogeneous) is not simply connected, it does have intrinsically projective representations. However, because the double loop that goes twice from 1 to Λ to $\Lambda\bar{\Lambda}$ and then back to 1 *can* be contracted to a point, we must have

$$\left[U(\Lambda)U(\bar{\Lambda})U^{-1}(\Lambda\bar{\Lambda})\right]^2 = 1$$

and hence the phase $e^{i\phi(\Lambda,\bar{\Lambda})}$ is just a sign

$$U(\Lambda)U(\bar{\Lambda}) = \pm U(\Lambda\bar{\Lambda})\,. \tag{2.7.43}$$

Likewise, for the inhomogeneous Lorentz group

$$U(\Lambda,a)U(\bar{\Lambda},\bar{a}) = \pm U(\Lambda\bar{\Lambda},\Lambda\bar{a}+a)\,. \tag{2.7.44}$$

These 'representations up to a sign' are familiar; they are just the states of integer spin, for which the signs in Eqs. (2.7.43) and (2.7.44) are always $+1$, and the states of half-integer spin, for which these signs are $+1$ or -1 according to whether the path from 1 to Λ to $\Lambda\bar{\Lambda}$ and then back to 1 is or is not contractible to a point. This difference arises because a rotation of 2π around the three-axis acting on a state with angular-momentum three-component σ produces a phase $e^{2i\pi\sigma}$, and thus has no effect on a state of integer spin and produces a sign change when acting on a state of half-integer spin. (These two cases correspond to the two irreducible representations of the first homotopy group, Z_2.) Thus Eq. (2.7.43) or Eq. (2.7.44) imposes a superselection rule: we must not mix states of integer and half-integer spin.

For finite mass, the limitation to integer or half-integer spin was previously derived by purely algebraic means from the well-known representations of the generators of the little group, which here are just the angular-momentum matrices $\vec{J}^{(j)}$ with j integer or half-integer. On the other hand, for zero mass the action of the little group on physical one-particle states is just a rotation around the momentum, and here there is no *algebraic* reason for a limitation to integer or half-integer helicity. There is, however, a *topological* reason: a rotation by an angle 4π around the momentum can be continuously deformed into no rotation at all, so the factor $\exp(4\pi i\sigma)$ must be unity, and hence σ must be an integer or half-integer.

Instead of working with projective representations and imposing a superselection rule, we can just as well expand the Lorentz group, taking it as $SL(2,C)$ itself, instead of $SL(2,C)/Z_2$ as before. Ordinary rotation invariance forbids transitions between states of integer and half-integer total spin, so the only difference is that now the group is simply-connected, and it therefore has only ordinary representations, not projective representations, so that we cannot infer a superselection rule. This does not mean that we actually can prepare physical systems in linear combinations of states of integer and half-integer spin, but only that the observed Lorentz invariance of nature cannot be used to show that such superpositions are impossible.

Similar remarks apply to any symmetry group. If its Lie algebra involves central charges, then we can always expand the algebra to include generators that commute with anything, and whose eigenvalues are the central charges, just as we did when we added a mass operator to the Lie algebra of the Galilean group at the end of Section 2.4. The expanded Lie algebra is then, of course, free of central charges, so the part of the group near the identity has only ordinary representations, and does not require any superselection rule. Likewise, even though a Lie group G may not be simply connected, it can always be expressed as C/H, where C is a simply connected group known as the 'universal covering group' of G, and H is an invariant subgroup† of C. In general, we may just as well take the symmetry group as C instead of G, because there is no difference in their consequences, except that G implies a superselection rule, while C does not. In short, the issue of superselection rules is a bit of a red herring; *it may or it may not be possible to prepare physical systems in arbitrary superpositions of states, but one cannot settle the question by*

---

† The first homotopy group of C/H is H. We have seen that the covering group of the homogeneous Lorentz group is $SL(2,C)$, and the covering group of the three-dimensional rotation group is $SU(2)$. This connection with SL and SU groups is special to the case of three, four, or six dimensions; for general dimensions d the covering group of $SO(d)$ is given a special name, '$Spin(d)$'.

reference to symmetry principles, because whatever one thinks the symmetry group of nature may be, there is always another group whose consequences are identical except for the absence of superselection rules.

Appendix A The Symmetry Representation Theorem

This appendix presents the proof of the fundamental theorem of Wigner[2] that any symmetry transformation can be represented on the Hilbert space of physical states by an operator that is either linear and unitary or antilinear and antiunitary. For our present purposes, the property of symmetry transformations on which we chiefly rely is that they are ray transformations T that preserve transition probabilities, in the sense that if Ψ_1 and Ψ_2 are state-vectors belonging to rays $\mathscr{R}_1$ and $\mathscr{R}_2$ then any state-vectors Ψ'_1 and Ψ'_2 belonging to the transformed rays $T\mathscr{R}_1$ and $T\mathscr{R}_2$ satisfy

$$|(\Psi'_1, \Psi'_2)|^2 = |(\Psi_1, \Psi_2)|^2. \tag{2.A.1}$$

We also require that a symmetry transformation should have an inverse that preserves transition probabilities in the same sense.

To start, consider some complete orthonormal set of state-vectors Ψ_k belonging to rays $\mathscr{R}_k$, with

$$(\Psi_k, \Psi_l) = \delta_{kl}, \tag{2.A.2}$$

and let Ψ'_k be some arbitrary choice of state-vectors belonging to the transformed rays $T\mathscr{R}_k$. From Eq. (2.A.1), we have

$$|(\Psi'_k, \Psi'_l)|^2 = |(\Psi_k, \Psi_l)|^2 = \delta_{kl}.$$

But (Ψ'_k, Ψ'_k) is automatically real and positive, so this requires that it should have the value unity, and therefore

$$(\Psi'_k, \Psi'_l) = \delta_{kl}. \tag{2.A.3}$$

It is easy to see that these transformed states Ψ'_k also form a complete set, for if there were any non-zero state-vector Ψ' that was orthogonal to all of the Ψ'_k, then the inverse transform of the ray to which Ψ' belongs would consist of non-zero state-vectors Ψ'' for which, for all k:

$$|(\Psi_k, \Psi'')|^2 = |(\Psi'_k, \Psi')|^2 = 0,$$

which is impossible since the Ψ_k were assumed to form a complete set.

We must now establish a phase convention for the states Ψ'_k. For this purpose, we single out one of the Ψ_k, say Ψ_1, and consider the state-vectors

$$\Upsilon_k \equiv \frac{1}{\sqrt{2}}[\Psi_1 + \Psi_k] \tag{2.A.4}$$

belonging to some ray $\mathscr{S}_k$, with $k \neq 1$. Any state-vector Υ'_k belonging to the transformed ray $T\mathscr{S}_k$ may be expanded in the state-vectors Ψ'_l,

$$\Upsilon'_k = \sum\nolimits_l c_{kl} \Psi'_l \, .$$

From Eq. (2.A.1) we have

$$|c_{kk}| = |c_{k1}| = \frac{1}{\sqrt{2}}$$

and for $l \neq k$ and $l \neq 1$:

$$c_{kl} = 0.$$

For any given k, by an appropriate choice of phase of the two state-vectors Υ'_k and Ψ'_k we can clearly adjust the phases of the two non-zero coefficients c_{kk} and c_{k1} so that both coefficients are just $1/\sqrt{2}$. From now on, the state-vectors Υ'_k and Ψ'_k chosen in this way will be denoted $U\Upsilon_k$ and $U\Psi_k$. As we have seen,

$$U\frac{1}{\sqrt{2}}[\Psi_k + \Psi_1] = U\Upsilon_k = \frac{1}{\sqrt{2}}[U\Psi_k + U\Psi_1] \, . \tag{2.A.5}$$

However, it still remains to define $U\Psi$ for general state-vectors Ψ.

Now consider an arbitrary state-vector Ψ belonging to an arbitrary ray $\mathscr{R}$, and expand it in the Ψ_k:

$$\Psi = \sum_k C_k \Psi_k \, . \tag{2.A.6}$$

Any state Ψ' that belongs to the transformed ray $T\mathscr{R}$ may similarly be expanded in the complete orthonormal set $U\Psi_k$:

$$\Psi' = \sum_k C'_k U\Psi_k \, . \tag{2.A.7}$$

The equality of $|(\Psi_k, \Psi)|^2$ and $|(U\Psi_k, \Psi')|^2$ tells us that for all k (including $k = 1$):

$$|C_k|^2 = |C'_k|^2, \tag{2.A.8}$$

while the equality of $|(\Upsilon_k, \Psi)|^2$ and $|(U\Upsilon_k, \Psi')|^2$ tells us that for all $k \neq 1$:

$$|C_k + C_1|^2 = |C'_k + C'_1|^2. \tag{2.A.9}$$

The ratio of Eqs. (2.A.9) and (2.A.8) yields the formula

$$\text{Re}\,(C_k/C_1) = \text{Re}\,(C'_k/C'_1) \tag{2.A.10}$$

which with Eq. (2.A.8) also requires

$$\text{Im}\,(C_k/C_1) = \pm\text{Im}\,(C'_k/C'_1) \tag{2.A.11}$$

and therefore either

$$C_k/C_1 = C'_k/C'_1, \tag{2.A.12}$$

or else

$$C_k/C_1 = (C'_k/C'_1)^*. \tag{2.A.13}$$

Furthermore, we can show that the same choice must be made for each k. (This step in the proof was omitted by Wigner.) To see this, suppose that for some k, we have $C_k/C_1 = C'_k/C'_1$, while for some $l \neq k$, we have instead $C_l/C_1 = (C'_l/C'_1)^*$. Suppose also that both ratios are complex, so that these are really different cases. (This incidentally requires that $k \neq 1$ and $l \neq 1$, as well as $k \neq l$.) We will show that this is impossible.

Define a state-vector $\Phi \equiv \frac{1}{\sqrt{3}}[\Psi_1 + \Psi_k + \Psi_l]$. Since all the ratios of the coefficients in this state-vector are real, we must get the same ratios in any state-vector Φ' belonging to the transformed ray:

$$\Phi' = \frac{\alpha}{\sqrt{3}}[U\Psi_1 + U\Psi_k + U\Psi_l]\,,$$

where α is a phase factor with $|\alpha| = 1$. But then the equality of the transition probabilities $|(\Phi, \Psi)|$ and $|(\Phi', \Psi')|$ requires that

$$\left|1 + \frac{C'_k}{C'_1} + \frac{C'_l}{C'_1}\right|^2 = \left|1 + \frac{C_k}{C_1} + \frac{C_l}{C_1}\right|^2$$

and hence

$$\left|1 + \frac{C_k}{C_1} + \frac{C^*_l}{C^*_1}\right|^2 = \left|1 + \frac{C_k}{C_1} + \frac{C_l}{C_1}\right|^2.$$

This is only possible if

$$\text{Re}\left(\frac{C_k}{C_1}\frac{C^*_l}{C^*_1}\right) = \text{Re}\left(\frac{C_k}{C_1}\frac{C_l}{C_1}\right)$$

or, in other words, if

$$\text{Im}\left(\frac{C_k}{C_1}\right)\,\text{Im}\left(\frac{C_l}{C_1}\right) = 0\,.$$

Hence either C_k/C_1 or C_l/C_1 must be real for any pair k, l, in contradiction with our assumptions. We see then that for a given symmetry transformation T applied to a given state-vector $\sum_k C_k\Psi_k$, we must have either Eq. (2.A.12) for all k, or else Eq. (2.A.13) for all k.

Wigner ruled out the second possibility, Eq. (2.A.13), because as he showed any symmetry transformation for which this possibility is realized would have to involve a reversal in the time coordinate, and in the proof he presented he was considering only symmetries like rotations that do not affect the direction of time. Here we are treating symmetries involving time-reversal on the same basis as all other symmetries, so we will have to consider that, for each symmetry T and state-vector $\sum_k C_k \Psi_k$, either Eq. (2.A.12) or Eq. (2.A.13) may apply. Depending on which of these alternatives is realized, we will now define $U\Psi$ to be the particular one of the state-vectors Ψ' belonging to the ray $T\mathscr{R}$ with phase chosen so that either $C_1 = C_1'$ or $C_1 = C_1'^*$, respectively. Then either

$$U(\sum_k C_k \Psi_k) = \sum_k C_k U \Psi_k \tag{2.A.14}$$

or else

$$U(\sum_k C_k \Psi_k) = \sum_k C_k^* U \Psi_k \,. \tag{2.A.15}$$

It remains to be proved that for a given symmetry transformation, we must make the same choice between Eqs. (2.A.14) and (2.A.15) for arbitrary values of the coefficients C_k. Suppose that Eq. (2.A.14) applies for a state-vector $\sum_k A_k \Psi_k$ while Eq. (2.A.15) applies for a state-vector $\sum_k B_k \Psi_k$. Then the invariance of transition probabilities requires that

$$|\sum_k B_k^* A_k|^2 = |\sum_k B_k A_k|^2$$

or equivalently

$$\sum_{kl} \text{Im}\,(A_k^* A_l)\,\text{Im}\,(B_k^* B_l) = 0\,. \tag{2.A.16}$$

We cannot rule out the possibility that Eq. (2.A.16) may be satisfied for a pair of state-vectors $\sum_k A_k \Psi_k$ and $\sum_k B_k \Psi_k$ belonging to different rays. However, for any pair of such state-vectors, with neither A_k nor B_k all of the same phase (so that Eqs. (2.A.14) and (2.A.15) are not the same), we can always find a third state-vector $\sum_k C_k \Psi_k$ for which*

$$\sum_{kl} \text{Im}\,(C_k^* C_l)\,\text{Im}\,(A_k^* A_l) \neq 0 \tag{2.A.17}$$

and also

$$\sum_{kl} \text{Im}\,(C_k^* C_l)\,\text{Im}\,(B_k^* B_l) \neq 0\,. \tag{2.A.18}$$

---

* If for some pair k,l both $A_k^* A_l$ and $B_k^* B_l$ are complex, then choose all Cs to vanish except for C_k and C_l, and choose these two coefficients to have different phases. If $A_k^* A_l$ is complex but $B_k^* B_l$ is real for some pair k,l, then there must be some other pair m,n (possibly with either m or n but not both equal to k or l) for which $B_m^* B_n$ is complex. If also $A_m^* A_n$ is complex, then choose all Cs to vanish except for C_m and C_n, and choose these two coefficients to have different phase. If $A_m^* A_n$ is real, then choose all Cs to vanish except for C_k, C_l, C_m, and C_n, and choose these four coefficients all to have different phases. The case where $B_k^* B_l$ is complex but $A_k^* A_l$ is real is handled in just the same way.

As we have seen, it follows from Eq. (2.A.17) that the same choice between Eqs. (2.A.14) and (2.A.15) must be made for $\sum_k A_k \Psi_k$ and $\sum_k C_k \Psi_k$, and it follows from Eq. (2.A.18) that the same choice between Eqs. (2.A.14) and (2.A.15) must be made for $\sum_k B_k \Psi_k$ and $\sum_k C_k \Psi_k$, so the same choice between Eqs. (2.A.14) and (2.A.15) must also be made for the two state-vectors $\sum_k A_k \Psi_k$ and $\sum_k B_k \Psi_k$ with which we started. We have thus shown that for a given symmetry transformation T either all state-vectors satisfy Eq. (2.A.14) or else they all satisfy Eq. (2.A.15).

It is now easy to show that as we have defined it, the quantum mechanical operator U is either linear and unitary or antilinear and antiunitary. First, suppose that Eq. (2.A.14) is satisfied for all state-vectors $\sum_k C_k \Psi_k$. Any two state-vectors Ψ and Φ may be expanded as

$$\Psi = \sum_k A_k \Psi_k \,, \quad \Phi = \sum_k B_k \Psi_k$$

and so, using Eq. (2.A.14),

$$\begin{aligned} U(\alpha\Psi + \beta\Phi) &= U \sum_k (\alpha A_k + \beta B_k)\Psi_k = \sum_k (\alpha A_k + \beta B_k) U\Psi_k \\ &= \alpha \sum_k A_k U\Psi_k + \beta \sum_k B_k U\Psi_k \,. \end{aligned}$$

Using Eq. (2.A.14) again, this gives

$$U(\alpha\Psi + \beta\Phi) = \alpha U\Psi + \beta U\Phi, \tag{2.A.19}$$

so U is *linear*. Also, using Eqs. (2.A.2) and (2.A.3), the scalar product of the transformed states is

$$(U\Psi, U\Phi) = \sum_{kl} A_k^* B_l (U\Psi_k, U\Psi_l) = \sum_k A_k^* B_k,$$

and hence

$$(U\Psi, U\Phi) = (\Psi, \Phi), \tag{2.A.20}$$

so U is *unitary*.

The case of a symmetry that satisfies Eq. (2.A.15) for all state-vectors may be dealt with in much the same way. The reader can probably supply the arguments without help, but since antilinear operators may be unfamiliar, we shall give the details here anyway. Suppose that Eq. (2.A.15) is satisfied for all state-vectors $\sum_k C_k \Psi_k$. Any two state-vectors Ψ and Φ may be expanded as before, and so:

$$\begin{aligned} U(\alpha\Psi + \beta\Phi) &= U \sum_k (\alpha A_k + \beta B_k)\Psi_k \\ &= \sum_k (\alpha^* A_k^* + \beta^* B_k^*) U\Psi_k = \alpha^* \sum_k A_k^* U\Psi_k + \beta^* \sum_k B_k^* U\Psi_k . \end{aligned}$$

Using Eq. (2.A.15) again, this gives

$$U(\alpha\Psi + \beta\Phi) = \alpha^* U\Psi + \beta^* U\Phi, \tag{2.A.21}$$

so U is *antilinear*. Also, using Eqs. (2.A.2) and (2.A.3), the scalar product of the transformed states is

$$(U\Psi, U\Phi) = \sum_{kl} A_k B_l^* (U\Psi_k, U\Psi_l) = \sum_k A_k B_k^*$$

and hence

$$(U\Psi, U\Phi) = (\Psi, \Phi)^*, \tag{2.A.22}$$

so U is *antiunitary*.

Appendix B Group Operators and Homotopy Classes

In this appendix we shall prove the theorem stated in Section 2.7, that the phases of the operators $U(T)$ for finite symmetry transformations T may be chosen so that these operators form a representation of the symmetry group, rather than a projective representation, provided (a) the generators of the group can be defined so that there are no central charges in the Lie algebra, and (b) the group is simply connected. We shall also comment on the projective representations encountered for groups that are not simply connected, and their relation to the homotopy classes of the group.

To prove this theorem, let us recall the method by which we construct the operators corresponding to symmetry transformations. As described in Section 2.2, we introduce a set of real variables θ^a to parameterize these transformations, in such a way that the transformations satisfy the composition rule (2.2.15):

$$T(\bar{\theta})T(\theta) = T\Big(f(\bar{\theta}, \theta)\Big).$$

We want to construct operators $U(T(\theta)) \equiv U[\theta]$ that satisfy the corresponding condition*

$$U[\bar{\theta}]U[\theta] = U\Big[f(\bar{\theta}, \theta)\Big]. \tag{2.B.1}$$

To do this, we lay down arbitrary 'standard' paths $\Theta^a_\theta(s)$ in group parameter space, running from the origin to each point θ, with $\Theta^a_\theta(0) = 0$ and $\Theta^a_\theta(1) = \theta^a$, and define $U_\theta(s)$ along each such path by the differential

* Square brackets are used here to distinguish U operators constructed as functions of the group parameters from those expressed as functions of the group transformations themselves.

equation

$$\frac{d}{ds}U_\theta(s) = it_a U_\theta(s) h^a_{\ b}\left(\Theta_\theta(s)\right)\frac{d\Theta^b_\theta(s)}{ds} \tag{2.B.2}$$

with the initial condition

$$U_\theta(0) = 1, \tag{2.B.3}$$

where

$$[h^{-1}]^a_{\ b}(\theta) \equiv \left[\frac{\partial f^a(\bar{\theta},\theta)}{\partial \bar{\theta}^b}\right]_{\bar{\theta}=0}. \tag{2.B.4}$$

We are eventually going to identify the operators $U[\theta]$ with $U_\theta(1)$, but first we must establish some of the properties of $U_\theta(s)$.

In order to check the composition rule, consider two points θ_1 and θ_2, and define a path $\mathscr{P}$ that runs from 0 to θ_1 and thence to $f(\theta_2,\theta_1)$:

$$\Theta^a_{\mathscr{P}}(s) \equiv \begin{cases} \Theta^a_{\theta_1}(2s) & 0 \le s \le \frac{1}{2}, \\ f^a\left(\Theta_{\theta_2}(2s-1),\theta_1\right) & \frac{1}{2} \le s \le 1. \end{cases} \tag{2.B.5}$$

At the end of the first segment, we are at $U_{\mathscr{P}}(\frac{1}{2}) = U_{\theta_1}(1)$. To evaluate $U_{\mathscr{P}}(s)$ along the second segment, we need the derivative of $f^a(\Theta_{\theta_2}(2s-1),\theta_1)$. For this purpose, we use the fundamental associativity condition:

$$f^a\left(f(\theta_3,\theta_2),\theta_1\right) = f^a\left(\theta_3, f(\theta_2,\theta_1)\right). \tag{2.B.6}$$

Matching the coefficients of θ^c_3 in the limit $\theta_3 \to 0$ yields the result:

$$\frac{\partial f^a(\theta_2,\theta_1)}{\partial \theta^b_2} h^c_{\ a}\left(f(\theta_2,\theta_1)\right) = h^c_{\ b}(\theta_2). \tag{2.B.7}$$

Along the second segment the differential equation (2.B.2) for $U_{\mathscr{P}}(s)$ is thus the same as the differential equation for $U_{\theta_2}(2s-1)$. They satisfy different initial conditions, but $U_{\mathscr{P}}(s)U^{-1}_{\theta_1}(1)$ also satisfies the same differential equation as $U_{\theta_2}(2s-1)$, and in addition the same initial condition: at $s=\frac{1}{2}$, both are unity. We therefore conclude that for $\frac{1}{2} \le s \le 1$,

$$U_{\mathscr{P}}(s)\, U^{-1}_{\theta_1}(s) = U_{\theta_2}(2s-1)$$

and in particular

$$U_{\mathscr{P}}(1) = U_{\theta_2}(1)\, U_{\theta_1}(1). \tag{2.B.8}$$

However, this does *not* say that $U_\theta(1)$ satisfies the desired composition rule (2.B.1), because although the path $\Theta_{\mathscr{P}}(s)$ runs from $\theta^a = 0$ to $\theta^a = f^a(\theta_2,\theta_1)$, in general it will not be the same as whatever 'standard' path $\Theta_{f(\theta_2,\theta_1)}$ we have chosen to run directly from $\theta^a = 0$ to $\theta^a = f^a(\theta_2,\theta_1)$. We need to show that $U_\theta(1)$ is independent of the path from 0 to θ in order to be able to identify $U[\theta]$ as $U_\theta(1)$.

For this purpose, consider the variation δU of $U_\theta(s)$ produced by a variation $\delta\Theta(s)$ in the path from 0 to θ. Taking the variation of Eq. (2.B.2) gives the differential equation

$$\frac{d}{ds}\delta U = it_a\delta U\, h^a{}_b(\Theta)\frac{d\Theta^b}{ds} + it_a U h^a{}_{b,c}(\Theta)\delta\Theta^c\frac{d\Theta^b}{ds} + it_a U h^a{}_b(\Theta)\frac{d\delta\Theta^b}{ds}$$

where $h^a{}_{b,c} \equiv \partial h^a{}_b/\partial\Theta^c$. Using the Lie commutation relations (2.2.22) (without central charges) and rearranging a bit, this gives

$$\frac{d}{ds}\left(U^{-1}\delta U\right) = \frac{d}{ds}\left(i\,U^{-1}t_a U h^a{}_b\delta\Theta^b\right)$$
$$+i\,U^{-1}t_a U\delta\Theta^b\frac{d\Theta^c}{ds}\left(h^a{}_{c,b} - h^a{}_{b,c} + C^a{}_{ed}h^e{}_b h^d{}_c\right)\,. \tag{2.B.9}$$

However, by taking the limit $\theta_3, \theta_2 \to 0$ in the associativity condition (2.B.6), we find for all θ:

$$h(\theta)^a{}_{b,c} = -f^a{}_{de}h(\theta)^d{}_b h(\theta)^e{}_c\,, \tag{2.B.10}$$

where $f^a{}_{de}$ is the coefficient defined by (2.2.19). Antisymmetrizing in b and c shows that the last term in Eq. (2.B.9) vanishes

$$h^a{}_{c,b} - h^a{}_{b,c} + C^a{}_{ed}h^e{}_b h^d{}_c = 0\,. \tag{2.B.11}$$

Eq. (2.B.9) thus tells us that the quantity

$$U^{-1}\delta U - iU^{-1}t_a U h^a{}_b\delta\Theta^b$$

is constant along the path $\theta(s)$. It follows that $U_\theta(1)$ is stationary under any infinitesimal variation of the path that leaves the endpoints $\Theta(0) = 0$ and $\Theta(1) = \theta$ (and $U_\theta(0) = 1$) fixed. But assumption (b) tells us that any path from $\Theta(0) = 0$ to $\Theta(1) = \theta$ can be continuously deformed into any other, so we may now regard $U_\theta(1)$ as a path-independent function of θ alone:

$$U_\theta(1) \equiv U[\theta]. \tag{2.B.12}$$

In particular, since the path $\mathscr{P}$ leads from 0 to $f(\theta_2, \theta_1)$, we have

$$U_{\mathscr{P}}(1) = U[f(\theta_2, \theta_1)] \tag{2.B.13}$$

so that Eq. (2.B.8) shows that $U[\theta]$ satisfies the group multiplication law (2.B.1), as was to be proved.

Now that we have constructed a non-projective representation $U[\theta]$, it remains to prove that any projective representation $\tilde{U}[\theta]$ of the same group with the same representation generators t_a can only differ from $U[\theta]$ by a phase:

$$\tilde{U}[\theta] = e^{i\alpha(\theta)}U[\theta]$$

so that the phase ϕ in the multiplication law for $\tilde{U}[\theta]$:

$$\tilde{U}[\theta']\tilde{U}[\theta] = e^{i\phi(\theta',\theta)}\tilde{U}[f(\theta',\theta)]$$

can be removed by a simple change of phase of $\tilde{U}[\theta]$. To see this, consider the operator

$$U[\theta]^{-1}U[\theta']^{-1}\tilde{U}[\theta']\tilde{U}[\theta] = U[f(\theta',\theta)]^{-1}\tilde{U}[f(\theta',\theta)]e^{i\phi(\theta',\theta)}.$$

Because $U[\theta]$ and $\tilde{U}[\theta]$ have the same generators, the derivative of the left-hand side with respect to θ'^a vanishes at $\theta' = 0$, and so

$$0 = \frac{\partial}{\partial\theta^b}\Big\{U[\theta]^{-1}\tilde{U}[\theta]\Big\} + i\phi_b(\theta)U[\theta]^{-1}\tilde{U}[\theta],$$

where

$$\phi_b(\theta) \equiv h^a{}_b(\theta)\left[\frac{\partial}{\partial\theta'^b}\phi(\theta',\theta)\right]_{\theta'=0}.$$

Differentiating this result with respect to θ^c and antisymmetrizing in b and c gives immediately

$$0 = \frac{\partial\phi_b(\theta)}{\partial\theta^c} - \frac{\partial\phi_c(\theta)}{\partial\theta^b}.$$

A familiar theorem[13] tells us that in a simply connected space, this requires that ϕ_b is just a gradient of some function β:

$$\phi_b(\theta) = \frac{\partial\beta(\theta)}{\partial\theta^b}.$$

Thus the quantity $U[\theta]^{-1}\tilde{U}[\theta]e^{i\beta(\theta)}$ is actually constant in θ. Setting it equal to its value at $\theta = 0$, we see that $\tilde{U}$ is just proportional to U:

$$\tilde{U}[\theta] = U[\theta]\exp(-i\beta(\theta) + i\beta(0))$$

as claimed above.

* * *

The above analysis provides some information about the nature of the phase factors that can appear in the group multiplication law when the Lie algebra is free of central charges but the group is not simply connected. Suppose that the path $\mathscr{P}$ from zero to θ to $f(\bar{\theta},\theta)$ cannot be deformed into the standard path we have chosen to go from zero to $f(\bar{\theta},\theta)$, or in other words, that the loop from zero to θ to $f(\bar{\theta},\theta)$ and then back to zero is not continuously deformable to a point. Then $U^{-1}(f(\theta_2,\theta_1))U(\theta_2)U(\theta_1)$ can be a phase factor $\exp(i\phi(\theta_2,\theta_1)) \neq 1$, but ϕ will be the same for all other loops into which this can be continuously deformed. The set consisting of all loops that start and end at the origin and that can be continuously deformed into a given loop is known as the *homotopy class*[14] of that loop, we have thus seen that $\phi(\theta_2,\theta_1)$ depends only on the homotopy class of the loop from zero to θ to $f(\bar{\theta}, \theta)$ and then back to zero. The set of homotopy classes forms a group; the 'product' of the homotopy class for loops $\mathscr{L}_1$ and $\mathscr{L}_2$ is the homotopy class of the loop formed by going around $\mathscr{L}_1$ and then $\mathscr{L}_2$; the 'inverse' of the homotopy class of the loop $\mathscr{L}$ is the homotopy class of the loop obtained by going around $\mathscr{L}$ in the opposite direction; and the 'identity' is the homotopy class of loops that can be deformed into a point at the origin. This group is known as the *first homotopy group* or *fundamental group* of the space in question. It is easy to see that the phase factors form a representation of this group: if going around loop $\mathscr{L}$ gives a phase factor $e^{i\phi}$, and going around loop $\bar{\mathscr{L}}$ gives a phase factor $e^{i\bar{\phi}}$, then going around both loops gives a phase factor $e^{i\phi}e^{i\bar{\phi}}$. Hence we can catalog all the possible types of projective representations of a given group $\mathscr{G}$ (with no central charges) if we know the one-dimensional representations of the first homotopy group of the parameter space of $\mathscr{G}$. Homotopy groups will be discussed in greater detail in Volume II.

Appendix C Inversions and Degenerate Multiplets

It is usually assumed that the inversions T and P take one-particle states into other one-particle states of the same species, perhaps with phase factors that depend on the particle species. In Section 2.6 we noted in passing that inversions might act in a more complicated way than this on degenerate multiplets of one-particle states, a possibility that seems to have been first suggested by Wigner[15] in 1964. This appendix will explore generalized versions of the inversion operators, in which finite matrices appear in place of the inversion phases, but without making some of Wigner's limiting assumptions.

Let us start with time-reversal. Wigner limited the possible action of the inversion operators by assuming that their squares are proportional to the unit operator. Because T is antiunitary, it is easy to see that the corresponding proportionality factor for T^2 can only be ± 1, perhaps with different signs for subspaces separated by superselection rules. When the sign for T^2 on the space of states with even or odd values of $2j$ is opposite to the sign $(-1)^{2j}$ found in Section 2.6, the physical states involved must furnish representations of the operator T that are more complicated than that assumed so far. But if we are willing to admit this possibility, there does not seem to be any good reason to impose Wigner's condition that T^2 is proportional to unity. It is not convincing to appeal to the structure of the extended Poincaré group; the only useful definition of any of the inversion operators is one that makes the operator exactly or approximately conserved, and this may not be the definition that makes T^2 proportional to the unit operator.

To explore more general possibilities for time-reversal, let us assume that on a massive one-particle state it has the action

$$\mathsf{T}\Psi_{\mathbf{p},\sigma,n} = (-1)^{j-\sigma} \sum_m \mathscr{T}_{mn} \Psi_{-\mathbf{p},-\sigma,m} , \tag{2.C.1}$$

where $\mathbf{p}$, j, and σ are the particle's momentum, spin, and spin z-component, and n, m are indices labelling members of a degenerate multiplet of particle species. (The appearance of the factor $(-1)^{j-\sigma}$ and the reversal of $\mathbf{p}$ and σ are deduced in the same way as in Section 2.6.) The matrix $\mathscr{T}_{mn}$ is unknown, except that because T is antiunitary, $\mathscr{T}$ must be unitary.

Now let us see how we can simplify this transformation by an appropriate choice of basis for the one-particle states. Defining new states by the unitary transformation $\Psi'_{\mathbf{p},\sigma,n} = \sum_m \mathscr{U}_{mn}\Psi_{\mathbf{p},\sigma,m}$, we find the same transformation (2.C.1), with the matrix $\mathscr{T}_{mn}$ changed to

$$\mathscr{T}' = \mathscr{U}^{-1}\mathscr{T}\mathscr{U}^* . \tag{2.C.2}$$

We cannot in general make $\mathscr{T}'$ diagonal by such a choice of basis of the one-particle states, as we could if T were unitary. But we can instead make it block-diagonal, with the blocks either 1×1 phases, or 2×2 matrices of the form

$$\begin{pmatrix} 0 & e^{i\phi/2} \\ e^{-i\phi/2} & 0 \end{pmatrix} , \tag{2.C.3}$$

where the ϕ are various real phases.

(Here is the proof. First, note that Eq. (2.C.2) gives

$$\mathscr{T}'\mathscr{T}'^* = \mathscr{U}^{-1}\mathscr{T}\mathscr{T}^*\mathscr{U} .$$

This is a unitary transformation, so it can be chosen to diagonalize the unitary matrix $\mathscr{T}\mathscr{T}^*$. Assuming this to have been done, and dropping primes, we have

$$\mathscr{T} = D\mathscr{T}^T , \tag{2.C.4}$$

where D is a unitary diagonal matrix, say with phases $e^{i\phi_n}$ along the main diagonal. One immediate consequence is that the diagonal component $\mathscr{T}_{nn}$ vanishes unless $e^{i\phi_n} = 1$. Furthermore, if $e^{i\phi_n} = 1$ but $e^{i\phi_m} \neq 1$, then Eq. (2.C.4) tells us that $\mathscr{T}_{nm} = \mathscr{T}_{mn} = 0$. By listing first all rows and columns for which $e^{i\phi_n} = 1$, the matrix $\mathscr{T}$ is put in the form

$$\mathscr{T} = \begin{pmatrix} \mathscr{A} & 0 \\ 0 & \mathscr{B} \end{pmatrix} , \tag{2.C.5}$$

where $\mathscr{A}$ is symmetric as well as unitary, and the diagonal elements of $\mathscr{B}$ all vanish. Because $\mathscr{A}$ is symmetric, it can be expressed as the exponential of a symmetric anti-Hermitian matrix, so it can be diagonalized by a transformation (2.C.2) acting only on $\mathscr{A}$, with the corresponding submatrix of $\mathscr{U}$ real and hence orthogonal. It is therefore only necessary to consider the submatrix $\mathscr{B}$ that connects the rows and columns for which $e^{i\phi_n} \neq 0$. For $n \neq m$, Eq. (2.C.4) gives $\mathscr{T}_{nm} = e^{i\phi_n}\mathscr{T}_{mn}$ and $\mathscr{T}_{mn} = e^{i\phi_m}\mathscr{T}_{nm}$, so $\mathscr{T}_{nm} = e^{i\phi_n}e^{i\phi_m}\mathscr{T}_{nm}$ and also $\mathscr{T}_{mn} = e^{i\phi_n}e^{i\phi_m}\mathscr{T}_{mn}$. Hence $\mathscr{T}_{nm} = \mathscr{T}_{mn} = 0$ unless $e^{i\phi_n}e^{i\phi_m} = 1$. If we list first all rows and columns of $\mathscr{B}$ with a given phase $e^{i\phi_1} \neq 1$, and then all rows and columns with the opposite phase, and then all rows and columns with some other phase $e^{i\phi_2} \neq 1$ not equal to $e^{\pm i\phi_1}$, and then all rows and columns with opposite phase, and so on, the matrix $\mathscr{B}$ becomes of block diagonal form

$$\mathscr{B} = \begin{pmatrix} \mathscr{B}_1 & 0 & \cdots \\ 0 & \mathscr{B}_2 & \cdots \\ \cdots & \cdots & \cdots \end{pmatrix}, \tag{2.C.6}$$

where

$$\mathscr{B}_i = \begin{pmatrix} 0 & e^{i\phi_i/2}\mathscr{C}_i \\ e^{-i\phi_i/2}\mathscr{C}_i^T & 0 \end{pmatrix}. \tag{2.C.7}$$

Furthermore, the unitarity of $\mathscr{T}$ and hence of $\mathscr{B}$ requires that $\mathscr{C}_i\mathscr{C}_i^\dagger = \mathscr{C}_i^\dagger\mathscr{C}_i = 1$, and hence $\mathscr{C}_i$ is square and unitary. By applying a transformation (2.C.2) with $\mathscr{U}$ block-diagonal in the same sense as $\mathscr{T}$, and with the matrix in the ith block of form

$$\begin{pmatrix} V_i & 0 \\ 0 & W_i \end{pmatrix}$$

with V_i and W_i unitary, the submatrices $\mathscr{C}_i$ are subjected to the transformations $\mathscr{C}_i \to V_i^{-1}\mathscr{C}_i W_i^*$, so we can clearly choose this transformation to make $\mathscr{C}_i = 1$. This establishes a correspondence between pairs of individual rows and columns within each block with phases $e^{i\phi_i}$ and $e^{-i\phi_i}$. To put the matrix $\mathscr{B}$ into block-diagonal form with 2×2 blocks of form (2.C.3), it is now only necessary to rearrange the rows and columns so that within the ith block we list rows and columns with phase $e^{i\phi_i}$ alternating with the corresponding rows and columns with phase $e^{-i\phi_i}$.)

It is important to note that where $e^{i\phi} \neq 1$, it is not possible to choose states to diagonalize the time-reversal transformation. If we have a pair of states $\Psi_{\mathbf{p},\sigma,\pm}$ on which T acts with a matrix (2.C.3), then

$$\mathsf{T}\Psi_{\mathbf{p},\sigma,\pm} = e^{\pm i\phi/2}(-1)^{j-\sigma}\Psi_{-\mathbf{p},-\sigma,\mp}\,. \tag{2.C.8}$$

Then on an arbitrary linear combination of these states, time-reversal gives

$$\mathsf{T}\left(c_+\Psi_{\mathbf{p},\sigma,+} + c_-\Psi_{\mathbf{p},\sigma,-}\right) = (-)^{j-\sigma}\left(e^{i\phi/2}c_+^*\Psi_{-\mathbf{p},-\sigma,-} + e^{-i\phi/2}c_-^*\Psi_{-\mathbf{p},-\sigma,+}\right).$$

For $c_+\Psi_{\mathbf{p},\sigma,+} + c_-\Psi_{\mathbf{p},\sigma,-}$ to be transformed under T by a phase λ, it is necessary that

$$e^{i\phi/2}c_+^* = \lambda c_- \qquad e^{-i\phi/2}c_-^* = \lambda c_+ .$$

But combining these equations gives $e^{\pm i\phi/2}c_\pm^* = |\lambda|^2 c_\pm^* e^{\mp i\phi/2}$, which is impossible unless either $c_+ = c_- = 0$ or $e^{i\phi}$ is unity. Thus for $e^{i\phi} \neq 1$, time-reversal invariance imposes a two-fold degeneracy on these states, beyond that associated with their spin.

Of course, if there is an additional 'internal' symmetry operator S which subjects these states to the transformation

$$\mathsf{S}\Psi_{\mathbf{p},\sigma,\pm} = e^{\pm i\phi/2}\Psi_{-\mathbf{p},\sigma,\mp} ,$$

then we can redefine the time-reversal operator as $\mathsf{T}' \equiv \mathsf{S}^{-1}\mathsf{T}$, and this operator would not mix the states $\Psi_{\mathbf{p},\sigma,\pm}$ with one another. It is only in the case where no such internal symmetry exists that we can attribute the doubling of particle states to time-reversal itself.

Let's come back now to the question of the square of T. Repeating the transformation (2.C.8) gives

$$\mathsf{T}^2\Psi_{\mathbf{p},\sigma,\pm} = (-1)^{2j}e^{\mp i\phi}\Psi_{\mathbf{p},\sigma,\pm} . \tag{2.C.9}$$

If we were to assume with Wigner that T^2 is proportional to the unit operator, then we would have to have $e^{i\phi} = e^{-i\phi}$, and since the phase is then real it would have to be $+1$ or -1. The choice $e^{i\phi} = -1$ would still require a two-fold degeneracy of one-particle states beyond that associated with their spin, and under Wigner's assumptions *all* particles would show this doubling. But there is no reason not to take a general phase ϕ in Eq. (2.C.8), one that may vanish for some particles and not for others. Thus the fact that observed particles do not show the extra two-fold degeneracy does not rule out the possibility that others might.

We may also consider the possibility of more complicated representations of the parity operator P, with

$$\mathsf{P}\Psi_{\mathbf{p},\sigma,n} = \sum_m \mathscr{P}_{nm}\Psi_{-\mathbf{p},\sigma,m} \tag{2.C.10}$$

with a unitary but otherwise unconstrained matrix $\mathscr{P}$. Unlike the case of time-reversal, here we may always diagonalize this matrix by a choice of basis for the states. But this choice of basis may not be the one in which time-reversal acts simply, so, in principle, P and T together can impose additional degeneracies that would not be required by P or T alone.

As discussed in Chapter 5, any quantum field theory is expected to respect a symmetry known as CPT, which acts on one-particle states as

$$\mathsf{CPT}\Psi_{\mathbf{p},\sigma,n} = (-1)^{j-\sigma}\Psi_{\mathbf{p},-\sigma,n^c} , \tag{2.C.11}$$

where n^c denotes the antiparticle (or 'charge-conjugate') of particle n. No phases or matrices are allowed in this transformation (though of course we could always introduce such phases or matrices by combining CPT with good internal symmetries.) It follows that

$$(\mathsf{CPT})^2\Psi_{\mathbf{p},\sigma,n} = (-1)^{2j}\Psi_{\mathbf{p},-\sigma,n}\,, \tag{2.C.12}$$

so the possibility suggested by Wigner of a sign $-(-1)^{2j}$ in the action of $(\mathsf{CPT})^2$ does not arise in quantum field theory.

To the extent that T is a good symmetry of some class of phenomena, so is the inversion $\mathsf{CP} \equiv (\mathsf{CPT})\mathsf{T}^{-1}$. For the states that transform under T in the conventional way

$$\mathsf{T}\Psi_{\mathbf{p},\sigma,n} \propto \Psi_{-\mathbf{p},-\sigma,n}\,, \tag{2.C.13}$$

the CP operator also acts conventionally

$$\mathsf{CP}\Psi_{\mathbf{p},\sigma,n} \propto \Psi_{-\mathbf{p},\sigma,n^c}\,. \tag{2.C.14}$$

The operator $\mathsf{C} \equiv \mathsf{CPP}^{-1}$ then just interchanges particles and antiparticles

$$\mathsf{C}\Psi_{\mathbf{p},\sigma,n} \propto \Psi_{\mathbf{p},\sigma,n^c}\,. \tag{2.C.15}$$

On the other hand, where T has the unconventional representation (2.C.8), Eq. (2.C.11) gives

$$\mathsf{CP}\Psi_{\mathbf{p},\sigma,\pm} = e^{\mp i\phi/2}\Psi_{-\mathbf{p},\sigma,\mp^c} \tag{2.C.16}$$

In particular, it is possible that the degeneracy indicated by the label $\pm$ may be the same as the particle–antiparticle degeneracy, so that the antiparticle (as defined by CPT) of the state $\Psi_\pm$ is $\Psi_\mp$. In this case, CP would have the unconventional property of *not* interchanging particles and antiparticles. As far as these particles are concerned, CP and T would be what are usually called P and CT. But this is not merely a matter of definition; on other particles CP and T would still have their usual effect.

No examples are known of particles that furnish unconventional representations of inversions, so these possibilities will not be pursued further here. From now on, the inversions will be assumed to have the conventional action assumed in Section 2.6.

Problems

1. Suppose that observer $\mathcal{O}$ sees a W-boson (spin one and mass $m \neq 0$) with momentum $\mathbf{p}$ in the y-direction and spin z-component σ. A second observer $\mathcal{O}'$ moves relative to the first with velocity $\mathbf{v}$ in the z-direction. How does $\mathcal{O}'$ describe the W state?

2. Suppose that observer $\mathcal{O}$ sees a photon with momentum $\mathbf{p}$ in the y-direction and polarization vector in the z-direction. A second observer $\mathcal{O}'$ moves relative to the first with velocity $\mathbf{v}$ in the z-direction. How does $\mathcal{O}'$ describe the same photon?

3. Derive the commutation relations for the generators of the Galilean group directly from the group multiplication law (without using our results for the Lorentz group). Include the most general set of central charges that cannot be eliminated by redefinition of the group generators.

4. Show that the operators $P_\mu P^\mu$ and $W_\mu W^\mu$ commute with all Lorentz transformation operators $U(\Lambda, a)$, where $W_\mu \equiv \epsilon_{\mu\nu\rho\lambda} J^{\nu\rho} P^\lambda$.

5. Consider physics in two space and one time dimensions, assuming invariance under a 'Lorentz' group $SO(2,1)$. How would you describe the spin states of a single *massive* particle? How do they behave under Lorentz transformations? What about the inversions P and T?

6. As in Problem 5, consider physics in two space and one time dimensions, assuming invariance under a 'Lorentz' group $O(2,1)$. How would you describe the spin states of a single *massless* particle? How do they behave under Lorentz transformations? What about the inversions P and T?

References

1. P. A. M. Dirac, *The Principles of Quantum Mechanics*, 4th edn (Oxford University Press, Oxford, 1958).

2. E. P. Wigner, *Gruppentheorie und ihre Anwendung auf die Quantenmechanik der Atomspektren* (Braunschweig, 1931): pp. 251–3 (English translation, Academic Press, Inc, New York, 1959). For massless particles, see also E. P. Wigner, in *Theoretical Physics* (International Atomic Energy Agency, Vienna, 1963): p. 64.

3. G. C. Wick, A. S. Wightman, and E. P. Wigner, *Phys. Rev.* **88**, 101 (1952).

3*a*. See, e.g, S. Weinberg, *Gravitation and Cosmology* (Wiley, New York, 1972): Section 2.1.

4. E. Inönü and E. P. Wigner, *Nuovo Cimento* **IX**, 705 (1952).

5. E. P. Wigner, *Ann. Math.* **40**, 149 (1939).

6. G. W. Mackey, *Ann. Math.* **55**, 101 (1952); **58**, 193 (1953); *Acta. Math.* **99**, 265 (1958); *Induced Representations of Groups and Quantum Mechanics* (Benjamin, New York, 1968).

7. See, e.g., A. R. Edmonds, *Angular Momentum in Quantum Mechanics*, (Princeton University Press, Princeton, 1957): Chapter 4; M. E. Rose, *Elementary Theory of Angular Momentum* (John Wiley & Sons, New York, 1957): Chapter IV; L. D. Landau and E. M. Lifshitz, *Quantum Mechanics – Non Relativistic Theory*, 3rd edn. (Pergamon Press, Oxford, 1977): Section 58; Wu-Ki Tung, *Group Theory in Physics* (World Scientific, Singapore, 1985): Sections 7.3 and 8.1.

8. T. D. Lee and C. N. Yang, *Phys. Rev.* **104**, 254 (1956); C. S. Wu *et al.*, *Phys. Rev.* **105**, 1413 (1957); R. Garwin, L. Lederman, and M. Weinrich, *Phys. Rev.* **105**, 1415 (1957); J. I. Friedman and V. L. Telegdi, *Phys. Rev.* **105**, 1681 (1957).

9. J. H. Christenson, J. W. Cronin, V. L. Fitch, and R. Turlay, *Phys. Rev. Letters* **13**, 138 (1964).

10. H. A. Kramers, *Proc. Acad. Sci. Amsterdam* **33**, 959 (1930); also see F. J. Dyson, *J. Math. Phys.* **3**, 140 (1962).

11. V. Bargmann, *Ann. Math.* **59**, 1 (1954): Theorem 7.1.

12. See, e.g., H. W. Turnbull and A. C. Aitken, *An Introduction to the Theory of Canonical Matrices* (Dover Publications, New York, 1961): p. 194.

13. See e. g. H. Flanders, *Differential Forms* (Academic Press, New York, 1963): Section 3.6.

14. For an introduction to homotopy classes and groups, see, e.g., J. G. Hocking and G. S. Young, *Topology* (Addison-Wesley, Reading, MA, 1961): Chapter 4; C. Nash and S. Sen, *Topology and Geometry for Physicists* (Academic Press, London, 1983): Chapters 3 and 5.

15. E. P. Wigner, in *Group Theoretical Concepts and Methods in Elementary Particle Physics*, ed. by F. Gürsey (Gordon and Breach, New York, 1964): p. 37.

3
Scattering Theory

The general principles of relativistic quantum mechanics described in the previous chapter have so far been applied here only to states of a single stable particle. Such one-particle states by themselves are not very exciting — it is only when two or more particles interact with each other that anything interesting can happen. But experiments do not generally follow the detailed course of events in particle interactions. Rather, the paradigmatic experiment (at least in nuclear or elementary particle physics) is one in which several particles approach each other from a macroscopically large distance, and interact in a microscopically small region, after which the products of the interaction travel out again to a macroscopically large distance. The physical states before and after the collision consist of particles that are so far apart that they are effectively non-interacting, so they can be described as direct products of the one-particle states discussed in the previous chapter. In such an experiment, all that is measured is the probability distribution, or 'cross-sections', for transitions between the initial and final states of distant and effectively non-interacting particles. This chapter will outline the formalism[1] used for calculating these probabilities and cross-sections.

3.1 'In' and 'Out' States

A state consisting of several non-interacting particles may be regarded as one that transforms under the inhomogeneous Lorentz group as a direct product of one-particle states. To label the one-particle states we use their four-momenta p^μ, spin z-component (or, for massless particles, helicity) σ, and, since we now may be dealing with more than one species of particle, an additional discrete label n for the particle type, which includes a specification of its mass, spin, charge, etc. The general transformation rule is

$$U(\Lambda, a)\Psi_{p_1,\sigma_1,n_1;p_2,\sigma_2,n_2;\cdots} = \exp\Big(-ia_\mu((\Lambda p_1)^\mu + (\Lambda p_2)^\mu + \cdots)\Big)$$
$$\times \sqrt{\frac{(\Lambda p_1)^0(\Lambda p_2)^0\cdots}{p_1^0 p_2^0\cdots}} \sum_{\sigma_1'\sigma_2'\cdots} D^{(j_1)}_{\sigma_1'\sigma_1}\Big(W(\Lambda, p_1)\Big) D^{(j_2)}_{\sigma_2'\sigma_2}\Big(W(\Lambda, p_2)\Big)\cdots$$
$$\times \Psi_{\Lambda p_1,\sigma_1',n_1;\Lambda p_2,\sigma_2',n_2;\cdots} \tag{3.1.1}$$

where $W(\Lambda, p)$ is the Wigner rotation (2.5.10), and $D^{(j)}_{\sigma'\sigma}(W)$ are the conventional $(2j+1)$-dimensional unitary matrices representing the three-dimensional rotation group. (This is for massive particles; for any massless particle, the matrix $D^{(j)}_{\sigma'\sigma}(W(\Lambda, p))$ is replaced with $\delta_{\sigma'\sigma}\exp(i\sigma\theta(\Lambda, p))$, where θ is the angle defined by Eq. (2.5.43).) The states are normalized as in Eq. (2.5.19)

$$\Big(\Psi_{p_1',\sigma_1',n_1';p_2',\sigma_2',n_2';\cdots}\,,\ \Psi_{p_1,\sigma_1,n_1;p_2,\sigma_2,n_2;\cdots}\Big)$$
$$= \delta^3(\mathbf{p}_1' - \mathbf{p}_1)\delta_{\sigma_1'\sigma_1}\delta_{n_1'n_1}\delta^3(\mathbf{p}_2' - \mathbf{p}_2)\delta_{\sigma_2'\sigma_2}\delta_{n_2'n_2}\cdots$$
$$\pm \text{permutations} \tag{3.1.2}$$

with the term '$\pm$ permutations' included to take account of the possibility that it is some permutation of the particle types $n_1', n_2', \cdots$ that are of the same species as the particle types $n_1, n_2, \cdots$. (As discussed more fully in Chapter 4, its sign is -1 if this permutation includes an odd permutation of half-integer spin particles, and otherwise $+1$. This will not be important in the work of the present chapter.)

We often use an abbreviated notation, letting one Greek letter, say α, stand for the whole collection $p_1, \sigma_1, n_1; p_2, \sigma_2, n_2; \cdots$. In this notation, Eq. (3.1.2) is written simply

$$(\Psi_{\alpha'}, \Psi_\alpha) = \delta(\alpha' - \alpha) \tag{3.1.3}$$

with $\delta(\alpha' - \alpha)$ standing for the sum of products of delta functions and Kronecker deltas appearing on the right-hand side of Eq. (3.1.2). Also, in summing over states, we write

$$\int d\alpha \cdots \equiv \sum_{n_1\sigma_1 n_2\sigma_2\cdots} \int d^3p_1\, d^3p_2 \cdots\,, \tag{3.1.4}$$

it being understood that in such sums and integrals, we include only configurations that do not differ merely by the exchange of identical particles. In particular, the completeness relation for states normalized as in Eq. (3.1.3) reads

$$\Psi = \int d\alpha\ \Psi_\alpha(\Psi_\alpha, \Psi)\,. \tag{3.1.5}$$

The transformation rule (3.1.1) is only possible for particles that for one reason or another are not interacting. Setting $\Lambda^\mu{}_\nu = \delta^\mu{}_\nu$ and $a^\mu =$

$(0,0,0,\tau)$, for which $U(\Lambda, a) = \exp(iH\tau)$, Eq. (3.1.1) requires among other things that Ψ_α be an energy eigenstate

$$H\Psi_\alpha = E_\alpha \Psi_\alpha \tag{3.1.6}$$

with an energy equal to the sum of the one-particle energies

$$E_\alpha = p_1^0 + p_2^0 + \cdots \tag{3.1.7}$$

and with no interaction terms, terms that would involve more than one particle at a time.

On the other hand, the transformation rule (3.1.1) does apply in scattering processes at times $t \to \pm\infty$. As explained at the beginning of this chapter, in the typical scattering experiment we start with particles at time $t \to -\infty$ so far apart that they are not yet interacting, and end with particles at $t \to +\infty$ so far apart that they have ceased interacting. We therefore have not one but two sets of states that transform as in Eq. (3.1.1): *the 'in' and 'out' states** Ψ_α^+ *and* Ψ_α^- *will be found to contain the particles described by the label* α *if observations are made at* $t \to -\infty$ *or* $t \to +\infty$, *respectively.*

Note how this definition is framed. To maintain manifest Lorentz invariance, in the formalism we are using here, state-vectors do not change with time — a state-vector Ψ describes the whole spacetime history of a system of particles. (This is known as the *Heisenberg picture*, in distinction with the Schrödinger picture, where the operators are constant and the states change with time.) Thus we do *not* say that $\Psi_\alpha^\pm$ are the limits at $t \to \mp\infty$ of a time-dependent state-vector $\Psi(t)$.

However, implicit in the definition of the states is a choice of the inertial frame from which the observer views the system; different observers see *equivalent* state-vectors, but not the *same* state-vector. In particular, suppose that a standard observer $\mathcal{O}$ sets his or her clock so that $t = 0$ is at some time during the collision process, while some other observer $\mathcal{O}'$ at rest with respect to the first uses a clock set so that $t' = 0$ is at a time $t = \tau$; that is, the two observers' time coordinates are related by $t' = t - \tau$. Then if $\mathcal{O}$ sees the system to be in a state Ψ, $\mathcal{O}'$ will see the system in a state $U(1, -\tau)\Psi = \exp(-iH\tau)\Psi$. Thus the appearance of the state long before or long after the collision (in whatever basis is used by $\mathcal{O}$) is found by applying a time-translation operator $\exp(-iH\tau)$ with $\tau \to -\infty$ or $\tau \to +\infty$, respectively. Of course, if the state is really an energy eigenstate, then it cannot be localized in time — the operator $\exp(-iH\tau)$ yields an inconsequential phase factor $\exp(-iE_\alpha\tau)$. Therefore, we must consider wave-packets, superpositions $\int d\alpha\, g(\alpha)\Psi_\alpha$ of states, with an amplitude $g(\alpha)$

* The labels '+' and '−' for 'in' and 'out' states may seem backward, but they seem to have become traditional. They arise from the signs in Eq. (3.1.16).

that is non-zero and smoothly varying over some finite range ΔE of energies. The 'in' and 'out' states are defined so that the superposition

$$\exp(-iH\tau)\int d\alpha\, g(\alpha)\Psi_\alpha{}^\pm = \int d\alpha\, e^{-iE_\alpha\tau}g(\alpha)\Psi_\alpha{}^\pm$$

has the appearance of a corresponding superposition of free-particle states for $\tau \ll -1/\Delta E$ or $\tau \gg +1/\Delta E$, respectively.

To make this concrete, suppose we can divide the time-translation generator H into two terms, a free-particle Hamiltonian H_0 and an interaction V,

$$H = H_0 + V \tag{3.1.8}$$

in such a way that H_0 has eigenstates Φ_α that have the same appearance as the eigenstates Ψ_α^+ and Ψ_α^- of the complete Hamiltonian

$$H_0\Phi_\alpha = E_\alpha\Phi_\alpha\,, \tag{3.1.9}$$

$$(\Phi_{\alpha'}, \Phi_\alpha) = \delta(\alpha' - \alpha)\,. \tag{3.1.10}$$

Note that H_0 is assumed here to have the same spectrum as the full Hamiltonian H. This requires that the masses appearing in H_0 be the physical masses that are actually measured, which are not necessarily the same as the 'bare' mass terms appearing in H; the difference if there is any must be included in the interaction V, not H_0. Also, any relevant bound states in the spectrum of H should be introduced into H_0 as if they were elementary particles.**

The 'in' and 'out' states can now be defined as eigenstates of H, not H_0,

$$H\Psi_\alpha{}^\pm = E_\alpha\Psi_\alpha{}^\pm \tag{3.1.11}$$

which satisfy the condition

$$\int d\alpha\, e^{-iE_\alpha\tau}g(\alpha)\Psi_\alpha{}^\pm \to \int d\alpha\, e^{-iE_\alpha\tau}g(\alpha)\Phi_\alpha \tag{3.1.12}$$

for $\tau \to -\infty$ or $\tau \to +\infty$, respectively.

Eq. (3.1.12) can be rewritten as the requirement that:

$$\exp(-iH\tau)\int d\alpha\, g(\alpha)\Psi_\alpha{}^\pm \to \exp(-iH_0\tau)\int d\alpha\, g(\alpha)\Phi_\alpha$$

for $\tau \to -\infty$ or $\tau \to +\infty$, respectively. This is sometimes rewritten as a formula for the 'in' and 'out' states:

$$\Psi_\alpha{}^\pm = \Omega(\mp\infty)\Phi_\alpha\,, \tag{3.1.13}$$

** Alternatively, in non-relativistic problems we can include the binding potential in H_0. In the application of this method to 'rearrangement collisions,' where some bound states appear in the initial state but not the final state, or vice-versa, one must use a different split of H into H_0 and V in the initial and final states.

where

$$\Omega(\tau) \equiv \exp(+iH\tau)\exp(-iH_0\tau)\,. \tag{3.1.14}$$

However, it should be kept in mind that $\Omega(\mp\infty)$ in Eq. (3.1.13) gives meaningful results only when acting on a smooth superposition of energy eigenstates.

One immediate consequence of the definition (3.1.12) is that the 'in' and 'out' states are normalized just like the free-particle states. To see this, note that since the left-hand side of Eq. (3.1.12) is obtained by letting the unitary operator $\exp(-iH\tau)$ act on a time-independent state, its norm is independent of time, and therefore equals the norm of its limit for $\tau \to \infty$, i.e., the norm of the right-hand side of Eq. (3.1.12):

$$\int d\alpha\, d\beta\, \exp(-i(E_\alpha - E_\beta)\tau)g(\alpha)g^*(\beta)(\Psi_\beta{}^\pm, \Psi_\alpha{}^\pm)$$
$$= \int d\alpha\, d\beta\, \exp(-i(E_\alpha - E_\beta)\tau)g(\alpha)g^*(\beta)(\Phi_\beta, \Phi_\alpha)\,.$$

Since this is supposed to be true for all smooth functions $g(\alpha)$, the scalar products must be equal

$$(\Psi_\beta{}^\pm, \Psi_\alpha{}^\pm) = (\Phi_\beta, \Phi_\alpha) = \delta(\beta - \alpha)\,. \tag{3.1.15}$$

It is useful for some purposes to have an explicit though formal solution of the energy eigenvalue equation (3.1.11) satisfying the conditions (3.1.12). For this purpose, write Eq. (3.1.11) as

$$(E_\alpha - H_0)\Psi_\alpha{}^\pm = V\Psi_\alpha{}^\pm\,.$$

The operator $E_\alpha - H_0$ is not invertible; it annihilates not only the free-particle state Φ_α, but also the continuum of other free-particle states Φ_β of the same energy. Since the 'in' and 'out' states become just Φ_α for $V \to 0$, we tentatively write the formal solutions as Φ_α plus a term proportional to V:

$$\Psi_\alpha{}^\pm = \Phi_\alpha + (E_\alpha - H_0 \pm i\epsilon)^{-1}V\Psi_\alpha{}^\pm\,, \tag{3.1.16}$$

or, expanding in a complete set of free-particle states,

$$\Psi_\alpha{}^\pm = \Phi_\alpha + \int d\beta\, \frac{T_{\beta\alpha}{}^\pm\,\Phi_\beta}{E_\alpha - E_\beta \pm i\epsilon}\,, \tag{3.1.17}$$

$$T_{\beta\alpha}{}^\pm \equiv (\Phi_\beta, V\Psi_\alpha{}^\pm)\,, \tag{3.1.18}$$

with ϵ a positive infinitesimal quantity, inserted to give meaning to the reciprocal of $E_\alpha - H_0$. These are known as the *Lippmann-Schwinger equations.*[1a] We shall use Eq. (3.1.17) at the end of the next section to give a slightly less unrigorous proof of the orthonormality of the 'in' and 'out' states.

It remains to be shown that Eq. (3.1.17), with a $+i\epsilon$ or $-i\epsilon$ in the denominator, satisfies the condition (3.1.12) for an 'in' or an 'out' state, respectively. For this purpose, consider the superpositions

$$\Psi_g{}^{\pm}(t) \equiv \int d\alpha\, e^{-iE_\alpha t} g(\alpha) \Psi_\alpha{}^{\pm}\,, \tag{3.1.19}$$

$$\Phi_g(t) \equiv \int d\alpha\, e^{-iE_\alpha t} g(\alpha) \Phi_\alpha\,. \tag{3.1.20}$$

We want to show that $\Psi_g{}^{+}(t)$ and $\Psi_g{}^{-}(t)$ approach $\Phi_g(t)$ for $t \to -\infty$ and $t \to +\infty$, respectively. Using Eq. (3.1.17) in Eq. (3.1.19) gives

$$\Psi_g{}^{\pm}(t) = \Phi_g(t) + \int d\alpha \int d\beta\, \frac{e^{-iE_\alpha t} g(\alpha) T_{\beta\alpha}{}^{\pm} \Phi_\beta}{E_\alpha - E_\beta \pm i\epsilon}\,. \tag{3.1.21}$$

Let us recklessly interchange the order of integration, and consider first the integrals

$$\mathscr{I}_\beta{}^{\pm} \equiv \int d\alpha\, \frac{e^{-iE_\alpha t} g(\alpha) T_{\beta\alpha}{}^{\pm}}{E_\alpha - E_\beta \pm i\epsilon}\,.$$

For $t \to -\infty$, we can close the contour of integration for the energy variable E_α in the upper half-plane with a large semi-circle, with the contribution from this semi-circle killed by the factor $\exp(-iE_\alpha t)$, which is exponentially small for $t \to -\infty$ and $\operatorname{Im} E_\alpha > 0$. The integral is then given by a sum over the singularities of the integral in the upper half-plane. The functions $g(\alpha)$ and $T_{\beta\alpha}{}^{\pm}$ may, in general, be expected to have some singularities at values of E_α with finite positive imaginary parts, but just as for the large semi-circle, their contribution is exponentially damped for $t \to -\infty$. (Specifically, $-t$ must be much greater than both the time-uncertainty in the wave-packet $g(\alpha)$ and the duration of the collision, which respectively govern the location of the singularities of $g(\alpha)$ and $T_{\beta\alpha}{}^{\pm}$ in the complex E_α plane.) This leaves the singularity in $(E_\alpha - E_\beta \pm i\epsilon)^{-1}$, which is in the upper half-plane for $\mathscr{I}_\beta{}^{-}$ but not $\mathscr{I}_\beta{}^{+}$. We conclude then that $\mathscr{I}_\beta{}^{+}$ vanishes for $t \to -\infty$. In the same way, for $t \to +\infty$ we must close the contour of integration in the lower half-plane, and so $\mathscr{I}_\beta{}^{-}$ vanishes in this limit. We conclude that $\Psi_g{}^{\pm}(t)$ approaches $\Phi_g(t)$ for $t \to \mp\infty$, in agreement with the defining condition (3.1.12).

* * *

For future use, we note a convenient representation of the factor $(E_\alpha - E_\beta \pm i\epsilon)^{-1}$ in Eq. (3.1.17). In general, we can write

$$(E \pm i\epsilon)^{-1} = \frac{\mathscr{P}_\epsilon}{E} \mp i\pi\delta_\epsilon(E)\,, \tag{3.1.22}$$

where

$$\frac{\mathscr{P}_\epsilon}{E} \equiv \frac{E}{E^2+\epsilon^2}\,, \tag{3.1.23}$$

$$\delta_\epsilon(E) \equiv \frac{\epsilon}{\pi(E^2+\epsilon^2)}\,. \tag{3.1.24}$$

The function (3.1.23) is just $1/E$ for $|E| \gg \epsilon$, and vanishes for $E \to 0$, so for $\epsilon \to 0$ it behaves just like the 'principal value function' $\mathscr{P}/E$, which allows us to give meaning to integrals of $1/E$ times any smooth function of E, by excluding an infinitesimal interval around $E = 0$. The function (3.1.24) is of order ϵ for $|E| \gg \epsilon$, and gives unity when integrated over all E, so in the limit $\epsilon \to 0$ it behaves just like the familiar delta function $\delta(E)$. With this understanding, we can drop the label ϵ in Eq. (3.1.22), and write simply

$$(E \pm i\epsilon)^{-1} = \frac{\mathscr{P}}{E} \mp i\pi\delta(E)\,. \tag{3.1.25}$$

3.2 The S-matrix

An experimentalist generally prepares a state to have a definite particle content at $t \to -\infty$, and then measures what this state looks like at $t \to +\infty$. If the state is prepared to have a particle content α for $t \to -\infty$, then it is the 'in' state $\Psi_\alpha{}^+$, and if it is found to have the particle content β at $t \to +\infty$, then it is the 'out' state $\Psi_\beta{}^-$. The probability amplitude for the transition $\alpha \to \beta$ is thus the scalar product

$$S_{\beta\alpha} = (\Psi_\beta{}^-, \Psi_\alpha{}^+)\,. \tag{3.2.1}$$

This array of complex amplitudes is known as the *S-matrix*.[2] If there were no interactions then 'in' and 'out' states would be the same, and then $S_{\beta\alpha}$ would be just $\delta(\alpha - \beta)$. The rate for a reaction $\alpha \to \beta$ is thus proportional to $|S_{\beta\alpha} - \delta(\alpha-\beta)|^2$. We shall see in detail in Section 3.4 what $S_{\beta\alpha}$ has to do with measured rates and cross-sections.

Perhaps it should be stressed that 'in' and 'out' states do not inhabit two different Hilbert spaces. They differ only in how they are labelled: by their appearance either at $t \to -\infty$ or $t \to +\infty$. Any 'in' state can be expanded as a sum of 'out' states, with expansion coefficients given by the S-matrix (3.2.1).

Since $S_{\beta\alpha}$ is the matrix connecting two complete sets of orthonormal states, it must be unitary. To see this in greater detail, apply the completeness relation (3.1.5) to the 'out' states, and write

$$\int d\beta\; S^*_{\beta\gamma} S_{\beta\alpha} = \int d\beta\; (\Psi_\gamma{}^+, \Psi_\beta{}^-)(\Psi_\beta{}^-, \Psi_\alpha{}^+) = (\Psi_\gamma{}^+, \Psi_\alpha{}^+)\,.$$

Using (3.1.15), this gives

$$\int d\beta\ S^*_{\beta\gamma} S_{\beta\alpha} = \delta(\gamma - \alpha) \tag{3.2.2}$$

or in brief, $S^\dagger S = 1$. In the same way, completeness for the 'in' states gives*

$$\int d\beta\ S_{\gamma\beta} S_{\alpha\beta}{}^* = \delta(\gamma - \alpha) \tag{3.2.3}$$

or, in other words, $SS^\dagger = 1$.

It is often convenient instead of dealing with the S-matrix to work with an operator S, defined to have matrix elements between *free-particle* states equal to the corresponding elements of the S-matrix:

$$(\Phi_\beta, S\Phi_\alpha) \equiv S_{\beta\alpha}\ . \tag{3.2.4}$$

The explicit though highly formal expression (3.1.13) for the 'in' and 'out' states yields a formula for the S-operator:

$$S = \Omega(\infty)^\dagger \Omega(-\infty) = U(+\infty, -\infty)\,, \tag{3.2.5}$$

where

$$U(\tau, \tau_0) \equiv \Omega(\tau)^\dagger \Omega(\tau_0) = \exp(iH_0\tau)\exp(-iH(\tau - \tau_0))\exp(-iH_0\tau_0)\ . \tag{3.2.6}$$

This will be used in the next section to examine the Lorentz invariance of the S-matrix, and in Sec. 3.5 to derive a formula for the S-matrix in time-dependent perturbation theory.

The methods of the previous section can be used to derive a useful alternative formula for the S-matrix. Let's return to Eq. (3.1.21) for the 'in' state Ψ^+, but this time take $t \to +\infty$. We must now close the contour of integration for E_α in the *lower* half-E_α-plane, and although as before the singularities in $T_{\beta\alpha}{}^+$ and $g(\alpha)$ make no contribution for $t \to +\infty$, we now do pick up a contribution from the singular factor $(E_\alpha - E_\beta + i\epsilon)^{-1}$. The contour runs from $E_\alpha = -\infty$ to $E_\alpha = +\infty$, and then back to $E_\alpha = -\infty$ on a large semi-circle in the lower half-plane, so it circles the singularity in a *clockwise* direction. By the method of residues, this contribution to the integral over E_α is given by the value of the integrand at $E_\alpha = E_\beta - i\epsilon$, times a factor $-2i\pi$. That is, in the limit $\epsilon \to 0+$, for $t \to +\infty$ the integral over α in (3.1.21) has the asymptotic behavior

$$\mathscr{I}_\beta{}^+ \to -2i\pi e^{-iE_\beta t} \int d\alpha\ \delta(E_\alpha - E_\beta) g(\alpha) T_{\beta\alpha}{}^+$$

* An alternative proof is given at the end of this section. Note that for infinite 'matrices,' the unitarity conditions $S^\dagger S$=1 and $SS^\dagger = 1$ are not equivalent.

and hence, for $t \to +\infty$,

$$\Psi_g{}^+(t) \to \int d\beta\, e^{-iE_\beta t}\Phi_\beta\left[g(\beta) - 2i\pi \int d\alpha\, \delta(E_\alpha - E_\beta) g(\alpha) T_{\beta\alpha}{}^+\right].$$

But expanding (3.1.19) for $\Psi_g{}^+$ in a complete set of 'out' states gives

$$\Psi_g{}^+(t) = \int d\alpha\, e^{-iE_\alpha t} g(\alpha) \int d\beta\, \Psi_\beta{}^-\, S_{\beta\alpha}\,.$$

Since $S_{\beta\alpha}$ contains a factor $\delta(E_\beta - E_\alpha)$, this may be rewritten

$$\Psi_g{}^+(t) = \int d\beta\, \Psi_\beta{}^-\, e^{-iE_\beta t} \int d\alpha\, g(\alpha) S_{\beta\alpha}$$

and, using the defining property (3.1.12) for 'out' states, this has the asymptotic behavior for $t \to +\infty$

$$\Psi_g{}^+(t) \to \int d\beta\, \Phi_\beta\, e^{-iE_\beta t} \int d\alpha\, g(\alpha) S_{\beta\alpha}\,.$$

Comparing this with our previous result, we find

$$\int d\alpha\, g(\alpha) S_{\beta\alpha} = g(\beta) - 2i\pi \int d\alpha\, \delta(E_\alpha - E_\beta) g(\alpha) T_{\beta\alpha}{}^+$$

or in other words

$$S_{\beta\alpha} = \delta(\beta - \alpha) - 2i\pi\delta(E_\alpha - E_\beta) T_{\beta\alpha}{}^+\,. \tag{3.2.7}$$

This suggests a simple approximation for the S-matrix: for a weak interaction V, we can neglect the difference between 'in' and free-particle states in (3.1.18), in which case Eq. (3.2.7) gives

$$S_{\beta\alpha} \simeq \delta(\beta - \alpha) - 2i\pi\delta(E_\alpha - E_\beta)(\Phi_\beta, V\Phi_\alpha)\,. \tag{3.2.8}$$

This is known as the *Born approximation*.[3] Higher-order terms are discussed in Section 3.5.

* * *

We can use the Lippmann–Schwinger equations (3.1.16) for the 'in' and 'out' states to give a proof[4] of the orthonormality of these states and the unitarity of the S-matrix, as well as Eq. (3.2.7), without having to deal with limits at $t \to \mp\infty$. First, by using (3.1.16) on either the left- or right-hand side of the matrix element $(\Psi_\beta^\pm, V\Psi_\alpha^\pm)$ and equating the results, we find that

$$\begin{aligned}(\Psi_\beta{}^\pm, V\Phi_\alpha) &+ (\Psi_\beta{}^\pm, V(E_\alpha - H_0 \pm i\epsilon)^{-1} V\Psi_\alpha{}^\pm)\\ &= (\Phi_\beta, V\Psi_\alpha{}^\pm) + (\Psi_\beta{}^\pm, V(E_\beta - H_0 \mp i\epsilon)^{-1} V\Psi_\alpha{}^\pm).\end{aligned}$$

Summing over a complete set Φ_γ of intermediate states, this gives the equation:

$$T_{\alpha\beta}{}^{\pm *} - T_{\beta\alpha}{}^{\pm} = -\int d\gamma\; T_{\gamma\beta}{}^{\pm *} T_{\gamma\alpha}{}^{\pm} \times \Big([E_\alpha - E_\gamma \pm i\epsilon]^{-1} - [E_\beta - E_\gamma \mp i\epsilon]^{-1} \Big) . \tag{3.2.9}$$

To prove the orthonormality of the 'in' and 'out' states, divide Eq. (3.2.9) by $E_\alpha - E_\beta \pm 2i\epsilon$. This gives

$$\left(\frac{T_{\alpha\beta}{}^{\pm}}{E_\beta - E_\alpha \pm 2i\epsilon} \right)^{*} + \frac{T_{\beta\alpha}{}^{\pm}}{E_\alpha - E_\beta \pm 2i\epsilon} = -\int d\gamma \left(\frac{T_{\gamma\beta}{}^{\pm}}{E_\beta - E_\gamma \pm i\epsilon} \right)^{*} \frac{T_{\gamma\alpha}{}^{\pm}}{E_\alpha - E_\gamma \pm i\epsilon} .$$

The 2ϵs in the denominators on the left-hand side can be replaced with ϵs, since the only important thing is that these are positive infinitesimals. We see then that $\delta(\beta - \alpha) + T_{\beta\alpha}{}^{\pm}/(E_\alpha - E_\beta \pm i\epsilon)$ is unitary. With (3.1.17), this is just the statement that the $\Psi_\alpha{}^{\pm}$ form two orthonormal sets of state-vectors. The unitarity of the S-matrix can be proved in a similar fashion by multiplying (3.2.9) with $\delta(E_\beta - E_\alpha)$ instead of $(E_\alpha - E_\beta \pm 2i\epsilon)^{-1}$.

3.3 Symmetries of the S-Matrix

In this section we will consider both what is meant by the invariance of the S-matrix under various symmetries, and what are the conditions on the Hamiltonian that will ensure such invariance properties.

Lorentz Invariance

For any proper orthochronous Lorentz transformation $x \to \Lambda x + a$, we may define a unitary operator $U(\Lambda, a)$ by specifying that it acts as in Eq. (3.1.1) on either the 'in' *or* the 'out' states. When we say that a theory is Lorentz-invariant, we mean that the same operator $U(\Lambda, a)$ acts as in (3.1.1) on *both* 'in' *and* 'out' states. Since the operator $U(\Lambda, a)$ is unitary, we may write

$$S_{\beta\alpha} = \left(\Psi_\beta{}^{-}, \Psi_\alpha{}^{+} \right) = \left(U(\Lambda, a)\Psi_\beta{}^{-}, U(\Lambda, a)\Psi_\alpha{}^{+} \right)$$

so using (3.1.1), we obtain the Lorentz invariance (actually, covariance) property of the S-matrix: for arbitrary Lorentz transformations $\Lambda^\mu{}_\nu$ and translations a^μ,

$$S_{p_1',\sigma_1',n_1';\,p_2',\sigma_2',n_2';\cdots,\;\; p_1,\sigma_1,n_1;\;p_2,\sigma_2,n_2;\cdots}$$
$$= \exp\Big(ia_\mu \Lambda^\mu{}_\nu (p_1'^\nu + p_2'^\nu + \ldots - p_1^\nu - p_2^\nu - \ldots)\Big)$$
$$\times \sqrt{\frac{(\Lambda p_1)^0 (\Lambda p_2)^0 \cdots (\Lambda p_1')^0 (\Lambda p_2')^0 \cdots}{p_1^0 p_2^0 \cdots \;\; p_1'^0 p_2'^0 \cdots}}$$
$$\times \sum_{\bar\sigma_1 \bar\sigma_2 \cdots} D^{(j_1)}_{\bar\sigma_1 \sigma_1}\Big(W(\Lambda, p_1)\Big) D^{(j_2)}_{\bar\sigma_2 \sigma_2}\Big(W(\Lambda, p_2)\Big) \cdots$$
$$\times \sum_{\bar\sigma_1' \bar\sigma_2' \cdots} D^{(j_1')*}_{\bar\sigma_1' \sigma_1'}\Big(W(\Lambda, p_1')\Big) D^{(j_2')*}_{\bar\sigma_2' \sigma_2'}\Big(W(\Lambda, p_2')\Big) \cdots$$
$$\times S_{\Lambda p_1',\bar\sigma_1',n_1';\,\Lambda p_2',\bar\sigma_2',n_2';\cdots,\;\; \Lambda p_1,\bar\sigma_1,n_1;\,\Lambda p_2,\bar\sigma_2,n_2;\cdots} \quad (3.3.1)$$

(Primes are used to distinguish final from initial particles; bars are used to distinguish summation variables.) In particular, since the left-hand side is independent of a^μ, so must be the right-hand side, and so the S-matrix vanishes unless the four-momentum is conserved. We can therefore write the part of the S-matrix that represents actual interactions among the particles in the form:

$$S_{\beta\alpha} - \delta(\beta - \alpha) = -2\pi i\, M_{\beta\alpha} \delta^4(p_\beta - p_\alpha). \quad (3.3.2)$$

(However, as we will see in the next chapter, the amplitude $M_{\beta\alpha}$ itself contains terms that involve further delta function factors.)

Eq. (3.3.1) should be regarded as a definition of what we mean by the Lorentz invariance of the S-matrix, rather than a theorem, because it is only for certain special choices of Hamiltonian that there exists a unitary operator that acts as in (3.1.1) on both 'in' and 'out' states. We need to formulate conditions on the Hamiltonian that would ensure the Lorentz invariance of the S-matrix. For this purpose, it will be convenient to work with the operator S defined by Eq. (3.2.4):

$$S_{\beta\alpha} = (\Phi_\beta, S\Phi_\alpha).$$

As we have defined the free-particle states Φ_α in Chapter 2, they furnish a representation of the inhomogeneous Lorentz group, so we can always define a unitary operator $U_0(\Lambda, a)$ that induces the transformation (3.1.1) on these states:

$$U_0(\Lambda, a)\Phi_{p_1,\sigma_1,n_1;\,p_2,\sigma_2,n_2;\cdots} = \exp\Big(-ia_\mu \Lambda^\mu{}_\nu (p_1^\nu + p_2^\nu + \ldots)\Big)$$
$$\times \sqrt{\frac{(\Lambda p_1)^0 (\Lambda p_2)^0 \cdots}{p_1^0 p_2^0 \cdots}} \sum_{\sigma_1' \sigma_2' \cdots} D^{(j_1)}_{\sigma_1' \sigma_1}\Big(W(\Lambda, p_1)\Big) D^{(j_2)}_{\sigma_2' \sigma_2}\Big(W(\Lambda, p_2)\Big) \cdots$$
$$\times \Phi_{\Lambda p_1,\sigma_1',n_1;\,\Lambda p_2,\sigma_2',n_2;\cdots}$$

Eq. (3.3.1) will thus hold if this unitary operator commutes with the

S-operator:

$$U_0(\Lambda, a)^{-1} S\, U_0(\Lambda, a) = S\ .$$

This condition can also be expressed in terms of infinitesimal Lorentz transformations. Just as in Section 2.4, there will exist a set of Hermitian operators, a momentum $\mathbf{P}_0$, an angular momentum $\mathbf{J}_0$, and a boost generator $\mathbf{K}_0$, that together with H_0 generate the infinitesimal version of inhomogeneous Lorentz transformations when acting on free-particle states. Eq. (3.3.1) is equivalent to the statement that the S-matrix is unaffected by such transformations, or in other words, that the S-operator commutes with these generators:

$$[H_0, S] = [\mathbf{P}_0, S] = [\mathbf{J}_0, S] = [\mathbf{K}_0, S] = 0. \tag{3.3.3}$$

Because the operators H_0, $\mathbf{P}_0$, $\mathbf{J}_0$, and $\mathbf{K}_0$ generate infinitesimal inhomogeneous Lorentz transformations on the Φ_α, they automatically satisfy the commutation relations (2.4.18)–(2.4.24):

$$[J_0^i, J_0^j] = i\,\epsilon_{ijk} J_0^k\ , \tag{3.3.4}$$

$$[J_0^i, K_0^j] = i\,\epsilon_{ijk} K_0^k\ , \tag{3.3.5}$$

$$[K_0^i, K_0^j] = -i\,\epsilon_{ijk} J_0^k\ , \tag{3.3.6}$$

$$[J_0^i, P_0^j] = i\,\epsilon_{ijk} P_0^k\ , \tag{3.3.7}$$

$$[K_0^i, P_0^j] = -iH_0\delta_{ij}\ , \tag{3.3.8}$$

$$[J_0^i, H_0] = [P_0^i, H_0] = [P_0^i, P_0^j] = 0\ , \tag{3.3.9}$$

$$[K_0^i, H_0] = -i\,P_0^i\ , \tag{3.3.10}$$

where i, j, k, etc. run over the values 1, 2, and 3, and ϵ_{ijk} is the totally antisymmetric quantity with $\epsilon_{123} = +1$.

In the same way, we may define a set of 'exact generators,' operators $\mathbf{P}$, $\mathbf{J}$, $\mathbf{K}$ that together with the full Hamiltonian H generate the transformations (3.1.1) on, say, the 'in' states. (As already mentioned, what is not obvious is that the same operators generate the same transformations on the 'out' states.) The group structure tells us that these exact generators satisfy the same commutation relations:

$$[J^i, J^j] = i\,\epsilon_{ijk} J^k\ , \tag{3.3.11}$$

$$[J^i, K^j] = i\,\epsilon_{ijk} K^k\ , \tag{3.3.12}$$

$$[K^i, K^j] = -i\,\epsilon_{ijk} J^k\ , \tag{3.3.13}$$

$$[J^i, P^j] = i\,\epsilon_{ijk} P^k\ , \tag{3.3.14}$$

$$[K^i, P^j] = -iH\delta_{ij}\ , \tag{3.3.15}$$

$$[J^i, H] = [P^i, H] = [P^i, P^j] = 0\ , \tag{3.3.16}$$

$$[K^i, H] = -i\,P^i. \tag{3.3.17}$$

In virtually all known field theories, the effect of interactions is to add an interaction term V to the Hamiltonian, while leaving the momentum and angular momentum unchanged:

$$H = H_0 + V \ , \qquad \mathbf{P} = \mathbf{P}_0 \ , \qquad \mathbf{J} = \mathbf{J}_0 \, . \tag{3.3.18}$$

(The only known exceptions are theories with topologically twisted fields, such as those with magnetic monopoles, where the angular momentum of states depends on the interactions.) Eq. (3.3.18) implies that the commutation relations (3.3.11), (3.3.14), and (3.3.16) are satisfied provided that the interaction commutes with the free-particle momentum and angular-momentum operators

$$[V, \mathbf{P}_0] = [V, \mathbf{J}_0] = 0. \tag{3.3.19}$$

It is easy to see from the Lippmann–Schwinger equation (3.1.16) or equivalently from (3.1.13) that the operators that generate translations and rotations when acting on the 'in' (and 'out') states are indeed simply $\mathbf{P}_0$ and $\mathbf{J}_0$. Also we easily see that $\mathbf{P}_0$ and $\mathbf{J}_0$ commute with the operator $U(t, t_0)$ defined by Eq.(3.2.6), and hence with the S-operator $U(\infty, -\infty)$. Further, we already know that the S-operator commutes with H_0, because there are energy-conservation delta functions in both terms in (3.2.7). This leaves just the boost generator $\mathbf{K}_0$ which we need to show commutes with the S-operator.

On the other hand, it is *not* possible to set the boost generator $\mathbf{K}$ equal to its free-particle counterpart $\mathbf{K}_0$, because then Eqs. (3.3.15) and (3.3.8) would give $H = H_0$, which is certainly not true in the presence of interactions. Thus when we add an interaction V to H_0, we must also add a correction $\mathbf{W}$ to the boost generator:

$$\mathbf{K} = \mathbf{K}_0 + \mathbf{W}. \tag{3.3.20}$$

Of the remaining commutation relations, let us concentrate on Eq. (3.3.17), which may now be put in the form

$$[\mathbf{K}_0, V] = -[\mathbf{W}, H]. \tag{3.3.21}$$

By itself, the condition (3.3.21) is empty, because for any V we could always define $\mathbf{W}$ by giving its matrix elements between H-eigenstates Ψ_α and Ψ_β as $-(\Psi_\beta, [\mathbf{K}_0, V]\Psi_\alpha)/(E_\beta - E_\alpha)$. Recall that the crucial point in the Lorentz invariance of a theory is not that there should exist a set of exact generators satisfying Eqs. (3.3.11)–(3.3.17), but rather that these operators should act the same way on 'in' and 'out' states; merely finding an operator $\mathbf{K}$ that satisfies Eq. (3.3.21) is not enough. Eq. (3.3.21) does become significant if we add the requirement that matrix elements of $\mathbf{W}$ should be smooth functions of the energies, and in particular should not have singularities of the form $(E_\beta - E_\alpha)^{-1}$. We shall now show that

Eq. (3.3.21), together with an appropriate smoothness condition on $\mathbf{W}$, does imply the remaining Lorentz invariance condition $[\mathbf{K}_0, S] = 0$.

To prove this, let us consider the commutator of $\mathbf{K}_0$ with the operator $U(t, t_0)$ defined by Eq. (3.2.6) for finite t and t_0. Using Eq. (3.3.10) and the fact that $\mathbf{P}_0$ commutes with H_0 yields:

$$[\mathbf{K}_0, \exp(iH_0t)] = t\mathbf{P}_0 \exp(iH_0t)$$

while Eq. (3.3.21) (which is equivalent to Eq. (3.3.17)) yields

$$[\mathbf{K}, \exp(iHt)] = t\mathbf{P} \exp(iHt) = t\mathbf{P}_0 \exp(iHt)\,.$$

The momentum operators then cancel in the commutator of $\mathbf{K}_0$ with U, and we find:

$$[\mathbf{K}_0, U(\tau, \tau_0)] = -\mathbf{W}(\tau)U(\tau, \tau_0) + U(\tau, \tau_0)\mathbf{W}(\tau_0)\,, \tag{3.3.22}$$

where

$$\mathbf{W}(t) \equiv \exp(iH_0t)\mathbf{W} \exp(-iH_0t). \tag{3.3.23}$$

If the matrix elements of $\mathbf{W}$ between H_0-eigenstates are sufficiently smooth functions of energy, then matrix elements of $\mathbf{W}(t)$ between smooth superpositions of energy eigenstates vanish for $t \to \pm\infty$, so Eq. (3.3.22) gives in effect:

$$0 = [\mathbf{K}_0, U(\infty, -\infty)] = [\mathbf{K}_0, S], \tag{3.3.24}$$

as was to be shown. This is the essential result: Eq. (3.3.21), together with the smoothness condition on matrix elements of $\mathbf{W}$ that ensures that $\mathbf{W}(t)$ vanishes for $t \to \pm\infty$, provides a sufficient condition for the Lorentz invariance of the S-matrix. This smoothness condition is a natural one, because it is much like the condition on matrix elements of V that is needed to make $V(t)$ vanish for $t \to \pm\infty$, as required in order to justify the very idea of an S-matrix.

We can also use Eq.(3.3.22) with $\tau = 0$ and $\tau_0 = \mp\infty$ to show that

$$\mathbf{K}\Omega(\mp\infty) = \Omega(\mp\infty)\mathbf{K}_0\,, \tag{3.3.25}$$

where $\Omega(\mp\infty)$ is according to (3.1.13) the operator that converts a free-particle state Φ_α into the corresponding 'in' or 'out' state $\Psi_\alpha{}^\pm$. Also, it follows trivially from Eqs. (3.3.18) and (3.3.19) that the same is true for the momentum and angular momentum:

$$\mathbf{P}\Omega(\mp\infty) = \Omega(\mp\infty)\mathbf{P}_0\,, \tag{3.3.26}$$

$$\mathbf{J}\Omega(\mp\infty) = \Omega(\mp\infty)\mathbf{J}_0\,. \tag{3.3.27}$$

Finally, since all Φ_α and $\Psi_\alpha{}^\pm$ are eigenstates of H_0 and H respectively with the same eigenvalue E_α, we have

$$H\Omega(\mp\infty) = \Omega(\mp\infty)H_0\,. \tag{3.3.28}$$

Eqs. (3.3.25)–(3.3.28) show that with our assumptions, 'in' and 'out' states do transform under inhomogeneous Lorentz transformations just like the free-particle states. Also, since these are similarity transformations, we now see that the exact generators **K**, **P**, **J**, and H satisfy the same commutation relations as $\mathbf{K}_0$, $\mathbf{P}_0$, $\mathbf{J}_0$, and H_0. This is why it turned out to be unnecessary in proving the Lorentz invariance of the S-matrix to use the other commutation relations (3.3.12), (3.3.13), and (3.3.15) that involve **K**.

Internal Symmetries

There are various symmetries, like the symmetry in nuclear physics under interchange of neutrons and protons, or the 'charge-conjugation' symmetry between particles and antiparticles, that have nothing directly to do with Lorentz invariance, and further appear the same in all inertial frames. Such a symmetry transformation T acts on the Hilbert space of physical states as a unitary operator $U(T)$ that induces linear transformations on the indices labelling particle species

$$U(T)\Psi_{p_1\sigma_1 n_1;\, p_2\sigma_2 n_2;\, \cdots} = \sum_{\bar{n}_1 \bar{n}_2 \cdots} \mathscr{D}_{\bar{n}_1 n_1}(T)\, \mathscr{D}_{\bar{n}_2 n_2}(T) \cdots \times \Psi_{p_1\sigma_1 \bar{n}_1;\, p_2\sigma_2 \bar{n}_2;\, \cdots} \,. \tag{3.3.29}$$

In accordance with the general discussion in Chapter 2, the $U(T)$ must satisfy the group multiplication rule

$$U(\bar{T})\, U(T) = U(\bar{T}T)\,, \tag{3.3.30}$$

where $\bar{T}T$ is the transformation obtained by first performing the transformation T, then some other transformation $\bar{T}$. Acting on Eq. (3.3.29) with $U(\bar{T})$, we see that the matrices $\mathscr{D}$ satisfy the same rule

$$\mathscr{D}(\bar{T})\mathscr{D}(T) = \mathscr{D}(\bar{T}T). \tag{3.3.31}$$

Also, taking the scalar product of the states obtained by acting with $U(T)$ on two different 'in' states or two different 'out' states, and using the normalization condition (3.1.2), we see that $\mathscr{D}(T)$ must be unitary

$$\mathscr{D}^\dagger(T) = \mathscr{D}^{-1}(T). \tag{3.3.32}$$

Finally, taking the scalar product of the states obtained by acting with $U(T)$ on one 'out' state and one 'in' state shows that $\mathscr{D}$ commutes with the S-matrix, in the sense that

$$\sum_{\bar{N}_1 \bar{N}_2 \cdots} \sum_{\bar{N}'_1 \bar{N}'_2 \cdots} \mathscr{D}^*_{\bar{N}'_1 n'_1}(T) \mathscr{D}^*_{\bar{N}'_2 n'_2}(T) \cdots \mathscr{D}_{\bar{N}_1 n_1}(T)\, \mathscr{D}_{\bar{N}_2 n_2}(T) \cdots \times S_{p'_1\sigma'_1 \bar{N}'_1;\, p'_2\sigma'_2 \bar{N}'_2;\, \cdots\, ,\; p_1\sigma_1 \bar{N}_1;\, p_2\sigma_2 \bar{N}_2;\, \cdots} = S_{p'_1\sigma'_1 n'_1;\, p'_2\sigma'_2 n'_2;\, \cdots\, ,\; p_1\sigma_1 n_1;\, p_2\sigma_2 n_2;\, \cdots} \,. \tag{3.3.33}$$

Again, this is a definition of what we mean by a theory being invariant under the internal symmetry T, because to derive Eq. (3.3.33) we still need to show that the *same* unitary operator $U(T)$ will induce the transformation (3.3.29) on both 'in' and 'out' states. This will be the case if there is an 'unperturbed' transformation operator $U_0(T)$ that induces these transformations on free-particle states,

$$U_0(T)\Phi_{p_1\sigma_1n_1;p_2\sigma_2n_2;\cdots} = \sum_{\bar{N}_1\bar{N}_2\cdots} \mathscr{D}_{\bar{N}_1n_1}(T)\,\mathscr{D}_{\bar{N}_2n_2}(T)\cdots\Phi_{p_1\sigma_1\bar{N}_1;p_2\sigma_2\bar{N}_2;\cdots} \tag{3.3.34}$$

and that commutes with both the free-particle and interaction parts of the Hamiltonian

$$U_0^{-1}(T)H_0U_0(T) = H_0\,, \tag{3.3.35}$$

$$U_0^{-1}(T)VU_0(T) = V. \tag{3.3.36}$$

From either the Lippmann–Schwinger equation (3.1.17) or from (3.1.13), we see that the operator $U_0(T)$ will induce the transformations (3.3.29) on 'in' and 'out' states as well as free-particle states, so that we can derive Eq. (3.3.29) taking $U(T)$ as $U_0(T)$.

A special case of great physical importance is that of a one-parameter Lie group, where T is a function of a single parameter θ, with

$$T(\bar{\theta})T(\theta) = T(\bar{\theta}+\theta). \tag{3.3.37}$$

As shown in Section 2.2, in this case the corresponding Hilbert-space operators must take the form

$$U(T(\theta)) = \exp(iQ\theta) \tag{3.3.38}$$

with Q a Hermitian operator. Likewise the matrices $\mathscr{D}(T)$ take the form

$$\mathscr{D}_{n'n}(T(\theta)) = \delta_{n'n}\exp(iq_n\theta)\,, \tag{3.3.39}$$

where q_n are a set of real species-dependent numbers. Here Eq. (3.3.33) simply tells us that the qs are conserved: $S_{\beta\alpha}$ vanishes unless

$$q_{n'_1} + q_{n'_2} + \cdots = q_{n_1} + q_{n_2} + \cdots \quad . \tag{3.3.40}$$

The classic example of such a conservation law is that of conservation of electric charge. Also, all known processes conserve baryon number (the number of baryons, such as protons, neutrons, and hyperons, minus the number of their antiparticles) and lepton number (the number of leptons, such as electrons, muons, τ particles, and neutrinos, minus the number of their antiparticles) but as we shall see in Volume II, these conservation laws are believed to be only very good approximations. There are other conservation laws of this type that are definitely only approximate, such as the conservation of the quantity known as strangeness, which was introduced to explain the relatively long life of a class of particles discovered by Rochester and Butler[5] in cosmic rays in 1947. For instance, the mesons now called* K^+ and K^0 are assigned strangeness +1 and the hyperons Λ^0, Σ^+, Σ^0, Σ^- are assigned strangeness −1, while the more familiar protons, neutrons, and π mesons (or pions) are taken to have strangeness zero. The conservation of strangeness in strong interactions explains why strange particles are always produced in association with one another, as in reactions like $\pi^+ + n \rightarrow K^+ + \Lambda^0$, while the relatively slow decays of strange particles into non-strange ones such as $\Lambda^0 \rightarrow p + \pi^-$ and $K^+ \rightarrow \pi^+ + \pi^0$ show that the interactions that do not conserve strangeness are very weak.

The classic example of a 'non-Abelian' symmetry whose generators do not commute with one another is isotopic spin symmetry, which was suggested[6] in 1937 on the basis of an experiment[7] that showed the existence of a strong proton–proton force similar to that between protons and neutrons. Mathematically, the group is $SU(2)$, like the covering group of the group of three-dimensional rotations; its generators are denoted t_i, with $i = 1, 2, 3$, and satisfy a commutation relation like (2.4.18):

$$[t_i, t_j] = i\epsilon_{ijk} t_k .$$

To the extent that isotopic spin symmetry is respected, it requires particles to form degenerate multiplets labelled with an integer or half-integer T and with $2T + 1$ components distinguished by their t_3 values, just like the degenerate spin multiplets required by rotational invariance. These include the nucleons p and n with $T = \frac{1}{2}$ and $t_3 = \frac{1}{2}, -\frac{1}{2}$; the pions π^+, π^0, and π^- with $T = 1$ and $t_3 = +1, 0, -1$; and the Λ^0 hyperon with $T = 0$ and $t_3 = 0$. These examples illustrate the relation between electric charge Q, the third component of isotopic spin t_3, the baryon number B, and the strangeness S:

$$Q = t_3 + (B + S)/2 .$$

This relation was originally inferred from observed selection rules, but it was interpreted[8] by Gell-Mann and Ne'eman in 1960 to be a consequence of the embedding of both the isospin $\vec{T}$ and the 'hypercharge' $Y \equiv B + S$ in the Lie algebra of a larger but more badly broken non-Abelian internal symmetry, based on the non-Abelian group $SU(3)$. As we will see in Volume II, today both isospin and $SU(3)$ symmetry are understood as incidental consequences of the small masses of the two or three lightest quarks in the modern theory of strong interactions, quantum chromodynamics.

* Superscripts denote electric charges in units of the absolute value of the electronic charge. A 'hyperon' is any particle carrying non-zero strangeness and unit baryon number.

The implications of isotopic spin symmetry for reactions among strongly interacting particles can be worked out by the same familiar methods that were invented for deriving the implications of rotational invariance. In particular, for a two-body reaction $A + B \to C + D$, Eq. (3.3.33) requires that the S-matrix may be put in the form (suppressing all but isospin labels):

$$S_{t_{C3}t_{D3},t_{A3}t_{B3}} = \sum_{T,t_3} C_{T_C T_D}(Tt_3; t_{C3}t_{D3}) C_{T_A T_B}(Tt_3; t_{A3}t_{B3}) S_T \ ,$$

where $C_{j_1 j_2}(j\sigma; \sigma_1\sigma_2)$ is the usual Clebsch–Gordan coefficient[9] for forming a spin j with three-component σ from spins j_1 and j_2 with three-components σ_1 and σ_2, respectively; and S_T is a 'reduced' S-matrix depending on T and on all the suppressed momentum and spin variables, but not on the isospin three-components $t_{A3}, t_{B3}, t_{C3}, t_{D3}$. Of course, this, like all of the consequences of isotopic spin invariance, is only approximate, because this symmetry is not respected by electromagnetic (and other) interactions, as shown for instance by the fact that different members of the same isospin multiplet like p and n have different electric charges and slightly different masses.

Parity

To the extent that the symmetry under the transformation $\mathbf{x} \to -\mathbf{x}$ is really valid, there must exist a unitary operator P under which both 'in' and 'out' states transform as a direct product of single-particle states:

$$\mathsf{P}\Psi^{\pm}_{p_1\sigma_1 n_1; p_2\sigma_2 n_2; \cdots} = \eta_{n_1}\eta_{n_2}\cdots \Psi^{\pm}_{\mathscr{P}p_1\sigma_1 n_1; \mathscr{P}p_2\sigma_2 n_2; \cdots} \tag{3.3.41}$$

where η_n is the intrinsic parity of particles of species n, and $\mathscr{P}$ reverses the space components of p^μ. (This is for massive particles; the modification for massless particles is obvious.) The parity conservation condition for the S-matrix is then:

$$S_{p'_1\sigma'_1 n'_1; p'_2\sigma'_2 n'_2; \cdots\ ,\ p_1\sigma_1 n_1; p_2\sigma_2 n_2; \cdots} = \eta^*_{n'_1}\eta^*_{n'_2}\cdots\eta_{n_1}\eta_{n_2}\cdots$$
$$\times\ S_{\mathscr{P}p'_1\sigma'_1 n'_1; \mathscr{P}p'_2\sigma'_2 n'_2; \cdots\ ,\ \mathscr{P}p_1\sigma_1 n_1; \mathscr{P}p_2\sigma_2 n_2; \cdots}\ . \tag{3.3.42}$$

Just as for internal symmetries, an operator P satisfying Eq. (3.3.41) will actually exist if the operator P_0 which is defined to act this way on free-particle states commutes with V as well as H_0.

The phases η_n may be inferred either from dynamical models or from experiment, but neither can provide a unique determination of the ηs. This is because we are always free to redefine P by combining it with any conserved internal symmetry operator. For instance, if P is conserved, then so is

$$\mathsf{P}' \equiv \mathsf{P}\exp(i\alpha B + i\beta L + i\gamma Q),$$

where B, L, and Q are respectively baryon number, lepton number, and electric charge, and α, β, and γ are arbitrary real phases; hence either P or P' could be called the parity operator. The neutron, proton, and electron have different combinations of values for B, L, and Q, so by judicious choice of the phases α, β, and γ we can define the intrinsic parities of all three particles to be $+1$. However, once we have done this the intrinsic parities of other particles like the charged pion (which can be emitted in a transition $n \to p + \pi^-$) are no longer arbitrary. Also, the intrinsic parity of any particle like the neutral pion π^0 which carries no conserved quantum numbers is always meaningful.

The foregoing remarks help to clarify the question of whether intrinsic parities must always have the values ± 1. It is easy to say that space inversion P has the group multiplication law $\mathsf{P}^2 = 1$; however, the parity operator that is conserved may not be this one, but rather may differ from it by a phase transformation of some sort. In any case, whether or not $\mathsf{P}^2 = 1$, the operator P^2 behaves just like an internal symmetry transformation:

$$\mathsf{P}^2\Psi^{\pm}_{p_1\sigma_1 n_1;\,p_2\sigma_2 n_2;\cdots} = \eta^2_{n_1}\eta^2_{n_2}\cdots\Psi^{\pm}_{p_1\sigma_1 n_1;\,p_2\sigma_2 n_2;\cdots}\,.$$

If this internal symmetry is part of a continuous symmetry group of phase transformations, such as the group of multiplication by the phases $\exp(i\alpha B + i\beta L + i\gamma Q)$ with arbitrary values of α, β, and γ, then its inverse square root must also be a member of this group, say I_P, with $I_P^2\mathsf{P}^2 = 1$ and $[I_P, \mathsf{P}] = 0$. (For instance, if $\mathsf{P}^2 = \exp(i\alpha B + \cdots)$, then take $I_P = \exp(-\frac{1}{2}i\alpha B + \cdots)$.) We can then define a new parity operator $\mathsf{P}' \equiv \mathsf{P}I_P$ with $\mathsf{P}'^2 = 1$. This is conserved to the same extent as P, so there is no reason why we should not call this *the* parity operator, in which case the intrinsic parities can only take the values ± 1.

The only sort of theory in which it is not necessarily possible to define parity so that all intrinsic parities have the values ± 1 is one in which there is some discrete internal symmetry which is not a member of any continuous symmetry group of phase transformations.[10] For instance, it is a consequence of angular-momentum conservation that the total number F of all particles of half-integer spin can only change by even numbers, so the internal symmetry operator $(-1)^F$ is conserved. All known particles of half-integer spin have odd values of the sum $B + L$ of baryon number and lepton number, so as far as we know, $(-1)^F = (-1)^{B+L}$. If this is true, then $(-1)^F$ is part of a continuous symmetry group, consisting of the operators $\exp(i\alpha(B + L))$ with arbitrary real α, and has an inverse square root $\exp(-i\alpha(B + L)/2)$. In this case, if $\mathsf{P}^2 = (-1)^F$ then P can be redefined so that all intrinsic parities are ± 1. However, if there were to be discovered a particle of half-integer spin and an even value of $B+L$ (such as a so-called Majorana neutrino, with $j = \frac{1}{2}$ and $B + L = 0$), then it would be possible to have $\mathsf{P}^2 = (-1)^F$ without our being able to redefine the parity operator itself to have eigenvalues ± 1. In this case, of course, we would have $\mathsf{P}^4 = 1$, so all particles would have intrinsic parities either ± 1 or (like the Majorana neutrino) $\pm i$.

It follows from Eq. (3.3.42) that if the product of intrinsic parities in the final state is equal to the product of intrinsic parities in the initial state, or equal to minus this product, then the S-matrix must be respectively even or odd overall in the three-momenta. For instance, it was observed[11] in 1951 that a pion can be absorbed by a deuteron from the $\ell = 0$ ground state of the $\pi^- d$ atom, in the reaction $\pi^- + d \to n + n$. (As discussed in Section 3.7, the orbital angular-momentum quantum number ℓ can be used in relativistic physics in the same way as in non-relativistic wave mechanics.) The initial state has total angular-momentum $j = 1$ (the pion and deuteron having spins zero and one, respectively), so the final state must have orbital angular-momentum $\ell = 1$ and total neutron spin $s = 1$. (The other possibilities $\ell = 1$, $s = 0$; $\ell = 0$, $s = 1$; and $\ell = 2$, $s = 1$, which are allowed by angular-momentum conservation, are forbidden by the requirement that the final state be antisymmetric in the two neutrons.) Because the final state has $\ell = 1$, the matrix element is odd under reversal of the direction of all three-momenta, so we can conclude that the intrinsic parities of the particles in this reaction must be related by:

$$\eta_d \eta_{\pi^-} = -\eta_n^2 .$$

The deuteron is known to be a bound state of a proton and neutron with even orbital angular-momentum (chiefly $\ell = 0$), and as we have seen we can take the neutron and proton to have the same intrinsic parity, so $\eta_d = \eta_n^2$, and we can conclude that $\eta_{\pi^-} = -1$; that is, the negative pion is a *pseudoscalar* particle. The π^+ and π^0 have also been found to have negative parity, as would be expected from the symmetry (isospin invariance) among these three particles.

The negative parity of the pion has some striking consequences. A spin zero particle that decays into three pions must have intrinsic parity $\eta_\pi^3 = -1$, because in the Lorentz frame in which the decaying particle is at rest, rotational invariance only allows the matrix element to depend on scalar products of the pion momenta with each other, all of which are even under reversal of all momenta. (The triple scalar product $\mathbf{p}_1 \cdot (\mathbf{p}_2 \times \mathbf{p}_3)$ formed from the three pion momenta vanishes because $\mathbf{p}_1 + \mathbf{p}_2 + \mathbf{p}_3 = 0$.) For the same reason, a spin zero particle that decays into two pions must have intrinsic parity $\eta_\pi^2 = +1$. In particular, among the strange particles discovered in the late 1940s there seemed to be two different particles of zero spin (inferred from the angular distribution of their decay products): one, the τ, was identified by its decay into three pions, and hence was assigned a parity -1, while the other, the θ, was identified by its decay into two pions and was assigned a parity $+1$. The trouble with all this was that as the τ and θ were studied in greater detail, they seemed increasingly to have identical masses and lifetimes. After many suggested solutions of this puzzle, Lee and Yang in 1956 finally cut the Gordian knot, and proposed that the τ and θ are the same particle, (now known as the $K^{\pm}$) and that parity is simply not conserved in the weak interactions that lead to its decay.[12]

As we shall see in detail in the next section of this chapter, the rate for a physical process $\alpha \to \beta$ (with $\alpha \neq \beta$) is proportional to $|S_{\beta\alpha}|^2$, with proportionality factors that are invariant under reversal of all three-momenta. As long as the states α and β contain definite numbers of particles of each type, the phase factors in Eq. (3.3.42) have no effect on $|S_{\beta\alpha}|^2$, so Eq. (3.3.42) would imply that the rate for $\alpha \to \beta$ is invariant under the reversal of direction of all three-momenta. As we have seen, this is a trivial consequence of rotational invariance for the decays of a K meson into two or three pions, but it is a non-trivial restriction on rates in more complicated processes. For example, following theoretical suggestions by Lee and Yang, Wu together with a group at the National Bureau of Standards measured the angular distribution of the electron in the final state of the beta decay $Co^{60} \to Ni^{60} + e^- + \bar{\nu}$ with a polarized cobalt source.[13] (No attempt was made in this experiment to measure the momentum of the antineutrino or nickel nucleus.) The electrons were found to be preferentially emitted in a direction opposite to that of the spin of the decaying nucleus, which would, of course, be impossible if the decay rate were invariant under a reversal of all three-momenta. A similar result was found in the decay of a positive muon (polarized in its production in the process $\pi^+ \to \mu^+ + \nu$) into a positron, neutrino, and antineutrino.[14] In this way, it became clear that parity is indeed not conserved in the weak interactions responsible for these decays. Nevertheless, for reasons discussed in Section 12.5, parity *is* conserved in the strong and electromagnetic interactions, and therefore continues to play an important part in theoretical physics.

Time-Reversal

We saw in Section 2.6 that the time-reversal operator T acting on a one-particle state $\Psi_{p,\sigma,n}$ gives a state $\Psi_{\mathscr{P}p,-\sigma,n}$ with reversed spin and momentum, times a phase $\zeta_n(-1)^{j-\sigma}$. A multi-particle state transforms as usual as a direct product of one-particle states, except that since this is a time-reversal transformation, we expect 'in' and 'out' states to be interchanged:

$$\mathsf{T}\Psi^{\pm}_{p_1\sigma_1 n_1;\,p_2\sigma_2 n_2\cdots} = \zeta_{n_1}(-1)^{j_1-\sigma_1}\zeta_{n_2}(-1)^{j_2-\sigma_2}\dots\Psi^{\mp}_{\mathscr{P}p_1\,-\sigma_1 n_1;\,\mathscr{P}p_2\,-\sigma_2 n_2;\,\cdots}\,. \tag{3.3.43}$$

(Again, this is for massive particles, with obvious modifications being required for massless particles.) It will be convenient to abbreviate this assumption as

$$\mathsf{T}\Psi^{\pm}_{\alpha} = \Psi^{\mp}_{\mathscr{T}\alpha}\,, \tag{3.3.44}$$

where $\mathscr{T}$ indicates a reversal of sign of three-momenta and spins as well as multiplication by the phase factors shown in Eq. (3.3.43). Because T is antiunitary, we have

$$(\Psi^{-}_{\beta},\Psi^{+}_{\alpha}) = (\mathsf{T}\Psi^{+}_{\alpha},\mathsf{T}\Psi^{-}_{\beta})\,, \tag{3.3.45}$$

so the time-reversal invariance condition for the S-matrix is

$$S_{\beta,\alpha} = S_{\mathscr{T}\alpha,\,\mathscr{T}\beta} \tag{3.3.46}$$

or in more detail

$$\begin{aligned} &S_{p'_1\sigma'_1 n'_1;\,p'_2\sigma'_2 n'_2;\,\cdots\,,\;p_1\sigma_1 n_1;\,p_2\sigma_2 n_2;\,\cdots} \\ &= \zeta_{n'_1}(-1)^{j'_1-\sigma'_1}\zeta_{n'_2}(-1)^{j'_2-\sigma'_2}\cdots\zeta^{*}_{n_1}(-1)^{j_1-\sigma_1}\zeta^{*}_{n_2}(-1)^{j_2-\sigma_2}\cdots \\ &\quad\times S_{\mathscr{P}p_1\,-\sigma_1\,n_1;\,\mathscr{P}p_2\,-\sigma_2\,n_2;\,\cdots\,,\;\mathscr{P}p'_1\,-\sigma'_1\,n'_1;\,\mathscr{P}p'_2\,-\sigma'_2\,n'_2;\,\cdots}\,. \end{aligned} \tag{3.3.47}$$

Note that in addition to the reversal of momenta and spins, the role of initial and final states is interchanged, as would be expected for a symmetry involving the reversal of time.

The S-matrix will satisfy this transformation rule if the operator T_0 that induces time-reversal transformations on free-particle states

$$\mathsf{T}_0\Phi_\alpha \equiv \Phi_{\mathscr{T}\alpha} \tag{3.3.48}$$

commutes not only with the free-particle Hamiltonian (which is automatic) but also with the interaction:

$$\mathsf{T}_0^{-1}H_0\mathsf{T}_0 = H_0\,, \tag{3.3.49}$$

$$\mathsf{T}_0^{-1}V\mathsf{T}_0 = V. \tag{3.3.50}$$

In this case we can take $\mathsf{T} = \mathsf{T}_0$, and use either (3.1.13) or (3.1.16) to show that time-reversal transformations do act as stated in Eq. (3.3.44). For instance, operating on the Lippmann–Schwinger equation (3.1.16) with T and using Eqs. (3.3.48)–(3.3.50), we have

$$\mathsf{T}\Psi^{\pm}_{\alpha} = \Phi_{\mathscr{T}\alpha} + [E_\alpha - H_0 \mp i\epsilon]^{-1}V\mathsf{T}\Psi^{\pm}_{\alpha}\,,$$

with the sign of $\pm i\epsilon$ reversed because T is antiunitary. This is just the Lippmann–Schwinger equation for $\Psi^{\mp}_{\mathscr{T}\alpha}$, thus justifying Eq. (3.3.44). Similarly, because T is antiunitary it changes the sign of the i in the exponent of $\Omega(t)$, so that

$$\mathsf{T}\Omega(-\infty)\Phi_\alpha = \Omega(\infty)\Phi_{\mathscr{T}\alpha}\,,$$

again leading to Eq. (3.3.44).

In contrast with the case of parity conservation, the time-reversal invariance condition (3.3.46) does *not* in general tell us that the rate for the process $\alpha \to \beta$ is the same as for the process $\mathscr{T}\alpha \to \mathscr{T}\beta$. However, something like this is true in cases where the S-matrix takes the form

$$S_{\beta\alpha} = S^{(0)}_{\beta\alpha} + S^{(1)}_{\beta\alpha}\,, \tag{3.3.51}$$

where $S^{(1)}$ is small, while $S^{(0)}$ happens to have matrix element zero for some particular process of interest, though it generally has much larger matrix elements than $S^{(1)}$. (For instance, the process might be nuclear beta decay, $N \to N' + e^- + \bar{\nu}$, with $S^{(0)}$ the S-matrix produced by the strong nuclear and electromagnetic interactions alone, and $S^{(1)}$ the correction to the S-matrix produced by the weak interactions. Section 3.5 shows how the use of the 'distorted-wave Born approximation' leads to an S-matrix of the form (3.3.51) in cases of this type. In some cases $S^{(0)}$ is simply the unit operator.) To first order in $S^{(1)}$, the unitarity condition for the S-operator reads

$$1 = S^\dagger S = S^{(0)\dagger}S^{(0)} + S^{(0)\dagger}S^{(1)} + S^{(1)\dagger}S^{(0)}.$$

Using the zeroth-order relation $S^{(0)\dagger}S^{(0)} = 1$, this gives a reality condition for $S^{(1)}$:

$$S^{(1)} = -S^{(0)}S^{(1)\dagger}S^{(0)}. \tag{3.3.52}$$

If $S^{(1)}$ as well as $S^{(0)}$ satisfies the time-reversal condition (3.3.46), then this can be put in the form

$$S^{(1)}_{\beta\alpha} = -\int d\gamma \int d\gamma'\; S^{(0)}_{\beta\gamma'}\; S^{(1)*}_{\mathscr{T}\gamma',\mathscr{T}\gamma}\; S^{(0)}_{\gamma\alpha}\,. \tag{3.3.53}$$

Since $S^{(0)}$ is unitary, the rates for the processes $\alpha \to \beta$ and $\mathscr{T}\alpha \to \mathscr{T}\beta$ are thus the same if summed over sets $\mathscr{I}$ and $\mathscr{F}$ of final and initial states that are complete with respect to $S^{(0)}$. (By being 'complete' here is meant that if $S^{(0)}_{\alpha'\alpha}$ is non-zero, and either α or α' are in $\mathscr{I}$, then both states are in $\mathscr{I}$; and similarly for $\mathscr{F}$.) In the simplest case we have 'complete' sets $\mathscr{I}$ and $\mathscr{F}$ consisting of just one state each; that is, both the initial and the final states are eigenvectors of $S^{(0)}$ with eigenvalues $e^{2i\delta_\alpha}$ and $e^{2i\delta_\beta}$, respectively. (The δ_α and δ_β are called 'phase shifts'; they are real because

$S^{(0)}$ is unitary.) In this case, Eq. (3.3.53) becomes simply:

$$S^{(1)}_{\beta\alpha} = -e^{2i(\delta_\alpha+\delta_\beta)} S^{(1)*}_{\mathscr{T}\beta\,\mathscr{T}\alpha}, \tag{3.3.54}$$

and it is clear that the absolute value of the S-matrix for the process $\alpha \to \beta$ is the same as for the process $\mathscr{T}\alpha \to \mathscr{T}\beta$. This is the case for instance in nuclear beta decay (in the approximation in which we ignore the relatively weak Coulomb interaction between the electron and nucleus in the final state), because both the initial and final states are eigenstates of the strong interaction S-matrix (with $\delta_\alpha = \delta_\beta = 0$). Thus if time-reversal invariance is respected, the differential rate for a beta decay process should be unchanged if we reverse both the momenta and the spin z-components σ of all particles. This prediction was *not* contradicted in the 1956 experiments[13,14] that discovered the non-conservation of parity; for instance, time-reversal invariance is consistent with the observation that electrons from the decay $\text{Co}^{60} \to \text{Ni}^{60} + e^- + \bar{\nu}$ are preferentially emitted in a direction opposite to that of the Co^{60} spin. As described below, indirect evidence against time-reversal invariance did emerge in 1964, but it remains a useful approximate symmetry in weak as well as strong and electromagnetic interactions.

In some cases we can use a basis of states for which $\mathscr{T}\alpha = \alpha$ and $\mathscr{T}\beta = \beta$, for which Eq. (3.3.54) reads

$$S^{(1)}_{\beta\alpha} = -e^{2i(\delta_\alpha+\delta_\beta)} S^{(1)*}_{\beta\alpha}, \tag{3.3.55}$$

which just says that $iS^{(1)}_{\beta,\alpha}$ has the phase $\delta_\alpha + \delta_\beta$ mod π. This is known as *Watson's theorem.*[15] The phases in Eqs. (3.3.54) or (3.3.55) may be measured in processes where there is interference between different final states. For instance, in the decay of the spin 1/2 hyperon Λ into a nucleon and a pion, the final state can only have orbital angular-momentum $\ell = 0$ or $\ell = 1$; the angular distribution of the pion relative to the Λ spin involves the interference between these states, and hence according to Watson's theorem depends on the difference $\delta_s - \delta_p$ of their phase shifts.

PT

Although the 1957 experiments on parity violation did not rule out time-reversal invariance, they did show immediately that the product PT is not conserved. If conserved this operator would have to be antiunitary for the same reasons as for T, so in processes like nuclear beta decay its consequences would take the form of relations like Eq. (3.3.54):

$$S^{(1)}_{\beta\alpha} = -e^{2i(\delta_\alpha+\delta_\beta)} S^{(1)*}_{\mathscr{P}\mathscr{T}\beta\,\mathscr{P}\mathscr{T}\alpha},$$

where $\mathscr{PT}$ reverses the signs of all spin z-components but *not* any momenta. Neglecting the final-state Coulomb interaction, it would then follow that there could be no preference for the electron in the decay $Co^{60} \to Ni^{60} + e^- + \bar{\nu}$ to be emitted in the same or the opposite direction to the Co^{60} spin, in contradiction with what was observed.

C, CP, *and* CPT

As already mentioned, there is an internal symmetry transformation, known as charge-conjugation, which interchanges particles and antiparticles. Formally, this entails the existence of a unitary operator C, whose effect on multi-particle states is:

$$\mathsf{C}\Psi^{\pm}_{p_1\sigma_1 n_1;\,p_2\sigma_2 n_2;\,\cdots} = \xi_{n_1}\xi_{n_2}\cdots\Psi^{\pm}_{p_1\sigma_1 n_1^c;\,p_2\sigma_2 n_2^c;\,\cdots}\,, \tag{3.3.56}$$

where n^c is the antiparticle of particle type n, and ξ_n is yet another phase. If this is true for both 'in' and 'out' states then the S-matrix satisfies the invariance conditions

$$S_{p_1'\sigma_1'n_1';\,p_2'\sigma_2'n_2';\,\cdots\,,\;p_1\sigma_1 n_1;\,p_2\sigma_2 n_2;\,\cdots}$$
$$= \xi^*_{n_1'}\xi^*_{n_2'}\cdots\xi_{n_1}\xi_{n_2}\cdots S_{p_1'\sigma_1'n_1^{c\prime};\,p_2'\sigma_2'n_2^{c\prime};\,\cdots\,,\;p_1\sigma_1 n_1^c;\,p_2\sigma_2 n_2^c;\,\cdots}\,. \tag{3.3.57}$$

As with other internal symmetries, the S-matrix will satisfy this condition if the operator C_0 that is defined to act as stated in Eq. (3.3.56) on free-particle states commutes with the interaction V as well as H_0; in this case, we take $\mathsf{C} = \mathsf{C}_0$.

The phases ξ_n are called charge-conjugation parities. Just as for the ordinary parities η_n, the ξ_n are in general not uniquely defined, because for any operator C that is defined to satisfy Eq. (3.3.56), we can find another such operator with different ξ_n by multiplying C with any internal symmetry phase transformation, such as $\exp(i\alpha B + i\beta L + i\gamma Q)$; the only particles whose charge-conjugation parities are individually measurable are those completely neutral particles like the photon or the neutral pion that carry no conserved quantum numbers and are their own antiparticles. In reactions involving only completely neutral particles, Eq. (3.3.57) tells us that the product of the charge-conjugation parities in the initial and final states must be equal; for instance, as we shall see the photon is required by quantum electrodynamics to have charge-conjugation parity $\eta_\gamma = -1$, so the observation of the neutral pion decay $\pi^0 \to 2\gamma$ requires that $\eta_{\pi^0} = +1$; it then follows that the process $\pi^0 \to 3\gamma$ should be forbidden, as is in fact known to be the case. For these two particles, the charge-conjugation parities are real, either $+1$ or -1. Just as for ordinary parity, this will always be the case if all internal phase transformation symmetries are members of continuous groups of phase transformations, because then we can redefine C by multiplying by the inverse square root of the internal symmetry equal to C^2, with the result that the new C satisfies $\mathsf{C}^2 = 1$.

For general reactions, Eq. (3.3.57) requires that the rate for a process equals the rate for the same process with particles replaced with their corresponding antiparticles. This was not directly contradicted by the 1957 experiments on parity non-conservation (it will be a long time before anyone is able to study the beta decay of anticobalt), but these experiments showed that C is not conserved in the *theory* of weak interactions as modified by Lee and Yang[12] to take account of parity non-conservation. (As we shall see below, the observed violation of TP conservation would imply a violation of C conservation in *any* field theory of weak interactions, not just in the particular theory considered by Lee and Yang.) It is understood today that C as well as P is not conserved in the weak interactions responsible for processes like beta decay and the decay of the pion and muon, though both C and P are conserved in the strong and electromagnetic interactions.

Although the early experiments on parity non-conservation indicated that neither C nor P are conserved in the weak interactions, they left open the possibility that their product CP is universally conserved. For some years it was expected (though not with complete confidence) that CP would be found to be generally conserved. This had particularly important consequences for the properties of the neutral K mesons. In 1954 Gell-Mann and Pais[16] had pointed out that because the K^0 meson is not its own antiparticle (the K^0 carries a non-zero value for the approximately conserved quantity known as strangeness) the particles with definite decay rates would be not K^0 or $\overline{K^0}$, but the linear combinations $K^0 \pm \overline{K^0}$. This was originally explained in terms of C conservation, but with C not conserved in the weak interactions, the argument may be equally well be based on CP conservation. If we arbitrarily define the phases in the CP operator and in the K^0 and $\overline{K^0}$ states so that

$$\mathsf{CP}\Psi_{K^0} = \Psi_{\overline{K^0}}$$

and

$$\mathsf{CP}\Psi_{\overline{K^0}} = \Psi_{K^0}$$

then we can define self-charge-conjugate one-particle states

$$\Psi_{K_1^0} \equiv \frac{1}{\sqrt{2}}[\Psi_{K^0} + \Psi_{\overline{K^0}}]$$

and

$$\Psi_{K_2^0} \equiv \frac{1}{\sqrt{2}}[\Psi_{K^0} - \Psi_{\overline{K^0}}] ,$$

which have CP eigenvalues +1 and −1, respectively. The fastest available decay mode of these particles is into two-pion states, but CP conservation would allow this only** for the K_1, not the K_2. The K_2^0 would thus be expected to decay only by slower modes, into three pions or into a pion, muon or electron, and neutrino. Nevertheless, it was found by Fitch and Cronin in 1964 that the long-lived neutral K-meson does have a small probability for decaying into two pions.[17] The conclusion was that CP is not exactly conserved in the weak interactions, although it seems more nearly conserved than C or P individually.

As we shall see in Chapter 5, there are good reasons to believe that although neither C nor CP is strictly conserved, the product CPT is exactly conserved in all interactions, at least in any quantum field theory. It is CPT that provides a precise correspondence between particles and antiparticles, and in particular it is the fact that CPT commutes with the Hamiltonian that tells us that stable particles and antiparticles have exactly the same mass. Because CPT is antiunitary, it relates the S-matrix for an arbitrary process to the S-matrix for the *inverse* process with all spin three-components reversed and particles replaced with antiparticles. However, in cases where the S-matrix can be divided into a weak term $S^{(1)}$ that produces a given reaction and a strong term $S^{(0)}$ that acts in the initial and final states, we can use the same arguments that were used above in studying the implications of T conservation to show that the rate for any process is equal to the rate for the same process with particles replaced with antiparticles and spin three-components reversed, provided that we sum over sets of initial and final states that are complete with respect to $S^{(0)}$. In particular, although the partial rates for decay of the particle into a pair of final states β_1, β_2 with $S^{(0)}_{\beta_1\beta_2} \neq 0$ may differ from the partial rates for the decay of the antiparticle into the corresponding final states $\mathscr{CPT}\beta_1$ and $\mathscr{CPT}\beta_2$, we shall see in Section 3.5 that (without any approximations) the total decay rate of any particle is equal to that of its antiparticle.

We can now understand why the 1957 experiments on parity violation could be interpreted in the context of the existing theory of weak interactions as evidence that C conservation as well as P conservation are badly violated but CP is not. These theories were field theories, and therefore automatically conserved CPT. Since the experiments showed that PT conservation but not T conservation is badly violated in nuclear beta decay, *any* theory that was consistent with these experiments and in which CPT is conserved would have to also incorporate C but not CP non-conservation.

** The neutral K-mesons have spin zero, so the two-pion final state has $\ell = 0$, and hence P = +1. Further, C = +1 for two π^0s because the π^0 has C = +1, and also for an $\ell = 0$ π^+-π^- state because C interchanges the two pions.

Similarly, the observation in 1964 of small violations of CP conservation in the weak interactions together with the assumed invariance of all interactions under CPT allowed the immediate inference that the weak interactions also do not exactly conserve T. This has since been verified[18] by more detailed studies of the K^0–$\overline{K^0}$ system, but it has so far been impossible to find other direct evidence of the failure of invariance under time-reversal.

3.4 Rates and Cross-Sections

The S-matrix $S_{\beta\alpha}$ is the probability amplitude for the transition $\alpha \to \beta$, but what does this have to do with the transition rates and cross-sections measured by experimentalists? In particular, (3.3.2) shows that $S_{\beta\alpha}$ has a factor $\delta^4(p_\beta - p_\alpha)$, which ensures the conservation of the total energy and momentum, so what are we to make of the factor $[\delta^4(p_\beta - p_\alpha)]^2$ in the transition probability $|S_{\beta\alpha}|^2$? The proper way to approach these problems is by studying the way that experiments are actually done, using wave packets to represent particles localized far from each other before a collision, and then following the time-history of these superpositions of multi-particle states. In what follows we will instead give a quick and easy derivation of the main results, actually more a mnemonic than a derivation, with the excuse that (as far as I know) no interesting open problems in physics hinge on getting the fine points right regarding these matters.

We consider our whole system of physical particles to be enclosed in a large box with a macroscopic volume V. For instance, we can take this box as a cube, but with points on opposite sides identified, so that the single-valuedness of the spatial wave function requires the momenta to be quantized

$$\mathbf{p} = \frac{2\pi}{L}(n_1, n_2, n_3)\,, \tag{3.4.1}$$

where the n_i are integers, and $L^3 = V$. Then all three-dimensional delta-functions become

$$\delta^3_V(\mathbf{p}' - \mathbf{p}) \equiv \frac{1}{(2\pi)^3}\int_V d^3x\, e^{i(\mathbf{p}-\mathbf{p}')\cdot\mathbf{x}} = \frac{V}{(2\pi)^3}\delta_{\mathbf{p}',\mathbf{p}}\,, \tag{3.4.2}$$

where $\delta_{\mathbf{p}',\mathbf{p}}$ is an ordinary Kronecker delta symbol, equal to one if the subscripts are equal and zero otherwise. The normalization condition (3.1.2) thus implies that the states we have been using have scalar products in a box which are not just sums of products of Kronecker deltas, but also contain a factor $[V/(2\pi)^3]^N$, where N is the number of particles in the state. To calculate transition probabilities we should use states of unit norm, so let us introduce states normalized approximately for our box

$$\Psi_\alpha^{\text{Box}} \equiv \left[(2\pi)^3/V\right]^{N_\alpha/2} \Psi_\alpha \tag{3.4.3}$$

with norm

$$\left(\Psi_\beta^{\text{Box}},\ \Psi_\alpha^{\text{Box}}\right) = \delta_{\beta\alpha} \tag{3.4.4}$$

where $\delta_{\beta\alpha}$ is a product of Kronecker deltas, one for each three-momentum, spin, and species label, plus terms with particles permuted. Correspondingly, the S-matrix may be written

$$S_{\beta\alpha} = \left[V/(2\pi)^3\right]^{(N_\beta+N_\alpha)/2} S_{\beta\alpha}^{\text{Box}}\,, \tag{3.4.5}$$

where $S_{\beta\alpha}^{\text{Box}}$ is calculated using the states (3.4.3).

Of course, if we just leave our particles in the box forever, then every possible transition will occur again and again. To calculate a meaningful transition probability we also have to put our system in a 'time box'. We suppose that the interaction is turned on for only a time T. One immediate consequence is that the energy-conservation delta function is replaced with

$$\delta_T(E_\alpha - E_\beta) = \frac{1}{2\pi}\int_{-T/2}^{T/2} \exp\left(i(E_\alpha - E_\beta)t\right) dt\,. \tag{3.4.6}$$

The probability that a multi-particle system, which is in a state α before the interaction is turned on, is found in a state β after the interaction is turned off, is

$$P(\alpha \to \beta) = \left|S_{\beta\alpha}^{\text{Box}}\right|^2 = \left[(2\pi)^3/V\right]^{(N_\alpha+N_\beta)} \left|S_{\beta\alpha}\right|^2. \tag{3.4.7}$$

This is the probability for a transition into one specific box state β. The number of one-particle box states in a momentum-space volume d^3p is $Vd^3p/(2\pi)^3$, because this is the number of triplets of integers n_1, n_2, n_3 for which the momentum (3.4.1) lies in the momentum-space volume d^3p around $\mathbf{p}$. We shall define the final-state interval $d\beta$ as a product of $d^3\mathbf{p}$ for each final particle, so the total number of states in this range is

$$d\mathcal{N}_\beta = \left[V/(2\pi)^3\right]^{N_\beta} d\beta\,. \tag{3.4.8}$$

Hence the total probability for the system to wind up in a range $d\beta$ of final states is

$$dP(\alpha \to \beta) = P(\alpha \to \beta)d\mathcal{N}_\beta = \left[(2\pi)^3/V\right]^{N_\alpha} |S_{\beta\alpha}|^2\, d\beta\,. \tag{3.4.9}$$

We will restrict our attention throughout this section to final states β that are not only different (however slightly) from the initial state α, but that also satisfy the more stringent condition, that no subset of the particles in the state β (other than the whole state itself) has precisely the same four-momentum as some corresponding subset of the particles in the state α. (In the language to be introduced in the next chapter, this means that we are considering only the connected part of the S-matrix.) For such states, we may define a delta function-free matrix element $M_{\beta\alpha}$:

$$S_{\beta\alpha} \equiv -2i\pi\delta_V^3(\mathbf{p}_\beta - \mathbf{p}_\alpha)\delta_T(E_\beta - E_\alpha)M_{\beta\alpha}\,. \tag{3.4.10}$$

Our introduction of the box allows us to interpret the squares of delta functions in $|S_{\beta\alpha}|^2$ for $\beta \neq \alpha$ as

$$\left[\delta_V^3(\mathbf{p}_\beta - \mathbf{p}_\alpha)\right]^2 = \delta_V^3(\mathbf{p}_\beta - \mathbf{p}_\alpha)\delta_V^3(0) = \delta_V^3(\mathbf{p}_\beta - \mathbf{p}_\alpha)V/(2\pi)^3\,,$$

$$\left[\delta_T(E_\beta - E_\alpha)\right]^2 = \delta_T(E_\beta - E_\alpha)\delta_T(0) = \delta_T(E_\beta - E_\alpha)T/2\pi\,,$$

so Eq. (3.4.9) gives a differential transition probability

$$dP(\alpha \to \beta) = (2\pi)^2\left[(2\pi)^3/V\right]^{N_\alpha - 1}(T/2\pi)|M_{\beta\alpha}|^2$$
$$\times\ \delta_V^3(\mathbf{p}_\beta - \mathbf{p}_\alpha)\delta_T(E_\beta - E_\alpha)d\beta\,.$$

If we let V and T be very large, the delta function product here may be interpreted as an ordinary four-dimensional delta function $\delta^4(p_\beta - p_\alpha)$. In this limit, the transition probability is simply proportional to the time T during which the interaction is acting, with a coefficient that may be interpreted as a differential transition *rate*:

$$d\Gamma(\alpha \to \beta) \equiv dP(\alpha \to \beta)/T$$
$$= (2\pi)^{3N_\alpha - 2}V^{1-N_\alpha}|M_{\beta\alpha}|^2\delta^4(p_\beta - p_\alpha)d\beta\,, \tag{3.4.11}$$

where now

$$S_{\beta\alpha} \equiv -2\pi i\delta^4(p_\beta - p_\alpha)M_{\beta\alpha}\,. \tag{3.4.12}$$

This is the master formula which is used to interpret calculations of S-matrix elements in terms of predictions for actual experiments. We will come back to the interpretation of the factor $\delta^4(p_\alpha - p_\beta)d\beta$ later in this section.

There are two cases of special importance:

$\boldsymbol{N_\alpha = 1}$**:**

Here the volume V cancels in Eq. (3.4.11), which gives the transition rate for a single-particle state α to decay into a general multi-particle state β as

$$d\Gamma(\alpha \to \beta) = 2\pi|M_{\beta\alpha}|^2\delta^4(p_\beta - p_\alpha)d\beta\,. \tag{3.4.13}$$

Of course, this makes sense only if the time T during which the interaction acts is much less than the mean lifetime τ_α of the particle α, so we cannot pass to the limit $T \to \infty$ in $\delta_T(E_\alpha - E_\beta)$. There is an unremovable width $\Delta E \simeq 1/T \gtrsim 1/\tau_\alpha$ in this delta function, so Eq. (3.4.13) is only useful if the total decay rate $1/\tau_\alpha$ is much less than any of the characteristic energies of the process.

$N_\alpha = 2$ **:**

Here the rate (3.4.11) is proportional to $1/V$, or in other words, to the density of either particle at the position of the other one. Experimentalists generally report not the transition rate per density, but the *rate per flux*, also known as the *cross-section.* The flux of either particle at the position of the other particle is defined as the product of the density $1/V$ and the relative velocity u_α:

$$\Phi_\alpha = u_\alpha/V \; . \tag{3.4.14}$$

(A general definition of u_α is given below; for the moment we will content ourselves with specifying that if either particle is at rest then u_α is defined as the velocity of the other.) Thus the differential cross-section is

$$d\sigma(\alpha \to \beta) \equiv d\Gamma(\alpha \to \beta)/\Phi_\alpha = (2\pi)^4 u_\alpha^{-1} |M_{\beta\alpha}|^2 \delta^4(p_\beta - p_\alpha) d\beta \; . \tag{3.4.15}$$

Even though the cases $N_\alpha = 1$ and $N_\alpha = 2$ are the most important, transition rates for $N_\alpha \geq 3$ are all measurable in principle, and some of them are very important in chemistry, astrophysics, etc. (For instance, in one of the main reactions that release energy in the sun, two protons and an electron turn into a deuteron and a neutrino.) Section 3.6 presents an application of the master transition rate formula (3.4.11) for general numbers N_α of initial particles.

We next take up the question of the Lorentz transformation properties of rates and cross-sections, which will help us to give a more general definition of the relative velocity u_α in Eq. (3.4.15). The Lorentz transformation rule (3.3.1) for the S-matrix is complicated by the momentum-dependent matrices associated with each particle's spin. To avoid this complication, consider the absolute-value squared of (3.3.1) (after factoring out the Lorentz-invariant delta function in Eq. (3.4.12)), and sum over all spins. The unitarity of the matrices $D^{(j)}_{\bar{\sigma}\sigma}(W)$ (or their analogs for zero mass) then shows that, apart from the energy factors in (3.3.1), the sum is Lorentz-invariant. That is, the quantity

$$\sum_{\text{spins}} |M_{\beta\alpha}|^2 \prod_\beta E \prod_\alpha E \equiv R_{\beta\alpha} \tag{3.4.16}$$

is a scalar function of the four-momenta of the particles in states α and β. (By $\prod_\alpha E$ and $\prod_\beta E$ is meant the product of all the single-particle energies $p^0 = \sqrt{\mathbf{p}^2 + m^2}$ for the particles in the states α and β.)

We can now write the spin-summed single-particle decay rate (3.4.13) as

$$\sum_{\text{spins}} d\Gamma(\alpha \to \beta) = 2\pi E_\alpha^{-1} R_{\beta\alpha} \delta^4(p_\beta - p_\alpha) d\beta \Big/ \prod_\beta E \,.$$

The factor $d\beta/\prod_\beta E$ may be recognized as the product of the Lorentz-invariant momentum-space volume elements (2.5.15), so it is Lorentz-invariant. So also are $R_{\beta\alpha}$ and $\delta^4(p_\beta - p_\alpha)$, leaving just the non-invariant factor $1/E_\alpha$, where E_α is the energy of the single initial particle. Our conclusion then is that the decay rate has the same Lorentz transformation property as $1/E_\alpha$. This is, of course, just the usual special-relativistic time dilation — the faster the particle, the slower it decays.

Similarly, our result (3.4.15) for the spin-summed cross-section may be written as

$$\sum_{\text{spins}} d\sigma(\alpha \to \beta) = (2\pi)^4 u_\alpha^{-1} E_1^{-1} E_2^{-1} R_{\beta\alpha} \delta^4(p_\alpha - p_\beta) d\beta \Big/ \prod_\beta E \,,$$

where E_1 and E_2 are the energies of the two particles in the initial state α. It is conventional to define the cross-section to be (when summed over spins) a Lorentz-invariant function of four-momenta. The factors $R_{\beta\alpha}$, $\delta^4(p_\beta - p_\alpha)$, and $d\beta/\prod_\beta E$ are already Lorentz-invariant, so this means that we must define the relative velocity u_α in arbitrary inertial frames so that $u_\alpha E_1 E_2$ is a scalar. We also mentioned earlier that in the Lorentz frame in which one particle (say, particle 1) is at rest, u_α is the velocity of the other particle. This uniquely determines u_α to have the value in general Lorentz frames*

$$u_\alpha = \sqrt{(p_1 \cdot p_2)^2 - m_1^2 m_2^2} \Big/ E_1 E_2 \tag{3.4.17}$$

where p_1, p_2 and m_1, m_2 are the four-momenta and mass of the two particles in the initial state α.

As a bonus, we note that in the 'center-of-mass' frame, where the total three-momentum vanishes, we have

$$p_1 = (\mathbf{p}, E_1) \,, \qquad p_2 = (-\mathbf{p}, E_2)$$

and here Eq. (3.4.17) gives

$$u_\alpha = \frac{|\mathbf{p}|(E_1 + E_2)}{E_1 E_2} = \left| \frac{\mathbf{p}_1}{E_1} - \frac{\mathbf{p}_2}{E_2} \right| \tag{3.4.18}$$

* Eq. (3.4.17) makes it obvious that $E_1 E_2 u_\alpha$ is a scalar. Also, when particle 1 is at rest, we have $\mathbf{p}_1 = 0$, $E_1 = m_1$, so $p_1 \cdot p_2 = -m_1 E_2$, and so Eq. (3.4.17) gives

$$u_\alpha = \sqrt{E_2^2 - m_2^2}/E_2 = |\mathbf{p}_2|/E_2 \,,$$

which is just the velocity of particle 2.

as might have been expected for a relative velocity. However, in this frame u_α is not really a physical velocity; in particular, Eq. (3.4.18) shows that for extremely relativistic particles, it can take values as large as 2.

We now turn to the interpretation of the so-called phase-space factor $\delta^4(p_\beta - p_\alpha)d\beta$, which appears in the general formula (3.4.11) for transition rates, and also in Eqs. (3.4.13) and (3.4.15) for decay rates and cross-sections. We here specialize to the case of the 'center-of-mass' Lorentz frame, where the total three-momentum of the initial state vanishes

$$\mathbf{p}_\alpha = 0\,. \tag{3.4.19}$$

(For $N_\alpha = 1$, this is just the case of a particle decaying at rest.) If the final state consists of particles with momenta $\mathbf{p}'_1, \mathbf{p}'_2, \cdots$, then

$$\delta^4(p_\beta - p_\alpha)d\beta = \delta^3(\mathbf{p}'_1 + \mathbf{p}'_2 + \cdots)\delta(E'_1 + E'_2 + \cdots - E)d^3p'_1 d^3p'_2 \cdots\,, \tag{3.4.20}$$

where $E \equiv E_\alpha$ is the total energy of the initial state. Any one of the $\mathbf{p}'_k$ integrals, say over $\mathbf{p}'_1$, can be done trivially by just dropping the momentum delta function

$$\delta^4(p_\beta - p_\alpha)d\beta \to \delta(E'_1 + E'_2 + \cdots - E)d^3p'_2 \cdots \tag{3.4.21}$$

with the understanding that wherever $\mathbf{p}'_1$ appears (as in E'_1) it must be replaced with

$$\mathbf{p}'_1 = -\mathbf{p}'_2 - \mathbf{p}'_3 - \cdots\,. \tag{3.4.22}$$

We can similarly use the remaining delta function to eliminate any *one* of the remaining integrals.

In the simplest case, there are just two particles in the final state. Here (3.4.21) gives

$$\delta^4(p_\beta - p_\alpha)d\beta \to \delta(E'_1 + E'_2 - E)d^3\mathbf{p}'_2\,.$$

In more detail, this is

$$\delta^4(p_\beta - p_\alpha)d\beta \to \delta\left(\sqrt{|\mathbf{p}_1{}'|^2 + m_1'^2} + \sqrt{|\mathbf{p}_1{}'|^2 + m_2'^2} - E\right)|\mathbf{p}_1{}'|^2 d|\mathbf{p}_1{}'|d\Omega\,, \tag{3.4.23}$$

where

$$\mathbf{p}'_2 = -\mathbf{p}'_1$$

and $d\Omega \equiv \sin\theta\, d\theta\, d\phi$ is the solid-angle differential for $\mathbf{p}_1{}'$. This can be simplified by using the standard formula

$$\delta(f(x)) = \delta(x - x_0)/\left|f'(x_0)\right|\,,$$

where $f(x)$ is an arbitrary real function with a single simple zero at $x = x_0$. In our case, the argument $E'_1 + E'_2 - E$ of the delta function in Eq. (3.4.23) has a unique zero at $|\mathbf{p}'_1| = k'$, where

$$k' = \sqrt{(E^2 - m_1'^2 - m_2'^2)^2 - 4m_1'^2 m_2'^2}/2E\,, \tag{3.4.24}$$

$$E'_1 = \sqrt{k'^2 + m_1'^2} = \frac{E^2 - m_2'^2 + m_1'^2}{2E}\,, \tag{3.4.25}$$

$$E'_2 = \sqrt{k'^2 + m_2'^2} = \frac{E^2 - m_1'^2 + m_2'^2}{2E}\,, \tag{3.4.26}$$

with derivative

$$\left[\frac{d}{d|\mathbf{p}_1{}'|}\left(\sqrt{|\mathbf{p}_1{}'|^2 + m_1'^2} + \sqrt{|\mathbf{p}_1{}'|^2 + m_2'^2} - E\right)\right]_{|\mathbf{p}'_1|=k'}$$

$$= \frac{k'}{E'_1} + \frac{k'}{E'_2} = \frac{k'E}{E'_1 E'_2}\,. \tag{3.4.27}$$

We can thus drop the delta function and the differential $d|\mathbf{p}_1{}'|$ in Eq. (3.4.23), by dividing by (3.4.27),

$$\delta^4(p_\beta - p_\alpha)d\beta \to \frac{k'E'_1E'_2}{E}d\Omega \tag{3.4.28}$$

with the understanding that k', E'_1, and E'_2 are given everywhere by Eqs. (3.4.24)–(3.4.26). In particular, the differential rate (3.4.13) for decay of a one-particle state of zero momentum and energy E into two particles is

$$\frac{d\Gamma(\alpha \to \beta)}{d\Omega} = \frac{2\pi k'E'_1E'_2}{E}|M_{\beta\alpha}|^2 \tag{3.4.29}$$

and the differential cross-section for the two-body scattering process $1\,2 \to 1'2'$ is given by Eq. (3.4.15) as

$$\frac{d\sigma(\alpha \to \beta)}{d\Omega} = \frac{(2\pi)^4 k'E'_1E'_2}{Eu_\alpha}|M_{\beta\alpha}|^2 = \frac{(2\pi)^4 k'E'_1E'_2E_1E_2}{E^2 k}|M_{\beta\alpha}|^2\,, \tag{3.4.30}$$

where $k \equiv |\mathbf{p}_1| = |\mathbf{p}_2|$.

The above case $N_\beta = 2$ is particularly simple, but there is one nice result for $N_\beta = 3$ that is also worth recording. For $N_\beta = 3$, Eq. (3.4.21) gives

$$\delta^4(p_\beta - p_\alpha)d\beta \to d^3\mathbf{p}'_2\, d^3\mathbf{p}'_3$$
$$\times\,\delta\left(\sqrt{(\mathbf{p}'_2 + \mathbf{p}'_3)^2 + m_1'^2} + \sqrt{\mathbf{p}_2'^2 + m_2'^2} + \sqrt{\mathbf{p}_3'^2 + m_3'^2} - E\right)\,.$$

We write the momentum-space volume as

$$d^3p'_2\, d^3p'_3 = |\mathbf{p}'_2|^2\, d|\mathbf{p}'_2|\, |\mathbf{p}'_3|^2\, d|\mathbf{p}'_3|\, d\Omega_3\, d\phi_{23}\, d\cos\theta_{23}\,,$$

where $d\Omega_3$ is the differential element of solid angle for $\mathbf{p}'_3$, and θ_{23} and ϕ_{23} are the polar and azimuthal angles of $\mathbf{p}'_2$ relative to the $\mathbf{p}'_3$ direction. The orientation of the plane spanned by $\mathbf{p}'_2$ and $\mathbf{p}'_3$ is specified by ϕ_{23} and the direction of $\mathbf{p}'_3$, with the remaining angle θ_{23} fixed by the energy-conservation condition

$$\sqrt{\mathbf{p}'^2_2 + 2|\mathbf{p}'_2|\;|\mathbf{p}'_3|\cos\theta_{23} + |\mathbf{p}'_3|^2 + m'^2_1} + \sqrt{|\mathbf{p}'_2|^2 + m'^2_2} + \sqrt{|\mathbf{p}'^2_3| + m'^2_3} = E\,.$$

The derivative of the argument of the delta function with respect to $\cos\theta_{23}$ is

$$\frac{\partial E'_1}{\partial \cos\theta_{23}} = \frac{|\mathbf{p}'_2|\,|\mathbf{p}'_3|}{E'_1}\,,$$

so we can do the integral over $\cos\theta_{23}$ by just dropping the delta function and dividing by this derivative

$$\delta^4(p_\beta - p_\alpha)d\beta \to |\mathbf{p}'_2|\,d\,|\mathbf{p}'_2|\;|\mathbf{p}'_3|\,d\,|\mathbf{p}'_3|\;E'_1\,d\Omega_3\,d\phi_{23}\,.$$

Replacing momenta with energies, this is finally

$$\delta^4(p_\beta - p_\alpha)d\beta \to E'_1\,E'_2\,E'_3\;dE'_2 dE'_3\;d\Omega_3\,d\phi_{23}\,. \tag{3.4.31}$$

But recall that the quantity (3.4.16), obtained by summing $|M_{\beta\alpha}|^2$ over spins and multiplying with the product of energies, is a scalar function of four-momenta. If we approximate this scalar as a constant, then Eq. (3.4.31) tells us that for a fixed initial state, the distribution of events plotted in the E'_2, E'_3 plane is uniform. Any departure from a uniform distribution of events in this plot thus provides a useful clue to the dynamics of the decay process, including possible centrifugal barriers or resonant intermediate states. This is known as a Dalitz plot,[19] because of its use by Dalitz in 1953 to analyze the decay $K^+ \to \pi^+ + \pi^+ + \pi^-$.

3.5 Perturbation Theory

The technique that has historically been most useful in calculating the S-matrix is perturbation theory, an expansion in powers of the interaction term V in the Hamiltonian $H = H_0 + V$. Eqs. (3.2.7) and (3.1.18) give the S-matrix as

$$S_{\beta\alpha} = \delta(\beta - \alpha) - 2i\pi\delta(E_\beta - E_\alpha)T_{\beta\alpha}{}^+$$
$$T_{\beta\alpha}{}^+ = (\Phi_\beta, V\Psi_\alpha{}^+),$$

where $\Psi_\alpha{}^+$ satisfies the Lippmann–Schwinger equation (3.1.17):

$$\Psi_\alpha{}^+ = \Phi_\alpha + \int d\gamma \, \frac{T_{\gamma\alpha}{}^+ \Phi_\gamma}{E_\alpha - E_\gamma + i\epsilon}.$$

Operating on this equation with V and taking the scalar product with Φ_β yields an integral equation for T^+

$$T_{\beta\alpha}{}^+ = V_{\beta\alpha} + \int d\gamma \, \frac{V_{\beta\gamma} T_{\gamma\alpha}{}^+}{E_\alpha - E_\gamma + i\epsilon}, \tag{3.5.1}$$

where

$$V_{\beta\alpha} \equiv (\Phi_\beta, V\Phi_\alpha). \tag{3.5.2}$$

The perturbation series for $T_{\beta\alpha}{}^+$ is obtained by iteration from Eq. (3.5.1)

$$T_{\beta\alpha}{}^+ = V_{\beta\alpha} + \int d\gamma \, \frac{V_{\beta\gamma} V_{\gamma\alpha}}{E_\alpha - E_\gamma + i\epsilon} + \int d\gamma d\gamma' \, \frac{V_{\beta\gamma} V_{\gamma\gamma'} V_{\gamma'\alpha}}{(E_\alpha - E_\gamma + i\epsilon)(E_\alpha - E_{\gamma'} + i\epsilon)} + \cdots. \tag{3.5.3}$$

The method of calculation based on Eq. (3.5.3), which dominated calculations of the S-matrix in the 1930s, is today known as *old-fashioned perturbation theory*. Its obvious drawback is that the energy denominators obscure the underlying Lorentz invariance of the S-matrix. It still has some uses, however, in clarifying the way that singularities of the S-matrix arise from various intermediate states. For the most part in this book, we will rely on a rewritten version of Eq. (3.5.3), known as *time-dependent perturbation theory*, which has the virtue of making Lorentz invariance much more transparent, while somewhat obscuring the contribution of individual intermediate states.

The easiest way to derive the time-ordered perturbation expansion is to use Eq. (3.2.5), which gives the S-operator as

$$S = U(\infty, -\infty) \, ,$$

where

$$U(\tau, \tau_0) \equiv \exp(iH_0\tau) \exp(-iH(\tau - \tau_0)) \exp(-iH_0\tau_0).$$

Differentiating this formula for $U(\tau, \tau_0)$ with respect to τ gives the differential equation

$$i\frac{d}{d\tau} U(\tau, \tau_0) = V(\tau) U(\tau, \tau_0), \tag{3.5.4}$$

where

$$V(t) \equiv \exp(iH_0 t) V \exp(-iH_0 t). \tag{3.5.5}$$

(Operators with this sort of time-dependence are said to be defined in the *interaction picture*, to distinguish their time-dependence from the time-dependence $O_H(t) = \exp(iHt)O_H \exp(-iHt)$ required in the Heisenberg picture of quantum mechanics.) Eq. (3.5.4) as well as the initial condition $U(\tau_0, \tau_0) = 1$ is obviously satisfied by the solution of the integral equation

$$U(\tau, \tau_0) = 1 - i \int_{\tau_0}^{\tau} dt\, V(t) U(t, \tau_0). \tag{3.5.6}$$

By iteration of this integral equation, we obtain an expansion for $U(\tau, \tau_0)$ in powers of V

$$\begin{aligned} U(\tau, \tau_0) = 1 - i \int_{\tau_0}^{\tau} dt_1\, V(t_1) + (-i)^2 \int_{\tau_0}^{\tau} dt_1 \int_{\tau_0}^{t_1} dt_2\, V(t_1)\, V(t_2) \\ + (-i)^3 \int_{\tau_0}^{\tau} dt_1 \int_{\tau_0}^{t_1} dt_2 \int_{\tau_0}^{t_2} dt_3\, V(t_1)\, V(t_2)\, V(t_3) + \cdots . \end{aligned} \tag{3.5.7}$$

Setting $\tau = \infty$ and $\tau_0 = -\infty$ then gives the perturbation expansion for the S-operator:

$$\begin{aligned} S = 1 - i \int_{-\infty}^{\infty} dt_1 V(t_1) + (-i)^2 \int_{-\infty}^{\infty} dt_1 \int_{-\infty}^{t_1} dt_2\, V(t_1) V(t_2) \\ + (-i)^3 \int_{-\infty}^{\infty} dt_1 \int_{-\infty}^{t_1} dt_2 \int_{-\infty}^{t_2} dt_3\, V(t_1)\, V(t_2)\, V(t_3) + \cdots . \end{aligned} \tag{3.5.8}$$

This can also be derived directly from the old-fashioned perturbation expansion (3.5.3), by using the Fourier representation of the energy factors in Eq. (3.5.3):

$$(E_\alpha - E_\gamma + i\epsilon)^{-1} = -i \int_0^{\infty} d\tau \exp(i(E_\alpha - E_\gamma)\tau) \tag{3.5.9}$$

with the understanding that such integrals are to be evaluated by inserting a convergence factor $e^{-\epsilon\tau}$ in the integrand, with $\epsilon \to 0+$.

There is a way of rewriting Eq. (3.5.8) that proves very useful in carrying out manifestly Lorentz-invariant calculations. Define the *time-ordered product* of any time-dependent operators as the product with factors arranged so that the one with the latest time-argument is placed leftmost, the next-latest next to the leftmost, and so on. For instance,

$$\begin{aligned} T\{V(t)\} &= V(t)\,, \\ T\{V(t_1)V(t_2)\} &= \theta(t_1 - t_2)V(t_1)V(t_2) + \theta(t_2 - t_1)V(t_2)V(t_1), \end{aligned}$$

and so on, where $\theta(\tau)$ is the step function, equal to $+1$ for $\tau > 0$ and to zero for $\tau < 0$. The time-ordered product of n Vs is a sum over all $n!$ permutations of the Vs, each of which gives the same integral over all

$t_1 \cdots t_n$, so Eq. (3.5.8) may be written

$$S = 1 + \sum_{n=1}^{\infty} \frac{(-i)^n}{n!} \int_{-\infty}^{\infty} dt_1 dt_2 \dots dt_n \, T\Big\{V(t_1) \cdots V(t_n)\Big\}. \tag{3.5.10}$$

This is sometimes known as the *Dyson series.*[20] This series can be summed if the $V(t)$ at different times all commute; the sum is then

$$S = \exp\left(-i \int_{-\infty}^{\infty} dt \, V(t)\right) .$$

Of course, this is not usually the case; in general (3.5.10) does not even converge, and is at best an asymptotic expansion in whatever coupling constant factors appear in V. However Eq. (3.5.10) is sometimes written in the general case as

$$S = T \exp\left(-i \int_{-\infty}^{\infty} dt \, V(t)\right)$$

with T indicating here that the expression is to be evaluated by time-ordering each term in the series expansion for the exponential.

We can now readily find one large class of theories for which the S-matrix is manifestly Lorentz-invariant. Since the elements of the S-matrix are the matrix elements of the S-operator between free-particle states Φ_α, Φ_β, etc., what we want is that the S-operator should commute with the operator $U_0(\Lambda, a)$ that produces Lorentz transformations on these free-particle states. Equivalently, the S-operator must commute with the generators of $U_0(\Lambda, a)$: H_0, $\mathbf{P}_0$ $\mathbf{J}_0$, and $\mathbf{K}_0$. To satisfy this requirement, let's try the hypothesis that $V(t)$ is an integral over three-space

$$V(t) = \int d^3x \, \mathscr{H}(\mathbf{x}, t) \tag{3.5.11}$$

with $\mathscr{H}(x)$ a scalar in the sense that

$$U_0(\Lambda, a)\mathscr{H}(x)U_0^{-1}(\Lambda, a) = \mathscr{H}(\Lambda x + a). \tag{3.5.12}$$

(By equating the coefficients of a^0 for infinitesimal transformations it can be checked that $\mathscr{H}(x)$ has a time-dependence consistent with Eq. (3.5.5).) Then S may be written as a sum of four-dimensional integrals

$$S = 1 + \sum_{n=1}^{\infty} \frac{(-i)^n}{n!} \int d^4x_1 \cdots d^4x_n \, T\Big\{\mathscr{H}(x_1) \cdots \mathscr{H}(x_n)\Big\}. \tag{3.5.13}$$

Everything is now manifestly Lorentz invariant, except for the time-ordering of the operator product.

Now, the time-ordering of two spacetime points x_1, x_2 is Lorentz-invariant unless $x_1 - x_2$ is space-like, i.e., unless $(x_1 - x_2)^2 > 0$, so the time-ordering in Eq. (3.5.13) introduces no special Lorentz frame if (though not only if) the $\mathscr{H}(x)$ all commute at space-like or light-like* separations:

$$[\mathscr{H}(x), \mathscr{H}(x')] = 0 \quad \text{for} \quad (x - x')^2 \geq 0. \tag{3.5.14}$$

We can use the results of Section 3.3 to give a formal non-perturbative proof that an interaction (3.5.11) satisfying Eqs. (3.5.12) and (3.5.14) does lead to an S-matrix with the correct Lorentz transformation properties. For an infinitesimal boost, Eq. (3.5.12) gives

$$-i[\mathbf{K}_0, \mathscr{H}(\mathbf{x}, t)] = t\nabla\mathscr{H}(\mathbf{x}, t) + \mathbf{x}\frac{\partial}{\partial t}\mathscr{H}(\mathbf{x}, t)\,, \tag{3.5.15}$$

so integrating over $\mathbf{x}$ and setting $t = 0$,

$$[\mathbf{K}_0, V] = \left[\mathbf{K}_0, \int d^3x\, \mathscr{H}(\mathbf{x}, 0)\right] = [H_0, \mathbf{W}]\,, \tag{3.5.16}$$

where

$$\mathbf{W} \equiv -\int d^3x\, \mathbf{x}\, \mathscr{H}(\mathbf{x}, 0). \tag{3.5.17}$$

If (as is usually the case) the matrix elements of $\mathscr{H}(\mathbf{x}, 0)$ between eigenstates of H_0 are smooth functions of the energy eigenvalues, then the same is true of V, as is necessary for the validity of scattering theory, and also true of $\mathbf{W}$, which is necessary in the proof of Lorentz invariance. The other condition for Lorentz invariance, the commutation relation (3.3.21), is also valid if and only if

$$0 = [\mathbf{W}, V] = \int d^3x \int d^3y\, \mathbf{x}\, [\mathscr{H}(\mathbf{x}, 0), \mathscr{H}(\mathbf{y}, 0)]\,. \tag{3.5.18}$$

This condition would follow from the 'causality' condition (3.5.14), but provides a somewhat less restrictive sufficient condition for Lorentz invariance of the S-matrix.

Theories of this class are not the only ones that are Lorentz invariant, but the most general Lorentz invariant theories are not very different. In particular, there is always a commutation condition something like (3.5.14) that needs to be satisfied. This condition has no counterpart for nonrelativistic systems, for which time-ordering is always Galilean-invariant. *It is this condition that makes the combination of Lorentz invariance and quantum mechanics so restrictive.*

* * *

---

* We write the condition on x and x' here as $(x - x')^2 \geq 0$ instead of $(x - x')^2 > 0$, because as we shall see in Chapter 6, Lorentz invariance can be disturbed by troublesome singularities at $x = x'$.

The methods described so far in this section are useful when the interaction operator V is sufficiently small. There is also a modified version of this approximation, known as the *distorted-wave Born approximation*, that is useful when the interaction contains two terms

$$V = V_s + V_w \tag{3.5.19}$$

with V_w weak but V_s strong. We can define $\Psi_{s\alpha}{}^{\pm}$ as what the 'in' and 'out' states would be if V_s were the whole interaction

$$\Psi_{s\alpha}{}^{\pm} = \Phi_\alpha + (E_\alpha - H_0 \pm i\epsilon)^{-1} V_s \Psi_{s\alpha}{}^{\pm}. \tag{3.5.20}$$

We can then write (3.1.16) as

$$\begin{aligned}
T_{\beta\alpha}{}^{+} &= (\Phi_\beta, V\Psi_\alpha{}^{+}) \\
&= \left(\left[\Psi_{s\beta}{}^{-} - (E_\beta - H_0 - i\epsilon)^{-1} V_s \Psi_{s\beta}{}^{-}\right], (V_s + V_w)\Psi_\alpha{}^{+}\right) \\
&= (\Psi_{s\beta}{}^{-}, V_w \Psi_\alpha{}^{+}) \\
&\quad + \left(\Psi_{s\beta}{}^{-}, \left[V_s - V_s(E_\beta - H_0 + i\epsilon)^{-1}(V_s + V_w)\right] \Psi_\alpha{}^{+}\right)
\end{aligned}$$

and so

$$T_{\beta\alpha}{}^{+} = (\Psi_{s\beta}{}^{-}, V_w \Psi_\alpha{}^{+}) + (\Psi_{s\beta}{}^{-}, V_s \Phi_\alpha). \tag{3.5.21}$$

The second term on the right-hand side is just what $T_{\beta\alpha}{}^{+}$ would be in the presence of strong interactions alone

$$T_{s\ \beta\alpha}{}^{+} \equiv (\Phi_\beta, V_s \Psi_{s\alpha}{}^{+}) = (\Psi_{s\beta}{}^{-}, V_s \Phi_\alpha). \tag{3.5.22}$$

(For a proof of Eq. (3.5.22), just drop V_w everywhere in the derivation of Eq. (3.5.21).) Eq. (3.5.21) is most useful when this second term vanishes: that is, when the process $\alpha \to \beta$ cannot be produced by the strong interactions alone. (For example, in nuclear beta decay we need a weak nuclear force to turn neutrons into protons, even though we cannot ignore the presence of the strong nuclear force acting in the initial and final nuclear states.) For such processes, the matrix element (3.5.22) vanishes, so Eq. (3.5.21) reads

$$T_{\beta\alpha}{}^{+} = (\Psi_{s\beta}{}^{-},\ V_w \Psi_\alpha{}^{+}). \tag{3.5.23}$$

So far, this is all exact. However, this way of rewriting the T-matrix becomes worthwhile when V_w is so weak that we may neglect its effect on the state $\Psi_\alpha{}^{+}$ in Eq. (3.5.23), and hence replace $\Psi_\alpha{}^{+}$ with the state $\Psi_{s\alpha}{}^{+}$, which takes account only of the strong interaction V_s. In this approximation, Eq. (3.5.23) becomes

$$T_{\beta\alpha}{}^{+} \simeq (\Psi_{s\beta}{}^{-},\ V_w \Psi_{s\alpha}{}^{+}). \tag{3.5.24}$$

This is valid to first order in V_w, but to all orders in V_s. This approximation is ubiquitous in physics; for instance, the S-matrix element for nuclear beta or gamma decay is calculated using Eq. (3.5.24) with V_s the strong nuclear interaction and V_w respectively either the weak nuclear interaction or the electromagnetic interaction, and with $\Psi_{s\beta}{}^-$ and $\Psi_{s\alpha}{}^+$ the final and initial nuclear states.

3.6 Implications of Unitarity

The unitarity of the S-matrix imposes an interesting and useful condition relating the amplitude $M_{\alpha\alpha}$ for forward scattering in an arbitrary multi-particle state α to the total rate for all reactions in that state. Recall that in the general case, where the state β may or may not be the same as the state α, the S-matrix may be written as in (3.3.2):

$$S_{\beta\alpha} = \delta(\beta - \alpha) - 2\pi i \delta^4(p_\beta - p_\alpha) M_{\beta\alpha}.$$

The unitarity condition then gives

$$\delta(\gamma - \alpha) = \int d\beta\, S^*_{\beta\gamma} S_{\beta\alpha} = \delta(\gamma - \alpha) - 2\pi i \delta^4(p_\gamma - p_\alpha) M_{\gamma\alpha}$$
$$+ 2\pi i \delta^4(p_\gamma - p_\alpha) M^*_{\alpha\gamma} + 4\pi^2 \int d\beta\, \delta^4(p_\beta - p_\gamma) \delta^4(p_\beta - p_\alpha) M^*_{\beta\gamma} M_{\beta\alpha}\, .$$

Cancelling the term $\delta(\gamma - \alpha)$ and a factor $2\pi\delta^4(p_\gamma - p_\alpha)$, we find that for $p_\gamma = p_\alpha$

$$0 = -iM_{\gamma\alpha} + iM^*_{\alpha\gamma} + 2\pi \int d\beta\, \delta^4(p_\beta - p_\alpha) M^*_{\beta\gamma} M_{\beta\alpha}\, . \tag{3.6.1}$$

This is most useful in the special case $\alpha = \gamma$, where it reads

$$\operatorname{Im} M_{\alpha\alpha} = -\pi \int d\beta\, \delta^4(p_\beta - p_\alpha) |M_{\beta\alpha}|^2. \tag{3.6.2}$$

Using Eq. (3.4.11), this can be expressed as a formula for the total rate for all reactions produced by an initial state α in a volume V

$$\Gamma_\alpha \equiv \int d\beta\, \frac{d\Gamma(\alpha \to \beta)}{d\beta}$$
$$= (2\pi)^{3N_\alpha - 2} V^{1-N_\alpha} \int d\beta\, \delta^4(p_\beta - p_\alpha) |M_{\beta\alpha}|^2$$
$$= -\frac{1}{\pi}(2\pi)^{3N_\alpha - 2} V^{1-N_\alpha} \operatorname{Im} M_{\alpha\alpha}. \tag{3.6.3}$$

In particular, where α is a two-particle state, this can be written

$$\operatorname{Im} M_{\alpha\alpha} = -u_\alpha \sigma_\alpha / 16\pi^3\, , \tag{3.6.4}$$

where u_α is the relative velocity (3.4.17) in state α, and σ_α is the *total* cross-section in this state, given by (3.4.15) as

$$\sigma_\alpha \equiv \int d\beta \; d\sigma(\alpha \to \beta)/d\beta = (2\pi)^4 u_\alpha^{-1} \int d\beta \, |M_{\beta\alpha}|^2 \delta^4(p_\beta - p_\alpha). \qquad (3.6.5)$$

This is usually expressed in a slightly different way, in terms of a *scattering amplitude* $f(\alpha \to \beta)$. Eq. (3.4.30) shows that the differential cross-section for *two-body* scattering in the center-of-mass frame is

$$\frac{d\sigma(\alpha \to \beta)}{d\Omega} = \frac{(2\pi)^4 k' E_1' E_2' E_1 E_2}{k E^2} |M_{\beta\alpha}|^2 \,, \qquad (3.6.6)$$

where k' and k are the magnitudes of the momenta in the final and initial states. We therefore define the scattering amplitude as*

$$f(\alpha \to \beta) \equiv -\frac{4\pi^2}{E}\sqrt{\frac{k' E_1' E_2' E_1 E_2}{k}}\, M_{\beta\alpha} \,, \qquad (3.6.7)$$

so that the differential cross-section is simply

$$\frac{d\sigma(\alpha \to \beta)}{d\Omega} = |f(\alpha \to \beta)|^2. \qquad (3.6.8)$$

In particular, for *elastic* two-body scattering, we have

$$f(\alpha \to \beta) \equiv -\frac{4\pi^2 E_1 E_2}{E} M_{\beta\alpha} \,. \qquad (3.6.9)$$

Using (3.4.18) for the relative velocity u_α, the unitarity prediction (3.6.3) now reads

$$\mathrm{Im}\, f(\alpha \to \alpha) = \frac{k}{4\pi}\sigma_\alpha. \qquad (3.6.10)$$

This form of the unitarity condition (3.6.3) is known as the *optical theorem.*[22]

There is a pretty consequence of the optical theorem that tells much about the pattern of scattering at high energy. The scattering amplitude f may be expected to be a smooth function of angle, so there must be some solid angle $\Delta\Omega$ within which $|f|^2$ has nearly the same value (say, within a factor 2) as in the forward direction. The total cross-section is then bounded by

$$\sigma_\alpha \geq \int |f|^2 d\Omega \geq \frac{1}{2}\, |f(\alpha \to \alpha)|^2 \Delta\Omega \geq \frac{1}{2}\, |\mathrm{Im}\, f(\alpha \to \alpha)|^2 \Delta\Omega \,.$$

* The phase of f is conventional, and is motivated by the wave mechanical interpretation[21] of f as the coefficient of the outgoing wave in the solution of the time-independent Schrödinger equation. The normalization of f used here is slightly unconventional for inelastic scattering; usually f is defined so that a ratio of final and initial velocities appears in the formula for the differential cross-section.

Using Eq. (3.6.10) then yields an upper bound on $\Delta\Omega$

$$\Delta\Omega \leq 32\pi^2 \Big/ k^2\sigma_\alpha \,. \tag{3.6.11}$$

As we shall see in the next section, total cross-sections are usually expected to approach constants or grow slowly at high energy, so Eq. (3.6.11) shows that the solid angle around the forward direction within which the differential cross-section is roughly constant shrinks at least as fast as $1/k^2$ for $k \to \infty$. This increasingly narrow peak in the forward direction at high energies is known as a *diffraction peak.*

Returning now to the general case of reactions involving arbitrary numbers of particles, we can use Eq. (3.6.2) together with CPT invariance to say something about the relations for total interaction rates of particles and antiparticles. Because CPT is antiunitary, its conservation does not in general imply any simple relation between a process $\alpha \to \beta$ and the process with particles replaced with their antiparticles. Instead, it provides a relation between a process and the *inverse* process involving antiparticles: we can use the same arguments that allowed us to deduce (3.3.46) from time-reversal invariance to show that CPT invariance requires the S-matrix to satisfy the condition

$$S_{\beta,\alpha} = S_{\mathscr{CPT}\alpha\,,\,\mathscr{CPT}\beta}\,, \tag{3.6.12}$$

where $\mathscr{CPT}$ indicates that we must reverse all spin z-components, change all particles into their corresponding antiparticles, and multiply the matrix element by various phase factors for the particles in the initial state and by their complex conjugates for the particles in the final state. Since CPT invariance also requires particles to have the same masses as their corresponding antiparticles, the same relation holds for the coefficient of $\delta^4(p_\alpha - p_\beta)$ in $S_{\beta\alpha}$:

$$M_{\beta\,,\,\alpha} = M_{\mathscr{CPT}\alpha\,,\,\mathscr{CPT}\beta}\,. \tag{3.6.13}$$

In particular, when the initial and final states are the same the phase factors all cancel, and Eq. (3.6.13) says that

$$\begin{aligned} &M_{p_1\sigma_1 n_1;\,p_2\sigma_2 n_2;\,\cdots\,,\,p_1\sigma_1 n_1;\,p_2\sigma_2 n_2;\,\cdots} \\ &\quad = M_{p_1 -\sigma_1 n_1^c;\,p_2 -\sigma_2 n_2^c;\,\cdots\,,\,p_1 -\sigma_1 n_1^c;\,p_2 -\sigma_2 n_2^c;\,\cdots}\,, \end{aligned} \tag{3.6.14}$$

where a superscript c on n indicates the antiparticle of n. The generalized optical theorem (3.6.2) then tells us that *the total reaction rate from an initial state consisting of some set of particles is the same as for an initial state consisting of the corresponding antiparticles with spins reversed*:

$$\Gamma_{p_1\sigma_1 n_1;\,p_2\sigma_2 n_2;\,\cdots} = \Gamma_{p_1 -\sigma_1 n_1^c;\,p_2 -\sigma_2 n_2^c;\,\cdots}\,. \tag{3.6.15}$$

In particular, applying this to one-particle states, we see that the decay rate of any particle equals the decay rate of the antiparticle with reversed spin. Rotational invariance does not allow particle decay rates to depend on the spin z-component of the decaying particle, so a special case of the general result (3.6.15) is that unstable particles and their corresponding antiparticles have precisely the same lifetimes.

* * *

The same argument that led from the unitarity condition $S^\dagger S = 1$ to our result (3.6.2) also allows us to use the other unitarity relation $SS^\dagger = 1$ to derive the result

$$\text{Im}\, M_{\alpha\alpha} = -\pi \int d\beta\, \delta^4(p_\beta - p_\alpha)|M_{\alpha\beta}|^2. \tag{3.6.16}$$

Putting this together with Eq. (3.6.2) then yields the reciprocity relation

$$\int d\beta\, \delta^4(p_\beta - p_\alpha)|M_{\beta\alpha}|^2 = \int d\beta\, \delta^4(p_\beta - p_\alpha)|M_{\alpha\beta}|^2 \tag{3.6.17}$$

or in other words

$$\int d\beta\, c_\alpha \frac{d\Gamma(\alpha \to \beta)}{d\beta} = \int d\beta\, c_\beta \frac{d\Gamma(\beta \to \alpha)}{d\alpha}\,, \tag{3.6.18}$$

where $c_\alpha \equiv [V/(2\pi)^3]^{N_\alpha}$. This result can be used in deriving some of the most important results of kinetic theory.[23] If $P_\alpha d\alpha$ is the probability of finding the system in a volume $d\alpha$ of the space of multi-particle states Φ_α, then the rate of decrease in P_α due to transitions to all other states is $P_\alpha \int d\beta\, d\Gamma(\alpha \to \beta)/d\beta$, while the rate of increase of P_α due to transitions from all other states is $\int d\beta\, P_\beta\, d\Gamma(\beta \to \alpha)/d\alpha$; the rate of change of P_α is then

$$\frac{dP_\alpha}{dt} = \int d\beta\, P_\beta \frac{d\Gamma(\beta \to \alpha)}{d\alpha} - P_\alpha \int d\beta \frac{d\Gamma(\alpha \to \beta)}{d\beta}\,. \tag{3.6.19}$$

It follows immediately that $\int P_\alpha d\alpha$ is time-independent. (Just interchange the labelling of the integration variables in the integral of the second term in Eq. (3.6.19).) On the other hand, the rate of change of the entropy $-\int d\alpha\, P_\alpha \ln(P_\alpha/c_\alpha)$ is

$$-\frac{d}{dt}\int d\alpha\, P_\alpha \ln(P_\alpha/c_\alpha) = -\int d\alpha \int d\beta\, \Big(\ln(P_\alpha/c_\alpha) + 1\Big) \times \left[P_\beta \frac{d\Gamma(\beta \to \alpha)}{d\alpha} - P_\alpha \frac{d\Gamma(\alpha \to \beta)}{d\beta}\right].$$

Interchanging the labelling of the integration variables in the second term, this may be written

$$-\frac{d}{dt}\int d\alpha\, P_\alpha \ln(P_\alpha/c_\alpha) = \int d\alpha \int d\beta\, P_\beta \ln\left(\frac{P_\beta c_\alpha}{P_\alpha c_\beta}\right) \frac{d\Gamma(\beta \to \alpha)}{d\alpha}\,.$$

Now, for any positive quantities x and y, the function $y \ln(y/x)$ satisfies the inequality**

$$y \ln\left(\frac{y}{x}\right) \geq y - x \,.$$

The rate of change of the entropy is thus bounded by

$$-\frac{d}{dt}\int d\alpha\, P_\alpha \ln(P_\alpha/c_\alpha) \geq \int d\alpha \int d\beta \left[\frac{P_\beta}{c_\beta} - \frac{P_\alpha}{c_\alpha}\right] c_\beta \frac{d\Gamma(\beta \to \alpha)}{d\alpha}$$

or interchanging variables of integration in the second term

$$-\frac{d}{dt}\int d\alpha\, P_\alpha \ln(P_\alpha/c_\alpha) \geq \int d\alpha \int d\beta\, \frac{P_\beta}{c_\beta}\left[c_\beta \frac{d\Gamma(\beta \to \alpha)}{d\alpha} - c_\alpha \frac{d\Gamma(\alpha \to \beta)}{d\beta}\right] .$$

But the unitarity relation (3.6.18) (with α and β interchanged) tells us that the integral over α on the right-hand side of this inequality vanishes, so we may conclude that the entropy always increases:

$$-\frac{d}{dt}\int d\alpha\, P_\alpha \ln(P_\alpha/c_\alpha) \geq 0 \,. \tag{3.6.20}$$

This is the 'Boltzmann H-theorem.' This theorem is often derived in statistical mechanics textbooks either by using the Born approximation, for which $|M_{\beta\alpha}|^2$ is symmetric in α and β so that $c_\beta d\Gamma(\beta \to \alpha)/d\alpha = c_\alpha d\Gamma(\alpha \to \beta)/d\beta$, or by assuming time-reversal invariance, which would tell us that $|M_{\beta\alpha}|^2$ is unchanged if we interchange α and β and also reverse all momenta and spins. Of course, neither the Born approximation nor time-reversal invariance are exact, so it is a good thing that the unitarity result (3.6.18) is all we need in order to derive the H-theorem.

The increase of the entropy stops when the probability P_α becomes a function only of conserved quantities such as the total energy and charge, times the factor c_α. In this case the conservation laws require $d\Gamma(\beta \to \alpha)/d\alpha$ to vanish unless $P_\alpha/c_\alpha = P_\beta/c_\beta$, so we can replace P_β with $P_\alpha c_\beta/c_\alpha$ in the first term of Eq. (3.6.19). Using Eq. (3.6.18) again then shows that in this case, P_α is time-independent. Here again, we need only the unitarity relation (3.6.18), not the Born approximation or time-reversal invariance.

3.7 Partial-Wave Expansions*

It is often convenient to work with the S-matrix in a basis of free-particle states in which all variables are discrete, except for the total momentum

** The difference between the left- and right-hand sides approaches the positive quantity $(x-y)^2/2y$ for $x \to y$, and has a derivative with respect to x that is positive- or negative-definite for all $x > y$ or $x < y$, respectively.

* This section lies somewhat out of the book's main line of development, and may be omitted in a first reading.

and energy. This is possible because the components of the momenta $\mathbf{p}_1, \cdots, \mathbf{p}_n$ in an n-particle state of definite total momentum $\mathbf{p}$ and total energy E form a $(3n-4)$-dimensional *compact* space; for instance, for $n = 2$ particles in the center-of-mass frame with $\mathbf{p} = 0$, this space is a two-dimensional spherical surface. Any function on such a compact space may be expanded in a series of generalized 'partial waves', such as the spherical harmonics that are commonly used in representing functions on the two-sphere. We may thus define a basis for these n-particle states that apart from the continuous variables $\mathbf{p}$ and E is discrete: we label the free-particle states in such a basis as $\Phi_{E\,\mathbf{p}\,N}$, with the index N incorporating all spin and species labels as well as whatever indices are used to label the generalized partial waves. These states may conveniently be chosen to be normalized so that their scalar products are:

$$\left(\Phi_{E'\,\mathbf{p}'\,N'}, \Phi_{E\,\mathbf{p}\,N}\right) = \delta(E'-E)\delta^3(\mathbf{p}'-\mathbf{p})\delta_{N',N}. \tag{3.7.1}$$

The S-operator then has matrix elements in this basis of the form

$$\left(\Phi_{E'\,\mathbf{p}'N'}\,,\ S\Phi_{E\,\mathbf{p}\,N}\right) = \delta(E'-E)\delta^3(\mathbf{p}'-\mathbf{p})S_{N',N}(E,\mathbf{p}), \tag{3.7.2}$$

where $S_{N',N}$ is a unitary matrix. Similarly, the T-operator, whose free-particle matrix elements $(\Phi_\beta, T\Phi_\alpha)$ are defined to be the quantities $T_{\beta\alpha}{}^+$ defined by Eq. (3.1.18), may be expressed in our new basis (in accordance with Eq. (3.4.12)) as

$$\left(\Phi_{E\,\mathbf{p}'\,N'},\ T\Phi_{E\,\mathbf{p}\,N}\right) = \delta^3(\mathbf{p}'-\mathbf{p})M_{N',N}(E,\mathbf{p}) \tag{3.7.3}$$

and the relation (3.2.7) is now an ordinary matrix equation:

$$S_{N',N}(E,\mathbf{p}) = \delta_{N',N} - 2i\pi M_{N',N}(E,\mathbf{p}). \tag{3.7.4}$$

We shall use this general formalism in the following section; for the present we will concentrate on reactions in which the initial state involves just two particles.

For example, consider a state consisting of two non-identical particles of species n_1, n_2 with non-zero masses M_1, M_2 and arbitrary spins s_1, s_2. In this case, the states may be labelled by their total momentum $\mathbf{p} = \mathbf{p}_1 + \mathbf{p}_2$, the energy E, the species labels n_1, n_2, the spin z-components σ_1, σ_2, and a pair of integers ℓ, m (with $|m| \leq \ell$) that specify the dependence of the state on the directions of, say, $\mathbf{p}_1$. Alternatively, we can form a convenient discrete basis by using Clebsch–Gordan coefficients[9] to combine the two spins to give a total spin s with z-component μ, and then using Clebsch–Gordan coefficients again to combine this with the orbital angular momentum ℓ with three-component m to form a total angular momentum j with three-component σ. This gives a basis of states $\Phi_{E\,\mathbf{p}\,j\sigma\ell s n}$ (with n a 'channel index' labelling the two particle species' n_1, n_2), defined by their scalar products with the states of definite individual momenta and spin three-components:

$$(\Phi_{\mathbf{p}_1\sigma_1\mathbf{p}_2\sigma_2 n'}\,,\,\Phi_{E\,\mathbf{p}\,j\sigma\ell sn}) \equiv (|\mathbf{p}_1|E_1E_2/E)^{-1/2}\delta^3(\mathbf{p}-\mathbf{p}_1-\mathbf{p}_2)$$
$$\times\,\delta\left(E-\sqrt{\mathbf{p}_1^2+M_1^2}-\sqrt{\mathbf{p}_2^2+M_2^2}\right)\delta_{n',n}$$
$$\times\sum_{m,\mu} C_{s_1s_2}(s,\mu;\sigma_1,\sigma_2)\,C_{\ell s}(j,\sigma;m,\mu)Y_\ell^m(\hat{\mathbf{p}}_1)\,. \tag{3.7.5}$$

where Y_ℓ^m are the usual spherical harmonics.[24] The factor $(|\mathbf{p}_1|E_1E_2/E)^{-1/2}$ is inserted so that in the center-of-mass frame these states will be properly normalized:

$$\left(\Phi_{E'\,\mathbf{p}'\,j'\sigma'\ell' s' n'},\,\Phi_{E\,0\,j\sigma\ell sn}\right) = \delta^3(\mathbf{p}')\delta(E'-E)\delta_{j',j}\delta_{\sigma',\sigma}\delta_{\ell',\ell}\delta_{s',s}\delta_{n',n}\,. \tag{3.7.6}$$

For identical particles to avoid double counting we must integrate only over half of the two-particle momentum space, so an extra factor $\sqrt{2}$ should appear in the scalar product (3.7.6).

In the center-of-mass frame the matrix elements of any momentum-conserving and rotationally-invariant operator O must take the form:

$$\left(\Phi_{E\,\mathbf{p}'\,j'\sigma'\ell' s' n'}\,,\,O\,\Phi_{E\,0\,j\sigma\ell sn}\right) = \delta^3(\mathbf{p}')O^j_{\ell' s' n',\ell sn}(E)\delta_{jj'}\delta_{\sigma\sigma'}. \tag{3.7.7}$$

(The fact that this is diagonal in j and σ follows from the commutation of O with $\mathbf{J}^2$ and J_3, and the further fact that the coefficient of $\delta_{\sigma\sigma'}$ is independent of σ follows from the commutation of O with $J_1 \pm iJ_2$. This is a special case of a general result known as the Wigner–Eckart theorem.[25]) Applying this to the operator M whose matrix elements are the quantities $M_{\beta\alpha}$, it follows that the scattering amplitude (3.6.7) in the center-of-mass system takes the form

$$f(\mathbf{k}\,\sigma_1,-\mathbf{k}\,\sigma_2,n\;\rightarrow\;\mathbf{k}'\,\sigma_1',-\mathbf{k}'\,\sigma_2',n')$$
$$\equiv -4\pi^2\sqrt{\frac{k'E_1'E_2'E_1E_2}{E^2k}}\,M_{\mathbf{k}'\sigma_1'\,-\mathbf{k}'\sigma_2'\,n',\,\mathbf{k}\,\sigma_1\,-\mathbf{k}\,\sigma_2\,n}$$
$$= -\frac{4\pi^2}{k}\sum_{j\sigma\ell' m' s'\mu'\ell m s\mu} C_{s_1s_2}(s,\mu;\sigma_1,\sigma_2)C_{\ell s}(j,\sigma;m,\mu)$$
$$\times\,C_{s_1's_2'}(s',\mu';\sigma_1',\sigma_2')C_{\ell' s'}(j,\sigma;m',\mu')$$
$$\times\,Y_{\ell'}^{m'}(\hat{\mathbf{k}}')Y_\ell^{m*}(\hat{\mathbf{k}})M^j_{\ell' s' n',\ell sn}(E). \tag{3.7.8}$$

The differential scattering cross-section is $|f|^2$. We will take the direction of the initial momentum $\mathbf{k}$ to be along the three-direction, in which case

$$Y_\ell^m(\hat{\mathbf{k}}) = \delta_{m0}\sqrt{\frac{2\ell+1}{4\pi}}. \tag{3.7.9}$$

Integrating $|f|^2$ over the direction of the final momentum $\mathbf{k}'$ and respectively summing and averaging over final and initial spin three-components gives the total cross-section** for transitions from channel n to n':

$$\sigma(n \to n'; E) = \frac{\pi}{k^2(2s_1+1)(2s_2+1)} \sum_{j\ell s\ell' s'} (2j+1) \times \left|\delta_{\ell'\ell}\delta_{s's}\delta_{n'n} - S^j_{\ell' s' n', \ell s n}(E)\right|^2 . \tag{3.7.10}$$

Summing (3.7.10) over all two-body channels gives the total cross-section for all elastic or inelastic two-body reactions:

$$\sum_{n'} \sigma(n \to n'; E) = \frac{\pi}{k^2(2s_1+1)(2s_2+1)} \sum_{j\ell s}(2j+1) \times \left[(1 - S^j(E))^\dagger (1 - S^j(E))\right]_{\ell s n, \ell s n} . \tag{3.7.11}$$

For comparison, Eqs. (3.7.8), (3.7.9), (3.7.4), and the Clebsch–Gordan sum rules** give the spin-averaged forward scattering amplitude as

$$f(n;E) = \frac{i}{2k(2s_1+1)(2s_2+1)} \sum_{j\ell s}(2j+1)[1 - S^j_{\ell sn, \ell sn}]$$

The optical theorem (3.6.10) then gives the total cross-section as

$$\sigma_{\text{total}}(n;E) = \frac{2\pi}{k^2(2s_1+1)(2s_2+1)} \sum_{j\ell s}(2j+1)\,\text{Re}\,[1 - S^j(E)]_{\ell sn, \ell sn} . \tag{3.7.12}$$

If only two-body channels can be reached from the channel n at energy E, then the matrix $S^j(E)$ (or at least some submatrix that includes channel n) is unitary, and thus

$$\left[(1 - S^j(E))^\dagger (1 - S^j(E))\right]_{\ell s n, \ell s n} = 2\,\text{Re}\,[1 - S^j(E)]_{\ell sn, \ell sn} , \tag{3.7.13}$$

so (3.7.12) and (3.7.11) are equal. On the other hand, if channels involving three or more particles are open, then the difference of (3.7.12) and (3.7.11)

** In deriving this result, we use standard sum rules[9] for the Clebsch–Gordan coefficients: first

$$\sum_{\sigma_1, \sigma_2} C_{s_1, s_2}(s, \mu; \sigma_1, \sigma_2) C_{s_1, s_2}(\bar{s}, \bar{\mu}; \sigma_1, \sigma_2) = \delta_{s\bar{s}}\delta_{\mu\bar{\mu}}$$

and the same with primes; then

$$\sum_{m\tilde{\sigma}} C_{\ell s}(j, \sigma; m, \tilde{\sigma}) C_{\ell s}(\bar{j}, \bar{\sigma}; m, \tilde{\sigma}) = \delta_{j\bar{j}}\delta_{\sigma\bar{\sigma}}$$

and finally

$$\sum_{\sigma\mu} C_{\ell s}(j, \sigma; 0, \mu) C_{\bar{\ell} s}(j, \sigma; 0, \mu) = \frac{2j+1}{2\ell+1}\delta_{\bar{\ell}\ell} .$$

gives the total cross-section for producing extra particles:

$$\sigma_{\text{production}}(n;E) = \frac{\pi}{k^2(2s_1+1)(2s_2+1)} \sum_{j\ell s}(2j+1) \times \left[1 - S^j(E)^\dagger S^j(E)\right]_{\ell s n, \ell s n}, \tag{3.7.14}$$

and this must be positive.

The partial wave expansion is particularly useful when applied to processes where the relevant part of the S-matrix is diagonal. This is the case, for instance, if the initial channel n contains just two spinless particles, and no other channels are open at this energy, as in $\pi^+ - \pi^+$ or $\pi^+ - \pi^0$ scattering at energies below the threshold for producing extra pions (provided one ignores weak and electromagnetic interactions). For a pair of spinless particles we have $j = \ell$, and angular momentum conservation keeps the S-matrix diagonal. It is also possible for the S-matrix to be diagonal in certain processes involving particles with spin; for instance in pion–nucleon scattering we can have $j = \ell + \frac{1}{2}$ or $j = \ell - \frac{1}{2}$, but for a given j these two states have opposite parity, so they cannot be connected by non-zero S-matrix elements. In any case, if for some n and E the S-matrix elements $S_{N',j\ell sn}(E,0)$ all vanish unless N' is the two-body state j, ℓ, s, n, then unitarity requires that

$$S^j_{\ell' s' n', \ell s n}(E) = \exp\left[2i\delta_{j\ell sn}(E)\right]\delta_{\ell'\ell}\delta_{s's}\delta_{n'n}\,, \tag{3.7.15}$$

where $\delta_{j\ell sn}(E)$ is a real phase, commonly known as the *phase shift*. This formula is also often used where the two-body part of the S-matrix is diagonal but channels containing three or more particles are also open; in such cases the phase shift must have a positive imaginary part, to keep (3.7.14) positive. For real phase shifts, the elastic and total cross-sections are then given by Eq. (3.7.10) or Eq. (3.7.12) as:

$$\sigma(n \to n; E) = \sigma_{\text{total}}(n;E) = \frac{4\pi}{k^2(2s_1+1)(2s_2+1)} \sum_{j\ell s}(2j+1)\sin^2\delta_{j\ell sn}(E)\,. \tag{3.7.16}$$

This familiar result is usually derived in non-relativistic quantum mechanics by studying the coordinate space wave function for a particle in a potential. The derivation given here is offered both to show that the partial wave expansion applies for elastic scattering even at relativistic velocities, and also to emphasize that it depends on no particular dynamical assumptions, only on unitarity and invariance principles.

It is also often useful to introduce phase shifts in dealing with problems where several channels are open, forming a few irreducible representations of some internal symmetry group. The classic example of such an internal symmetry is isotopic spin symmetry, for which the channel index n includes a specification of the isospins T_1, T_2 of the two particles together with their three-components t_1, t_2; the states in channel n may be expressed as linear combinations of the tth components of irreducible representations T, with coefficients given by the familiar Clebsch–Gordan coefficients $C_{T_1T_2}(T,t : t_1,t_2)$. Suppose that for the channels and energy of interest the S-matrix is diagonal in ℓ and s as well as j, T, and t. Unitarity and isospin symmetry then allow us to write the S-matrix as

$$S^j_{\ell' s' T' t', \ell s T t} = \exp[2i\delta_{j\ell s T}(E)]\delta_{\ell'\ell}\delta_{s's}\delta_{T'T}\delta_{t't}\,, \tag{3.7.17}$$

with $\delta_{j\ell sT}(E)$ a real phase shift, t-independent according to the Wigner–Eckart theorem. The partial cross-sections can again be calculated from Eq. (3.7.10), and the total cross-section is given by Eq. (3.7.12) as

$$\sigma_{\text{total}}(t_1,t_2;E) = \frac{4\pi}{k^2(2s_1+1)(2s_2+1)} \times \sum_{j\ell sTt}(2j+1)C_{T_1,T_2}(T,t;t_1,t_2)^2\sin^2\delta_{j\ell sT}(E). \tag{3.7.18}$$

For instance, in pion–pion scattering we have phase shifts $\delta_{\ell\ell 0T}(E)$ with $T=0$ or $T=2$ for each even ℓ and $T=1$ for each odd ℓ, while for pion–nucleon scattering we have phase shifts $\delta_{j\,j\pm\frac{1}{2}\,\frac{1}{2}\,T}$ with $T=\frac{1}{2}$ or $T=\frac{3}{2}$.

We can gain some useful insight about the threshold behavior of the scattering amplitudes and phase shifts from considerations of analyticity that are nearly independent of any dynamical assumptions. Unless there are special circumstances that would produce singularities in momentum space, we would expect the matrix element $M_{\mathbf{k}'\sigma_1'\,-\mathbf{k}'\sigma_2' n',\,\mathbf{k}\sigma_1\,-\mathbf{k}\sigma_2 n}$ to be an analytic function[†] of the three-momenta $\mathbf{k}$ and $\mathbf{k}'$ near $k=0$ or $k'=0$ or (for elastic scattering) $k=k'=0$. Turning to the partial wave expansion (3.7.8) for M, we note that $k^\ell Y_\ell^m(\hat{\mathbf{k}})$ is a simple polynomial function of the three-vector $\mathbf{k}$, so in order for $M_{\mathbf{k}'\sigma_1'\,-\mathbf{k}'\sigma_2' n',\,\mathbf{k}\sigma_1\,-\mathbf{k}\sigma_2 n}$ to be an analytic function of the three-momenta $\mathbf{k}$ and $\mathbf{k}'$ near $k=0$ or $k'=0$, the coefficients $M^j_{\ell' s' n', \ell s n}$ or equivalently $\delta_{\ell'\ell}\delta_{s's}\delta_{n'n} - S^j_{\ell' s' n', \ell s n}$ must go as $k^{\ell+\frac{1}{2}}k'^{\ell'+\frac{1}{2}}$ when k and/or k' go to zero. Hence for small k and/or k', it is only the lowest partial waves in the initial and/or final state that contribute appreciably to the scattering amplitude. We have three possible cases:

† For instance, in the Born approximation (3.2.8), M is proportional to the Fourier transform of the coordinate space matrix elements of the interaction, and hence is analytic at zero momenta as long as these matrix elements fall off sufficiently rapidly at large separations. The chief exception is for scattering involving long-distance forces, such as the Coulomb force.

Exothermic reactions

Here k' approaches a finite value as $k \to 0$, and in this limit $\delta_{\ell'\ell}\delta_{s's}\delta_{n'n} - S^j_{\ell's'n',\ell sn}$ goes as $k^{\ell+\frac{1}{2}}$. The cross-section (3.7.11) in this case goes as $k^{2\ell-1}$, where ℓ is here the *lowest* orbital angular momentum that can lead to the reaction. In the most usual case $\ell = 0$, so the reaction cross-section goes as $1/k$. (This is the case, for instance, in the absorption of slow neutrons by complex nuclei, or for the annihilation of electron-positron pairs into photons at low energy, aside from the higher-order effects of Coulomb forces.) The reaction rate is the cross-section times the flux, which goes like k, so the rate for an exothermic reaction behaves like a constant for $k \to 0$. However, it is the cross-section rather than the reaction rate that determines the probability of absorption when a beam crosses a given thickness of target material, and the factor $1/k$ makes this probability very high for slow neutrons in an absorbing material like boron.

Endothermic Reactions

Here the reaction is forbidden until k reaches a finite threshold value, where $k' = 0$. Just above this threshold $\delta_{\ell'\ell}\delta_{s's}\delta_{n'n} - S^j_{\ell's'n',\ell sn}$ goes as $(k')^{\ell'+\frac{1}{2}}$. The cross-section (3.7.11) in this case goes as $(k')^{2\ell'+1}$, where ℓ' is here the *lowest* orbital angular momentum that can be produced at threshold. In the most usual case $\ell' = 0$, so the reaction cross-section rises above threshold like k', and hence like $\sqrt{E - E_{\text{threshold}}}$. (This is the case, for instance, in the associated production of strange particles, or the production of electron–positron pairs in the scattering of photons.)

Elastic Reactions

Here $k = k'$, so k and k' go to zero together. (This is the case where $n' = n$, or where n' consists of particles in the same isotopic spin multiplets as are those in n.) In elastic scattering the partial waves with $\ell = \ell' = 0$ are always present, so in the limit $k \to 0$ the scattering amplitude (3.7.8) becomes a constant:

$$f(\mathbf{k}, \sigma_1, -\mathbf{k}, \sigma_2, n \;\to\; \mathbf{k}', \sigma_1', -\mathbf{k}', \sigma_2', n') \quad\to$$
$$\sum_{s\sigma} C_{s_1 s_2}(s, \sigma; \sigma_1, \sigma_2) C_{s_1' s_2'}(s, \sigma; \sigma_1', \sigma_2')\, a_s(n \to n')\,, \qquad (3.7.19)$$

where a is a constant, known as the *scattering length*, defined by the limit

$$S^s_{0sn',0sn} \to \delta_{n',n} + 2ik a_s(n \to n') \qquad (3.7.20)$$

for $k = k' \to 0$. Summing $4\pi|f|^2$ over final spins and averaging over initial spins gives the total cross-section for the transition $n \to n'$ at $k = k' = 0$:

$$\sigma(n \to n'; k = 0) = \frac{4\pi}{(2s_1+1)(2s_2+1)} \sum_s (2s+1) a_s^2(n \to n') . \quad (3.7.21)$$

The classic instance of the use of this formula is in neutron–proton scattering, where there are two scattering lengths, with the spin singlet length a_0 considerably larger than the spin triplet length a_1.

The partial wave expansion can also be used to make a crude guess about the behavior of cross-sections at high energy. With decreasing wavelengths, we may expect scattering to be described more or less classically: a particle of momentum k and orbital angular momentum ℓ would have an impact parameter ℓ/k, and will therefore strike a disk of radius R if $\ell \leq kR$. This can be interpreted as a statement about S-matrix elements:

$$S^j_{\ell\, s\, n, \ell\, s\, n} \to \begin{cases} 0 & \ell \ll kR_n \\ 1 & \ell \gg kR_n \end{cases} , \quad (3.7.22)$$

where R_n is some sort of interaction radius for channel n. For a given $\ell \gg s$, there are $2s+1$ values of j, all close enough to ℓ to approximate $2j + 1 \simeq 2\ell + 1$, so the sum over j and s in Eq. (3.7.12) merely gives a factor of order

$$\sum_{j\,s} (2j+1) = (2\ell+1) \sum_s (2s+1) = (2\ell+1)(2s_1+1)(2s_2+1).$$

The total cross-section is then given for $k \gg 1/R_n$ by Eq. (3.7.12) as

$$\sigma_{\text{total}}(n; E) \to \frac{2\pi}{k^2} \sum_{\ell \leq kR_n} (2\ell+1) \to 2\pi R_n^2 . \quad (3.7.23)$$

In exactly the same way, Eq. (3.7.10) gives the elastic scattering cross-section

$$\sigma(n \to n; E) \to \pi R_n^2 . \quad (3.7.24)$$

The difference between Eqs. (3.7.23) and (3.7.24) gives an inelastic cross-section πR_n^2, which is what we would expect for collisions with an opaque disk of radius R_n. (The somewhat surprising elastic scattering cross-section πR_n^2 may be attributed to diffraction by the disk.) On the other hand, if we assume along with Eq. (3.7.22) that $S^j_{\ell\, s\, n, \ell\, s\, n}$ is complex only for impact parameters ℓ/k within a small range of width $\Delta_n \ll R_n$ around $\ell/k = R_n$, then using the inequality $|\text{Im}\,(1 - S^j_{\ell\, s\, n, \ell\, s\, n})| \leq 2$, the same analysis gives a bound on the real part of the forward scattering amplitude

$$|\text{Re}\, f(n; E)| \leq 2k\, R_n \Delta_n \ll |\text{Im}\, f(n; E)| . \quad (3.7.25)$$

The smallness of the real part of the forward scattering amplitude at high energy is confirmed by experiment.

So far we have not said anything about whether the interaction radius R_n itself may depend on energy. As a very crude guess, we may take R_n as the distance at which the factor $\exp(-\mu r)$ in the Yukawa potential (1.2.74) takes a value proportional to some unknown power of E, in which case R_n goes as $\log E$ for $E \to \infty$, and the cross-sections go as $(\log E)^2$. As it happens, it has been rigorously shown[26] on the basis of very general assumptions that the total cross-section can grow no faster than $(\log E)^2$ for $E \to \infty$, and in fact the observed proton–proton total cross-section rises something like $(\log E)^2$ at high energies, so this rough picture of high-energy scattering does seem to have some correspondence with reality.

3.8 Resonances*

It often happens that the particles participating in a multi-particle collision can form an intermediate state consisting of a *single* unstable particle R, that eventually decays into the particles observed as the final state. If the total decay rate of R is small the cross-section exhibits a rapid variation (usually a peak), known as a *resonance*, at the energy of the intermediate state R.

We shall see that the behavior of the cross-sections near a resonance is pretty much prescribed by the unitarity condition alone, which is a good thing since there are a number of very different mechanisms that can produce a nearly stable state:

(a) The simplest possibility is that the Hamiltonian can be decomposed into two terms, a 'strong' Hamiltonian $H_0 + V_s$, which has the particle R as an eigenstate, plus a weak perturbation V_w, which allows R to decay into various states, including the initial and final states α, β of our collision process. For instance there is a neutral particle, the Z^0, with $j = 1$ and mass 91 GeV, that would be stable in the absence of the electroweak interactions. These interactions allow the Z^0 to decay into electron–positron pairs, muon–antimuon pairs, etc., but with a total decay rate that is much less than the Z^0 mass. In 1989 the Z^0 particle was seen as a resonance** in electron–positron collisions at CERN and Stanford,

* This section lies somewhat out of the book's main line of development, and may be omitted in a first reading.

** Incidentally, this example shows that a resonant state only needs to decay *relatively* slowly; the Z^0 lifetime is 2.6×10^{-25} seconds, which is not long enough for a Z^0 travelling near the speed of light to cross an atomic nucleus. What is important is that the decay rate is 36 times smaller than the rate $\hbar/M_Z$ of oscillation of the Z^0 wave function in its rest frame.

in the reactions $e^+ + e^- \to Z^0 \to e^+ + e^-$, $e^+ + e^- \to Z^0 \to \mu^+ + \mu^-$, etc.

(b) In some cases a particle is long-lived because there is a potential barrier that nearly prevents its constituents from escaping. The classic example is nuclear alpha decay: it may be energetically possible for a nucleus to emit an alpha particle (a He^4 nucleus) but the strong electrostatic repulsion between the alpha particle and the nucleus creates a barrier region around the daughter nucleus, which the alpha particle is classically forbidden to enter. The decay then can proceed only by quantum mechanical barrier penetration, and is exponentially slow. Such an unstable state shows up as a resonance in the scattering of the alpha particle on the daughter nucleus. For instance, the lowest-energy state of the Be^8 nucleus is unstable against decay into two alpha particles, and is seen as a resonance in He^4–He^4 scattering. (In addition to Coulomb barriers, there are also centrifugal barriers that help lengthen the life of alpha-, beta-, and gamma-unstable nuclei of high spin.)

(c) It is possible for complicated systems to be nearly stable for statistical reasons, without the presence of any potential barriers or weak interactions. For instance, an excited state of a heavy nucleus may be able to decay only if, through a statistical fluctuation, a large part of its energy is concentrated on a single neutron. This state will then show up as a resonance in the scattering of a neutron on the daughter nucleus.

These mechanisms for producing long-lived states are so different that it is truly fortunate that most of the properties of resonances follow from unitarity alone, without regard to the dynamical mechanism that produces the resonance.

First, let's consider the energy dependence of the matrix element for a reaction near a resonance. A wave packet $\int d\alpha\, g(\alpha)\Psi_\alpha{}^+ \exp(-iE_\alpha t)$ of 'in' states has a time-dependence given by (3.1.19)

$$\int d\alpha\, g(\alpha)\Psi_\alpha^+ e^{-iE_\alpha t} = \int d\alpha\, g(\alpha)\Phi_\alpha e^{-iE_\alpha t} + \int d\beta\, \Phi_\beta \int d\alpha\, \frac{e^{-iE_\alpha t}\, g(\alpha)\, T_{\beta\alpha}{}^+}{E_\alpha - E_\beta + i\epsilon} \,.$$

As mentioned in Section 3.1, a pole in the function $T_{\beta\alpha}{}^+$ in the lower-half complex E_α plane would make a contribution to the second term that decays exponentially as $t \to \infty$. Specifically, a pole at $E_\alpha = E_R - i\Gamma/2$ yields a term in the *amplitude* that behaves like $\exp(-iE_R t - \Gamma t/2)$, so it corresponds to a state whose *probability* decays like $\exp(-\Gamma t)$. We conclude then that a long-lived state of energy E_R with a slow decay rate

Γ produces a term in the scattering amplitude that varies as

$$T_{\beta\alpha}{}^{+} \sim (E_\alpha - E_R + i\Gamma/2)^{-1} + \text{ constant} . \tag{3.8.1}$$

To go further, it will be convenient to adopt as a basis the orthonormal discrete multi-particle states $\Phi_{\mathbf{p}EN}$ discussed in the previous section; $\mathbf{p}$ and E are the total momentum and energy, and N is an index that takes only discrete (though infinitely many) values. In this basis, the S-matrix may be written

$$S_{\mathbf{p}'E'N',\,\mathbf{p}EN} = \delta^3(\mathbf{p}' - \mathbf{p})\delta(E' - E)S_{N'N}(\mathbf{p}, E) . \tag{3.8.2}$$

Near a resonance, we expect the center-of-mass frame amplitude $S(0, E) \equiv \mathscr{S}(E)$ to have the form

$$\mathscr{S}_{N'N}(E) \equiv S_{N'N}(0, E) = \mathscr{S}_{0N'N} + \frac{\mathscr{R}_{N'N}}{E - E_R + i\Gamma/2} , \tag{3.8.3}$$

where $\mathscr{S}_0$ and $\mathscr{R}$ are approximately constant at least over the relatively small range of energies $|E - E_R| \lesssim \Gamma$.

In this basis, the unitarity of the S-matrix is an ordinary matrix equation

$$\mathscr{S}(E)^\dagger \mathscr{S}(E) = 1 . \tag{3.8.4}$$

Applied to Eq. (3.8.3), this tells us that the non-resonant background S-matrix is unitary

$$\mathscr{S}_0{}^\dagger \mathscr{S}_0 = 1 , \tag{3.8.5}$$

and also that the residue matrix $\mathscr{R}$ satisfies the two conditions

$$\mathscr{S}_0{}^\dagger \mathscr{R} + \mathscr{R}^\dagger \mathscr{S}_0 = 0 , \tag{3.8.6}$$

$$-\frac{i}{2}\Gamma \mathscr{S}_0{}^\dagger \mathscr{R} + \frac{i}{2}\Gamma \mathscr{R}^\dagger \mathscr{S}_0 + \mathscr{R}^\dagger \mathscr{R} = 0 . \tag{3.8.7}$$

These conditions can be put in a more transparent form by setting

$$\mathscr{R} \equiv -i\Gamma \mathscr{A} \mathscr{S}_0 . \tag{3.8.8}$$

The unitarity conditions on the matrix $\mathscr{A}$ are then simply

$$\mathscr{A}^\dagger = \mathscr{A} , \qquad \mathscr{A}^2 = \mathscr{A} . \tag{3.8.9}$$

Any such Hermitian idempotent matrix is called a *projection matrix*. Such matrices can always be expressed as a sum of dyads of orthonormal vectors $u^{(r)}$:

$$\mathscr{A}_{N'N} = \sum_r u_{N'}^{(r)} u_N^{(r)*} , \qquad \sum_N u_N^{(r)*} u_N^{(s)} = \delta_{rs} . \tag{3.8.10}$$

The discrete part of the S-matrix is then

$$\mathscr{S}_{N'N}(E) = \sum_{N''} \left[\delta_{N'N''} - i \frac{\Gamma}{E - E_R + i\Gamma/2} \sum_r u_{N'}^{(r)} u_{N''}^{(r)*} \right] \mathscr{S}_{0N''N} . \tag{3.8.11}$$

Each term in the sum over r can be thought of as arising from a different resonant state, all these states having the same values for E_R and Γ.

What has this to do with rates and cross-sections? For simplicity let's now ignore the non-resonant background scattering, setting $\mathscr{S}_{0\,N'N}$ equal to $\delta_{N'N}$; we will come back to the more general case a little later. Then for the two-body discrete center-of-mass states described in the previous section, Eq. (3.8.11) reads:

$$\mathscr{S}_{j'\sigma'\ell's'n',\,j\sigma\ell sn}(E) = \delta_{j'j}\delta_{\sigma'\sigma}\delta_{\ell'\ell}\delta_{s's}\delta_{n'n} \\ -i\frac{\Gamma}{E - E_R + i\Gamma/2}\sum_r u^{(r)}_{j'\sigma'\ell's'n'}\, u^{(r)*}_{j\sigma\ell sn} \,. \tag{3.8.12}$$

In all cases the label r will include an index σ_R giving the z-component of the total angular momentum of the resonant state; for a resonant state of total angular momentum j_R, σ_R takes $2j_R + 1$ values. If there is no other degeneracy, then r just labels the value of σ_R, and

$$u^{(\sigma_R)}_{j\sigma\ell sn} = \delta_{j_R,j}\delta_{\sigma_R,\sigma}u_{\ell sn} \,, \tag{3.8.13}$$

where $u_{\ell sn}$ are a set of complex amplitudes that (because of the Wigner–Eckart theorem) are independent of σ. Now Eq. (3.8.12) gives the amplitude S^j defined by Eq. (3.7.7) as

$$S^j_{\ell's'n',\ell sn}(E) = \delta_{\ell',\ell}\delta_{s's}\delta_{n'n} - i\delta_{j,j_R}\frac{\Gamma}{E - E_R + i\Gamma/2}u_{\ell's'n'}\, u^*_{\ell sn} \,. \tag{3.8.14}$$

Also, Eq. (3.8.10) now reads

$$\sum_{\ell sn} |u_{\ell sn}|^2 \; + \; \cdots = 1 \tag{3.8.15}$$

with the dots representing the positive contribution of any states containing three or more particles. As we shall see, the quantities $|u_{\ell sn}|^2$ have the interpretation of branching ratios for the decay of the resonant state into the various accessible two-body states.

Eq. (3.7.12) now gives the total cross-section for all reactions in channel n:

$$\sigma_{\rm total}(n;E) = \frac{\pi(2j_R+1)}{k^2(2s_1+1)(2s_2+1)}\frac{\Gamma\Gamma_n}{(E-E_R)^2+\Gamma^2/4} \,, \tag{3.8.16}$$

where

$$\Gamma_n \equiv \Gamma\sum_{\ell s} |u_{\ell sn}|^2 \,. \tag{3.8.17}$$

This is a version of the celebrated Breit–Wigner single-level formula.[27] We can also use these results to calculate the cross-section for resonant scattering from an initial two-body channel n to a final two-body channel

n'. Using Eq. (3.8.14) in Eq. (3.7.10) gives

$$\sigma(n \to n'; E) = \frac{\pi(2j_R + 1)}{k^2(2s_1 + 1)(2s_2 + 1)} \frac{\Gamma_n \Gamma_{n'}}{(E - E_R)^2 + \Gamma^2/4} . \tag{3.8.18}$$

This shows that the probabilities that the resonant state will decay into any one of the final two-body channels n' are proportional to the $\Gamma_{n'}$. According to Eq. (3.8.15), the sum of the Γ_n (including contributions from final states containing three or more particles) is just equal to the total decay rate Γ, so we can conclude that Γ_n is just the rate for the decay of the resonant state into channel n.

We see in Eqs. (3.8.16) and (3.8.18) the characteristic resonant peak at energy E_R, with a width (the full width at half maximum) equal to the decay rate Γ. (The individual Γ_n are often called partial widths.) Since $\Gamma_n \leq \Gamma$, the total cross-section at the peak of the resonance is roughly bounded by one square wavelength, $(2\pi/k)^2$. This rule, that cross-sections at a single resonance are roughly bounded by a square wavelength, is universally applicable even in classical physics (where energy conservation plays the role played here by unitarity), as for instance in the resonant interaction of sound waves with bubbles in the sea, or gravitational waves with gravitational wave antennae. (In the latter case, the branching ratio for oscillations in any laboratory mass to lose their energy through gravitational radiation is tiny, so the cross-section even at a resonance peak is vastly less then a square wavelength.[28])

Incidentally, it often happens that a resonance is detected, but energy measurements are insufficiently precise to resolve its width. In this case, what is measured experimentally is the integral of the cross-section over the resonant peak. For the total cross-section (3.8.16), this is

$$\int \sigma_{\text{total}}(n; E) dE = \frac{2\pi^2(2j_R + 1)\Gamma_n}{k_R^2(2s_1 + 1)(2s_2 + 1)} . \tag{3.8.19}$$

Such experiments can reveal only the partial width for decay of the resonant state into the initial particles, not the total width or branching ratios.

This formalism can also be applied when the resonant states with a given spin z-component form a multiplet related by some symmetry group. For instance, to the extent that isospin symmetry is respected, for a resonance of total isospin T_R the index r labelling resonant states includes a specification not only of the angular-momentum z-component σ_R, but also of an isospin three-component t_R, taking values $-T_R, -T_R+1, \cdots T_R$. In this case there is no change in the above results for the total and partial cross-sections, because each two-body channel n has definite values t_1, t_2 of the isospin z-components of the two particles, and hence can only couple to the resonant state with the single value t_1+t_2 for t_R. The partial widths Γ_n here depend on t_1 and t_2 only through factors $C_{T_1,T_2}(T_R,t_R;t_1,t_2)^2$.

The presence of a resonance shows itself in a characteristic behavior of phase shifts near the resonance. Returning to the general formula (3.8.11) (but still taking $\mathscr{S}_0=1$), we see from Eq. (3.8.10) that for each individual resonant state r, there is an eigenvector $u_N^{(r)}$ of $\mathscr{S}_{N'N}(E)$ with eigenvalue

$$\exp(2i\delta^{(r)}(E)) = 1 - i\frac{\Gamma}{E-E_R+i\Gamma/2}$$

or, in other words,

$$\tan\delta^{(r)}(E) = -\frac{\Gamma/2}{E-E_R}\,. \tag{3.8.20}$$

We see that over an energy range of order Γ centered around the resonant energy, the the 'eigenphase' $\delta^{(r)}(E)$ jumps from a value $\nu\pi$ (with ν a positive or negative integer) below the resonance to $(\nu+1)\pi$ above it. However, in order to use this result to say something about reaction rates, we need to know the eigenvectors $u_N^{(r)}$, which, in general, have components with arbitrary numbers of particles with various momenta, spin, and species.

These results are much more useful in those special cases when the particles in a particular channel N are forbidden (usually by conservation laws) from making a transition to any other channel. With this assumption, it is not difficult to include the effects of a non-resonant background scattering matrix $\mathscr{S}_0$ in the general result (3.8.11). In order for $\mathscr{S}_{N'N}$ to vanish for some particular N and all $N'\neq N$, it is necessary that the same is true of $\mathscr{S}_{0N'N}$, and also true of $u_{N'}^{(r)}$ for any r for which $u_N^{(r)}\neq 0$. The unitarity requirement (3.8.5) then requires that for this N

$$\mathscr{S}_{0N'N} = \exp(2i\delta_{0N})\delta_{N'N}$$

and Eq. (3.8.10) requires that

$$u_N^{(r)*}u_N^{(s)} = \delta_{rs}\,,$$

so that there can be only one term r in Eq. (3.8.11) for which $u_N^{(r)}\neq 0$. In this case, Eq. (3.8.11) gives

$$\begin{aligned}\mathscr{S}_{N'N}(E) &= \delta_{N'N}\left[1-\frac{i\Gamma}{E-E_R+i\Gamma/2}\right]\exp(2i\delta_{0N})\\ &\equiv \delta_{N'N}\exp(2i\delta_N(E))\end{aligned}$$

with total phase shift

$$\delta_N(E) = \delta_{0N} - \arctan\left(\frac{\Gamma/2}{E-E_R}\right)\,. \tag{3.8.21}$$

We see that over an energy range of order Γ centered around the resonant energy E_R, the phase shift $\delta_N(E)$ jumps from a value δ_{0N} below the resonance to $\delta_{0N} + \pi$ above it. For instance, as we saw in the previous section, these assumptions are satisfied in various two-body reactions such as pion–pion and pion–nucleon scattering at energies below the threshold for producing extra pions, with N incorporating the total and orbital angular-momenta j, ℓ (with $j = \ell$ for pion–pion scattering) and the total angular-momentum z-component σ, as well as the total isospin T and its three-component t. The Wigner–Eckart theorem allows the phase shifts to depend only on j, ℓ, and T, not on t or σ. There are famous resonances in these channels: in pion–pion scattering there is a resonance at 770 MeV called ρ with $j = \ell = 1$, $T = 1$, and $\Gamma = 150$ MeV; in pion–nucleon scattering there is a resonance at 1232 MeV called Δ with $j = \frac{3}{2}$, $\ell = 1$, $T = \frac{3}{2}$ and $\Gamma = 110$ to 120 MeV.

Inspection of Eq. (3.7.12) or Eq. (3.7.18) shows that the total cross-section reaches a peak when the resonant phase shift passes through $\pi/2$ (or odd-integer multiples of $\pi/2$.) The non-resonant phase shifts are typically rather small, so as we saw earlier, σ_{total} will exhibit a sharp peak when the phase shift δ_ℓ goes through $\pi/2$, at an energy close to E_R. However, it sometimes happens that the non-resonant background phase shift δ_{0N} is near $\pi/2$, in which case the cross-section will exhibit a sharp *dip* as the phase shift rises through π near E_R, due to destructive interference between the resonance and the non-resonant background amplitude. Such dips were first observed by Ramsauer and Townsend[29] in 1922, in the scattering of electrons by noble gas atoms.

Problems

1. Consider a theory with a separable interaction; that is,

$$(\Phi_\beta, V\Phi_\alpha) = g\, u_\beta\, u_\alpha^* \,,$$

where g is a real coupling constant, and u_α is a set of complex quantities with

$$\sum_\alpha |u_\alpha|^2 = 1.$$

Use the Lippmann–Schwinger equation (3.1.16) to find explicit solutions for the 'in' and 'out' states and the S-matrix.

2. Suppose that a resonance of spin one is discovered in e^+–e^- scattering at a total energy of 150 GeV and with a cross-section (in the center-of-mass frame, averaged over initial spins, and summed over final spins) for elastic e^+–e^- scattering at the peak of the resonance equal to 10^{-31} cm^2. What is the branching ratio for the decay of the resonant state R by the mode $R \to e^- + e^+$? What is the total

cross-section for e^+–e^- scattering at the peak of the resonance? (In answering both questions, ignore the non-resonant background scattering.)

3. Express the differential cross-section for two-body scattering in the *laboratory* frame, in which one of the two particles is initially at rest, in terms of kinematic variables and the matrix element $M_{\beta\alpha}$. (Derive the result directly, without using the results derived in this chapter for the differential cross-section in the center-of-mass frame.)

4. Derive the perturbation expansion (3.5.8) directly from the expansion (3.5.3) of old-fashioned perturbation theory.

5. We can define 'standing wave' states $\Psi_\alpha{}^0$ by a modified version of the Lippmann–Schwinger equation

$$\Psi_\alpha{}^0 = \Phi_\alpha + \frac{\mathscr{P}}{E_\alpha - H_0} V \Psi_\alpha{}^0 \, .$$

Show that the matrix $K_{\beta\alpha} \equiv \pi\delta(E_\beta - E_\alpha)(\Phi_\beta, V\Psi_\alpha{}^0)$ is Hermitian. Show how to express the S-matrix in terms of the K-matrix.

6. Express the differential cross-section for elastic π^+–proton and π^-–proton scattering in terms of the phase shifts for states of definite total angular momentum, parity, and isospin.

7. Show that the states $\Phi_{E\mathbf{p}j\sigma sn}$ defined by Eq. (3.7.5) are correctly normalized to have the scalar products (3.7.6).

References

1. For more details, see M. L. Goldberger and K. M. Watson, *Collision Theory* (John Wiley & Sons, New York, 1964); R. G. Newton, *Scattering Theory of Waves and Particles*, 2nd edn (Springer-Verlag, New York, 1982).

1*a*. B. Lippmann and J. Schwinger, *Phys. Rev.* **79**, 469 (1950).

2. J. A. Wheeler, *Phys. Rev.* **52**, 1107 (1937); W. Heisenberg, *Z. Phys.* **120**, 513, 673 (1943).

3. M. Born, *Z. Phys.*. **37**, 863 (1926); **38**, 803 (1926).

4. C. Møller, *Kgl. Danske Videnskab. Mat. Fys. Medd.* **23**, No. 1 (1945); **22**, No. 19 (1946).

5. G. D. Rochester and C. C. Butler, *Nature* **160**, 855 (1947). For a historical review, see G. D. Rochester, in *Pions to Quarks – Particle Physics in the 1950s*, ed. by L. M. Brown, M. Dresden, and L. Hoddeson (Cambridge University Press, Cambridge, UK, 1989).

6. G. Breit, E. U. Condon, and R. S. Present, *Phys. Rev.* **50**, 825 (1036); B. Cassen and E. U. Condon, *Phys. Rev.* **50**, 846 (1936); G. Breit and E. Feenberg, *Phys. Rev.* **50**, 850 (1936).

7. M. A. Tuve, N. Heydenberg, and L. R. Hafstad, *Phys. Rev.* **50**, 806 (1936).

8. M. Gell-Mann, Cal. Tech. Synchotron Laboratory Report CTSL-20 (1961); *Phys. Rev.* **125**, 1067 (1962); Y. Ne'eman *Nucl. Phys.* **26**, 222 (1961).

9. See e.g. A. R. Edmonds, *Angular Momentum in Quantum Mechanics* (Princeton University Press, Princeton, 1957): Chapter 3 (where $C_{j_1 j_2}(jm; m_1 m_2)$ is denoted $(j_1\, j_2\, j\, m | j_1\, m_1\, j_2\, m_2)$); M. E. Rose, *Elementary Theory of Angular Momentum* (John Wiley & Sons, New York, 1957): Chapter III (where $C_{j_1 j_2}(jm; m_1 m_2)$ is denoted $C(j_1\, j_2\, j\,;\, m_1\, m_2\, m)$).

10. G. Feinberg and S. Weinberg, *Nuovo Cimento* Serie X, **14**, 571 (1959).

11. W. Chinowsky and J. Steinberger, *Phys. Rev.* **95**, 1561 (1954); also see B. Ferretti, *Report of an International Conference on Fundamental Particles and Low Temperatures, Cambridge, 1946* (The Physical Society, London, 1947).

12. T. D. Lee and C. N. Yang, *Phys. Rev.* **104**, 254 (1956).

13. C. S. Wu *et al.*, *Phys. Rev.* **105**, 1413 (1957).

14. R. Garwin, L. Lederman, and M. Weinrich, *Phys. Rev.* **105**, 1415 (1957); J. I. Friedman and V. L. Telegdi, *Phys. Rev.* **105**, 1681 (1957).

15. K. M. Watson, *Phys. Rev.* **88**, 1163 (1952).

16. M. Gell-Mann and A. Pais, *Phys. Rev.* **97**, 1387 (1955); also see A. Pais and O. Piccioni, *Phys. Rev.* **100**, 1487 (1955).

17. J. H. Christenson, J. W. Cronin, V. L. Fitch, and R. Turlay, *Phys. Rev. Letters* **13**, 138 (1964).

18. K. R. Schubert *et al.*, *Phys. Lett.* **31B**, 662 (1970). This reference analyzes neutral kaon data without the assumption of CPT invariance, and finds that the part of the CP-violating amplitude that conserves

CPT and violates T has both real and imaginary parts five standard deviations from zero, while the part that conserves T and violates CPT is within a standard deviation of zero.

19. R. H. Dalitz, *Phil. Mag.* **44**, 1068 (1953); also see E. Fabri, *Nuovo Cimento* **11**, 479 (1954).

20. F. J. Dyson, *Phys. Rev.* **75**, 486, 1736 (1949).

21. See, e.g., L. I. Schiff, *Quantum Mechanics*, 1st edn (McGraw-Hill, New York, 1949): Section 19.

22. This was first proved in classical electrodynamics. See, e. g., H. A. Kramers, *Atti Congr. Intern. Fisici, Como, 1927*; reprinted in H. A. Kramers, *Collected Scientific Papers* (North-Holland, Amsterdam, 1956). For the proof in quantum mechanics, see E. Feenberg, *Phys. Rev.* **40**, 40 (1932); N. Bohr, R. E. Peierls, and G. Placzek, *Nature* **144**, 200 (1939).

23. A general version of this argument was given in the late 1960s in unpublished work of C. N. Yang and C. P. Yang. Also see A. Aharony, in *Modern Developments in Thermodynamics* (Wiley, New York, 1973): pp. 95–114, and references therein.

24. See, e.g., A. R. Edmonds, *Angular Momentum in Quantum Mechanics*, (Princeton University Press, Princeton, 1957): Chapter 2; M. E. Rose, *Elementary Theory of Angular Momentum* (John Wiley & Sons, New York, 1957): Appendix III; L. D. Landau and E. M. Lifshitz, *Quantum Mechanics — Non Relativistic Theory*, 3rd edn (Pergamon Press, Oxford, 1977): Section 28.

25. E. P. Wigner, *Gruppentheorie* (Friedrich Vieweg und Sohn, Braunschweig, 1931); C. Eckart, *Rev. Mod. Phys.* **2**, 305 (1930).

26. M. Froissart, *Phys. Rev.* **123**, 1053 (1961).

27. G. Breit and E. P. Wigner, *Phys. Rev.* **49**, 519 (1936).

28. See, e.g., S. Weinberg, *Gravitation and Cosmology* (Wiley, New York, 1972): Section 10.7.

29. R. Kollath, *Phys. Zeit.* **31**, 985 (1931).

4

The Cluster Decomposition Principle

Up to this point we have not had much to say about the detailed structure of the Hamiltonian operator H. This operator can be defined by giving all its matrix elements between states with arbitrary numbers of particles. Equivalently, as we shall show here, any such operator may be expressed as a function of certain operators that create and destroy single particles. We saw in Chapter 1 that such creation and annihilation operators were first encountered in the canonical quantization of the electromagnetic field and other fields in the early days of quantum mechanics. They provided a natural formalism for theories in which massive particles as well as photons can be produced and destroyed, beginning in the early 1930s with Fermi's theory of beta decay.

However, there is a deeper reason for constructing the Hamiltonian out of creation and annihilation operators, which goes beyond the need to quantize any pre-existing field theory like electrodynamics, and has nothing to do with whether particles can actually be produced or destroyed. The great advantage of this formalism is that if we express the Hamiltonian as a sum of products of creation and annihilation operators, with suitable non-singular coefficients, then the S-matrix will automatically satisfy a crucial physical requirement, the cluster decomposition principle,[1] which says in effect that distant experiments yield uncorrelated results. Indeed, it is for this reason that the formalism of creation and annihilation operators is widely used in non-relativistic quantum statistical mechanics, where the number of particles is typically fixed. In relativistic quantum theories, the cluster decomposition principle plays a crucial part in making field theory inevitable. There have been many attempts to formulate a relativistically invariant theory that would not be a local field theory, and it is indeed possible to construct theories that are not field theories and yet yield a Lorentz-invariant S-matrix for two-particle scattering,[2] but such efforts have always run into trouble in sectors with more than two particles: either the three-particle S-matrix is not Lorentz-invariant, or else it violates the cluster decomposition principle.

In this chapter we will first discuss the basis of states containing arbitrary numbers of bosons and fermions, then define the creation and annihilation operators, and finally show how their use facilitates the construction of Hamiltonians that yield S-matrices satisfying the cluster decomposition condition.

4.1 Bosons and Fermions

The Hilbert space of physical states is spanned by states containing 0, 1, 2, $\cdots$ free particles. These can be free-particle states, or 'in' states, or 'out' states; for definiteness we shall deal here with the free-particle states $\Phi_{\mathbf{p}_1\,\sigma_1\,n_1,\mathbf{p}_2\,\sigma_2\,n_2,\cdots}$, but all our results will apply equally to 'in' or 'out' states. As usual, σ labels spin z-components (or helicities, for massless particles) and n labels particle species.

We must now go into a matter that has been passed over in Chapter 3; the symmetry properties of these states. As far as we know, all particles are either *bosons* or *fermions*, the difference being that a state is unchanged by the interchange of two identical bosons, and changes sign under the interchange of two identical fermions. That is

$$\Phi_{\cdots\,\mathbf{p}\,\sigma\,n\,\cdots\,\mathbf{p}'\,\sigma'\,n\,\cdots} = \pm\,\Phi_{\cdots\,\mathbf{p}'\,\sigma'\,n\,\cdots\,\mathbf{p}\,\sigma\,n\,\cdots} \tag{4.1.1}$$

with an upper or lower sign if n is a boson or a fermion, respectively, and dots representing other particles that may be present in the state. (Equivalently, this could be stated as a condition on the 'wave functions,' the coefficients of these multi-particle basis vectors in physically allowable state-vectors.) These two cases are often referred to as Bose or Fermi 'statistics'. We will see in the next chapter that Bose and Fermi statistics are only possible for particles that have integer or half-integer spins, respectively, but we shall not need this information in the present chapter. In this section we shall offer a non-rigorous argument that all particles must be either bosons or fermions, and then set up normalization conditions for multi-boson or multi-fermion states.

First note that if two particles with momenta and spins $\mathbf{p}, \sigma$ and $\mathbf{p}', \sigma'$ belong to identical species n, then the state-vectors $\Phi_{\cdots\,\mathbf{p}\,\sigma\,n\,\cdots\,\mathbf{p}'\,\sigma'\,n\,\cdots}$ and $\Phi_{\cdots\,\mathbf{p}'\,\sigma'\,n\,\cdots\,\mathbf{p}\,\sigma\,n\,\cdots}$ represent the same physical state; if this were not the case then the particles would be distinguished by their order in the labelling of the state-vector, and the first listed would not be identical with the second. Since the two state-vectors are physically indistinguishable, they must belong to the same ray, and so

$$\Phi_{\cdots\,\mathbf{p}\,\sigma\,n\,\cdots\,\mathbf{p}'\,\sigma'\,n\,\cdots} = \alpha_n\,\Phi_{\cdots\,\mathbf{p}'\,\sigma'\,n\,\cdots\,\mathbf{p}\,\sigma\,n\,\cdots}\,, \tag{4.1.2}$$

where α_n is a complex number of unit absolute value. We may regard this as part of the definition of what we mean by identical particles.

The crux of the matter is to decide on what the phase factor α_n may depend. *If* it depends only on the species index n, then we are nearly done. Interchanging the two particles in Eq. (4.1.2) again, we find

$$\Phi_{\cdots \mathbf{p}\sigma n \cdots \mathbf{p}'\sigma' n \cdots} = \alpha_n^2 \Phi_{\cdots \mathbf{p}\sigma n \cdots \mathbf{p}'\sigma' n \cdots}$$

so that $\alpha_n^2 = 1$, yielding Eq. (4.1.1) as the only two possibilities.

On what else could α_n depend? It might depend on the numbers and species of the other particles in the state (indicated by dots in Eqs. (4.1.1) and (4.1.2)), but this would lead to the uncomfortable result that the symmetry of state-vectors under interchange of particles here on earth may depend on the presence of particles elsewhere in the universe. This is the sort of thing that is ruled out by the cluster decomposition principle, to be discussed later in this chapter. The phase α_n cannot have any non-trivial dependence on the spins of the two particles that are interchanged, because then these spin-dependent phase factors would have to furnish a representation of the rotation group, and there are no non-trivial representations of the three-dimensional rotation group that are one-dimensional – that is, by phase factors. The phase α_n might conceivably depend on the momenta of the two particles that are interchanged, but Lorentz invariance would require α_n to depend only on the scalar $p_1^\mu p_{2\mu}$; this is symmetric under interchange of particles 1 and 2, and therefore such dependence would not change the argument leading to the conclusion that $\alpha_n^2 = 1$.

The logical gap in the above argument is that (although our notation hides the fact) the states $\Phi_{\mathbf{p}_1 \sigma_1 n, \mathbf{p}_2 \sigma_2 n, \cdots}$ may carry a phase factor that depends on the path through momentum space by which the momenta of the particles are brought to the values $\mathbf{p}_1$, $\mathbf{p}_2$, etc. In this case the interchange of two particles twice might change the state by a phase factor, so that $\alpha_n^2 \neq 1$. We will see in Section 9.7 that this is a real possibility in two-dimensional space, but not for three or more spatial dimensions.

What about interchanges of particles belonging to different species? If we like, we can avoid this question by simply agreeing from the beginning to label the state-vector by listing all photon momenta and helicities first, then all electron momenta and spin z-components, and so on through the table of elementary particle types. Alternatively, we can allow the particle labels to appear in any order, and *define* the state-vectors with particle labels in an arbitrary order as equal to the state-vector with particle labels in some standard order times phase factors, whose dependence on the interchange of particles of different species can be anything we like. In order to deal with symmetries like isospin invariance that relate particles of different species, it is convenient to adopt a convention that generalizes Eq. (4.1.1): the state-vector will be taken to be symmetric under interchange of any bosons with each other, or any bosons with any fermions, and antisymmetric with respect to interchange of any two fermions with each other, in all cases, whether the particles are of the same species or not.*

The normalization of these states must be defined in consistency with these symmetry conditions. To save writing, we will use a label q to denote all the quantum numbers of a single particle: its momentum, $\mathbf{p}$, spin z-component (or, for massless particles, helicity) σ, and species n. The N-particle states are thus labelled $\Phi q_1 \ldots q_N$ (with $N = 0$ for the vacuum state Φ_0.) For $N = 0$ and $N = 1$ the question of symmetry does not arise: here we have

$$(\Phi_0, \Phi_0) = 1 \tag{4.1.3}$$

and

$$(\Phi_{q'}, \Phi_q) = \delta(q' - q)\,, \tag{4.1.4}$$

where $\delta(q' - q)$ is a product of all the delta functions and Kronecker deltas for the particle's quantum numbers,

$$\delta(q' - q) \equiv \delta^3(\mathbf{p}' - \mathbf{p})\,\delta_{\sigma'\sigma}\delta_{n'n}. \tag{4.1.5}$$

On the other hand, for $N = 2$ the states $\Phi_{q_1'q_2'}$ and $\Phi_{q_2'q_1'}$ are physically the same, so here we must take

$$\left(\Phi_{q_1'q_2'}\,,\, \Phi_{q_1q_2}\right) = \delta(q_1' - q_1)\,\delta(q_2' - q_2) \pm\, \delta(q_2' - q_1)\,\delta(q_1' - q_2) \tag{4.1.6}$$

the sign $\pm$ being $-$ if both particles are fermions and $+$ otherwise. This obviously is consistent with the above stated symmetry properties of the states. More generally,

$$\left(\Phi_{q_1'q_2'\cdots q_M'}, \Phi_{q_1q_2\cdots q_N}\right) = \delta_{NM} \sum_{\mathscr{P}} \delta_{\mathscr{P}} \prod_i \delta(q_i - q'_{\mathscr{P}i}). \tag{4.1.7}$$

The sum here is over all permutations $\mathscr{P}$ of the integers $1, 2, \cdots, N$. (For instance, in the first term in Eq. (4.1.6), $\mathscr{P}$ is the identity, $\mathscr{P}1 = 1$, $\mathscr{P}2 = 2$, while in the second term $\mathscr{P}1 = 2$, $\mathscr{P}2 = 1$.) Also, $\delta_{\mathscr{P}}$ is a sign factor equal to -1 if $\mathscr{P}$ involves an odd permutation of fermions (an odd number of fermion interchanges) and $+1$ otherwise. It is easy to see that Eq. (4.1.7) has the desired symmetry or antisymmetry properties under interchange of the q_i, and also under interchange of the q_j'.

* In fact, by the same reasoning, the symmetry or antisymmetry of the state-vector under interchange of particles of the same species but different helicities or spin z-components is purely conventional, because we could have agreed from the beginning to list first the momenta of photons of helicity $+1$, then the momenta of all photons of helicity -1, then the momenta of all electrons of spin z-component $+\frac{1}{2}$, and so on. We adopt the *convention* that the state-vector is symmetric or antisymmetric under interchange of identical bosons or fermions of different helicities or spin z-components in order to facilitate the use of rotational invariance.

4.2 Creation and Annihilation Operators

Creation and annihilation operators may be defined in terms of their effect on the normalized multi-particle states discussed in the previous section. The *creation operator* $a^\dagger(q)$ (or in more detail, $a^\dagger(\mathbf{p},\sigma,n)$) is defined as the operator that simply adds a particle with quantum numbers q at the front of the list of particles in the state

$$a^\dagger(q)\Phi_{q_1q_2\cdots q_N} \equiv \Phi_{qq_1q_2\cdots q_N}\,. \tag{4.2.1}$$

In particular, the N-particle state can be obtained by acting on the vacuum with N creation operators

$$a^\dagger(q_1)a^\dagger(q_2)\cdots a^\dagger(q_N)\Phi_0 = \Phi_{q_1\cdots q_N}\,. \tag{4.2.2}$$

It is conventional for this operator to be called $a^\dagger(q)$; *its* adjoint, which is then called $a(q)$, may be calculated from Eq. (4.1.7). As we shall now show, $a(q)$ removes a particle from any state on which it acts, and is therefore known as an *annihilation operator.* In particular, when the particles $q\,q_1\cdots q_N$ are either all bosons or all fermions, we have

$$a(q)\Phi_{q_1q_2\cdots q_N} = \sum_{r=1}^{N}(\pm)^{r+1}\delta(q-q_r)\Phi_{q_1\cdots q_{r-1}q_{r+1}\cdots q_N}\,, \tag{4.2.3}$$

with a $+1$ or -1 sign for bosons or fermions, respectively . (Here is the proof. We want to calculate the scalar product of $a(q)\Phi_{q_1q_2\cdots q_N}$ with an arbitrary state $\Phi_{q_1'\cdots q_M'}$. Using Eq. (4.2.1), this is

$$\left(\Phi_{q_1'\cdots q_M'},\, a(q)\Phi_{q_1\cdots q_N}\right) \equiv \left(a^\dagger(q)\Phi_{q_1'\cdots q_M'},\, \Phi_{q_1\cdots q_N}\right) = \left(\Phi_{qq_1'\cdots q_M'},\, \Phi_{q_1\cdots q_N}\right).$$

We now use Eq. (4.1.7). The sum over permutations $\mathscr{P}$ of $1,2,\cdots,N$ can be written as a sum over the integer r that is permuted into the first place, i.e. $\mathscr{P}r=1$, and over mappings $\bar{\mathscr{P}}$ of the remaining integers $1,\cdots,r-1,\,r+1,\cdots,N$ into $1,\cdots,N-1$. Furthermore, the sign factor is

$$\delta_{\mathscr{P}} = (\pm)^{r-1}\delta_{\bar{\mathscr{P}}}$$

with upper and lower signs for bosons and fermions, respectively. Hence, using Eq. (4.1.7) twice,

$$\begin{aligned}\left(\Phi_{q_1'\cdots q_M'},\, a(q)\Phi_{q_1\cdots q_N}\right) &= \delta_{N,M+1}\\ &\times \sum_{r=1}^{N}\sum_{\bar{\mathscr{P}}}(\pm)^{r-1}\delta_{\bar{\mathscr{P}}}\delta(q-q_r)\prod_{i=1}^{M}\delta(q_i'-q_{\mathscr{P}i})\\ &= \delta_{N,M+1}\sum_{r=1}^{N}(\pm)^{r-1}\delta(q-q_r)\left(\Phi_{q_1'\cdots q_M'},\, \Phi_{q_1\cdots q_{r-1}q_{r+1}\cdots q_N}\right).\end{aligned}$$

Both sides of Eq. (4.2.3) thus have the same matrix element with any state $\Phi_{q'_1\cdots q'_M}$, and are therefore equal, as was to be shown.) As a special case of Eq. (4.2.3), we note that for both bosons and fermions, $a(q)$ annihilates the vacuum

$$a(q)\Phi_0 = 0\,. \tag{4.2.4}$$

As defined here, the creation and annihilation operators satisfy an important commutation or anticommutation relation. Applying the operator $a(q')$ to Eq. (4.2.1) and using Eq. (4.2.3) gives

$$a(q')a^\dagger(q)\Phi_{q_1\cdots q_N} = \delta(q'-q)\Phi_{q_1\cdots q_N} + \sum_{r=1}^{N}(\pm)^{r+2}\delta(q'-q_r)\Phi_{qq_1\cdots q_{r-1}q_{r+1}\cdots q_N}\,.$$

(The sign in the second term is $(\pm)^{r+2}$ because q_r is in the $(r+1)$-th place in $\Phi_{qq_1\cdots q_N}$.) On the other hand, applying the operator $a^\dagger(q)$ to Eq. (4.2.3) gives

$$a^\dagger(q)a(q')\Phi_{q_1\cdots q_N} = \sum_{r=1}^{N}(\pm)^{r+1}\delta(q'-q_r)\Phi_{qq_1\cdots q_{r-1}q_{r+1}\cdots q_N}\,.$$

Subtracting or adding, we have then

$$\Big[a(q')a^\dagger(q) \mp a^\dagger(q)a(q')\Big]\;\Phi_{q_1\cdots q_N} = \delta(q'-q)\Phi_{q_1\cdots q_N}\,.$$

But this holds for all states $\Phi_{q_1\cdots q_N}$ (and may easily be seen to hold also for states containing both bosons and fermions) and therefore implies the operator relation

$$a(q')a^\dagger(q) \mp a^\dagger(q)a(q') = \delta(q'-q)\,. \tag{4.2.5}$$

In addition, Eq. (4.2.2) gives immediately

$$a^\dagger(q')a^\dagger(q) \mp a^\dagger(q)a^\dagger(q') = 0 \tag{4.2.6}$$

and so also

$$a(q')a(q) \mp a(q)a(q') = 0\,. \tag{4.2.7}$$

As always, the top and bottom signs apply for bosons and fermions, respectively. According to the conventions discussed in the previous section, the creation and/or annihilation operators for particles of two different species commute if either particle is a boson, and anticommute if both are fermions.

The above discussion could have been presented in reverse order (and in most textbooks usually is). That is, we could have started with the commutation or anticommutation relations Eqs. (4.2.5)–(4.2.7), derived from the canonical quantization of some given field theory. Multi-particle states would have then been defined by Eq. (4.2.2), and their scalar products Eq. (4.1.7) derived from the commutation or anticommutation relations. In fact, as discussed in Chapter 1, such a treatment would be much closer to the way that this formalism developed historically. We have followed an unhistorical approach here because we want to free ourselves from any dependence on pre-existing field theories, and rather wish to understand why field theories are the way they are.

We will now prove the fundamental theorem quoted at the beginning of this chapter: *any* operator $\mathscr{O}$ may be expressed as a sum of products of creation and annihilation operators

$$\mathscr{O} = \sum_{N=0}^{\infty} \sum_{M=0}^{\infty} \int dq'_1 \cdots dq'_N dq_1 \cdots dq_M \\ \times a^\dagger(q'_1) \cdots a^\dagger(q'_N) a(q_M) \cdots a(q_1) \\ \times C_{NM}(q'_1 \cdots q'_N q_1 \cdots q_M) \,. \tag{4.2.8}$$

That is, we want to show that the C_{NM} coefficients can be chosen to give the matrix elements of this expression any desired values. We do this by mathematical induction. First, it is trivial that by choosing C_{00} properly, we can give $(\Phi_0, \mathscr{O}\Phi_0)$ any desired value, irrespective of the values of C_{NM} with $N > 0$ and/or $M > 0$. We need only use Eq. (4.2.4) to see that Eq. (4.2.8) has the vacuum expectation value

$$(\Phi_0, \mathscr{O}\Phi_0) = C_{00} \,.$$

Now suppose that the same is true for all matrix elements of $\mathscr{O}$ between N- and M-particle states, with $N < L, M \le K$ or $N \le L, M < K$; that is, that these matrix elements have been given some desired values by an appropriate choice of the corresponding coefficients C_{NM}. To see that the same is then also true of matrix elements of $\mathscr{O}$ between any L- and K-particle states, use Eq. (4.2.8) to evaluate

$$\left(\Phi_{q'_1 \cdots q'_L}, \mathscr{O}\Phi_{q_1 \cdots q_K}\right) = L!K!\,C_{LK}(q'_1 \cdots q'_L q_1 \cdots q_K) \\ + \text{terms involving } C_{NM} \text{ with } N < L, M \le K \text{ or } N \le L, M < K \,.$$

Whatever values have already been given to C_{NM} with $N < L, M \le K$ or $N \le L, M < K$, there is clearly some choice of C_{LK} which gives this matrix element any desired value.

Of course, an operator need not be expressed in the form (4.2.8), with all creation operators to the left of all annihilation operators. (This is often called the 'normal' order of the operators.) However, if the formula for some operator has the creation and annihilation operators in some other order, we can always bring the creation operators to the left of the annihilation operators by repeated use of the commutation or anticommutation relations, picking up new terms from the delta function in Eq. (4.2.5).

For instance, consider any sort of additive operator F (like momentum, charge, etc.) for which

$$F\Phi_{q_1\cdots q_N} = \left(f(q_1) + \cdots + f(q_N)\right)\Phi_{q_1\cdots q_N} \,. \tag{4.2.9}$$

Such an operator can be written as in Eq. (4.2.8), but using only the term with $N = M = 1$:

$$F = \int dq\, a^\dagger(q) a(q) f(q) \,. \tag{4.2.10}$$

In particular, the free-particle Hamiltonian is always

$$H_0 = \int dq\, a^\dagger(q) a(q) E(q) \tag{4.2.11}$$

where $E(q)$ is the single-particle energy

$$E(\mathbf{p}, \sigma, n) = \sqrt{\mathbf{p}^2 + m_n^2} \,.$$

We will need the transformation properties of the creation and annihilation operators for various symmetries. First, let's consider inhomogeneous proper orthochronous Lorentz transformations. Recall that the N-particle states have the Lorentz transformation property

$$\begin{aligned} U_0(\Lambda, \alpha)\Phi_{\mathbf{p}_1\sigma_1 n_1, \mathbf{p}_2\sigma_2 n_2, \cdots} &= e^{-i(\Lambda p_1)\cdot\alpha}\, e^{-i(\Lambda p_2)\cdot\alpha} \\ &\cdots \sqrt{\frac{(\Lambda p_1)^0 (\Lambda p_2)^0 \cdots}{p_1^0 p_2^0 \cdots}} \\ &\times \sum_{\bar\sigma_1 \bar\sigma_2 \cdots} D^{(j_1)}_{\bar\sigma_1 \sigma_1}\Big(W(\Lambda, p_1)\Big) D^{(j_2)}_{\bar\sigma_2 \sigma_2}\Big(W(\Lambda, p_2)\Big) \\ &\cdots \Phi_{\mathbf{p}_{1\Lambda}\bar\sigma_1 n_1, \mathbf{p}_{2\Lambda}\bar\sigma_2 n_2, \cdots} \,. \end{aligned}$$

Here $\mathbf{p}_\Lambda$ is the three-vector part of Λp, $D^{(j)}_{\bar\sigma\sigma}(R)$ is the same unitary spin-j representation of the three-dimensional rotation group as used in Section 2.5, and $W(\Lambda, p)$ is the particular rotation

$$W(\Lambda, p) \equiv L^{-1}(\Lambda p) \Lambda L(p),$$

where $L(p)$ is the standard 'boost' that takes a particle of mass m from rest to four-momentum p^μ. (Of course, m and j depend on the species label n. This is all for $m \neq 0$; we will return to the massless particle case in the following chapter.) Now, these states can be expressed as in Eq. (4.2.2)

$$\Phi_{\mathbf{p}_1\sigma_1 n_1, \mathbf{p}_2\sigma_2 n_2, \cdots} = a^\dagger(\mathbf{p}_1\sigma_1 n_1) a^\dagger(\mathbf{p}_2\sigma_2 n_2) \cdots \Phi_0 \,,$$

where Φ_0 is the Lorentz-invariant vacuum state

$$U_0(\Lambda, \alpha)\Phi_0 = \Phi_0 \,.$$

In order that the state (4.2.2) should transform properly, it is necessary and sufficient that the creation operator have the transformation rule

$$U_0(\Lambda,\alpha)a^\dagger(\mathbf{p}\sigma n)U_0^{-1}(\Lambda,\alpha) = e^{-i(\Lambda p)\cdot\alpha}\sqrt{(\Lambda p)^0/p^0} \times \sum_{\bar\sigma} D^{(j)}_{\bar\sigma\sigma}\Big(W(\Lambda,p)\Big)a^\dagger(\mathbf{p}_\Lambda\,\bar\sigma\, n)\,. \quad (4.2.12)$$

In the same way, the operators C, P, and T, that induce charge-conjugation, space inversion, and time-reversal transformations on free particle states* transform the creation operators as:

$$\mathsf{C}a^\dagger(\mathbf{p}\,\sigma\, n)\mathsf{C}^{-1} = \xi_n\, a^\dagger(\mathbf{p}\,\sigma\, n^c)\,, \quad (4.2.13)$$

$$\mathsf{P}a^\dagger(\mathbf{p}\,\sigma\, n)\mathsf{P}^{-1} = \eta_n\, a^\dagger(-\mathbf{p}\,\sigma\, n)\,, \quad (4.2.14)$$

$$\mathsf{T}a^\dagger(\mathbf{p}\,\sigma\, n)\mathsf{T}^{-1} = \zeta_n(-1)^{j-\sigma}\, a^\dagger(-\mathbf{p}\;-\sigma\, n)\,, \quad (4.2.15)$$

As mentioned in the previous section, although we have been dealing with operators that create and annihilate particles in free-particle states, the whole formalism can be applied to 'in' and 'out' states, in which case we would introduce operators $a_{\rm in}$ and $a_{\rm out}$ defined in the same way by their action on these states. These operators satisfy a Lorentz transformation rule just like Eq. (4.2.12), but with the true Lorentz transformation operator $U(\Lambda,\alpha)$ instead of the free-particle operator $U_0(\Lambda,\alpha)$.

4.3 Cluster Decomposition and Connected Amplitudes

It is one of the fundamental principles of physics (indeed, of all science) that experiments that are sufficiently separated in space have unrelated results. The probabilities for various collisions measured at Fermilab should not depend on what sort of experiments are being done at CERN at the same time. If this principle were not valid, then we could never make any predictions about any experiment without knowing everything about the universe.

In S-matrix theory, the cluster decomposition principle states that if multi-particle processes $\alpha_1 \to \beta_1, \alpha_2 \to \beta_2, \cdots, \alpha_{\mathcal{N}} \to \beta_{\mathcal{N}}$ are studied in $\mathcal{N}$ very distant laboratories, then the S-matrix element for the overall process

* We omit the subscript '0' on these operators, because in virtually all cases where C, P, and/or T are conserved, the operators that induce these transformations on 'in' and 'out' states are the same as those defined by their action on free-particle states. This is not the case for continuous Lorentz transformations, for which it is necessary to distinguish between the operators $U(\Lambda,a)$ and $U_0(\Lambda,a)$.

factorizes. That is,*

$$S_{\beta_1+\beta_2+\cdots+\beta_{\mathcal{N}},\,\alpha_1+\alpha_2+\cdots+\alpha_{\mathcal{N}}} \to S_{\beta_1\alpha_1}\,S_{\beta_2\alpha_2}\cdots S_{\beta_{\mathcal{N}}\alpha_{\mathcal{N}}} \tag{4.3.1}$$

if for all $i \neq j$, *all* of the particles in states α_i and β_i are at a great spatial distance from *all* of the particles in states α_j and β_j. This factorization of S-matrix elements will ensure a factorization of the corresponding transition probabilities, corresponding to uncorrelated experimental results.

There is a combinatoric trick that allows us to rewrite Eq. (4.3.1) in a more transparent way. Suppose we define the *connected* part of the S-matrix, $S^{\rm C}_{\beta\alpha}$, by the formula**

$$S_{\beta\alpha} = \sum_{\rm PART} (\pm)\, S^{\rm C}_{\beta_1\alpha_1} S^{\rm C}_{\beta_2\alpha_2}\cdots\,. \tag{4.3.2}$$

Here the sum is over all different ways of partitioning the particles in the state α into clusters $\alpha_1, \alpha_2, \cdots$, and likewise a sum over all ways of partitioning the particles in the state β into clusters $\beta_1, \beta_2, \cdots$, not counting as different those that merely arrange particles within a given cluster or permute whole clusters. The sign is $+$ or $-$ according to whether the rearrangements $\alpha \to \alpha_1\alpha_2\cdots$ and $\beta \to \beta_1\beta_2\cdots$ involve altogether an even or an odd number of fermion interchanges, respectively. The term 'connected' is used because of the interpretation of $S^{\rm C}_{\beta\alpha}$ in terms of diagrams representing different contributions in perturbation theory, to be discussed in the next section.

This is a recursive definition. For each α and β, the sum on the right-hand side of Eq. (4.3.2) consists of a term $S^{\rm C}_{\beta\alpha}$, plus a sum Σ' over products of two or more $S^{\rm C}$-matrix elements, with a total number of particles in each of the states α_j and β_j that is *less* than the number of particles in

* We are here returning to the notation used in Chapter 3; Greek letters α or β stand for a collection of particles, including for each particle a specification of its momentum, spin, and species. Also, $\alpha_1 + \alpha_2 + \cdots + \alpha_{\mathcal{N}}$ is the state formed by combining all the particles in the states $\alpha_1, \alpha_2, \cdots$, and $\alpha_{\mathcal{N}}$, and likewise for $\beta_1 + \beta_2 + \cdots + \beta_{\mathcal{N}}$.

** This decomposition has been used in classical statistical mechanics by Ursell, Mayer, and others, and in quantum statistical mechanics by Lee and Yang and others.[3] It has also been used to calculate many-body ground state energies by Goldstone[4] and Hugenholtz.[5] In all of these applications the purpose of isolating the connected parts of Green's functions, partition functions, resolvents, etc., is to deal with objects with a simple volume dependence. This is essentially our purpose too, because as we shall see, the crucial property of the connected parts of the S-matrix is that they are proportional to a single momentum-conservation delta function, and in a box the delta function becomes a Kronecker delta times the volume. The cluster decomposition is also the same formal device as that used in the theory of noise[6] to decompose the correlation function of several random variables into its 'cumulants'; if the random variable receives contributions from a large number N of independent fluctuations, then each cumulant is proportional to N.

the states α and β

$$S_{\beta\alpha} = S^{\rm C}_{\beta\alpha} + \sum_{\rm PART}{}' (\pm)\, S^{\rm C}_{\beta_1\alpha_1} S^{\rm C}_{\beta_2\alpha_2} \cdots .$$

Suppose that the $S^{\rm C}$-matrix elements in this sum have already been chosen in such a way that Eq. (4.3.2) is satisfied for states β, α containing together fewer than, say, N particles. Then no matter what values are found in this way for the S-matrix elements appearing in the sum Σ', we can always choose the remaining term $S^{\rm C}_{\beta\alpha}$ so that Eq. (4.3.2) is also satisfied for states α, β containing a total of N particles.† Thus Eq. (4.3.2) contains no information in itself; it is merely a definition of $S^{\rm C}$.

If the states α and β each consist of just a single particle, say with quantum numbers q and q' respectively, then the only term on the right-hand-side of Eq. (4.3.2) is just $S^{\rm C}_{\beta\alpha}$ itself, so for one-particle states

$$S^{\rm C}_{q'q} \equiv S_{q'q} = \delta(q' - q) \,. \tag{4.3.3}$$

(Apart from possible degeneracies, the fact that $S_{q'q}$ is proportional to $\delta(q' - q)$ follows from conservation laws. The absence of any proportionality factor in Eq. (4.3.3) is based on a suitable choice of the relative phase of 'in' and 'out' states.) We are here assuming that single-particle states are stable, so that there are no transitions between single-particle states and any others, such as the vacuum.

For transitions between two-particle states, Eq. (4.3.2) reads

$$S_{q'_1 q'_2, q_1 q_2} = S^{\rm C}_{q'_1 q'_2, q_1 q_2} + \delta(q'_1 - q_1)\delta(q'_2 - q_2) \pm \delta(q'_1 - q_2)\delta(q'_2 - q_1) \,. \tag{4.3.4}$$

(We are here using Eq. (4.3.3).) The sign $\pm$ is $-$ if both particles are fermions, and otherwise $+$. We recognize that the two delta function terms just add up to the norm (4.1.6), so here $S^{\rm C}_{\beta\alpha}$ is just $(S-1)_{\beta\alpha}$. But the general case is more complicated.

For transitions between three-particle or four-particle states, Eq. (4.3.2) reads

$$\begin{aligned} S_{q'_1 q'_2 q'_3, q_1 q_2 q_3} &= S^{\rm C}_{q'_1 q'_2 q'_3, q_1 q_2 q_3} \\ &+ \delta(q'_1 - q_1) S^{\rm C}_{q'_2 q'_3, q_2 q_3} \pm \text{permutations} \\ &+ \delta(q'_1 - q_1)\delta(q'_2 - q_2)\delta(q'_3 - q_3) \pm \text{permutations} \end{aligned} \tag{4.3.5}$$

---

† A technicality should be mentioned here. This argument works only if we neglect the possibility that for one or more of the connected S-matrix elements in Eq. (4.3.2), the states α_j and β_j both contain no particles at all. We must therefore define the connected vacuum–vacuum element $S^{\rm C}_{0,0}$ to be zero. We do not use Eq. (4.3.2) for the vacuum–vacuum S-matrix $S_{0,0}$, which in the absence of time-varying external fields is simply defined to be unity, $S_{0,0} = 1$. We will have more to say about the vacuum–vacuum amplitude in the presence of external fields in Volume II.

and

$$\begin{aligned}
S_{q_1'q_2'q_3'q_4',q_1q_2q_3q_4} &= S^{\rm C}_{q_1'q_2'q_3'q_4',q_1q_2q_3q_4} \\
&+ S^{\rm C}_{q_1'q_2',q_1q_2}S^{\rm C}_{q_3'q_4',q_3q_4} \pm \text{permutations} \\
&+ \delta(q_1'-q_1)S^{\rm C}_{q_2'q_3'q_4',q_2q_3q_4} \pm \text{permutations} \\
&+ \delta(q_1'-q_1)\delta(q_2'-q_2)S^{\rm C}_{q_3'q_4',q_3q_4} \pm \text{permutations} \\
&+ \delta(q_1'-q_1)\delta(q_2'-q_2)\delta(q_3'-q_3)\delta(q_4'-q_4) \pm \text{permutations}\,. \quad (4.3.6)
\end{aligned}$$

(Taking account of all permutations, there are a total of $1+9+6=16$ terms in Eq. (4.3.5) and $1+18+16+72+24=131$ terms in Eq. (4.3.6). If we had not assumed that one-particle states are stable, there would be even more terms.) As explained previously, the definition of $S^{\rm C}_{\beta\alpha}$ is recursive: we use Eq. (4.3.4) to define $S^{\rm C}_{\beta\alpha}$ for two-particle states, then use this definition in Eq. (4.3.5) when we define $S^{\rm C}_{\beta\alpha}$ for three-particle states, then use both of these definitions in Eq. (4.3.6) to obtain the definition of $S^{\rm C}_{\beta\alpha}$ for four-particle states, and so on.

The point of this definition of the connected part of the S-matrix is that the cluster decomposition principle is equivalent to the requirement that $S^{\rm C}_{\beta\alpha}$ must vanish when any one or more of the particles in the states β and/or α are far away in space from the others.‡ To see this, suppose that the particles in the states β and α are grouped into clusters $\beta_1, \beta_2, \cdots$ and $\alpha_1, \alpha_2, \cdots$, and that all particles in the set $\alpha_i + \beta_i$ are far from all particles in the set $\alpha_j + \beta_j$ for any $j \neq i$. Then if $S^{\rm C}_{\beta'\alpha'}$ vanishes if any particles in β' or α' are far from the others, it vanishes if any particles in these states are in different clusters, so the definition (4.3.2) yields

$$S_{\beta\alpha} \to \sum{}^{(1)}(\pm)S^{\rm C}_{\beta_{11}\alpha_{11}}S^{\rm C}_{\beta_{12}\alpha_{12}}\cdots \times \sum{}^{(2)}(\pm)S^{\rm C}_{\beta_{21}\alpha_{21}}S^{\rm C}_{\beta_{22}\alpha_{22}}\cdots \times \cdots\,, \quad (4.3.7)$$

where $\Sigma^{(j)}$ is a sum over all different ways of partitioning the clusters β_j and α_j into subclusters $\beta_{j1}, \beta_{j2}, \cdots$ and $\alpha_{j1}, \alpha_{j2}, \cdots$. But referring back to Eq. (4.3.2), this is just the desired factorization property (4.3.1).

For instance, suppose that in the four-particle reaction $1234 \to 1'2'3'4'$, we let particles $1, 2, 1'$, and $2'$ be very far from $3, 4, 3'$, and $4'$. Then if $S^{\rm C}_{\beta\alpha}$ vanishes when any particles in β and/or α are far from the others, the only terms in Eq. (4.3.6) that survive (in an even more abbreviated notation) are

‡ In order to give a meaning to 'far', we will have to Fourier transform $S^{\rm C}$, so that each three-momentum label $\mathbf{p}$ is replaced with a spatial coordinate three-vector $\mathbf{x}$.

$$\begin{aligned} S_{1'2'3'4',1234} \to & S^{\rm C}_{1'2',12}S^{\rm C}_{3'4',34} \\ & + (\delta_{1'1}\delta_{2'2} \pm \delta_{1'2}\delta_{2'1})S^{\rm C}_{3'4',34} \\ & + (\delta_{3'3}\delta_{4'4} \pm \delta_{3'4}\delta_{4'3})S^{\rm C}_{1'2',12} \\ & + (\delta_{1'1}\delta_{2'2} \pm \delta_{1'2}\delta_{2'1})(\delta_{3'3}\delta_{4'4} \pm \delta_{3'4}\delta_{4'3}) . \end{aligned}$$

Comparison with Eq. (4.3.4) shows that this is just the required factorization condition (4.3.1)

$$S_{1'2'3'4',1234} \to S_{1'2',12}S_{3'4',34} .$$

We have formulated the cluster decomposition principle in coordinate space, as the condition that $S^{\rm C}_{\beta\alpha}$ vanishes if any particles in the states β or α are far from any others. It is convenient for us to reexpress this in momentum space. The coordinate space matrix elements are defined as a Fourier transform

$$\begin{aligned} S^{\rm C}_{{\bf x}'_1{\bf x}'_2\cdots,{\bf x}_1{\bf x}_2\cdots} \equiv & \int d^3{\bf p}'_1 d^3{\bf p}'_2 \cdots d^3{\bf p}_1 d^3{\bf p}_2 \cdots \; S^{\rm C}_{{\bf p}'_1{\bf p}'_2\cdots{\bf p}_1{\bf p}_2\cdots} \\ & \times e^{i{\bf p}'_1\cdot{\bf x}'_1} e^{i{\bf p}'_2\cdot{\bf x}'_2} \cdots e^{-i{\bf p}_1\cdot{\bf x}_1} e^{-i{\bf p}_2\cdot{\bf x}_2 \cdots} . \end{aligned} \tag{4.3.8}$$

(We are here temporarily dropping spin and species labels, which just go along with the momentum or coordinate labels.) If $|S^{\rm C}_{{\bf p}'_1{\bf p}'_2\cdots{\bf p}_1{\bf p}_2\cdots}|$ were sufficiently well behaved (to be specific, if it were Lebesgue integrable) then according to the Riemann–Lebesgue theorem[7] the integral (4.3.8) would vanish when any combination of spatial coordinates goes to infinity. Now, this is certainly too strong a requirement. Translational invariance tells us that the connected part of the S-matrix, like the S-matrix itself, can only depend on differences of coordinate vectors, and therefore does not change at all if all of the x_i and x'_j vary together, with their differences held constant. This requires that the elements of $S^{\rm C}$ in a momentum basis must, like those of S, be proportional to a three-dimensional delta function that ensures momentum conservation (and makes $|S^{\rm C}_{{\bf p}'_1{\bf p}'_2\cdots{\bf p}_1{\bf p}_2\cdots}|$ *not* Lebesgue integrable), as well as the energy-conservation delta function required by scattering theory. That is, we can write

$$\begin{aligned} S^{\rm C}_{{\bf p}'_1{\bf p}'_2\cdots,{\bf p}_1{\bf p}_2\cdots} = & \, \delta^3({\bf p}'_1 + {\bf p}'_2 + \cdots - {\bf p}_1 - {\bf p}_2 - \cdots) \\ & \times \delta(E'_1 + E'_2 + \cdots - E_1 - E_2 - \cdots) C_{{\bf p}'_1{\bf p}'_2\cdots,{\bf p}_1{\bf p}_2\cdots} . \end{aligned} \tag{4.3.9}$$

This is no problem: the cluster decomposition principle only requires that Eq. (4.3.8) vanish when the *differences* among some of the ${\bf x}_i$ and/or ${\bf x}'_i$ become large. However, if C itself in Eq. (4.3.9) contained additional delta functions of linear combinations of the three-momenta, then this principle would not be satisfied. For instance, suppose that there were a delta function in C that required that the sum of the ${\bf p}'_i$ and $-{\bf p}_j$ for some subset of the particles vanished. Then Eq. (4.3.8) would not vary if all of the ${\bf x}'_i$ and

$\mathbf{x}_j$ for the particles in that subset moved together (with constant differences) away from all the other $\mathbf{x}'_k$ and $\mathbf{x}_\ell$, in contradiction to the cluster decomposition principle. Loosely speaking then, the cluster decomposition principle simply says that *the connected part of the S-matrix, unlike the S-matrix itself, contains just a single momentum-conservation delta function.*

In order to put this a bit more precisely, we can say that the coefficient function $C_{\mathbf{p}'_1\mathbf{p}'_2\cdots,\mathbf{p}_1\mathbf{p}_2\cdots}$ in Eq. (4.3.9) is a smooth function of its momentum labels. But how smooth? It would be most straightforward if we could simply require that $C_{\mathbf{p}'_1\mathbf{p}'_2\cdots,\mathbf{p}_1\mathbf{p}_2\cdots}$ be analytic in all of the momenta at $\mathbf{p}'_1 = \mathbf{p}'_2 = \cdots = \mathbf{p}_1 = \mathbf{p}_2 = \cdots = 0$. This requirement would indeed guarantee that $S^C_{\mathbf{x}'_1\mathbf{x}'_2\cdots,\mathbf{x}_1\mathbf{x}_2\cdots}$ vanishes exponentially fast when any of the $\mathbf{x}$ and $\mathbf{x}'$ is very distant from any of the other $\mathbf{x}$ and $\mathbf{x}'$. However, an exponential fall-off of S^C is not an essential part of the cluster decomposition principle, and, in fact, the requirement of analyticity is not met in all theories. Most notably, in theories with massless particles, S^C can have poles at certain values of the $\mathbf{p}$ and $\mathbf{p}'$. For instance, as we will see in Chapter 10, if a massless particle can be emitted in the transition $1 \to 3$ and absorbed in the transition $2 \to 4$, then $S^C_{34,12}$ will have a term proportional to $1/(p_1 - p_3)^2$. After Fourier transforming, such poles yield terms in $S^C_{\mathbf{x}'_1\mathbf{x}'_2\cdots,\mathbf{x}_1\mathbf{x}_2\cdots}$ that fall off only as negative powers of coordinate differences.[1] There is no need to formulate the cluster decomposition principle so stringently that such behavior is ruled out. Thus the 'smoothness' condition on S^C should be understood to allow various poles and branch-cuts at certain values of the $\mathbf{p}$ and $\mathbf{p}'$, but not singularities as severe as delta functions.

4.4 Structure of the Interaction

We now ask, what sort of Hamiltonian will yield an S-matrix that satisfies the cluster decomposition principle? It is here that the formalism of creation and annihilation operators comes into its own. The answer is contained in the theorem that the S-matrix satisfies the cluster decomposition principle if (and as far as I know, only if) the Hamiltonian can be expressed as in Eq. (4.2.8):

$$H = \sum_{N=0}^{\infty} \sum_{M=0}^{\infty} \int dq'_1 \cdots dq'_N \, dq_1 \cdots dq_M \\ \times a^\dagger(q'_1) \cdots a^\dagger(q'_N) a(q_M) \cdots a(q_1) \\ \times h_{NM}(q'_1 \cdots q'_N \,,\, q_1 \cdots q_M) \tag{4.4.1}$$

with coefficient functions h_{NM} that contain just a *single* three-dimensional momentum-conservation delta function (returning here briefly to a more explicit notation)

$$h_{NM}(\mathbf{p}'_1\sigma'_1n'_1\cdots\mathbf{p}'_N\sigma'_Nn'_N\,,\,\mathbf{p}_1\sigma_1n_1\cdots\mathbf{p}_M\sigma_Mn_M)$$
$$=\delta^3(\mathbf{p}'_1+\cdots+\mathbf{p}'_N-\mathbf{p}_1-\cdots-\mathbf{p}_M)$$
$$\times\,\tilde{h}_{NM}(\mathbf{p}'_1\sigma'_1n'_1\cdots\mathbf{p}'_N\sigma'_Nn'_N\,,\,\mathbf{p}_1\sigma_1n_1\cdots\mathbf{p}_M\sigma_Mn_M)\,, \qquad (4.4.2)$$

where $\tilde{h}_{NM}$ contains no delta function factors. Note that Eq. (4.4.1) by itself has no content — we saw in Section 4.2 that *any* operator can be put in this form. It is only Eq. (4.4.1) combined with the requirement that h_{NM} has only the single delta function shown in Eq. (4.4.2) that guarantees that the S-matrix satisfies the cluster decomposition principle.

The validity of this theorem in perturbation theory will become obvious when we develop the Feynman diagram formalism in Chapter 6. The trusting reader may prefer to skip the rest of the present chapter, and move on to consider the implications of this theorem in Chapter 5. However, the proof has some instructive features, and will help to clarify in what sense the field theory of the next chapter is inevitable.

To prove this theorem, we make use of perturbation theory in its time-dependent form. (One of the advantages of time-dependent perturbation theory is that it makes the combinatorics underlying the cluster decomposition principle much more transparent; if E is a sum of one-particle energies then e^{-iEt} is a product of functions of the individual energies, while $[E-E_\alpha+i\epsilon]^{-1}$ is not.) The S-matrix is given by Eq. (3.5.10) as*

$$S_{\beta\alpha}=\sum_{n=0}^{\infty}\frac{(-i)^n}{n!}\int_{-\infty}^{\infty}dt_1\cdots dt_n\Big(\Phi_\beta,T\Big\{V(t_1)\cdots V(t_n)\Big\}\Phi_\alpha\Big)\,, \qquad (4.4.3)$$

where the Hamiltonian is split into a free-particle part H_0 and an interaction V, and

$$V(t)\equiv\exp(iH_0t)V\exp(-iH_0t)\,. \qquad (4.4.4)$$

Now, the states Φ_α and Φ_β may be expressed as in Eq. (4.2.2) as products of creation operators acting on the vacuum Φ_0, and $V(t)$ is itself a sum of products of creation and annihilation operators, so each term in the sum (4.4.3) may be written as a sum of vacuum expectation values of products of creation and annihilation operators. By using the commutation or anticommutation relations (4.2.5) we may move each annihilation operator in turn to the right past all the creation operators. For each annihilation operator moved to the right past a creation operator we have two terms, as shown by writing Eq. (4.2.5) in the form

$$a(q')a^\dagger(q)=\pm a^\dagger(q)a(q')+\delta(q'-q)\,.$$

---

* We are now adopting the convention that for $n=0$, the time-ordered product in Eq. (4.4.3) is taken as the unit operator, so the $n=0$ term in the sum just yields the term $\delta(\beta-\alpha)$ in $S_{\beta\alpha}$.

Moving other creation operators past the annihilation operator in the first term generates yet more terms. But Eq. (4.2.4) shows that any annihilation operator that moves all the way to the right and acts on Φ_0 gives zero, so in the end all we have left is the delta functions. In this way, the vacuum expectation value of a product of creation and annihilation operators is given by a sum of different terms, each term equal to a product of delta functions and $\pm$ signs from the commutators or anticommutators. It follows that each term in Eq. (4.4.3) may be expressed as a sum of terms, each term equal to a product of delta functions and $\pm$ signs from the commutators or anticommutators and whatever factors are contributed by $V(t)$, integrated over all the times and integrated and summed over the momenta, spins, and species in the arguments of the delta functions.

Each of the terms generated in this way may be symbolized by a diagram. (This is not yet the full Feynman diagram formalism, because we are not yet going to associate numerical quantities with the ingredients in the diagrams; we are using the diagrams here only as a way of keeping track of three-momentum delta functions.) Draw n points, called *vertices*, one for each $V(t)$ operator. For each delta function produced when an annihilation operator in one of these $V(t)$ operators moves past a creation operator in the initial state Φ_α, draw a line coming into the diagram from below that ends at the corresponding vertex. For each delta function produced when an annihilation operator in the adjoint of the final state Φ_β moves past a creation operator in one of the $V(t)$, draw a line from the corresponding vertex upwards out of the diagram. For each delta function produced when an annihilation operator in one $V(t)$ moves past a creation operator in another $V(t)$ draw a line between the two corresponding vertices. Finally, for each delta function produced when an annihilation operator in the adjoint of the final state moves past a creation operator in the initial state, draw a line from bottom to top, right through the diagrams. Each of the delta functions associated with one of these lines enforces the equality of the momentum arguments of the pair of creation and annihilation operators represented by the line. There is also at least one delta function contributed by each of the vertices, which enforces the conservation of the total three-momentum at the vertex.

Such a diagram may be connected (every point connected to every other by a set of lines) and if not connected, it breaks up into a number of connected pieces. The $V(t)$ operator associated with a vertex in one connected component effectively commutes with the $V(t)$ associated with any vertex in any other connected component, because for this diagram, we are not including any terms in which an annihilation operator in one vertex destroys a particle that is produced by a creation operator in the other vertex — if we did, the two vertices would be in the same connected component. Thus the matrix element in Eq. (4.4.3) can be expressed as a sum over *products* of contributions, one from each connected component:

$$\left(\Phi_\beta, T\left\{V(t_1)\cdots V(t_n)\right\}\Phi_\alpha\right)$$
$$= \sum_{\text{clusterings}} (\pm) \prod_{j=1}^{\nu} \left(\Phi_{\beta_j}, T\left\{V(t_{j1})\cdots V(t_{jn_j})\right\}\Phi_{\alpha_j}\right)_{\mathrm{C}} . \tag{4.4.5}$$

Here the sum is over all ways of splitting up the incoming and outgoing particles and $V(t)$ operators into ν clusters (including a sum over ν from 1 to n) with the n_j operators $V(t_{j_1})\cdots V(t_{jn_j})$ and the subsets of initial particles α_j and final particles β_j all in the jth cluster. Of course, this means that

$$n = n_1 + \cdots + n_\nu$$

and also the set α is the union of all the particles in the subsets $\alpha_1, \alpha_2, \cdots \alpha_\nu$, and likewise for the final state. Some of the clusters in Eq. (4.4.5) may contain no vertices at all, i.e., $n_j = 0$; for these factors, we must take the matrix element factor in Eq. (4.4.5) to vanish unless β_j and α_j are both one-particle states (in which case it is just a delta function $\delta(\alpha_j - \beta_j)$), because the only connected diagrams without vertices consist of a single line running through the diagram from bottom to top. Most important, the subscript C in Eq. (4.4.5) means that we exclude any contributions corresponding to disconnected diagrams, that is, any contributions in which any $V(t)$ operator or any initial or final particle is not connected to every other by a sequence of particle creations and annihilations.

Now let us use Eq. (4.4.5) in the sum (4.4.3). Every time variable is integrated from $-\infty$ to $+\infty$, so it makes no difference which of the $t_1, \cdots t_n$ are sorted out into each cluster. The sum over clusterings therefore yields a factor $n!/n_1!n_2!\cdots n_\nu!$, equal to the number of ways of sorting out n vertices into ν clusters, each containing $n_1, n_2, \cdots$ vertices:

$$\int_{-\infty}^{\infty} dt_1 \cdots dt_n \left(\Phi_\beta,\ T\left\{V(t_1)\cdots V(t_n)\right\}\Phi_\alpha\right)$$
$$= \sum_{\text{PART}} (\pm) \sum_{\substack{n_1\cdots n_\nu \\ n_1+\cdots+n_\nu=n}} \frac{n!}{n_1!n_2!\cdots n_\nu!} \prod_{j=1}^{\nu} \int_{-\infty}^{\infty} dt_{j_1}\cdots dt_{jn_j}$$
$$\times \quad \left(\Phi_{\beta_j}, T\left\{V(t_{j_1})\cdots V(t_{jn_j})\right\}\Phi_{\alpha_j}\right)_{\mathrm{C}} .$$

The first sum here is over all ways of partitioning the particles in the initial and final states into clusters $\alpha_1\cdots\alpha_\nu$ and $\beta_1\cdots\beta_\nu$ (including a sum over the number ν of clusters). The factor $n!$ here cancels the $1/n!$ in Eq. (4.4.3), and the factor $(-i)^n$ in the perturbation series for (4.4.5) can be written as a product $(-i)^{n_1}\cdots(-i)^{n_\nu}$, so instead of summing over n and then summing separately over $n_1, \cdots n_\nu$ constrained by $n_1 + \cdots + n_\nu = n$, we can simply sum independently over each $n_1, \cdots n_\nu$. This gives finally

$$S_{\beta\alpha} = \sum_{\text{PART}} (\pm) \prod_{j=1}^{\nu} \sum_{n_j=0}^{\infty} \frac{(-i)^{n_j}}{n_j!} \int_{-\infty}^{\infty} dt_{j_1} \cdots dt_{jn_j}$$
$$\times \left(\Phi_{\beta_j}, T\left\{V(t_{j_1}) \cdots V(t_{jn_j})\right\}\Phi_{\alpha_j}\right)_{\text{C}} .$$

Comparing this with the definition (4.3.2) of the connected matrix elements $S^{\text{C}}_{\beta\alpha}$, we see that these matrix elements are just given by the factors in the product here

$$S^{\text{C}}_{\beta\alpha} = \sum_{n=0}^{\infty} \frac{(-i)^n}{n!} \int_{-\infty}^{\infty} dt_1 \cdots dt_n \left(\Phi_\beta, T\left\{V(t_1) \cdots V(t_n)\right\}\Phi_\alpha\right)_{\text{C}} . \tag{4.4.6}$$

(The subscript j is dropped on all the ts and ns, as these are now mere integration and summation variables.) We see that $S^{\text{C}}_{\beta\alpha}$ is calculated by a very simple prescription: *$S^{\text{C}}_{\beta\alpha}$ is the sum of all contributions to the S-matrix that are connected, in the sense that we drop all terms in which any initial or final particle or any operator $V(t)$ is not connected to all the others by a sequence of particle creations and annihilations.* This justifies the adjective 'connected' for S^{C}.

As we have seen, momentum is conserved at each vertex and along every line, so the connected parts of the S-matrix individually conserve momentum: $S^{\text{C}}_{\beta\alpha}$ contains a factor $\delta^3(\mathbf{p}_\beta - \mathbf{p}_\alpha)$. What we want to prove is that $S^{\text{C}}_{\beta\alpha}$ contains no other delta functions.

We now make the assumption that the coefficient fractions h_{NM} in the expansion (4.4.1) of the Hamiltonian in terms of creation and annihilation operators are proportional to a *single* three-dimensional delta function, that ensures momenta conservation. This is automatically true for the free-particle Hamiltonian H_0, so it is also then true separately for the interaction V. Returning to the graphical interpretation of the matrix elements that we have been using, this means that each vertex contributes one three-dimensional delta function. (The other delta functions in matrix elements $V_{\gamma\delta}$ simply keep the momentum of any particle that is not created or annihilated at the corresponding vertex unchanged.) Now, most of these delta functions simply go to fix the momentum of intermediate particles. The only momenta that are left unfixed by such delta functions are those that circulate in loops of internal lines. (Any line which if cut leaves the diagram disconnected carries a momentum that is fixed by momentum conservation as some linear combination of the momenta of the lines coming into or going out of the diagram. If the diagram has L lines that can all be cut at the same time without the diagram becoming disconnected, then we say it has L independent loops, and there are L

momenta that are not fixed by momentum conservation.) With V vertices, I internal lines, and L loops, there are V delta functions, of which $I-L$ go to fix internal momenta, leaving $V-I+L$ delta functions relating the momenta of incoming or outgoing particles. But a well-known topological identity** tells that for any graph consisting of C connected pieces, the numbers of vertices, internal lines, and loops are related by

$$V-I+L=C\,. \tag{4.4.7}$$

Hence for a connected matrix element like $S^{C}_{\beta\alpha}$, which arises from graphs with $C=1$, we find just a single three-dimensional delta function $\delta^3(\mathbf{p}_\beta-\mathbf{p}_\alpha)$, as was to be proved.

It was not important in the above argument that the time variables were integrated from $-\infty$ to $+\infty$. Thus exactly the same arguments can be used to show that if the coefficients $h_{N,M}$ in the Hamiltonian contain just single delta functions, then $U(t,t_0)$ can also be decomposed into connected parts, each containing a single momentum-conservation delta function factor. On the other hand, the connected part of the S-matrix also contains an energy-conservation delta function, and when we come to Feynman diagrams in Chapter 6 we shall see that $S^{C}_{\beta\alpha}$ contains only a single energy-conservation delta function, $\delta(E_\beta-E_\alpha)$, while $U(t,t_0)$ contains no energy-conservation delta functions at all.

It should be emphasized that the requirement that h_{NM} in Eq. (4.4.1) should have only a single three-dimensional momentum conservation delta function factor is very far from trivial, and has far-reaching implications. For instance, assume that V has non-vanishing matrix elements between two-particle states. Then Eq. (4.4.1) must contain a term with $N=M=2$, and coefficient

$$v_{2,2}(\mathbf{p}'_1\mathbf{p}'_2,\mathbf{p}_1\,\mathbf{p}_2)=V_{\mathbf{p}'_1\mathbf{p}'_2,\mathbf{p}_1\,\mathbf{p}_2}\,. \tag{4.4.8}$$

(We are here temporarily dropping spin and species labels.) But then the matrix element of the interaction between three-particle states is

$$\begin{aligned}V_{\mathbf{p}'_1\mathbf{p}'_2\mathbf{p}'_3,\mathbf{p}_1\,\mathbf{p}_2\,\mathbf{p}_3} &= v_{3,3}(\mathbf{p}'_1\mathbf{p}'_2\mathbf{p}'_3,\mathbf{p}_1\,\mathbf{p}_2\,\mathbf{p}_3)\\ &\quad+v_{2,2}(\mathbf{p}'_1\mathbf{p}'_2,\mathbf{p}_1\mathbf{p}_2)\,\delta^3(\mathbf{p}'_3-\mathbf{p}_3)\pm\text{permutations.}\end{aligned} \tag{4.4.9}$$

** A graph consisting of a single vertex has $V=1$, $L=0$, and $C=1$. If we add $V-1$ vertices with just enough internal lines to keep the graph connected, we have $I=V-1$, $L=0$, and $C=1$. Any additional internal lines attached (without new vertices) to the same connected graph produce an equal number of loops, so $I=V+L-1$ and $C=1$. If a disconnected graph consists of C such connected parts, the sums of I, V, and L in each connected part will than satisfy $\sum I=\sum V+\sum L-C$.

As mentioned at the beginning of this chapter, we might try to make a relativistic quantum theory that is not a field theory by choosing $v_{2,2}$ so that the two–body S-matrix is Lorentz-invariant, and adjusting the rest of the Hamiltonian so that there is no scattering in states containing three or more particles. We would then have to take $v_{3,3}$ to cancel the other terms in Eq. (4.4.9)

$$v_{3,3}(\mathbf{p}'_1\mathbf{p}'_2\mathbf{p}'_3, \mathbf{p}_1\,\mathbf{p}_2\,\mathbf{p}_3) = -v_{2,2}(\mathbf{p}'_1\mathbf{p}'_2, \mathbf{p}_1\mathbf{p}_2)\,\delta^3(\mathbf{p}'_3 - \mathbf{p}_3) \mp \text{permutations.} \tag{4.4.10}$$

However, this would mean that each term in $v_{3,3}$ contains *two* delta function factors (recall that $v_{2,2}(\mathbf{p}'_1\mathbf{p}'_2, \mathbf{p}_1\mathbf{p}_2)$ has a factor $\delta^3(\mathbf{p}'_1 + \mathbf{p}'_2 - \mathbf{p}_1 - \mathbf{p}_2)$) and this would violate the cluster decomposition principle. Thus in a theory satisfying the cluster decomposition principle, the existence of scattering processes involving two particles makes processes involving three or more particles inevitable.

* * *

When we set out to solve three-body problems in quantum theories that satisfy the cluster decomposition principle, the term $v_{3,3}$ in Eq. (4.4.9) gives no particular trouble, but the extra delta function in the other terms makes the Lippmann–Schwinger equation difficult to solve directly. The problem is that these delta functions make the kernel $[E_\alpha - E_\beta + i\epsilon]^{-1} V_{\beta\alpha}$ of this equation not square-integrable, even after we factor out an overall momentum conservation delta function. In consequence, it cannot be approximated by a finite matrix, even one of very large rank. To solve problems involving three or more particles, it is necessary to replace the Lippmann–Schwinger equation with one that has a connected right-hand side. Such equations have been developed for the scattering of three or more particles,[8,9] and in non-relativistic scattering problems they can be solved recursively, but they have not turned out to be useful in relativistic theories and so will not be described in detail here.

However, recasting the Lippmann–Schwinger equation in this manner is useful in another way. Our arguments in this section have so far relied on perturbation theory. I do not know of any non-perturbative proof of the main theorem of this section, but it has been shown[9] that these reformulated non-perturbative dynamical equations are *consistent* with the requirement that $U^C(t, t_0)$ (and hence S^C) should also contain only a single momentum-conservation delta function, as required by the cluster decomposition principle, provided that the Hamiltonian satisfies our condition that the coefficient functions $h_{N,M}$ each contain only a single momentum-conservation delta function.

Problems

1. Define generating functionals for the S-matrix and its connected part:

$$F[v] \equiv 1 + \sum_{N=1}^{\infty}\sum_{M=1}^{\infty} \frac{1}{N!M!} \int v^*(q_1') \cdots v^*(q_N') v(q_1) \cdots v(q_M)$$
$$\times\, S_{q_1' \cdots q_N', q_1 \cdots q_M}\, dq_1' \cdots dq_N'\, dq_1 \cdots dq_M$$
$$F^C[v] \equiv \sum_{N=1}^{\infty}\sum_{M=1}^{\infty} \frac{1}{N!M!} \int v^*(q_1') \cdots v^*(q_N') v(q_1) \cdots v(q_M)$$
$$\times\, S^C_{q_1' \cdots q_N', q_1 \cdots q_M}\, dq_1' \cdots dq_N'\, dq_1 \cdots dq_M .$$

Derive a formula relating $F[v]$ and $F^C[v]$. (You may consider the purely bosonic case.)

2. Consider an interaction

$$V = g \int d^3\mathbf{p}_1\, d^3\mathbf{p}_2\, d^3\mathbf{p}_3\, d^3\mathbf{p}_4\, \delta^3(\mathbf{p}_1 + \mathbf{p}_2 - \mathbf{p}_3 - \mathbf{p}_4)$$
$$\times\; a^\dagger(\mathbf{p}_1)\, a^\dagger(\mathbf{p}_2)\, a(\mathbf{p}_3)\, a(\mathbf{p}_4)\,,$$

where g is a real constant and $a(\mathbf{p})$ is the annihilation operator of a spinless boson of mass $M > 0$. Use perturbation theory to calculate the S-matrix element for scattering of these particles in the center-of-mass frame to order g. What is the corresponding differential cross-section?

3. A *coherent state* Φ_λ is defined to be an eigenstate of the annihilation operators $a(q)$ with eigenvalues $\lambda(q)$. Construct such a state as a superposition of the multi-particle states $\Phi_{q_1 q_2 \cdots q_N}$.

References

1. The cluster decomposition principle seems to have been first stated explicitly in quantum field theory by E. H. Wichmann and J. H. Crichton, *Phys. Rev.* **132**, 2788 (1963).

2. See, e.g., B. Bakamijian and L. H. Thomas, *Phys. Rev.* **92**, 1300 (1953).

3. For references, see T. D. Lee and C. N. Yang, *Phys. Rev.* **113**, 1165 (1959).

4. J. Goldstone, *Proc. Roy. Soc. London* **A239**, 267 (1957)

5. N. M. Hugenholtz, *Physica* **23**, 481 (1957).

6. See, e.g., R. Kubo, *J. Math. Phys.* **4**, 174 (1963).

7. E. C. Titchmarsh, *Introduction to the Theory of Fourier Integrals* (Oxford University Press, Oxford, 1937): Section 1.8.

8. L. D. Faddeev. *Zh. Ekxper. i Teor. Fiz.* **39**. 1459 (1961) (translation: *Soviet Phys — JETP* **12**, 1014 (1961)); *Dokl. Akad. Nauk. SSSR* **138**, 565 (1961) and **145**, 30 (1962) (translations *Soviet Physics — Doklady* **6**, 384 (1961) and **7**, 600 (1963)).

9. S. Weinberg, *Phys. Rev.* **133**, B232 (1964)

5

Quantum Fields and Antiparticles

We now have all the pieces needed to motivate the introduction of quantum fields.[1] In the course of this construction, we shall encounter some of the most remarkable and universal consequences of the union of relativity with quantum mechanics: the connection between spin and statistics, the existence of antiparticles, and various relationships between particles and antiparticles, including the celebrated CPT theorem.

5.1 Free Fields

We have seen in Chapter 3 that the S-matrix will be Lorentz-invariant if the interaction can be written as

$$V(t) = \int d^3x \, \mathscr{H}(\mathbf{x}, t) \,, \tag{5.1.1}$$

where $\mathscr{H}$ is a scalar, in the sense that

$$U_0(\Lambda, a)\mathscr{H}(x)U_0^{-1}(\Lambda, a) = \mathscr{H}(\Lambda x + a) \,, \tag{5.1.2}$$

and satisfies the additional condition:

$$[\mathscr{H}(x), \mathscr{H}(x')] = 0 \quad \text{for} \quad (x - x')^2 \geq 0 \,. \tag{5.1.3}$$

As we shall see, there are more general possibilities, but none of them are very different from this. (For the present we are leaving it as an open question whether Λ here is restricted to a proper orthochronous Lorentz transformation, or can also include space inversions.) In order to facilitate also satisfying the cluster decomposition principle we are going to construct $\mathscr{H}(x)$ out of creation and annihilation operators, but here we face a problem: as shown by Eq. (4.2.12), under Lorentz transformations each such operator is multiplied by a matrix that depends on the momentum carried by that operator. How can we couple such operators together to make a scalar? The solution is to build $\mathscr{H}(x)$ out of *fields* — both annihilation fields $\psi^+_\ell(x)$ and creation fields $\psi^-_\ell(x)$:

$$\psi^+_\ell(x) = \sum_{\sigma n} \int d^3p \; u_\ell(x;\mathbf{p},\sigma,n) a(\mathbf{p},\sigma,n) \,, \tag{5.1.4}$$

$$\psi^-_\ell(x) = \sum_{\sigma n} \int d^3p \; v_\ell(x;\mathbf{p},\sigma,n) a^\dagger(\mathbf{p},\sigma,n). \tag{5.1.5}$$

with coefficients* $u_\ell(x;\mathbf{p},\sigma,n)$ and $v_\ell(x;\mathbf{p},\sigma,n)$ chosen so that under Lorentz transformations each field is multiplied with a position-independent matrix:

$$U_0(\Lambda,a)\psi^+_\ell(x)U_0^{-1}(\Lambda,a) = \sum_{\bar\ell} D_{\ell\bar\ell}(\Lambda^{-1})\psi^+_{\bar\ell}(\Lambda x + a) \,, \tag{5.1.6}$$

$$U_0(\Lambda,a)\psi^-_\ell(x)U_0^{-1}(\Lambda,a) = \sum_{\bar\ell} D_{\ell\bar\ell}(\Lambda^{-1})\psi^-_{\bar\ell}(\Lambda x + a) \,. \tag{5.1.7}$$

(We might, in principle, have different transformation matrices $D^\pm$ for the annihilation and creation fields, but as we shall see, it is always possible to choose the fields so that these matrices are the same.) By applying a second Lorentz transformation $\bar\Lambda$, we find that

$$D(\Lambda^{-1})D(\bar\Lambda^{-1}) = D((\bar\Lambda\Lambda)^{-1}) \,,$$

so taking $\Lambda_1 = (\Lambda)^{-1}$ and $\Lambda_2 = (\bar\Lambda)^{-1}$, we see that the D-matrices furnish a *representation* of the homogeneous Lorentz group:

$$D(\Lambda_1)D(\Lambda_2) = D(\Lambda_1\Lambda_2) \,. \tag{5.1.8}$$

There are many such representations, including the scalar $D(\Lambda) = 1$, the vector $D(\Lambda)^\mu{}_\nu = \Lambda^\mu{}_\nu$, and a host of tensor and spinor representations. These particular representations are irreducible, in the sense that it is not possible by a choice of basis to reduce all $D(\Lambda)$ to the same block-diagonal form, with two or more blocks. However, we do not require at this point that $D(\Lambda)$ be irreducible; in general it is a block-diagonal matrix with an arbitrary array of irreducible representations in the blocks. That is, the index ℓ here includes a label that runs over the types of particle described and the irreducible representations in the different blocks, as well as another that runs over the components of the individual irreducible representations. Later we will separate these fields into irreducible fields that each describe only a single particle species (and its antiparticle) and transform irreducibly under the Lorentz group.

---

* A reminder: the labels n and σ run over all different particle species and spin z-components, respectively.

Once we have learned how to construct fields satisfying the Lorentz transformation rules (5.1.6) and (5.1.7), we will be able to construct the interaction density as

$$\mathscr{H}(x) = \sum_{NM} \sum_{\ell'_1 \cdots \ell'_N} \sum_{\ell_1 \cdots \ell_M} g_{\ell'_1 \cdots \ell'_N, \, \ell_1 \cdots \ell_M} \times \; \psi^-_{\ell'_1}(x) \cdots \psi^-_{\ell'_N}(x) \, \psi^+_{\ell_1}(x) \cdots \psi^+_{\ell_M}(x) \tag{5.1.9}$$

and this will be a scalar in the sense of Eq. (5.1.2) if the constant coefficients $g_{\ell'_1 \cdots \ell'_N, \, \ell_1 \cdots \ell_M}$ are chosen to be Lorentz covariant, in the sense that for all Λ:

$$\sum_{\ell'_1 \cdots \ell'_N} \sum_{\ell_1 \cdots \ell_M} D_{\ell'_1 \bar{\ell}'_1}(\Lambda^{-1}) \cdots D_{\ell'_N \bar{\ell}'_N}(\Lambda^{-1}) D_{\ell_1 \bar{\ell}_1}(\Lambda^{-1}) \cdots D_{\ell_M \bar{\ell}_M}(\Lambda^{-1}) \times g_{\ell'_1 \cdots \ell'_N \ell_1 \cdots \ell_M} = g_{\bar{\ell}'_1 \cdots \bar{\ell}'_N \bar{\ell}_1 \cdots \bar{\ell}_M} . \tag{5.1.10}$$

(Note that we do not include derivatives here, because we regard the derivatives of components of these fields as just additional sorts of field components.) The task of finding coefficients $g_{\ell'_1 \cdots \ell'_N, \, \ell_1 \cdots \ell_M}$ that satisfy Eq. (5.1.10) is no different in principle (and not much more difficult in practice) than that of using Clebsch–Gordan coefficients to couple together various representations of the three-dimensional rotation group to form rotational scalars. Later we will be able to combine creation and annihilation fields so that this density also commutes with itself at space-like separations.

Now, what shall we take as the coefficient functions $u_\ell(x; \mathbf{p}, \sigma, n)$ and $v_\ell(x; \mathbf{p}, \sigma, n)$? Eq. (4.2.12) and its adjoint give the transformation rules** for the annihilation and creation operators

$$U_0(\Lambda, b) a(\mathbf{p}, \sigma, n) U_0^{-1}(\Lambda, b) = \exp\Big(i(\Lambda p) \cdot b\Big) \sqrt{(\Lambda p)^0 / p^0} \times \sum_{\bar{\sigma}} D^{(j_n)}_{\sigma \bar{\sigma}}\Big(W^{-1}(\Lambda, p)\Big) a(\mathbf{p}_\Lambda, \bar{\sigma}, n) , \tag{5.1.11}$$

$$U_0(\Lambda, b) a^\dagger(\mathbf{p}, \sigma, n) U_0^{-1}(\Lambda, b) = \exp\Big(-i(\Lambda p) \cdot b\Big) \sqrt{(\Lambda p)^0 / p^0} \times \sum_{\bar{\sigma}} D^{(j_n)*}_{\sigma \bar{\sigma}}\Big(W^{-1}(\Lambda, p)\Big) a^\dagger(\mathbf{p}_\Lambda, \bar{\sigma}, n) \tag{5.1.12}$$

where j_n is the spin of particles of species n, and $\mathbf{p}_\Lambda$ is the three-vector part of Λp. (We have used the unitarity of the rotation matrices $D^{(j_n)}_{\sigma \bar{\sigma}}$ to put both Eqs. (5.1.11) and (5.1.12) in the form shown here.) Also, as we saw in Section 2.5 the volume element d^3p/p^0 is Lorentz-invariant, so we

** This is for massive particles. The case of zero mass will be taken up in Section 5.9.

can replace d^3p in Eqs. (5.1.4) and (5.1.5) with $d^3(\Lambda p)p^0/(\Lambda p)^0$. Putting this all together, we find

$$U_0(\Lambda,b)\psi^+_{\bar\ell}(x)U_0^{-1}(\Lambda,b) = \sum_{\sigma\bar\sigma n}\int d^3(\Lambda p)\; u_\ell(x;\mathbf{p},\sigma,n)$$
$$\times\;\exp\Big(i(\Lambda p)\cdot b\Big)D^{(j_n)}_{\sigma\bar\sigma}\Big(W^{-1}(\Lambda,p)\Big)\sqrt{p^0/(\Lambda p)^0}\;a(\mathbf{p}_\Lambda,\bar\sigma,n)$$

and

$$U_0(\Lambda,b)\psi^-_{\bar\ell}(x)U_0^{-1}(\Lambda,b) = \sum_{\sigma\bar\sigma n}\int d^3(\Lambda p)\; v_\ell(x;\mathbf{p},\sigma,n)$$
$$\times\;\exp\Big(-i(\Lambda p)\cdot b\Big)D^{(j_n)*}_{\sigma\bar\sigma}\Big(W^{-1}(\Lambda,p)\Big)\sqrt{p^0/(\Lambda p)^0}\;a^\dagger(\mathbf{p}_\Lambda,\bar\sigma,n)\,.$$

We see that in order for the fields to satisfy the Lorentz transformation rules (5.1.6) and (5.1.7), it is necessary and sufficient that

$$\sum_{\bar\ell} D_{\ell\bar\ell}(\Lambda^{-1})u_{\bar\ell}(\Lambda x+b;\mathbf{p}_\Lambda,\sigma,n) = \sqrt{p^0\big/(\Lambda p)^0}$$
$$\times\sum_{\bar\sigma} D^{(j_n)}_{\sigma\bar\sigma}\Big(W^{-1}(\Lambda,p)\Big)\exp\Big(+i(\Lambda p)\cdot b\Big)u_\ell(x;\mathbf{p},\bar\sigma,n)$$

and

$$\sum_{\bar\ell} D_{\ell\bar\ell}(\Lambda^{-1})v_{\bar\ell}(\Lambda x+b;\mathbf{p}_\Lambda,\sigma,n) = \sqrt{p^0\big/(\Lambda p)^0}$$
$$\times\sum_{\bar\sigma} D^{(j_n)*}_{\sigma\bar\sigma}\Big(W^{-1}(\Lambda,p)\Big)\exp\Big(-i(\Lambda p)\cdot b\Big)v_\ell(x;\mathbf{p},\bar\sigma,n)$$

or somewhat more conveniently

$$\sum_{\bar\sigma} u_{\bar\ell}(\Lambda x+b;\mathbf{p}_\Lambda,\bar\sigma,n)D^{(j_n)}_{\bar\sigma\sigma}\Big(W(\Lambda,p)\Big) = \sqrt{p^0\big/(\Lambda p)^0}$$
$$\times\sum_{\ell} D_{\bar\ell\ell}(\Lambda)\exp\Big(i(\Lambda p)\cdot b\Big)u_\ell(x;\mathbf{p},\sigma,n) \tag{5.1.13}$$

and

$$\sum_{\bar\sigma} v_{\bar\ell}(\Lambda x+b;\mathbf{p}_\Lambda,\bar\sigma,n)D^{(j_n)*}_{\bar\sigma\sigma}\Big(W(\Lambda,p)\Big) = \sqrt{p^0\big/(\Lambda p)^0}$$
$$\times\sum_{\ell} D_{\bar\ell\ell}(\Lambda)\exp\Big(-i(\Lambda p)\cdot b\Big)v_\ell(x;\mathbf{p},\sigma,n)\,. \tag{5.1.14}$$

These are the fundamental requirements that will allow us to calculate the u_ℓ and v_ℓ coefficient functions in terms of a finite number of free parameters.

We will use Eqs. (5.1.13) and (5.1.14) in three steps, considering in turn three different types of proper orthochronous Lorentz transformation:

Translations

First we consider Eqs. (5.1.13) and (5.1.14) with $\Lambda = 1$ and b arbitrary. We see immediately that $u_\ell(x; \mathbf{p}, \sigma, n)$ and $v_\ell(x; \mathbf{p}, \sigma, n)$ must take the form

$$u_\ell(x; \mathbf{p}, \sigma, n) = (2\pi)^{-3/2} e^{ip\cdot x} u_\ell(\mathbf{p}, \sigma, n) , \tag{5.1.15}$$

$$v_\ell(x; \mathbf{p}, \sigma, n) = (2\pi)^{-3/2} e^{-ip\cdot x} v_\ell(\mathbf{p}, \sigma, n) , \tag{5.1.16}$$

so the fields are Fourier transforms:

$$\psi_\ell^+(x) = \sum_{\sigma,n} (2\pi)^{-3/2} \int d^3p \; u_\ell(\mathbf{p}, \sigma, n) e^{ip\cdot x} a(\mathbf{p}, \sigma, n) , \tag{5.1.17}$$

and

$$\psi_\ell^-(x) = \sum_{\sigma,n} (2\pi)^{-3/2} \int d^3p \; v_\ell(\mathbf{p}, \sigma, n) e^{-ip\cdot x} a^\dagger(\mathbf{p}, \sigma, n) . \tag{5.1.18}$$

(The factors $(2\pi)^{-3/2}$ could be absorbed into the definition of u_ℓ and v_ℓ, but it is conventional to show them explicitly in these Fourier integrals.) Using Eqs. (5.1.15) and (5.1.16), we see that Eqs. (5.1.13) and (5.1.14) are satisfied if and only if

$$\sum_{\bar\sigma} u_{\bar\ell}(\mathbf{p}_\Lambda, \bar\sigma, n) D^{(j_n)}_{\bar\sigma\sigma}\Big(W(\Lambda, p)\Big) = \sqrt{\frac{p^0}{(\Lambda p)^0}} \sum_\ell D_{\bar\ell\ell}(\Lambda) u_\ell(\mathbf{p}, \sigma, n) \tag{5.1.19}$$

and

$$\sum_{\bar\sigma} v_{\bar\ell}(\mathbf{p}_\Lambda, \bar\sigma, n) D^{(j_n)*}_{\bar\sigma\sigma}\Big(W(\Lambda, p)\Big) = \sqrt{\frac{p^0}{(\Lambda p)^0}} \sum_\ell D_{\bar\ell\ell}(\Lambda) v_\ell(\mathbf{p}, \sigma, n) . \tag{5.1.20}$$

for arbitrary homogeneous Lorentz transformations Λ.

Boosts

Next take $\mathbf{p} = 0$ in Eqs. (5.1.19) and (5.1.20), and let Λ be the standard boost $L(q)$ that takes a particle of mass m from rest to some four-momentum q^μ. Then $L(p) = 1$, and

$$W(\Lambda, p) \equiv L^{-1}(\Lambda p)\Lambda L(p) = L^{-1}(q)L(q) = 1 .$$

Hence in this special case, Eqs. (5.1.19) and (5.1.20) give

$$u_{\bar\ell}(\mathbf{q}, \sigma, n) = (m/q^0)^{1/2} \sum_\ell D_{\bar\ell\ell}(L(q)) \, u_\ell(0, \sigma, n) \tag{5.1.21}$$

and

$$v_{\bar{\ell}}(\mathbf{q},\sigma,n) = (m/q^0)^{1/2} \sum_\ell D_{\bar{\ell}\ell}(L(q))\, v_\ell(0,\sigma,n)\,. \tag{5.1.22}$$

In other words, if we know the quantities $u_\ell(0,\sigma,n)$ and $v_\ell(0,\sigma,n)$ for zero momentum, then for a given representation $D(\Lambda)$ of the homogeneous Lorentz group, we know the functions $u_\ell(\mathbf{p},\sigma,n)$ and $v_\ell(\mathbf{p},\sigma,n)$ for all $\mathbf{p}$. (Explicit formulas for the matrices $D_{\bar{\ell}\ell}(L(q))$ will be given for arbitrary representations of the homogeneous Lorentz group in Section 5.7.)

Rotations

Next, take $\mathbf{p} = 0$, but this time let Λ be a Lorentz transformation with $\mathbf{p}_\Lambda = 0$; that is, take Λ as a rotation R. Here obviously $W(\Lambda, p) = R$, and so Eqs. (5.1.19) and (5.1.20) read

$$\sum_{\bar{\sigma}} u_{\bar{\ell}}(0,\bar{\sigma},n)\, D^{(j_n)}_{\bar{\sigma}\sigma}(R) = \sum_\ell D_{\bar{\ell}\ell}(R) u_\ell(0,\sigma,n) \tag{5.1.23}$$

and

$$\sum_{\bar{\sigma}} v_{\bar{\ell}}(0,\bar{\sigma},n)\, D^{(j_n)*}_{\bar{\sigma}\sigma}(R) = \sum_\ell D_{\bar{\ell}\ell}(R) v_\ell(0,\sigma,n)\,, \tag{5.1.24}$$

or equivalently

$$\sum_{\bar{\sigma}} u_{\bar{\ell}}(0,\bar{\sigma},n) \mathbf{J}^{(j_n)}_{\bar{\sigma}\sigma} = \sum_\ell \boldsymbol{\mathscr{J}}_{\bar{\ell}\ell} u_\ell(0,\sigma,n) \tag{5.1.25}$$

and

$$\sum_{\bar{\sigma}} v_{\bar{\ell}}(0,\bar{\sigma},n) \mathbf{J}^{(j_n)*}_{\bar{\sigma}\sigma} = -\sum_\ell \boldsymbol{\mathscr{J}}_{\bar{\ell}\ell} v_\ell(0,\sigma,n), \tag{5.1.26}$$

where $\mathbf{J}^{(j)}$ and $\boldsymbol{\mathscr{J}}$ are the angular-momentum matrices in the representations $D^{(j)}(R)$ and $D(R)$, respectively. Any representation $D(\Lambda)$ of the homogeneous Lorentz group obviously yields a representation of the rotation group when Λ is restricted to rotations R; Eqs. (5.1.25) and (5.1.26) tell us that if the field $\psi^\pm_\ell(x)$ is to describe particles of some particular spin j, then this representation $D(R)$ must contain among its irreducible components the spin-j representation $D^{(j)}(R)$, with the coefficients $u_\ell(0,\sigma,n)$ and $v_\ell(0,\sigma,n)$ simply describing how the spin-j representation of the rotation group is embedded in $D(R)$. We shall see in Section 5.6 that each *irreducible* representation of the proper orthochronous Lorentz group contains any given irreducible representation of the rotation group at most once, so that if the fields $\psi^+_\ell(x)$ and $\psi^-_\ell(x)$ transform irreducibly, then they are unique up to overall scale. More generally, the number of free parameters in the annihilation or creation fields (including their overall scales) is equal to the number of irreducible representations in the field.

It is straightforward to show that coefficient functions $u_\ell(\mathbf{p},\sigma,n)$ and $v_\ell(\mathbf{p},\sigma,n)$ given by Eqs. (5.1.21) and (5.1.22), with $u_\ell(0,\sigma,n)$ and $v_\ell(0,\sigma,n)$ satisfying Eqs. (5.1.23) and (5.1.24), will automatically satisfy the more general requirements (5.1.19) and (5.1.20). This is left as an exercise for the reader.

Let us now return to the cluster decomposition principle. Inserting Eqs. (5.1.17) and (5.1.18) in Eq. (5.1.9) and integrating over $\mathbf{x}$, the interaction Hamiltonian is

$$V = \sum_{NM} \int d^3\mathbf{p}'_1 \cdots d^3\mathbf{p}'_N \, d^3\mathbf{p}_1 \cdots d^3\mathbf{p}_M \sum_{\sigma'_1\cdots\sigma'_N} \sum_{\sigma_1\cdots\sigma_M} \sum_{n'_1\cdots n'_N} \sum_{n_1\cdots n_M}$$
$$\times a^\dagger(\mathbf{p}'_1\,\sigma'_1\,n'_1)\cdots a^\dagger(\mathbf{p}'_N\,\sigma'_N\,n'_N)\, a(\mathbf{p}_M\sigma_M n_M)\cdots a(\mathbf{p}_1\sigma_1 n_1)$$
$$\times \mathscr{V}_{NM}(\mathbf{p}'_1\,\sigma'_1\,n'_1\cdots \mathbf{p}'_N\,\sigma'_N\,n'_N\,,\ \mathbf{p}_1\sigma_1 n_1\cdots \mathbf{p}_M\sigma_M n_M) \tag{5.1.27}$$

with coefficient functions given by

$$\mathscr{V}_{NM}(\mathbf{p}'_1\,\sigma'_1\,n'_1\cdots,\mathbf{p}_1\sigma_1 n_1\cdots) = \delta^3(\mathbf{p}'_1+\cdots-\mathbf{p}_1-\cdots)$$
$$\times \tilde{\mathscr{V}}_{NM}(\mathbf{p}'_1\,\sigma'_1\,n'_1\cdots,\mathbf{p}_1\sigma_1 n_1\cdots)\,, \tag{5.1.28}$$

where

$$\tilde{\mathscr{V}}_{NM}(\mathbf{p}'_1\,\sigma'_1\,n'_1\cdots \mathbf{p}'_N\,\sigma'_N\,n'_N\,,\ \mathbf{p}_1\sigma_1 n_1\cdots \mathbf{p}_M\sigma_M n_M) = (2\pi)^{3-3N/2-3M/2}$$
$$\times \sum_{\ell'_1\cdots\ell'_N} \sum_{\ell_1\cdots\ell_M} g_{\ell'_1\cdots\ell'_N,\ell_1\cdots\ell_M}\, v_{\ell'_1}(\mathbf{p}'_1\,\sigma'_1\,n'_1)\cdots v_{\ell'_N}(\mathbf{p}'_N\,\sigma'_N\,n'_N)$$
$$\times u_{\ell_1}(\mathbf{p}_1\sigma_1 n_1)\cdots u_{\ell_M}(\mathbf{p}_M\sigma_M n_M)\,. \tag{5.1.29}$$

This interaction is manifestly of the form that will guarantee that the S-matrix satisfies the cluster decomposition principle: $\mathscr{V}_{NM}$ has a single delta function factor, with a coefficient $\tilde{\mathscr{V}}_{NM}$ that (at least for a finite number of field types) has at most branch point singularities at zero particle momenta. In fact, we could turn this argument around; any operator can be written as in Eq. (5.1.27), and the cluster decomposition principle requires that the coefficient $\mathscr{V}_{NM}$ may be written as in Eq. (5.1.28) as the product of a single momentum-conservation delta function times a smooth coefficient function. Any sufficiently smooth function (but *not* one containing additional delta functions) can be expressed as in Eq. (5.1.29).† *The cluster decomposition principle together with Lorentz invariance thus makes it natural that the interaction density should be constructed out of the annihilation and creation fields.*

---

† For general functions the indices ℓ and ℓ' may have to run over an infinite range. The reasons for restricting ℓ and ℓ' to a finite range have to do with the principle of renormalizability, discussed in Chapter 12.

If all we needed were to construct a scalar interaction density that satisfied the cluster decomposition principle, then we could combine annihilation and creation operators in arbitrary polynomials (5.1.9), with coupling coefficients $g_{\ell'_1 \cdots \ell'_N, \ell_1 \cdots \ell_M}$ subject only to the invariance condition (5.1.10) (and a suitable reality condition). However, for the Lorentz invariance of the S-matrix it is necessary also that the interaction density satisfy the commutation condition (5.1.3). This condition is not satisfied for arbitrary functions of the creation and annihilation fields because

$$[\psi_\ell^+(x), \psi_{\bar\ell}^-(y)]_\mp = (2\pi)^{-3} \sum_{\sigma\, n} \int d^3p\; u_\ell(\mathbf{p},\sigma,n) v_{\bar\ell}(\mathbf{p},\sigma,n) e^{ip\cdot(x-y)} \qquad (5.1.30)$$

(with the sign $\mp$ indicating a commutator or anticommutator if the particles destroyed and created by the components ψ_ℓ^+ and $\psi_{\bar\ell}^-$ are bosons or fermions, respectively,) and in general this does not vanish even for $x-y$ space-like. It is obviously not possible to avoid this problem by making the interaction density out of creation or annihilation fields alone, for then the interaction could not be Hermitian. The only way out of this difficulty is to combine annihilation and creation fields in linear combinations:

$$\psi_\ell(x) \equiv \kappa_\ell \psi_\ell^+(x) + \lambda_\ell \psi_\ell^-(x)\,, \qquad (5.1.31)$$

with the constants κ and λ and any other arbitrary constants in the fields adjusted so that for $x-y$ space-like

$$[\psi_\ell(x), \psi_{\ell'}(y)]_\mp = [\psi_\ell(x), \psi_{\ell'}^\dagger(y)]_\mp = 0. \qquad (5.1.32)$$

We will see in subsequent sections of this chapter how to do this for various irreducibly transforming fields. (By including explicit constants κ and λ in Eq. (5.1.31) we are leaving ourselves free to choose the overall scale of the annihilation and creation fields in any way that seems convenient.) The Hamiltonian density $\mathscr{H}(x)$ will satisfy the commutation condition (5.1.3) if it is constructed out of such fields and their adjoints, with an even number of any field components that destroy and create fermions.

The condition (5.1.32) is often described as a *causality* condition, because if $x-y$ is space-like then no signal can reach y from x, so that a measurement of ψ_ℓ at point x should not be able to interfere with a measurement of $\psi_{\ell'}$ or $\psi_{\ell'}^\dagger$ at point y. Such considerations of causality are plausible for the electromagnetic field, any one of whose components may be measured at a given spacetime point, as shown in a classic paper of Bohr and Rosenfeld.[2] However, we will be dealing here with fields like the Dirac field of the electron that do not seem in any sense measurable. The point of view taken here is that Eq. (5.1.32) is needed for the Lorentz invariance of the S-matrix, without any ancillary assumptions about measurability or causality.

There is an obstacle to the construction of fields (5.1.31) satisfying

(5.1.32). It may be that the particles that are destroyed and created by these fields carry non-zero values of one or more conserved quantum numbers like the electric charge. For instance, if particles of species n carry a value $q(n)$ for the electric charge Q, then

$$[Q, a(\mathbf{p}, \sigma, n)] = -q(n)a(\mathbf{p}, \sigma, n) \,,$$

$$[Q, a^\dagger(\mathbf{p}, \sigma, n)] = +q(n)a^\dagger(\mathbf{p}, \sigma, n) \,.$$

In order that $\mathscr{H}(x)$ should commute with the charge operator Q (or some other symmetry generator) it is necessary that it be formed out of fields that have simple commutation relations with Q:

$$[Q, \psi_\ell(x)] = -q_\ell \psi_\ell(x) \tag{5.1.33}$$

for then we can make $\mathscr{H}(x)$ commute with Q by constructing it as a sum of products of fields $\psi_{\ell_1}\psi_{\ell_2}\cdots$ and adjoints $\psi^\dagger_{m_1}\psi^\dagger_{m_2}\cdots$ such that

$$q_{\ell_1} + q_{\ell_2} + \cdots - q_{m_1} - q_{m_2} - \cdots = 0.$$

Now, Eq. (5.1.33) is satisfied for one particular component $\psi^+_\ell(x)$ of the annihilation field if and only if all particle species n that are destroyed by the field carry the same charge $q(n) = q_\ell$, and it is satisfied for one particular component $\psi^-_\ell(x)$ of the creation field if and only if all particle species $\bar{n}$ that are created by the field carry the charge $q(\bar{n}) = -q_\ell$. We see that in order for such a theory to conserve quantum numbers like electric charge, there must be a doubling of particle species carrying non-zero values of such quantum numbers: if a particular component of the annihilation field destroys a particle of species n, then the same component of the creation field must create particles of a species $\bar{n}$, known as the *antiparticles* of the particles of species n, which have opposite values of all conserved quantum numbers. *This is the reason for antiparticles.*

If the representation $D(\Lambda)$ is not irreducible, then we can adopt a basis for the fields in which $D(\Lambda)$ breaks up into blocks along the main diagonal, such that fields that belong to different blocks do not transform into each other under Lorentz transformations. Also, Lorentz transformations have no effect on the particle species. Therefore, instead of considering one big field, including many irreducible components and many particle species, we shall from now on restrict our attention to fields that destroy only a single type of particle (dropping the label n) and create only the corresponding antiparticle, and that transform irreducibly under the Lorentz group (which as mentioned above may or may not be supposed to include space inversion), with the understanding that, in general, we shall have to consider many different such fields, some perhaps formed as the derivatives of other fields. In the following sections we are going to finish the determination of the coefficient functions $u_\ell(\mathbf{p}, \sigma)$ and $v_\ell(\mathbf{p}, \sigma)$, fix the relative values of the constants κ and λ, and deduce the relations between the properties of particles and antiparticles first for fields that belong to the simplest irreducible representations of the Lorentz group, the scalar, vector, and Dirac spinor representations. After that we will repeat the analysis for a completely general irreducible representation.

A word about field equations. Inspection of Eqs. (5.1.31), (5.1.17), and (5.1.18) shows that all the components of a field of definite mass m satisfy the Klein–Gordon equation:

$$(\Box - m^2)\psi_\ell(x) = 0\,. \tag{5.1.34}$$

Some fields satisfy other field equations as well, depending on whether or not there are more field components than independent particle states. Traditionally in quantum field theory one begins with such field equations, or with the Lagrangian from which they are derived, and then uses them to derive the expansion of the fields in terms of one-particle annihilation and creation operators. In the approach followed here, we start with the particles, and derive the fields according to the dictates of Lorentz invariance, with the field equations arising almost incidentally as a byproduct of this construction.

* * *

A technicality must be mentioned here. According to the theorem proved in Section 4.4, the condition that guarantees that a theory will satisfy the cluster decomposition principle is that the interaction can be expressed as a sum of products of creation and annihilation operators, with all creation operators to the left of all annihilation operators, and with coefficients that contain only a single momentum-conservation delta function. For this reason, we should write the interaction in the 'normal ordered' form

$$V = \int d^3x \;\; :\mathscr{F}(\psi(x), \psi^\dagger(x)): \tag{5.1.35}$$

the colons indicating that the enclosed expression is to be rewritten (ignoring non-vanishing commutators or anticommutators, but including minus signs for permutations of fermionic operators) so that all creation operators stand to the left of all annihilation operators. By using the commutation or anticommutation relations of the fields, any such normal-ordered function of the fields can just as well be written as a sum of ordinary products of the fields with c-number coefficients. Rewriting $:\mathscr{F}:$ in this way makes it obvious that despite the normal ordering, $:\mathscr{F}(\psi(x), \psi^\dagger(x)):$ will commute with $:\mathscr{F}(\psi(y), \psi^\dagger(y)):$ when $x - y$ is space-like, if it is constructed out of fields that satisfy Eq. (5.1.32), with even numbers of any fermionic field components.

5.2 Causal Scalar Fields

We first consider one-component annihilation and creation fields $\phi^+(x)$ and $\phi^-(x)$ that transform as the simplest of all representations of the Lorentz group, the scalar, with $D(\Lambda) = 1$. Restricted to rotations, this is just the scalar representation of the rotation group, for which $\mathscr{J} = 0$, so Eqs. (5.1.25) and (5.1.26) have no solutions except for $j = 0$, in which case $\sigma, \bar{\sigma}$ take only the value zero. Thus a scalar field can only describe particles of zero spin. Assuming also for the moment that the field describes only a single species of particle, with no distinct antiparticle (and dropping the species label n as well as the spin label σ and the field label ℓ), the quantities $u_\ell(0\sigma n)$ and $v_\ell(0\sigma n)$ are here just the numbers $u(0)$ and $v(0)$. It is conventional to adjust the overall scales of the annihilation and creation fields so that these constants both have the values $(2m)^{-1/2}$. Eqs. (5.1.21) and (5.1.22) then give simply

$$u(\mathbf{p}) = (2p^0)^{-1/2} \tag{5.2.1}$$

and

$$v(\mathbf{p}) = (2p^0)^{-1/2}\,. \tag{5.2.2}$$

The fields (5.1.17) and (5.1.18) are then, in the scalar case,

$$\phi^+(x) = \int d^3p\,(2\pi)^{-3/2}(2p^0)^{-1/2}a(\mathbf{p})e^{ip\cdot x} \tag{5.2.3}$$

and

$$\phi^-(x) = \int d^3p\,(2\pi)^{-3/2}(2p^0)^{-1/2}a^\dagger(\mathbf{p})e^{-ip\cdot x} = \phi^{+\dagger}(x). \tag{5.2.4}$$

A Hamiltonian density $\mathscr{H}(x)$ that is formed as a polynomial in $\phi^+(x)$ and $\phi^-(x)$ will automatically satisfy the requirement (5.1.9), that it transform as a scalar. It remains to satisfy the other condition for the Lorentz invariance of the S-matrix, that $\mathscr{H}(x)$ commute with $\mathscr{H}(y)$ at space-like separations $x-y$. If $\mathscr{H}(x)$ were a polynomial in $\phi^+(x)$ alone, there would be no problem. All annihilation operators commute or anticommute, so $\phi^+(x)$ either commutes or anticommutes with $\phi^+(y)$ for all x and y, according to whether the particle is a boson or fermion, respectively:

$$[\phi^+(x), \phi^+(y)]_\mp = 0\,. \tag{5.2.5}$$

Hence any $\mathscr{H}(x)$ formed as a polynomial in $\phi^+(x)$ (or, for fermions, any such even polynomial) will commute with $\mathscr{H}(y)$ for all x and y. The problem, of course, is that, in order to be Hermitian, $\mathscr{H}(x)$ must involve $\phi^{+\dagger}(x) = \phi^-(x)$ as well as $\phi^+(x)$, and $\phi^+(x)$ does not commute or anticommute with $\phi^-(y)$ for general space-like separations. Using the commutation (for bosons) or anticommutation (for fermions) relations

(4.2.5), we have

$$[\phi^+(x), \phi^-(y)]_\mp = \int \frac{d^3p\, d^3p'}{(2\pi)^3(2p^0 \cdot 2p'^0)^{1/2}}\, e^{ip\cdot x}\, e^{-ip'\cdot y} \delta^3(\mathbf{p} - \mathbf{p}')\,,$$

which collapses to the single integral

$$[\phi^+(x), \phi^-(y)]_\mp = \Delta_+(x-y)\,, \tag{5.2.6}$$

where Δ_+ is a standard function:

$$\Delta_+(x) \equiv \frac{1}{(2\pi)^3} \int \frac{d^3p}{2p^0}\, e^{ip\cdot x}\,. \tag{5.2.7}$$

This is manifestly Lorentz-invariant, and therefore for space-like x it can depend only on the invariant square $x^2 > 0$. We can thus evaluate $\Delta_+(x)$ for space-like x by choosing the coordinate system so that

$$x^0 = 0\,, \quad |\mathbf{x}| = \sqrt{x^2}\,.$$

Eq. (5.2.7) then gives

$$\begin{aligned}\Delta_+(x) &= \frac{1}{(2\pi)^3} \int \frac{d^3p}{2\sqrt{\mathbf{p}^2 + m^2}}\, e^{i\mathbf{p}\cdot\mathbf{x}} \\ &= \frac{4\pi}{(2\pi)^3} \int_0^\infty \frac{p^2 dp}{2\sqrt{\mathbf{p}^2 + m^2}}\, \frac{\sin(p\sqrt{x^2})}{p\sqrt{x^2}}\,.\end{aligned}$$

Changing the variable of integration to $u \equiv p/m$, this is

$$\Delta_+(x) = \frac{m}{4\pi^2\sqrt{x^2}} \int_0^\infty \frac{u\,du}{\sqrt{u^2+1}}\, \sin(m\sqrt{x^2}u) \tag{5.2.8}$$

or, in terms of a standard Hankel function,

$$\Delta_+(x) = \frac{m}{4\pi^2\sqrt{x^2}}\, K_1\left(m\sqrt{x^2}\right)\,. \tag{5.2.9}$$

This isn't zero, so what are we to do with it? Note that even though $\Delta_+(x)$ is not zero, for $x^2 > 0$ it is even in x^μ. Instead of using only $\phi^+(x)$, suppose we try to construct $\mathscr{H}(x)$ out of a linear combination

$$\phi(x) \equiv \kappa\phi^+(x) + \lambda\phi^-(x).$$

Using Eq. (5.2.6), we have then for $x - y$ space-like

$$\begin{aligned}[\phi(x), \phi^\dagger(y)]_\mp &= |\kappa|^2[\phi^+(x), \phi^-(y)]_\mp + |\lambda|^2[\phi^-(x), \phi^+(y)]_\mp \\ &= (|\kappa|^2 \mp |\lambda|^2)\Delta_+(x-y)\end{aligned}$$

$$\begin{aligned}[\phi(x), \phi(y)]_\mp &= \kappa\lambda([\phi^+(x), \phi^-(y)]_\mp + [\phi^-(x), \phi^+(y)]_\mp) \\ &= \kappa\lambda(1 \mp 1)\Delta_+(x-y)\,.\end{aligned}$$

Both of these will vanish if and only if the particle is a *boson* (i.e., it is the top sign that applies) and κ and λ are equal in magnitude

$$|\kappa| = |\lambda| .$$

We can change the relative phase of κ and λ by redefining the phases of the states so that $a(\mathbf{p}) \to e^{i\alpha}a(\mathbf{p})$, $a^\dagger(\mathbf{p}) \to e^{-i\alpha}a^\dagger(\mathbf{p})$, and hence $\kappa \to \kappa e^{i\alpha}$, $\lambda \to \lambda e^{-i\alpha}$. Taking $\alpha = \frac{1}{2}\text{Arg}(\lambda/\kappa)$, we can in this way make κ and λ equal in phase, and hence equal.

Redefining $\phi(x)$ to absorb the overall factor $\kappa = \lambda$, we have then

$$\phi(x) = \phi^+(x) + \phi^{+\dagger}(x) = \phi^\dagger(x) . \tag{5.2.10}$$

The interaction density $\mathscr{H}(x)$ will commute with $\mathscr{H}(y)$ at space-like separations $x-y$ if formed as a normal-ordered polynomial in the self-adjoint scalar field $\phi(x)$.

Even though the choice of the relative phase of the two terms in Eq. (5.2.10) is a matter of convention, it is a convention that once adopted must be used wherever a scalar field for this particle appears in the interaction Hamiltonian density. For instance, suppose that the interaction density involved not only the field (5.2.10), but also another scalar field for the same particle

$$\tilde{\phi}(x) = e^{i\alpha}\phi^+(x) + e^{-i\alpha}\phi^{+\dagger}(x)$$

with α an arbitrary phase. This $\tilde{\phi}$, like ϕ, would be causal in the sense that $\tilde{\phi}(x)$ commutes with $\tilde{\phi}(y)$ when $x-y$ is space-like, but $\tilde{\phi}(x)$ would not commute with $\phi(y)$ at space-like separations, and therefore we cannot have both of these fields appearing in the same theory.

If the particles that are destroyed and created by $\phi(x)$ carry some conserved quantum number like electric charge, then $\mathscr{H}(x)$ will conserve the quantum number if and only if each term in $\mathscr{H}(x)$ contains equal numbers of operators $a(\mathbf{p})$ and $a(\mathbf{p})^\dagger$. But this is impossible if $\mathscr{H}(x)$ is formed as a polynomial in $\phi(x) = \phi^+(x) + \phi^{+\dagger}(x)$. To put this another way, in order that $\mathscr{H}(x)$ should commute with the charge operator Q (or some other symmetry generator) it is necessary that it be formed out of fields that have simple commutation relations with Q. This is true for $\phi^+(x)$ and its adjoint, for which

$$[Q, \phi^+(x)]_- = -q\phi^+(x) ,$$
$$[Q, \phi^{+\dagger}(x)]_- = +q\phi^{+\dagger}(x) ,$$

but not for the self-adjoint field (5.2.10).

In order to deal with this problem, we must suppose that there are *two* spinless bosons, with the same mass m, but charges $+q$ and $-q$, respectively. Let $\phi^+(x)$ and $\phi^{+c}(x)$ denote the annihilation fields for these two particles, so that*

$$[Q, \phi^+(x)]_- = -q\phi^+(x) ,$$
$$[Q, \phi^{+c}(x)]_- = +q\phi^{+c}(x) .$$

Define $\phi(x)$ as the linear combination

$$\phi(x) = \kappa\phi^+(x) + \lambda\phi^{+c\dagger}(x) ,$$

which manifestly has the same commutator with Q as $\phi^+(x)$ alone

$$[Q, \phi(x)]_- = -q\phi(x) .$$

The commutator or anticommutator of $\phi(x)$ with its adjoint is then, at space-like separation

$$\begin{aligned}[\phi(x), \phi^\dagger(y)]_\mp &= |\kappa|^2[\phi^+(x), \phi^{+\dagger}(y)] + |\lambda|^2[\phi^{+c\dagger}(x), \phi^{+c}(y)]_\mp \\ &= (|\kappa|^2 \mp |\lambda|^2)\Delta_+(x-y) ,\end{aligned}$$

while $\phi(x)$ and $\phi(y)$ automatically commute or anticommute with each other for all x and y because ϕ^+ and $\phi^{+c\dagger}$ destroy and create different particles. In deriving this result, we have tacitly assumed that the particle and antiparticle have the same mass, so that the commutators or anticommutators involve the same function $\Delta_+(x-y)$. Fermi statistics is again ruled out here, because it is not possible that $\phi(x)$ should anticommute with $\phi^\dagger(y)$ at space-like separations unless $\kappa = \lambda = 0$, in which case the fields simply vanish. So a spinless particle must be a boson.

For Bose statistics, in order that a complex $\phi(x)$ should commute with $\phi^\dagger(y)$ at space-like separations, it is necessary and sufficient that $|\kappa|^2 = |\lambda|^2$, as well as for the particle and antiparticle to have the same mass. By redefining the relative phase of states of these two particles, we can again give κ and λ the same phase, in which case $\kappa = \lambda$. This common factor can again be eliminated by a redefinition of the field ϕ, so that

$$\phi(x) = \phi^+(x) + \phi^{c+\dagger}(x)$$

or in more detail

$$\phi(x) = \int \frac{d^3p}{(2\pi)^{3/2}(2p^0)^{1/2}} \left[a(\mathbf{p})e^{ip\cdot x} + a^{c\dagger}(\mathbf{p})e^{-ip\cdot x}\right] . \tag{5.2.11}$$

This is the essentially unique causal scalar field. This formula can be used both for purely neutral spinless particles that are their own antiparticles (in which case we take $a^c(\mathbf{p}) = a(\mathbf{p})$), and for particles with distinct antiparticles (for which $a^c(\mathbf{p}) \neq a(\mathbf{p})$).

* The label 'c' denotes 'charge conjugate'. It should be kept in mind that a particle that carries no conserved quantum numbers may or may not be its own antiparticle, with $a^c(\mathbf{p}) = a(\mathbf{p})$.

For future use, we note here that the commutator of the complex scalar field with its adjoint is

$$[\phi(x), \phi^\dagger(y)] = \Delta(x-y)\,, \tag{5.2.12}$$

where

$$\Delta(x-y) \equiv \Delta_+(x-y) - \Delta_+(y-x) = \int \frac{d^3p}{2p^0(2\pi)^3}\left[e^{ip\cdot(x-y)} - e^{-ip\cdot(x-y)}\right]. \tag{5.2.13}$$

Let's now consider the effect of the various inversion symmetries on this field. First, from the results of Section 4.2, we can readily see that the effect of the space-inversion operator on the annihilation and creation operators is:**

$$\mathsf{P}a(\mathbf{p})\mathsf{P}^{-1} = \eta^* a(-\mathbf{p})\,, \tag{5.2.14}$$

$$\mathsf{P}a^{c\dagger}(\mathbf{p})\mathsf{P}^{-1} = \eta^c a^{c\dagger}(-\mathbf{p})\,, \tag{5.2.15}$$

where η and η^c are the intrinsic parities of the particle and antiparticle, respectively. Applying these results to the annihilation field (5.2.3) and the charge-conjugate of the creation field (5.2.4), and changing the variable of integration from $\mathbf{p}$ to $-\mathbf{p}$, we see that

$$\mathsf{P}\phi^+(x)\mathsf{P}^{-1} = \eta^*\phi^+(\mathscr{P}x) \tag{5.2.16}$$

$$\mathsf{P}\phi^{+c\dagger}(x)\mathsf{P}^{-1} = \eta^c\phi^{+c\dagger}(\mathscr{P}x)\,, \tag{5.2.17}$$

where as before $\mathscr{P}x = (-\mathbf{x}, x^0)$. We see that in general applying the space inversion to the scalar field $\phi(x) = \phi^+(x) + \phi^{+c\dagger}(x)$ would give a different field $\phi_P = \eta^*\phi^+ + \eta^c\phi^{+c\dagger}$. Both fields are separately causal, but if ϕ and $\phi_P^\dagger$ appear in the same interaction then we are in trouble, because in general they do not commute at space-like separations. The only way to preserve Lorentz invariance as well as parity conservation and the hermiticity of the interaction is to require that ϕ_P be proportional to ϕ, and hence that

$$\eta^c = \eta^*. \tag{5.2.18}$$

That is, *the intrinsic parity $\eta\eta^c$ of a state containing a spinless particle and its antiparticle is even.* We have now simply

$$\mathsf{P}\phi(x)\mathsf{P}^{-1} = \eta^*\phi(\mathscr{P}x)\,. \tag{5.2.19}$$

** We are omitting the subscript 0 on inversion operators P, C, and T, because in virtually all cases where these inversions are good symmetries, the same operators induce inversion transformations on 'in' and 'out' states and on free-particle states.

These results also apply when the spinless particle is its own antiparticle, for which $\eta^c = \eta$, and imply that the intrinsic parity of such a particle is real: $\eta = \pm 1$.

Charge-conjugation can be handled in much the same way. From the results of Section 4.2, we have

$$\mathsf{C}a(\mathbf{p})\mathsf{C}^{-1} = \xi^* a^c(\mathbf{p})\,, \tag{5.2.20}$$

$$\mathsf{C}a^{c\dagger}(\mathbf{p})\mathsf{C}^{-1} = \xi^c a^\dagger(\mathbf{p})\,, \tag{5.2.21}$$

where ξ and ξ^c are the phases associated with the operation of charge-conjugation on one-particle states. It follows then that

$$\mathsf{C}\phi^+(x)\mathsf{C}^{-1} = \xi^* \phi^{+c}(x)\,, \tag{5.2.22}$$

$$\mathsf{C}\phi^{+c\dagger}(x)\mathsf{C}^{-1} = \xi^c \phi^{+\dagger}(x)\,. \tag{5.2.23}$$

In order that $\mathsf{C}\phi(x)\mathsf{C}^{-1}$ should be proportional to the field $\phi^\dagger(x)$ with which it commutes at space-like separations, it is evidently necessary that

$$\xi^c = \xi^*\,. \tag{5.2.24}$$

Just as for ordinary parity, the intrinsic charge-conjugation parity $\xi\xi^c$ of a state consisting of a spinless particle and its antiparticle is even. We now have simply

$$\mathsf{C}\phi(x)\mathsf{C}^{-1} = \xi^* \phi^\dagger(x)\,. \tag{5.2.25}$$

Again, these results apply also in the case where the particle is its own antiparticle, where $\xi^c = \xi$. In this case the charge-conjugation parity like the ordinary parity must be real, $\xi = \pm 1$.

Finally we come to time-reversal. From Section 4.2 we have

$$\mathsf{T}a(\mathbf{p})\mathsf{T}^{-1} = \zeta^* a(-\mathbf{p})\,, \tag{5.2.26}$$

$$\mathsf{T}a^{c\dagger}(\mathbf{p})\mathsf{T}^{-1} = \zeta^c a^{c\dagger}(-\mathbf{p})\,. \tag{5.2.27}$$

Recalling that T is antiunitary, and again changing the variable of integration from $\mathbf{p}$ to $-\mathbf{p}$, we find that

$$\mathsf{T}\phi^+(x)\mathsf{T}^{-1} = \zeta^* \phi^+(-\mathscr{P}x) \tag{5.2.28}$$

$$\mathsf{T}\phi^{+c\dagger}(x)\mathsf{T}^{-1} = \zeta^c \phi^{+c\dagger}(-\mathscr{P}x)\,. \tag{5.2.29}$$

In order for $\mathsf{T}\phi(x)\mathsf{T}^{-1}$ to be simply related to the field ϕ at the time-reversed point $-\mathscr{P}x$, we must have

$$\zeta^c = \zeta^* \tag{5.2.30}$$

and then

$$\mathsf{T}\phi(x)\mathsf{T}^{-1} = \zeta^* \phi(-\mathscr{P}x)\,. \tag{5.2.31}$$

5.3 Causal Vector Fields

We now take up the next simplest kind of field, which transforms as a four-vector, the simplest non-trivial representation of the homogeneous Lorentz group. There are massive particles, the $W^{\pm}$ and Z^0, that at low energies are described by such fields and that play an increasing role in modern elementary particle physics, so this example is not merely of pedagogical interest. (Also, although we are here considering only massive particles, one approach to quantum electrodynamics is to describe the photon in terms of a massive vector field in the limit of very small mass). For the moment we will suppose that only one species of particle is described by this field (dropping the species label n); then we shall consider the possibility that the field describes both a particle and a distinct antiparticle.

In the four-vector representation of the Lorentz group, the rows and columns of the representation matrices $D(\Lambda)$ are labelled with four-component indices μ, ν, etc., with

$$D(\Lambda)^{\mu}{}_{\nu} \equiv \Lambda^{\mu}{}_{\nu}. \tag{5.3.1}$$

The annihilation and creation parts of the vector field are written:

$$\phi^{+\mu}(x) = \sum_{\sigma}(2\pi)^{-3/2}\int d^3p\; u^{\mu}(\mathbf{p},\sigma)\, a(\mathbf{p},\sigma)\, e^{ip\cdot x}\,, \tag{5.3.2}$$

$$\phi^{-\mu}(x) = \sum_{\sigma}(2\pi)^{-3/2}\int d^3p\; v^{\mu}(\mathbf{p},\sigma)\, a^{\dagger}(\mathbf{p},\sigma)\, e^{-ip\cdot x}\,. \tag{5.3.3}$$

The coefficient functions $u^{\mu}(\mathbf{p},\sigma)$ and $v^{\mu}(\mathbf{p},\sigma)$ for arbitrary momentum are given in terms of those for zero momentum by Eqs. (5.1.21) and (5.1.22), which here read:

$$u^{\mu}(\mathbf{p},\sigma) = (m/p^0)^{1/2} L(p)^{\mu}{}_{\nu} u^{\nu}(0,\sigma)\,, \tag{5.3.4}$$

$$v^{\mu}(\mathbf{p},\sigma) = (m/p^0)^{1/2} L(p)^{\mu}{}_{\nu} v^{\nu}(0,\sigma)\,. \tag{5.3.5}$$

(We are using the usual summation convention for spacetime indices μ, ν, etc.) Also, the coefficient functions at zero momentum are subject to the conditions (5.1.25) and (5.1.26):

$$\sum_{\bar{\sigma}} u^{\mu}(0,\bar{\sigma})\mathbf{J}^{(j)}_{\bar{\sigma}\sigma} = \mathscr{J}^{\mu}{}_{\nu} u^{\nu}(0,\sigma) \tag{5.3.6}$$

and

$$-\sum_{\bar{\sigma}} v^{\mu}(0,\bar{\sigma})\mathbf{J}^{(j)*}_{\bar{\sigma}\sigma} = \mathscr{J}^{\mu}{}_{\nu} v^{\nu}(0,\sigma)\,. \tag{5.3.7}$$

The rotation generators $\mathscr{J}^\mu{}_\nu$ in the four-vector representation are given by Eq. (5.3.1) as

$$(\mathscr{J}_k)^0{}_0 = (\mathscr{J}_k)^0{}_i = (\mathscr{J}_k)^i{}_0 = 0\,, \tag{5.3.8}$$

$$(\mathscr{J}_k)^i{}_j = -i\epsilon_{ijk}\,, \tag{5.3.9}$$

with i, j, k here running over the values 1, 2, and 3. We note in particular that $\mathscr{J}^2$ takes the form

$$(\mathscr{J}^2)^0{}_0 = (\mathscr{J}^2)^0{}_i = (\mathscr{J}^2)^i{}_0 = 0\,, \tag{5.3.10}$$

$$(\mathscr{J}^2)^i{}_j = 2\delta^i{}_j\,. \tag{5.3.11}$$

From Eqs. (5.3.6) and (5.3.7) it follows then that

$$\sum_{\bar\sigma} u^0(0,\bar\sigma)(\mathbf{J}^{(j)})^2_{\bar\sigma\sigma} = 0\,, \tag{5.3.12}$$

$$\sum_{\bar\sigma} u^i(0,\bar\sigma)(\mathbf{J}^{(j)})^2_{\bar\sigma\sigma} = 2u^i(0,\sigma) \tag{5.3.13}$$

and

$$\sum_{\bar\sigma} v^0(0,\bar\sigma)(\mathbf{J}^{(j)*})^2_{\bar\sigma\sigma} = 0\,, \tag{5.3.14}$$

$$\sum_{\bar\sigma} v^i(0,\bar\sigma)(\mathbf{J}^{(j)*})^2_{\bar\sigma\sigma} = 2v^i(0,\sigma)\,. \tag{5.3.15}$$

Also, we recall the familiar result that $(\mathbf{J}^{(j)})^2_{\bar\sigma\sigma} = j(j+1)\delta_{\bar\sigma\sigma}$. From Eqs. (5.3.12)–(5.3.15) we see that there are just two possibilities for the spin of the particle described by the vector field: either $j = 0$, for which at $\mathbf{p} = 0$ only u^0 and v^0 are non-zero, or else $j = 1$ (so that $j(j+1) = 2$), for which at $\mathbf{p} = 0$ only the space-components u^i and v^i are non-zero. Let us look in a little more detail at each of these two possibilities.

Spin Zero

By an appropriate choice of normalization of the fields, we can take the only non-vanishing component of $u^\mu(0)$ and $v^\mu(0)$ to have the conventional values:

$$u^0(0) = i(m/2)^{1/2}$$
$$v^0(0) = -i(m/2)^{1/2}\,.$$

(The label σ here takes only the single value zero, and is therefore dropped.) Then Eqs. (5.3.4) and (5.3.5) yield for general momenta

$$u^\mu(\mathbf{p}) = ip^\mu(2p^0)^{-1/2} \tag{5.3.16}$$

and

$$v^\mu(\mathbf{p}) = -ip^\mu(2p^0)^{-1/2}. \tag{5.3.17}$$

The vector annihilation and creation fields here are nothing but the derivatives of the scalar annihilation and creation fields $\phi^\pm$ for a spinless particle that were defined in the previous section:

$$\phi^{+\mu}(x) = \partial^\mu\phi^+(x) \qquad \phi^{-\mu}(x) = \partial^\mu\phi^-(x)\,. \tag{5.3.18}$$

It is obvious that the causal vector field for a spinless particle is also just the derivative of the causal scalar field:

$$\phi^\mu(x) = \phi^{+\mu}(x) + \phi^{-\mu}(x) = \partial^\mu\phi(x)\,. \tag{5.3.19}$$

Hence we need not explore this case any further here.

Spin One

From Eqs. (5.3.6) and (5.3.7) we see immediately that the vectors $u^i(0,0)$ and $v^i(0,0)$ for $\sigma = 0$ are in the 3-direction. By a suitable normalization of the fields, we can take these vectors to have the values

$$u^\mu(0,0) = v^\mu(0,0) = (2m)^{-1/2}\begin{bmatrix} 0 \\ 0 \\ 1 \\ 0 \end{bmatrix} \tag{5.3.20}$$

with four-vector components listed always in the order 1, 2, 3, 0. To find the other components, we use Eqs. (5.3.6), (5.3.7), and (5.3.9) to calculate the effect of the raising and lowering operators $J_1^{(1)} \pm iJ_2^{(1)}$ on u and v. This gives:

$$u^\mu(0,+1) = -v^\mu(0,-1) = -\frac{1}{\sqrt{2}}(2m)^{-1/2}\begin{bmatrix} 1 \\ +i \\ 0 \\ 0 \end{bmatrix}, \tag{5.3.21}$$

$$u^\mu(0,-1) = -v^\mu(0,+1) = \frac{1}{\sqrt{2}}(2m)^{-1/2}\begin{bmatrix} 1 \\ -i \\ 0 \\ 0 \end{bmatrix}. \tag{5.3.22}$$

Applying Eqs. (5.3.4) and (5.3.5) now yields

$$u^\mu(\mathbf{p},\sigma) = v^{\mu *}(\mathbf{p},\sigma) = (2p^0)^{-1/2}e^\mu(\mathbf{p},\sigma) \tag{5.3.23}$$

where

$$e^\mu(\mathbf{p},\sigma) \equiv L^\mu{}_\nu(\mathbf{p})e^\nu(0,\sigma) \tag{5.3.24}$$

with

$$e^\mu(0,0) = \begin{bmatrix} 0 \\ 0 \\ 1 \\ 0 \end{bmatrix}, \quad e^\mu(0,+1) = -\frac{1}{\sqrt{2}} \begin{bmatrix} 1 \\ +i \\ 0 \\ 0 \end{bmatrix}, \quad e^\mu(0,-1) = \frac{1}{\sqrt{2}} \begin{bmatrix} 1 \\ -i \\ 0 \\ 0 \end{bmatrix}. \tag{5.3.25}$$

The annihilation and creation fields (5.3.2) and (5.3.3) here are

$$\phi^{+\mu}(x) = \phi^{-\mu\dagger}(x) = (2\pi)^{-3/2} \sum_\sigma \int \frac{d^3p}{\sqrt{2p^0}} \, e^\mu(\mathbf{p},\sigma)\, a(\mathbf{p},\sigma)\, e^{ip\cdot x} \,. \tag{5.3.26}$$

The fields $\phi^{+\mu}(x)$ and $\phi^{+\nu}(y)$ of course commute (or anticommute) for all x and y, but $\phi^{+\mu}(x)$ and $\phi^{-\nu}(y)$ do not. Their commutator (for bosons) or anticommutator (for fermions) is

$$[\phi^{+\mu}(x), \phi^{-\nu}(y)]_\mp = \int \frac{d^3p}{(2\pi)^3 2p^0} \, e^{ip\cdot(x-y)} \, \Pi^{\mu\nu}(\mathbf{p}) \tag{5.3.27}$$

where

$$\Pi^{\mu\nu}(\mathbf{p}) \equiv \sum_\sigma e^\mu(\mathbf{p},\sigma) e^{\nu *}(\mathbf{p},\sigma) \,. \tag{5.3.28}$$

A straightforward calculation using Eq. (5.3.25) shows that $\Pi^{\mu\nu}(0)$ is the projection matrix on the space orthogonal to the time-direction, and Eq. (5.3.24) then shows that $\Pi^{\mu\nu}(\mathbf{p})$ is the projection matrix on the space orthogonal to the four-vector p^μ:

$$\Pi^{\mu\nu}(\mathbf{p}) = \eta^{\mu\nu} + p^\mu p^\nu / m^2 \,. \tag{5.3.29}$$

The commutator (or anticommutator) (5.3.27) may then be written in terms of the Δ_+ function defined in the previous section, as

$$[\phi^{+\mu}(x), \phi^{-\nu}(y)]_\mp = \left[\eta^{\mu\nu} - \frac{\partial^\mu \partial^\nu}{m^2} \right] \Delta_+(x-y) \,. \tag{5.3.30}$$

For our present purposes, the important thing about this expression is that for $x-y$ space-like it does not vanish and is *even* in $x-y$. We can therefore repeat the reasoning of the previous section in seeking to construct a causal field: we form a linear combination of annihilation and creation fields

$$v^\mu(x) \equiv \kappa \phi^{+\mu}(x) + \lambda \phi^{-\mu}(x)$$

for which, for $x-y$ spacelike,

$$[v^\mu(x), v^\nu(y)]_\mp = \kappa\lambda \, [1 \mp 1] \left[\eta^{\mu\nu} - \frac{\partial^\mu \partial^\nu}{m^2} \right] \Delta_+(x-y)$$

and

$$[v^\mu(x), v^{\nu\dagger}(y)]_\mp = (|\kappa|^2 \mp |\lambda|^2)\left[\eta^{\mu\nu} - \frac{\partial^\mu \partial^\nu}{m^2}\right]\Delta_+(x-y).$$

In order for both to vanish for space-like $x-y$, it is necessary and sufficient that the spin one particles be *bosons* and that $|\kappa| = |\lambda|$. By a suitable choice of phase of the one-particle states we can give κ and λ the same phase, so that $\kappa = \lambda$, and then drop the common factor κ by redefining the overall normalization of the field. After all this, we find that the causal vector field for a massive particle of spin one is

$$v^\mu(x) = \phi^{+\mu}(x) + \phi^{+\mu\dagger}(x)\,. \tag{5.3.31}$$

We note that this is real:

$$v^\mu(x) = v^{\mu\dagger}(x)\,. \tag{5.3.32}$$

However, if the particles it describes carry a non-zero value of some conserved quantum number Q, then we cannot construct an interaction that conserves Q out of such a field. Instead, we must suppose that there is another boson of the same mass and spin which carries an opposite value of Q, and construct the causal field as

$$v^\mu(x) = \phi^{+\mu}(x) + \phi^{+c\mu\dagger}(x) \tag{5.3.33}$$

or in more detail

$$v^\mu(x) = (2\pi)^{-3/2}\sum_\sigma \int \frac{d^3p}{\sqrt{2p^0}}$$
$$\times \left[e^\mu(\mathbf{p},\sigma)a(\mathbf{p},\sigma)e^{ip\cdot x} + e^{\mu *}(\mathbf{p},\sigma)a^{c\dagger}(\mathbf{p},\sigma)e^{-ip\cdot x}\right]\,, \tag{5.3.34}$$

where the superscript c indicates operators that create the antiparticle that is charge-conjugate to the particle annihilated by $\phi^{+\mu}(x)$. This again is a causal field, but no longer real. We can also use this formula for the case of a purely neutral, spin one particle that is its own antiparticle, by simply setting $a^c(\mathbf{p}) = a(\mathbf{p})$. In either case, the commutator of a vector field with its adjoint is

$$[v^\mu(x), v^{\nu\dagger}(y)] = \left[\eta^{\mu\nu} - \frac{\partial^\mu \partial^\nu}{m^2}\right]\Delta(x-y)\,, \tag{5.3.35}$$

where $\Delta(x-y)$ is the function (5.2.13).

The real and complex fields we have constructed for a massive, spin one particle satisfy interesting field equations. First, since p^μ in the exponential in Eq. (5.3.26) satisfies $p^2 = -m^2$, the field satisfies the Klein–Gordon equation:

$$(\Box - m^2)v^\mu(x) = 0\,, \tag{5.3.36}$$

just as for the scalar field. In addition, since Eq. (5.3.24) shows that

$$e^\mu(\mathbf{p},\sigma)\,p_\mu = 0\,, \tag{5.3.37}$$

we now have another equation

$$\partial_\mu v^\mu(x) = 0\,. \tag{5.3.38}$$

In the limit of small mass, Eqs. (5.3.36) and (5.3.38) are just the equations for the potential four-vector of electrodynamics in what is called Lorentz gauge.

However, we cannot obtain electrodynamics from just any theory of massive spin one particles by letting the mass go to zero. The trouble can be seen by considering the rate of production of a spin one particle by an interaction density $\mathscr{H} = J_\mu v^\mu$, where J_μ is an arbitrary four-vector current. Squaring the matrix element and summing over the spin z-components of the spin one particle gives a rate proportional to

$$\sum_\sigma |< J_\mu > e^\mu(\mathbf{p},\sigma)^*|^2 =< J_\mu >< J_\nu >^*\,\Pi^{\mu\nu}(\mathbf{p})\,,$$

where $\mathbf{p}$ is the momentum of the emitted spin one particle, and $< J_\mu >$ is the matrix element of the current (say, at $x = 0$) between the initial and final states of all other particles. The term $p^\mu p^\nu/m^2$ in $\Pi^{\mu\nu}(\mathbf{p})$ will, in general, cause the emission rate to blow up when $m \to 0$. The only way to avert this catastrophe is to suppose that $< J_\mu > p^\mu$ vanishes, which in coordinate space is just the statement that the current J^μ must be *conserved*, in the sense that $\partial_\mu J^\mu = 0$. Indeed, the need for conservation of the current can be seen by simply counting states. A massive spin one particle has three spin states, which can be taken as the states with helicity +1, 0, and −1, while any massless, spin one particle like the photon can only have helicities +1 and −1: the current conservation condition just ensures that the helicity zero states of the spin one particle are not emitted in the limit of zero mass.

The inversions can be dealt with in much the same way as for the scalar field discussed in the previous section. To evaluate the effect of space inversion, we need a formula for $e^\mu(-\mathbf{p},\sigma)$. Using $L^\mu{}_\nu(-\mathbf{p}) = \mathscr{P}^\mu{}_\rho L^\rho{}_\tau(\mathbf{p})\mathscr{P}^\tau{}_\nu$ and Eq. (5.3.24), we have

$$e^\mu(-\mathbf{p},\sigma) = -\mathscr{P}^\mu{}_\nu e^\nu(\mathbf{p},\sigma). \tag{5.3.39}$$

Also, to evaluate the effect of time-reversal we need a formula for $(-1)^{1+\sigma}e^{\mu*}(-\mathbf{p},-\sigma)$. Using $(-1)^{1+\sigma}e^{\mu\,*}(0,-\sigma) = -e^\mu(0,\sigma)$ and the above formula for $L^\mu{}_\nu(-\mathbf{p})$, we find

$$(-1)^{1+\sigma}e^{\mu*}(-\mathbf{p},-\sigma) = \mathscr{P}^\mu{}_\nu\, e^\nu(\mathbf{p},\sigma). \tag{5.3.40}$$

Using these results and the transformation properties of the annihilation and creation operators given in Section 4.2, it is straightforward to work out the inversion transformation properties of the annihilation and creation fields. Once again we find that, in order for causal fields to be transformed into other fields with which they commute at space-like separations, it is necessary that the intrinsic space inversion, charge-conjugation, and time-reversal phases for spin one particles and their antiparticles be related by

$$\eta^c = \eta^* \,, \tag{5.3.41}$$

$$\xi^c = \xi^* \,, \tag{5.3.42}$$

$$\zeta^c = \zeta^* \,. \tag{5.3.43}$$

(In particular all phases must be real if the spin one particle is its own antiparticle.) With these phase conditions satisfied, our causal vector field (5.3.34) has the inversion transformation properties

$$\mathsf{P}v^\mu(x)\mathsf{P}^{-1} = -\eta^* \mathscr{P}^\mu{}_\nu v^\nu(\mathscr{P}x) \,, \tag{5.3.44}$$

$$\mathsf{C}v^\mu(x)\mathsf{C}^{-1} = \xi^* v^{\mu\dagger}(x) \,, \tag{5.3.45}$$

$$\mathsf{T}v^\mu(x)\mathsf{T}^{-1} = \zeta^* \mathscr{P}^\mu{}_\nu v^\nu(-\mathscr{P}x) \,. \tag{5.3.46}$$

In particular, the minus sign in Eq. (5.3.44) means that a vector field that transforms as a polar vector, with no extra phases or signs accompanying the matrix $\mathscr{P}^\mu{}_\nu$, describes a spin one particle with intrinsic parity $\eta = -1$.

5.4 The Dirac Formalism

Among all the representations of the homogeneous Lorentz group, there is one that plays a special role in physics. As we saw in Section 1.1, this representation was introduced into the theory of the electron by Dirac,[3] but as so often happens it was already known to mathematicians,[4] because it provides the basis of one of the two broad classes of representations of the rotation or Lorentz groups (actually, of their covering groups — see Section 2.7) in any number of dimensions. From the point of view we are following here, the structure and properties of any quantum field are dictated by the representation of the homogeneous Lorentz group under which it transforms, so it will be natural for us to describe the Dirac formalism as it first appeared in mathematics, rather than as it was introduced by Dirac.

By a representation of the homogeneous Lorentz group, we mean a set of matrices $D(\Lambda)$ satisfying the group multiplication law

$$D(\bar{\Lambda})D(\Lambda) = D(\bar{\Lambda}\Lambda) \,.$$

Just as for the unitary operators $U(\Lambda)$, we can study the properties of these matrices by considering the infinitesimal case,

$$\Lambda^\mu{}_\nu = \delta^\mu{}_\nu + \omega^\mu{}_\nu\,, \tag{5.4.1}$$

$$\omega_{\mu\nu} = -\omega_{\nu\mu}\,, \tag{5.4.2}$$

for which

$$D(\Lambda) = 1 + \frac{i}{2}\,\omega_{\mu\nu}\mathscr{J}^{\mu\nu} \tag{5.4.3}$$

with $\mathscr{J}^{\mu\nu} = -\mathscr{J}^{\nu\mu}$ a set of matrices satisfying the commutation relations (2.4.12):

$$i[\mathscr{J}^{\mu\nu}, \mathscr{J}^{\rho\sigma}] = \eta^{\nu\rho}\mathscr{J}^{\mu\sigma} - \eta^{\mu\rho}\mathscr{J}^{\nu\sigma} - \eta^{\sigma\mu}\mathscr{J}^{\rho\nu} + \eta^{\sigma\nu}\mathscr{J}^{\rho\mu}\,. \tag{5.4.4}$$

To find such a set of matrices, suppose we first construct matrices γ^μ that satisfy the *anti*commutation relations

$$\{\gamma^\mu, \gamma^\nu\} = 2\eta^{\mu\nu} \tag{5.4.5}$$

and tentatively define

$$\mathscr{J}^{\mu\nu} = -\frac{i}{4}[\gamma^\mu, \gamma^\nu]\,, \tag{5.4.6}$$

It is elementary, using Eq. (5.4.5), to show that

$$[\mathscr{J}^{\mu\nu}, \gamma^\rho] = -i\gamma^\mu\eta^{\nu\rho} + i\gamma^\nu\eta^{\mu\rho} \tag{5.4.7}$$

and from this we easily see that Eq. (5.4.6) does indeed satisfy the desired commutation relation Eq. (5.4.4). We shall further assume that the matrices γ_μ are *irreducible*; that is, that there is no proper subspace that is left invariant by all these matrices. Otherwise we could choose some smaller set of field components, which would transform as in Eqs. (5.4.3) and (5.4.6), with an irreducible set of γ_μs.

Any set of matrices satisfying a relation like Eq. (5.4.5) (or its Euclidean analog, with $\eta_{\mu\nu}$ replaced with a Kronecker delta) is called a *Clifford algebra.* The importance in mathematics of this particular representation of the homogeneous Lorentz group (or, more accurately, its covering group) arises from the fact (shown in Section 5.6) that the most general irreducible representation of the Lorentz group is either a tensor, or a spinor transforming as in Eqs. (5.4.3) and (5.4.6), or a direct product of a spinor and a tensor.

The commutation relation (5.4.7) can be summarized by saying that γ^ρ is a *vector*, in the sense that Eq. (5.4.3) satisfies

$$D(\Lambda)\gamma^\rho D^{-1}(\Lambda) = \Lambda_\sigma{}^\rho\gamma^\sigma\,. \tag{5.4.8}$$

In the same sense, the unit matrix is trivially a *scalar*

$$D(\Lambda)\,1\,D^{-1}(\Lambda) = 1 \tag{5.4.9}$$

and Eq. (5.4.4) shows that $\mathscr{J}^{\rho\sigma}$ is an antisymmetric *tensor*

$$D(\Lambda)\mathscr{J}^{\rho\sigma}D^{-1}(\Lambda) = \Lambda_\mu{}^\rho\Lambda_\nu{}^\sigma\mathscr{J}^{\mu\nu} . \tag{5.4.10}$$

The matrices γ^μ can be used to construct other totally antisymmetric tensors

$$\mathscr{A}^{\rho\sigma\tau} \equiv \gamma^{[\rho}\gamma^\sigma\gamma^{\tau]} , \tag{5.4.11}$$

$$\mathscr{P}^{\rho\sigma\tau\eta} = \gamma^{[\rho}\gamma^\sigma\gamma^\tau\gamma^{\eta]} . \tag{5.4.12}$$

The brackets here are a standard notation, indicating that we are to sum over all permutations of the indices within the brackets, with a plus or minus sign for even or odd permutations, respectively. For instance, Eq. (5.4.11) is shorthand for

$$\mathscr{A}^{\rho\sigma\tau} \equiv \gamma^\rho\gamma^\sigma\gamma^\tau - \gamma^\rho\gamma^\tau\gamma^\sigma - \gamma^\sigma\gamma^\rho\gamma^\tau + \gamma^\tau\gamma^\rho\gamma^\sigma + \gamma^\sigma\gamma^\tau\gamma^\rho - \gamma^\tau\gamma^\sigma\gamma^\rho .$$

By repeated use of Eq. (5.4.5) we can write any product of γs as a sum of antisymmetrized products of γs times a product of metric tensors, so the totally antisymmetric tensors form a complete basis for the set of all matrices that can be constructed from the Dirac matrices.

This formalism automatically contains a parity transformation, conventionally taken as

$$\beta \equiv i\gamma^0 . \tag{5.4.13}$$

Applied to the Dirac matrices, this gives

$$\beta\gamma^i\beta^{-1} = -\gamma^i , \qquad \beta\gamma^0\beta^{-1} = +\gamma^0 . \tag{5.4.14}$$

(We here label indices so that μ runs over values $0, 1, 2, \cdots$.) The same similarity transformation, applied to any product of γ-matrices, then yields just a plus or minus sign, according to whether the product contains an even or an odd number of γs with space-like indices, respectively. In particular,

$$\beta\mathscr{J}^{ij}\beta^{-1} = \mathscr{J}^{ij} , \tag{5.4.15}$$

$$\beta\mathscr{J}^{i0}\beta^{-1} = -\mathscr{J}^{i0} . \tag{5.4.16}$$

Everything so far in this section applies in any number of spacetime dimensions and for any 'metric' $\eta_{\mu\nu}$. In four spacetime dimensions, however, there is a special feature, that no totally antisymmetric tensor can have more than four indices, so the sequence of tensors $\mathbf{1}, \gamma^\rho, \mathscr{J}^{\rho\sigma}, \mathscr{A}^{\rho\sigma\tau}, \cdots$ terminates with the tensor (5.4.12). Furthermore, each of these tensors transforms differently under Lorentz and/or parity transformations so they are all linearly independent.* The number of linearly independent components of these tensors is one for **1**, four for γ^ρ, six for $\mathcal{J}^{\rho\sigma}$, four for $\mathcal{A}^{\rho\sigma\tau}$, and one for $\mathcal{P}^{\mu\nu\rho\sigma}$, or 16 independent components in all. (The general rule is that a totally antisymmetric tensor with n indices in d dimensions has a number of independent components equal to the binomial coefficient $d!/n!(d-n)!$) There are at most ν^2 independent $\nu \times \nu$ matrices, so they must have at least $\sqrt{16} = 4$ rows and columns. Dirac matrices of the minimum dimensionality are necessarily irreducible; if reducible, the subspace left invariant by these matrices would furnish a representation of lower dimensionality. We shall therefore take the γ^μ to be 4×4 matrices.

(More generally, in any even number d of spacetime dimensions, one can form antisymmetric tensors with $0, 1, \cdots, d$ indices, which contain altogether a number of independent components equal to

$$\sum_{n=0}^{d} \frac{d!}{n!(d-n)!} = 2^d ,$$

so the γ-matrices must have at least $2^{d/2}$ rows and columns. In spaces or spacetimes with odd dimensionality, the totally antisymmetric tensors of rank n and $d-n$ can be linearly related by the conditions**

$$\gamma^{[\mu_1}\gamma^{\mu_2}\cdots\gamma^{\mu_r]} \propto \epsilon^{\mu_1\mu_2\cdots\mu_d}\,\gamma_{[\mu_{r+1}}\gamma_{\mu_{r+2}}\cdots\gamma_{\mu_d]},$$

for $r = 0, 1, 2, ..., d-1$, with $\epsilon^{\mu_1\mu_2\cdots\mu_d}$ totally antisymmetric, and the left-hand side taken as the unit matrix for $r = 0$. Under these conditions there are only 2^{d-1} independent tensors, requiring γ-matrices of dimensionality at least $2^{(d-1)/2}$.)

Returning now to four spacetime dimensions, we shall choose an explicit set of 4×4 γ-matrices. One very convenient choice is

$$\gamma^0 = -i\begin{bmatrix} 0 & 1 \\ 1 & 0 \end{bmatrix}, \qquad \boldsymbol{\gamma} = -i\begin{bmatrix} 0 & \boldsymbol{\sigma} \\ -\boldsymbol{\sigma} & 0 \end{bmatrix}, \tag{5.4.17}$$

where **1** is the unit 2×2 matrix, and the components of $\boldsymbol{\sigma}$ are the usual

---

* Alternatively, these matrices can be shown to be linearly independent by noting that they form an orthogonal set, with the scalar product of two matrices defined by the trace of their product. Note that none of these matrices can vanish, because each component of each of these tensors is proportional to a product of different γ-matrices, and such a product has a square equal to plus or minus the product of the corresponding squares, and hence equal to ± 1.

** This constraint does not interfere with the inclusion of space inversion in the Dirac representation of the Lorentz group in odd-dimensional spacetime, because here the tensor $\epsilon^{\mu_1\mu_2\cdots\mu_d}$ is even under inversion of space coordinates. If we don't care about space inversion, we can also construct $2^{(d-1)/2}$-dimensional irreducible representations of the proper orthochronous Lorentz group in even spacetime dimensions by imposing the above condition relating antisymmetrized products of r and $d-r$ Dirac matrices. An example is provided by the submatrices in Eqs. (5.4.19) and (5.4.20) below.

Pauli matrices

$$\sigma_1 = \begin{pmatrix} 0 & 1 \\ 1 & 0 \end{pmatrix}, \quad \sigma_2 = \begin{pmatrix} 0 & -i \\ i & 0 \end{pmatrix}, \quad \sigma_3 = \begin{pmatrix} 1 & 0 \\ 0 & -1 \end{pmatrix}. \tag{5.4.18}$$

(The σ_i are just the 2×2 γ-matrices in three dimensions.) It can be shown[5] that any other irreducible set of γ-matrices are related to these by a similarity transformation. From Eq. (5.4.17), we can easily calculate the Lorentz group generators (5.4.6):

$$\mathscr{J}^{ij} = \frac{1}{2}\,\epsilon_{ijk} \begin{bmatrix} \sigma_k & 0 \\ 0 & \sigma_k \end{bmatrix} \tag{5.4.19}$$

$$\mathscr{J}^{i0} = +\frac{i}{2} \begin{bmatrix} \sigma_i & 0 \\ 0 & -\sigma_i \end{bmatrix}. \tag{5.4.20}$$

(Here ϵ_{ijk} is the totally antisymmetric tensor in three dimensions, with $\epsilon_{123} \equiv +1$.) We note that these are block-diagonal, so the Dirac matrices provide a *reducible* representation of the proper orthochronous Lorentz group, the direct sum of two irreducible representation with $\mathscr{J}^{ij} = \pm i\,\epsilon_{ijk}\mathscr{J}^{k0}$.

It is convenient to write the totally antisymmetric tensors (5.4.11) and (5.4.12) in a somewhat simpler way. The matrix (5.4.12) is totally antisymmetric, and therefore proportional to the pseudotensor $\epsilon^{\rho\sigma\tau\eta}$, defined as a totally antisymmetric quantity with $\epsilon^{0123} = +1$. Setting ρ, σ, τ, η equal to 0,1,2,3, respectively, we see that

$$\mathscr{P}^{\rho\sigma\tau\eta} = 4!\, i\, \epsilon^{\rho\sigma\tau\eta}\gamma_5\,, \tag{5.4.21}$$

where

$$\gamma_5 \equiv -i\gamma^0\gamma^1\gamma^2\gamma^3\,. \tag{5.4.22}$$

The matrix γ_5 is a pseudoscalar in the sense that

$$[\mathscr{J}^{\rho\sigma}, \gamma_5] = 0\,, \tag{5.4.23}$$

$$\beta\gamma_5\beta^{-1} = -\gamma_5\,. \tag{5.4.24}$$

Similarly, $\mathscr{A}^{\rho\sigma\tau}$ must be proportional to $\epsilon^{\rho\sigma\tau\eta}$ contracted with some matrix $\mathscr{A}_\eta$, and by setting ρ, σ, τ equal in turn to 0,1,2 or 0,1,3 or 0,2,3 or 1,2,3, we find

$$\mathscr{A}^{\rho\sigma\tau} = 3!\, i\, \epsilon^{\rho\sigma\tau\eta}\gamma_5\gamma_\eta\,. \tag{5.4.25}$$

The 16 independent 4×4 matrices can therefore be taken as the components of the scalar 1, the vector γ^ρ, the antisymmetric tensor $\mathscr{J}^{\rho\sigma}$, the 'axial' vector $\gamma_5\gamma_\eta$, and the pseudoscalar γ_5. It is easy to see that the matrix γ_5 has unit square

$$\gamma_5{}^2 = 1 \tag{5.4.26}$$

and anticommutes with all γ^μ

$$\{\gamma_5, \gamma^\mu\} = 0\,. \tag{5.4.27}$$

The notation γ_5 is particularly appropriate, because the anticommutation relations (5.4.26) and (5.4.27), together with Eq. (5.4.5) show that $\gamma^0, \gamma^1, \gamma^2, \gamma^3, \gamma_5$ provide a Clifford algebra in five spacetime dimensions. For the particular 4×4 representation (5.4.17) of the γ-matrices, the matrix γ_5 is

$$\gamma_5 = \begin{pmatrix} 1 & 0 \\ 0 & -1 \end{pmatrix}. \tag{5.4.28}$$

This representation is convenient because it reduces $\mathscr{J}^{\rho\sigma}$ and γ_5 to block-diagonal form. As we shall see, this makes it particularly useful for dealing with particles in the ultra-relativistic limit, $v \to c$. (It is not, however, the representation described in Section 1.1 that was originally introduced by Dirac, because Dirac was mostly interested in electrons in atoms where $v << c$, and in this case it is more convenient to adopt a representation for which γ^0 rather than γ_5 is diagonal.)

The representation of the homogeneous Lorentz group we have constructed here is not unitary, because the generators $\mathscr{J}^{\rho\sigma}$ are not all represented by Hermitian matrices. In particular, in the representation (5.4.17) we have $\mathscr{J}^{ij}$ Hermitian, but $\mathscr{J}^{i0}$ is anti-Hermitian. Such reality conditions can conveniently be written in a manifestly Lorentz-invariant fashion by introducing the matrix $\beta \equiv i\gamma^0$ of Eq. (5.4.13), which in the representation (5.4.17) takes the form

$$\beta = \begin{bmatrix} 0 & 1 \\ 1 & 0 \end{bmatrix}. \tag{5.4.29}$$

Inspection of Eq. (5.4.17) shows that

$$\beta \gamma^{\mu\dagger} \beta = -\gamma^\mu \tag{5.4.30}$$

and it follows then that

$$\beta \mathscr{J}^{\rho\sigma\dagger} \beta = \mathscr{J}^{\rho\sigma}\,. \tag{5.4.31}$$

Hence, though not unitary, the matrices $D(\Lambda)$ satisfy the pseudounitarity relation

$$\beta\, D(\Lambda)^\dagger \beta = D(\Lambda)^{-1}\,. \tag{5.4.32}$$

Also, γ_5 is Hermitian and anticommutes with β so

$$\beta {\gamma_5}^\dagger \beta = -\gamma_5 \tag{5.4.33}$$

and it follows that

$$\beta(\gamma_5\gamma_\mu)^\dagger \beta = -\gamma_5\gamma_\mu. \tag{5.4.34}$$

The Dirac and related matrices also have important symmetry properties. Inspection of Eqs. (5.4.17) and (5.4.18) shows that γ_μ is symmetric for $\mu = 0, 2$ and antisymmetric for $\mu = 1, 3$, so

$$\gamma_\mu^{\rm T} = -\mathscr{C}\gamma_\mu\mathscr{C}^{-1}\,, \tag{5.4.35}$$

where T denotes a transpose, and

$$\mathscr{C} \equiv \gamma_2\beta = -i\begin{bmatrix} \sigma_2 & 0 \\ 0 & -\sigma_2 \end{bmatrix}\,. \tag{5.4.36}$$

It follows immediately that

$$\mathscr{J}_{\mu\nu}^{\rm T} = -\mathscr{C}\mathscr{J}_{\mu\nu}\mathscr{C}^{-1}\,, \tag{5.4.37}$$

$$\gamma_5^{\rm T} = +\mathscr{C}\gamma_5\mathscr{C}^{-1}\,, \tag{5.4.38}$$

$$(\gamma_5\gamma_\mu)^{\rm T} = +\mathscr{C}\gamma_5\gamma_\mu\mathscr{C}^{-1}\,. \tag{5.4.39}$$

These signs will prove significant when we consider the charge-conjugation properties of various currents in the next section. Of course, we can combine our results for adjoints and transposes to obtain the complex conjugates of the Dirac and allied matrices:

$$\gamma_\mu^* = \beta\mathscr{C}\gamma_\mu\mathscr{C}^{-1}\beta\,, \tag{5.4.40}$$

$$\mathscr{J}_{\mu\nu}^* = -\beta\mathscr{C}\mathscr{J}_{\mu\nu}\mathscr{C}^{-1}\beta\,, \tag{5.4.41}$$

$$\gamma_5^* = -\beta\mathscr{C}\gamma_5\mathscr{C}^{-1}\beta\,, \tag{5.4.42}$$

$$(\gamma_5\gamma_\mu)^* = -\beta\mathscr{C}\gamma_5\gamma_\mu\mathscr{C}^{-1}\beta. \tag{5.4.43}$$

5.5 Causal Dirac Fields

We now want to construct particle annihilation and antiparticle creation fields that transform under the Lorentz group according to the Dirac representation of this group, discussed in the previous section. In general these take the form given in Eqs. (5.1.17) and (5.1.18):

$$\psi_\ell^+(x) = (2\pi)^{-3/2}\sum_\sigma\int d^3p\; u_\ell(\mathbf{p},\sigma)e^{ip\cdot x}a(\mathbf{p},\sigma) \tag{5.5.1}$$

and

$$\psi_\ell^{-c}(x) = (2\pi)^{-3/2}\sum_\sigma\int d^3p\; v_\ell(\mathbf{p},\sigma)e^{-ip\cdot x}a^{c\dagger}(\mathbf{p},\sigma). \tag{5.5.2}$$

with the particle species label omitted here. In order to calculate the coefficient functions $u_\ell(\mathbf{p},\sigma)$ and $v_\ell(\mathbf{p},\sigma)$ appearing in these formulas, we must first use Eqs. (5.1.25) and (5.1.26) to find u_ℓ and v_ℓ for zero momentum, and then apply Eqs. (5.1.21) and (5.1.22) to calculate them for arbitrary momenta, with $D_{\bar\ell\ell}(\Lambda)$ in both cases taken as the 4×4 Dirac representation of the homogeneous Lorentz group discussed in the previous section.

Using Eq. (5.4.19), the zero-momentum conditions (5.1.25) and (5.1.26) read*

$$\sum_{\bar\sigma} u_{\bar m\pm}(0,\bar\sigma)\mathbf{J}^{(j)}_{\bar\sigma\sigma} = \sum_m \tfrac{1}{2}\boldsymbol{\sigma}_{\bar m m}u_{m\pm}(0,\sigma)$$

and

$$-\sum_{\bar\sigma} v_{\bar m\pm}(0,\bar\sigma)\mathbf{J}_{\bar\sigma\sigma}{}^{(j)*} = \sum_m \tfrac{1}{2}\boldsymbol{\sigma}_{\bar m m}v_{m\pm}(0,\sigma)\,.$$

In other words, if we regard $u_{m\pm}(0,\sigma)$ and $v_{m\pm}(0,\sigma)$ as the m,σ elements of matrices $U_\pm$ and $V_\pm$, we have in matrix notation

$$U_\pm\mathbf{J}^{(j)} = \tfrac{1}{2}\boldsymbol{\sigma}U_\pm \tag{5.5.3}$$

and

$$-V_\pm\mathbf{J}^{(j)*} = \tfrac{1}{2}\boldsymbol{\sigma}V_\pm. \tag{5.5.4}$$

Now, the $(2j+1)$-dimensional matrices $\mathbf{J}^{(j)}$ and $-\mathbf{J}^{(j)*}$ and the 2×2 matrices $\frac{1}{2}\boldsymbol{\sigma}$ all provide irreducible representations of the Lie algebra of the rotation group. A general theorem of group theory known as Schur's lemma[6] tells us that when a matrix like $U_\pm$ or $V_\pm$ connects two such representations as in Eqs. (5.5.3) and (5.5.4), the matrix must either vanish (a possibility of no interest here) or else be square and non-singular. Hence the Dirac field can only describe particles of spin $j=\frac{1}{2}$ (so that $2j+1=2$) and the matrices $\mathbf{J}^{(1/2)}$ and $-\mathbf{J}^{(1/2)*}$ must be the same as $\frac{1}{2}\boldsymbol{\sigma}$ up to a similarity transformation. In fact, in the standard representation (2.5.21), (2.5.22) of the rotation generators, we have $\mathbf{J}^{(1/2)}=\frac{1}{2}\boldsymbol{\sigma}$ and $-\mathbf{J}^{(1/2)*}=\frac{1}{2}\sigma_2\boldsymbol{\sigma}\sigma_2$. It follows then that $U_\pm$ and $V_\pm\sigma_2$ must commute with $\boldsymbol{\sigma}$, and hence must be proportional to the unit matrix:

$$u_{m,\pm}(0,\sigma) = c_\pm\delta_{m\sigma}\,, \qquad v_{m,\pm}(0,\sigma) = -id_\pm(\sigma_2)_{m\sigma}\,. \tag{5.5.5}$$

---

* We are here dropping the species label n, and replacing the four-component index ℓ with a pair of indices, one 2-valued index m labelling the rows and columns of the submatrices in Eqs. (5.4.19) and (5.4.20), and a second index taking values $\pm$, labelling the rows and columns of the supermatrix in Eqs. (5.4.19) and (5.4.20).

In other words

$$u(0,\,\tfrac{1}{2}) = \begin{bmatrix} c_+ \\ 0 \\ c_- \\ 0 \end{bmatrix}, \qquad u(0,-\tfrac{1}{2}) = \begin{bmatrix} 0 \\ c_+ \\ 0 \\ c_- \end{bmatrix},$$

$$v(0,\,\tfrac{1}{2}) = \begin{bmatrix} 0 \\ d_+ \\ 0 \\ d_- \end{bmatrix}, \qquad v(0,-\tfrac{1}{2}) = -\begin{bmatrix} d_+ \\ 0 \\ d_- \\ 0 \end{bmatrix}$$

and the spinors at finite momentum are

$$u(\mathbf{p},\sigma) = \sqrt{m/p^0}D\Big(L(p)\Big)u(0,\sigma)\,, \tag{5.5.6}$$

$$v(\mathbf{p},\sigma) = \sqrt{m/p^0}D\Big(L(p)\Big)v(0,\sigma)\,. \tag{5.5.7}$$

It now only remains to say something about the constants $c_\pm$ and $d_\pm$. In general, these are quite arbitrary — we could even choose c_- and d_- or c_+ and d_+ to be zero if we liked, so that the Dirac field would have only *two* non-vanishing components. The only physical principle that could tell us anything about the relative values of the $c_\pm$ or the $d_\pm$ is the conservation of parity. We recall that under a space inversion, the particle annihilation and antiparticle creation operators undergo the transformations:

$$\mathsf{P}a(\mathbf{p},\sigma)\mathsf{P}^{-1} = \eta^*a(-\mathbf{p},\sigma) \tag{5.5.8}$$

$$\mathsf{P}a^{c\dagger}(\mathbf{p},\sigma)\mathsf{P}^{-1} = \eta^c a^{c\dagger}(-\mathbf{p},\sigma) \tag{5.5.9}$$

and so

$$\mathsf{P}\psi_\ell^+(x)\mathsf{P}^{-1} = \eta^*(2\pi)^{-3/2}\sum_\sigma\int d^3p\, u_\ell(-\mathbf{p},\sigma)e^{ip\cdot\mathscr{P}x}a(\mathbf{p},\sigma)\,, \tag{5.5.10}$$

$$\mathsf{P}\psi_\ell^{-c}(x)\mathsf{P}^{-1} = \eta^c(2\pi)^{-3/2}\sum_\sigma\int d^3p\, v_\ell(-\mathbf{p},\sigma)e^{-ip\cdot\mathscr{P}x}a^{c\dagger}(\mathbf{p},\sigma). \tag{5.5.11}$$

Also, Eqs. (5.4.16), (5.1.21), and (5.1.22) give

$$u(-\mathbf{p},\sigma) = \sqrt{m/p^0}\,\beta D(L(\mathbf{p}))\beta u(0,\sigma) \tag{5.5.12}$$

$$v(-\mathbf{p},\sigma) = \sqrt{m/p^0}\,\beta D(L(\mathbf{p}))\beta v(0,\sigma)\,. \tag{5.5.13}$$

(Since $\beta^2 = 1$, we are no longer making a distinction between β and β^{-1}.) In order that the parity operator should transform the annihilation and creation fields at the point x into something proportional to these fields at $\mathscr{P}x$, it is necessary that $\beta u(0,\sigma)$ and $\beta v(0,\sigma)$ be proportional to $u(0,\sigma)$ and $v(0,\sigma)$, respectively:

$$\beta u(0,\sigma) = b_u u(0,\sigma)\,, \qquad \beta v(0,\sigma) = b_v v(0,\sigma)\,, \tag{5.5.14}$$

where b_u and b_v are sign factors, $b_u^2 = b_v^2 = 1$. In this case, the fields have the simple space-inversion properties:

$$\mathsf{P}\psi^+(x)\mathsf{P}^{-1} = \eta^* b_u \beta \psi^+(\mathscr{P}x)\,, \tag{5.5.15}$$

$$\mathsf{P}\psi^{-c}(x)\mathsf{P}^{-1} = \eta^c b_v \beta \psi^{-c}(\mathscr{P}x). \tag{5.5.16}$$

By adjusting the overall scales of the fields, we can choose the coefficient functions at zero momentum to have the form:

$$u(0,\tfrac{1}{2}) = \frac{1}{\sqrt{2}}\begin{bmatrix}1\\0\\b_u\\0\end{bmatrix}, \qquad u(0,-\tfrac{1}{2}) = \frac{1}{\sqrt{2}}\begin{bmatrix}0\\1\\0\\b_u\end{bmatrix}, \tag{5.5.17}$$

$$v(0,\tfrac{1}{2}) = \frac{1}{\sqrt{2}}\begin{bmatrix}0\\1\\0\\b_v\end{bmatrix}, \qquad v(0,-\tfrac{1}{2}) = \frac{-1}{\sqrt{2}}\begin{bmatrix}1\\0\\b_v\\0\end{bmatrix}. \tag{5.5.18}$$

Now let's try to put together the annihilation and creation fields in a linear combination

$$\psi(x) = \kappa\psi^+(x) + \lambda\psi^{-c}(x) \tag{5.5.19}$$

that commutes or anticommutes with itself and its adjoint at space-like separations. A straightforward calculation gives

$$[\psi_\ell(x), \psi^\dagger_{\bar{\ell}}(y)]_\mp = (2\pi)^{-3}\int d^3p\;[|\kappa|^2 N_{\ell\bar{\ell}}(\mathbf{p})e^{ip\cdot(x-y)} \mp |\lambda|^2 M_{\ell\bar{\ell}}(\mathbf{p})e^{-ip\cdot(x-y)}]\,, \tag{5.5.20}$$

where

$$N_{\ell\bar{\ell}}(\mathbf{p}) \equiv \sum_\sigma u_\ell(\mathbf{p},\sigma)u^*_{\bar{\ell}}(\mathbf{p},\sigma)\,, \tag{5.5.21}$$

$$M_{\ell\bar{\ell}}(\mathbf{p}) \equiv \sum_\sigma v_\ell(\mathbf{p},\sigma)v^*_{\bar{\ell}}(\mathbf{p},\sigma)\,. \tag{5.5.22}$$

By using either the eigenvalue conditions (5.5.14) or the explicit formulas (5.5.17) and (5.5.18), we find at zero momentum:

$$N(0) = \frac{1+b_u\beta}{2}\,, \qquad M(0) = \frac{1+b_v\beta}{2}\,. \tag{5.5.23}$$

From Eqs. (5.5.6) and (5.5.7) we have then

$$N(\mathbf{p}) = \frac{m}{2p^0} D\Big(L(p)\Big)[1 + b_u\beta]D^\dagger\Big(L(p)\Big) , \tag{5.5.24}$$

$$M(\mathbf{p}) = \frac{m}{2p^0} D\Big(L(p)\Big)[1 + b_v\beta]D^\dagger\Big(L(p)\Big) . \tag{5.5.25}$$

The pseudounitarity condition (5.4.32) yields

$$D\Big(L(p)\Big)\beta D^\dagger\Big(L(p)\Big) = \beta$$

and

$$D\Big(L(p)\Big)D^\dagger\Big(L(p)\Big) = D\Big(L(p)\Big)\beta D^{-1}\Big(L(p)\Big)\beta .$$

We also recall that $\beta = i\gamma^0$, so by using the Lorentz transformation rule (5.4.8) we have

$$D\Big(L(p)\Big)\beta D^{-1}\Big(L(p)\Big) = iL_\mu{}^0(p)\gamma^\mu = -ip_\mu\gamma^\mu/m. \tag{5.5.26}$$

Putting this together, we find**

$$N(\mathbf{p}) = \frac{1}{2p^0}[-ip^\mu\gamma_\mu + b_u m]\beta , \tag{5.5.27}$$

$$M(\mathbf{p}) = \frac{1}{2p^0}[-ip^\mu\gamma_\mu + b_v m]\beta . \tag{5.5.28}$$

Using this in Eq. (5.5.20) yields finally

$$\begin{aligned}[\psi_\ell(x), \psi^\dagger_{\bar\ell}(y)]_\mp = \Big(|\kappa|^2[-\gamma^\mu\partial_\mu + b_u m]\beta\Delta_+(x-y) \\ \mp|\lambda|^2[-\gamma^\mu\partial_\mu + b_v m]\beta\Delta_+(y-x)\Big)_{\ell\bar\ell},\end{aligned} \tag{5.5.29}$$

where Δ_+ is the function introduced in Section 5.2:

$$\Delta_+(x) \equiv \int \frac{d^3p}{2p^0(2\pi)^3}e^{ip\cdot x} .$$

We saw in Section 5.2 that for $x - y$ space-like $\Delta_+(x-y)$ is an even function of $x - y$ and, of course, this implies that its first derivatives are odd functions of $x - y$. Hence, in order that both the derivative and the non-derivative terms in the commutator or anticommutator should vanish at space-like separations, it is necessary and sufficient that

$$|\kappa|^2 = \mp|\lambda|^2 \tag{5.5.30}$$

** Sometimes an extra factor $\sqrt{p^0/m}$ is included in the Dirac spinors, so that m appears in place of p^0 in the denominators of the spin sums (5.5.27) and (5.5.28). The normalization convention used here has the advantage that it goes smoothly over to the case $m = 0$.

and

$$|\kappa|^2 b_u = \pm|\lambda|^2 b_v \,. \tag{5.5.31}$$

Clearly Eq. (5.5.30) is only possible if we choose the bottom sign, $\mp = +$; that is, *the particles described by a Dirac field must be fermions.* It is also then necessary that $|\kappa|^2 = |\lambda|^2$ and $b_u = -b_v$. Just as for scalars we have the freedom to redefine the relative phase of the creation and annihilation operators to make the ratio κ/λ real and positive, in which case $\kappa = \lambda$, and by adjusting the overall scale and phase of the field ψ we may then take

$$\kappa = \lambda = 1. \tag{5.5.32}$$

Finally if we like we can replace ψ with $\gamma_5\psi$, which changes the sign of both b_u and b_v, so we can always take

$$b_u = -b_v = +1. \tag{5.5.33}$$

For future use, we record here that the Dirac field is now

$$\psi_\ell(x) = (2\pi)^{-3/2}\sum_\sigma \int d^3p\ [u_\ell(\mathbf{p},\sigma)e^{ip\cdot x}a(\mathbf{p},\sigma) + v_\ell(\mathbf{p},\sigma)e^{-ip\cdot x}a^{c\dagger}(\mathbf{p},\sigma)] \tag{5.5.34}$$

while the coefficient functions at zero momentum are

$$u(0,\ \tfrac{1}{2}) = \frac{1}{\sqrt{2}}\begin{bmatrix}1\\0\\1\\0\end{bmatrix}, \qquad u(0,-\ \tfrac{1}{2}) = \frac{1}{\sqrt{2}}\begin{bmatrix}0\\1\\0\\1\end{bmatrix}, \tag{5.5.35}$$

$$v(0,\ \tfrac{1}{2}) = \frac{1}{\sqrt{2}}\begin{bmatrix}0\\1\\0\\-1\end{bmatrix}, \qquad v(0,-\ \tfrac{1}{2}) = \frac{1}{\sqrt{2}}\begin{bmatrix}-1\\0\\1\\0\end{bmatrix}. \tag{5.5.36}$$

The spin sums are

$$N(\mathbf{p}) = \frac{1}{2p^0}[-ip^\mu\gamma_\mu + m]\beta\,, \tag{5.5.37}$$

$$M(\mathbf{p}) = \frac{1}{2p^0}[-ip^\mu\gamma_\mu - m]\beta, \tag{5.5.38}$$

so the anticommutator is given by Eq. (5.5.20) as

$$[\psi_\ell(x), \psi^\dagger_{\bar\ell}(y)]_+ = \{[-\gamma^\mu\partial_\mu + m]\beta\}_{\ell\bar\ell}\ \Delta(x-y)\,. \tag{5.5.39}$$

Now let's return to the requirement that under a space inversion the field $\psi(x)$ must transform into something proportional to $\psi(\mathscr{P}x)$. For this to be possible the phases in Eqs. (5.5.15) and (5.5.16) must be equal, and so the intrinsic parities of particles and their antiparticles must be related by

$$\eta^c = -\eta^* \,. \tag{5.5.40}$$

That is, *the intrinsic parity $\eta\eta^c$ of a state consisting of a spin $\frac{1}{2}$ particle and its antiparticle is odd.* It is for this reason that negative parity mesons like the ρ^0 and J/ψ can be interpreted as s-wave bound states of quark–antiquark pairs. Eqs. (5.5.15) and (5.5.16) now give the transformation of the causal Dirac field under space inversion as

$$\mathsf{P}\psi(x)\mathsf{P}^{-1} = \eta^*\beta\psi(\mathscr{P}x) \,. \tag{5.5.41}$$

Before going on to the other inversions, this is a good place to mention that Eqs. (5.5.14), (5.5.33), and (5.5.26) show that $u(\mathbf{p},\sigma)$ and $v(\mathbf{p},\sigma)$ are eigenvectors of $-ip^\mu\gamma_\mu/m$ with eigenvalues $+1$ and -1, respectively:

$$(ip^\mu\gamma_\mu + m)u(\mathbf{p},\sigma) = 0 \,, \qquad (-ip^\mu\gamma_\mu + m)v(\mathbf{p},\sigma) = 0 \,. \tag{5.5.42}$$

It follows then that the field (5.5.34) satisfies the differential equation

$$(\gamma^\mu\partial_\mu + m)\psi(x) = 0 \,. \tag{5.5.43}$$

This is the celebrated Dirac equation for a free particle of spin $\frac{1}{2}$. From the point of view adopted here, the free-particle Dirac equation is nothing but a Lorentz-invariant record of the convention that we have used in putting together the two irreducible representations of the proper orthochronous Lorentz group to form a field that transforms simply also under space inversion.

In order to work out the charge-conjugation and time-reversal properties of the Dirac field, we will need expressions for the complex-conjugates of the u and v coefficient functions. These functions are real for zero momentum, but to obtain the coefficient functions at finite momentum we have to multiply with the complex matrix $D(L(p))$. From Eq. (5.4.41) we see that for general real $\omega_{\mu\nu}$:

$$[\exp(\tfrac{1}{2}i\mathscr{J}^{\mu\nu}\omega_{\mu\nu})]^* = \beta\mathscr{C}\exp(\tfrac{1}{2}i\mathscr{J}^{\mu\nu}\omega_{\mu\nu})\mathscr{C}^{-1}\beta$$

and so in particular

$$D(L(p))^* = \beta\mathscr{C}D(L(p))\mathscr{C}^{-1}\beta \,.$$

We also note that $\mathscr{C}^{-1}\beta u(0,\sigma) = -v(0,\sigma)$ and $\mathscr{C}^{-1}\beta v(0,\sigma) = -u(0,\sigma)$, so

$$u^*(\mathbf{p},\sigma) = -\beta\mathscr{C}v(\mathbf{p},\sigma) \,, \tag{5.5.44}$$

$$v^*(\mathbf{p},\sigma) = -\beta\mathscr{C}u(\mathbf{p},\sigma) \,. \tag{5.5.45}$$

In order for the field to transform under charge-conjugation into another field with which it commutes at space-like separations, it is necessary again that the charge conjugation parities of the particle and antiparticle be related by

$$\xi^c = \xi^* \, . \tag{5.5.46}$$

In this case, the field transforms as

$$\mathsf{C}\psi(x)\mathsf{C}^{-1} = -\xi^* \beta \mathscr{C} \psi^*(x) \, . \tag{5.5.47}$$

(We are calling the Hermitian adjoint of the field on the right-hand side ψ^* instead of $\psi^\dagger$ to emphasize that this is still a column vector, not a row.)

Although we have been distinguishing particles from their antiparticles, we have not ruled out the possibility that the two are actually identical. Such spin $\frac{1}{2}$ particles are called *Majorana fermions.* Following the same reasoning that led to Eq. (5.5.47), the Dirac field of such a particle must satisfy the reality condition

$$\psi(x) = -\beta \mathscr{C} \psi^*(x) \, . \tag{5.5.48}$$

For Majorana fermions the intrinsic space-inversion parity must be imaginary, $\eta = \pm i$, while the charge-conjugation parity must be real, $\xi = \pm 1$.

There is an important difference between fermions and bosons in the intrinsic charge-conjugation phase of states consisting of a particle and its antiparticle. Such a state may be written

$$\Phi \equiv \sum_{\sigma,\sigma'} \int d^3p \int d^3p' \; \chi(\mathbf{p},\sigma;\mathbf{p}',\sigma') a^\dagger(\mathbf{p},\sigma)\, a^{c\dagger}(\mathbf{p}',\sigma')\Phi_0 \, ,$$

where Φ_0 is the vacuum state. Under charge-conjugation, this state is transformed into

$$\mathsf{C}\Phi = \xi\xi^c \sum_{\sigma,\sigma'} \int d^3p \int d^3p' \; \chi(\mathbf{p},\sigma;\mathbf{p}',\sigma') a^{c\dagger}(\mathbf{p},\sigma)\, a^{\dagger}(\mathbf{p}',\sigma')\Phi_0 \, .$$

Interchanging the variables of integration and summation and using the anticommutation of the creation operators and Eq. (5.5.46), we can rewrite this as

$$\mathsf{C}\Phi = -\sum_{\sigma,\sigma'} \int d^3p \int d^3p' \; \chi(\mathbf{p}',\sigma';\mathbf{p},\sigma) a^\dagger(\mathbf{p},\sigma)\, a^{c\dagger}(\mathbf{p}',\sigma')\Phi_0 \, .$$

That is, *the intrinsic charge-conjugation parity of a state consisting of a particle described by a Dirac field and its antiparticle is odd,* in the sense that if the wave function χ of the state is even or odd under interchange of the momenta and spins of the particle and antiparticle, then the charge-conjugation operator applied to such a state gives a sign -1 or $+1$, respectively. The classic example here is positronium, the bound state of an electron and a positron. The two lowest states are a pair of nearly degenerate s-wave states with total spin $s = 0$ and $s = 1$, known respectively as para- and ortho-positronium. The wave function for these two states is even under interchange of momenta and odd or even respectively under interchange of spin z-components, so para- and ortho-positronium have $\mathsf{C} = +1$ and $\mathsf{C} = -1$, respectively. These values are dramatically confirmed in the decay modes of positronium: para-positronium decays rapidly into a pair of photons (each of which has $\mathsf{C} = -1$), while ortho-positronium can only decay much more slowly into three or more photons. In the same way, single ρ^0 and ω^0 mesons are produced as resonances in high-energy electron–positron annihilation through a one-photon intermediate state, so they must have $\mathsf{C} = -1$, which is consistent with their interpretation as quark–antiquark bound states with orbital angular momentum zero and total quark spin one.

Now we come to time-reversal. Recall the transformation properties of the particle annihilation and antiparticle creation operators given by Eq. (4.2.15):

$$\mathsf{T}a(\mathbf{p},\sigma)\mathsf{T}^{-1} = \zeta^*(-1)^{\frac{1}{2}-\sigma}a(-\mathbf{p},-\sigma)\,, \tag{5.5.49}$$

$$\mathsf{T}a^{c\dagger}(\mathbf{p},\sigma)\mathsf{T}^{-1} = \zeta^c(-1)^{\frac{1}{2}-\sigma}a^{c\dagger}(-\mathbf{p},-\sigma)\,. \tag{5.5.50}$$

Time-reversal of the field (5.5.34) thus gives

$$\mathsf{T}\psi_\ell(x)\mathsf{T}^{-1} = (2\pi)^{-3/2}\sum_\sigma\int d^3p\,(-1)^{\frac{1}{2}-\sigma}$$
$$\times\,[\zeta^* u_\ell^*(\mathbf{p},\sigma)e^{-ip\cdot x}a(-\mathbf{p},-\sigma) + \zeta^c v_\ell^*(\mathbf{p},\sigma)e^{ip\cdot x}a^{c\dagger}(-\mathbf{p},-\sigma)]$$

In order to put this back in the form given for ψ, we shall redefine the variables of integration and summation as $-\mathbf{p}$ and $-\sigma$, so we need formulas for $u_\ell^*(-\mathbf{p},-\sigma)$ and $v_\ell^*(-\mathbf{p},-\sigma)$ in terms of $u_\ell(\mathbf{p},\sigma)$ and $v_\ell(\mathbf{p},\sigma)$, respectively. For this purpose, we can use the fact that $\mathscr{J}^{i0}$ anticommutes with β and commutes with γ_5 together with our former result for $D(L(\mathbf{p}))^*$ to write

$$D^*(L(-\mathbf{p})) = \gamma_5\beta D^*(L(\mathbf{p}))\beta\gamma_5 = \gamma_5\mathscr{C}D(L(\mathbf{p}))\mathscr{C}^{-1}\gamma_5\,.$$

Also, Eqs. (5.4.36) and (5.5.35)–(5.5.36) give

$$\gamma_5\mathscr{C}^{-1}u(0,-\sigma) = (-1)^{\frac{1}{2}-\sigma}u(0,\sigma)\,,$$

$$\gamma_5\mathscr{C}^{-1}v(0,-\sigma) = (-1)^{\frac{1}{2}-\sigma}v(0,\sigma)\,,$$

so

$$(-1)^{\frac{1}{2}-\sigma}u^*(-\mathbf{p},-\sigma) = -\gamma_5\mathscr{C}\,u(\mathbf{p},\sigma)\,, \tag{5.5.51}$$

$$(-1)^{\frac{1}{2}-\sigma}v^*(-\mathbf{p},-\sigma) = -\gamma_5\mathscr{C}\,v(\mathbf{p},\sigma)\,. \tag{5.5.52}$$

We see then that in order for time-reversal to take the Dirac field into something proportional to itself at the time-reversed point (with which it would anticommute at space-like separations) it is necessary that the intrinsic time-reversal phases be related by

$$\zeta^c = \zeta^* \tag{5.5.53}$$

and in this case

$$\mathsf{T}\psi(x)\mathsf{T}^{-1} = -\zeta^*\gamma_5\mathscr{C}\psi(-\mathscr{P}x). \tag{5.5.54}$$

Now let us consider how to construct scalar interaction densities out of the Dirac fields and their adjoints. As already mentioned the Dirac representation is not unitary, so $\psi^\dagger\psi$ is not a scalar. To deal with this complication it is convenient to define a new sort of adjoint:

$$\bar{\psi} \equiv \psi^\dagger\beta. \tag{5.5.55}$$

Using the pseudounitarity condition (5.4.32), we see that the fermion bilinears constructed with $\bar{\psi}$ have the Lorentz transformation property

$$U_0(\Lambda)[\bar{\psi}(x)M\psi(x)]U_0^{-1}(\Lambda) = \bar{\psi}(\Lambda x)D(\Lambda)\,M\,D^{-1}(\Lambda)\psi(\Lambda x). \tag{5.5.56}$$

Also, under a space inversion

$$\mathsf{P}[\bar{\psi}(x)M\psi(x)]\mathsf{P}^{-1} = \bar{\psi}(\mathscr{P}x)\beta M\beta\psi(\mathscr{P}x). \tag{5.5.57}$$

Taking the matrix M as $\mathbf{1}$, γ^μ, $\mathscr{J}^{\mu\nu}$, $\gamma_5\gamma^\mu$, or γ_5 yields a bilinear $\bar{\psi}M\psi$ that transforms as a scalar, vector, tensor, axial vector, and pseudoscalar, respectively. (The terms 'axial' and 'pseudo' indicate that these have space-inversion properties opposite to those of ordinary vectors and scalars: a pseudoscalar has negative parity, while the space and time components of an axial vector have positive and negative parity, respectively.) These results apply also when the two fermion fields in the bilinear refer to different particle species, except that in this case a space inversion also yields a ratio of the intrinsic parities.

For instance, the original Fermi theory of beta decay involved an interaction density proportional to $\bar{\psi}_p\gamma^\mu\psi_n\,\bar{\psi}_e\gamma_\mu\psi_\nu$. Later it was realized that the most general Lorentz-invariant and parity-conserving non-derivative beta decay interaction takes the form of a linear combination of products like this, with γ_μ replaced with any one of the five covariant types of 4×4 matrices 1, γ^μ, $\mathscr{J}^{\mu\nu}$, $\gamma_5\gamma^\mu$, or γ_5. (As discussed in Chapter 2, we are defining the space-inversion operator so that the proton, neutron, and electron all have parity +1. If the neutrino is massless then its parity may also be defined as +1, if necessary by replacing the neutrino field with $\gamma_5\psi_\nu$.) When Lee and Yang[7] called parity conservation into question in 1956, they expanded the list of possible non-derivative interactions to include ten terms proportional to $\bar{\psi}_pM\psi_n\,\bar{\psi}_eM\psi_\nu$ and also $\bar{\psi}_pM\psi_n\,\bar{\psi}_eM\gamma_5\psi_\nu$, with M running over the matrices 1, γ^μ, $\mathscr{J}^{\mu\nu}$, $\gamma_5\gamma^\mu$, or γ_5.

It is also of some interest to study the charge-conjugation properties of these bilinears. Using Eqs. (5.5.47) and (5.4.35)–(5.4.39), we have

$$\begin{aligned}\mathsf{C}(\bar{\psi}M\psi)\mathsf{C}^{-1} &= (\beta\mathscr{C}\psi)^{\mathrm{T}}\beta M(\beta\mathscr{C}\psi^*) = -(\beta\mathscr{C}\psi^*)^{\mathrm{T}}M^{\mathrm{T}}\mathscr{C}\psi \\ &= \bar{\psi}\mathscr{C}^{-1}M^{\mathrm{T}}\mathscr{C}\psi = \pm\bar{\psi}M\psi \end{aligned} \tag{5.5.58}$$

the sign in the last expression being + for the matrices **1**, $\gamma_5\gamma_\mu$, and γ_5, and − for γ_μ and $\mathscr{J}_{\mu\nu}$. (The minus sign in the first line arises from Fermi statistics. We ignore a c-number anticommutator.) A boson field that interacts with the current $\bar{\psi}M\psi$ must therefore have $\mathsf{C} = +1$ for scalars, pseudoscalars, or axial vectors, and $\mathsf{C} = -1$ for vectors or antisymmetric tensors. This is one way of seeing that the π^0 (which couples to pseudoscalar or axial-vector nucleon currents) has $\mathsf{C} = +1$, while the photon has $\mathsf{C} = -1$.

5.6 General Irreducible Representations of the Homogeneous Lorentz Group*

We shall now generalize from the special cases of vector and Dirac fields to the case of a field that transforms according to a general irreducible representation of the homogeneous Lorentz group. All fields may be constructed as direct sums of these irreducible fields.

A general representation of the proper orthochronous homogeneous Lorentz group (or, more properly, its infinitesimal part) is provided by a set of matrices $\mathscr{J}_{\mu\nu}$ satisfying the same commutation relations (5.4.4) as the generators of the group

$$[\mathscr{J}_{\mu\nu}, \mathscr{J}_{\rho\sigma}] = i\left(\mathscr{J}_{\rho\nu}\eta_{\sigma\mu} + \mathscr{J}_{\mu\rho}\eta_{\nu\sigma} - \mathscr{J}_{\sigma\nu}\eta_{\rho\mu} - \mathscr{J}_{\mu\sigma}\eta_{\nu\rho}\right) , \tag{5.6.1}$$

(Of course, $\mathscr{J}_{\mu\nu} = -\mathscr{J}_{\nu\mu}$, and indices on $\mathscr{J}_{\mu\nu}$ are as usual raised or lowered by contraction with $\eta^{\mu\nu}$ or $\eta_{\mu\nu}$.) To see how to construct such matrices, first divide the six independent components of $\mathscr{J}_{\mu\nu}$ into two three-vectors: an angular momentum matrix

$$\mathscr{J}_1 = \mathscr{J}_{23} , \qquad \mathscr{J}_2 = \mathscr{J}_{31} , \qquad \mathscr{J}_3 = \mathscr{J}_{12} \tag{5.6.2}$$

and a boost

$$\mathscr{K}_1 = \mathscr{J}_{10} , \qquad \mathscr{K}_2 = \mathscr{J}_{20} , \qquad \mathscr{K}_3 = \mathscr{J}_{30} . \tag{5.6.3}$$

Eq. (5.6.1) then reads

$$[\mathscr{J}_i, \mathscr{J}_j] = i\epsilon_{ijk}\mathscr{J}_k , \tag{5.6.4}$$

* This section lies somewhat out of the book's main line of development, and may be omitted in a first reading.

$$[\mathscr{J}_i, \mathscr{K}_j] = i\epsilon_{ijk}\mathscr{K}_k\,, \tag{5.6.5}$$

$$[\mathscr{K}_i, \mathscr{K}_j] = -i\epsilon_{ijk}\mathscr{J}_k\,, \tag{5.6.6}$$

where i, j, k run over the values $1, 2, 3$, and ϵ_{ijk} is the totally antisymmetric quantity with $\epsilon_{123} \equiv +1$. Eq. (5.6.4) just says that the matrices $\mathscr{J}$ generate a representation of the rotation subgroup of the Lorentz group, and Eq. (5.6.5) just represents the fact that $\mathscr{K}$ is a three-vector. The minus sign in the right-hand side of Eq. (5.6.6) arises from the fact that $\eta_{00} = -1$, and plays a crucial role in what follows.

It is very convenient to replace the matrices $\mathscr{J}$ and $\mathscr{K}$ with two decoupled spin three-vectors, writing

$$\mathscr{A} \equiv \frac{1}{2}(\mathscr{J} + i\mathscr{K})\,, \tag{5.6.7}$$

$$\mathscr{B} \equiv \frac{1}{2}(\mathscr{J} - i\mathscr{K})\,. \tag{5.6.8}$$

It is easy to see that the commutation relations (5.6.4)–(5.6.6) are equivalent to

$$\left[\mathscr{A}_i, \mathscr{A}_j\right] = i\epsilon_{ijk}\;\mathscr{A}_k\,, \tag{5.6.9}$$

$$\left[\mathscr{B}_i, \mathscr{B}_j\right] = i\epsilon_{ijk}\;\mathscr{B}_k\,, \tag{5.6.10}$$

$$\left[\mathscr{A}_i, \mathscr{B}_j\right] = 0\,, \tag{5.6.11}$$

We find matrices satisfying Eqs. (5.6.9)–(5.6.11) in the same way that we find matrices representing the spins of a pair of uncoupled particles — as a direct sum. That is, we label the rows and columns of these matrices with a pair of integers and/or half-integers a, b, running over the values

$$a = -A, -A+1, \cdots, +A\,, \tag{5.6.12}$$

$$b = -B, -B+1, \cdots, +B \tag{5.6.13}$$

and take**

$$(\mathscr{A})_{a'b',ab} = \delta_{b'b}\;\mathbf{J}^{(A)}_{a'a}\,, \tag{5.6.14}$$

$$(\mathscr{B})_{a'b',ab} = \delta_{a'a}\;\mathbf{J}^{(B)}_{b'b}\,, \tag{5.6.15}$$

where $\mathbf{J}^{(A)}$ and $\mathbf{J}^{(B)}$ are the standard spin matrices for spins A or B:

$$\left(\mathbf{J}^{(A)}_3\right)_{a'a} = a\delta_{a'a}\,, \tag{5.6.16}$$

$$\left(\mathbf{J}^{(A)}_1 \pm i\mathbf{J}^{(A)}_2\right)_{a'a} = \delta_{a',a\pm 1}\sqrt{(A \mp a)(A \pm a + 1)}\,, \tag{5.6.17}$$

** There is an alternative formalism,[8] based on the fact that the spin j representation of the rotation group can be written as the symmetrized direct product of $2j$ spin $1/2$ representations — i.e., as a symmetric $SU(2)$ tensor with $2j$ two-valued indices. We can therefore write fields belonging to the (A, B) representation with $2A$ two-valued $(1/2, 0)$ indices and $2B$ two-valued $(0, 1/2)$ indices, the latter written with dots to distinguish them from the former.

and likewise for $\mathbf{J}^{(B)}$. The representation is labelled by the values of the positive integers and/or half-integers A and B. We see that the (A, B) representation has dimensionality $(2A+1)(2B+1)$.

The finite-dimensional representations of the homogeneous Lorentz group are not unitary, because $\mathscr{A}$ and $\mathscr{B}$ are Hermitian, and therefore $\mathscr{J}$ is Hermitian but $\mathscr{K}$ is *anti-Hermitian.* This is because of the i in Eqs. (5.6.7) and (5.6.8), which is required by the minus sign in (5.6.6), and hence stems from the fact that the homogeneous Lorentz group is not the same as the four-dimensional rotation group $SO(4)$, a compact group, but instead is the non-compact group known as $SO(3,1)$. It is only compact groups that can have finite-dimensional unitary representations (aside from representations in which the non-compact part is represented trivially, by the identity). There is no problem in working with non-unitary representations, because the objects we are now concerned with are fields, not wave functions, and do not need to have a Lorentz-invariant positive norm.

In contrast, the rotation group is represented unitarily, with its generators represented by the Hermitian matrices

$$\mathscr{J} = \mathscr{A} + \mathscr{B}\,, \tag{5.6.18}$$

By the usual rules of vector addition, we can see that a field that transforms according to the (A, B) representation of the homogeneous Lorentz group has components that rotate like objects of spin j, with

$$j = A+B,\ A+B-1, \cdots, |A-B|\,.$$

This is enough to identify the (A, B) representations with the perhaps more familiar tensors and spinors. For instance, a $(0,0)$ field is obviously scalar, with only a single $j=0$ component. A $(\frac{1}{2},0)$ or $(0,\frac{1}{2})$ field can only have $j=+\frac{1}{2}$; these are the top (i.e., $\gamma_5=+1$) and bottom ($\gamma_5=-1$) two components of the Dirac spinor. A $(\frac{1}{2},\frac{1}{2})$ field has components with $j=1$ and $j=0$, corresponding to the spatial part $\mathbf{v}$ and time-component v^0 of a four-vector v^μ. More generally, an (A, A) field contains terms with only integer spins $2A, 2A-1, \cdots, 0$, and corresponds to a traceless symmetric tensor of rank $2A$. (Note that the number of independent components of a symmetric tensor of rank $2A$ in four dimensions is

$$\frac{4\cdot 5\cdots(4+2A-1)}{(2A)!} = \frac{(3+2A)!}{6(2A)!}$$

and the tracelessness condition reduces this to

$$\frac{(3+2A)!}{6(2A)!} - \frac{(1+2A)!}{6(2A-2)!} = (2A+1)^2$$

as expected for an (A, A) field.) One more example: a $(1,0)$ or $(0,1)$ field can only have $j=1$, and corresponds to an antisymmetric tensor $F^{\mu\nu}$ that satisfies the further irreducibility 'duality' conditions

$$F^{\mu\nu} = \pm \frac{i}{2} \epsilon^{\mu\nu\lambda\rho} F_{\lambda\rho}$$

for (1,0) and (0,1) fields, respectively. Of course, it is only in four dimensions that an antisymmetric two-index tensor $F^{\mu\nu}$ can be divided into such 'self-dual' and 'anti-self-dual' parts.

A general tensor of rank N transforms as the direct product of N $(\frac{1}{2},\frac{1}{2})$ four-vector representations. It can therefore be decomposed (by suitable symmetrizations and antisymmetrizations and extracting traces) into irreducible terms (A,B) with $A = N/2, N/2-1, \cdots$ and $B = N/2, N/2-1, \cdots$. In this way, we can construct any irreducible representation (A,B) for which $A+B$ is an integer. The spin representations, for which $A+B$ is half an odd integer, can similarly be constructed from the direct product of these tensor representations and the Dirac representation $(\frac{1}{2},0)\oplus(0,\frac{1}{2})$. For instance, taking the direct product of the vector $(\frac{1}{2},\frac{1}{2})$ representation and the Dirac $(\frac{1}{2},0)\oplus(0,\frac{1}{2})$ representation gives a spinor-vector ψ^μ, that transforms according to the reducible representation

$$(\tfrac{1}{2},\tfrac{1}{2})\otimes[(\tfrac{1}{2},0)\oplus(0,\tfrac{1}{2})] = (\tfrac{1}{2},1)\oplus(\tfrac{1}{2},0)\oplus(1,\tfrac{1}{2})\oplus(0,\tfrac{1}{2})\,.$$

The quantity $\gamma_\mu\psi^\mu$ would transform as an ordinary $(\frac{1}{2},0)\oplus(0,\frac{1}{2})$ Dirac field, so we can isolate the $(\frac{1}{2},1)\oplus(1,\frac{1}{2})$ representation† by requiring that $\gamma_\mu\psi^\mu = 0$. This is the *Rarita–Schwinger field*.[9]

So far in this section we have only considered the representations of the proper orthochronous Lorentz group. In any representation of the Lorentz group including space inversion, there must be a matrix β which reverses the signs of tensors with odd numbers of space indices, and in particular

$$\beta\mathscr{J}\beta^{-1} = +\mathscr{J}\,, \qquad \beta\mathscr{K}\beta^{-1} = -\mathscr{K}\,. \tag{5.6.19}$$

In terms of the matrices (5.6.7) and (5.6.8), this is

$$\beta\mathscr{A}\beta^{-1} = \mathscr{B}\,, \qquad \beta\mathscr{B}\beta^{-1} = \mathscr{A}\,. \tag{5.6.20}$$

Thus an irreducible (A,B) representation of the proper orthochronous homogeneous Lorentz group does not provide a representation of the Lorentz group including space inversion unless $A = B$. As we have seen, these (A,A) representations are the scalar, the vector, and the symmetric traceless tensors. For $A \neq B$, the irreducible representations of the Lorentz

† According to Eq. (5.6.18), such a field transforms under ordinary rotations as a direct sum of two $j = 3/2$ and two $j = \frac{1}{2}$ components. The doubling is eliminated by imposing the Dirac equation $[\gamma^\nu\partial_\nu + m]\psi^\mu = 0$, and the remaining $j = \frac{1}{2}$ component is eliminated by requiring that $\partial_\mu\psi^\mu = 0$. With these conditions, the field describes a single particle of spin $j = 3/2$.

group including space inversion are the direct sums $(A,B) \oplus (B,A)$, of dimensionality $2(2A+1)(2B+1)$. One of these is the $(\frac{1}{2},0) \oplus (0,\frac{1}{2})$ Dirac representation discussed in Section 5.4. The 4×4 matrix (5.4.29) provides the β-matrix for this representation. Another familiar example is the $(1,0)\oplus(0,1)$ representation, which as we have seen is just the antisymmetric tensor of second rank, including both self-dual and anti-self-dual parts.

5.7 General Causal Fields*

We now proceed to construct causal fields that transform according to the general irreducible (A,B) representations described in the previous section. The index ℓ is replaced here with a pair of indices a, b, running over the ranges (5.6.12), (5.6.13), so the fields are now written as

$$\psi_{ab}(x) = (2\pi)^{-3/2} \sum_\sigma \int d^3p \left[\kappa\, a(\mathbf{p},\sigma) e^{ip\cdot x} u_{ab}(\mathbf{p},\sigma) \right.$$
$$\left. + \lambda\, a^{c\dagger}(\mathbf{p},\sigma) e^{-ip\cdot x} v_{ab}(\mathbf{p},\sigma)\right] \tag{5.7.1}$$

with κ and λ arbitrary constants. We are here leaving open the possibility that this particle is its own antiparticle, in which case $a^c(\mathbf{p},\sigma) = a(\mathbf{p},\sigma)$.

Our first task is to find the zero-momentum coefficient functions $u_{ab}(0,\sigma)$ and $v_{ab}(0,\sigma)$. The fundamental conditions (5.1.25)–(5.1.26) on $u(0,\sigma)$ and $v(0,\sigma)$ read here

$$\sum_{\bar\sigma} u_{\bar a\bar b}(0,\bar\sigma) \mathbf{J}^{(j)}_{\bar\sigma\sigma} = \sum_{a,b} \mathscr{J}_{\bar a\bar b,ab} u_{ab}(0,\sigma)\,,$$

$$-\sum_{\bar\sigma} v_{\bar a\bar b}(0,\bar\sigma) \mathbf{J}^{(j)*}_{\bar\sigma\sigma} = \sum_{a,b} \mathscr{J}_{\bar a\bar b,ab} v_{ab}(0,\sigma)\,,$$

or using Eqs. (5.6.14)–(5.6.15)

$$\sum_{\bar\sigma} u_{\bar a\bar b}(0,\bar\sigma) \mathbf{J}^{(j)}_{\bar\sigma\sigma} = \sum_a \mathbf{J}^{(A)}_{\bar a a} u_{a\bar b}(0,\sigma) + \sum_b \mathbf{J}^{(B)}_{\bar b b} u_{\bar a b}(0,\sigma)\,, \tag{5.7.2}$$

$$-\sum_{\bar\sigma} v_{\bar a\bar b}(0,\bar\sigma) \mathbf{J}^{(j)*}_{\bar\sigma\sigma} = \sum_a \mathbf{J}^{(A)}_{\bar a a} v_{a\bar b}(0,\sigma) + \sum_b \mathbf{J}^{(B)}_{\bar b b} v_{\bar a b}(0,\sigma)\,. \tag{5.7.3}$$

But Eq. (5.7.2) is the defining condition for the Clebsch–Gordan coefficients $C_{AB}(j\sigma;ab)$! These coefficients are defined by the requirement that

* This section lies somewhat out of the book's main line of development, and may be omitted in a first reading.

if Ψ_{ab} are states that under an infinitesimal rotation transform as

$$\delta\Psi_{ab} = i\sum_{\bar{a}} \boldsymbol{\theta}\cdot\mathbf{J}^{(A)}_{\bar{a}a}\Psi_{\bar{a}b} + i\sum_{\bar{b}} \boldsymbol{\theta}\cdot\mathbf{J}^{(B)}_{\bar{b}b}\Psi_{a\bar{b}}$$

then, under the same rotation, the state

$$\Psi^j{}_\sigma \equiv \sum_{ab} C_{AB}(j\sigma;ab)\Psi_{ab}$$

transforms as

$$\delta\Psi^j{}_\sigma = i\sum_{\bar{\sigma}} \boldsymbol{\theta}\cdot\mathbf{J}^{(j)}_{\bar{\sigma}\sigma}\Psi^j{}_{\bar{\sigma}}\ .$$

Inspection of Eq. (5.7.2) shows that this requirement is satisfied by the coefficients $u_{ab}(0,\sigma)$, and therefore, up to a possible proportionality factor, $u_{ab}(0,\sigma)$ is just $C_{AB}(j\sigma;ab)$. This constant is conventionally chosen so that

$$u_{ab}(0,\sigma) = (2m)^{-1/2}C_{AB}(j\sigma;ab)\ . \tag{5.7.4}$$

This result is unique because each irreducible (A,B) representation of the homogeneous Lorentz group contains a given spin j representation of the rotation group at most *once*. Similarly, inspection of Eqs. (5.6.16)–(5.6.17) shows that the complex conjugates of the angular momentum matrices are

$$-\mathbf{J}^{(j)*}_{\sigma\sigma'} = (-1)^{\sigma-\sigma'}\mathbf{J}^{(j)}_{-\sigma,-\sigma'}\ . \tag{5.7.5}$$

Therefore if we write Eq. (5.7.3) in terms of $(-1)^{j-\sigma}v_{ab}(\mathbf{p},-\sigma)$, it takes the same form as Eq. (5.7.2). With a suitable adjustment of a constant factor, the unique solution for $v(0,\sigma)$ is

$$v_{ab}(0,\sigma) = (-1)^{j+\sigma}u_{ab}(0,-\sigma)\ . \tag{5.7.6}$$

We must now perform a boost to calculate the coefficient functions for finite momentum. For a fixed direction $\hat{\mathbf{p}} \equiv \mathbf{p}/|\mathbf{p}|$, we can write the boost (2.5.24) as a function of a parameter θ defined by

$$\cosh\theta = \sqrt{\mathbf{p}^2+m^2}/m\ , \qquad \sinh\theta = |\mathbf{p}|/m \tag{5.7.7}$$

and write $L^\mu{}_\nu(\theta)$ in place of $L^\mu{}_\nu(p)$, where

$$\begin{aligned} L^i{}_k(\theta) &= \delta_{ik} + (\cosh\theta - 1)\hat{p}_i\hat{p}_k, \\ L^i{}_0(\theta) &= L^0{}_i(p) = \hat{p}_i\sinh\theta, \\ L^0{}_0(\theta) &= \cosh\theta\ . \end{aligned} \tag{5.7.8}$$

The advantage of this parameterization is that

$$L(\bar{\theta})L(\theta) = L(\bar{\theta}+\theta). \tag{5.7.9}$$

For infinitesimal θ, we have $[L(\theta)]^\mu{}_\nu \to \delta^\mu{}_\nu + \omega^\mu{}_\nu$, where $\omega^i{}_0 = \omega^0{}_i = \hat{p}_i\theta$ and $\omega^i{}_j = \omega^0{}_0 = 0$. Following the same reasoning that led from Eq.

(2.2.24) to Eq. (2.2.26), it follows then that

$$D\Big(L(p)\Big) = \exp(-i\hat{\mathbf{p}}\cdot\mathscr{K}\theta)\,. \tag{5.7.10}$$

This is for any representation of the homogeneous Lorentz group; for the irreducible (A,B) representations, Eqs. (5.6.7) and (5.6.8) give

$$i\mathscr{K} = \mathscr{A} - \mathscr{B} \tag{5.7.11}$$

and since $\mathscr{A}$ and $\mathscr{B}$ are commuting matrices

$$D\Big(L(p)\Big) = \exp(-\hat{\mathbf{p}}\cdot\mathscr{A}\theta)\exp(+\hat{\mathbf{p}}\cdot\mathscr{B}\theta)\,. \tag{5.7.12}$$

In more detail, using Eqs. (5.6.14) and (5.6.15)

$$D\Big(L(p)\Big)_{a'b',ab} = \Big(\exp\big(-\hat{\mathbf{p}}\cdot\mathbf{J}^{(A)}\theta\big)\Big)_{a'a}\Big(\exp\big(+\hat{\mathbf{p}}\cdot\mathbf{J}^{(B)}\theta\big)\Big)_{b'b}\,. \tag{5.7.13}$$

Eqs. (5.7.4) and (5.7.6) then give the coefficient functions at finite momentum as

$$u_{ab}(\mathbf{p},\sigma) = \frac{1}{\sqrt{2p^0}}\sum_{a'b'}\Big(\exp\big(-\hat{\mathbf{p}}\cdot\mathbf{J}^{(A)}\theta\big)\Big)_{aa'}\Big(\exp\big(+\hat{\mathbf{p}}\cdot\mathbf{J}^{(B)}\theta\big)\Big)_{bb'} \times C_{AB}(j\sigma;a'b') \tag{5.7.14}$$

and

$$v_{ab}(\mathbf{p},\sigma) = (-1)^{j+\sigma}u_{ab}(\mathbf{p},-\sigma)\,. \tag{5.7.15}$$

These results give the field explicitly for a given transformation type (A,B), so the field (5.1.31) of this type is unique up to the choice of the constant factors κ and λ.

It is very easy in this formalism to construct Lorentz scalar interaction densities. The (A,B) representation of the homogeneous Lorentz group is just the direct product of the $(A,0)$ and $(0,B)$ representations, so the general Lorentz transformation rules (5.1.6), (5.1.7) read here

$$U_0(\Lambda)\psi_{ab}(x)U_0^{-1}(\Lambda) = \sum_{a'b'} D^{A0}_{a,a'}(\Lambda^{-1})D^{0B}_{b,b'}(\Lambda^{-1})\psi_{a'b'}(\Lambda x)\,. \tag{5.7.16}$$

Furthermore, Eqs. (5.6.14) and (5.6.15) show that the matrix generators of the $(A,0)$ and $(0,B)$ representations are just the spin matrices for spin A and B, respectively. Thus we can construct scalars of the form

$$\sum_{a_1a_2\cdots a_n}\sum_{b_1b_2\cdots b_n} g_{a_1a_2\cdots a_n;b_1b_2\cdots b_n}\,\psi^{(1)}_{a_1b_1}(x)\,\psi^{(2)}_{a_2b_2}(x)\cdots\psi^{(n)}_{a_nb_n}(x) \tag{5.7.17}$$

by simply taking $g_{a_1a_2\cdots a_n;b_1b_2\cdots b_n}$ as the product of a coefficient for coupling spins $A_1, A_2, \cdots A_n$ to make a scalar and a coefficient for coupling spins $B_1, B_2, \cdots B_n$ to make a scalar. (Even though we do not explicitly consider interactions involving derivatives, we will in this way obtain the most general interaction involving n fields, because the derivative of a field of type (A, B) can always be decomposed into fields of other types without derivatives.) For instance, the most general Lorentz scalar formed from a product of three fields of transformation types (A_1, B_1), (A_2, B_2), and (A_3, B_3) is

$$g \sum_{a_1 a_2 a_3} \sum_{b_1 b_2 b_3} \begin{pmatrix} A_1 & A_2 & A_3 \\ a_1 & a_2 & a_3 \end{pmatrix} \begin{pmatrix} B_1 & B_2 & B_3 \\ b_1 & b_2 & b_3 \end{pmatrix} \psi^{(1)}_{a_1 b_1} \psi^{(2)}_{a_2 b_2} \psi^{(3)}_{a_3 b_3} \tag{5.7.18}$$

with a single free parameter g. This is the most general three-field interaction. (The brackets in (5.7.18) denote the Wigner "three-j" symbols:[10]

$$\begin{pmatrix} j_1 & j_2 & j_3 \\ m_1 & m_2 & m_3 \end{pmatrix} \equiv \sum_{m'_3} C_{j_1 j_2}(j_3 m'_3, m_1 m_2) C_{j_3 j_3}(00, m'_3 m_3)$$

which describe the coupling of three spins to make a rotational scalar.)

For the S-matrix to be Lorentz-invariant it is not enough that the interaction density $\mathscr{H}(x)$ be a scalar like (5.7.18); it is also necessary that $\mathscr{H}(x)$ should commute with $\mathscr{H}(y)$ at space-like separations $x - y$. To see how to satisfy this condition, consider the commutator or anticommutator of two fields for the same particle species, a field ψ of type (A, B), and the adjoint $\tilde{\psi}^\dagger$ of a field $\tilde{\psi}$ of type $(\tilde{A}, \tilde{B})$. We find

$$\Big[\psi_{ab}(x), \tilde{\psi}^\dagger_{\tilde{a}\tilde{b}}(y)\Big]_\mp = (2\pi)^{-3} \int d^3p \; (2p^0)^{-1} \pi_{ab,\tilde{a}\tilde{b}}(\mathbf{p}) \times \Big[\kappa\tilde{\kappa}^* e^{ip\cdot(x-y)} \mp \lambda\tilde{\lambda}^* e^{-ip\cdot(x-y)}\Big] , \tag{5.7.19}$$

where $\pi(\mathbf{p})$ is the spin sum

$$(2p^0)^{-1} \pi_{ab,\tilde{a}\tilde{b}}(\mathbf{p}) \equiv \sum_\sigma u_{ab}(\mathbf{p}, \sigma) \tilde{u}^*_{\tilde{a}\tilde{b}}(\mathbf{p}, \sigma) = \sum_\sigma v_{ab}(\mathbf{p}, \sigma) \tilde{v}^*_{\tilde{a}\tilde{b}}(\mathbf{p}, \sigma) \tag{5.7.20}$$

and as usual, the top and bottom signs are for bosons and fermions, respectively. (We allow here for different coefficients $\tilde{\kappa}$ and $\tilde{\lambda}$ in the $\tilde{\psi}$ field.) In more detail

$$\pi_{ab,\tilde{a}\tilde{b}}(\mathbf{p}) = \sum_{a'b'} \sum_{\tilde{a}'\tilde{b}'} \sum_\sigma C_{AB}\Big(j\sigma; a'b'\Big) \; C_{\tilde{A}\tilde{B}}\Big(j\sigma; \tilde{a}'\tilde{b}'\Big) \times \Big(\exp\Big(-\hat{\mathbf{p}} \cdot \mathbf{J}^{(A)}\theta\Big)\Big)_{aa'} \Big(\exp\Big(\hat{\mathbf{p}} \cdot \mathbf{J}^{(B)}\theta\Big)\Big)_{bb'} \times \Big(\exp\Big(-\hat{\mathbf{p}} \cdot \mathbf{J}^{(\tilde{A})}\theta\Big)\Big)^*_{\tilde{a}\tilde{a}'} \Big(\exp\Big(\hat{\mathbf{p}} \cdot \mathbf{J}^{(\tilde{B})}\theta\Big)\Big)^*_{\tilde{b}\tilde{b}'} . \tag{5.7.21}$$

The function $\pi(\mathbf{p})$ has been calculated explicitly.[11] What concerns us here is the fact that it turns out to be the mass-shell value of a *polynomial* function P of $\mathbf{p}$ and p^0:

$$\pi_{ab,\tilde{a}\tilde{b}}(\mathbf{p}) = P_{ab,\tilde{a}\tilde{b}}\Big(\mathbf{p}, \sqrt{\mathbf{p}^2 + m^2}\Big) \tag{5.7.22}$$

and that P is even or odd according to whether $2A + 2\tilde{B}$ is an even or odd integer

$$P(-\mathbf{p}, -p^0) = (-)^{2A+2\tilde{B}}\ P(\mathbf{p}, p^0)\,. \tag{5.7.23}$$

We shall check this here for just one particular direction of $\mathbf{p}$. Taking $\mathbf{p}$ in the three-direction, (5.7.21) gives

$$\pi_{ab,\tilde{a}\tilde{b}}(\mathbf{p}) = \sum_\sigma C_{AB}(j\sigma\,;ab)\ C_{\tilde{A}\tilde{B}}(j\sigma\,;\tilde{a}\tilde{b})\exp\Big([-a+b-\tilde{a}+\tilde{b}]\,\theta\Big)$$

The Clebsch–Gordan coefficients vanish unless $\sigma = a + b$ and $\sigma = \tilde{a} + \tilde{b}$, so we can replace

$$-a+b-\tilde{a}+\tilde{b} = -2a+\sigma+2\tilde{b}-\sigma = 2\tilde{b}-2a\,.$$

We can write $\exp(\pm\theta)$ as $(p^0 \pm p^3)/m$, so here

$$\pi_{ab,\tilde{a}\tilde{b}}(\mathbf{p}) = \sum_\sigma C_{AB}(j\sigma\,;ab)C_{\tilde{A}\tilde{B}}(j\sigma, \tilde{a}\tilde{b})$$
$$\times \begin{cases} \Big[(p^0+p^3)/m\Big]^{2\tilde{b}-2a} & (\tilde{b} \geq a) \\ \Big[(p^0-p^3)/m\Big]^{2a-2\tilde{b}} & (a \geq \tilde{b}) \end{cases}$$

where $p^0 \equiv \sqrt{\mathbf{p}^{\,2}+m^2}$. We see that $\pi(\mathbf{p})$ can indeed be written as the mass-shell value of a polynomial $P(\mathbf{p}, p^0)$. Also, $2\tilde{b} - 2a$ equals $2\tilde{B} + 2A$ minus an even integer, so the polynomial satisfies the reflection condition (5.7.23).

Any polynomial in $\mathbf{p}$ and $\sqrt{\mathbf{p}^{\,2}+m^2}$ can be written in a form *linear* in $\sqrt{\mathbf{p}^{\,2}+m^2}$ (by expressing even powers of $\sqrt{\mathbf{p}^{\,2}+m^2}$ in terms of $\mathbf{p}$) so $\pi(\mathbf{p})$ can be written

$$\pi_{ab,\tilde{a}\tilde{b}}(\mathbf{p}) = P_{ab,\tilde{a}\tilde{b}}(\mathbf{p}) + 2\sqrt{\mathbf{p}^{\,2}+m^2}\,Q_{ab,\tilde{a}\tilde{b}}(\mathbf{p})\,, \tag{5.7.24}$$

where P and Q are now polynomials in $\mathbf{p}$ alone, with

$$P(-\mathbf{p}) = (-)^{2A+2\tilde{B}}P(\mathbf{p}) \tag{5.7.25}$$

$$Q(-\mathbf{p}) = -(-)^{2A+2\tilde{B}}Q(\mathbf{p})\,. \tag{5.7.26}$$

For $x - y$ space-like, we can adopt a Lorentz frame in which $x^0 = y^0$, and write Eq. (5.7.19) as

$$[\psi_{ab}(x), \psi^\dagger_{\tilde{a}\tilde{b}}(y)]_\mp = [\kappa\tilde{\kappa}^* \mp (-)^{2A+2\tilde{B}}\lambda\tilde{\lambda}^*]P_{ab,\tilde{a}\tilde{b}}(-i\nabla)\Delta_+(\mathbf{x}-\mathbf{y}, 0)$$
$$+[\kappa\tilde{\kappa}^* \pm (-)^{2A+2\tilde{B}}\lambda\tilde{\lambda}^*]Q_{ab,\tilde{a}\tilde{b}}(-i\nabla)\delta^3(\mathbf{x}-\mathbf{y})\,.$$

In order that this should vanish when $\mathbf{x} \neq \mathbf{y}$, we must have

$$\kappa\tilde{\kappa}^* = \pm(-1)^{2A+2\tilde{B}}\lambda\tilde{\lambda}^*. \tag{5.7.27}$$

Now let us consider the special case where ψ and $\tilde{\psi}$ are the same, so in particular $A = \tilde{A}$ and $B = \tilde{B}$. (It is unavoidable that such commutators or anticommutators will appear in $[\mathscr{H}(x), \mathscr{H}(y)]$, because the hermiticity of the Hamiltonian requires that if $\mathscr{H}(x)$ involves ψ, it also involves $\psi^\dagger$.) In this case, Eqs. (5.7.27) gives

$$|\kappa|^2 = \pm(-)^{2A+2B}|\lambda|^2 \, .$$

This is possible if and only if

$$\pm(-1)^{2A+2B} = +1 \tag{5.7.28}$$

and

$$|\kappa|^2 = |\lambda|^2 \, . \tag{5.7.29}$$

Of course, $2A+2B$ differs from $2j$ by an *even* integer, so Eq. (5.7.28) says that *our particle is a boson or fermion according to whether* $2j$ *is even or odd.* This is the general relation between spin and statistics,[12] of which we have already seen special examples for particles described by scalar, vector, or Dirac fields.

Now let's return to the general case, where the fields ψ and $\tilde{\psi}$ may be different. Using Eq. (5.7.27), and dividing both sides by $|\tilde{\kappa}|^2 = |\tilde{\lambda}|^2$, we have

$$\frac{\kappa}{\tilde{\kappa}} = (-1)^{2B+2\tilde{B}} \, \frac{\lambda}{\tilde{\lambda}} \, .$$

It follows that, for any field,

$$\lambda = (-)^{2B} c \, \kappa \, , \tag{5.7.30}$$

where c is the same factor for all fields of a given particle. Furthermore, Eq. (5.7.29) shows that c is just a phase, $|c| = 1$. We can therefore eliminate c for all fields by a redefinition of the relative phase of the operators $a(\mathbf{p}, \sigma)$ and $a^{c\dagger}(\mathbf{p}, \sigma)$, so that $c = 1$, and hence $\lambda = (-)^{2B}\kappa$. Also, the factor κ for each field type may be eliminated by a redefinition of the over-all scale of the field. We emerge from all this with a formula for the (A, B) field of a given particle, that is unique up to overall scale

$$\psi_{ab}(x) = (2\pi)^{-3/2} \sum_\sigma \int d^3p \, \Big[u_{ab}(\mathbf{p}, \sigma) a(\mathbf{p}, \sigma) e^{ip\cdot x} + (-)^{2B} v_{ab}(\mathbf{p}, \sigma) a^{c\dagger}(\mathbf{p}, \sigma) e^{-ip\cdot x} \Big] \, . \tag{5.7.31}$$

The different fields for a given particle do not really represent possibilities that are physically distinct. For instance, the possible fields for $j = 0$ are those of type (A, A) (because the triangle inequality $|A - B| \leq j \leq A + B$ here requires $A = B$). Starting with a $(0, 0)$ scalar field ϕ, we can easily construct such (A, A) fields from the $2A$th derivative

$$\{\partial_{\mu_1} \cdots \partial_{\mu_{2A}}\}\phi\,, \tag{5.7.32}$$

where $\{\}$ here denotes the traceless part; for instance

$$\{\partial_\mu \partial_\nu\} \equiv \frac{\partial^2}{\partial x^\mu \partial x^\nu} - \tfrac{1}{4}\,\eta_{\mu\nu}\,\Box\,.$$

(Recall that a traceless symmetric tensor of rank N transforms according to the $(N/2, N/2)$ representation.) But Eq. (5.7.31) represents the *unique* causal (A, B) field for a given particle of spin j, so the (A, A) fields (5.7.31) for $j = 0$ can be nothing but linear combinations of the $2A$th derivatives (5.7.32) of a scalar field.

More generally, *any* field (A, B) for a given particle of spin j can be expressed as a differential operator of rank $2B$ acting on the field[13] $\varphi_\sigma(x)$ of type $(j, 0)$ (or a differential operator of rank $2A$ acting on the field of type $(0, j)$). To see this, consider the field

$$\{\partial_{\mu_1} \cdots \partial_{\mu_{2B}}\}\ \varphi_\sigma\,. \tag{5.7.33}$$

This transforms as the direct product of the representations (B, B) and $(j, 0)$, and hence by the usual rules of vector addition, it can be decomposed into fields transforming according to all the irreducible representations (A, B) with $|j - B| \leq A \leq j + B$, or equivalently $|A - B| \leq j \leq A + B$. Since Eq. (5.7.31) represents the unique field of type (A, B) for a given particle of spin j, it can be nothing** but the (A, B) field obtained from the derivatives (5.7.33).

Now let us consider the behavior of these fields under inversions, beginning with space inversion. Using the results of Section 4.2, the space-inversion properties of the particle annihilation and antiparticle creation operators are:

$$\mathsf{P}a(\mathbf{p}, \sigma)\mathsf{P}^{-1} = \eta^* a(-\mathbf{p}, \sigma)\,, \tag{5.7.34}$$

$$\mathsf{P}a^{c\dagger}(\mathbf{p}, \sigma)\mathsf{P}^{-1} = \eta^{c} a^{c\dagger}(-\mathbf{p}, \sigma)\,, \tag{5.7.35}$$

where η and η^c are the intrinsic parities for the particle and antiparticle, respectively. The general causal (A, B) field (5.7.31) thus transforms under

** The only possible flaw in this argument would be if some of the (A, B) fields obtained in this way actually vanished. But in this case, the $(j, 0)$ field φ_σ would satisfy a field equation $\sum_\sigma M_\sigma(\partial/\partial x)\varphi_\sigma(x) = 0$ and hence, for *each* $\bar\sigma$, $\sum_\sigma M_\sigma(ip)u_\sigma(\mathbf{p}, \bar\sigma) = 0$. For the (j, σ) representation the Clebsch–Gordan coefficient $C_{j0}(j\bar\sigma; \sigma 0)$ is just the Kronecker symbol $\delta_{\bar\sigma\sigma}$, so this would require $\sum_\sigma M_\sigma(ip)D_{\sigma\sigma'}(L(p)) = 0$, which is impossible unless all the $M_\sigma(ip)$ vanish, since $D(\Lambda)$ has an inverse $D(\Lambda^{-1})$. The $(j, 0)$ fields $\varphi_\sigma(x)$ thus satisfy no field equation other than the Klein–Gordon equation $(\Box - m^2)\ \varphi_\sigma(x) = 0$, and therefore none of the (A, B) fields obtained from (5.7.33) can vanish.

the parity operator P into

$$\mathsf{P}\psi_{ab}^{AB}(x)\mathsf{P}^{-1} = (2\pi)^{-3/2}\sum_\sigma\int d^3p\,\Big[\eta^* a(-\mathbf{p},\sigma)e^{ip\cdot x}\,u_{ab}^{AB}(\mathbf{p},\sigma)$$
$$+\eta^{c}(-)^{2B}a^{c\dagger}(-\mathbf{p},\sigma)e^{-ip\cdot x}\,v_{ab}^{AB}(\mathbf{p},\sigma)\Big]. \tag{5.7.36}$$

We want to change the integration variable from $\mathbf{p}$ to $-\mathbf{p}$, and for this purpose we will need to evaluate $u_{ab}(-\mathbf{p},\sigma)$ and $v_{ab}(-\mathbf{p},\sigma)$. To do this, we need only glance back at (5.7.14) and (5.7.15), and use the symmetry property of the Clebsch–Gordan coefficient[14]

$$C_{AB}(j\sigma;ab) = (-)^{A+B-j}C_{BA}(j\sigma;ba). \tag{5.7.37}$$

This gives

$$u_{ab}^{AB}(-\mathbf{p},\sigma) = (-)^{A+B-j}\,u_{ba}^{BA}(\mathbf{p},\sigma)\,, \tag{5.7.38}$$

$$v_{ab}^{AB}(-\mathbf{p},\sigma) = (-)^{A+B-j}\,v_{ba}^{BA}(\mathbf{p},\sigma)\,, \tag{5.7.39}$$

so

$$\mathsf{P}\psi_{ab}^{AB}(x)\mathsf{P}^{-1} = (2\pi)^{-3/2}\sum_\sigma\int d^3p\,(-1)^{A+B-j}$$
$$\times\Big[\eta^* a(\mathbf{p},\sigma)e^{ip\cdot\mathscr{P}x}u_{ba}^{BA}(\mathbf{p},\sigma) + \eta^{c}(-)^{2B}a^{c\dagger}(\mathbf{p},\sigma)e^{-ip\cdot\mathscr{P}x}v_{ba}^{BA}(\mathbf{p},\sigma)\Big]\,, \tag{5.7.40}$$

where, as before, $\mathscr{P}x \equiv (-\mathbf{x}, x^0)$. This is the causal field ψ_{ba}^{BA} evaluated at $\mathscr{P}x$, except that the coefficients of the annihilation and creation terms may not be the same as called for in Eq. (5.7.31). But these coefficients *must* be the same up to an overall constant factor as in Eq. (5.7.31) because, aside from scale, Eq. (5.7.31) is the unique causal field of any type. Hence the ratio of the coefficients of the two terms in Eq. (5.7.40) must be the same as in Eq. (5.7.31) (but with B replaced with A because this is supposed to be a (B,A) field):

$$\eta^{c}(-)^{2B}/\eta^* = (-)^{2A}\,. \tag{5.7.41}$$

But $A-B$ differs from the spin j by only an integer, so this gives

$$\eta^{c} = \eta^*(-)^{2j}\,. \tag{5.7.42}$$

We saw special cases of this result in Sections 5.2, 5.3, and 5.5, where $j=0$, $j=1$, and $j=\frac{1}{2}$, respectively. We now see that the result is general; *the intrinsic parity $\eta^{c}\eta$ of a particle–antiparticle pair is $+1$ for bosons, and -1 for fermions.* Using Eq. (5.7.42) in Eq. (5.7.40), our final result for space inversion is

$$\mathsf{P}\psi_{ab}^{AB}(x)\mathsf{P}^{-1} = \eta^*(-)^{A+B-j}\psi_{ba}^{BA}(-\mathbf{x},x^0)\,. \tag{5.7.43}$$

Let's see how this applies to the Dirac field. For the top $(\frac{1}{2},0)$ and bottom $(0,\frac{1}{2})$ components of the Dirac field, the sign $(-1)^{A+B-j}$ is just

+1, so the parity operator simply takes $\mathbf{x}$ into $-\mathbf{x}$; reverses the top and bottom components; and multiplies the field with η^*. The reversal of the top and bottom components of the Dirac field is accomplished by the matrix β in (5.5.41).

Now let us consider charge-conjugation. Its effect on the particle annihilation and antiparticle creation operators is

$$\mathsf{C}a(\mathbf{p},\sigma)\,\mathsf{C}^{-1} = \xi^*\, a^c(\mathbf{p},\sigma), \tag{5.7.44}$$

$$\mathsf{C}a^{c\dagger}(\mathbf{p},\sigma)\,\mathsf{C}^{-1} = \xi^c a^\dagger(\mathbf{p},\sigma)\,, \tag{5.7.45}$$

where ξ and ξ^c are the charge-conjugation parities of the particle and antiparticle, respectively. Applying this transformation to the field (5.7.31), we find

$$\mathsf{C}\psi_{ab}^{AB}(x)\mathsf{C}^{-1} = (2\pi)^{-3/2}\sum_\sigma \int d^3p\; u_{ab}^{AB}(\mathbf{p},\sigma)$$
$$\times\left[\xi^* a^c(\mathbf{p},\sigma)e^{ip\cdot x} + \xi^c(-)^{2B}a^\dagger(\mathbf{p},-\sigma)(-)^{j-\sigma}e^{-ip\cdot x}\right]. \tag{5.7.46}$$

It is useful to compare this formula for the charge-conjugate of an (A,B) field with the adjoint of the (B,A) field for the same particle:

$$\psi_{ba}^{BA\dagger}(x) = (2\pi)^{-3/2}\sum_\sigma \int d^3p\, u_{ba}^{BA*}\,(\mathbf{p},\sigma)$$
$$\times\left[(-1)^{2A}(-1)^{j-\sigma}a^c(\mathbf{p},-\sigma)e^{ip\cdot x} + a^\dagger(\mathbf{p},\sigma)e^{-ip\cdot x}\right]. \tag{5.7.47}$$

To calculate the u^*, we use our previous result

$$\mathbf{J}^{(j)*} = -\mathscr{C}\mathbf{J}^{(j)}\mathscr{C}^{-1}\,, \qquad \mathscr{C}_{\bar\sigma\sigma} \propto (-1)^{j-\sigma}\delta_{\bar\sigma,-\sigma}\,.$$

The Clebsch–Gordan coefficient in Eq. (5.7.14) is real, so

$$u_{ba}^{BA}(\mathbf{p},\sigma)^* = \frac{1}{\sqrt{2p^0}}\sum_{a'b'}\Big(\exp(-\hat{\mathbf{p}}\cdot\mathbf{J}^{(A)}\theta)\Big)_{-a,-a'}\Big(\exp\,(\hat{\mathbf{p}}\cdot\mathbf{J}^{(B)}\theta)\Big)_{-b,-b'}$$
$$\times(-)^{a'-a}(-)^{b'-b}C_{BA}(j\sigma;b'a')\,.$$

We use the reflection property of the Clebsch–Gordan coefficients[14]

$$C_{BA}(j,-\sigma;-b',-a') = C_{AB}(j\sigma;a'b') \tag{5.7.48}$$

and the fact that those coefficients vanish unless $a'+b'=\sigma$, to write

$$u_{-b,-a}^{BA}(\mathbf{p},-\sigma)^* = (-)^{a+b-\sigma}\,u_{ab}^{AB}(\mathbf{p},\sigma)\,. \tag{5.7.49}$$

The field adjoint (5.7.47) is then (replacing $a\to-a$, $b\to-b$, $\sigma\to-\sigma$)

$$\psi_{-b,-a}^{BA\dagger}(x) = (2\pi)^{-3/2}\sum_\sigma \int d^3p\;(-)^{a+b-\sigma}u_{ab}^{AB}(\mathbf{p},\sigma)$$
$$\times\left[(-)^{2A}(-)^{j+\sigma}\;a^c(\mathbf{p},\sigma)e^{ip\cdot x} + a^\dagger(\mathbf{p},\sigma)e^{-ip\cdot x}\right].$$

Using the sign relation $(-)^{-2A-j} = (-)^{2B+j}$, this is

$$(-)^{-2A-a-b-j}\ \psi^{BA^\dagger}_{-b,-a}(x) = (2\pi)^{-3/2}\sum_\sigma \int d^3p\ u^{AB}_{ab}(\mathbf{p},\sigma)$$
$$\times\left[a^c(\mathbf{p},\sigma)e^{ip\cdot x} + (-)^{j-\sigma+2B}a^\dagger(\mathbf{p},-\sigma)e^{-ip\cdot x}\right]. \quad (5.7.50)$$

In order that $\mathsf{C}\ \psi^{AB}_{ab}(x)\ \mathsf{C}^{-1}$ should commute or anticommute with all ordinary fields at space-like separations, it is necessary that it be proportional to $\psi^{BA^\dagger}_{-b,-a}(x)$, because this is the adjoint of the unique causal field of transformation type (B, A). Comparing Eq. (5.7.50) with Eq. (5.7.46), we see that this is only possible if the charge-conjugation parities are related by

$$\xi^* = \xi^c\,, \quad (5.7.51)$$

in which case

$$\mathsf{C}\ \psi^{AB}_{ab}(x)\ \mathsf{C}^{-1} = \xi^*(-)^{-2A-a-b-j}\ \psi^{BA^\dagger}_{-b-a}(x). \quad (5.7.52)$$

We have already encountered the relation (5.7.51) for spins 0, 1, and $\frac{1}{2}$ in Sections 5.2, 5.3, and 5.5, and noted some of its implications for electron–positron and quark–antiquark states in Section 5.5.

In particular, for a particle that is its own antiparticle, Eq. (5.7.52) is satisfied without any charge-conjugation operator on the left-hand side or phase ξ^* on the right:

$$\psi^{AB}_{ab}(x) = (-)^{-2A-a-b-j}\ \psi^{BA\dagger}_{-b-a}(x)\,. \quad (5.7.53)$$

We have already seen an example of this sort of reality condition for Majorana spin $\frac{1}{2}$ particles in Section 5.5.

Finally we come to time-reversal. Applied to particle annihilation and antiparticle creation operators, this gives

$$\mathsf{T}a(\mathbf{p},\sigma)\mathsf{T}^{-1} = \zeta^*(-1)^{j-\sigma}a(-\mathbf{p},-\sigma)\,, \quad (5.7.54)$$

$$\mathsf{T}a^{c\dagger}(\mathbf{p},\sigma)\mathsf{T}^{-1} = \zeta^c(-1)^{j-\sigma}a^{c\dagger}(-\mathbf{p},-\sigma)\,. \quad (5.7.55)$$

The irreducible field (5.7.31) thus has the transformation property

$$\mathsf{T}\psi^{AB}_{ab}(x)\mathsf{T}^{-1} = (2\pi)^{-3/2}\sum_\sigma \int d^3p\, u^{AB*}_{ab}(\mathbf{p},\sigma)(-1)^{j-\sigma}$$
$$\times\left[\zeta^* a(-\mathbf{p},-\sigma)e^{-ip\cdot x} + \zeta^c(-1)^{2B}a^{c\dagger}(-\mathbf{p},-\sigma)e^{ip\cdot x}\right]. \quad (5.7.56)$$

To calculate the complex conjugate of the coefficient function, we use Eq. (5.7.14) and the standard formula[14]

$$C_{AB}(j,\sigma;a,b) = (-)^{A+B-j}C_{AB}(j,-\sigma;-a,-b) \quad (5.7.57)$$

and find:

$$u_{ab}^{AB*}(-\mathbf{p},-\sigma) = (-)^{a+b+\sigma+A+B-j}u_{-a,-b}^{AB}(\mathbf{p},\sigma)\,. \tag{5.7.58}$$

Changing the variables of integration and summation in Eq. (5.7.56) to $-\mathbf{p}$ and $-\sigma$, we find that in order for an (A, B) field to be transformed by time-reversal into something proportional to another (A, B) field, it is necessary that

$$\zeta^c = \zeta^*\,, \tag{5.7.59}$$

in which case

$$\mathsf{T}\psi_{ab}^{AB}(x)\mathsf{T}^{-1} = (-)^{a+b+A+B-2j}\,\zeta^*\,\psi_{-a,-b}^{AB}(\mathbf{x},-x^0). \tag{5.7.60}$$

* * *

It should be mentioned that from time to time various difficulties have been reported[15] in the field theory of particles with spin $j \geq 3/2$. Generally, these are encountered in the study of the propagation of a higher spin field in the presence of c-number external field. Depending on the details of the theory, the difficulties encountered include non-causality, inconsistency, unphysical mass states, and violation of unitarity. I will not go into details about these problems here, because it seems to me that they are not relevant to the calculational scheme described in this chapter, for the following reasons:

(1) The fields $\psi_{ab}(x)$ have been constructed here directly from the creation and annihilation operators for physical particles, so no question of inconsistency or unphysical mass states can arise. These are free fields, but by incorporating them into an interaction Hamiltonian density in the interaction picture, we can use perturbation theory to calculate S-matrix elements that automatically satisfy the cluster decomposition principle. As long as the interaction Hamiltonian is Hermitian, there can be no difficulty with unitarity. Lorentz invariance is guaranteed in perturbation theory as long as we add appropriate local but non-covariant terms in the Hamiltonian density; though a rigorous proof is lacking, there is no reason to doubt that this is always possible. Thus any difficulties with higher spin can only arise when we try to go beyond perturbation theory.
(2) As discussed in Section 13.6, the solution of field equations in the presence of a c-number background field (the context where all the problems with higher spin have been found) does go beyond perturbation theory, in that the results correspond to summing an infinite subset of terms in the perturbation series. This partial summation is justified, even for weak external fields, if the fields are sufficiently slowly varying, the smallness of energy denominators making up for the weakness of the fields. But the results obtained in this way depend on all the details of the interaction

of the high-spin particle with the external fields: not only the multipole moments of the particle but also possible terms in the interaction that are non-linear in the external fields. The problems reported[15] with higher spin have been encountered only for higher-spin particles that have been arbitrarily assumed to have only very simple interactions with external fields. No one has shown that the problems persist for arbitrary interactions, and as we shall see in Chapter 12, particles of higher spin are expected to have interactions of all possible types allowed by symmetry principles.

(3) In fact, there are good reasons to believe that the problems with higher spin disappear if the interaction with external fields is sufficiently complicated. For one thing, there is no doubt about the existence of higher-spin *particles*, including various stable nuclei and hadronic resonances. If there is any problem with higher spin, it can only be for 'point' particles, that is, those whose interactions with external fields are particularly simple. It should be kept in mind that the requirement of simplicity depends on the choice of which field we choose to represent the higher-spin particle. Remember that any free field types for a given particle can be expressed as a derivative operator acting on any other field type, so in the interaction picture any interaction with external fields may be written in terms of any field types we like, but interactions that are simple when expressed in terms of a field of one type may look complicated when expressed in terms of a field of another type. So the requirement of simplicity does not seem to have any objective content.

(4) Also, both higher-dimensional 'Kaluza–Klein' theories and string theories provide examples of consistent theories of a charged massive particles of spin two interacting with an electromagnetic background field.[16] (It was found that the consistency of the theory depends on the assumption of realistic external fields that satisfy the field equations, a point generally neglected in earlier work.) Reformulating this work in the interaction picture, the spin two particle is represented by a $(1,1)$ free field, but as mentioned above, the interactions may be reexpressed in the interaction picture in terms of any field type (A, B) that contains the $j = 2$ representation of the rotation group.

5.8 The CPT Theorem

We have seen that the demands of relativity combined with quantum mechanics require the existence of antiparticles. Not only is it necessary that every particle have an antiparticle (which may for a purely neutral particle be itself); there is a precise relation between the properties of particles and antiparticles, that can be summarized in the statement that *for an appropriate choice of inversion phases, the product* CPT *of all the inversions is conserved.* This is the celebrated CPT theorem.*

As a first step in the proof, let us work out the effect of the product CPT on free fields of various types. For a scalar, vector, or Dirac field the results of Sections 5.2, 5.3, and 5.5 give

$$\mathsf{CPT}\,\phi(x)\,[\mathsf{CPT}]^{-1} = \zeta^*\xi^*\eta^*\phi^\dagger(-x)\,, \tag{5.8.1}$$

$$\mathsf{CPT}\,\phi_\mu(x)\,[\mathsf{CPT}]^{-1} = -\zeta^*\xi^*\eta^*\phi_\mu^\dagger(-x)\,, \tag{5.8.2}$$

$$\mathsf{CPT}\,\psi(x)\,[\mathsf{CPT}]^{-1} = -\zeta^*\xi^*\eta^*\gamma_5\psi^*(-x)\,. \tag{5.8.3}$$

(Of course, the phases ζ, ξ, and η depend on the species of particle described by each field.) We are going to choose the phases so that for all particles

$$\zeta\,\xi\,\eta = 1\,. \tag{5.8.4}$$

Then any tensor $\phi_{\mu_1\ldots\mu_n}$ formed from any set of scalar and vector fields and their derivatives transforms into

$$\mathsf{CPT}\,\phi_{\mu_1\ldots\mu_n}(x)\,[\mathsf{CPT}]^{-1} = (-)^n\phi^\dagger_{\mu_1\ldots\mu_n}(-x)\,. \tag{5.8.5}$$

(Any complex numerical coefficient appearing in these tensors is transformed into its complex conjugate because CPT is antiunitary.) We can easily see that the same transformation rule applies to tensors formed from bilinear combinations of Dirac fields. Applying Eq. (5.8.3) to such a bilinear gives

$$\begin{aligned}\mathsf{CPT}[\bar\psi_1(x)M\psi_2(x)][\mathsf{CPT}]^{-1} &= \psi_1^T(-x)\gamma_5\beta M^*\gamma_5\psi_2^*(-x)\\ &= [\bar\psi_1(-x)\gamma_5 M\gamma_5\psi_2(-x)]^\dagger\,.\end{aligned} \tag{5.8.6}$$

(A minus sign from the anticommutation of β and γ_5 is cancelled by the minus sign from the anticommutation of fermionic operators.) If the bilinear is a tensor of rank n, then M is a product of n modulo 2 Dirac matrices, so $\gamma_5 M\gamma_5 = (-1)^n M$, and the bilinear therefore satisfies Eq. (5.8.5).

A Hermitian scalar interaction density $\mathscr{H}(x)$ must be formed from

* The original proofs of this theorem were by Lüders and Pauli.[17] It has been proved rigorously in axiomatic field theory,[18] by using commutativity assumptions to extend the Lorentz invariance of the theory to the complex Lorentz group, then using complex Lorentz transformations to prove a reflection property of vacuum expectation values of products of fields, and then using this reflection property to infer the existence of an antiunitary operator that induces CPT transformations on the fields.

tensors with an *even* total number of spacetime indices, and therefore

$$\mathsf{CPT}\, \mathscr{H}(x)\, [\mathsf{CPT}]^{-1} = \mathscr{H}(-x)\,. \tag{5.8.7}$$

More generally (and somewhat more easily) we can see that the same is true for Hermitian scalars formed from the fields $\psi^{AB}_{ab}(x)$ belonging to one or more of the general irreducible representations of the homogeneous Lorentz group. Putting together our results in the previous section for the effects of inversions on such fields, we find

$$\mathsf{CPT}\, \psi^{AB}_{ab}(x)\, [\mathsf{CPT}]^{-1} = (-1)^{2B} \psi^{AB\dagger}_{ab}(-x)\,. \tag{5.8.8}$$

(For the Dirac field the factor $(-1)^{2B}$ is supplied by the matrix γ_5 in Eq. (5.8.3).) In order to couple together a product $\psi^{A_1B_1}_{a_1b_1}(x)\,\psi^{A_2B_2}_{a_2b_2}(x)\cdots$ to form a scalar $\mathscr{H}(x)$, it is necessary that both $A_1+A_2+\cdots$ and $B_1+B_2+\cdots$ be integers, so $(-1)^{2B_1+2B_2+\cdots} = 1$, and so a Hermitian scalar $\mathscr{H}(x)$ will automatically satisfy Eq. (5.8.7).

From Eq. (5.8.7) it follows immediately that CPT commutes with the interaction $V \equiv \int d^3x\, \mathscr{H}(\vec{x},0)$:

$$\mathsf{CPT}\, V\, [\mathsf{CPT}]^{-1} = V\,. \tag{5.8.9}$$

Also, in any theory CPT commutes with the free-particle Hamiltonian H_0. Thus the operator CPT, which has been defined here by its operation on free-particle operators, acts on 'in' and 'out' states in the way described in Section 3.3. The physical consequences of this symmetry principle have already been discussed in Sections 3.3 and 3.6.

5.9 Massless Particle Fields

Up to this point we have dealt only with the fields of massive particles. For some of these fields, such as the scalar and Dirac fields discussed in Sections 5.2 and 5.5, there is no special problem in passing to the limit of zero mass. On the other hand, we saw in Section 5.3 that there *is* a difficulty in taking the zero-mass limit of the vector field for a particle of spin one: at least one of the polarization vectors blows up in this limit. In fact, we shall see in this section that the creation and annihilation operators for physical massless particles of spin $j \geq 1$ cannot be used to construct all of the irreducible (A, B) fields that can be constructed for finite mass. This peculiar limitation on field types will lead us naturally to the introduction of gauge invariance.

Just as we did for massive particles, let us attempt to construct a general free field for a massless particle as a linear combination of the annihilation operators $a(\mathbf{p}, \sigma)$ for particles of momentum $\mathbf{p}$ and helicity σ, and the corresponding creation operators $a^{c\dagger}(\mathbf{p},\sigma)$ for the antiparticles:*

$$\psi_\ell(x) = (2\pi)^{-3/2} \int d^3p \sum_\sigma \Big[\kappa\, a(\mathbf{p},\sigma) u_\ell(\mathbf{p},\sigma)\, e^{ip\cdot x} + \lambda\, a^{c\dagger}(\mathbf{p},\sigma) v_\ell(\mathbf{p},\sigma)\, e^{-ip\cdot x}\Big] \tag{5.9.1}$$

where now $p^0 \equiv |\mathbf{p}|$. The creation operators transform just like the one-particle states in Eq. (2.5.42)

$$U(\Lambda) a^\dagger(\mathbf{p},\sigma) U^{-1}(\Lambda) = \sqrt{\frac{(\Lambda p)^0}{p^0}}\, \exp\Big(i\sigma\theta(p,\Lambda)\Big) a^\dagger(\mathbf{p}_\Lambda,\sigma)\,, \tag{5.9.2}$$

$$U(\Lambda) a^{c\dagger}(\mathbf{p},\sigma) U^{-1}(\Lambda) = \sqrt{\frac{(\Lambda p)^0}{p^0}}\, \exp\Big(i\sigma\theta(p,\Lambda)\Big) a^{c\dagger}(\mathbf{p}_\Lambda,\sigma)\,, \tag{5.9.3}$$

and hence also

$$U(\Lambda) a(\mathbf{p},\sigma) U^{-1}(\Lambda) = \sqrt{\frac{(\Lambda p)^0}{p^0}}\, \exp\Big(-i\sigma\theta(p,\Lambda)\Big) a(\mathbf{p}_\Lambda,\sigma)\,, \tag{5.9.4}$$

where $p_\Lambda \equiv \Lambda p$, and θ is the angle defined by Eqs. (2.5.43). Hence if we want the field to transform according to some representation $D(\Lambda)$ of the homogeneous Lorentz group

$$U(\Lambda)\psi_\ell(x) U^{-1}(\Lambda) = \sum_{\bar\ell} D_{\ell\bar\ell}(\Lambda^{-1})\psi_{\bar\ell}(\Lambda x)\,, \tag{5.9.5}$$

then we must take the coefficient functions u and v to satisfy the relations

$$u_{\bar\ell}(\mathbf{p}_\Lambda,\sigma)\, \exp\Big(i\sigma\theta(p,\Lambda)\Big) = \sqrt{\frac{p^0}{(\Lambda p)^0}} \sum_\ell D_{\bar\ell\ell}(\Lambda) u_\ell(\mathbf{p},\sigma)\,, \tag{5.9.6}$$

$$v_{\bar\ell}(\mathbf{p}_\Lambda,\sigma)\, \exp\Big(-i\sigma\theta(p,\Lambda)\Big) = \sqrt{\frac{p^0}{(\Lambda p)^0}} \sum_\ell D_{\bar\ell\ell}(\Lambda) v_\ell(\mathbf{p},\sigma) \tag{5.9.7}$$

in place of Eqs. (5.1.19) and (5.1.20). (Again, $p_\Lambda \equiv \Lambda p$.) As in the massive particle case, we can satisfy these requirements by setting (in place of

* We deal here with only a single species of particle, and drop the species label n. Also, κ and λ are constant coefficients to be determined by the requirement of causality with some convenient choice of normalization of the coefficient functions u_ℓ and v_ℓ.

Eqs. (5.1.21) and (5.1.22))

$$u_{\bar{\ell}}(\mathbf{p},\sigma) = \sqrt{\frac{|\mathbf{k}|}{p^0}} \sum_{\ell} D_{\bar{\ell}\ell}\left(\mathscr{L}(p)\right) u_{\ell}(\mathbf{k},\sigma) , \tag{5.9.8}$$

$$v_{\bar{\ell}}(\mathbf{p},\sigma) = \sqrt{\frac{|\mathbf{k}|}{p^0}} \sum_{\ell} D_{\bar{\ell}\ell}\left(\mathscr{L}(p)\right) v_{\ell}(\mathbf{k},\sigma) , \tag{5.9.9}$$

where $\mathbf{k}$ is a standard momentum, say $(0, 0, k)$, and $\mathscr{L}(p)$ is a standard Lorentz transformation that takes a massless particle from momentum $\mathbf{k}$ to momentum $\mathbf{p}$. Also, in place of Eqs. (5.1.23) and (5.1.24), the coefficient functions at the standard momentum must satisfy

$$u_{\bar{\ell}}(\mathbf{k},\sigma)\ \exp\Big(i\sigma\theta(k, W)\Big) = \sum_{\ell} D_{\bar{\ell}\ell}(W) u_{\ell}(\mathbf{k},\sigma) \tag{5.9.10}$$

$$v_{\bar{\ell}}(\mathbf{k},\sigma)\ \exp\Big(-i\sigma\theta(k, W)\Big) = \sum_{\ell} D_{\bar{\ell}\ell}(W) v_{\ell}(\mathbf{k},\sigma) \tag{5.9.11}$$

where $W^{\mu}{}_{\nu}$ is an arbitrary element of the 'little group' for four-momentum $k = (\mathbf{k}, |\mathbf{k}|)$, i.e., an arbitrary Lorentz transformation that leaves this four-momentum invariant.

We can extract the content of Eqs. (5.9.10) and (5.9.11) by considering separately the two kinds of little-group elements in Eq. (2.5.28). For a rotation $R(\theta)$ by an angle θ around the z-axis, given by Eq. (2.5.27),

$$R^{\mu}{}_{\nu}(\theta) = \begin{bmatrix} \cos\theta & \sin\theta & 0 & 0 \\ -\sin\theta & \cos\theta & 0 & 0 \\ 0 & 0 & 1 & 0 \\ 0 & 0 & 0 & 1 \end{bmatrix} ,$$

we find from Eqs. (5.9.10) and (5.9.11)

$$u_{\bar{\ell}}(\mathbf{k},\sigma) e^{i\sigma\theta} = \sum_{\ell} D_{\bar{\ell}\ell}\Big(R(\theta)\Big) u_{\ell}(\mathbf{k},\sigma) \tag{5.9.12}$$

$$v_{\bar{\ell}}(\mathbf{k},\sigma) e^{-i\sigma\theta} = \sum_{\ell} D_{\bar{\ell}\ell}\Big(R(\theta)\Big) v_{\ell}(\mathbf{k},\sigma)\ . \tag{5.9.13}$$

For combined rotations and boosts $S(\alpha, \beta)$ in the $x - y$ plane, given by (2.5.26),

$$S^{\mu}{}_{\nu}(\alpha,\beta) = \begin{bmatrix} 1 & 0 & -\alpha & \alpha \\ 0 & 1 & -\beta & \beta \\ \alpha & \beta & 1-\gamma & \gamma \\ \alpha & \beta & -\gamma & 1+\gamma \end{bmatrix} ,$$

$$\gamma \equiv (\alpha^2 + \beta^2)/2 ,$$

Eqs. (5.9.10) and (5.9.11) give

$$u_{\bar{\ell}}(\mathbf{k},\sigma) = \sum_{\ell} D_{\bar{\ell}\ell}\Big(S(\alpha,\beta)\Big)u_{\ell}(\mathbf{k},\sigma) \ , \tag{5.9.14}$$

$$v_{\bar{\ell}}(\mathbf{k},\sigma) = \sum_{\ell} D_{\bar{\ell}\ell}\Big(S(\alpha,\beta)\Big)v_{\ell}(\mathbf{k},\sigma) \ . \tag{5.9.15}$$

Eqs. (5.9.12)–(5.9.15) are the conditions that determine the coefficient functions u and v at the standard momentum $\mathbf{k}$; Eqs. (5.9.8) and (5.9.9) then give them at arbitrary momenta. The equations for v are just the complex conjugates of the equations for u, so with a suitable adjustment of the constants κ and λ we may normalize the coefficient functions so that

$$v_{\ell}(\mathbf{p},\sigma) = u_{\ell}(\mathbf{p},\sigma)^* \ . \tag{5.9.16}$$

The problem is that we cannot find a u_ℓ that satisfies Eq. (5.9.14) for general representations of the homogeneous Lorentz group, even for those representations for which it is possible to construct fields for particles of a given helicity in the case $m \neq 0$.

To see what goes wrong here, let's try to construct the four-vector $[(\frac{1}{2},\frac{1}{2})]$ field for a massless particle of helicity ± 1. In the four-vector representation, we have simply

$$D^{\mu}{}_{\nu}(\Lambda) = \Lambda^{\mu}{}_{\nu} \ .$$

It is conventional to write the coefficient function u_μ here in terms of a 'polarization vector' e_μ:

$$u_{\mu}(\mathbf{p},\sigma) \equiv (2p^0)^{-1/2}\, e_{\mu}(\mathbf{p},\sigma) \ , \tag{5.9.17}$$

so that Eq. (5.9.8) gives

$$e^{\mu}(\mathbf{p},\sigma) = \mathscr{L}(\mathbf{p})^{\mu}{}_{\nu} e^{\nu}(\mathbf{k},\sigma) \ . \tag{5.9.18}$$

Also, Eqs. (5.9.12) and (5.9.14) read here

$$e^{\mu}(\mathbf{k},\sigma)\, e^{i\sigma\theta} = R(\theta)^{\mu}{}_{\nu} e^{\nu}(\mathbf{k},\sigma) \ , \tag{5.9.19}$$

$$e^{\mu}(\mathbf{k},\sigma) = S(\alpha,\beta)^{\mu}{}_{\nu} e^{\nu}(\mathbf{k},\sigma) \ . \tag{5.9.20}$$

Eq. (5.9.19) requires that (up to a constant which can be absorbed into the coefficients κ and λ),

$$e^{\mu}(\mathbf{k},\pm 1) = (1,\pm i,0,0)/\sqrt{2} \ . \tag{5.9.21}$$

But then Eq. (5.9.20) would require also that $\alpha \pm i\beta = 0$, which is impossible for general real α,β. We therefore cannot satisfy the fundamental requirement (5.9.14) or (5.9.10); instead, we have here

$$D^\mu{}_\nu\Big(W(\theta,\alpha,\beta)\Big)e^\nu(\mathbf{k},\pm 1) = S^\mu{}_\lambda(\alpha,\beta)R^\lambda{}_\nu(\theta)e^\nu(\mathbf{k},\pm 1)$$
$$= \exp\,(\pm i\theta)\left\{e^\mu(\mathbf{k},\pm 1) + \frac{(\alpha \pm i\beta)}{\sqrt{2}|\mathbf{k}|}k^\mu\right\}. \quad (5.9.22)$$

We have thus come to the conclusion that no four-vector field can be constructed from the annihilation and creation operators for a particle of mass zero and helicity ± 1.

Let's temporarily close our eyes to this difficulty, and go ahead anyway, using Eqs. (5.9.18) and (5.9.21) to define a polarization vector for arbitrary momentum, and take the field as

$$a_\mu(x) = \int d^3p\,(2\pi)^{-3/2}(2p^0)^{-1/2}$$
$$\times \sum_{\sigma=\pm 1}\left[e_\mu(\mathbf{p},\sigma)e^{ip\cdot x}a(\mathbf{p},\sigma) + e_\mu(\mathbf{p},\sigma)^* e^{-ip\cdot x}a^{c\dagger}(\mathbf{p},\sigma)\right]. \quad (5.9.23)$$

We will come back later to consider how such a field can be used as an ingredient in a physical theory.

The field (5.9.23) of course satisfies

$$\Box a^\mu(x) = 0\,. \quad (5.9.24)$$

Other properties of the field follow from those of the polarization vector. (We shall need these properties of the polarization vector later when we come to quantum electrodynamics.). Note that the Lorentz transformation $\mathscr{L}(p)$ that takes a massless particle momentum from $\mathbf{k}$ to $\mathbf{p}$ may be written as a 'boost' $\mathscr{B}(|\mathbf{p}|)$ along the z-axis which takes the particle from energy $|\mathbf{k}|$ to energy $|\mathbf{p}|$, followed by a standardized rotation $R(\hat{\mathbf{p}})$ that takes the z-axis into the direction of $\mathbf{p}$. Since $e^\nu(\mathbf{k},\pm 1)$ is a purely spatial vector with only x and y components, it is unaffected by the boost along the z-axis, and so

$$e^\mu(\mathbf{p},\pm 1) = R(\hat{\mathbf{p}})^\mu{}_\nu\, e^\nu(\mathbf{k},\pm 1)\,. \quad (5.9.25)$$

In particular, $e^0(\mathbf{k},\pm 1) = 0$ and $\mathbf{k}\cdot\mathbf{e}(\mathbf{k},\pm 1) = 0$ so

$$e^0(\mathbf{p},\pm 1) = 0 \quad (5.9.26)$$

and

$$\mathbf{p}\cdot\mathbf{e}(\mathbf{p},\pm 1) = 0\,. \quad (5.9.27)$$

It follows that

$$a^0(x) = 0 \quad (5.9.28)$$

and

$$\nabla\cdot\mathbf{a}(x) = 0\,. \quad (5.9.29)$$

As we shall see in Chapter 9, these are the conditions satisfied by the vacuum vector potential of electrodynamics in what is called Coulomb or radiation gauge.

The fact that a^0 vanishes in all Lorentz frames shows vividly that a^μ cannot be a four-vector. Instead, Eq. (5.9.22) shows that for a general momentum $\mathbf{p}$ and a general Lorentz transformation Λ, in place of Eq. (5.9.6) we have

$$e^\mu(\mathbf{p}_\Lambda, \pm 1)\exp(\pm i\theta(\mathbf{p}, \Lambda)) = D^\mu{}_\nu(\Lambda)\, e^\nu(\mathbf{p}, \pm 1) + p^\mu \Omega_\pm(\mathbf{p}, \Lambda)\,, \tag{5.9.30}$$

so that under a general Lorentz transformation

$$U(\Lambda) a_\mu(x) U^{-1}(\Lambda) = \Lambda^\nu{}_\mu a_\nu(\Lambda x) + \partial_\mu \Omega(x, \Lambda)\,, \tag{5.9.31}$$

where $\Omega(x, \Lambda)$ is a linear combination of annihilation and creation operators, whose precise form will not concern us here. As we will see in more detail in Chapter 8, we will be able to use a field like $a^\mu(x)$ as an ingredient in Lorentz-invariant physical theories if the couplings of $a^\mu(x)$ are not only formally Lorentz-invariant (that is, invariant under formal Lorentz transformations under which $a^\mu \to \Lambda^\mu{}_\nu\, a^\nu$), but are also invariant under the 'gauge' transformations $a_\mu \to a_\mu + \partial_\mu \Omega$. This is accomplished by taking the couplings of a_μ to be of the form $a_\mu j^\mu$, where j^μ is a four-vector current with $\partial_\mu j^\mu = 0$.

Although there is no ordinary four-vector field for massless particles of helicity ± 1, there is no problem in constructing an antisymmetric tensor field for such particles. From Eq. (5.9.22) and the invariance of k^μ under the little group we see immediately that

$$D^\mu{}_\rho\Big(W(\theta, \alpha, \beta)\Big) D^\nu{}_\sigma\Big(W(\theta, \alpha, \beta)\Big)\Big(k^\rho e^\sigma(\mathbf{k}, \pm 1) - k^\sigma e^\rho(\mathbf{k}, \pm 1)\Big)$$
$$= e^{\pm i\theta}\Big(k^\mu e^\nu(\mathbf{k}, \pm 1) - k^\nu e^\mu(\mathbf{k}, \pm 1)\Big)\,. \tag{5.9.32}$$

This shows that the coefficient function that satisfies Eq. (5.9.6) for the antisymmetric tensor representation of the homogeneous Lorentz group is (with an appropriate choice of normalization)

$$u^{\mu\nu}(\mathbf{p}, \pm 1) = i(2\pi)^{-3/2}(2p^0)^{-3/2}\,[p^\mu e^\nu(\mathbf{p}, \pm 1) - p^\nu e^\mu(\mathbf{p}, \pm 1)\;]\,, \tag{5.9.33}$$

where $e^\mu(\mathbf{p}, \pm 1)$ is given by Eq. (5.9.25). Using this together with Eq. (5.9.23) gives the general antisymmetric tensor field for massless particles of helicity ± 1 in the form

$$f_{\mu\nu} = \partial_\mu a_\nu - \partial_\nu a_\mu\,. \tag{5.9.34}$$

Note that this is a tensor even though a^μ is not a four-vector, because the extra term in Eq. (5.9.31) drops out in Eq. (5.9.34). Note also that Eqs. (5.9.34), (5.9.24), (5.9.28), and (5.9.29) show that $f^{\mu\nu}$ satisfies the vacuum Maxwell equations:

$$\partial_\mu f^{\mu\nu} = 0\,, \tag{5.9.35}$$

$$\epsilon^{\rho\sigma\mu\nu}\partial_\sigma f_{\mu\nu} = 0\,. \tag{5.9.36}$$

To calculate the commutation relations for the tensor fields we need sums over helicities of the bilinears $e^\mu e^{\nu *}$. The explicit formula (5.9.21) gives

$$\sum_{\sigma=\pm 1} e^i(\mathbf{k},\sigma)e^j(\mathbf{k},\sigma)^* = \delta_{ij} - \frac{k^i k^j}{|\mathbf{k}|^2}$$

and so, using Eq. (5.9.25),

$$\sum_{\sigma=\pm 1} e^i(\mathbf{p},\sigma)e^j(\mathbf{p},\sigma)^* = \delta_{ij} - \frac{p^i p^j}{|\mathbf{p}|^2}\,. \tag{5.9.37}$$

A straightforward calculation gives then

$$[f_{\mu\nu}(x), f_{\rho\sigma}(y)^\dagger] = (2\pi)^{-3}\left[-\eta_{\mu\rho}\partial_\nu\partial_\sigma + \eta_{\nu\rho}\partial_\mu\partial_\sigma + \eta_{\mu\sigma}\partial_\nu\partial_\rho - \eta_{\nu\sigma}\partial_\mu\partial_\rho\right]$$
$$\times \int d^3p(2p^0)^{-1}\left[|\kappa|^2 e^{ip\cdot(x-y)} - |\lambda|^2 e^{-ip\cdot(x-y)}\right]\,. \tag{5.9.38}$$

This clearly vanishes for $x^0 = y^0$ if and only if

$$|\kappa|^2 = |\lambda|^2 \tag{5.9.39}$$

in which case since $f_{\mu\nu}$ is a tensor the commutator also vanishes for all space-like separations. Eq. (5.9.39) also implies that the commutator of the a^μ vanishes at equal times, and as we shall see in Chapter 8 this is enough to yield a Lorentz-invariant S-matrix. The relative phase of the creation and annihilation operators can be adjusted so that $\kappa = \lambda$; the fields are then Hermitian if the particles are their own charge-conjugates, as is the case for the photon.

Why should we want to use fields like $a^\mu(x)$ in constructing theories of massless particles of spin one, rather than being content with fields like $f^{\mu\nu}(x)$ with simple Lorentz transformation properties? The presence of the derivatives in Eq. (5.9.34) means that an interaction density constructed solely from $f_{\mu\nu}$ and its derivatives will have matrix elements that vanish more rapidly for small massless particle energy and momentum than one that uses the vector field a_μ. Interactions in such a theory will have a correspondingly rapid fall-off at large distances, faster than the usual inverse-square law. This is perfectly possible, but gauge-invariant theories that use vector fields for massless spin one particles represent a more general class of theories, including those that are actually realized in nature.

Parallel remarks apply to gravitons, massless particles of helicity ± 2. From the annihilation and creation operators for such particles we can construct a tensor $R_{\mu\nu\rho\sigma}$ with the algebraic properties of the Riemann–Christoffel curvature tensor: antisymmetric within the pairs μ,ν and ρ,σ, and symmetric between the pairs. However, in order to incorporate the usual inverse-square gravitational interactions we need to introduce a field $h_{\mu\nu}$ that transforms as a symmetric tensor, up to gauge transformations of the sort associated in general relativity with general coordinate transformations. Thus in order to construct a theory of massless particles of helicity ± 2 that incorporates long-range interactions, it is necessary for it to have a symmetry something like general covariance. As in the case of electromagnetic gauge invariance, this is achieved by coupling the field to a conserved 'current' $\theta^{\mu\nu}$, now with two spacetime indices, satisfying $\partial_\mu \theta^{\mu\nu} = 0$. The only such conserved tensor is the energy-momentum tensor, aside from possible total derivative terms that do not affect the long-range behavior of the force produced.** The fields of massless particles of spin $j \geq 3$ would have to couple to conserved tensors with three or more spacetime indices, but aside from total derivatives there are none, so *high-spin massless particles cannot produce long-range forces.*

* * *

The problems we have encountered in constructing four-vector fields for helicities ± 1 or symmetric tensor fields for helicity ± 2 are just special cases of a more general limitation. To see this, let's consider how to construct fields for massless particles belonging to arbitrary representations of the homogeneous Lorentz group. As we saw in Section 5.6, any representation $D(\Lambda)$ of the homogeneous Lorentz group can be decomposed into $(2A+1)(2B+1)$-dimensional representations (A, B), for which the generators of the homogeneous Lorentz group are represented by

$$(\mathscr{J}_{ij})_{a'b',ab} = \epsilon_{ijk} \left[(J_k^{(A)})_{a'a}\, \delta_{b'b} + (J_k{}^{(B)})_{b'b}\, \delta_{a'a} \right],$$

$$(\mathscr{J}_{k0})_{a'b',ab} = -i \left[(J_k^{(A)})_{a'a}\, \delta_{b'b} - (J_k{}^{(B)})_{b'b}\, \delta_{a'a} \right],$$

where $\mathbf{J}^{(j)}$ are the angular-momentum matrices for spin j. For θ infinites-

** If $\theta^{\mu_1\cdots\mu_N}$ is a tensor current satisfying $\partial_{\mu_1}\theta^{\mu_1\cdots\mu_N} = 0$, then $\int d^3x\, \theta^{0\mu_2\cdots\mu_N}$ is a conserved quantity that transforms like a tensor of rank $N-1$. The only such conserved tensors are the scalar 'charges' associated with various continuous symmetries, and the energy-momentum four-vector. The conservation of any other four-vector, or any tensor of higher rank, would forbid all but forward collisions.

imal, $D(R(\theta)) = 1 + i\mathcal{J}_{12}\theta$, so Eqs. (5.9.12) and (5.9.13) give

$$\sigma u_{ab}(\mathbf{k},\sigma) = (a+b)u_{ab}(\mathbf{k},\sigma)\,,$$

$$-\sigma v_{ab}(\mathbf{k},\sigma) = (a+b)v_{ab}(\mathbf{k},\sigma)\,,$$

and so $u_{ab}(\mathbf{k},\sigma)$ and $v_{ab}(\mathbf{k},\sigma)$ must vanish unless $\sigma = a+b$ and $\sigma = -a-b$, respectively. Also, letting α and β become infinitesimal in Eq. (5.9.14) gives

$$\begin{aligned}
0 &= (\mathcal{J}_{31} + \mathcal{J}_{01})_{ab,a'b'}\; u_{a'b'}(\mathbf{k},\sigma) \\
&= (J_2^{(A)} + iJ_1^{(A)})_{aa'}u_{a'b}(\mathbf{k},\sigma) + (J_2^{(B)} - iJ_1^{(B)})_{bb'}u_{ab'}(\mathbf{k},\sigma)\,, \\
0 &= (\mathcal{J}_{32} + \mathcal{J}_{02})_{ab,a'b'}\; u_{a'b'}(\mathbf{k},\sigma) \\
&= (-J_1^{(A)} + iJ_2^{(A)})_{aa'}u_{a'b}(\mathbf{k},\sigma) + (-J_1^{(B)} - iJ_2^{(B)})_{bb'}u_{ab'}(\mathbf{k},\sigma)\,,
\end{aligned}$$

or more simply

$$\left(J_1^{(A)} - i\,J_2^{(A)}\right)_{aa'} u_{a'b}(\mathbf{k},\sigma) = 0\,,$$

$$\left(J_1^{(B)} + i\,J_2^{(B)}\right)_{bb'} u_{ab'}(\mathbf{k},\sigma) = 0\;\;.$$

These require that $u_{ab}(\mathbf{k},\sigma)$ vanishes unless

$$a = -A,\quad b = \;+B \tag{5.9.40}$$

and the same is obviously also true of $v_{ab}(\mathbf{k},\sigma)$. Putting this together, we see that a field of type (A,B) can be formed only from the annihilation operators for a massless particle of helicity σ and the creation operators for the antiparticle of helicity $-\sigma$, where

$$\sigma = B - A\,. \tag{5.9.41}$$

For instance, the $(\frac{1}{2},0)$ and $(0,\frac{1}{2})$ parts of the Dirac field for a massless particle can only destroy particles of helicity $-\frac{1}{2}$ and $+\frac{1}{2}$ respectively, and create antiparticles of helicity $+\frac{1}{2}$ and $-\frac{1}{2}$, respectively. In the 'two-component' theory of the neutrino, there is only a $(\frac{1}{2},0)$ field and its adjoint, so neutrinos have helicity $-\frac{1}{2}$ and antineutrinos helicity $+\frac{1}{2}$ in this theory.

By the same methods as in Section 5.7, it can be shown that the $(j,0)$ and $(0,j)$ fields for massless particles of spin j (i.e., helicity $\mp j$) commute with each other and their adjoints at space-like separations if the coefficients of the annihilation and creation terms in Eq. (5.9.1) satisfy Eq. (5.9.39). The relative phase of the annihilation and creation operators may then be adjusted so that these coefficients are equal. It is easy to see that the fields for a massless particle of spin j of type $(A, A+j)$ or $(B+j, B)$ are just the $2A$th or $2B$th derivatives of fields of type $(0,j)$ or $(j,0)$, respectively, so these more general fields do not need to be considered separately here.

We can now see why it was impossible to construct a vector field for massless particles of helicity ± 1. A vector field transforms according to the $(\frac{1}{2}, \frac{1}{2})$ representation, and hence according to Eq. (5.9.41) can only describe helicity zero. (It *is*, of course, possible to construct a vector field for helicity zero — just take the derivative $\partial_\mu \phi$ of a massless scalar field ϕ.) The simplest covariant massless field for helicity ± 1 has the Lorentz transformation type $(1,0) \oplus (0,1)$; that is, it is an antisymmetric tensor $f_{\mu\nu}$. Similarly, the simplest covariant massless field for helicity ± 2 has the Lorentz transformation type $(2,0) \oplus (0,2)$: a fourth rank tensor which like the Riemann–Christoffel curvature tensor is antisymmetric within each pair of indices and symmetric between the two pairs.

The discussion of the inversions P, C, T given in the previous section can be carried over to the case of zero mass with only obvious modifications.

Problems

1. Show that if the zero-momentum coefficient functions satisfy the conditions (5.1.23) and (5.1.24), then the coefficient functions (5.1.21) and (5.1.22) for arbitrary momentum satisfy the defining conditions Eqs. (5.1.19) and (5.1.20).

2. Consider a free field $\psi^\mu_\ell(x)$ which annihilates and creates a self-charge-conjugate particle of spin $\frac{3}{2}$ and mass $m \neq 0$. Show how to calculate the coefficient functions $u^\mu_\ell(\mathbf{p}, \sigma)$, which multiply the annihilation operators $a(\mathbf{p}, \sigma)$ in this field, in such a way that the field transforms under Lorentz transformations like a Dirac field ψ_ℓ with an extra four-vector index μ. What field equations and algebraic and reality conditions does this field satisfy? Evaluate the matrix $P^{\mu\nu}(p)$, defined (for $p^2 = -m^2$) by

$$\sum_\sigma u^\mu_\ell(\mathbf{p}, \sigma) u^{\nu\,*}_m(\mathbf{p}, \sigma) \equiv (2p^0)^{-1} P^{\mu\nu}_{\ell m}(p).$$

   What are the commutation relations of this field? How does the field transform under the inversions P, C, T?

3. Consider a free field $h^{\mu\nu}(x)$ satisfying $h^{\mu\nu}(x) = h^{\nu\mu}(x)$ and $h^\mu{}_\mu(x) = 0$, which annihilates and creates a particle of spin two and mass $m \neq 0$. Show how to calculate the coefficient functions $u^{\mu\nu}(\mathbf{p}, \sigma)$, which multiply the annihilation operators $a(\mathbf{p}, \sigma)$ in this field, in such a way that the field transforms under Lorentz transformations like a tensor. What field equations does this field satisfy? Evaluate the function $P^{\mu\nu,\kappa\lambda}(p)$, defined by

$$\sum_\sigma u^{\mu\nu}(\mathbf{p}, \sigma) u^{\kappa\lambda\,*}(\mathbf{p}, \sigma) \equiv (2p^0)^{-1} P^{\mu\nu,\kappa\lambda}(p).$$

What are the commutation relations of this field? How does the field transform under the inversions P, C, T?

4. Show that the fields for a massless particle of spin j of type $(A, A+j)$ or $(B+j, B)$ are the $2A$th or $2B$th derivatives of fields of type $(0, j)$ or $(j, 0)$, respectively.

5. Work out the transformation properties of fields of transformation type $(j, 0) + (0, j)$ for massless particles of helicity $\pm j$ under the inversions P, C, T.

6. Consider a generalized Dirac field ψ that transforms according to the $(j, 0) + (0, j)$ representation of the homogeneous Lorentz group. List the tensors that can be formed from products of the components of ψ and $\psi^\dagger$. Check your result against what we found for $j = \frac{1}{2}$.

7. Consider a general field ψ_{ab} describing particles of spin j and mass $m \neq 0$, that transforms according to the (A, B) representation of the homogeneous Lorentz group. Suppose it has an interaction Hamiltonian of the form

$$V = \int d^3x \, [\psi_{ab}(x) J^{ab}(x) + J^{ab\dagger}(x) \psi^\dagger_{ab}(x)] \, ,$$

where J^{ab} is an external c-number current. What is the asymptotic behavior of the matrix element for emitting these particles for energy $E \gg m$ and definite helicity? (Assume that the Fourier transform of the current has values for different a, b that are of the same order of magnitude, and that do not depend strongly on E.)

References

1. The point of view adopted in this chapter was presented in a series of papers: S. Weinberg, *Phys. Rev.* **133**, B1318 (1964); **134**, B882 (1964); **138**, B988 (1965); **181**, 1893 (1969). A similar approach has been followed in unpublished lectures by E. Wichmann.

2. N. Bohr and L. Rosenfeld, *Kgl. Danske Vidensk. Selskab Mat.-Fys. Medd.*, No. 12 (1933) (translation in *Selected Papers of Leon Rosenfeld*, ed. by R. S. Cohen and J. Stachel (Reidel, Dordrecht, 1979)); *Phys. Rev.* **78**, 794 (1950).

3. P. A. M. Dirac, *Proc. Roy. Soc.* (London) **A117**, 610 (1928).

4. E. Cartan, *Bull. Soc. Math. France* **41**, 53 (1913).

5. See, e.g., J. M. Jauch and F. Rohrlich, *The Theory of Photons and Electrons* (Addison-Wesley, Cambridge, MA, 1955): Appendix A2; H. Georgi, *Lie Algebras in Particle Physics* (Benjamin–Cummings, Reading, MA, 1982): pp. 15, 198. The original reference is I. Schur, *Sitz. Preuss. Akad.*, p. 406 (1905).

6. See, e.g., H. Georgi, *Lie Algebras in Particle Physics* (Benjamin/Cummings, Reading, MA, 1982): pp. 15, 198. The original reference is I. Schur, *Sitz. Preuss. Akad.*, p. 406 (1905).

7. T. D. Lee and C. N. Yang, *Phys. Rev.* **104**, 254 (1956).

8. See, e.g., B. L. van der Waerden, *Die gruppentheoretische Methode in der Quantenmechanik* (Springer Verlag, Berlin, 1932); G. Ya. Lyubarski, *The Applications of Group Theory in Physics*, translated by S. Dedijer (Pergamon Press, New York, 1960).

9. W. Rarita and J. Schwinger, *Phys. Rev.* **60**, 61 (1941).

10. See, e.g., A. R. Edmonds, *Angular Momentum in Quantum Mechanics*, (Princeton University Press, Princeton, 1957): Chapter 3.

11. S. Weinberg, *Phys. Rev.* **181**, 1893 (1969), Section V.

12. M. Fierz, *Helv. Phys. Acta* **12**, 3 (1939); W. Pauli, *Phys. Rev.* **58**, 716 (1940). Non-perturbative proofs in axiomatic field theory were given by G. Lüders and B. Zumino, *Phys. Rev.* **110**, 1450 (1958) and N. Burgoyne, *Nuovo Cimento* **8**, 807 (1958). Also see R. F. Streater and A. S. Wightman, *PCT, Spin & Statistics, and All That* (Benjamin, New York, 1968).

13. Fields in the $(j,0)+(0,j)$ representation were introduced by H. Joos, *Fortschr. Phys.* **10**, 65 (1962); S. Weinberg, *Phys. Rev.* **133**, B1318 (1964).

14. A. R. Edmonds, Ref. 10, or M. E. Rose, *Elementary Theory of Angular Momentum* (John Wiley & Sons, New York, 1957): Chapter III .

15. G. Velo and D. Zwanziger, *Phys. Rev.* **186**, 1337 (1969); **188**, 2218 (1969); A. S. Wightman, in *Proceedings of the Fifth Coral Gables Conference on Symmetry Principles at High Energy*, ed. by T. Gudehus, G. Kaiser, and A. Perlmutter (Gordon and Breach, New York, 1969); B. Schroer, R. Seiler, and J. A. Swieca, *Phys. Rev.* D **2**, 2927 (1970); and other references quoted therein.

16. C. R. Nappi and L. Witten, *Phys. Rev.* D **40**, 1095 (1989); P. C. Argyres and C. R. Nappi, *Phys. Lett.* **B224**, 89 (1989). For the

derivation of a consistent theory of $j = 3/2$ particles in external fields from a Kaluza–Klein theory, see S. D. Rindani and M. Sivakumar, *J. Phys. G: Nucl. Phys.* **12**, 1335 (1986); *J. Phys. C: Particles & Fields* **49**, 601 (1991).

17. G. Lüders, *Kong. Dansk. Vid. Selskab, Mat.-Fys. Medd.* **28**, 5 (1954); *Ann. Phys.* **2**, 1 (1957); W. Pauli, *Nuovo Cimento* **6**, 204 (1957). When Lüders first considered how the inversions are related, it was still taken for granted that P is conserved, so his theorem stated that C conservation is equivalent to T invariance.

18. R. Jost, *Helv. Phys. Acta* **30**, 409 (1957); F. J. Dyson, *Phys. Rev.* **110**, 579 (1958). Also see Streater and Wightman, Ref. 12.

6

The Feynman Rules

In previous chapters the use of covariant free fields in the construction of the Hamiltonian density has been motivated by the requirement that the S-matrix satisfy Lorentz invariance and cluster decomposition conditions. With the Hamiltonian density constructed in this way, it makes no difference which form of perturbation theory we use to calculate the S-matrix; the results will automatically satisfy these invariance and clustering conditions in each order in the interaction density. Nevertheless, there are obvious practical advantages in using a version of perturbation theory in which the Lorentz invariance and cluster decomposition properties of the S-matrix are kept manifest at every stage in the calculation. This was not true for the perturbation theory used in the 1930s, now known as 'old-fashioned perturbation theory', described at the beginning of Section 3.5. The great achievement of Feynman, Schwinger, and Tomonaga in the late 1940s was to develop perturbative techniques for calculating the S-matrix, in which Lorentz invariance and cluster decomposition properties are transparent throughout. This chapter will outline the diagrammatic calculational technique first described by Feynman at the Poconos Conference in 1948. Feynman was led to these diagrammatic rules in part through his development of a path-integral approach, which will be the subject of Chapter 9. In this chapter, we shall use the approach described by Dyson[1] in 1949, which until the 1970s was the basis of almost all analyses of perturbation theory in quantum field theory, and still provides a particularly transparent introduction to the Feynman rules.

6.1 Derivation of the Rules

Our starting point is a formula for the S-matrix, obtained by putting together the Dyson series (3.5.10) with expression (4.2.2) for the free-particle states:

$$S_{\mathbf{p}'_1\sigma'_1n'_1;\mathbf{p}'_2\sigma'_2n'_2;\cdots,\ \mathbf{p}_1\sigma_1n_1;\mathbf{p}_2\sigma_2n_2;\cdots}$$
$$=\sum_{N=0}^{\infty}\frac{(-i)^N}{N!}\int d^4x_1\cdots d^4x_N\Big(\Phi_0,\cdots a(\mathbf{p}'_2\sigma'_2n'_2)a(\mathbf{p}'_1\sigma'_1n'_1)$$
$$\times\, T\Big\{\mathscr{H}(x_1)\cdots\mathscr{H}(x_N)\Big\}a^\dagger(\mathbf{p}_1\sigma_1n_1)a^\dagger(\mathbf{p}_2\sigma_2n_2)\cdots\Phi_0\Big)\ . \tag{6.1.1}$$

As a reminder: $\mathbf{p}$, σ, and n label particle momenta, spin, and species; primes denote labels for particles in the final state; Φ_0 is the free-particle vacuum state; a and $a^\dagger$ are annihilation and creation operators; T indicates a time-ordering, which puts the $\mathscr{H}(x)$ in an order in which the arguments x^0 decrease from left to right; and $\mathscr{H}(x)$ is the interaction Hamiltonian density, taken as a polynomial in the fields and their adjoints

$$\mathscr{H}(x)=\sum_i g_i\mathscr{H}_i(x)\,, \tag{6.1.2}$$

each term $\mathscr{H}_i$ being a product of definite numbers of fields and field adjoints of each type. The field of a particle of species n that transforms under a particular representation of the homogeneous Lorentz group (with or without space inversions) is given by

$$\psi_\ell(x)=\sum_\sigma(2\pi)^{-3/2}\int d^3p\,\Big[u_\ell(\mathbf{p},\sigma,n)\,a(\mathbf{p},\sigma,n)\,e^{ip\cdot x}$$
$$+\,v_\ell(\mathbf{p},\sigma,n)a^\dagger(\mathbf{p},\sigma,n^c)\,e^{-ip\cdot x}\Big]\ . \tag{6.1.3}$$

Here n^c denotes the antiparticle of the species n, and $\exp(\pm ip\cdot x)$ is calculated with p^0 set equal to $\sqrt{\mathbf{p}^2+m_n^2}$. The coefficient functions u_ℓ and v_ℓ depend on the Lorentz transformation properties of the field and the spin of the particle it describes; they were calculated in Chapter 5. (For instance, in the scalar field the u_ℓ for a particle of energy E is simply $(2E)^{-1/2}$, while in a Dirac field u_ℓ and v_ℓ are the normalized Dirac spinors introduced in Section 5.5.) The index ℓ on the field should here be understood to indicate the particle type and the representation of the Lorentz group by which the field transforms, as well as including a running index labelling the components in this representation. There is no need to deal separately with interactions that involve derivatives of fields; from our point of view, the derivative of a field (6.1.3) is just another field described by (6.1.3), with different u_ℓ and v_ℓ. We will here make a distinction between some particle species that we arbitrarily call 'particles', for instance electrons, protons, etc., and those we call 'antiparticles', such as positrons and antiprotons. The field operators that destroy particles and create antiparticles are called simply 'fields'; their adjoints, which destroy antiparticles and create particles, are called 'field adjoints'. Of course, some particle species like the photon and π^0 are their own antiparticles; for these the field adjoints are proportional to the fields.

We now proceed to move all annihilation operators to the right in Eq. (6.1.1), repeatedly using for this purpose the commutation or anti-commutation relations:

$$a(\mathbf{p}\,\sigma\,n)a^\dagger(\mathbf{p}'\sigma'n') = \pm a^\dagger(\mathbf{p}'\sigma'n')a(\mathbf{p}\,\sigma\,n) + \delta^3(\mathbf{p}'-\mathbf{p})\delta_{\sigma'\sigma}\delta_{n'n} \tag{6.1.4}$$

$$a(\mathbf{p}\,\sigma\,n)a(\mathbf{p}'\sigma'n') = \pm a(\mathbf{p}'\sigma'n')a(\mathbf{p}\,\sigma\,n) \tag{6.1.5}$$

$$a^\dagger(\mathbf{p}\,\sigma\,n)a^\dagger(\mathbf{p}'\sigma'n') = \pm a^\dagger(\mathbf{p}'\sigma'n')a^\dagger(\mathbf{p}\,\sigma\,n) \tag{6.1.6}$$

(and likewise for antiparticles), the $\pm$ sign on the right being $-$ if both particles n, n' are fermions, and $+$ if either or both are bosons. Whenever an annihilation operator appears on the extreme right (or a creation operator on the extreme left), the corresponding contribution to Eq. (6.1.1) vanishes, because these operators annihilate the vacuum state:

$$a(\mathbf{p}\,\sigma\,n)\,\Phi_0 = 0\,, \tag{6.1.7}$$

$$\Phi_0^\dagger\, a^\dagger(\mathbf{p}\,\sigma\,n) = 0. \tag{6.1.8}$$

The remaining contributions to Eq. (6.1.1) are those arising from the delta function terms on the right-hand side of Eq. (6.1.4), with every creation and annihilation operator in the initial or final states or in the interaction Hamiltonian density paired in this way with some other annihilation or creation operator.

In this way, the contribution to Eq. (6.1.1) of a given order in each of the terms $\mathscr{H}_i$ in the polynomial $\mathscr{H}(\psi(x),\psi^\dagger(x))$ is given by a sum, over all ways of pairing creation and annihilation operators,[2] of the integrals of products of factors, as follows:

(a) Pairing of a final particle having quantum numbers $\mathbf{p}', \sigma', n'$ with a field adjoint $\psi_\ell^\dagger(x)$ in $\mathscr{H}_i(x)$ yields a factor

$$\Big[a(\mathbf{p}'\,\sigma'n'),\,\psi_\ell^\dagger(x)\Big]_\mp = (2\pi)^{-3/2}e^{-ip'\cdot x}u_\ell^*(\mathbf{p}'\sigma'n')\,. \tag{6.1.9}$$

(b) Pairing of a final antiparticle having quantum numbers $\mathbf{p}', \sigma', n'^c$ with a field $\psi_\ell(x)$ in $\mathscr{H}_i(x)$ yields a factor

$$\Big[a(\mathbf{p}'\sigma'\,n'^c),\psi_\ell(x)\Big]_\mp = (2\pi)^{-3/2}e^{-ip'\cdot x}v_\ell(\mathbf{p}'\sigma'n')\,. \tag{6.1.10}$$

(c) Pairing of an initial particle having quantum numbers $\mathbf{p}, \sigma, n$ with a field $\psi_\ell(x)$ in $\mathscr{H}_i(x)$ yields a factor

$$\Big[\psi_\ell(x), a^\dagger(\mathbf{p}\sigma n)\Big]_\mp = (2\pi)^{-3/2}e^{ip\cdot x}u_\ell(\mathbf{p}\sigma n)\,. \tag{6.1.11}$$

(d) Pairing of an initial antiparticle having quantum numbers $\mathbf{p}, \sigma, n^c$ with a field adjoint $\psi_\ell^\dagger(x)$ in $\mathscr{H}_i(x)$ yields a factor

$$\Big[\psi_\ell^\dagger(x), a^\dagger(\mathbf{p}\,\sigma\,n^c)\Big]_\mp = (2\pi)^{-3/2} e^{ip\cdot x} v_\ell^*(\mathbf{p}\,\sigma\,n)\,. \qquad (6.1.12)$$

(e) Pairing of a final particle (or antiparticle) having numbers $\mathbf{p}', \sigma', n'$ with an initial particle (or antiparticle) having quantum numbers $\mathbf{p}, \sigma, n$ yields a factor

$$\Big[a(\mathbf{p}'\sigma' n'), a^\dagger(\mathbf{p}\,\sigma\,n)\Big]_\mp = \delta^3(\mathbf{p}' - \mathbf{p})\delta_{\sigma'\sigma}\delta_{n'n}\,. \qquad (6.1.13)$$

(f) Pairing of a field $\psi_\ell(x)$ in $\mathscr{H}_i(x)$ with a field adjoint $\psi_m^\dagger(y)$ in $\mathscr{H}_j(y)$ yields a factor*

$$\theta(x-y)\Big[\psi_\ell^+(x), \psi_m^{+\dagger}(y)\Big]_\mp \pm \theta(y-x)\Big[\psi_m^{-\dagger}(y), \psi_\ell^-(x)\Big]_\mp$$
$$\equiv -i\,\Delta_{\ell m}(x,y)\,, \qquad (6.1.14)$$

where ψ^+ and ψ^- are the terms in ψ that destroy particles and create antiparticles, respectively:

$$\psi_\ell^+(x) = (2\pi)^{-3/2} \int d^3p \sum_\sigma u_\ell(\mathbf{p}\,\sigma\,n)\, e^{ip\cdot x} a(\mathbf{p}\,\sigma\,n)\,, \qquad (6.1.15)$$

$$\psi_\ell^-(x) = (2\pi)^{-3/2} \int d^3p \sum_\sigma v_\ell(\mathbf{p}\,\sigma\,n)\, e^{-ip\cdot x} a^\dagger(\mathbf{p}\,\sigma\,n^c)\,. \qquad (6.1.16)$$

Recall that $\theta(x-y)$ is a step function, equal to $+1$ for $x^0 > y^0$ and zero for $x^0 < y^0$. These step functions appear in Eq. (6.1.14) because of the time-ordering in Eq. (6.1.1); we can encounter a pairing of an annihilation field $\psi^+(x)$ in $\mathscr{H}(x)$ with a creation field $\psi^{+\dagger}(y)$ in $\mathscr{H}(y)$ only if $\mathscr{H}(x)$ was initially to the left of $\mathscr{H}(y)$ in Eq. (6.1.1), i.e., if $x^0 > y^0$; similarly, we encounter a pairing of an annihilation field $\psi^{-\dagger}(y)$ in $\mathscr{H}(y)$ with a creation field $\psi^-(x)$ in $\mathscr{H}(x)$ only if $\mathscr{H}(y)$ was initially to the left of $\mathscr{H}(x)$ in Eq. (6.1.1), i.e., if $y^0 > x^0$. (The $\pm$ sign in the second term in (6.1.14) will be explained a little later.) The quantity (6.1.14) is known as a *propagator*; it is calculated in the following section.

The S-matrix is obtained by multiplying these factors together, along with additional numerical factors to be discussed below, then integrating over $x_1 \cdots x_N$, then summing over all pairings, and then over the numbers of interactions of each type. Before filling in all the details, it will be convenient first to describe a diagrammatic formalism for keeping track of all these pairings.

* If the interaction $\mathscr{H}(x)$ is written in the normal-ordered form, as in Eq. (5.1.33), then there is no pairing of fields and field adjoints in the *same* interaction. Otherwise some sort of regularization is needed to give meaning to $\Delta_{\ell m}(0)$.

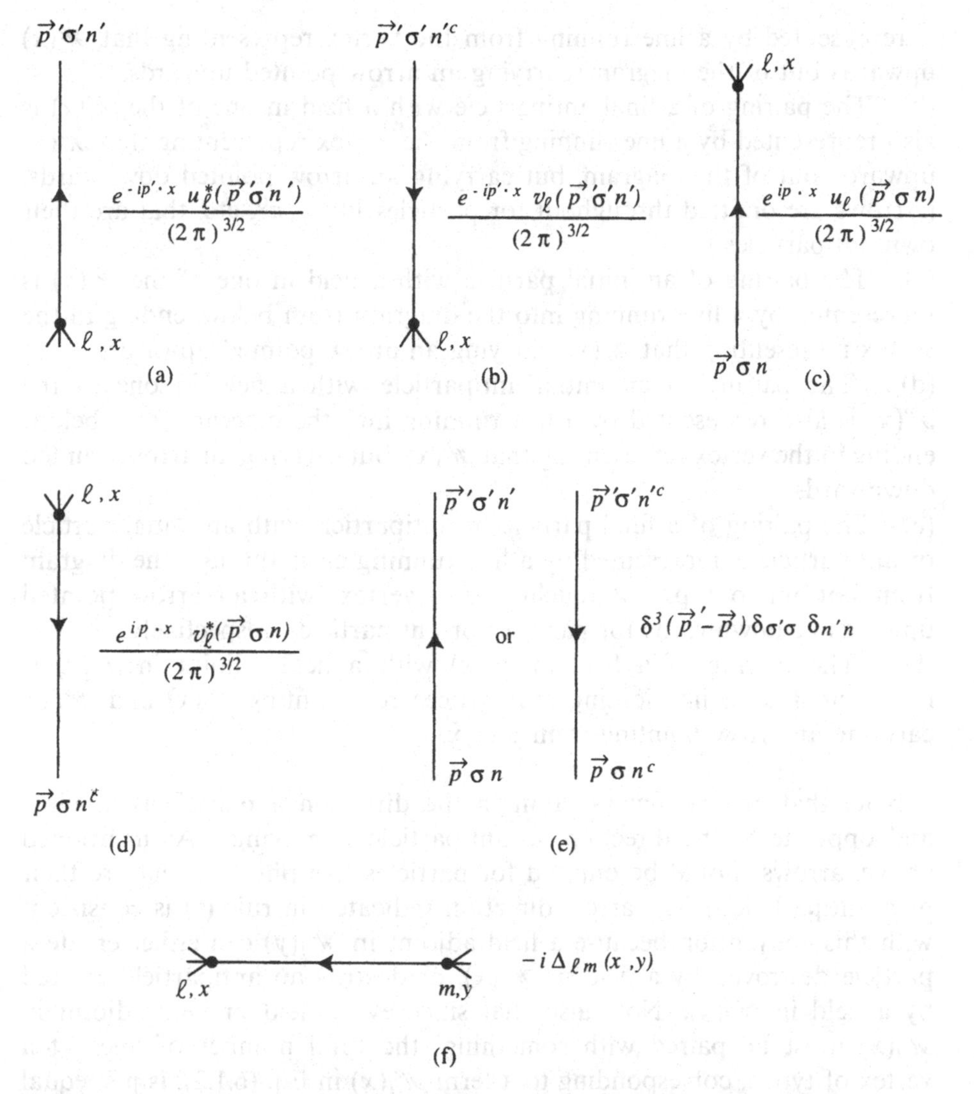

Figure 6.1. Graphical representation of pairings of operators arising in the coordinate-space evaluation of the S-matrix. The expressions on the right are the factors that must be included in the coordinate-space integrand of the S-matrix for each line of the Feynman diagram.

The rules for calculating the S-matrix are conveniently summarized in terms of *Feynman diagrams.* (See Figure 6.1.) The diagrams consist of points called *vertices*, each representing one of the $\mathscr{H}_i(x)$, and *lines*, each representing the pairing of a creation with an annihilation operator. More specifically:

(a) The pairing of a final particle with a field adjoint in one of the $\mathscr{H}(x)$

is represented by a line running from the vertex representing that $\mathscr{H}(x)$ upwards out of the diagram, carrying an arrow pointed upwards.
(b) The pairing of a final antiparticle with a field in one of the $\mathscr{H}(x)$ is also represented by a line running from the vertex representing that $\mathscr{H}(x)$ upwards out of the diagram, but carrying an arrow pointed downwards. (Arrows are omitted throughout for particles like γ, π^0, etc. that are their own antiparticles.)
(c) The pairing of an initial particle with a field in one of the $\mathscr{H}(x)$ is represented by a line running into the diagram from below, ending in the vertex representing that $\mathscr{H}(x)$, carrying an arrow pointed upwards.
(d) The pairing of an initial antiparticle with a field in one of the $\mathscr{H}(x)$ is also represented by a line running into the diagram from below, ending in the vertex representing that $\mathscr{H}(x)$, but carrying an arrow pointed downwards.
(e) The pairing of a final particle or antiparticle with an initial particle or antiparticle is represented by a line running clear through the diagram from bottom to top, not touching any vertex, with an arrow pointed upwards or downwards for particles or antiparticles, respectively.
(f) The pairing of a field in $\mathscr{H}(x)$ with a field adjoint in $\mathscr{H}(y)$ is represented by a line joining the vertices representing $\mathscr{H}(x)$ and $\mathscr{H}(y)$, carrying an arrow pointing from y to x.

Note that arrows always point in the direction a particle is moving, and opposite to the direction an antiparticle is moving. (As mentioned above, arrows should be omitted for particles like photons that are their own antiparticles.) The arrow direction indicated in rule (f) is consistent with this convention because a field adjoint in $\mathscr{H}_j(y)$ can either create a particle destroyed by a field in $\mathscr{H}_i(x)$, or destroy an antiparticle created by a field in $\mathscr{H}_i(x)$. Note also that since every field or field adjoint in $\mathscr{H}_i(x)$ must be paired with something, the total number of lines at a vertex of type i, corresponding to a term $\mathscr{H}_i(x)$ in Eq. (6.1.2), is just equal to the total number of field or field adjoint factors in $\mathscr{H}_i(x)$. Of these lines, the number with arrows pointed into the vertex or out of it equals the number of fields or field adjoints respectively in the corresponding interaction term.

To calculate the contribution to the S-matrix for a given process, of a given order N_i in each of the interaction terms $\mathscr{H}_i(x)$ in Eq. (6.1.2), we must carry out the following steps:

(i) Draw all Feynman diagrams containing N_i vertices of each type i, and containing a line coming into the diagrams from below for each particle or antiparticle in the initial state, and a line going upwards out of the diagram for every particle or antiparticle in the final state, together with any number of internal lines running from one vertex to another, as required to give each vertex the proper number of attached lines. The lines carry arrows as described above, each of which may point upwards or downwards. Each vertex is labelled with an interaction type i and spacetime coordinate x^μ. Each internal or external line is labelled at the end where it runs into a vertex with a field type ℓ (corresponding to the field $\psi_\ell(x)$ or $\psi_\ell^\dagger(x)$ that creates or destroys the particle or antiparticle at that vertex), and each external line where it enters or leaves the diagram is labelled with the quantum numbers $\mathbf{p}, \sigma, n$ or $\mathbf{p}', \sigma', n'$ of the initial or final particle (or antiparticle).

(ii) For each vertex of type i, include a factor $-i$ (from the $(-i)^N$ in Eq. (6.1.1)) and a factor g_i (the coupling constant multiplying the product of fields in $\mathscr{H}_i(x)$). For each line running upwards out of the diagram, include a factor (6.1.9) or (6.1.10), depending on whether the arrow is pointing up or down. For each line running from below into the diagram, include a factor (6.1.11) or (6.1.12), again depending on the arrow direction. For each line running straight through the diagram include a factor (6.1.13). For each internal line connecting two vertices include a factor (6.1.14).

(iii) Integrate the product of all these factors over the coordinates $x_1, x_2, \cdots$ of each vertex.

(iv) Add up the results obtained in this way from each Feynman diagram. The complete perturbation series for the S-matrix is obtained by adding up the contributions of each order in each interaction type, up to whatever order our strength permits.

Note that we have not included the factor $1/N!$ from Eq. (6.1.1) in these rules, because the time-ordered product in Eq. (6.1.1) is a sum over the $N!$ permutations of $x_1 x_2 \cdots x_N$, each permutation giving the same contribution to the final result. To put this another way, a Feynman diagram with N vertices is one of $N!$ identical diagrams, which differ only in permutations of the labels on the vertices, and this yields a factor of $N!$ which cancels the $1/N!$ in Eq. (6.1.1). (There are exceptions to this rule, discussed below.) For this reason, henceforth we do *not* include more than one of a set of Feynman diagrams that differ only by relabelling the vertices.

In some cases, there are additional combinatoric factors or signs that must be included in the contribution of individual Feynman diagrams:

(v) Suppose that an interaction $\mathscr{H}_i(x)$ contains (among other fields and field adjoints) M factors of the *same* field. Suppose that each of these fields is paired with a field adjoint in a different interaction (different for each one), or in the initial or final state. The first of these field adjoints

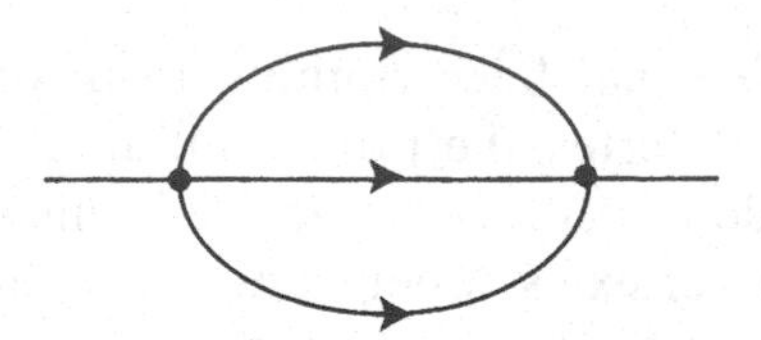

Figure 6.2. Example of a graph requiring extra combinatoric factors in the S-matrix. For an interaction involving, say, three factors of some field (as well as other fields) we usually include a factor $1/3!$ in the interaction Hamiltonian, to cancel factors arising from sums over ways of pairing these fields with their adjoints in other interactions. But in this diagram there are two such factors of $1/3!$ and only $3!$ different pairings, so we are left with an extra factor of $1/3!$.

can be paired with any one of the M identical fields in $\mathscr{H}_i(x)$; the second with any one of the remaining $M-1$ identical fields; and so on, yielding an extra factor of $M!$. To compensate for this, it is conventional to define the coupling constants g_i so that an explicit factor $1/M!$ appears in any $\mathscr{H}_i(x)$ containing M identical fields (or field adjoints.) For instance, the interaction of Mth order in a scalar field $\phi(x)$ would be written $g\phi^M/M!$. (More generally, one often also displays an explicit factor of $1/M!$ when the interaction involves a sum of M factors of fields from the same symmetry multiplet, or when for this or any other reason the coupling coefficient is totally symmetric or antisymmetric under permutations of M boson or fermion fields.)

However, this cancellation of $M!$ factors is not always complete. For instance, consider a Feynman diagram in which the M identical fields in one interaction $\mathscr{H}_i(x)$ are paired with M corresponding field adjoints in a *single* other interaction $\mathscr{H}_j(y)$. (See Figure 6.2.) Then by following the above analysis, we find only $M!$ different pairings (since it makes no difference which of the field adjoints we call the first, second, ...), cancelling only one of the two factors of $1/M!$ in the two different interactions. In this case, we would have to insert an extra factor of $1/M!$ 'by hand' into the contribution of such a Feynman diagram.

Other combinatoric factors arise when some of the permutations of vertices have no effect on the Feynman diagram. We noted earlier that the factor $1/N!$ in the series (6.1.1) is usually cancelled by the sum over the $N!$ diagrams that differ only in the labelling of the N vertices. However, this cancellation is incomplete when relabelling the vertices does not yield a new diagram. This happens most commonly in the calculation of vacuum-to-vacuum S-matrix elements in a theory with a quadratic interaction $\mathscr{H} = \psi^\dagger_\ell M_{\ell\ell'}\psi_{\ell'}$, where M may depend on external fields. (The physical significance of such vacuum fluctuation diagrams is discussed in detail in Volume II.) The Feynman diagram of Nth order in $\mathscr{H}$ is a ring with N corners. (See Figure 6.3.) There are only $(N-1)!$ different diagrams here because a permutation of labels that moves each label to the next vertex around the ring yields the same diagram. Hence such a graph is accompanied with a factor

$$\frac{(N-1)!}{N!} = \frac{1}{N}. \tag{6.1.17}$$

(vi) In theories involving fermion fields, the use of Eqs. (6.1.4)–(6.1.6) to move annihilation and creation operators to the right and left introduces minus signs into the contribution of various pairings. To be specific, we get a minus sign wherever the permutation of the operators in Eq. (6.1.1) that is required to put all paired operators adjacent to one another (with annihilation operators just to the left of the paired creation operators) involves an odd number of interchanges of fermion operators. (This is because to compute the contribution of a certain pairing, we can first permute all operators in Eq. (6.1.1) so that each annihilation operator is just to the left of the creation operator with which it is paired, ignoring all commutators and anticommutators of unpaired operators, and then replace each product of paired operators with their commutators or anticommutators.) One immediate consequence is to produce the minus sign in the relative sign of the two terms in Eq. (6.1.14) for the fermion propagator. Whatever permutation puts the annihilation part $\psi^+(x)$ of a field in $\mathscr{H}(x)$ just to the left of the creation part $\psi^{+\dagger}(y)$ of a field adjoint in $\mathscr{H}(y)$, the permutation that puts the annihilation part $\psi^{-\dagger}(y)$ of the field adjoint just to the left of the creation part $\psi^-(x)$ of the field involves one extra interchange of fermion operators, yielding the minus sign in the second term of Eq. (6.1.14) for fermions.

In addition, minus signs can arise in the contribution of whole Feynman diagrams. As an example, let us take up a theory in which the sole interaction of fermions takes the form

$$\mathscr{H}(x) = \sum_{\ell mk} g_{\ell mk} \psi^\dagger_\ell(x) \psi_m(x) \phi_k(x) , \tag{6.1.18}$$

where $g_{\ell mk}$ are general constants, $\psi_\ell(x)$ are a set of complex fermion fields, and $\phi_m(x)$ are a set of real bosonic (but not necessarily scalar) fields. (Not only quantum electrodynamics, but the whole 'standard model' of weak, electromagnetic, and strong interactions, has fermionic interactions that can all be put in his form.) Let us first take up the process of fermion–fermion scattering, $12 \to 1'2'$, to second order in $\mathscr{H}$. The fermion operators in the second-order term in Eq. (6.1.1) appear in the order (with obvious abbreviations)

$$a(2')a(1')\psi^\dagger(x)\psi(x)\psi^\dagger(y)\psi(y)a^\dagger(1)a^\dagger(2) \ . \tag{6.1.19}$$

There are two connected diagrams to this order, corresponding to the

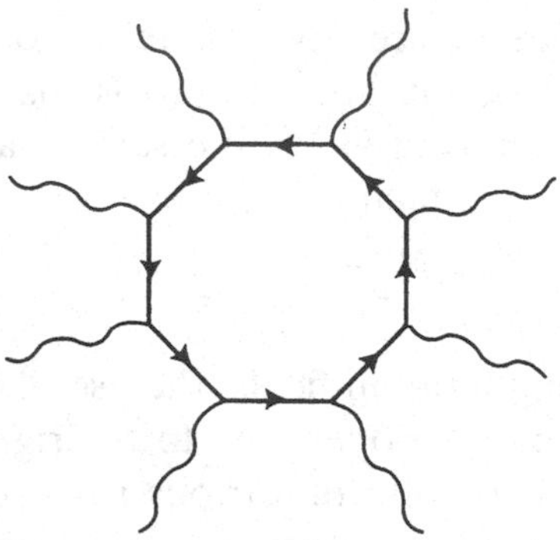

Figure 6.3. An eighth-order graph for the vacuum-to-vacuum amplitude with particles interacting only with an external field. In this diagram the external field is represented by wiggly lines. There are 7! such diagrams, differing only by relabelling the vertices, and not counting as different those labellings that simply rotate the ring. The factor 1/8! from the Dyson formula (6.1.1) is therefore not entirely cancelled here, leaving us with an extra factor 1/8.

pairings

$$[a(2')\psi^\dagger(x)]\ [a(1')\psi^\dagger(y)]\ [\psi(y)a^\dagger(1)]\ [\psi(x)a^\dagger(2)] \tag{6.1.20}$$

and

$$[a(1')\psi^\dagger(x)]\ [a(2')\psi^\dagger(y)]\ [\psi(y)a^\dagger(1)]\ [\psi(x)a^\dagger(2)]\ . \tag{6.1.21}$$

(See Figure 6.4.) To go from (6.1.19) to (6.1.20) requires an *even* permutation of fermionic operators. (For instance, move $\psi(x)$ past three operators to the right, and then move $a(1')$ past one operator to the right.) Thus there is no extra minus sign in the contribution of the pairing (6.1.20). This in itself is not so important; the *overall* sign of the S-matrix does not matter in transition rates, and in any case depends on sign conventions for the initial and final states. What is important is that the contributions of pairings (6.1.20) and (6.1.21) have opposite sign, as can be seen most easily by noting that the only difference between these two pairings is the interchange of two fermionic operators, $a(1')$ and $a(2')$. In fact, this relative minus sign is just what is required by Fermi statistics: it makes the scattering amplitude antisymmetric under the interchange of particles $1'$ and $2'$ (or 1 and 2).

However, it must not be thought that all sign factors can be related in such a simple way to the antisymmetry of the final or initial states, even in the lowest order of perturbation theory. To illustrate this point, let's now consider fermion–antifermion scattering, $1\,2^c \to 1'2'^c$, to second order in the same interaction (6.1.18). The fermionic operators in the second-order

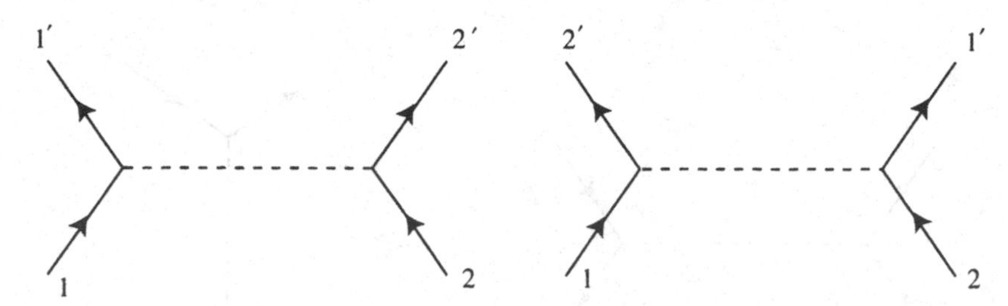

Figure 6.4. The connected second-order diagrams for fermion–fermion scattering in a theory with interaction (6.1.18). Here straight lines represent fermions; dotted lines are neutral bosons. There is a minus sign difference in the contributions of these two diagrams, arising from an extra interchange of fermion operators in the pairings represented by the second diagram.

term in Eq. (6.1.1) appear in the order:

$$a(2'^c)a(1')\psi^\dagger(x)\psi(x)\psi^\dagger(y)\psi(y)a^\dagger(1)a^\dagger(2^c) \; . \qquad (6.1.22)$$

Here again there are two Feynman diagrams to this order, corresponding to the pairings

$$[a(2'^c)\psi(x)]\;[a(1')\psi^\dagger(x)]\;[\psi(y)a^\dagger(1)]\;[\psi^\dagger(y)a^\dagger(2^c)] \qquad (6.1.23)$$

and

$$[a(2'^c)\psi(x)]\;[a(1')\psi^\dagger(y)]\;[\psi(y)a^\dagger(1)]\;[\psi^\dagger(x)a^\dagger(2^c)] \; . \qquad (6.1.24)$$

(See Figure 6.5.) To go from (6.1.22) to (6.1.23) requires an even permutation of fermionic operators (for instance, move $\psi(x)$ past two operators to the left and move $\psi^\dagger(y)$ past two operators to the right) so there is no extra minus sign in the contribution of the pairing (6.1.23). On the other hand, to go from (6.1.22) to (6.1.24) requires an odd permutation of fermionic operators (the same as for (6.1.23), *plus* the interchange of $\psi^\dagger(x)$ and $\psi^\dagger(y)$) so the contribution of this pairing does come with an extra minus sign.**

Additional signs are encountered when we consider contributions of higher order. In theories of the type considered here, in which the interactions of fermions all take the form (6.1.18), the fermion lines in

** Actually, this sign is not wholly unrelated to the requirements of Fermi statistics. The same field can destroy a particle and create an antiparticle, so there is a relation, known as 'crossing symmetry', between processes in which initial particles or antiparticles are exchanged with final antiparticles or particles. In particular the amplitudes for the process $1\,2^c \to 1'2'^c$ are related to those for the 'crossed' process $1\,2' \to 1'2$; the two pairings (6.1.23) and (6.1.24) just correspond to the two diagrams for this process, which differ by an interchange of 1 and 2′ (or 1′ and 2), so the antisymmetry of the scattering amplitude under interchange of initial (or final) particles naturally requires a minus sign in the relative contribution of these two pairings. However, crossing symmetry is not an ordinary symmetry (it involves an analytic continuation in kinematic variables) and it is difficult to use it with any precision for general processes.

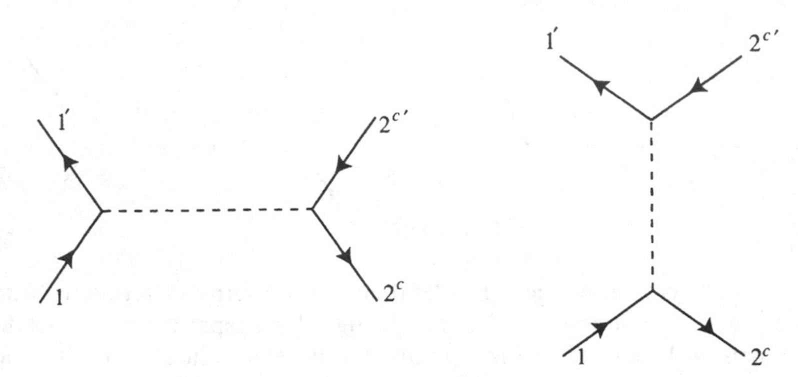

Figure 6.5. The connected second-order diagrams for fermion–antifermion scattering in a theory with interaction (6.1.18). Here straight lines represent fermions or antifermions, depending on the arrow direction; dotted lines are neutral bosons. There is again a minus sign difference in the contributions of these two diagrams, arising from an extra interchange of fermion operators in the pairings represented by the second diagram.

general Feynman diagrams form either chains of lines that pass through the diagram with arbitrary numbers of interactions with the boson fields, as in Figure 6.6, or else fermionic *loops*, like that shown in Figure 6.7. Consider the effect of adding a fermionic loop with M corners to the Feynman diagram for any process. This corresponds to the pairing of fermionic operators

$$[\psi(x_1)\bar{\psi}(x_2)]\ [\psi(x_2)\bar{\psi}(x_3)]\cdots[\psi(x_M)\bar{\psi}(x_1)]\,. \tag{6.1.25}$$

On the other hand, these operators appear in Eq. (6.1.1) in the order

$$\bar{\psi}(x_1)\psi(x_1)\bar{\psi}(x_2)\psi(x_2)\cdots\bar{\psi}(x_M)\psi(x_M)\,. \tag{6.1.26}$$

To go from (6.1.26) to (6.1.25) requires an odd permutation of fermionic operators (move $\bar{\psi}(x_1)$ to the right past $2M-1$ operators) so the contribution of each such fermionic loop is accompanied with a minus sign.

These rules yield the full S-matrix, including contributions from processes in which various clusters of particles interact in widely separated regions of spacetime. As discussed in Chapter 4, to calculate the part of the S-matrix that excludes such contributions, we should include only *connected* Feynman diagrams. In particular, this excludes lines passing clean through the diagram without interacting, which would yield the factors (6.1.13).

To make the Feynman rules perfectly clear, we will calculate the low-order contributions to the S-matrix for particle scattering in two different theories.

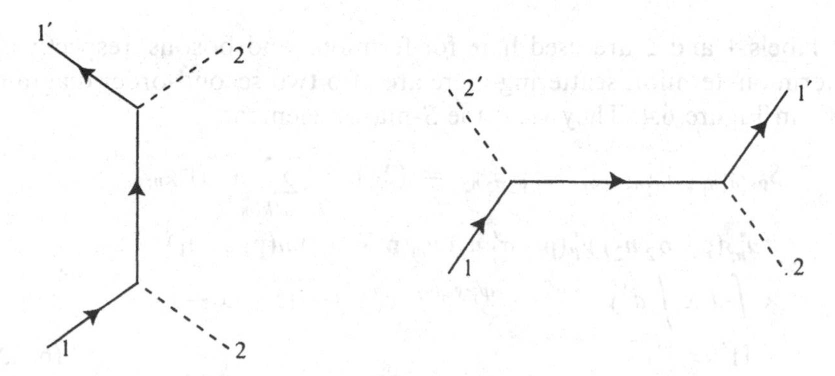

Figure 6.6. The connected second-order diagrams for boson–fermion scattering in a theory with interaction (6.1.18). Straight lines are fermions; dashed lines are neutral bosons.

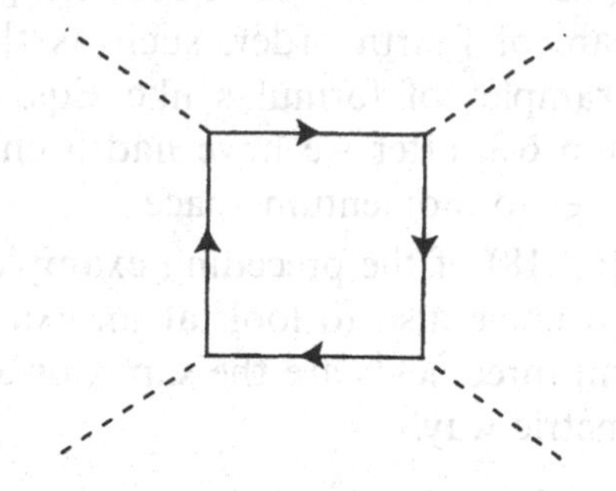

Figure 6.7. The lowest-order connected diagram for boson–boson scattering in a theory with interaction (6.1.18). Such fermion loop graphs yield an extra minus sign, arising from permutations of the paired fermion fields.

Theory I

Consider the theory of fermions and self-charge-conjugate bosons with interaction (6.1.18). The lowest-order connected diagrams for fermion–boson scattering are shown in Figure 6.6. Following the rules outlined in Figure 6.1, the corresponding S-matrix element is

$$
\begin{aligned}
&S_{\mathbf{p}_1{}'\sigma_1'n_1'\,\mathbf{p}_2{}'\sigma_2'n_2',\ \mathbf{p}_1\sigma_1n_1\,\mathbf{p}_2\sigma_2n_2} = \\
&(2\pi)^{-6}\sum_{k'l'm'\,klm}(-i)^2 g_{l'm'k'}\,g_{mlk}\,u^*_{l'}(\mathbf{p}_1{}'\sigma_1'n_1')\,u_l(\mathbf{p}_1\sigma_1n_1) \\
&\times\int d^4x\int d^4y\,\Big(-i\Delta_{m'm}(y-x)\Big)e^{-ip_1'\cdot y}e^{ip_1\cdot x} \\
&\times\Big[e^{-ip_2'\cdot y}u^*_{k'}(\mathbf{p}_2{}'\sigma_2'n_2')e^{ip_2\cdot x}u_k(\mathbf{p}_2\sigma_2n_2) \\
&+\ e^{\;ip_2'\cdot x}u^*_k(\mathbf{p}_2{}'\sigma_2'n_2')e^{ip_2\cdot y}u_{k'}(\mathbf{p}_2\sigma_2n_2)\Big]\,.
\end{aligned}
\qquad (6.1.27)
$$

(The labels 1 and 2 are used here for fermions and bosons, respectively.) For fermion–fermion scattering there are also two second-order diagrams, shown in Figure 6.4. They yield the S-matrix element

$$
\begin{aligned}
S_{\mathbf{p}_1{}'\sigma_1'n_1'\,\mathbf{p}_2{}'\sigma_2'n_2',\,\mathbf{p}_1\sigma_1n_1\,\mathbf{p}_2\sigma_2n_2} &= (2\pi)^{-6}\sum_{k'l'm'klm}(-i)^2 g_{m'mk'}\,g_{l'lk} \\
&\times u^*_{m'}(\mathbf{p}_2{}'\,\sigma_2'\,n_2')\,u^*_{l'}(\mathbf{p}_1{}'\,\sigma_1'\,n_1')\,u_m(\mathbf{p}_2\,\sigma_2\,n_2)u_l(\mathbf{p}_1\,\sigma_1\,n_1) \\
&\times\int d^4x\int d^4y\; e^{-ip_2'\cdot x}e^{-ip_1'\cdot y}e^{ip_2\cdot x}e^{ip_1\cdot y}(-i)\Delta_{k'k}(x-y) \\
&-[1'\rightleftharpoons 2']
\end{aligned}
\qquad (6.1.28)
$$

with the last term indicating subtraction of the preceding term with interchange of particles 1′ and 2′ (or equivalently 1 and 2). There are no second-order graphs for boson–boson scattering in this theory; the lowest-order graphs are of fourth order, such as that shown in Figure 6.7. More specific examples of formulas like Eqs. (6.1.27) and (6.1.28) will be given in Section 6.3, after we have had a chance to evaluate the propagators and go over to momentum space.

In the interaction (6.1.18) of the preceding example, the three fields are all different. It is instructive also to look at an example with a trilinear interaction in which all three fields are the same, or at least enter into the interaction in a symmetric way.

Theory II

Now take the interaction density to be a sum of terms that are trilinear in a set of *real* bosonic fields $\phi_\ell(x)$:

$$
\mathscr{H}(x) = \frac{1}{3!}\sum_{\ell mn} g_{\ell mn}\phi_\ell(x)\phi_m(x)\phi_n(x) \qquad (6.1.29)
$$

with $g_{\ell mn}$ a real totally symmetric coupling coefficient. Suppose we want to consider a scattering process $1\ 2 \to 1'2'$ to second order in this interaction. Each of the two vertices must have two of the four external lines attached to it. (The only other possibility is that one of the external lines is attached to one vertex and three to the other vertex, but the vertex with three external lines attached to it would have no remaining lines to connect it to the other vertex so this would be a disconnected contribution.) The additional line required at each vertex must then just serve to connect the two vertices to each other. There are three graphs of this type, differing in whether the other external line that is attached to the same vertex as line 1 is line 2 or 1′ or 2′. (See Figure 6.8.) Following the rules given above, the contribution to the S-matrix from those three

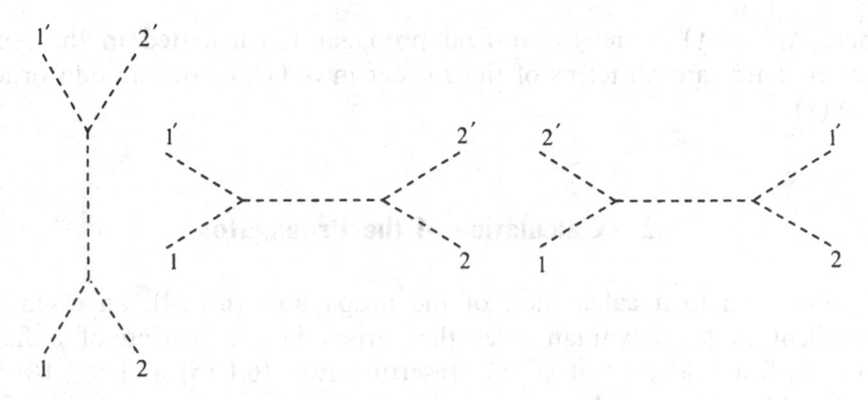

Figure 6.8. The connected-second order diagrams for boson–boson scattering in a theory with interaction (6.1.29).

diagrams is

$$S_{\mathbf{p}'_1\sigma'_1n'_1\ \mathbf{p}'_2\sigma'_2n'_2\ ,\ \mathbf{p}_1\sigma_1n_1\ \mathbf{p}_2\sigma_2n_2}$$

$$= (-i)^2(2\pi)^{-6} \sum_{\ell\ell'\ell''mm'm''} g_{\ell\ell'\ell''}\, g_{mm'm''} \int d^4x \int d^4y \Big(-i\Delta_{\ell''m''}(x,y)\Big)$$

$$\times \Big[u^*_{\ell}(\mathbf{p}'_1\sigma'_1n'_1)\, e^{-ip'_1\cdot x}\, u^*_{\ell'}(\mathbf{p}'_2\sigma'_2n'_2)\, e^{-ip'_2\cdot x}$$

$$\times\, u_m(\mathbf{p}_1\sigma_1n_1) e^{ip_1\cdot y}\, u_{m'}(\mathbf{p}_2\sigma_2n_2)\, e^{ip_2\cdot y}$$

$$+\, u^*_{\ell'}(\mathbf{p}'_1\sigma'_1n'_1)\, e^{-ip'_1\cdot x}\, u_{\ell}(\mathbf{p}_1\sigma_1n_1) e^{ip_1\cdot x}$$

$$\times\, u^*_{m'}(\mathbf{p}'_2\sigma'_2n'_2)\, e^{-ip'_2\cdot y}\, u_m(\mathbf{p}_2\sigma_2n_2)\, e^{ip_2\cdot y}$$

$$+\, u^*_{\ell'}(\mathbf{p}'_2\sigma'_2n'_2)\, e^{-ip'_2\cdot x}\, u_{\ell}(\mathbf{p}_1\sigma_1n_1) e^{ip_1\cdot x}$$

$$\times\, u^*_{m'}(\mathbf{p}'_1\sigma'_1n'_1)\, e^{-ip'_1\cdot y}\, u_m(\mathbf{p}_2\sigma_2n_2)\, e^{ip_2\cdot y}\Big]. \tag{6.1.30}$$

To be even more specific, if the bosons in this theory are spinless particles of a single species, then we write the interaction (6.1.29) in the form

$$\mathscr{H} = g\phi^3/3! \tag{6.1.31}$$

and the S-matrix element (6.1.30) for scalar–scalar scattering is

$$S_{\mathbf{p}'_1\ \mathbf{p}'_2\ ,\ \mathbf{p}_1\ \mathbf{p}_2} =$$

$$\frac{ig^2}{(2\pi)^6\sqrt{16E'_1E'_2E_1E_2}} \int d^4x \int d^4y\ \Delta_F(x-y)$$

$$\times \Big[\exp(-i(p'_1+p'_2)\cdot x)\exp(i(p_1+p_2)\cdot y)$$

$$+ \exp(i(p_1-p'_1)\cdot x)\exp(i(p_2-p'_2)\cdot y)$$

$$+ \exp(i(p_1-p'_2)\cdot x)\exp(i(p_2-p'_1)\cdot y)\Big],$$

where $\Delta_F(x-y)$ is the scalar field propagator, calculated in the next section. There are no terms of third order in $\mathscr{H}(x)$, or of any odd order in $\mathscr{H}(x)$.

6.2 Calculation of the Propagator

We now turn to a calculation of the propagator (6.1.14), an essential ingredient in the Feynman rules that arises in the pairing of a field $\psi_\ell(x)$ with a field adjoint $\psi_m^\dagger(y)$. Inserting Eqs. (6.1.15) and (6.1.16) in Eq. (6.1.14), and using the commutation or anticommutation relations for annihilation and creation operators, we have immediately

$$-i\Delta_{\ell m}(x,y) = \theta(x-y)(2\pi)^{-3}\int d^3p \sum_\sigma u_\ell(\mathbf{p}\,\sigma\,n)u_m^*(\mathbf{p}\,\sigma\,n)e^{ip\cdot(x-y)}$$
$$\pm\,\theta(y-x)(2\pi)^{-3}\int d^3p \sum_\sigma v_m^*(\mathbf{p}\,\sigma\,n)v_\ell(\mathbf{p}\,\sigma\,n)e^{ip\cdot(y-x)}\,. \tag{6.2.1}$$

In the course of calculating commutators and anticommutators in Chapter 5, we showed that

$$\sum_\sigma u_\ell(\mathbf{p}\,\sigma\,n)u_m^*(\mathbf{p}\,\sigma\,n) = \left(2\sqrt{\mathbf{p}^2+m_n^2}\right)^{-1} P_{\ell m}\left(\mathbf{p},\sqrt{\mathbf{p}^2+m_n^2}\right)\,, \tag{6.2.2}$$

$$\sum_\sigma v_\ell(\mathbf{p}\,\sigma\,n)v_m^*(\mathbf{p}\,\sigma\,n) = \pm\left(2\sqrt{\mathbf{p}^2+m_n^2}\right)^{-1} P_{\ell m}\left(-\mathbf{p},-\sqrt{\mathbf{p}^2+m_n^2}\right)\,, \tag{6.2.3}$$

where $P_{\ell m}(\mathbf{p},\omega)$ is a polynomial in $\mathbf{p}$ and ω. (Here as in Eq. (6.2.1), the top and bottom signs refer to bosonic and fermionic fields, respectively.) For instance, if $\psi_\ell(x)$ and $\psi_m(y)$ are scalar fields $\phi(x)$ and $\phi(y)$ for a particle of spin zero, then we have simply

$$P(p) = 1\,. \tag{6.2.4}$$

If $\psi_\ell(x)$ and $\psi_m(y)$ are Dirac fields for a particle of spin $\frac{1}{2}$, then

$$P_{\ell m}(p) = \Big[(-i\gamma_\mu p^\mu + m)\beta\Big]_{\ell m}\,, \tag{6.2.5}$$

where ℓ and m are here four-valued Dirac indices. (The matrix β appears here because we are considering the pairing of $\psi_\ell(x)$ with $\psi_m^\dagger(y)$. It is absent in the pairing of $\psi(x)$ with $\bar\psi(y) \equiv \psi^\dagger(y)\beta$.) If $\psi_\ell(x)$ and $\psi_m(y)$ are vector fields $V_\mu(x)$ and $V_\nu(y)$ for a particle of spin one, then

$$P_{\mu\nu}(p) = \eta_{\mu\nu} + m^{-2}p_\mu p_\nu\,. \tag{6.2.6}$$

More generally, if $\psi_\ell(x)$ and $\psi_m(y)$ are components of fields $\psi_{ab}(x)$ and $\psi_{\tilde{a}\tilde{b}}(y)$ for a particle of spin j, in the irreducible (A,B) and $(\tilde{A},\tilde{B})$ representations of the homogeneous Lorentz group, then

$$\begin{aligned}P_{ab,\tilde{a}\tilde{b}}(p) = &\sum_{a'b'}\sum_{\tilde{a}'\tilde{b}'}\sum_{\sigma} C_{AB}(j\sigma, a'b')C_{\tilde{A}\tilde{B}}(j\sigma, \tilde{a}'\tilde{b}') \\ &\times\left[\exp(-\theta\hat{p}\cdot\mathbf{J}^{(A)})\right]_{aa'}\left[\exp(+\theta\hat{p}\cdot\mathbf{J}^{(B)})\right]_{bb'} \\ &\times\left[\exp(-\theta\hat{p}\cdot\mathbf{J}^{(\tilde{A})})\right]_{\tilde{a}\tilde{a}'}\left[\exp(+\theta\hat{p}\cdot\mathbf{J}^{(\tilde{B})})\right]_{\tilde{b}\tilde{b}'}, \end{aligned} \quad (6.2.7)$$

where $\sinh\theta = |\mathbf{p}|/m$, while $a, b, \tilde{a}, \tilde{b}$ run by unit steps from $-A$ to $+A$, $-B$ to $+B$, $-\tilde{A}$ to $+\tilde{A}$, and $-\tilde{B}$ to $+\tilde{B}$, respectively, and likewise for the running indices $a', b', \tilde{a}'$, and $\tilde{b}'$.

Inserting Eqs. (6.2.2) and (6.2.3) in Eq. (6.2.1) yields

$$\begin{aligned}-i\Delta_{\ell m}(x,y) = &\,\theta(x-y)P_{\ell m}\left(-i\frac{\partial}{\partial x}\right)\Delta_+(x-y) \\ &+\theta(y-x)P_{\ell m}\left(-i\frac{\partial}{\partial x}\right)\Delta_+(y-x)\,, \end{aligned} \quad (6.2.8)$$

where $\Delta_+(x)$ is the function introduced in Chapter 5

$$\Delta_+(x) \equiv (2\pi)^{-3}\int d^3p\,(2p^0)^{-1}e^{ip\cdot x} \quad (6.2.9)$$

in which p^0 is taken as $+\sqrt{\mathbf{p}^2+m^2}$.

To go further, we must say a bit about how to extend the definition of the polynomial $P(p)$. Eqs. (6.2.2) and (6.2.3) only define $P(p)$ for four-momenta 'on the mass shell', i.e., with $p^0 = \pm\sqrt{\mathbf{p}^2+m^2}$. Any polynomial function of such four-momentum can always be taken as linear in p^0, because any power $(p^0)^{2\nu}$ or $(p^0)^{2\nu+1}$ can be written as $(\mathbf{p}^2+m^2)^\nu$ or $p^0(\mathbf{p}^2+m^2)^\nu$, respectively. Thus we can define a polynomial $P^{(L)}(q)$ by the conditions that

$$P^{(L)}(p) = P(p) \qquad \left(\text{for } p^0 = \sqrt{\mathbf{p}^2+m^2}\right),$$

$$P^{(L)}(q) = P^{(0)}(\mathbf{q}) + q^0P^{(1)}(\mathbf{q}) \quad (\text{for general } q^\mu)\,, \quad (6.2.10)$$

where $P^{(0,1)}$ are polynomials depending only on $\mathbf{q}$. We can now use the relations

$$\frac{\partial}{\partial x^0}\theta(x^0-y^0) = -\frac{\partial}{\partial x^0}\theta(y^0-x^0) = \delta(x^0-y^0) \quad (6.2.11)$$

(recall that $\theta(x)$ has a unit step at x^0, and is otherwise constant) to move the derivative operators to the left of the θ functions in Eq. (6.2.8)

$$\Delta_{\ell m}(x,y) = P^{(L)}_{\ell m}\left(-i\frac{\partial}{\partial x}\right)\Delta_F(x-y) + \delta(x^0-y^0)P^{(1)}_{\ell m}(-i\nabla)\Big[\Delta_+(x-y)-\Delta_+(y-x)\Big]\,, \tag{6.2.12}$$

where Δ_F is the 'Feynman propagator'

$$-i\Delta_F(x) \equiv \theta(x)\Delta_+(x) + \theta(-x)\Delta_+(-x)\,. \tag{6.2.13}$$

However, for $x^0 = 0$ the function $\Delta_+(x)$ is even in $\mathbf{x}$, since a change $\mathbf{x} \to -\mathbf{x}$ in Eq. (6.2.9) can be compensated by a change $\mathbf{p} \to -\mathbf{p}$ in the integration variable. We can therefore drop the second term in Eq. (6.2.12), and write simply

$$\Delta_{\ell m}(x,y) = P^{(L)}_{\ell m}\left(-i\frac{\partial}{\partial x}\right)\Delta_F(x-y)\,. \tag{6.2.14}$$

It will be most useful to use the expression of the Feynman propagator as a Fourier integral. The step functions in Eq. (6.2.13) have the Fourier representation*

$$\theta(t) = \frac{-1}{2\pi i}\int_{-\infty}^{\infty}\frac{\exp(-ist)}{s+i\epsilon}\,ds\,. \tag{6.2.15}$$

This can be combined with the Fourier integral (6.2.9) for $\Delta_+(x)$. We introduce new integration variables, $\mathbf{q} \equiv \mathbf{p}$, $q^0 = p^0 + s$ in the first term of Eq. (6.2.13), yielding

$$-i\Delta_F(x) = -\frac{1}{2\pi i}\int d^3q\int_{-\infty}^{\infty} dq^0\,\frac{\exp(i\mathbf{q}\cdot\mathbf{x} - iq^0x^0)}{(2\pi)^3 2\sqrt{\mathbf{q}^2+m^2}} \times\left[\left(q^0-\sqrt{\mathbf{q}^2+m^2}+i\epsilon\right)^{-1} + \left(-q^0-\sqrt{\mathbf{q}^2+m^2}+i\epsilon\right)^{-1}\right].$$

Combining denominators and adopting a four-dimensional notation, we have simply

$$\Delta_F(x) = (2\pi)^{-4}\int d^4q\,\frac{\exp(iq\cdot x)}{q^2+m^2-i\epsilon}\,, \tag{6.2.16}$$

where $q^2 \equiv \mathbf{q}^2 - (q^0)^2$. (In the denominator we have replaced $2\epsilon\sqrt{\mathbf{q}^2+m^2}$ with ϵ, because the only important thing about this quantity is that it

* To prove this, note that if $t > 0$ then the contour of integration can be closed with a large clockwise semi-circle in the lower half-plane, so the integral picks up a contribution of $-2\pi i$ from the pole at $s = -i\epsilon$. If $t < 0$ then the contour can be closed with a large counter-clockwise semi-circle in the upper half-plane, where the integrand is analytic, giving an integral equal to zero.

is a positive infinitesimal.) This shows incidentally that Δ_F is a Green's function for the Klein–Gordon differential operator, in the sense that

$$(\Box - m^2)\Delta_F(x) = -\delta^4(x) \tag{6.2.17}$$

with boundary conditions specified by the $-i\epsilon$ in the denominator: as shown by Eq. (6.2.13), $\Delta_F(x)$ for $x^0 \to +\infty$ or $x^0 \to -\infty$ involves only positive or negative frequency terms, $\exp(-ix^0\sqrt{\mathbf{p}^2+m^2})$ or $\exp(+ix^0\sqrt{\mathbf{p}^2+m^2})$, respectively.

Inserting Eq. (6.2.16) in Eq. (6.2.14) now gives the propagator as

$$\Delta_{\ell m}(x,y) = (2\pi)^{-4}\int d^4q\,\frac{P^{(L)}_{\ell m}(q)e^{iq\cdot(x-y)}}{q^2+m^2-i\epsilon}\,. \tag{6.2.18}$$

There is one obvious problem with this expression. The polynomial $P(p)$ is Lorentz-covariant when p is on the mass shell, $p^2 = -m^2$, but in Eq. (6.2.18) we integrate over all q^μ, not restricted to the mass shell. The polynomial $P^{(L)}(q)$ is defined for general q^μ to be linear in q^0, a condition that clearly does not respect Lorentz covariance unless the polynomial is also linear in each spatial component q^i as well. We can instead always define our extension of the polynomial $P(p)$ to general four-momenta q^μ, which we shall call simply $P(q)$, in such a way that $P(q)$ *is* Lorentz-covariant for general q^μ, in the sense that

$$P_{\ell m}(\Lambda q) = D_{\ell\ell'}(\Lambda)D^*_{mm'}(\Lambda)P_{\ell' m'}(q)\,,$$

where $\Lambda^\mu{}_\nu$ is a general Lorentz transformation, and $D(\Lambda)$ is the appropriate representation of the Lorentz group. For instance, for scalar, Dirac, and four-vector fields, these covariant extensions are obviously provided by just replacing p^μ with a general four-momentum q^μ in Eqs. (6.2.4), (6.2.5), and (6.2.6). For the scalar and Dirac fields, these are already linear in q^0, so here there is no difference between $P^{(L)}(q)$ and $P(q)$:

$$P^{(L)}_{\ell m}(q) = P_{\ell m}(q) \quad \text{(scalar, Dirac fields)}\,. \tag{6.2.19}$$

On the other hand, for the vector field of a spin one particle, the 00 components of the covariant polynomial $P_{\mu\nu}(q) \equiv \eta_{\mu\nu} + m^{-2}q_\mu q_\nu$ are quadratic in q^0, so here there *is* a difference:

$$\begin{aligned}P^{(L)}_{\mu\nu}(q) &= \eta_{\mu\nu} + m^{-2}\left[q_\mu q_\nu - \delta^0_\mu\delta^0_\nu(q_0^2 - \mathbf{q}^2 - m^2)\right]\\ &= P_{\mu\nu}(q) + m^{-2}(q^2+m^2)\delta^0_\mu\delta^0_\nu\,.\end{aligned} \tag{6.2.20}$$

(The extra term here is fixed by the two conditions that it must cancel the $(q_0)^2$ term in $P_{00}(q)$, and must vanish when q^μ is on the mass shell.) Inserting this in Eq. (6.2.18) gives the propagator of a vector field as

$$\Delta_{\mu\nu}(x,y) = (2\pi)^{-4}\int d^4q\,\frac{P_{\mu\nu}(q)e^{iq\cdot(x-y)}}{q^2+m^2-i\epsilon} + m^{-2}\delta^4(x-y)\delta^0_\mu\delta^0_\nu\,. \tag{6.2.21}$$

The first term is manifestly covariant, and the second term, though not covariant, is local, so it can be cancelled by adding a local non-covariant term to the Hamiltonian density. Specifically, if $V_\mu(x)$ interacts with other fields through a term $V_\mu(x)J^\mu(x)$ in $\mathscr{H}(x)$, then the effect of the second term in Eq. (6.2.21) is to produce an effective interaction

$$-i\mathscr{H}_{eff}(x) = \tfrac{1}{2}\Big[-iJ^\mu(x)\Big]\Big[-iJ^\nu(x)\Big]\Big[-im^{-2}\delta^0_\mu\delta^0_\nu\Big] .$$

(The factors $-i$ are the usual ones which always accompany vertices and propagators. The factor $\frac{1}{2}$ is needed because there are two ways to pair other fields with $\mathscr{H}_{eff}(x)$, differing in the interchange of J^μ and J^ν.) Thus the effect of the non-covariant second term in Eq. (6.2.21) can be cancelled by adding to $\mathscr{H}(x)$ the non-covariant term

$$\mathscr{H}_{NC}(x) = -\mathscr{H}_{eff}(x) = \frac{1}{2m^2}\Big[J^0(x)\Big]^2 . \tag{6.2.22}$$

It is the singularity of the equal-time commutators of vector fields at zero separation that requires us to employ a wider class of interactions than those with a scalar density. A detailed non-perturbative proof of the Lorentz invariance of the S-matrix in this theory will be given in the next chapter.

It should not be thought that this is solely a phenomenon associated with spins $j \geq 1$. For instance, consider the vector field associated with a particle of spin $j = 0$, equal (as discussed in Chapter 5) to the derivative $\partial_\lambda\phi(x)$ of a scalar field. For the pairing of this field with a scalar $\phi^\dagger(y)$, the polynomial $P(p)$ on the mass shell is

$$P_\lambda(p) = ip_\lambda , \tag{6.2.23}$$

while the pairing of $\partial_\lambda\phi(x)$ with $\partial_\eta\phi^\dagger(y)$ yields a polynomial

$$P_{\lambda,\eta}(p) = p_\lambda p_\eta . \tag{6.2.24}$$

The covariant polynomials for general off-shell four-momenta q^μ are again obtained by just substituting q^μ for p^μ in Eqs. (6.2.23) and (6.2.24). Eq. (6.2.23) shows that $P_\lambda(q)$ is already linear in q_0, so here there is no difference between $P_\lambda(q)$ and $P^{(L)}_\lambda(q)$. However, for Eq. (6.2.24) there is a difference:

$$\begin{aligned} P^{(L)}_{\lambda,\eta}(q) &= q_\lambda q_\eta - (q_0^2 - \mathbf{q}^2 - m^2)\delta^0_\lambda\delta^0_\eta \\ &= P_{\lambda,\eta}(q) + (q^2 + m^2)\delta^0_\lambda\delta^0_\eta, \end{aligned} \tag{6.2.25}$$

so here the propagator is

$$\Delta_{\lambda,\eta}(x,y) = (2\pi)^{-4}\int d^4q\, \frac{q_\lambda q_\eta e^{iq\cdot x}}{q^2 + m^2 - i\epsilon} + \delta^0_\lambda\delta^0_\eta\delta^4(x-y) . \tag{6.2.26}$$

Just as before, the non-covariant effects of the second term may be removed by adding to the interaction a non-covariant term

$$\mathscr{H}_{NC}(x) = \tfrac{1}{2}\left[J^0(x)\right]^2 , \qquad (6.2.27)$$

where $J^\mu(x)$ is here the current which multiplies $\partial_\mu \phi(x)$ in the covariant part of $\mathscr{H}(x)$.

It should be clear that (at least for massive particles) the effects of non-covariant parts of the propagator can always be cancelled in this way by adding non-covariant local terms to the Hamiltonian density. This is because the numerator $P^{(L)}_{\ell m}(q)$ in the propagator must equal the covariant polynomial $P_{\ell m}(q)$ when q^μ is on the mass shell, so the difference between $P^{(L)}_{\ell m}(q)$ and $P_{\ell m}(q)$ must contain a factor $q^2 + m^2$. This factor cancels the denominator $(q^2 + m^2 - i\epsilon)$ in the contribution of this difference to Eq. (6.2.18), so Eq. (6.2.18) always equals a covariant term plus a term proportional to the delta function $\delta^4(x - y)$ or its derivatives. The effect of the latter term may be cancelled by adding to the interaction a term quadratic in the currents to which the paired fields couple, or in their derivatives. In what follows, it will be assumed tacitly that such a term has been included in the interaction, and in consequence we shall use the *covariant* polynomial $P_{\ell m}(q)$ in the propagator (6.2.18), and will thus henceforth drop the label 'L'.

It may seem that this is a rather *ad hoc* procedure. Fortunately, in the canonical formalism discussed in the following chapter, the non-covariant term in the Hamiltonian density needed to cancel non-covariant terms in the propagator arises automatically. This, in fact, forms part of the motivation for introducing the canonical formalism.

* * *

Before closing this section, it may be useful to mention some other definitions of the propagator, equivalent to Eq. (6.2.1), that appear commonly in the literature. First, taking the vacuum expectation value of Eq. (6.1.14) gives

$$\begin{aligned}-i\Delta_{\ell m}(x,y) = \theta(x-y)\left\langle \left[\psi^+_\ell(x), \psi^{+\dagger}_m(y)\right]_\mp \right\rangle_0 \\ \pm\, \theta(y-x)\left\langle \left[\psi^{-\dagger}_m(y), \psi^-_\ell(x)\right]_\mp \right\rangle_0 . \qquad (6.2.28)\end{aligned}$$

(Here $\langle AB\cdots\rangle_0$ denotes the vacuum expectation value $(\Phi_0, AB\cdots\Phi_0)$.) Both $\psi^+_\ell(x)$ and $\psi^{-\dagger}_m(y)$ annihilate the vacuum, so only one term in each commutator or anticommutator in Eq. (6.2.28) actually contributes to the propagator:

$$-i\Delta_{\ell m}(x,y) = \theta(x-y)\langle\psi^+_\ell(x)\psi^{+\dagger}_m(y)\rangle_0 \pm \theta(y-x)\langle\psi^{-\dagger}_m(y)\psi^-_\ell(x)\rangle_0 . \quad (6.2.29)$$

Also, $\psi^{-\dagger}$ and ψ^{+} would annihilate a vacuum state on the right, and ψ^{-} and $\psi^{+\dagger}$ would annihilate the vacuum state on the left, so ψ^{+} and ψ^{-} may be replaced everywhere in Eq. (6.2.29) with the complete field $\psi = \psi^{+} + \psi^{-}$:

$$-i\Delta_{\ell m}(x,y) = \theta(x-y)\langle \psi_\ell(x)\psi^\dagger_m(y)\rangle_0 \pm \theta(y-x)\langle \psi^\dagger_m(y)\psi_\ell(x)\rangle_0 \,. \quad (6.2.30)$$

This is often written

$$-i\Delta_{\ell m}(x,y) = \langle T\{\psi_\ell(x)\psi^\dagger_m(y)\}\rangle_0 \,, \quad (6.2.31)$$

where T is a time-ordered product, whose definition is now extended** to all fields, with a minus sign for any odd permutation of fermionic operators.

6.3 Momentum Space Rules

The Feynman rules outlined in Section 6.1 specify how to calculate the contribution to the S-matrix of a given Nth order diagram, as the integral over N spacetime coordinates of a product of spacetime-dependent factors. For a final particle (or antiparticle) line with momentum p'^μ leaving a vertex with spacetime coordinate x^μ, we get a factor proportional to $\exp(-ip'\cdot x)$, and for an initial particle line with momentum p^μ entering a vertex with spacetime coordinate x^μ, we get a factor proportional to $\exp(+ip\cdot x)$. In Section 6.2 we saw that the factor associated with an internal line running from y to x can be expressed as a Fourier integral, over off-shell four-momenta q^μ, of an integrand proportional to $\exp(iq\cdot(x-y))$. We can think of q^μ as the four-momentum flowing along the internal line in the direction of the arrow from y to x. Hence the integral over each vertex's spacetime position merely yields a factor

$$(2\pi)^4\,\delta^4\left(\sum p + \sum q - \sum p' - \sum q'\right) \,, \quad (6.3.1)$$

where $\sum p'$ and $\sum p$ denote the total four-momentum of all the final or initial particles leaving or entering the vertex, and $\sum q'$ and $\sum q$ denote the total four-momentum of all the internal lines with arrows leaving or entering the vertex, respectively. Of course, in place of these integrals over x^μs, we now have to do integrals over the Fourier variables q^μ, one for each internal line.

These considerations can be encapsulated in a new set of Feynman rules

** This is not inconsistent with our previous definition of the time-ordered product of Hamiltonian densities in Chapter 3, because the Hamiltonian density can only contain even numbers of fermionic field factors.

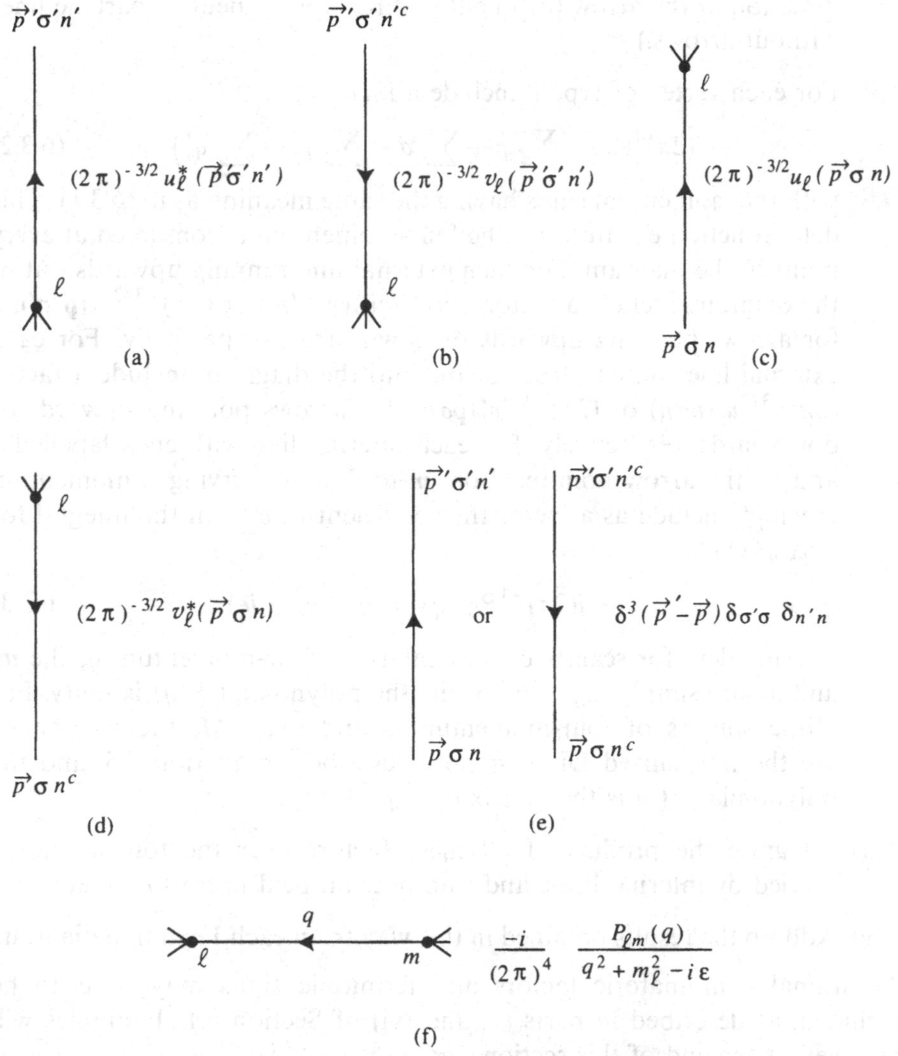

Figure 6.9. Graphical representation of pairings of operators arising in the momentum-space evaluation of the S-matrix. The expressions on the right are the factors that must be included in the momentum space integrand of the S-matrix for each line of the Feynman diagram.

(see Figure 6.9) for calculating contribution to the S-matrix as integrals over momentum variables:

(i) Draw all Feynman diagrams of the desired order, just as described in Section 6.1. However, instead of labelling each vertex with a spacetime coordinate, each internal line is now labelled with an off-mass-shell four-momentum, considered conventionally to flow in the direction of the arrow (or in either direction for neutral particle lines without arrows.)

(ii) For each vertex of type i, include a factor

$$-i(2\pi)^4 g_i\,\delta^4\left(\sum p + \sum q - \sum p' - \sum q'\right) \tag{6.3.2}$$

with the momentum sums having the same meaning as in (6.3.1). This delta function ensures that the four-momentum is conserved at every point in the diagram. For each external line running upwards out of the diagram, include a factor $(2\pi)^{-3/2}u^*_\ell(\mathbf{p}'\sigma'n')$ or $(2\pi)^{-3/2}v_\ell(\mathbf{p}'\sigma'n')$, for arrows pointing upwards or downwards, respectively. For each external line running from below into the diagram, include a factor $(2\pi)^{-3/2}u_\ell(\mathbf{p}\sigma n)$ or $(2\pi)^{-3/2}v^*_\ell(\mathbf{p}\sigma n)$, for arrows pointing upwards or downwards, respectively. For each internal line with ends labelled ℓ and m, the arrow pointing from m to ℓ, and carrying a momentum label q^μ, include as a factor the coefficient of $e^{iq\cdot x}$ in the integral for $-i\Delta_{\ell m}(x)$:

$$-i(2\pi)^{-4}P_{\ell m}(q)\Big/(q^2+m_\ell^2-i\epsilon)\,. \tag{6.3.3}$$

A reminder: for scalars or antiscalars of four-momentum q, the us and vs are simply $(2q^0)^{-1/2}$, while the polynomial $P(q)$ is unity. For Dirac spinors of four-momentum p and mass M, the us and vs are the normalized Dirac spinors described in Section 5.5, and the polynomial $P(p)$ is the matrix $(-i\gamma_\mu p^\mu + M)\beta$.

(iii) Integrate the product of all these factors over the four-momenta carried by internal lines, and sum over all field indices ℓ, m, etc.

(iv) Add up the results obtained in this way from each Feynman diagram.

Additional combinatoric factors and fermionic signs may need to be included, as described in parts (v) and (vi) of Section 6.1. Examples will be given at the end of this section.

We have a four-momentum integration variable for every internal line, but many of these are eliminated by the delta functions associated with vertices. Since energy and momentum are separately conserved for each connected part of a Feynman diagram, there will be C delta functions left over in a graph with C connected parts. Hence in a diagram with I internal lines and V vertices, the number of independent four-momenta that are *not* fixed by the delta functions is $I-[V-C]$. This is clearly also the number L of independent loops:

$$L = I - V + C\,, \tag{6.3.4}$$

which is defined as the maximum number of internal lines that can be cut without disconnecting the diagram, because any such and only such internal lines can be assigned an independent four-momentum. We can think of the independent momentum variables as characterizing the momenta that circulate in each loop. In particular a *tree* graph is one without loops; after taking the delta functions into account there are no momentum-space integrals left for such graphs.

For instance, in a theory with interaction (6.1.18), the S-matrix (6.1.27) for fermion–boson scattering is given by the momentum-space Feynman rules as

$$\begin{aligned}
&S_{\mathbf{p}_1{}'\sigma_1'n_1'\,\mathbf{p}_2{}'\sigma_2'n_2',\,\mathbf{p}_1\sigma_1n_1\,\mathbf{p}_2\sigma_2n_2} = \\
&\sum_{k'l'm'klm}(-i)^2(2\pi)^8 g_{l'm'k'}\,g_{mlk}\,u^*_{l'}(\mathbf{p}_1{}'\sigma_1'n_1')\,u_l(\mathbf{p}_1\sigma_1n_1)\\
&\times\int d^4q\left(-i(2\pi)^{-4}\frac{P_{m'm}(q)}{q^2+m_m^2-i\epsilon}\right)\\
&\times(2\pi)^{-6}\Big[u^*_{k'}(\mathbf{p}_2{}'\sigma_2'n_2')u_k(\mathbf{p}_2\sigma_2n_2)\delta^4(p_1+p_2-q)\delta^4(q-p_{1'}-p_{2'})\\
&\quad+\;u^*_k(\mathbf{p}_2{}'\sigma_2'n_2')u_{k'}(\mathbf{p}_2\sigma_2n_2)\delta^4(p_2-p_{1'}+q)\delta^4(p_1-p_{2'}-q)\Big]\,,
\end{aligned}$$

with labels 1 and 2 here denoting fermions and bosons, respectively. The momentum-space integral here is trivial, and gives

$$\begin{aligned}
&S_{\mathbf{p}_1{}'\sigma_1'n_1'\,\mathbf{p}_2{}'\sigma_2'n_2',\,\mathbf{p}_1\sigma_1n_1\,\mathbf{p}_2\sigma_2n_2} = i(2\pi)^{-2}\delta^4(p_1+p_2-p_1'-p_2')\\
&\times\sum_{k'l'm'klm} g_{l'm'k'}\,g_{mlk}\,u^*_{l'}(\mathbf{p}_1{}'\sigma_1'n_1')\,u_l(\mathbf{p}_1\sigma_1n_1)\\
&\times\Bigg[\frac{P_{m'm}(p_1+p_2)}{(p_1+p_2)^2+m_m^2-i\epsilon}\,u^*_{k'}(\mathbf{p}_2{}'\sigma_2'n_2')u_k(\mathbf{p}_2\sigma_2n_2)\\
&\quad+\;\frac{P_{m'm}(p_{2'}-p_1)}{(p_{2'}-p_1)^2+m_m^2-i\epsilon}\,u^*_k(\mathbf{p}_2{}'\sigma_2'n_2')u_{k'}(\mathbf{p}_2\sigma_2n_2)\Bigg]\,. \qquad (6.3.5)
\end{aligned}$$

In the same way, the S-matrix element (6.1.28) for fermion–fermion scattering in the same theory is

$$\begin{aligned}
&S_{\mathbf{p}_1{}'\sigma_1'n_1'\,\mathbf{p}_2{}'\sigma_2'n_2',\,\mathbf{p}_1\sigma_1n_1\,\mathbf{p}_2\sigma_2n_2} = i(2\pi)^{-2}\delta^4(p_1+p_2-p_1'-p_2')\\
&\times\sum_{k'l'm'klm} g_{m'mk'}\,g_{l'lk}\frac{P_{k'k}(p_{1'}-p_1)}{(p_{1'}-p_1)^2+m_k^2-i\epsilon}\\
&\times u^*_{m'}(\mathbf{p}_2{}'\,\sigma_2'\,n_2')\,u^*_{l'}(\mathbf{p}_1{}'\,\sigma_1'\,n_1')\,u_m(\mathbf{p}_2\,\sigma_2\,n_2)u_l(\mathbf{p}_1\,\sigma_1\,n_1)\\
&-[1'\rightleftharpoons 2']\,. \qquad (6.3.6)
\end{aligned}$$

These results illustrate the need for a more compact notation. We may define a fermion–boson coupling matrix

$$[\Gamma_k]_{lm} \equiv g_{lmk}\,. \qquad (6.3.7)$$

The matrix elements (6.3.5) and (6.3.6) for fermion–boson and fermion–

fermion scattering can then be rewritten in matrix notation as

$$S_{\mathbf{p}_1'\sigma_1'n_1',\,\mathbf{p}_2'\sigma_2'n_2',\,\mathbf{p}_1\sigma_1n_1\,\mathbf{p}_2\sigma_2n_2} = i(2\pi)^{-2}\delta^4(p_1+p_2-p_1'-p_2')\sum_{k'k}$$
$$\times\Bigg[\left(u^\dagger(\mathbf{p}_1'\sigma_1'n_1')\,\Gamma_{k'}\frac{P(p_1+p_2)}{(p_1+p_2)^2+M^2-i\epsilon}\Gamma_k u(\mathbf{p}_1\sigma_1n_1)\right)$$
$$\times u^*_{k'}(\mathbf{p}_2'\sigma_2'n_2')u_k(\mathbf{p}_2\sigma_2n_2)$$
$$+\left(u^\dagger(\mathbf{p}_1'\sigma_1'n_1')\,\Gamma_{k'}\frac{P(p_1-p_2')}{(p_1-p_2')^2+M^2-i\epsilon}\Gamma_k u(\mathbf{p}_1\sigma_1n_1)\right)$$
$$\times u^*_{k}(\mathbf{p}_2'\sigma_2'n_2')u_{k'}(\mathbf{p}_2\sigma_2n_2)\Bigg] \tag{6.3.8}$$

and

$$S_{\mathbf{p}_1'\sigma_1'n_1',\,\mathbf{p}_2'\sigma_2'n_2',\,\mathbf{p}_1\sigma_1n_1\,\mathbf{p}_2\sigma_2n_2} = i(2\pi)^{-2}\delta^4(p_1+p_2-p_1'-p_2')$$
$$\times\sum_{k'k}\frac{P_{k'k}(p_{1'}-p_1)}{(p_{1'}-p_1)^2+m_k^2-i\epsilon}$$
$$\times\left(u^\dagger(\mathbf{p}_2'\sigma_2'n_2')\,\Gamma_{k'}u(\mathbf{p}_2\sigma_2n_2)\right)\left(u^\dagger(\mathbf{p}_1'\sigma_1'n_1')\Gamma_k\,u(\mathbf{p}_1\sigma_1n_1)\right)$$
$$-[1'\rightleftharpoons 2']\,, \tag{6.3.9}$$

where M^2 and m^2 are the diagonal mass matrices of the fermions and bosons in Eqs. (6.3.8) and (6.3.9), respectively. The general rule is that in using matrix notation, one writes coefficient functions, coupling matrices, and propagators in an order dictated by following lines *backwards* from the order indicated by the arrows. In the same notation, the S-matrix for boson–boson scattering in the same theory would be given by a sum of one-loop diagrams, shown in Figure 6.7:

$$S_{\mathbf{p}_1'\sigma_1'n_1',\,\mathbf{p}_2'\sigma_2'n_2',\,\mathbf{p}_1\sigma_1n_1\,\mathbf{p}_2\sigma_2n_2} = -(2\pi)^{-6}\delta^4(p_1+p_2-p_1'-p_2')$$
$$\times\sum_{k_1k_2k_1'k_2'}u^*_{k_1'}(\mathbf{p}_1',\sigma_1',n_1')u^*_{k_2'}(\mathbf{p}_2',\sigma_2',n_2')u_{k_1}(\mathbf{p}_1,\sigma_1,n_1)u_{k_2}(\mathbf{p}_2,\sigma_2,n_2)$$
$$\times\int d^4q\,\mathrm{Tr}\Bigg\{\Gamma_{k_2'}\frac{P(q)}{q^2+M^2-i\epsilon}\Gamma_{k_1'}\frac{P(q+p_1')}{(q+p_1')^2+M^2-i\epsilon}$$
$$\times\Gamma_{k_1}\frac{P(q+p_1'-p_1)}{(q+p_1'-p_1)^2+M^2-i\epsilon}\Gamma_{k_2}\frac{P(q-p_2')}{(q-p_2')^2+M^2-i\epsilon}\Bigg\}$$
$$+\ldots\,, \tag{6.3.10}$$

where the ellipsis in the last line indicates terms obtained by permuting bosons 1′, 2′, and 2. The minus sign at the beginning of the right-hand side is the extra minus sign associated with fermionic loops. Note that after elimination of delta functions there is just one momentum-space integral here, as appropriate for a diagram with one loop. We shall see how to do this sort of momentum-space integral in Chapter 11.

To make this more specific, consider a theory with a Dirac spinor field $\psi(x)$ of mass M and a pseudoscalar field $\phi(x)$ of mass m, interacting through the interaction $-ig\phi\bar{\psi}\gamma_5\psi$. (The factor $-i$ is inserted to make this interaction Hermitian for real coupling constants g.) Recall that the polynomial $P(q)$ for the scalar is just unity, while for the spinor it is $[-i\gamma_\mu q^\mu+M]\beta$. Also, the u for a scalar of energy E is $(2E)^{-1/2}$, while for the spinor u is the conventionally normalized Dirac spinor discussed in Section 5.5. Eqs. (6.3.8), (6.3.9), and (6.3.10) give the lowest-order connected S-matrix elements for fermion–boson scattering, fermion–fermion scattering, and boson–boson scattering:

$$S_{\mathbf{p}'_1\sigma'_1\,\mathbf{p}'_2\,,\,\mathbf{p}_1\sigma_1\,\mathbf{p}_2} = -i(2\pi)^{-2}g^2(4E'_2E_2)^{-1/2}\delta^4(p_1+p_2-p'_1-p'_2)$$
$$\times\Bigg[\left(\bar{u}(\mathbf{p}_1{}'\sigma'_1)\,\gamma_5\frac{-i\gamma_\mu(p_1+p_2)^\mu+M}{(p_1+p_2)^2+M^2-i\epsilon}\gamma_5 u(\mathbf{p}_1\sigma_1)\right)$$
$$+\left(\bar{u}(\mathbf{p}_1{}'\sigma'_1)\,\gamma_5\frac{-i\gamma_\mu(p_1-p'_2)^\mu+M}{(p_1-p'_2)^2+M^2-i\epsilon}\gamma_5\, u(\mathbf{p}_1\,\sigma_1)\right)\Bigg]\,,$$

$$S_{\mathbf{p}_1{}'\sigma'_1\,\mathbf{p}_2{}'\sigma'_2\,,\,\mathbf{p}_1\sigma_1\,\mathbf{p}_2\sigma_2} = -i(2\pi)^{-2}g^2\delta^4(p_1+p_2-p'_1-p'_2)$$
$$\times\left(\bar{u}(\mathbf{p}'_2\,\sigma'_2)\,\gamma_5\,u(\mathbf{p}_2\,\sigma_2)\right)\,\left(\bar{u}(\mathbf{p}'_1\,\sigma'_1)\gamma_5\,u(\mathbf{p}_1\,\sigma_1)\right)$$
$$\times\frac{1}{(p_{1'}-p_1)^2+m^2-i\epsilon}$$
$$-[1'\rightleftharpoons 2']\,,$$

$$S_{\mathbf{p}'_1\,\mathbf{p}'_2\,,\,\mathbf{p}_1\,\mathbf{p}_2} = -(2\pi)^{-6}\,g^4\,(16E_1E_2E'_1E'_2)^{-1/2}\,\delta^4(p_1+p_2-p'_1-p'_2)$$
$$\times\int d^4q\;\mathrm{Tr}\Bigg\{\gamma_5\frac{-i\gamma_\mu q^\mu+M}{q^2+M^2-i\epsilon}\gamma_5\frac{-i\gamma_\mu(q+p'_1)^\mu+M}{(q+p'_1)^2+M^2-i\epsilon}$$
$$\times\gamma_5\frac{-i\gamma_\mu(q+p'_1-p_1)^\mu+M}{(q+p'_1-p_1)^2+M^2-i\epsilon}\gamma_5\frac{-i\gamma_\mu(q-p'_2)^\mu+M}{(q-p'_2)^2+M^2-i\epsilon}\Bigg\}+\ldots\,,$$

where in the last formula the ellipsis indicates a sum over permutations of particles 2, 1′, 2′. The factors β in the fermion propagator numerators have been used to replace $u^\dagger$ with $\bar{u}$.

* * *

Another useful topological result expresses a sort of conservation law of lines. For the moment we can think of all internal and external lines as being created at vertices and destroyed in pairs at the centers of internal lines or when external lines leave the diagram. (This has nothing to do with the directions of the arrows carried by these lines.) Equating the numbers of lines that are created and destroyed then gives

$$2I + E = \sum_i n_i V_i \,, \tag{6.3.11}$$

where I and E are the numbers of internal and external lines, V_i are the numbers of vertices of various types labelled i, and n_i are the number of lines attached to each vertex. (This also holds separately for fields of each type.) In particular, if all interactions involve the same number $n_i = n$ of fields, then this reads

$$2I + E = nV \,, \tag{6.3.12}$$

where V is the total number of all vertices. In this case, we can eliminate I from Eqs. (6.3.4) and (6.3.11), and find that for a connected (i.e., $C = 1$) graph the number of vertices is given by

$$V = \frac{2L + E - 2}{n - 2} \,. \tag{6.3.13}$$

For instance for a trilinear interaction the diagrams for a scattering process ($E = 4$) with $L = 0, 1, 2 \cdots$ has $V = 2, 4, 6 \cdots$ vertices. In general, the expansion in powers of the coupling constants is an expansion in increasing numbers of loops.

6.4 Off the Mass Shell

In the Feynman diagrams for any S-matrix element all external lines are 'on the mass shell'; that is, the four-momentum associated with an external line for a particle of mass m is constrained to satisfy $p_\mu p^\mu = -m^2$. It is often important also to consider Feynman diagrams 'off the mass shell', for which the external line energies like the energies associated with internal lines are free variables, unrelated to any three-momenta. For one thing, these arise as parts of larger Feynman diagrams; for instance, a loop appearing as an insertion in some internal line of a diagram could be regarded as a Feynman diagram with two external lines, both off the mass shell.

Of course, once we calculate the contribution of a given Feynman diagram off the mass shell, it is easy to calculate the associated S-matrix elements by going to the mass shell, taking the four-momentum p^μ flowing along the line *into* the diagram to have $p^0 = \sqrt{\mathbf{p}^2 + m^2}$ for particles in the initial state and $p^0 = -\sqrt{\mathbf{p}^2 + m^2}$ for particles in the final state, and including the appropriate external line factors $(2\pi)^{-3/2} u_\ell$ or $(2\pi)^{-3/2} v_\ell^*$ for initial particles or antiparticles and $(2\pi)^{-3/2} u_\ell^*$ or $(2\pi)^{-3/2} v_\ell$ for final particles or antiparticles. Indeed, when we come to the path integral approach in Chapter 9 we shall find it easiest first to derive the Feynman rules for diagrams with *all* external lines off the mass shell, and then obtain S-matrix elements by letting the momenta associated with external lines approach their appropriate mass shells.

Feynman graphs with lines off the mass shell are just a special case of a wider generalization of the Feynman rules that takes into account the effects of various possible external fields. Suppose we add a sum of terms involving external fields $\epsilon_a(x)$ to the Hamiltonian, so that the interaction $V(t)$ that is used in the Dyson series (3.5.10) for the S-matrix is replaced with

$$V_\epsilon(t) = V(t) + \sum_a \int d^3x \, \epsilon_a(\mathbf{x}, t) \, o_a(\mathbf{x}, t) . \tag{6.4.1}$$

The 'currents' $o_a(t)$ have the usual time-dependence of the interaction picture

$$o_a(t) = \exp(iH_0t) o_a(0) \exp(-iH_0t) , \tag{6.4.2}$$

but are otherwise quite arbitrary operators. The S-matrix for any given transition $\alpha \to \beta$ then becomes a functional $S_{\beta\alpha}[\epsilon]$ of the c-number functions $\epsilon_a(t)$. The Feynman rules for computing this functional are given by an obvious extension of the usual Feynman rules. In addition to the usual vertices obtained from $V(t)$, we must include additional vertices: if $o_a(x)$ is a product of n_a field factors, then any o_a vertex with position label x must have n_a lines of corresponding types attached, and makes a contribution to the position-space Feynman rules equal to $-i\epsilon_a(x)$ times whatever numerical factors appear in $o_a(x)$. It follows then that the rth variational derivative of $S_{\beta\alpha}[\epsilon]$ with respect to $\epsilon_a(x)$, $\epsilon_b(y) \cdots$ at $\epsilon = 0$ is given by position space diagrams with r additional vertices, to which are attached respectively $n_a, n_b \cdots$ internal lines, and no external lines. These vertices carry position labels $x, y \cdots$ over which we do *not* integrate; each such vertex makes a contribution equal to $-i$ times whatever numerical factors appear in the associated current o_a.

In particular, in the case where these currents are all single field factors, i.e.,

$$V_\epsilon(t) = V(t) + \sum_\ell \int d^3x \, \epsilon_\ell(\mathbf{x}, t) \, \psi_\ell(\mathbf{x}, t) ,$$

the rth variational derivative of $S_{\beta\alpha}[\epsilon]$ with respect to $\epsilon_\ell(x)$, $\epsilon_m(y) \cdots$ at $\epsilon = 0$ is given by position space diagrams with r additional vertices carrying spacetime labels $x, y \cdots$, to each of which is attached a single internal particle line of type $\ell, m \cdots$. These can be thought of as off-shell external lines, with the difference that their contribution to the matrix element is not a coefficient function like $(2\pi)^{-3/2} u_\ell(\mathbf{p}, \sigma) e^{ip\cdot x}$ or $(2\pi)^{-3/2} u_\ell^*(\mathbf{p}, \sigma) e^{-ip\cdot x}$ but a *propagator*, as well as a factor $-i$ from the vertex at the end of the line. We obtain a momentum space Feynman diagram with particles in states α and β on the mass shell plus r external lines of type $\ell, m\cdots$ carrying momenta $p, p'\cdots$ from the variational derivative

$$\left[\frac{\delta^r S_{\beta\alpha}[\epsilon]}{\delta\epsilon_\ell(x)\,\delta\epsilon_m(y)\cdots}\right]_{\epsilon=0}$$

by stripping away the propagators on each of the off-shell lines and then taking the appropriate Fourier transforms and multiplying with appropriate coefficient functions u_ℓ, u_ℓ^*, etc. and a factor $(-i)^r$.

It is very useful for a number of purposes to recognize that there is a simple relation between the sum of contributions from all perturbation theory diagrams for any off-shell amplitude and a matrix element, between eigenstates of the full Hamiltonian, of a time-ordered product of corresponding operators in the Heisenberg picture. This relation is provided by a theorem,[3] which states that to all orders of perturbation theory*

$$\left[\frac{\delta^r S_{\beta\alpha}[\epsilon]}{\delta\epsilon_a(x)\,\delta\epsilon_b(y)\cdots}\right]_{\epsilon=0} = \left(\Psi_\beta{}^-, T\Big\{-iO_a(x),\ -iO_b(y)\cdots\Big\}\Psi_\alpha{}^+\right), \tag{6.4.3}$$

where $O_a(x)$, etc. are the counterparts of $o_a(x)$ in the Heisenberg picture

$$O_a(\mathbf{x},t) = \exp(iHt)\,o_a(\mathbf{x},0)\exp(-iHt) = \Omega(t)\,o_a(\mathbf{x},t)\,\Omega^{-1}(t)\,, \tag{6.4.4}$$

$$\Omega(t) \equiv e^{iHt}\,e^{-iH_0t}\,, \tag{6.4.5}$$

and $\Psi_\beta{}^+$ and $\Psi_\beta{}^-$ are 'in' and 'out' eigenstates of the full Hamiltonian H, respectively.

Here is the proof. From Eq. (3.5.10), we see immediately that the left-hand side of Eq. (6.4.3) is

$$\left[\frac{\delta^r S[\epsilon]}{\delta\epsilon_{a_1}(x_1)\ldots\delta\epsilon_{a_r}(x_r)}\right]_{\epsilon=0} = \sum_{N=0}^{\infty}\frac{(-i)^{N+r}}{N!}\int_{-\infty}^{\infty} d\tau_1\cdots d\tau_N$$
$$\times\left(\Phi_\beta,\ T\Big\{V(\tau_1)\cdots V(\tau_N)o_{a_1}(x_1)\cdots o_{a_r}(x_r)\Big\}\Phi_\alpha\right). \tag{6.4.6}$$

For definiteness, suppose that $x_1^0 \geq x_2^0 \geq \cdots \geq x_r^0$. Then we can denote by $\tau_{01}\cdots\tau_{0N_0}$ all τs greater than x_1^0; by $\tau_{11}\cdots\tau_{1N_1}$ all τs between x_1^0 and x_2^0, and so on; finally denoting by $\tau_{r1}\cdots\tau_{rN_r}$ all τs that are less than x_r^0.

* For a single O operator, this is a version of the Schwinger action principle.[4]

Eq. (6.4.6) then becomes:

$$\left[\frac{\delta^r S[\epsilon]}{\delta\epsilon_{a_1}(x_1)\cdots\delta\epsilon_{a_r}(x_r)}\right]_{\epsilon=0} = \sum_{N=0}^{\infty}\frac{(-i)^{N+r}}{N!}\sum_{N_0 N_1\cdots N_r}\frac{N!\,\delta_{N,N_0+N_1+\cdots+N_r}}{N_0!N_1!\cdots N_r!}$$

$$\times\int_{x_1^0}^{\infty}d\tau_{01}\cdots d\tau_{0N_0}\int_{x_2^0}^{x_1^0}d\tau_{11}\cdots d\tau_{1N_1}\cdots\int_{-\infty}^{x_r^0}d\tau_{r1}\cdots d\tau_{rN_r}$$

$$\times\Bigg(\Phi_\beta,\ T\Big\{V(\tau_{01})\cdots V(\tau_{0N_0})\Big\}\,o_{a_1}(x_1)T\Big\{V(\tau_{11})\cdots V(\tau_{1N_1})\Big\}\,o_{a_2}(x_2)\cdots$$

$$\times\cdots o_{a_r}(x_r)\,T\Big\{V(\tau_{r1})\cdots V(\tau_{rN_r})\Big\}\Phi_\alpha\Bigg)\,.$$

The factor $N!/N_0!N_1!\cdots N_r!$ is the number of ways of sorting N τs into $r+1$ subsets, each containing $N_0, N_1,\cdots N_r$ of these τs. Instead of summing over $N_0, N_1,\cdots, N_r$, subject to the condition $N_0+N_1+\cdots+N_r = N$, and then summing over N, we can just sum independently over N_0, $N_1,\cdots, N_r$, setting N where it appears in $(-i)^N$ equal to $N_0+N_1+\cdots+N_r$. This gives

$$\left[\frac{\delta^r S[\epsilon]}{\delta\epsilon_{a_1}(x_1)\cdots\delta\epsilon_{a_r}(x_r)}\right]_{\epsilon=0} = (-i)^r\Bigg(\Phi_\beta, U(\infty, x_1^0)\,o_{a_1}(x_1)$$

$$\times\, U(x_1^0, x_2^0)\,o_{a_2}(x_2)\,U(x_2^0, x_3^0)\cdots o_{a_r}(x_r)\,U(x_r^0,-\infty)\Phi_\alpha\Bigg)\,, \quad (6.4.7)$$

where

$$U(t',t) = \sum_{N=0}^{\infty}\frac{(-i)^N}{N!}\int_t^{t'}d\tau_1\cdots d\tau_N\,T\Big\{V(\tau_1)\cdots V(\tau_N)\Big\}\,. \quad (6.4.8)$$

The operator $U(t',t)$ satisfies the differential equation

$$\frac{d}{dt'}\,U(t',t) = -iV(t')U(t',t) \quad (6.4.9)$$

with the obvious initial condition

$$U(t,t) = 1\,. \quad (6.4.10)$$

This has the solution

$$U(t',t) = \exp(iH_0t')\exp(-iH(t'-t))\exp(-iH_0t) = \Omega^{-1}(t')\Omega(t) \quad (6.4.11)$$

with Ω given by Eq. (6.4.5). Inserting Eq. (6.4.11) in Eq. (6.4.7) and using

Eq. (6.4.4), we have

$$\left[\frac{\delta S[\epsilon]}{\delta\epsilon_{a_1}(x_1)\cdots\delta\epsilon_{a_r}(x_r)}\right]_{\epsilon=0}$$
$$= (-i)^r\Big(\Omega(\infty)\Phi_\beta,\ O_{a_1}(x_1)\cdots O_{a_r}(x_r)\Omega(-\infty)\Phi_\alpha\Big)\,. \tag{6.4.12}$$

In deriving this result we supposed that $x_1^0 \geq x_2^0 \geq \cdots \geq x_r^0$, so we could just as well replace the product of operators on the right-hand-side with the time-ordered product:

$$\left[\frac{\delta S[\epsilon]}{\delta\epsilon_{a_1}(x_1)\cdots\delta\epsilon_{a_r}(x_r)}\right]_{\epsilon=0}$$
$$= (-i)^r\Big(\Omega(\infty)\Phi_\beta,\ T\Big\{O_{a_1}(x_1)\cdots O_{a_r}(x_r)\Big\}\Omega(-\infty)\Phi_\alpha\Big)\,. \tag{6.4.13}$$

But now both sides are entirely symmetric (or for fermions antisymmetric) in the as and xs, so this relation holds whatever the order of the times $x_1^0 \ldots x_r^0$. Also, we saw in Section 3.1 that (in the sense of Eq. (3.1.12))

$$\Psi_\beta{}^{\pm} = \Omega(\mp\infty)\Phi_\beta\,. \tag{6.4.14}$$

Hence Eq. (6.4.13) is the desired result (6.4.3).

Problems

1. Consider the theory of a real scalar field ϕ, with interaction (in the interaction picture) $V = g\int d^3x\,\phi(x)^3/3!$. Calculate the connected S-matrix element for scalar–scalar scattering to second order in g, doing all integrals. Use the results to calculate the differential cross-section for scalar–scalar scattering in the center-of-mass system.

2. Consider a theory involving a neutral scalar field $\phi(x)$ for a boson B and a complex Dirac field $\psi(x)$ for a fermion F, with interaction (in the interaction picture) $V = ig\int d^3x\,\bar\psi(x)\gamma_5\psi(x)\phi(x)$. Draw all the connected order-g^2 Feynman diagrams and calculate the corresponding S-matrix elements for the processes $F^c + B \to F^c + B$, $F + F^c \to F + F^c$, and $F^c + F \to B + B$ (where F^c is the antiparticle of F). Do all integrals.

3. Consider the theory of a real scalar field $\phi(x)$, with interaction $V = g\int d^3x\,\phi(x)^4/4!$. Calculate the S-matrix for scalar–scalar scattering to order g, and use the result to calculate the differential scattering cross-section. Calculate the correction terms in the S-matrix for scalar–scalar scattering to order g^2, expressing the result as an integral over a single four-momentum, but do all x-integrals.

4. What is the contribution in Feynman diagrams from the contraction of the derivative $\partial_\mu \psi_\ell(x)$ of a Dirac field with the adjoint $\psi_m^\dagger(y)$ of the field?

5. Use the theorem of Section 6.4 to give expressions for the vacuum expectation values of Heisenberg picture operators $(\Psi_0, \Phi(x)\Psi_0)$ and $(\Psi_0, T\{\Phi(x), \Phi(y)\}\Psi_0)$ in the theory of Problem 1, to orders g and g^2, respectively.

References

1. F. J. Dyson, *Phys. Rev.* **75**, 486, 1736 (1949).

2. The formal statement of this result is known as *Wick's theorem*; see G. C. Wick, *Phys. Rev.* **80**, 268 (1950).

3. I do not know who first proved this theorem. It was known in the early 1950s to several theorists, including M. Gell-Mann and F. E. Low.

4. J. Schwinger, *Phys. Rev.* **82**, 914 (1951).

7
The Canonical Formalism

Ever since the birth of quantum field theory in the papers of Born, Dirac, Fermi, Heisenberg, Jordan, and Pauli in the late 1920s, its development has been historically linked to the canonical formalism, so much so that it seems natural to begin any treatment of the subject today by postulating a Lagrangian and applying to it the rules of canonical quantization. This is the approach used in most books on quantum field theory. Yet historical precedent is not a very convincing reason for using this formalism. If we discovered a quantum field theory that led to a physically satisfactory S-matrix, would it bother us if it could not be derived by the canonical quantization of some Lagrangian?

To some extent this question is moot because, as we shall see in Section 7.1, all of the most familiar quantum field theories furnish canonical systems, and these can easily be put in a Lagrangian form. However, there is no proof that every conceivable quantum field theory can be formulated in this way. And even if it can, this does not in itself explain why we should *prefer* to use the Lagrangian formalism as a starting point in constructing various quantum field theories.

The point of the Lagrangian formalism is that it makes it easy to satisfy Lorentz invariance and other symmetries: a classical theory with a Lorentz-invariant Lagrangian density will when canonically quantized lead to a Lorentz-invariant quantum theory. That is, we shall see here that such a theory allows the construction of suitable quantum mechanical operators that satisfy the commutation relations of the Poincaré algebra, and therefore leads to a Lorentz-invariant S-matrix.

This is not so trivial. We saw in the previous chapter that in theories with derivative couplings or spins $j \geq 1$, it is not enough to take the interaction Hamiltonian as the integral over space of a scalar interaction density; we also need to add non-scalar terms to the interaction density to compensate for non-covariant terms in the propagators. The canonical formalism with a scalar Lagrangian density will automatically provide these extra terms. Later, when we come to non-Abelian gauge theories in Volume II, this extra convenience will become a necessity; it would be just about hopeless to try to guess at the form of the Hamiltonian in such theories without starting with a Lorentz-invariant and gauge-invariant Lagrangian density.

7.1 Canonical Variables

In this section we shall show that various quantum field theories that we have constructed so far satisfy the commutation rules and equations of motion of the Hamiltonian version of the canonical formalism. It is the Hamiltonian formalism that is needed to calculate the S-matrix (whether by operator or path-integral methods) but it is not always easy to choose Hamiltonians that yield a Lorentz-invariant S-matrix. In the balance of this chapter we shall take the Lagrangian version of the canonical formalism as our starting point, and use it to derive physically satisfactory Hamiltonians. The purpose of the present section is to identify the canonical fields and their conjugates in various field theories, to tell us how to separate the free-field terms in the Lagrangian, and incidentally to reassure us that the canonical formalism is indeed applicable to physically realistic theories.

We first show that the free fields constructed in Chapter 5 automatically provide a system of quantum operators $q^n(\mathbf{x}, t)$ and canonical conjugates $p_n(\mathbf{x}, t)$ that satisfy the familiar canonical commutation or anticommutation relations:

$$[q^n(\mathbf{x}, t), p_{\bar{n}}(\mathbf{y}, t)]_{\mp} = i\delta^3(\mathbf{x} - \mathbf{y})\delta^n_{\bar{n}} , \tag{7.1.1}$$

$$[q^n(\mathbf{x}, t), q^{\bar{n}}(\mathbf{y}, t)]_{\mp} = 0 , \tag{7.1.2}$$

$$[p_n(\mathbf{x}, t), p_{\bar{n}}(\mathbf{y}, t)]_{\mp} = 0 , \tag{7.1.3}$$

where the subscripts $\mp$ indicate that these are commutators if either of the particles created and destroyed by the two operators are bosons, and anticommutators if both particles are fermions. For instance, the real scalar field $\phi(\mathbf{x})$ for a self-charge-conjugate particle of zero spin was found in Section 5.2 to obey the commutation relation

$$[\phi(x), \phi(y)]_- = \Delta(x - y) ,$$

where Δ is the function

$$\Delta(x) \equiv \int \frac{d^3k}{2k^0(2\pi)^3} [e^{ik\cdot x} - e^{-ik\cdot x}]$$

with $k^0 \equiv \sqrt{\mathbf{k}^2 + m^2}$. We note that

$$\Delta(\mathbf{x}, 0) = 0 , \qquad \dot{\Delta}(\mathbf{x}, 0) = -i\delta^3(\mathbf{x}) .$$

(A dot denotes the derivative with respect to the time x^0.) It is easy then to see that the field and its time-derivative $\dot{\phi}$ obey the equal-time commutation relations:

$$[\phi(\mathbf{x},t), \dot{\phi}(\mathbf{y},t)]_- = i\delta^3(\mathbf{x}-\mathbf{y}) , \tag{7.1.4}$$

$$[\phi(\mathbf{x},t), \phi(\mathbf{y},t)]_- = 0 , \tag{7.1.5}$$

$$[\dot{\phi}(\mathbf{x},t), \dot{\phi}(\mathbf{y},t)]_- = 0 . \tag{7.1.6}$$

Therefore we may define canonical variables

$$q(\mathbf{x},t) \equiv \phi(\mathbf{x},t) , \qquad p(\mathbf{x},t) \equiv \dot{\phi}(\mathbf{x},t) \tag{7.1.7}$$

which satisfy the canonical commutation relations (7.1.1)–(7.1.3).

For the complex scalar field of a particle of spin zero with a distinct antiparticle, the commutation relations are

$$[\phi(x), \phi^\dagger(y)]_- = \Delta(x-y) , \qquad [\phi(x), \phi(y)]_- = 0 .$$

We may therefore define the free-particle canonical variables as the complex operators

$$q(\mathbf{x},t) \equiv \phi(\mathbf{x},t) , \tag{7.1.8}$$

$$p(\mathbf{x},t) \equiv \dot{\phi}^\dagger(\mathbf{x},t) . \tag{7.1.9}$$

Equivalently, writing $\phi \equiv (\phi_1 + i\phi_2)/\sqrt{2}$ with ϕ_k Hermitian for $k = 1, 2$, we have canonical variables

$$q^k(\mathbf{x},t) = \phi_k(\mathbf{x},t) , \tag{7.1.10}$$

$$p_k(\mathbf{x},t) = \dot{\phi}_k(\mathbf{x},t) , \tag{7.1.11}$$

and these satisfy the commutation relations (7.1.1)–(7.1.3).

For the real vector field of a particle of spin one, the commutation relations are given by Section 5.3 as

$$[v^\mu(x), v^\nu(y)]_- = \left[\eta^{\mu\nu} - \frac{\partial^\mu \partial^\nu}{m^2}\right] \Delta(x-y) .$$

(We are using v^μ rather than V^μ for the vector field because we want to reserve upper case letters here for the fields in the Heisenberg picture.) Here the free-particle canonical variables may be taken as

$$q^i(\mathbf{x},t) = v^i(\mathbf{x},t) , \tag{7.1.12}$$

$$p_i(\mathbf{x},t) = \dot{v}^i(\mathbf{x},t) + \frac{\partial v^0(\mathbf{x},t)}{\partial x^i} , \tag{7.1.13}$$

with $i = 1, 2, 3$. The reader may check that (7.1.12) and (7.1.13) satisfy the commutation relations (7.1.1)–(7.1.3). The field equations (5.3.36) and

(5.3.38) together with Eq. (7.1.13) allow us to express v^0 in terms of the other variables as

$$v^0 = m^{-2}\nabla \cdot \mathbf{p}\,, \tag{7.1.14}$$

so v^0 is not regarded as one of the qs. The extension of these results to complex vector fields may be handled just as for complex scalar fields.

For the Dirac field of a non-Majorana spin $\frac{1}{2}$ particle, Section 5.5 shows that the anticommutator is

$$[\psi_n(x), \psi^\dagger_{\bar{n}}(y)]_+ = \Big[(-\gamma^\mu \partial_\mu + m)\beta\Big]_{n,\bar{n}} \Delta(x-y)$$

and

$$[\psi_n(x), \psi_{\bar{n}}(y)]_+ = 0\,.$$

Here it would be inconsistent to take ψ_n and $\psi^\dagger_n$ to be independent canonical variables, because their anticommutator does not vanish at equal times. It is conventional instead to define

$$q^n(x) \equiv \psi_n(x)\,, \tag{7.1.15}$$

$$p_n(x) \equiv i\psi^\dagger_n(x)\,. \tag{7.1.16}$$

It is easy then to see that (7.1.15) and (7.1.16) satisfy the canonical anticommutation relations (7.1.1)–(7.1.3).

For any system of operators that satisfy commutation or anticommutation relations like (7.1.1)–(7.1.3), we may define a quantum mechanical functional derivative: for an arbitrary bosonic functional $F[q(t), p(t)]$ of $q^n(\mathbf{x}, t)$ and $p_n(\mathbf{x}, t)$ at a fixed time t, we define*

$$\frac{\delta F[q(t), p(t)]}{\delta q^n(\mathbf{x}, t)} \equiv i\Big[p_n(\mathbf{x}, t)\,,\; F[q(t), p(t)]\Big]\,, \tag{7.1.17}$$

$$\frac{\delta F[q(t), p(t)]}{\delta p_n(\mathbf{x}, t)} \equiv i\Big[F[q(t), p(t)]\,,\; q^n(\mathbf{x}, t)\Big]\,. \tag{7.1.18}$$

This definition is motivated by the fact that if $F[q(t), p(t)]$ is written with all qs to the left of all ps, then (7.1.17) and (7.1.18) are respectively just the left- and right-derivatives with respect to q^n and p_n. That is, for an

* We are here using a notation that will be adopted henceforth; if $f(x, y)$ is a function of two classes of variables collectively called x and y, then $F[f(y)]$ indicates a functional that depends on the values of $f(x, y)$ for all x at fixed y. By a bosonic functional we mean one in which each term contains only even numbers of fermionic fields.

arbitrary c-number** variation δq and δp of the qs and ps, we have

$$\delta F[q(t),p(t)] = \int d^3x \sum_n \left(\delta q^n(\mathbf{x},t) \frac{\delta F[q(t),p(t)]}{\delta q^n(\mathbf{x},t)} + \frac{\delta F[q(t),p(t)]}{\delta p_n(\mathbf{x},t)} \delta p_n(\mathbf{x},t) \right).$$

For more general functionals we need the definitions (7.1.17) and (7.1.18) to pin down various signs and equal-time commutators that may appear.

In particular, H_0 is the generator of time-translations on free-particle states in the sense that:

$$q^n(\mathbf{x},t) = \exp(iH_0 t) q^n(\mathbf{x},0) \exp(-iH_0 t) \,, \tag{7.1.19}$$

$$p_n(\mathbf{x},t) = \exp(iH_0 t) p_n(\mathbf{x},0) \exp(-iH_0 t) \,, \tag{7.1.20}$$

so the free-particle operators have the time-dependence

$$\dot{q}^n(\mathbf{x},t) = i[H_0, q^n(\mathbf{x},t)] = \frac{\delta H_0}{\delta p_n(\mathbf{x},t)} \,, \tag{7.1.21}$$

$$\dot{p}_n(\mathbf{x},t) = -i[p_n(\mathbf{x},t), H_0] = -\frac{\delta H_0}{\delta q^n(\mathbf{x},t)} \,. \tag{7.1.22}$$

We recognize these as the familiar dynamical equations in the Hamiltonian formalism.

The free-particle Hamiltonian is given as always by

$$H_0 = \sum_{n,\sigma} \int d^3k \; a^\dagger(\mathbf{k},\sigma,n)\, a(\mathbf{k},\sigma,n) \sqrt{\mathbf{k}^2 + m_n^2} \,. \tag{7.1.23}$$

This H_0 may be rewritten in terms of the qs and ps at time t. For instance, it is easy to see that for a real scalar field, Eq. (7.1.23) is equal up to a constant term to the functional

$$H_0 = \int d^3x \left[\tfrac{1}{2} p^2 + \tfrac{1}{2} (\nabla q)^2 + \tfrac{1}{2} m^2 q^2 \right] \,. \tag{7.1.24}$$

To be more precise, using (7.1.7) and the Fourier representation of the scalar field ϕ, we find that Eq. (7.1.24) becomes:

$$\begin{aligned} H_0 &= \tfrac{1}{2} \int d^3k \, k^0 \left[a(\mathbf{k}), a^\dagger(\mathbf{k}) \right]_+ \\ &= \int d^3k \, k^0 \left(a^\dagger(\mathbf{k}) a(\mathbf{k}) + \tfrac{1}{2} \delta^3(\mathbf{k} - \mathbf{k}) \right) . \end{aligned} \tag{7.1.25}$$

** Where q^n and p_n are bosonic or fermionic, δq^n and δp_n are understood to commute or anticommute with all fermionic operators, respectively, and to commute with all bosonic operators.

This is the same as Eq. (7.1.23), except for the infinite constant term. Such terms only affect the zero of energy, and have no physical significance in the absence of gravity.† Explicit forms for H_0 as a functional of the q and p variables for other fields will be given in Section 7.5.

It is usual in textbooks on quantum field theory to derive Eq. (7.1.25) as a consequence of Eq. (7.1.24), which in turn is derived from a Lagrangian density. This seems to me backward, for Eq. (7.1.25) *must* hold; if some assumed free-particle Lagrangian did not give Eq. (7.1.25) up to a constant term, we would conclude that it was the wrong Lagrangian. Rather, we should ask what free-field Lagrangian gives Eq. (7.1.25) for spinless particles, or more generally, gives the free-particle Hamiltonian (7.1.23). This question may be answered by the well-known Legendre transformation from the Hamiltonian to the Lagrangian; the free-field Lagrangian is given by

$$L_0[q(t), \dot{q}(t)] = \sum_n \int d^3x \, p_n(\mathbf{x}, t) \, \dot{q}^n(\mathbf{x}, t) \; - \; H_0 \, , \tag{7.1.26}$$

it being understood that p_n is replaced everywhere by its expression in terms of q^n and $\dot{q}^n$ (and, as we shall see, perhaps some auxiliary fields as well). For instance, from the Hamiltonian (7.1.24) and (7.1.7) we can derive the free-field Lagrangian for a scalar field:

$$\begin{aligned} L_0 &= \int d^3x \left[p\dot{q} - \tfrac{1}{2}p^2 - \tfrac{1}{2}(\nabla q)^2 - \tfrac{1}{2}m^2 q^2 \right] \\ &= \int d^3x \left[-\tfrac{1}{2}\partial_\mu \phi \, \partial^\mu \phi - \tfrac{1}{2}m^2 \phi^2 \right] . \end{aligned} \tag{7.1.27}$$

Whatever we suppose the complete Lagrangian of the scalar field may be, this is the term that must be separated out and treated as a term of zeroth order in perturbation theory. A similar exercise may be carried out for the other canonical systems described in this section, but from now on we shall content ourselves with guessing the form of the free-field Lagrangian and then confirming that it gives the correct free-particle Hamiltonian.

We have seen that various free-field theories can be formulated in canonical terms. It is then a short step to show that the same is true of the interacting fields. We can introduce canonical variables in what is called the 'Heisenberg picture', defined by

$$Q^n(\mathbf{x}, t) \equiv \exp(iHt) q^n(\mathbf{x}, 0) \exp(-iHt), \tag{7.1.28}$$

$$P_n(\mathbf{x}, t) \equiv \exp(iHt) p_n(\mathbf{x}, 0) \exp(-iHt), \tag{7.1.29}$$

† However, *changes* in such terms due to changes in the boundary conditions for the fields, as for instance quantizing in the space between parallel plates rather than in infinite space, are physically significant, and have even been measured.[1]

where H is the full Hamiltonian. Because this is a similarity transformation that commutes with H, the total Hamiltonian is the same functional of the Heisenberg picture operators as it was of the qs and ps:

$$H[Q,P] = e^{iHt}H[q,p]e^{-iHt} = H[q,p] \, .$$

Also, because Eqs. (7.1.28)–(7.1.29) define a similarity transformation, the Heisenberg picture operators again satisfy the canonical commutation or anticommutation relations:

$$[Q^n(\mathbf{x},t), P_{\bar{n}}(\mathbf{y},t)]_{\mp} = i\delta^3(\mathbf{x}-\mathbf{y})\delta^n_{\bar{n}} \, , \tag{7.1.30}$$

$$[Q^n(\mathbf{x},t), Q^{\bar{n}}(\mathbf{y},t)]_{\mp} = 0 \, , \tag{7.1.31}$$

$$[P_n(\mathbf{x},t), P_{\bar{n}}(\mathbf{y},t)]_{\mp} = 0 \, . \tag{7.1.32}$$

However, they now have the time-dependence

$$\dot{Q}^n(\mathbf{x},t) = i[H, Q^n(\mathbf{x},t)] = \frac{\delta H}{\delta P_n(\mathbf{x},t)} \, , \tag{7.1.33}$$

$$\dot{P}_n(\mathbf{x},t) = -i[P_n(\mathbf{x},t), H] = -\frac{\delta H}{\delta Q^n(\mathbf{x},t)} \, . \tag{7.1.34}$$

For instance, we might take the Hamiltonian for a real scalar field as the free-particle term (7.1.24) plus the integral of a scalar interaction density $\mathscr{H}$, so that in terms of Heisenberg-picture variables

$$H = \int d^3x \left[\tfrac{1}{2}P^2 + \tfrac{1}{2}(\nabla Q)^2 + \tfrac{1}{2}m^2 Q^2 + \mathscr{H}(Q) \right] \, . \tag{7.1.35}$$

In this case the canonical conjugate to Q is given by the same formula as for free fields:

$$P = \dot{Q} \, . \tag{7.1.36}$$

However, as we shall see, the relation between the canonical conjugates $P_n(x)$ and the field variables and their time-derivatives is in general not the same as for the free-particle operators, but must be inferred from Eqs. (7.1.33) and (7.1.34).

7.2 The Lagrangian Formalism

Having seen that various realistic theories may be cast in the canonical formalism, we must now face the question of how to choose the Hamiltonian. As we will see in the next section, the easiest way to enforce Lorentz invariance and other symmetries is to choose a suitable Lagrangian and use it to derive the Hamiltonian. There is not much loss of generality in this;

given a realistic Hamiltonian, we can generally reconstruct a Lagrangian from which it could be derived, by reversing the process that we are going to describe here of deriving Hamiltonians from Lagrangians. (The derivation of Eq. (7.1.26) gives one example of this reconstruction.) But although we can go from Hamiltonians to Lagrangians or Lagrangians to Hamiltonians, it is easier to explore physically satisfactory theories by listing possible Lagrangians, rather than Hamiltonians.

The Lagrangian is, in general, a functional* $L[\Psi(t), \dot{\Psi}(t)]$ of a set of generic fields $\Psi^\ell(\mathbf{x}, t)$ and their time-derivatives $\dot{\Psi}^\ell(\mathbf{x}, t)$. The conjugate fields $\Pi_\ell(\mathbf{x}, t)$ are defined as the variational derivatives**

$$\Pi_\ell(\mathbf{x}, t) \equiv \frac{\delta L[\Psi(t), \dot{\Psi}(t)]}{\delta \dot{\Psi}^\ell(\mathbf{x}, t)} , \tag{7.2.1}$$

The equations of motion are

$$\dot{\Pi}_\ell(\mathbf{x}, t) = \frac{\delta L[\Psi(t), \dot{\Psi}(t)]}{\delta \Psi^\ell(\mathbf{x}, t)} . \tag{7.2.2}$$

These field equations can be usefully reformulated as a variational principle. We define a functional of $\Psi^\ell(x)$ over all spacetime, known as the *action*

$$I[\Psi] \equiv \int_{-\infty}^{\infty} dt\, L[\Psi(t), \dot{\Psi}(t)] . \tag{7.2.3}$$

Under an arbitrary variation of $\Psi(x)$, the change in $I[\Psi]$ is

$$\delta I[\Psi] = \int_{-\infty}^{\infty} dt \int d^3x \left[\frac{\delta L}{\delta \Psi^\ell(x)}\, \delta \Psi^\ell(x) + \frac{\delta L}{\delta \dot{\Psi}^\ell(x)}\, \delta \dot{\Psi}^\ell(x) \right] .$$

Assuming that $\delta \Psi^\ell(x)$ vanishes for $t \to \pm\infty$, we may integrate by parts, and write

$$\delta I[\Psi] = \int d^4x \left[\frac{\delta L}{\delta \Psi^\ell(x)} - \frac{d}{dt} \frac{\delta L}{\delta \dot{\Psi}^\ell(x)} \right] \delta \Psi^\ell(x) . \tag{7.2.4}$$

We see that the action is stationary with respect to all variations $\delta \Psi^\ell$ that vanish at $t \to \pm\infty$ if and only if the fields satisfy the field equations (7.2.2).

---

* Recall that in the notation we use for functionals, a functional like L in which we display the variable t is understood to depend on the fields $\Psi^\ell(\mathbf{x}, t)$ and $\dot{\Psi}^\ell(\mathbf{x}, t)$, with the undisplayed variables ℓ and $\mathbf{x}$ running over all their values at a fixed value of the displayed variable t. We use upper case Ψs and Πs to indicate that these are interacting rather than free fields.

** Because the Ψs and $\dot{\Psi}$s do not in general satisfy simple commutation or anticommutation relations, we cannot give a simple definition of the functional derivatives occuring here as we did for functional derivatives with respect to the Qs and Ps in the previous section. Instead, we will simply specify that the variational derivatives are what they would be for c-number variables, with minus signs and equal-time commutators or anticommutators supplied as needed to make the formulas correct quantum mechanically. As far as I know, no important issues hinge on the details here.

Because the field equations are determined by the functional $I[\Psi]$, it is natural in trying to construct a Lorentz-invariant theory to make $I[\Psi]$ a scalar functional. In particular, since $I[\Psi]$ is a time-integral of $L[\Psi(t),\dot{\Psi}(t)]$, we guess that L should itself be a space-integral of an ordinary scalar function of $\Psi(x)$ and $\partial\Psi(x)/\partial x^\mu$, known as the *Lagrangian density* $\mathscr{L}$:

$$L[\Psi(t),\dot{\Psi}(t)] = \int d^3x\, \mathscr{L}\left(\Psi(\mathbf{x},t), \nabla\Psi(\mathbf{x},t),\ \dot{\Psi}(\mathbf{x},t)\right)\ , \tag{7.2.5}$$

so that the action is

$$I[\Psi] = \int d^4x\, \mathscr{L}\Big(\Psi(x), \partial\Psi(x)/\partial x^\mu\Big)\,. \tag{7.2.6}$$

All field theories used in current theories of elementary particles have Lagrangians of this form.

Varying $\Psi^\ell(x)$ by an amount $\delta\Psi^\ell(x)$, and integrating by parts, we find a variation in L:

$$\begin{aligned}\delta L &= \int d^3x \left[\frac{\partial\mathscr{L}}{\partial\Psi^\ell}\,\delta\Psi^\ell + \frac{\partial\mathscr{L}}{\partial(\nabla\Psi^\ell)}\nabla\delta\Psi^\ell + \frac{\partial\mathscr{L}}{\partial\dot{\Psi}^\ell}\,\delta\dot{\Psi}^\ell\right]\\ &= \int d^3x \left[\left(\frac{\partial\mathscr{L}}{\partial\Psi^\ell} - \nabla\cdot\frac{\partial\mathscr{L}}{\partial(\nabla\Psi^\ell)}\right)\delta\Psi^\ell + \frac{\partial\mathscr{L}}{\partial\dot{\Psi}^\ell}\,\delta\dot{\Psi}^\ell\right]\,,\end{aligned}$$

so (with obvious arguments suppressed)

$$\frac{\delta L}{\delta\Psi^\ell} = \frac{\partial\mathscr{L}}{\partial\Psi^\ell} - \nabla\cdot\frac{\partial\mathscr{L}}{\partial(\nabla\Psi^\ell)}\,, \tag{7.2.7}$$

$$\frac{\delta L}{\delta\dot{\Psi}^\ell} = \frac{\partial\mathscr{L}}{\partial\dot{\Psi}^\ell}\,. \tag{7.2.8}$$

The field equations (7.2.2) then read

$$\frac{\partial}{\partial x^\mu}\,\frac{\partial\mathscr{L}}{\partial(\partial\Psi^\ell/\partial x^\mu)} = \frac{\partial\mathscr{L}}{\partial\Psi^\ell}\,. \tag{7.2.9}$$

These are known as the *Euler–Lagrange equations.* As expected, if $\mathscr{L}$ is a scalar then these equations are Lorentz-invariant.

In addition to being Lorentz-invariant, the action I is required to be *real.* This is because we want just as many field equations as there are fields. By breaking up any complex fields into their real and imaginary parts, we can always think of I as being a functional only of a number of *real* fields, say N of them. If I were complex, with independent real and imaginary parts, then the real and imaginary parts of the conditions that I be stationary (the Euler–Lagrange equations) would yield $2N$ field equations for N fields, too many to be satisfied except in special cases. We will see in the next section that the reality of the action also ensures that the generators of various symmetry transformations are Hermitian operators.

Although the Lagrangian formalism makes it easy to construct theories that will satisfy Lorentz invariance and other symmetries, to calculate the S-matrix we need a formula for the interaction Hamiltonian. In general, the Hamiltonian is given by the *Legendre transformation*

$$H = \sum_\ell \int d^3x\, \Pi_\ell(\mathbf{x},t)\dot{\Psi}^\ell(\mathbf{x},t) - L[\Psi(t),\dot{\Psi}(t)]\,. \tag{7.2.10}$$

Although Eq. (7.2.1) does not in general allow $\dot{\Psi}^\ell$ to be expressed uniquely in terms of Ψ^ℓ and Π_ℓ, it is easy to see that Eq. (7.2.10) has vanishing variational derivative with respect to $\dot{\Psi}^\ell$ for any $\dot{\Psi}^\ell$ satisfying Eq. (7.2.1), so in general it is a functional only of Ψ^ℓ and Π_ℓ. Its variational derivatives with respect to these variables are

$$\left.\frac{\delta H}{\delta \Psi^\ell(\mathbf{x},t)}\right|_\Pi = \int d^3y \sum_{\ell'} \Pi_{\ell'}(\mathbf{y},t) \left.\frac{\delta \dot{\Psi}^{\ell'}(\mathbf{y},t)}{\delta \Psi^\ell(\mathbf{x},t)}\right|_\Pi - \left.\frac{\delta L}{\delta \Psi^\ell(\mathbf{x},t)}\right|_{\dot{\Psi}}$$
$$- \int d^3y \sum_{\ell'} \left.\frac{\delta L}{\delta \dot{\Psi}^{\ell'}(\mathbf{y},t)}\right|_\Psi \left.\frac{\delta \dot{\Psi}^{\ell'}(\mathbf{y},t)}{\delta \Psi^\ell(\mathbf{x},t)}\right|_\Pi ,$$

$$\left.\frac{\delta H}{\delta \Pi_\ell(\mathbf{x},t)}\right|_\Psi = \dot{\Psi}^\ell(\mathbf{x},t) + \int d^3y \sum_{\ell'} \Pi_{\ell'}(\mathbf{y},t) \left.\frac{\delta \dot{\Psi}^{\ell'}(\mathbf{y},t)}{\delta \Pi^\ell(\mathbf{x},t)}\right|_\Psi$$
$$- \int d^3y \sum_{\ell'} \left.\frac{\delta L}{\delta \dot{\Psi}^{\ell'}(\mathbf{y},t)}\right|_\Psi \left.\frac{\delta \dot{\Psi}^{\ell'}(\mathbf{y},t)}{\delta \Pi^\ell(\mathbf{x},t)}\right|_\Psi ,$$

where subscripts denote the quantities held fixed in these variational derivatives. Using the defining equation (7.2.1) for Π_ℓ, this simplifies to

$$\left.\frac{\delta H}{\delta \Psi^\ell(\mathbf{x},t)}\right|_\Pi = - \left.\frac{\delta L}{\delta \Psi^\ell(\mathbf{x},t)}\right|_{\dot{\Psi}} , \tag{7.2.11}$$

and

$$\left.\frac{\delta H}{\delta \Pi_\ell(\mathbf{x},t)}\right|_\Psi = \dot{\Psi}^\ell(\mathbf{x},t)\,. \tag{7.2.12}$$

The equations of motion (7.2.2) are then equivalent to

$$\left.\frac{\delta H}{\delta \Psi^\ell(\mathbf{x},t)}\right|_\Pi = -\dot{\Pi}_\ell(\mathbf{x},t)\,. \tag{7.2.13}$$

It is tempting now to identify the generic field variables Ψ^ℓ and their conjugates Π_ℓ with the canonical variables Q^n and P_n of the previous section, and impose on them the same canonical commutation relations (7.1.30)–(7.1.32), so that Eqs. (7.2.12) and (7.2.13) are the same as the Hamiltonian equations of motion (7.1.33) and (7.1.34). This is indeed the case for the simple example of the real scalar field Φ with non-derivative coupling. Consider the Lagrangian density[†]

$$\mathscr{L} = -\frac{1}{2}\,\partial_\mu\Phi\partial^\mu\Phi - \frac{m^2}{2}\Phi^2 - \mathscr{H}(\Phi)\,, \tag{7.2.14}$$

which can be obtained by adding a real function $-\mathscr{H}(\Phi)$ of Φ to the free-field Lagrangian density found in the previous section. The Euler–Lagrange equations here are

$$(\Box - m^2)\Phi = \mathscr{H}'(\Phi)\,. \tag{7.2.15}$$

From this Lagrangian density, we calculate a canonical conjugate to Φ:

$$\Pi = \frac{\partial\mathscr{L}}{\partial\dot{\Phi}} = \dot{\Phi}\,, \tag{7.2.16}$$

which is the same as Eq. (7.1.36) if we identify Φ and Π with the canonical variables Q and P. The Hamiltonian is now given by Eq. (7.2.10) as[‡]

$$\begin{aligned} H &= \int d^3x\,(\Pi\dot{\Phi} - \mathscr{L}) \\ &= \int d^3x \left[\,\tfrac{1}{2}\Pi^2 + \tfrac{1}{2}(\nabla\Phi)^2 + \tfrac{1}{2}m^2\Phi^2 + \mathscr{H}(\Phi)\right]\,, \end{aligned} \tag{7.2.17}$$

which we recognize as the Hamiltonian (7.1.35). This little exercise should not be regarded as another derivation of this Hamiltonian, but rather as a validation of the Lagrangian (7.2.14) as a possible theory of scalar fields.

Matters are not always so simple. We have already seen in the previous section that there are field variables, such as the time component of a vector field or the Hermitian conjugate of a Dirac field, that are not canonical field variables Q^n and do not have canonical conjugates; yet Lorentz invariance dictates that these must appear in the Lagrangians for the vector and Dirac fields.

From the point of view of the Lagrangian formalism, the special character of field variables like the time component of a vector field or the Hermitian conjugate of a Dirac field arises from the fact that although they appear in the Lagrangian, their time-derivatives do not. We shall denote the field variables Ψ^ℓ whose time-derivatives do not appear in the Lagrangian as C^r; the remaining independent field variables are the

---

† We do not include a free constant factor in the term $-\frac{1}{2}\,\partial_\mu\Phi\partial^\mu\Phi$, because any such constant if positive can be absorbed into the normalization of Φ. As we shall see, a negative constant here would lead to a Hamiltonian that is not bounded below. The constant m is known as the bare mass. The most general Lagrangian that satisfies the principle of renormalizability (discussed in Chapter 12) is of this form, with $\mathscr{H}(\Phi)$ a quartic polynomial in Φ.

‡ In order for H to be interpreted as an energy, it should be bounded below. The positivity of the first two terms shows that we guessed correctly as to the sign in the first term in Eq. (7.2.14). The remaining condition is that $\frac{1}{2}m^2\Phi^2 + \mathscr{H}(\Phi)$ must be bounded below as a function of Φ.

canonical variables Q^n. The Q^n have canonical conjugates

$$P_n(\mathbf{x},t) = \frac{\delta L[Q(t),\dot{Q}(t),C(t)]}{\delta \dot{Q}^n(\mathbf{x},t)}\,, \tag{7.2.18}$$

and satisfy the commutation relations (7.1.30)–(7.1.32), but there are no canonical conjugates for the C^r. Because $\delta L/\delta \dot{C}^r = 0$, the Hamiltonian (7.2.10) is in general

$$H = \sum_n \int d^3x\, P_n \dot{Q}^n - L[Q(t),\dot{Q}(t),C(t)]\,, \tag{7.2.19}$$

but this is not yet useful until we express the C^r and $\dot{Q}^\ell$ in terms of the Qs and Ps. The equations of motion of the C^r involve only fields and their first time-derivatives

$$0 = \frac{\delta L[Q(t),\dot{Q}(t),C(t)]}{\delta C^r(\mathbf{x},t)}\,. \tag{7.2.20}$$

In the simple cases to be discussed in this chapter, these equations together with Eq. (7.2.18) can be solved to give the C^r and $\dot{Q}^\ell$ in terms of the Qs and Ps. Section 7.6 shows how in such cases one can avoid the task of actually solving for the C^r and $\dot{Q}^\ell$. In gauge theories like electrodynamics other methods must be used: either choosing a particular gauge, as in Chapter 8, or the more modern covariant methods to be discussed in Volume II.

Once we have derived a Hamiltonian as a functional of the Heisenberg picture Qs and Ps, to use perturbation theory we must make a transition to the interaction picture. The Hamiltonian is time-independent, so it can be written in terms of the P_n and Q^n at $t = 0$, which are equal to the corresponding operators p_n and q^n in the interaction picture at $t = 0$. The Hamiltonian derived in this way may then be expressed in terms of the qs and ps of the interaction picture, and split into two parts, a suitable free-particle term H_0 and an interaction V. Finally, the time-dependence equations (7.1.21) and (7.1.22) and the commutation or anticommutation relations (7.1.1)–(7.1.3) are used to express the qs and ps in $V(t)$ as linear combinations of annihilation and creation operators.

We shall present a number of examples of this procedure in Section 7.5; for the moment we will give only one example of the simplest type, the scalar field with Hamiltonian (7.2.17). We split H into a free-particle term and an interaction

$$H = H_0 + V \tag{7.2.21}$$

$$H_0 = \int d^3x \left[\tfrac{1}{2}\Pi^2 + \tfrac{1}{2}(\nabla\Phi)^2 + \tfrac{1}{2}m^2\Phi^2\right] \tag{7.2.22}$$

$$V = \int d^3x\, \mathscr{H}(\Phi)\,. \tag{7.2.23}$$

Here Φ and Π are taken at the same time t, and H is independent of t, though H_0 and V usually are not.

We now pass to the interaction representation. Taking $t = 0$ in Eqs. (7.2.22) and (7.2.23), we can simply replace Φ, Π with the interaction picture variables ϕ, π, since they are defined by Eqs. (7.1.28) and (7.1.29) to be equal at that time. To calculate the interaction $V(t)$ in the interaction picture, we apply the similarity transformation (3.5.5)

$$V(t) = \exp(iH_0t)\, V\, \exp(-iH_0t) = \int d^3x\, \mathscr{H}\Big(\phi(\mathbf{x}, t)\Big) . \tag{7.2.24}$$

The same transformation applied to H_0 leaves it constant:

$$H_0 = \exp(iH_0t)\, H_0\, \exp(-iH_0(t)) = \int d^3x \left[\tfrac{1}{2}\pi^2(\mathbf{x}, t) + \tfrac{1}{2}\Big(\nabla\phi(\mathbf{x}, t)\Big)^2 + \tfrac{1}{2}m^2\phi^2 \right] . \tag{7.2.25}$$

The relation between π and $\dot{\phi}$ is dictated by Eq. (7.1.21)

$$\dot{\phi}(\mathbf{x}, t) = \frac{\delta H_0}{\delta \pi(\mathbf{x}, t)} = \pi(\mathbf{x}, t) . \tag{7.2.26}$$

(This happens to be the same relation as in Eq. (7.2.16), but as we shall see this is not to be expected in general.) Also, the equation of motion for ϕ is dictated by Eq. (7.1.22):

$$\dot{\pi}(\mathbf{x}, t) = -\frac{\delta H_0}{\delta \phi(\mathbf{x}, t)} = +\nabla^2\phi(\mathbf{x}, t) - m^2\phi(\mathbf{x}, t) , \tag{7.2.27}$$

which together with Eq. (7.2.26) yields the field equations

$$(\Box - m^2)\phi(x) = 0 . \tag{7.2.28}$$

The general real solution may be expressed as

$$\phi(x) = (2\pi)^{-3/2} \int d^3p (2p^0)^{-1/2} \left[e^{ip\cdot x} a(\mathbf{p}) + e^{-ip\cdot x} a^\dagger(\mathbf{p}) \right] \tag{7.2.29}$$

with $p^0 = \sqrt{\mathbf{p}^2 + m^2}$ understood, and $a(\mathbf{p})$ some as-yet-unknown operator function of $\mathbf{p}$. Eq. (7.2.26) then gives the canonical conjugate as

$$\pi(x) = -i(2\pi)^{-3/2} \int d^3p\, (p^0/2)^{1/2} \left[e^{ip\cdot x} a(\mathbf{p}) - e^{-ip\cdot x} a^\dagger(\mathbf{p}) \right] . \tag{7.2.30}$$

In order to get the desired commutation relations,

$$\Big[\phi(\mathbf{x}, t),\, \pi(\mathbf{y}, t)\Big]_- = i\delta^3(\mathbf{x} - \mathbf{y}) , \tag{7.2.31}$$

$$\Big[\phi(\mathbf{x}, t),\, \phi(\mathbf{y}, t)\Big]_- = 0 , \tag{7.2.32}$$

$$\Big[\pi(\mathbf{x}, t),\, \pi(\mathbf{y}, t)\Big]_- = 0 , \tag{7.2.33}$$

we must take the as to satisfy the familiar commutation relations

$$\left[a(\mathbf{p}),\ a^\dagger(\mathbf{p}')\right] = \delta^3(\mathbf{p}-\mathbf{p}'), \tag{7.2.34}$$

$$\left[a(\mathbf{p}),\ a(\mathbf{p}')\right] = 0\,. \tag{7.2.35}$$

Also, we have already shown in the previous section that using these expansions in Eq. (7.2.25) gives the usual formula (4.2.11) for the free-particle Hamiltonian, up to an inconsequential additive constant. As remarked before, these results should not be regarded so much as an alternative derivation of Eqs. (7.2.29), (7.2.34), and (7.2.35) (which were obtained in Chapter 5 on quite other grounds) but rather as a validation of the first two terms of Eq. (7.2.14) as the correct free-particle Lagrangian for a real scalar field. We can now proceed to use perturbation theory to calculate the S-matrix, taking (7.2.24) as $V(t)$, with the field $\phi(x)$ given by Eq. (7.2.29).

The procedures illustrated here will be carried out for examples that are more complicated and more interesting in Section 7.5.

* * *

In considering the various possible Lagrangian densities for physical theories it is common to apply integration by parts, treating as equivalent any Lagrangian densities that differ only by total derivatives $\partial_\mu \mathscr{F}^\mu$. It is obvious that such total derivative terms do not contribute to the action and hence do not affect the field equations. It is also obvious that a space-derivative term $\nabla \cdot \mathscr{F}$ in the Lagrangian density does not contribute to the Lagrangian and hence does not affect the quantum theory defined by the Lagrangian.¶ What is less obvious and worth noting here is that a *time*-derivative $\partial_0 \mathscr{F}^0$ in the Lagrangian density also does not affect the quantum structure of the theory. To see this, let's first consider the effect of adding a term to the Lagrangian of the more general form

$$\Delta L(t) = \int d^3x\, D_{n,\mathbf{x}}[Q(t)]\, \dot{Q}^n(\mathbf{x},t)\,, \tag{7.2.36}$$

where D is an arbitrary n- and $\mathbf{x}$-dependent functional of the values of Q at a given time. This changes the formula for the conjugate variables $P(t)$ as functionals of $Q(t)$ and $\dot{Q}(t)$ by the amount

$$\Delta P_n(\mathbf{x},t) = \frac{\delta\, \Delta L(t)}{\delta\, \dot{Q}^n(\mathbf{x},t)} = D_{n,\mathbf{x}}[Q(t)]\,. \tag{7.2.37}$$

It follows that there is no change in the Hamiltonian as expressed as a

---

¶ This is under the usual assumption, that the fields vanish at infinity. These results do not necessarily apply when we allow fields of different topology, as discussed in Volume II.

functional of the $Q(t)$ and $\dot{Q}(t)$:

$$\int d^3x\, \Delta P_n(\mathbf{x},t)\dot{Q}^n(\mathbf{x},t) - \Delta L(t) = 0. \tag{7.2.38}$$

Hence also there is no change in the Hamiltonian as expressed as a functional of the old canonical variables Q^n and P_n. However, the Hamiltonian is *not* the same functional of the *new* canonical variables Q^n and $P_n + \Delta P_n$ as it was of the Q^n and P_n, and in a theory described by the new Lagrangian $\mathscr{L} + \Delta\mathscr{L}$ it is the new canonical variables Q^n and $P_n + \Delta P_n$ rather than the Q^n and P_n that would satisfy the canonical commutation relations. The commutators of the Q^n with each other and of the Q^n with the P_m are given by the usual canonical relations, but the commutators of the P_n with each other are now

$$\begin{aligned}
[P_n(\mathbf{x},t), P_m(\mathbf{y},t)] &= [P_n(\mathbf{x},t) + \Delta P_n(\mathbf{x},t), P_m(\mathbf{y},t) + \Delta P_m(\mathbf{y},t)] \\
&\quad -[\Delta P_n(\mathbf{x},t), P_m(\mathbf{y},t) + \Delta P_m(\mathbf{y},t)] \\
&\quad -[P_n(\mathbf{x},t) + \Delta P_n(\mathbf{x},t), \Delta P_m(\mathbf{y},t)] \\
&\quad +[\Delta P_n(\mathbf{x},t), \Delta P_m(\mathbf{y},t)] \\
&= -i\frac{\delta D_{n,\mathbf{x}}[Q(t)]}{\delta Q^m(\mathbf{y},t)} + i\frac{\delta D_{m,\mathbf{y}}[Q(t)]}{\delta Q^n(\mathbf{x},t)}\,.
\end{aligned} \tag{7.2.39}$$

In general this doesn't vanish, but if the added term in the Lagrangian is a total time-derivative

$$\Delta L = \frac{dG}{dt} = \int d^3x\, \frac{\delta G[Q(t)]}{\delta Q^n(\mathbf{x},t)}\dot{Q}^n(\mathbf{x},t)\,, \tag{7.2.40}$$

then D in Eq. (7.2.36) is of the special form

$$D_{n,\mathbf{x}}[Q] = \frac{\delta G[Q(t)]}{\delta Q^n(\mathbf{x},t)}\,. \tag{7.2.41}$$

In this case the commutator (7.2.39) vanishes, so the variables Q^n and P_n satisfy the usual commutation relations. We have seen that a change of the form (7.2.36) in the Lagrangian does not change the form of the Hamiltonian as a functional of the Q^n and P_n, and since, as we have now shown, the commutation relations of these variables are also unchanged, the addition to the Lagrangian of the term (7.2.36) has no effect on the quantum structure of the theory. Different Lagrangian densities obtained from each other by partial integration may therefore be regarded as equivalent in quantum as well as classical field theory.

7.3 Global Symmetries

We now come to the real point of the Lagrangian formalism, that it provides a natural framework for the quantum mechanical implementation of symmetry principles. This is because the dynamical equations in the Lagrangian formalism take the form of a variational principle, the principle of stationary action. Consider any infinitesimal transformation of the fields

$$\Psi^{\ell}(x) \to \Psi^{\ell}(x) + i\epsilon\,\mathscr{F}^{\ell}(x) \tag{7.3.1}$$

that leaves the action (7.2.3) invariant:

$$0 = \delta I = i\epsilon \int d^4x\, \frac{\delta I[\Psi]}{\delta \Psi^{\ell}(x)} \mathscr{F}^{\ell}(x)\,. \tag{7.3.2}$$

(With ϵ a constant, such symmetries are known as *global* symmetries. In general, $\mathscr{F}^{\ell}$ depends on the fields and their derivatives at x.) Of course Eq. (7.3.2) is automatically satisfied for *all* infinitesimal variations of the fields if the fields satisfy the dynamical equations; by an infinitesimal symmetry transformation we mean one that leaves the action invariant even when the dynamical equations are *not* satisfied. If we now consider the same transformation with ϵ an arbitrary function of position in spacetime:

$$\Psi^{\ell}(x) \to \Psi^{\ell}(x) + i\epsilon(x)\,\mathscr{F}^{\ell}(x), \tag{7.3.3}$$

then, in general, the variation of the action will not vanish, but it will have to be of the form

$$\delta I = -\int d^4x\, J^{\mu}(x) \frac{\partial \epsilon(x)}{\partial x^{\mu}} \tag{7.3.4}$$

in order that it should vanish when $\epsilon(x)$ *is* constant. If we *now* take the fields in $I[\Psi]$ to satisfy the field equations then I is stationary with respect to arbitrary field variations that vanish at large spacetime distances, including variations of the form (7.3.3), so in this case (7.3.4) should vanish. Integrating by parts, we see that $J^{\mu}(x)$ must satisfy a conservation law:

$$0 = \frac{\partial J^{\mu}(x)}{\partial x^{\mu}}\,. \tag{7.3.5}$$

It follows immediately that

$$0 = \frac{dF}{dt}, \tag{7.3.6}$$

where

$$F \equiv \int d^3x\, J^0\,. \tag{7.3.7}$$

There is one such conserved current J^{μ} and one constant of the motion F for each independent infinitesimal symmetry transformation. This represents a general feature of the canonical formalism, often referred to as Noether's theorem: *symmetries imply conservation laws.*

Many symmetry transformations leave the Lagrangian and not just the action invariant. This is the case, for instance, for translations and rotations in space and also isospin transformations and other internal symmetry transformations, though not for general Lorentz transformations. When the Lagrangian is invariant we can go further, and write an explicit formula for the conserved quantities F. Consider a field variation (7.3.3) in which $\epsilon(x)$ depends on t but not $\mathbf{x}$. In this case the variation in the action is

$$\delta I = i \int dt \int d^3x \left[\frac{\delta L[\Psi(t), \dot{\Psi}(t)]}{\delta \Psi^\ell(\mathbf{x}, t)} \epsilon(t) \mathscr{F}^\ell(\mathbf{x}, t) + \frac{\delta L[\Psi(t), \dot{\Psi}(t)]}{\delta \dot{\Psi}^\ell(\mathbf{x}, t)} \frac{d}{dt} \Big(\epsilon(t) \mathscr{F}^\ell(\mathbf{x}, t) \Big) \right] . \tag{7.3.8}$$

The requirement that the Lagrangian be invariant under this transformation when ϵ is a constant yields

$$0 = \int d^3x \left[\frac{\delta L[\Psi(t), \dot{\Psi}(t)]}{\delta \Psi^\ell(\mathbf{x}, t)} \mathscr{F}^\ell(\mathbf{x}, t) + \frac{\delta L[\Psi(t), \dot{\Psi}(t)]}{\delta \dot{\Psi}^\ell(\mathbf{x}, t)} \frac{d}{dt} \mathscr{F}^\ell(\mathbf{x}, t) \right] , \tag{7.3.9}$$

so for general fields (whether or not the field equations are satisfied) the variation in the action is

$$\delta I = i \int dt \int d^3x \, \frac{\delta L[\Psi(t), \dot{\Psi}(t)]}{\delta \dot{\Psi}^\ell(\mathbf{x}, t)} \, \dot{\epsilon}(t) \, \mathscr{F}^\ell(\mathbf{x}, t). \tag{7.3.10}$$

Comparing this with Eq. (7.3.4) gives

$$F = -i \int d^3x \frac{\delta L[\Psi(t), \dot{\Psi}(t)]}{\delta \dot{\Psi}^\ell(\mathbf{x}, t)} \mathscr{F}^\ell(\mathbf{x}, t). \tag{7.3.11}$$

Using the symmetry condition (7.3.9), the reader can easily check that this F is indeed time-independent for any fields that satisfy the dynamical equations (7.2.2).

Other symmetry transformations such as isospin rotations leave not only the action and the Lagrangian invariant but also the Lagrangian density. In such cases we can go even further, and write an explicit formula for the current $J^\mu(x)$. Writing the action as in Eq. (7.2.6) as the integral of the Lagrangian density, its variation under the transformation (7.3.3) with a general infinitesimal parameter $\epsilon(x)$ is

$$\delta I[\Psi] = i \int d^4x \left[\frac{\partial \mathscr{L}(\Psi(x), \partial_\mu \Psi(x))}{\partial \Psi^\ell(x)} \mathscr{F}^\ell(x) \epsilon(x) + \frac{\partial \mathscr{L}(\Psi(x), \partial_\mu \Psi(x))}{\partial(\partial_\mu \Psi^\ell(x))} \partial_\mu \Big(\mathscr{F}^\ell(x) \epsilon(x) \Big) \right] . \tag{7.3.12}$$

The invariance of the Lagrangian density when ϵ is a constant requires that

$$0 = \frac{\partial \mathscr{L}(\Psi(x), \partial_\mu \Psi(x))}{\partial \Psi^\ell(x)} \mathscr{F}^\ell(x) + \frac{\partial \mathscr{L}(\Psi(x), \partial_\mu \Psi(x))}{\partial(\partial_\mu \Psi^\ell(x))} \partial_\mu \mathscr{F}^\ell(x) , \quad (7.3.13)$$

so for arbitrary fields the variation of the action is

$$\delta I[\Psi] = i \int d^4x \, \frac{\partial \mathscr{L}(\Psi(x), \partial_\mu \Psi(x))}{\partial(\partial_\mu \Psi^\ell(x))} \mathscr{F}^\ell(x) \partial_\mu \epsilon(x). \quad (7.3.14)$$

Comparison with Eq. (7.3.4) shows that

$$J^\mu = -i \frac{\partial \mathscr{L}}{\partial(\partial \Psi^\ell / \partial x^\mu)} \mathscr{F}^\ell . \quad (7.3.15)$$

Using the symmetry condition (7.3.13), it is easy to see directly that $\partial_\mu J^\mu$ vanishes when the fields satisfy the Euler–Lagrange equations (7.2.9). Note also that the integral of the time component of the current (7.3.15) has the previously derived value (7.3.11).

So far everything we have said would apply to classical as well as quantum mechanical field theories. The quantum properties of the conserved quantities F are most easily seen for symmetries of the Lagrangian (not necessarily the Lagrangian density) that transform the canonical fields $Q^n(\mathbf{x}, t)$ (that is, those of the Ψ^ℓ whose time derivatives appear in the Lagrangian) into **x**-dependent functionals of themselves at the same time. For such transformations, we have

$$\mathscr{F}^n(\mathbf{x}, t) = \mathscr{F}^n[Q(t); \mathbf{x}] . \quad (7.3.16)$$

As we shall see, infinitesimal spatial translations and rotations as well as all infinitesimal internal symmetry transformations are of the form (7.3.1), (7.3.16), with $\mathscr{F}^n$ a linear functional of the Q^m, but we will not need to assume here that the symmetry is linear. For all such symmetries the operator F is not only conserved; it also acts in quantum mechanics as a *generator* of this symmetry.

To see this, note first that when Ψ^ℓ is a canonical field Q^n, the functional derivative $\delta L / \delta \dot{\Psi}^\ell$ is equal to the canonical conjugate P_n, while when Ψ^ℓ is an auxiliary field C^r, this functional derivative vanishes; hence we may rewrite Eq. (7.3.11) in the form

$$F = -i \int d^3x \, P_n(\mathbf{x}, t) \mathscr{F}^n(\mathbf{x}, t) = -i \int d^3x \, P_n(\mathbf{x}, t) \, \mathscr{F}^n[Q(t), \mathbf{x}] . \quad (7.3.17)$$

To calculate the commutator (not anticommutator) of F with a canonical field $Q^m(\mathbf{x}, t)$ at an arbitrary time t, we can invoke Eq. (7.3.6) to evaluate F as a functional of the Qs and Ps at the time t, and then use the equal-time canonical commutation relations (7.1.30)–(7.1.32) to obtain*

$$[F\,,\,Q^n(\mathbf{x},t)]_- = -\mathscr{F}^n(\mathbf{x},t)\,. \tag{7.3.18}$$

It is in this sense that F is the generator of the transformation with Eq. (7.3.16). Eq. (7.3.17) and the canonical commutation rules give also

$$[F\,,\,P_n(\mathbf{x},t)]_- = \int d^3y\,P_m(\mathbf{y},t)\,\frac{\delta\mathscr{F}^m(Q(t);\mathbf{y})}{\delta Q^n(\mathbf{x},t)}\,. \tag{7.3.19}$$

Where F^m is linear, Eq. (7.3.19) tells us that P_n transforms contragrediently to Q^n.

As a first example, consider the symmetry transformation of spacetime translation:

$$\Psi^\ell(x) \to \Psi^\ell(x+\epsilon) = \Psi^\ell(x) + \epsilon^\mu\partial_\mu\Psi^\ell(x)\,. \tag{7.3.20}$$

This is of the form (7.3.1), with four independent parameters ϵ^μ and four corresponding transformation functions

$$\mathscr{F}^\ell_{\ \mu} = -i\partial_\mu\Psi^\ell\,. \tag{7.3.21}$$

In consequence we have four independent conserved currents, conventionally grouped together in the *energy–momentum tensor* $T^\mu{}_\nu$:

$$\partial_\mu T^\mu{}_\nu = 0 \tag{7.3.22}$$

from which we can derive time-independent quantities as the spatial integrals of the time components of the translation 'currents' (not to be confused with the canonical conjugate field variables $P_n(\mathbf{x},t)$):

$$P_\nu = \int d^3x\,T^0{}_\nu\,, \tag{7.3.23}$$

$$\frac{d}{dt}\,P_\nu = 0\,. \tag{7.3.24}$$

The Lagrangian is invariant under spatial translations, so in accordance with the above general results we can conclude that the spatial components of P_ν take the form

$$\mathbf{P} \equiv -\int d^3x\,P_n(\mathbf{x},t)\nabla Q^n(\mathbf{x},t)\,. \tag{7.3.25}$$

Using the equal-time commutation relations (7.1.30)–(7.1.32), we also find the commutator of this operator with the canonical fields and

* We are here assuming that for Q^n bosonic or fermionic the variation $\mathscr{F}^n$ is also respectively bosonic or fermionic, so that F is bosonic. The only exceptions are certain symmetries known as supersymmetries, for which F is fermionic and (7.3.18) is an anticommutator if Q^n is also fermionic.

conjugates:

$$[\mathbf{P}\,,\, Q^n(\mathbf{x},t)]_- = i\nabla Q^n(\mathbf{x},t)\,, \tag{7.3.26}$$

$$[\mathbf{P}\,,\, P_n(\mathbf{x},t)]_- = i\nabla P_n(\mathbf{x},t)\,. \tag{7.3.27}$$

It follows that for any function $\mathscr{G}$ of Qs and Ps that does not also depend explicitly on $\mathbf{x}$, we have

$$\left[\mathbf{P},\, \mathscr{G}(x)\right] = i\nabla\mathscr{G}(x)\,. \tag{7.3.28}$$

These results show that the operator $\mathbf{P}$ is indeed the generator of space translations.

In contrast, time-translations do not leave the Lagrangian $L(t)$ invariant. However, we already know the generator of time-translations; it is the Hamiltonian $P^0 \equiv H$, which as we know satisfies the commutation relation

$$[H\,,\, \mathscr{G}(\mathbf{x},t)] = -i\dot{\mathscr{G}}(\mathbf{x},t) \tag{7.3.29}$$

for any function $\mathscr{G}$ of Heisenberg picture operators.

If we further assume that the Lagrangian is the integral of a Lagrangian density, then we may also obtain an explicit formula for the energy–momentum tensor $T^\mu{}_\nu$. However, the Lagrangian density $\mathscr{L}(x)$ is not invariant under spacetime translations, so we cannot use Eq. (7.3.15) here. Instead, note that the change in the action under a spacetime-dependent translation

$$\Psi^\ell(x) \to \Psi^\ell(x+\epsilon(x)) = \Psi^\ell(x) + \epsilon^\mu(x)\partial_\mu\Psi^\ell(x) \tag{7.3.30}$$

is

$$\delta I[\Psi] = \int d^4x \left(\frac{\partial\mathscr{L}}{\partial\Psi^\ell}\epsilon^\mu\partial_\mu\Psi^\ell + \frac{\partial\mathscr{L}}{\partial(\partial_\nu\Psi^\ell)}\partial_\nu[\epsilon^\mu\partial_\mu\Psi^\ell]\right)\,. \tag{7.3.31}$$

The Euler–Lagrange equations (7.2.9) show that the terms proportional to ϵ add up to $\epsilon^\mu\partial_\mu\mathscr{L}$, so

$$\delta I[\Psi] = \int d^4x \left(\frac{\partial\mathscr{L}}{\partial x^\mu}\epsilon^\mu + \frac{\partial\mathscr{L}}{\partial(\partial_\nu\Psi^\ell)}\partial_\mu\Psi^\ell\,\partial_\nu\epsilon^\mu\right)\,. \tag{7.3.32}$$

Integrating by parts, we see that this takes the form of Eq. (7.3.4)

$$\delta I = -\int d^4x\, T^\nu{}_\mu\partial_\nu\epsilon^\mu \tag{7.3.33}$$

with 'currents'

$$T^\nu{}_\mu = \delta^\nu_\mu\mathscr{L} - \frac{\partial\mathscr{L}}{\partial(\partial_\nu\Psi^\ell)}\partial_\mu\Psi^\ell\,. \tag{7.3.34}$$

As a check, we may note that the spatial components of Eq. (7.3.23) are the same as our previous formula (7.3.25) for $\mathbf{P}$, while for $\mu = 0$

Eq. (7.3.23) gives the usual formula for the Hamiltonian:

$$H \equiv -P_0 = \int d^3x \left[\sum_n P_n \dot{Q}^n - \mathscr{L}\right] . \tag{7.3.35}$$

(A warning: the tensor $T^{\mu\nu}$ obtained by raising the second index in Eq. (7.3.34) is not in general symmetric, and therefore cannot be used as the right-hand side of the field equations of general relativity. The correct energy–momentum tensor to use as the source of the gravitational field is the symmetric tensor $\Theta^{\mu\nu}$ introduced in the next section.)

In many theories there are also one or more symmetry principles that state the invariance of the action, under a set of linear coordinate-independent transformations of the canonical fields

$$Q^n(x) \to Q^n(x) + i\epsilon^a \, (t_a)^n{}_m Q^m(x) \tag{7.3.36}$$

together with a set of suitable transformations on any auxiliary fields C^r:

$$C^r(x) \to C^r(x) + i\epsilon^a \, (\tau_a)^r{}_s C^s(x) . \tag{7.3.37}$$

Here t_a and τ_a are sets of Hermitian matrices furnishing some representations of the Lie algebra of the symmetry group, and we sum over repeated group indices a, b, etc. (For instance, in electrodynamics there is such a symmetry, for which the one matrix $t^n{}_m$ is diagonal, with the charges carried by each field on the main diagonal.) From any such symmetry, we can infer the existence of another set of conserved currents J^μ_a:

$$\partial_\mu J^\mu_a = 0 , \tag{7.3.38}$$

whose time components are the densities of a set of time-independent operators

$$T_a = \int d^3x \, J^0_a . \tag{7.3.39}$$

When the Lagrangian as well as the action is invariant under the transformation (7.3.36), Eq. (7.3.11) provides an explicit formula for the T_a:

$$T_a = -i \int d^3x \, P_n(\mathbf{x}, t)(t_a)^n{}_m Q^m(\mathbf{x}, t) . \tag{7.3.40}$$

The equal-time commutation relations here give

$$\left[T_a , \, Q^n(x)\right] = -(t_a)^n{}_m Q^m(x), \tag{7.3.41}$$

$$\left[T_a , \, P_n(x)\right] = +(t_a)^m{}_n P_m(x) . \tag{7.3.42}$$

(Where t_a is diagonal, this tells us that Q^n and P_n respectively lower and raise the value of T_a by an amount equal to the nth diagonal element of t_a.) Using these results, we can calculate the commutator of T_a with another generator T_b:

$$[T_a, T_b]_- = i\int d^3x\left[-P_m(t_a)^m{}_n(t_b)^n{}_kQ^k + P_n(t_b)^n{}_k(t_a)^k{}_mQ^m\right]. \tag{7.3.43}$$

Thus, if the matrices t_a form a Lie algebra with structure constants $f_{ab}{}^c$,

$$[t_a, t_b]_- = if_{ab}{}^ct_c, \tag{7.3.44}$$

then so do the quantum operators T_a:

$$[T_a, T_b]_- = if_{ab}{}^c\,T_c. \tag{7.3.45}$$

This confirms that the quantities (7.3.40) are correctly normalized to qualify as generators of the symmetry group.

Where the Lagrangian is the integral of a Lagrangian density which is invariant under (7.3.36) and (7.3.37) we can go further, and use Eq. (7.3.15) to provide an explicit formula for the currents associated with these global symmetries:

$$J^\mu_a \equiv -i\frac{\partial\mathscr{L}}{\partial(\partial Q^n/\partial x^\mu)}(t_a)^n{}_mQ^m - i\frac{\partial\mathscr{L}}{\partial(\partial C^r/\partial x^\mu)}(\tau_a)^r{}_sC^s. \tag{7.3.46}$$

As an illustration, suppose we have *two* real scalar fields of equal mass, with Lagrangian density

$$\mathscr{L} = -\tfrac{1}{2}\partial_\mu\Phi_1\partial^\mu\Phi_1 - \tfrac{1}{2}m^2\Phi_1^2 - \tfrac{1}{2}\partial_\mu\Phi_2\partial^\mu\Phi_2 - \tfrac{1}{2}m^2\Phi_2^2 - \mathscr{H}(\Phi_1^2+\Phi_2^2). \tag{7.3.47}$$

This is invariant under a linear transformation like (7.3.36):

$$\delta\Phi_1 = -\epsilon\Phi_2, \qquad \delta\Phi_2 = +\epsilon\Phi_1,$$

so there is a conserved current (7.3.46):

$$J^\mu = \Phi_2\partial^\mu\Phi_1 - \Phi_1\partial^\mu\Phi_2.$$

The explicit formula (7.3.46) for the current can be used to derive other useful commutation relations. In particular, since the Lagrangian density does not involve time-derivatives of the auxiliary fields, we have

$$J^0_a = -iP_n(t_a)^n{}_mQ^m. \tag{7.3.48}$$

We can then derive the equal-time commutators of general fields not only with the symmetry generators T_a, but also with the densities J^0_a:

$$[J^0_a(\mathbf{x},t), Q^n(\mathbf{y},t)] = -\delta^3(\mathbf{x}-\mathbf{y})(t_a)^n{}_mQ^m(\mathbf{x},t), \tag{7.3.49}$$

$$[J^0_a(\mathbf{x},t), P_m(\mathbf{y},t)] = \delta^3(\mathbf{x}-\mathbf{y})(t_a)^n{}_mP_n(\mathbf{x},t). \tag{7.3.50}$$

If the auxiliary fields are constructed as local functions of the Ps and Qs in such a way that they transform according to a representation of the symmetry algebra with generators τ_a, then also

$$[J_a^0(\mathbf{x},t), C^r(\mathbf{y},t)] = -\delta^3(\mathbf{x}-\mathbf{y})\,(\tau_a)^r{}_s C^s(\mathbf{x},t)\,. \tag{7.3.51}$$

We often summarize Eqs. (7.3.49) and (7.3.51) in the single commutation relation

$$[J_a^0(\mathbf{x},t)\,,\ \Psi^\ell(\mathbf{y},t)] = -\delta^3(\mathbf{x}-\mathbf{y})\,(t_a)^\ell{}_{\ell'}\Psi^{\ell'}(\mathbf{x},t)\,. \tag{7.3.52}$$

Commutation relations like (7.3.49)–(7.3.51) will be used in Chapter 10 to derive relations called Ward identities for matrix elements involving the current J^μ.

7.4 Lorentz Invariance

We are now going to show that the Lorentz invariance of the Lagrangian density implies the Lorentz invariance of the S-matrix. Consider an infinitesimal Lorentz transformation

$$\Lambda^\mu{}_\nu = \delta^\mu{}_\nu + \omega^\mu{}_\nu\,, \tag{7.4.1}$$

$$\omega_{\mu\nu} = -\omega_{\nu\mu}\,. \tag{7.4.2}$$

According to the analysis of the previous section, the invariance of the action under such transformations tells us immediately that there is a set of conserved 'currents' $\mathscr{M}^{\rho\mu\nu}$:

$$\partial_\rho \mathscr{M}^{\rho\mu\nu} = 0\,, \tag{7.4.3}$$

$$\mathscr{M}^{\rho\mu\nu} = -\mathscr{M}^{\rho\nu\mu}\,, \tag{7.4.4}$$

one current for each independent component of $\omega_{\mu\nu}$. The integrals of the time-components of these 'currents' then provide us with a set of time-independent tensors:

$$J^{\mu\nu} \equiv \int d^3x\, \mathscr{M}^{0\mu\nu}\,, \tag{7.4.5}$$

$$\frac{d}{dt} J^{\mu\nu} = 0\,. \tag{7.4.6}$$

The $J^{\mu\nu}$ will turn out to be the generators of the homogeneous Lorentz group.

We would like to have an explicit formula for the tensor $\mathscr{M}^{\rho\mu\nu}$, but Lorentz transformations act on the coordinates and hence cannot leave the Lagrangian density invariant, so we cannot immediately use the results of the previous section. However, translation invariance allows us to formulate Lorentz invariance as a symmetry of the Lagrangian density under a set of transformations on the fields and field-derivatives alone. The fields undergo the matrix transformation

$$\delta\Psi^\ell = \frac{i}{2}\,\omega^{\mu\nu}(\mathscr{J}_{\mu\nu})^\ell{}_m\Psi^m\,, \tag{7.4.7}$$

where $\mathscr{J}_{\mu\nu}$ are a set of matrices satisfying the algebra of the homogeneous Lorentz group

$$[\mathscr{J}_{\mu\nu},\,\mathscr{J}_{\rho\sigma}] = i\mathscr{J}_{\rho\nu}\eta_{\mu\sigma} - i\mathscr{J}_{\sigma\nu}\eta_{\mu\rho} - i\mathscr{J}_{\mu\sigma}\eta_{\nu\rho} + i\mathscr{J}_{\mu\rho}\eta_{\nu\sigma}\,. \tag{7.4.8}$$

For example, for a scalar field ϕ we have $\delta\phi = 0$, so $\mathscr{J}_{\mu\nu} = 0$, while for an irreducible field of type (A, B) we have

$$\mathscr{J}_{ij} = \epsilon_{ijk}(\mathscr{A}_k + \mathscr{B}_k)\,, \qquad \mathscr{J}_{k0} = -i(\mathscr{A}_k - \mathscr{B}_k)\,,$$

where $\mathscr{A}$ and $\mathscr{B}$ are spin matrices for spin A and B, respectively. We specially note that for a covariant vector field, $\delta V_\kappa = \omega_\kappa{}^\lambda V_\lambda$, so here

$$(\mathscr{J}_{\rho\sigma})_\kappa{}^\lambda = -i\eta_{\rho\kappa}\delta_\sigma{}^\lambda + i\eta_{\sigma\kappa}\delta_\rho{}^\lambda\,.$$

The derivative of a field that transforms as in Eq. (7.4.7) transforms like another such field, but with an extra vector index

$$\delta(\partial_\kappa\Psi_\ell) = \tfrac{1}{2}i\,\omega^{\mu\nu}(\mathscr{J}_{\mu\nu})_\ell{}^m\partial_\kappa\Psi_m + \omega_\kappa{}^\lambda\partial_\lambda\Psi_\ell\,. \tag{7.4.9}$$

The Lagrangian density is assumed to be invariant under the combined transformations Eqs. (7.4.7) and (7.4.9), so

$$0 = \frac{\partial\mathscr{L}}{\partial\Psi^\ell}\,\frac{i}{2}\,\omega^{\mu\nu}(\mathscr{J}_{\mu\nu})^\ell{}_m\Psi^m + \frac{\partial\mathscr{L}}{\partial(\partial_\kappa\Psi^\ell)}\,\frac{i}{2}\,\omega^{\mu\nu}(\mathscr{J}_{\mu\nu})^\ell{}_m\partial_\kappa\Psi^m$$
$$+ \frac{\partial\mathscr{L}}{\partial(\partial_\kappa\Psi^\ell)}\,\omega_\kappa{}^\lambda\partial_\lambda\Psi^\ell\,.$$

Setting the coefficient of $\omega^{\mu\nu}$ equal to zero gives

$$0 = \frac{i}{2}\,\frac{\partial\mathscr{L}}{\partial\Psi^\ell}\,(\mathscr{J}_{\mu\nu})^\ell{}_m\Psi^m + \frac{i}{2}\,\frac{\partial\mathscr{L}}{\partial(\partial_\kappa\Psi^\ell)}\,(\mathscr{J}_{\mu\nu})^\ell{}_m\partial_\kappa\Psi^m$$
$$+ \frac{1}{2}\,\frac{\partial\mathscr{L}}{\partial(\partial_\kappa\Psi^\ell)}\,(\eta_{\kappa\mu}\partial_\nu - \eta_{\kappa\nu}\partial_\mu)\Psi^\ell\,.$$

Using the Euler–Lagrange equations (7.2.9), and our formula (7.3.34) for the energy-momentum tensor $T_{\mu\nu}$, we may write this as

$$0 = \partial_\kappa\left[\frac{i}{2}\,\frac{\partial\mathscr{L}}{\partial(\partial_\kappa\Psi^\ell)}\,(\mathscr{J}_{\mu\nu})^\ell{}_m\Psi^m\right] - \frac{1}{2}(T_{\mu\nu} - T_{\nu\mu})\,. \tag{7.4.10}$$

This immediately suggests the definition of a new energy–momentum tensor, known as the *Belinfante tensor*:[2]

$$\Theta^{\mu\nu} = T^{\mu\nu} - \frac{i}{2}\partial_\kappa\left[\frac{\partial\mathscr{L}}{\partial(\partial_\kappa\Psi^\ell)}(\mathscr{J}^{\mu\nu})^\ell{}_m\Psi^m\right.$$
$$\left.-\frac{\partial\mathscr{L}}{\partial(\partial_\mu\Psi^\ell)}(\mathscr{J}^{\kappa\nu})^\ell{}_m\Psi^m - \frac{\partial\mathscr{L}}{\partial(\partial_\nu\Psi^\ell)}(\mathscr{J}^{\kappa\mu})^\ell{}_m\Psi^m\right] . \quad (7.4.11)$$

The quantity in square brackets is manifestly antisymmetric in μ and κ, so $\Theta^{\mu\nu}$ satisfies the same conservation law as $T^{\mu\nu}$,

$$\partial_\mu\Theta^{\mu\nu} = 0 . \quad (7.4.12)$$

For the same reason, when we set $\mu = 0$ in Eq. (7.4.11) the index κ runs over space components only, so the derivative term here drops out when we integrate over all space

$$\int \Theta^{0\nu} d^3x = \int T^{0\nu} d^3x = P^\nu , \quad (7.4.13)$$

where $P^0 \equiv H$. Thus $\Theta^{\mu\nu}$ can be regarded as *the* energy–momentum tensor, just as well as $T^{\mu\nu}$. However, Eq. (7.4.10) tells us that, unlike $T^{\mu\nu}$, the Belinfante tensor $\Theta^{\mu\nu}$ is not only conserved but also *symmetric*:

$$\Theta^{\mu\nu} = \Theta^{\nu\mu} . \quad (7.4.14)$$

It is $\Theta^{\mu\nu}$ rather than $T^{\mu\nu}$ that acts as a source of the gravitational field.[3] In consequence of the symmetry of $\Theta^{\mu\nu}$, we may construct one more conserved tensor density:

$$\mathscr{M}^{\lambda\mu\nu} \equiv x^\mu\Theta^{\lambda\nu} - x^\nu\Theta^{\lambda\mu} . \quad (7.4.15)$$

This is conserved, in the sense that

$$\partial_\lambda\mathscr{M}^{\lambda\mu\nu} = \Theta^{\mu\nu} - \Theta^{\nu\mu} = 0 . \quad (7.4.16)$$

Thus Lorentz invariance allows us to define one more time-independent tensor

$$J^{\mu\nu} = \int \mathscr{M}^{0\mu\nu} d^3x = \int d^3x(x^\mu\Theta^{0\nu} - x^\nu\Theta^{0\mu}) . \quad (7.4.17)$$

The rotation generator $J_k = \epsilon_{ijk}J^{ij}/2$ is not only time-independent, but also has no *explicit* time-dependence, so it commutes with the Hamiltonian

$$[H, \mathbf{J}] = 0 . \quad (7.4.18)$$

Also, applying Eq. (7.3.28) to the function $\Theta^{0\nu}$, we have

$$[P_j, J_i] = \frac{1}{2}\epsilon_{i\ell k}[P_j, J^{\ell k}] = \frac{i}{2}\epsilon_{i\ell k}\int d^3x\left(x^\ell\frac{\partial}{\partial x^j}\Theta^{0k} - x^k\frac{\partial}{\partial x^j}\Theta^{0\ell}\right)$$
$$= -i\,\epsilon_{ijk}\int d^3x\,\Theta^{0k}$$

and therefore

$$[P_j, J_i] = -i\,\epsilon_{ijk}\,P_k\,. \tag{7.4.19}$$

On the other hand, the 'boost' generator $K_k \equiv J^{k0}$, though time-independent, does explicitly involve the time coordinate

$$K_k = \int d^3x(x^k\Theta^{00} - x^0\Theta^{0k})\,,$$

or more explicitly

$$\mathbf{K} = -t\mathbf{P} + \int d^3x\,\mathbf{x}\,\Theta^{00}(\mathbf{x},t)\,. \tag{7.4.20}$$

Since this is a constant, we have $0 = \dot{\mathbf{K}} = -\mathbf{P} + i[H,\mathbf{K}]$, and therefore

$$[H,\mathbf{K}] = -i\,\mathbf{P}\,. \tag{7.4.21}$$

Also, applying Eq. (7.3.28) again gives

$$[P_j, K_k] = i\int d^3x\,x^k\,\frac{\partial}{\partial x^j}\,\Theta^{00} = -i\delta_{jk}\int d^3x\,\Theta^{00}$$

and therefore

$$[P_j, K_k] = -i\,\delta_{jk}\,H\,. \tag{7.4.22}$$

For any reasonable Lagrangian density, the operator (7.4.20) will be 'smooth' in the sense used in Section 3.3, i.e., the interaction terms in $e^{iH_0t}\int d^3x\,\mathbf{x}\,\Theta^{00}(\mathbf{x},0)e^{-iH_0t}$ vanish* for $t \to \pm\infty$. (Note that the interaction terms in $e^{iH_0t}\int d^3x\;\Theta^{00}(\mathbf{x},0)e^{-iH_0t}$ must vanish for $t \to \pm\infty$ in order to allow the introduction of 'in' and 'out' states and the S-matrix.) With this smoothness assumption and the commutation relation (7.4.21) in hand, we can repeat the arguments of Section 3.3, and conclude that the S-matrix is Lorentz-invariant.

* * *

The same arguments were also used in Section 3.3 to verify that the remaining commutation relations of the Lorentz group, those of the $J^{\mu\nu}$ with each other, take the proper form. This can also be shown directly for the commutators of the rotation generators, which here take the form

$$J^{ij} = \int d^3x\,\frac{\partial\mathscr{L}}{\partial\dot\Psi^\ell}\left(-x^i\partial_j\Psi^\ell + x^j\partial_i\Psi^\ell - i\left(\mathscr{J}^{ij}\right)^\ell{}_m\Psi^m\right)\,. \tag{7.4.23}$$

Since the Lagrangian density does not depend on the time-derivatives of the auxiliary fields, and the rotation generators do not mix canonical and

* When we say that some interaction-picture operator vanishes for $t \to \pm\infty$, we mean that its matrix elements between states that are smooth superpositions of energy eigenstates vanish in this limit.

auxiliary fields, this can also be written as a sum over canonical fields alone:

$$J^{ij} = \int d^3x \, P_n \left(-x^i \partial_j Q^n + x^j \partial_i Q^n - i \left(\mathscr{J}^{ij} \right)^n{}_{n'} Q^{n'} \right) . \tag{7.4.24}$$

It follows immediately then from the canonical commutation relations that

$$[J^{ij}, Q^n(x)]_- = -i(-x_i \partial_j + x_j \partial_i) Q^n(x) - (\mathscr{J}^{ij})^n{}_{n'} Q^{n'}(x) , \tag{7.4.25}$$

$$[J^{ij}, P_n(x)]_- = i(-x_i \partial_j + x_j \partial_i) P^n(x) + (\mathscr{J}^{ij})^{n'}{}_n P_{n'}(x) . \tag{7.4.26}$$

These results can be used to derive the usual commutation relations of the J^{ij} with each other and other generators.** If there are no auxiliary fields then the same arguments may be applied to the 'boost' generators to complete the demonstration that the P^μ and $J^{\mu\nu}$ satisfy the commutation relations of the inhomogeneous Lorentz group. However the 'boost' matrices $\mathscr{J}^{i0}$ will in general mix canonical and auxiliary fields (such as the components V^i and V^0 of a vector field), so the direct proof of the commutation relations of the J^{i0} with each other has to be given on a case by case basis. Fortunately, this is not needed for the proof of the Lorentz invariance of the S-matrix given in Section 3.3.

7.5 Transition to Interaction Picture: Examples

At the end of Section 7.2 we showed how to use the Lagrangian of a simple scalar field theory to derive the structure of the interaction and the free fields it contains in the interaction picture. We will now turn to somewhat more complicated and revealing examples.

Scalar Field, Derivative Coupling

First let's consider a neutral scalar field, but now with derivative coupling. We take the Lagrangian as

$$\mathscr{L} = -\tfrac{1}{2} \partial_\mu \Phi \partial^\mu \Phi - \tfrac{1}{2} m^2 \Phi^2 - J^\mu \partial_\mu \Phi - \mathscr{H}(\Phi) , \tag{7.5.1}$$

where J^μ is either a c-number external current (unrelated to currents J^μ introduced earlier), or a functional of various fields other than Φ (in which case terms involving these other fields need to be added to (7.5.1)). The

** Also, since J^{ij} commutes with H and $P_n \dot{Q}^n$, it commutes with L. The commutator of J^{ij} with the auxiliary fields must thus be consistent with the rotational invariance of L.

canonical conjugate to Φ is now

$$\Pi = \frac{\partial \mathscr{L}}{\partial \dot{\Phi}} = \dot{\Phi} - J^0 \tag{7.5.2}$$

and the Hamiltonian is

$$H = \int d^3x [\Pi\dot{\Phi} - \mathscr{L}]$$
$$= \int d^3x \Big[\Pi(\Pi + J^0) + \tfrac{1}{2}(\nabla\Phi)^2 - \tfrac{1}{2}(\Pi + J^0)^2$$
$$+ \tfrac{1}{2}m^2\Phi^2 + \mathbf{J}\cdot\nabla\Phi + J^0(\Pi + J^0) + \mathscr{H}(\Phi)\Big] .$$

Collecting terms, we can write this as

$$H = H_0 + V , \tag{7.5.3}$$

$$H_0 = \int d^3x \Big[\tfrac{1}{2}\Pi^2 + \tfrac{1}{2}(\nabla\Phi)^2 + \tfrac{1}{2}m^2\Phi^2 \Big] , \tag{7.5.4}$$

$$V = \int d^3x \Big[\Pi J^0 + \mathbf{J}\cdot\nabla\Phi + \tfrac{1}{2}(J^0)^2 + \mathscr{H}(\Phi)\Big] . \tag{7.5.5}$$

As explained in Section 7.2, we can pass to the interaction picture by simply replacing Π and Φ with π and ϕ (and likewise for any fields in the current J^μ, though we will not bother to indicate this explicitly):

$$H_0 = \int d^3x \left[\tfrac{1}{2}\pi^2(\mathbf{x},t) + \tfrac{1}{2}\Big(\nabla\phi(\mathbf{x},t)\Big)^2 + \tfrac{1}{2}m^2\phi^2(\mathbf{x},t) \right] , \tag{7.5.6}$$

$$V(t) = \int d^3x \Big[\pi(\mathbf{x},t)J^0(\mathbf{x},t) + \mathbf{J}(\mathbf{x},t)\cdot\nabla\phi(\mathbf{x},t)$$
$$+ \tfrac{1}{2}\Big[J^0(\mathbf{x},t)\Big]^2 + \mathscr{H}(\phi(\mathbf{x},t))\Big] . \tag{7.5.7}$$

The free-particle Hamiltonian is just the same as Eq. (7.2.25), and leads as in Section 7.2 to Eqs. (7.2.26)–(7.2.35). Indeed, whatever the total Hamiltonian may be, we *must* take (7.5.6) as the part we split off and call the free-particle part, with the remainder called the interaction, because as we have seen it is this form of the free-particle Hamiltonian that leads to the correct expansion (7.2.29) of the scalar field in terms of creation and annihilation operators that satisfy the commutation relations (7.2.34), (7.2.35). The last step is to replace π in the interaction Hamiltonian with its value $\dot{\phi}$ in the interaction picture (*not* its value $\dot{\phi} - J^0$ in the Heisenberg picture):

$$V(t) = \int d^3x \left[J^\mu(\mathbf{x},t)\partial_\mu\phi(\mathbf{x},t) + \tfrac{1}{2}\Big[J^0(\mathbf{x},t)\Big]^2 + \mathscr{H}(\phi(\mathbf{x},t))\right] . \tag{7.5.8}$$

The extra non-invariant term in Eq. (7.5.8) is just what we saw in Section 6.2 is needed to cancel a non-invariant term in the propagator of $\partial\phi$.

Vector Field, Spin One

Similar results are obtained in the canonical quantization of the vector field V_μ for a particle of spin one. Let's here keep an open mind, and write the Lagrangian density in a fairly general form

$$\mathscr{L} = -\tfrac{1}{2}\alpha\,\partial_\mu V_\nu\,\partial^\mu V^\nu - \tfrac{1}{2}\beta\,\partial_\mu V_\nu\,\partial^\nu V^\mu - \tfrac{1}{2}m^2\,V_\mu V^\mu + J_\mu V^\mu\,, \tag{7.5.9}$$

where α, β, and m^2 are so far arbitrary constants, and J_μ is either a c-number external current, or an operator depending on fields other than V^μ, in which case additional terms involving these fields must be added to $\mathscr{L}$. The Euler–Lagrange field equations for V_μ read

$$-\alpha\Box V_\nu - \beta\,\partial_\nu(\partial_\mu V^\mu) + m^2 V_\nu = J_\nu\,.$$

Taking the divergence gives

$$-(\alpha+\beta)\Box\partial_\lambda V^\lambda + m^2\partial_\lambda V^\lambda = \partial_\lambda J^\lambda\,. \tag{7.5.10}$$

This is the equation for an ordinary scalar field with mass $m^2/(\alpha+\beta)$ and source $\partial_\lambda J^\lambda/(\alpha+\beta)$. We want to describe a theory containing only particles of spin one, not spin zero, so to avoid the appearance of $\partial_\lambda V^\lambda$ as an independently propagating scalar field, we take $\alpha = -\beta$, in which case $\partial_\lambda V^\lambda$ can be expressed in terms of an external current or other fields, as $\partial_\lambda J^\lambda/m^2$. The constant α can be absorbed in the definition of V_μ, so we can take $\alpha = -\beta = 1$, and therefore

$$\mathscr{L} = -\tfrac{1}{4}F_{\mu\nu}F^{\mu\nu} - \tfrac{1}{2}m^2\,V_\mu V^\mu + J_\mu V^\mu\,, \tag{7.5.11}$$

where

$$F_{\mu\nu} \equiv \partial_\mu V_\nu - \partial_\nu V_\mu\,. \tag{7.5.12}$$

The derivative of the Lagrangian with respect to the time-derivative of the vector field is

$$\frac{\partial\mathscr{L}}{\partial\dot{V}_\mu} = -F^{0\mu}\,. \tag{7.5.13}$$

This is non-vanishing for μ a spatial index i, so the V^i are canonical fields, with conjugates

$$\Pi^i = F^{i0} = \dot{V}^i + \partial_i V^0\,. \tag{7.5.14}$$

On the other hand $F^{00} = 0$, so $\dot{V}^0$ does not appear in the Lagrangian, and V^0 is therefore an auxiliary field. This causes no serious difficulty: the fact that $\partial\mathscr{L}/\partial\dot{V}^0$ vanishes means that the field equation for V^0 involves no second time-derivatives, and can therefore be used as a constraint that eliminates a field variable. Specifically, the Euler–Lagrange equation for $\nu = 0$ is

$$\partial_i F^{i0} = m^2 V^0 - J^0 \tag{7.5.15}$$

or using Eq. (7.5.14)

$$V^0 = \frac{1}{m^2}(\nabla \cdot \boldsymbol{\Pi} + J^0)\,. \tag{7.5.16}$$

Now let us calculate the Hamiltonian $H = \int d^3x\,(\boldsymbol{\Pi}\cdot\dot{\mathbf{V}} - \mathscr{L})$ for this theory. Eq. (7.5.14) allows us to write $\dot{\mathbf{V}}$ in terms of $\boldsymbol{\Pi}$ and J^0:

$$\dot{\mathbf{V}} = -\nabla V^0 + \boldsymbol{\Pi} = \boldsymbol{\Pi} - \frac{1}{m^2}\nabla(\nabla\cdot\boldsymbol{\Pi} + J^0),$$

so

$$\begin{aligned} H = \int d^3x \Big[& \boldsymbol{\Pi}^2 + m^{-2}(\nabla\cdot\boldsymbol{\Pi})(\nabla\cdot\boldsymbol{\Pi} + J^0) \\ & - \tfrac{1}{2}\boldsymbol{\Pi}^2 + \tfrac{1}{2}(\nabla\times\mathbf{V})^2 + \tfrac{1}{2}m^2\mathbf{V}^2 \\ & - \tfrac{1}{2}m^{-2}(\nabla\cdot\boldsymbol{\Pi} + J^0)^2 - \mathbf{J}\cdot\mathbf{V} + m^{-2}J^0(\nabla\cdot\boldsymbol{\Pi} + J^0)\Big]\,. \end{aligned}$$

Again, we split this up into a free-particle term H_0 and interaction V:

$$H = H_0 + V\,, \tag{7.5.17}$$

and pass to the interaction picture by replacing the Heisenberg-picture quantities $\mathbf{V}$ and $\boldsymbol{\Pi}$ with their interaction-picture counterparts $\mathbf{v}$ and $\boldsymbol{\pi}$ (and, though not shown explicitly, likewise for whatever fields and conjugates are present in J^μ):

$$H_0 = \int d^3x \left[\frac{1}{2}\boldsymbol{\pi}^2 + \frac{1}{2m^2}(\nabla\cdot\boldsymbol{\pi})^2 + \frac{1}{2}(\nabla\times\mathbf{v})^2 + \frac{m^2}{2}\mathbf{v}^2\right], \tag{7.5.18}$$

$$V = \int d^3x \left[-\mathbf{J}\cdot\mathbf{v} + m^{-2}J^0\nabla\cdot\boldsymbol{\pi} + \frac{1}{2m^2}(J^0)^2\right]. \tag{7.5.19}$$

The relation between $\boldsymbol{\pi}$ and $\mathbf{v}$ is then

$$\dot{\mathbf{v}} = \frac{\delta H_0(\mathbf{v},\boldsymbol{\pi})}{\delta\boldsymbol{\pi}} = \boldsymbol{\pi} - m^{-2}\nabla(\nabla\cdot\boldsymbol{\pi}) \tag{7.5.20}$$

and the 'field equation' is

$$\dot{\boldsymbol{\pi}} = -\frac{\delta H_0(\mathbf{v},\boldsymbol{\pi})}{\delta\mathbf{v}} = +\nabla^2\mathbf{v} - \nabla(\nabla\cdot\mathbf{v}) - m^2\mathbf{v}\,. \tag{7.5.21}$$

Since V^0 is not an independent field variable, it is not related by a similarity transformation to any interaction-picture object v^0. Instead, we can *invent* a quantity

$$v^0 \equiv m^{-2}\nabla\cdot\boldsymbol{\pi}\,. \tag{7.5.22}$$

Eq. (7.5.20) then allows us to write π as

$$\pi = \dot{\mathbf{v}} + \nabla v^0 \,. \tag{7.5.23}$$

Inserting this in Eqs. (7.5.22) and (7.5.21) gives our field equations in the form

$$\nabla^2 v^0 + \nabla \cdot \dot{\mathbf{v}} - m^2 v^0 = 0 \,,$$
$$\nabla^2 \mathbf{v} - \nabla(\nabla \cdot \mathbf{v}) - \ddot{\mathbf{v}} - \nabla \dot{v}^0 - m^2 \mathbf{v} = 0 \,.$$

These can be combined in the covariant form

$$\Box\, v^\mu - \partial^\mu \partial_\nu v^\nu - m^2 v^\mu = 0 \,. \tag{7.5.24}$$

Taking the divergence gives

$$\partial_\mu v^\mu = 0 \tag{7.5.25}$$

and hence

$$(\Box - m^2)\, v^\mu = 0 \,. \tag{7.5.26}$$

A real vector field satisfying Eqs. (7.5.25) and (7.5.26) can be expressed as a Fourier transform

$$v^\mu(x) = (2\pi)^{-3/2} \sum_\sigma \int d^3p\, (2p^0)^{-1/2} \Big\{ e^\mu(\mathbf{p},\sigma) a(\mathbf{p},\sigma) e^{ip\cdot x} + e^{\mu *}(\mathbf{p},\sigma) a^\dagger(\mathbf{p},\sigma)\, e^{-ip\cdot x} \Big\} \,, \tag{7.5.27}$$

where $p^0 = \sqrt{\mathbf{p}^2 + m^2}$; the $e^\mu(\mathbf{p},\sigma)$ for $\sigma = +1, 0, -1$ are three independent vectors satisfying

$$p_\mu e^\mu(\mathbf{p},\sigma) = 0 \tag{7.5.28}$$

and normalized so that

$$\sum_\sigma e^\mu(\mathbf{p},\sigma) e^{\nu *}(\mathbf{p},\sigma) = \eta^{\mu\nu} + p^\mu p^\nu / m^2 \;\; ; \tag{7.5.29}$$

and the $a(\mathbf{p},\sigma)$ are operator coefficients. It is straightforward using Eqs. (7.5.23), (7.5.27), and (7.5.29) to calculate that $\mathbf{v}$ and π satisfy the correct commutation relations

$$\Big[v^i(\mathbf{x},t),\, \pi^j(\mathbf{y},t) \Big] = i\,\delta_{ij}\, \delta^3(\mathbf{x}-\mathbf{y}) \,,$$
$$\Big[v^i(\mathbf{x},t),\, v^j(\mathbf{y},t) \Big] = \Big[\pi^i(\mathbf{x},t),\, \pi^j(\mathbf{y},t) \Big] = 0 \,, \tag{7.5.30}$$

provided that $a(\mathbf{p},\sigma)$ and $a^\dagger(\mathbf{p},\sigma)$ satisfy the commutation relations

$$\Big[a(\mathbf{p},\sigma),\, a^\dagger(\mathbf{p}',\sigma') \Big] = \delta^3(\mathbf{p}'-\mathbf{p}) \delta_{\sigma'\sigma} \,, \tag{7.5.31}$$

$$\Big[a(\mathbf{p},\sigma),\, a(\mathbf{p}',\sigma') \Big] = 0 \,. \tag{7.5.32}$$

We already know that the vector field for a spin one particle must take the form (7.5.27), so our derivation of these results serves to verify that Eq. (7.5.18) gives the correct free-particle Hamiltonian for a massive particle of spin one. It is easy to check also that Eq. (7.5.18) may be written (up to a constant term) in the standard form of a free-particle energy, as $\sum_\sigma \int d^3p\, p^0\, a^\dagger(\mathbf{p},\sigma)a(\mathbf{p},\sigma)$. Finally, using Eq. (7.5.22) in Eq. (7.5.19) yields the interaction in the interaction picture

$$V(t) = \int d^3x \left[-J_\mu v^\mu + \frac{1}{2m^2}(J^0)^2 \right] . \tag{7.5.33}$$

The extra non-invariant term in Eq. (7.5.33) is just what we found in Chapter 6 is needed to cancel a non-invariant term in the propagator of the vector field.

Dirac field, Spin One Half

For the Dirac field of a particle of spin 1/2, we tentatively take the Lagrangian as

$$\mathscr{L} = -\bar{\Psi}(\gamma^\mu \partial_\mu + m)\Psi - \mathscr{H}(\bar{\Psi}, \Psi) \tag{7.5.34}$$

with $\mathscr{H}$ a real function of $\bar{\Psi}$ and Ψ. This is not real, but the action is, because

$$\bar{\Psi}\gamma^\mu\partial_\mu\Psi - (\bar{\Psi}\gamma^\mu\partial_\mu\Psi)^\dagger = \bar{\Psi}\gamma^\mu\partial_\mu\Psi + (\partial_\mu\bar{\Psi})\gamma^\mu\Psi = \partial_\mu(\bar{\Psi}\gamma^\mu\Psi) .$$

Hence the field equations obtained by requiring the action to be stationary with respect to $\bar{\Psi}$ are the adjoints of those obtained by requiring the action to be stationary with respect to Ψ, as necessary if we are to avoid having too many field equations. The canonical conjugate to Ψ is

$$\Pi = \frac{\partial \mathscr{L}}{\partial \dot{\Psi}} = -\bar{\Psi}\gamma^0 , \tag{7.5.35}$$

so we should not regard $\bar{\Psi}$ as a field like Ψ, but rather as proportional to the canonical conjugate of Ψ. The Hamiltonian is

$$H = \int d^3x [\Pi\dot{\Psi} - \mathscr{L}] = \int d^3x \left[\Pi\gamma^0[\boldsymbol{\gamma}\cdot\nabla + m]\Psi + \mathscr{H} \right] .$$

We write this as

$$H = H_0 + V , \tag{7.5.36}$$

where

$$H_0 = \int d^3x\, \Pi\gamma^0[\boldsymbol{\gamma}\cdot\nabla + m]\Psi , \tag{7.5.37}$$

$$V = \int d^3x\, \mathscr{H}(\bar{\Psi}, \Psi) . \tag{7.5.38}$$

We now pass to the interaction picture. Since Eq. (7.5.35) does not involve the time, the similarity transformation (7.1.28), (7.1.29) yields immediately

$$\pi = -\bar{\psi}\gamma^0 . \tag{7.5.39}$$

Likewise, H_0 and $V(t)$ can be calculated by replacing Ψ and Π with ψ and π in Eqs. (7.5.37) and (7.5.38). This gives the equation of motion

$$\dot{\psi} = \frac{\delta H_0}{\delta \pi} = \gamma^0(\boldsymbol{\gamma}\cdot\nabla + m)\psi \tag{7.5.40}$$

or more neatly

$$(\gamma^\mu \partial_\mu + m)\psi = 0 . \tag{7.5.41}$$

(The other equation of motion, $\dot{\pi} = -\delta H_0 / \delta\psi$, yields just the adjoint of this one.) Any field satisfying Eq. (7.5.41) can be written as a Fourier transform

$$\psi(x) = (2\pi)^{-3/2}\int d^3p \sum_\sigma \left\{u(\mathbf{p},\sigma)\, e^{ip\cdot x} a(\mathbf{p},\sigma) + \ v(\mathbf{p},\sigma) e^{-ip\cdot x} b^\dagger(\mathbf{p},\sigma)\right\} , \tag{7.5.42}$$

where $p^0 \equiv \sqrt{\mathbf{p}^2 + m^2}$; $a(\mathbf{p},\sigma)$ and $b^\dagger(\mathbf{p},\sigma)$ are operator coefficients; and $u(\mathbf{p}, \pm\frac{1}{2})$ are the two independent solutions of

$$(i\gamma^\mu p_\mu + m)\, u(\mathbf{p},\sigma) = 0 \tag{7.5.43}$$

and likewise

$$(-i\gamma^\mu p_\mu + m)\, v(\mathbf{p},\sigma) = 0 \tag{7.5.44}$$

normalized so that*

$$\sum_\sigma u(\mathbf{p},\sigma)\bar{u}(\mathbf{p},\sigma) = \frac{(-i\gamma^\mu p_\mu + m)}{2p^0} , \tag{7.5.45}$$

$$\sum_\sigma v(\mathbf{p},\sigma)\bar{v}(\mathbf{p},\sigma) = -\frac{(i\gamma^\mu p_\mu + m)}{2p^0} . \tag{7.5.46}$$

In order to obtain the desired anticommutators

$$\Big[\psi_\alpha(\mathbf{x},t),\, \bar{\psi}_\beta(\mathbf{y},t)\Big]_+ = \Big[\psi_\alpha(\mathbf{x},t),\, \pi_\gamma(\mathbf{y},t)\Big]_+ (\gamma^0)_{\gamma\beta}$$
$$= i\,(\gamma^0)_{\alpha\beta}\delta^3(\mathbf{x}-\mathbf{y}) , \tag{7.5.47}$$

$$\Big[\psi_\alpha(\mathbf{x},t),\, \psi_\beta(\mathbf{y},t)\Big]_+ = 0 , \tag{7.5.48}$$

* The matrix $i\gamma^\mu p_\mu$ has eigenvalues $\pm m$, so $\Sigma u\bar{u}$ and $\Sigma v\bar{v}$ must be proportional to the projection matrices $(-i\gamma^\mu p_\mu + m)/2m$ and $(i\gamma^\mu p_\mu + m)/2m$, respectively. The proportionality factor may be adjusted up to a sign by absorbing it in the definition of u and v. The overall sign is determined by positivity: Tr $\Sigma u\bar{u}\beta = \Sigma u^\dagger u$ and Tr $\Sigma v\bar{v}\beta = \Sigma v^\dagger v$ must be positive.

we must adopt the anticommutation relations

$$\Big[a(\mathbf{p},\sigma),\, a^\dagger(\mathbf{p}',\sigma')\Big]_+ = \Big[b(\mathbf{p},\sigma),\, b^\dagger(\mathbf{p}',\sigma')\Big]_+ = \delta^3(\mathbf{p}'-\mathbf{p})\delta_{\sigma'\sigma}\,, \qquad (7.5.49)$$

$$\Big[a(\mathbf{p},\sigma),\, a(\mathbf{p}',\sigma')\Big]_+ = \Big[b(\mathbf{p},\sigma),\, b(\mathbf{p}',\sigma')\Big]_+ =$$

$$\Big[a(\mathbf{p},\sigma),\, b(\mathbf{p}',\sigma')\Big]_+ = \Big[a(\mathbf{p},\sigma),\, b^\dagger(\mathbf{p}',\sigma')\Big]_+ = 0\,, \qquad (7.5.50)$$

and their adjoints. These agree with the results obtained in Chapter 5, thus verifying that (7.5.37) is the correct free-particle Hamiltonian for spin $\frac{1}{2}$. In terms of the as and bs, this Hamiltonian is

$$H_0 = \sum_\sigma \int d^3p\; p^0\Big(a^\dagger(\mathbf{p},\sigma)a(\mathbf{p},\sigma) - b(\mathbf{p},\sigma)b^\dagger(\mathbf{p},\sigma)\Big)\,. \qquad (7.5.51)$$

We can rewrite this as a more conventional free-particle Hamiltonian, plus another infinite c-number**

$$H_0 = \sum_\sigma \int d^3\mathbf{p}\; p^0\Big[a^\dagger(\mathbf{p},\sigma)a(\mathbf{p},\sigma) + b^\dagger(\mathbf{p},\sigma)b(\mathbf{p},\sigma) - \delta^3(\mathbf{p}-\mathbf{p})\Big]\,. \qquad (7.5.52)$$

The c-number term in Eq. (7.5.52) is only important if we worry about gravitational phenomena; otherwise here, as for the scalar field, we can throw it away, since it only affects the zero of energy with respect to which all energies are measured. With this understanding, H_0 is a positive operator, just as for bosons.

7.6 Constraints and Dirac Brackets

The chief obstacle to deriving the Hamiltonian from the Lagrangian is the occurrence of constraints. The standard analysis of this problem is that of Dirac,[5] whose terminology we will follow here. Dirac's analysis is not really needed for the simple theories discussed in this chapter, where it is easy to identify the unconstrained canonical variables. We shall use the theory of a real massive vector field for illustration here, returning to Dirac's approach in the next chapter, where it will be actually useful.

Primary constraints are either imposed on the system (as when in the next chapter we choose a gauge for the electromagnetic field) or arise from the structure of the Lagrangian itself. For an example of the latter type, consider the Lagrangian (7.5.11) of a massive vector field V^μ interacting

** Note the negative sign of the c-number term. The conjectured symmetry known as *supersymmetry*[4] connects the numbers of boson and fermion fields, in such a way that the c-numbers in H_0 all cancel.

with a current J_μ:

$$\mathscr{L} = -\tfrac{1}{4} F_{\mu\nu}F^{\mu\nu} - \tfrac{1}{2}m^2 V_\mu V^\mu + J_\mu V^\mu \tag{7.6.1}$$

where

$$F_{\mu\nu} \equiv \partial_\mu V_\nu - \partial_\nu V_\mu \,. \tag{7.6.2}$$

Suppose we try to treat all four components of V^μ on the same basis. We should then define the conjugates

$$\Pi^\mu \equiv \frac{\partial \mathscr{L}}{\partial(\partial_0 V_\mu)} = -F^{0\mu} \,. \tag{7.6.3}$$

We immediately find the primary constraint:

$$\Pi_0 = 0 \,. \tag{7.6.4}$$

More generally, we encounter primary constraints whenever the equations $\Pi_\ell = \delta L/\delta\partial_0\Psi^\ell$ cannot be solved to give all the $\partial_0\Psi^\ell$ (at least locally) in terms of Π_ℓ and Ψ^ℓ. This will be the case if and only if the matrix $\delta^2 L/\delta(\partial_0\Psi^\ell)\,\delta(\partial_0\Psi^m)$ has vanishing determinant. Such Lagrangians are called *irregular*.

Then there are *secondary constraints*, which arise from the requirement that the primary constraints be consistent with the equations of motion. For the massive vector field, this is just the Euler–Lagrange equation (7.5.16) for V^0:

$$\partial_i \Pi_i = m^2 V^0 - J^0 \,. \tag{7.6.5}$$

Here we are finished, but in other theories we might encounter further constraints by requiring consistency of the secondary constraints with the field equations, and so on. The distinction between primary, secondary, etc. constraints is not important; we will treat them all together here.

There is another distinction between certain types of constraint that is more important. The constraints we have found for the massive vector field are of a type known as *second class*, for which there is a universal prescription for the commutation relations. To explain the distinction between first and second class constraints, and the prescription used to deal with second class constraints, it is useful first to recall the definition of the Poisson brackets of classical mechanics.

Consider any Lagrangian $L(\Psi, \dot{\Psi})$ that depends on a set of variables $\Psi^a(t)$ and their time-derivatives $\dot{\Psi}^a(t)$. (The Lagrangians of quantum field theory are a special case, with the index a running over all pairs of ℓ and $\mathbf{x}$.) We can define canonical conjugates for *all* of these variables by

$$\Pi_a \equiv \frac{\partial L}{\partial \dot{\Psi}^a} \,. \tag{7.6.6}$$

The Πs and Ψs will in general not be independent variables, but may instead be related by various constraint equations, both primary and secondary. The Poisson bracket is then defined by

$$[A,B]_{\rm P} \equiv \frac{\partial A}{\partial \Psi^a}\frac{\partial B}{\partial \Pi_a} - \frac{\partial B}{\partial \Psi^a}\frac{\partial A}{\partial \Pi_a} \tag{7.6.7}$$

with the constraints ignored in calculating the derivatives with respect to Ψ^a and Π_a. In particular, we always have $[\Psi^a, \Pi_b]_{\rm P} = \delta^a_b$. (Here and below all fields are taken at the same time, and time arguments are everywhere dropped.) These brackets have the same algebraic properties as commutators:

$$[A,B]_{\rm P} = -[B,A]_{\rm P}, \tag{7.6.8}$$

$$[A,BC]_{\rm P} = [A,B]_{\rm P}C + B[A,C]_{\rm P}, \tag{7.6.9}$$

including the Jacobi identity

$$[A,[B,C]_{\rm P}]_{\rm P} + [B,[C,A]_{\rm P}]_{\rm P} + [C,[A,B]_{\rm P}]_{\rm P} = 0\,. \tag{7.6.10}$$

If we could adopt the usual commutation relations $[\Psi^a, \Pi_b] = i\delta^a_b$, $[\Psi^a, \Psi^b] = [\Pi_a, \Pi_b] = 0$, then the commutator of any two functions of the Ψs and Πs would be just $[A,B] = i[A,B]_{\rm P}$. But the constraints do not always allow this.

The constraints may in general be expressed in the form $\chi_N = 0$, where the χ_N are a set of functions of the Ψs and Πs. Because we are including secondary constraints along with the primary constraints, the set of all the constraints is necessarily consistent with the equations of motion $\dot{A} = [A,H]_{\rm P}$, and therefore

$$[\chi_N, H]_{\rm P} = 0 \tag{7.6.11}$$

when the constraint equations $\chi_N = 0$ are satisfied.

We call a constraint *first class* if its Poisson bracket with all the other constraints vanishes when (*after* calculating the Poisson brackets) we impose the constraints. We shall see a simple example of such a constraint in the quantization of the electromagnetic field in the next chapter, where the first class constraint arises from a symmetry of the action, electromagnetic gauge invariance. In fact, the set of first class constraints $\chi_N = 0$ is always associated with a group of symmetries, under which an arbitrary quantity A undergoes the infinitesimal transformation

$$\delta_N A \equiv \sum_N \epsilon_N\,[\chi_N, A]_{\rm P}\,. \tag{7.6.12}$$

(In field theory these are local transformations, because the index N contains a spacetime coordinate.) Eq. (7.6.11) shows that this transformation leaves the Hamiltonian invariant, and for first class constraints it also respects all other constraints. Such first class constraints can be eliminated by a choice of gauge.

After all of the first class constraints have been eliminated by a choice of gauge, the remaining constraint equations $\chi_N = 0$ are such that no linear combination $\sum_N u_N [\chi_N, \chi_M]_{\rm P}$ of the Poisson brackets of these constraints with each other vanishes. It follows that the matrix of the Poisson brackets of the remaining constraints is non-singular:

$$\operatorname{Det} C \neq 0\,, \tag{7.6.13}$$

where

$$C_{NM} \equiv [\chi_N, \chi_M]_{\rm P}\,. \tag{7.6.14}$$

Constraints of this sort are called *second class.* Note that there must always be an even number of second class constraints, because an antisymmetric matrix of odd dimensionality necessarily has vanishing determinant.

As we have seen, in the case of the massive real vector field the constraints are

$$\chi_{1\mathbf{x}} = \chi_{2\mathbf{x}} = 0\,, \tag{7.6.15}$$

where

$$\chi_{1\mathbf{x}} = \Pi_0(\mathbf{x})\,, \qquad \chi_{2\mathbf{x}} = \partial_i \Pi_i(\mathbf{x}) - m^2 V^0(\mathbf{x}) - J^0(\mathbf{x})\,. \tag{7.6.16}$$

The Poisson bracket of these constraints is

$$C_{1\mathbf{x},2\mathbf{y}} = -C_{2\mathbf{y},1\mathbf{x}} = [\chi_{1\mathbf{x}}, \chi_{2\mathbf{y}}]_{\rm P} = m^2 \delta^3(\mathbf{x} - \mathbf{y}) \tag{7.6.17}$$

and, of course,

$$C_{1\mathbf{x},1\mathbf{y}} = C_{2\mathbf{x},2\mathbf{y}} = 0\,. \tag{7.6.18}$$

This 'matrix' is obviously non-singular, so the constraints (7.6.15) are second class.

Dirac suggested that when all constraints are second class, the commutation relations will be given by

$$[A, B] = i[A, B]_{\rm D}\,, \tag{7.6.19}$$

where $[A, B]_{\rm D}$ is a generalization of the Poisson bracket known as the *Dirac bracket*:

$$[A, B]_{\rm D} \equiv [A, B]_{\rm P} - [A, \chi_N]_{\rm P} \, (C^{-1})^{NM} \, [\chi_M, B]_{\rm P}\,. \tag{7.6.20}$$

(Here N and M are compound indices including the position in space, taking values like $1, \mathbf{x}$ and $2, \mathbf{x}$ in the vector field example.) He noted that the Dirac bracket like the Poisson bracket satisfies the same algebraic relations as the commutators

$$[A, B]_{\rm D} = -[B, A]_{\rm D} \,, \tag{7.6.21}$$

$$[A, BC]_{\rm D} = [A, B]_{\rm D} C + B[A, C]_{\rm D} \,, \tag{7.6.22}$$

$$[A, [B, C]_{\rm D}]_{\rm D} + [B, [C, A]_{\rm D}]_{\rm D} + [C, [A, B]_{\rm D}]_{\rm D} = 0 \,, \tag{7.6.23}$$

and also the relations

$$[\chi_N, B]_{\rm D} = 0 \tag{7.6.24}$$

which make the commutation relations (7.6.19) consistent with the constraints $\chi_N = 0$. Also, the Dirac brackets are unchanged if we replace the χ_N with any functions χ'_N for which the equations $\chi'_N = 0$ and $\chi_N = 0$ define the same submanifold of phase space. But all these agreeable properties do not prove that the commutators are actually given by Eq. (7.6.19) in terms of the Dirac brackets.

This issue is illuminated if not settled by a powerful theorem proved by Maskawa and Nakajima.[6] They showed that for any set of canonical variables Ψ^a, Π_a governed by second class constraints, it is always possible by a canonical transformation* to construct two sets of variables Q^n, $\mathcal{Q}^r$ and their respective conjugates P_n, $\mathcal{P}_r$, such that the constraints read $\mathcal{Q}^r = \mathcal{P}_r = 0$. Using these coordinates to calculate Poisson brackets, and redefining the constraint functions as $\chi_{1r} = \mathcal{Q}^r$, $\chi_{2r} = \mathcal{P}_r$, we have

$$C_{1r,2s} = [\mathcal{Q}^r, \mathcal{P}_s]_P = \delta^r_s \,,$$

$$C_{1r,1s} = [\mathcal{Q}^r, \mathcal{Q}^s]_P = 0 \,, \qquad C_{2r,2s} = [\mathcal{P}_r, \mathcal{P}_s]_P = 0 \,,$$

and for any functions A, B

$$[A, \chi_{1r}]_P = -\frac{\partial A}{\partial \mathcal{P}_r} \,, \qquad [A, \chi_{2r}]_P = \frac{\partial A}{\partial \mathcal{Q}^r} \,,$$

This C-matrix has inverse $C^{-1} = -C$, so the Dirac brackets (7.6.20) are

---

* Recall that by a canonical transformation, we mean a transformation from a set of phase space coordinates Ψ^a, Π_a to some other phase space coordinates $\tilde{\Psi}^a$, $\tilde{\Pi}_a$, such that $[\tilde{\Psi}^a, \tilde{\Pi}_b]_P = \delta^a_b$ and $[\tilde{\Psi}^a, \tilde{\Psi}^b]_P = [\tilde{\Pi}_a, \tilde{\Pi}_b]_P = 0$, the Poisson brackets being calculated in terms of the Ψ^a and Π_a. It follows that the Poisson brackets for any functions A, B are the same whether calculated in terms of Ψ^a and Π_a or in terms of $\tilde{\Psi}^a$ and $\tilde{\Pi}_a$. It also follows that if Ψ^a and Π_a satisfy the Hamiltonian equations of motion, then so do $\tilde{\Psi}^a$ and $\tilde{\Pi}_a$, with the same Hamiltonian. The Lagrangian is changed by a canonical transformation, but only by a time-derivative, which does not affect the action.

here

$$[A,B]_D = [A,B]_P + [A,\chi_{1r}]_P[\chi_{2r},B]_P - [A,\chi_{2r}]_P[\chi_{1r},B]_P$$
$$= [A,B]_P - \frac{\partial A}{\partial \mathcal{Q}^r}\frac{\partial B}{\partial \mathcal{P}_r} + \frac{\partial B}{\partial \mathcal{Q}^r}\frac{\partial A}{\partial \mathcal{P}_r}$$
$$= \frac{\partial A}{\partial Q^n}\frac{\partial B}{\partial P_n} - \frac{\partial B}{\partial Q^n}\frac{\partial A}{\partial P_n} . \tag{7.6.25}$$

In other words, *the Dirac bracket is equal to the Poisson bracket calculated in terms of the reduced set of unconstrained canonical variables Q^n, P_n.* If we assume that these unconstrained variables satisfy the canonical commutation relations, then the commutators of general operators A, B are given by Eq. (7.6.19) in terms of the Dirac brackets.**

We now return to the massive vector field, to see how it can be quantized using Dirac brackets. This is a case where it is easy to express the constrained variables V^0 and Π_0 in terms of the unconstrained ones† V_i and Π_i; we have simply $\Pi_0 = 0$, and V^0 is given by Eq. (7.6.5). From Eqs. (7.6.17) and (7.6.18), we see that C_{NM} here has the inverse

$$(C^{-1})^{1\mathbf{x},2\mathbf{y}} = -(C^{-1})^{2\mathbf{y},1\mathbf{x}} = -m^{-2}\delta^3(\mathbf{x}-\mathbf{y}), \tag{7.6.26}$$

$$(C^{-1})^{1\mathbf{x},1\mathbf{y}} = (C^{-1})^{2\mathbf{x},2\mathbf{y}} = 0 . \tag{7.6.27}$$

Therefore the Dirac prescription (7.6.19), (7.6.20) yields the equal-time commutators

$$[A,B] = i[A,B]_\mathrm{P} + i m^{-2}\int d^3z \left([A,\Pi_0(\mathbf{z})]_\mathrm{P}\,[\partial_i\Pi_i(\mathbf{z}) - m^2 V^0(\mathbf{z}) - J^0(\mathbf{z}), B]_\mathrm{P} - A\leftrightarrow B\right) . \tag{7.6.28}$$

By definition, we have

$$[V^\mu(\mathbf{x}),\Pi_\nu(\mathbf{y})]_\mathrm{P} = \delta^3(\mathbf{x}-\mathbf{y})\delta^\mu_\nu\,, \quad [V^\mu(\mathbf{x}),V^\nu(\mathbf{y})]_\mathrm{P} = [\Pi_\mu(\mathbf{x}),\Pi_\nu(\mathbf{y})]_\mathrm{P} = 0 . \tag{7.6.29}$$

Hence

$$[V^i(\mathbf{x}),V^j(\mathbf{y})] = [V^0(\mathbf{x}),V^0(\mathbf{y})] = 0\,,$$

** It is still an open question whether we should adopt canonical commutation relations for the unconstrained variables Q^n, P_n constructed by the Maskawa-Nakajima canonical transformation. Ultimately, the test of such canonical commutation relations is their consistency with the free-field commutation relations derived in Chapter 5, but to apply this test we need to know what the Q^n and P_n are. In the Appendix to this chapter we display two large classes of theories in which we can identify a set of unconstrained Qs and Ps, such that the Dirac commutation relations (7.6.19) follow from the ordinary canonical commutation relations of the Qs and Ps. We shall also show that in these cases, the Hamiltonian defined in terms of the unconstrained Ψs and Πs may be written just as well in terms of the constrained variables.

† This is a special case of the theories discussed in Part A of the Appendix.

$$\begin{aligned}
&[V^i(\mathbf{x}), V^0(\mathbf{y})] = -im^{-2}\partial_i\delta^3(\mathbf{x}-\mathbf{y})\,, \\
&[V^i(\mathbf{x}), \Pi_j(\mathbf{y})] = i\delta^i_j\delta^3(\mathbf{x}-\mathbf{y})\,, \\
&[V^0(\mathbf{x}), \Pi_j(\mathbf{y})] = [V^\mu(\mathbf{x}), \Pi_0(\mathbf{y})] = 0\,, \\
&[\Pi^\mu(\mathbf{x}), \Pi^\nu(\mathbf{y}] = 0\,.
\end{aligned} \tag{7.6.30}$$

These are indeed just the commutation relations that we would find by assuming that the unconstrained variables satisfy the usual canonical commutation relations $[V^i(\mathbf{x}), \Pi_j(\mathbf{y})] = i\delta^i_j\delta^3(\mathbf{x}-\mathbf{y})$, $[V^i(\mathbf{x}), V^j(\mathbf{y})] = [\Pi_i(\mathbf{x}), \Pi_j(\mathbf{y})] = 0$, and using the constraints to evaluate the commutators involving Π_0 and V^0.

7.7 Field Redefinitions and Redundant Couplings*

Observables like masses and S-matrix elements are independent of some of the coupling parameters in any action, known as the *redundant* parameters. This is because changes in these parameters can be undone by simply redefining the field variables. A continuous redefinition of the fields, such as an infinitesimal local transformation $\Psi^\ell(x) \to \Psi^\ell(x) + \epsilon F^\ell(\Psi(x), \partial_\mu\Psi(x), \cdots)$, clearly cannot affect any *observable* of the theory,** though, of course, it would change the values of matrix elements of the fields themselves.

How can we tell whether some variation in the parameters of a theory can be cancelled by a field redefinition? A continuous local field redefinition will produce a change in the action of the form

$$\delta I[\Psi] = \epsilon \sum_\ell \int d^4x \, \frac{\delta I[\Psi]}{\delta \Psi^\ell(x)} \, F^\ell(\Psi(x), \partial\Psi(x), \cdots) \quad . \tag{7.7.1}$$

So any change δg_i in the coupling parameters g_i, for which the change in the action is of the form

$$\sum_i \frac{\partial I}{\partial g_i}\, \delta g_i = -\epsilon \sum_\ell \int d^4x \, \frac{\delta I[\Psi]}{\delta \Psi^\ell(x)} \, F^\ell(\Psi(x), \partial\Psi(x), \cdots)\,, \tag{7.7.2}$$

may be compensated by a field redefinition

$$\Psi^\ell(x) \to \Psi^\ell(x) + \epsilon F^\ell(\Psi(x), \partial_\mu\Psi(x), \cdots)\,,$$

* This section lies somewhat out of the book's main line of development, and may be omitted in a first reading.

** For instance, the theorem of Section 10.2 shows that as long as we multiply by the correct field renormalization constants, S-matrix elements can be obtained from the vacuum expectation value of a time-ordered product of *any* operators that have non-vanishing matrix elements between the vacuum and the one-particle states of the particles participating in the reaction.

and therefore can have no effect on any observables. In other words, *a coupling parameter is redundant if the change in the action when we vary this parameter vanishes when we use the field equations $\delta I/\delta\Psi^\ell = 0$.*

For example, suppose we write the Lagrangian density of a scalar field theory in the form

$$\mathscr{L} = -\tfrac{1}{2}Z\,(\partial^\mu\Phi\,\partial_\mu\Phi + m^2\Phi^2) - \tfrac{1}{24}gZ^2\Phi^4\,.$$

The constant Z is a redundant coupling, because

$$\frac{\partial I}{\partial Z} = \tfrac{1}{2}\int d^4x\,\Phi\,(\Box\Phi - m^2\Phi - \tfrac{1}{6}gZ\Phi^3)\,,$$

and this vanishes when we use the field equation

$$\Box\Phi - m^2\Phi = \tfrac{1}{6}gZ\Phi^3\,.$$

On the other hand, neither the bare mass m nor the coupling g are redundant, and no function of m and g is redundant.

In this example, the field redefinition needed to compensate for a change of Z is a simple rescaling, in which F is proportional to Φ. (For this reason Z is called a field renormalization constant.) This is the most general field transformation that leaves the general form of this action invariant. But for the more general actions considered in Sections 12.3 and 12.4, with arbitrary numbers of fields and derivatives, we would have to consider non-linear as well as linear field redefinitions, and an infinite subset of the parameters of the theory would be redundant.

Appendix Dirac Brackets from Canonical Commutators

In this Appendix we shall show, in theories of two types, that the formula giving commutators as Dirac brackets times i follows from the usual canonical commutation relations for a reduced set of variables.

A

Suppose (as in the case of a massive vector field V^μ) that the quantum variables Ψ^a and Π_a appearing in the Lagrangian L may be divided into two classes:* one set Q^n of independent canonical variables (like $V^i(\mathbf{x})$) with independent canonical conjugates $P_n = \partial L/\partial\dot{Q}^n$; and another set

* We are again using a compact notation, in which labels like a, n, and r include a space coordinate $\mathbf{x}$ as well as discrete indices. Repeated labels are summed and integrated. All quantum variables are understood to be evaluated at the same time, with the common time argument dropped everywhere. The quantities $\mathscr{Q}^r$ are the same as the C^r introduced in Section 7.2.

$\mathcal{Q}^r(\mathbf{x})$ (like V^0) whose time-derivatives do not appear in the Lagrangian. The primary constraints are the conditions $\chi_{1r} = 0$, where

$$\chi_{1r} = \mathscr{P}_r \tag{7.A.1}$$

are the variables conjugate to the $\mathcal{Q}^r$. The secondary constraints arise from the equations of motion $0 = \partial L/\partial \mathcal{Q}^r$ for the $\mathcal{Q}^r$; we suppose that these constraints can be 'solved' — i.e., they may be written in the form $\chi_{2r} = 0$, with χ_{2r} in the form

$$\chi_{2r} = \mathcal{Q}^r - f^r(Q, P)\,. \tag{7.A.2}$$

(An example is provided by Eq. (7.6.5), which gives V^0 in terms of the independent Ps (here, the Π_i) and Qs.) We assume that the independent Qs and Ps satisfy the usual canonical commutation rules:

$$[Q^n, P_m] = i\delta^n_m\,, \qquad [Q^n, Q^m] = [P_n, P_m] = 0\,. \tag{7.A.3}$$

The constraint $\chi_{2r} = 0$ yields the commutators involving $\mathcal{Q}$:

$$[\mathcal{Q}^r, Q^n] = -i\frac{\partial f^r}{\partial P_n}\,, \qquad [\mathcal{Q}^r, P_n] = i\frac{\partial f^r}{\partial Q^n}\,, \tag{7.A.4}$$

$$[\mathcal{Q}^r, \mathcal{Q}^s] = i\Gamma^{rs}\,, \tag{7.A.5}$$

where Γ^{rs} is the Poisson bracket

$$\Gamma^{rs} \equiv [f^r, f^s]_{\rm P}\,, \tag{7.A.6}$$

and, of course, all commutators involving $\mathscr{P}_r$ vanish:

$$[\mathscr{P}_r, Q^n] = [\mathscr{P}_r, P_n] = [\mathscr{P}_r, \mathcal{Q}^s] = [\mathscr{P}_r, \mathscr{P}_s] = 0\,. \tag{7.A.7}$$

Now let us compare these commutators with the Dirac brackets. The Poisson brackets of the constraint functions are

$$C_{1r,1s} \equiv [\chi_{1r}, \chi_{1s}]_{\rm P} = 0\,, \tag{7.A.8}$$

$$C_{1r,2s} \equiv -C_{2s,1r} \equiv [\chi_{1r}, \chi_{2s}]_{\rm P} = -\delta^s_r\,, \tag{7.A.9}$$

$$C_{2r,2s} \equiv [\chi_{2r}, \chi_{2s}]_{\rm P} = [f^r(Q,P), f^s(Q,P)]_{\rm P} \equiv \Gamma^{rs}\,. \tag{7.A.10}$$

(In the example of the massive vector field Γ^{rs} vanishes, but the discussion here will apply also for non-vanishing Γ^{rs}.) It is easy to see that the C-matrix has the inverse

$$(C^{-1})^{1r,1s} = \Gamma^{rs}\,, \qquad (C^{-1})^{2r,2s} = 0\,,$$
$$(C^{-1})^{1r,2s} = -(C^{-1})^{2s,1r} = \delta^r_s\,. \tag{7.A.11}$$

Also, the Poisson brackets of any function A with the constraint functions are

$$[A,\chi_{1r}]_{\rm P} = \frac{\partial A}{\partial \mathcal{Q}^r}\,, \qquad [A,\chi_{2r}]_{\rm P} = -\frac{\partial A}{\partial \mathcal{P}_r} - [A, f^r(Q,P)]_{\rm P}\,.$$

Hence the Dirac bracket is

$$[A,B]_{\rm D} = [A,B]_{\rm P} - \frac{\partial A}{\partial \mathcal{Q}^r}\frac{\partial B}{\partial \mathcal{P}_r} + \frac{\partial B}{\partial \mathcal{Q}^r}\frac{\partial A}{\partial \mathcal{P}_r}$$
$$+\frac{\partial A}{\partial \mathcal{Q}^r}\Gamma^{rs}\frac{\partial B}{\partial \mathcal{Q}^s} - \frac{\partial A}{\partial \mathcal{Q}^r}[B,f^r]_{\rm P} + [A,f^r]_{\rm P}\frac{\partial B}{\partial \mathcal{Q}^r}\,. \quad (7.A.12)$$

Now, if A and B are both functions only of the independent canonical variables Q^n and P_n, then $\partial A/\partial \mathcal{Q}^r = \partial B/\partial \mathcal{Q}^r = 0$, so the Dirac bracket is equal to the Poisson bracket. In particular,

$$[Q^n, P_m]_{\rm D} = \delta^n_m\,, \qquad [Q^n, Q^m]_{\rm D} = [P_n, P_m]_{\rm D} = 0\,. \quad (7.A.13)$$

Where A is $\mathcal{Q}^r$ and B is a function of Qs and Ps, it is only the fifth term on the right-hand side of Eq. (7.A.12) that contributes. In particular

$$[\mathcal{Q}^r, Q^n]_{\rm D} = -\frac{\partial f^r}{\partial P_n}\,, \qquad [\mathcal{Q}^r, P_n]_{\rm D} = +\frac{\partial f^r}{\partial Q^n}\,. \quad (7.A.14)$$

Where both A and B are $\mathcal{Q}$s, we have only the fourth term

$$[\mathcal{Q}^r, \mathcal{Q}^s]_{\rm D} = \Gamma^{rs}\,. \quad (7.A.15)$$

Finally, where A is $\mathcal{P}_r$ and B is anything, we have only the first and third terms, which cancel:

$$[\mathcal{P}_r, B]_{\rm D} = [\mathcal{P}_r, B]_{\rm P} + \frac{\partial B}{\partial \mathcal{Q}^r} = 0\,. \quad (7.A.16)$$

Comparison of Eqs. (7.A.13)–(7.A.16) with Eqs. (7.A.3)–(7.A.7) shows that in all cases, the commutators are equal to the Dirac brackets times i. This is only to be expected, because as remarked in Section 7.6 all Dirac brackets involving the constraint functions vanish, so the Dirac brackets involving $\mathcal{Q}^r$ and/or $\mathcal{P}_s$ are given by using the constraint equations to express $\mathcal{Q}^r$ and/or $\mathcal{P}_s$ in terms of the independent Qs and Ps.

B

Next consider the case where the constraints take the form of conditions $\chi_{1r}(\Psi) = 0$ on the Ψ^a, which can be solved by expressing them in terms of a smaller set of unconstrained variables Q^n, and an equal number of separate conditions $\chi_{2r}(\Pi) = 0$ on the Π_a, which can be solved by expressing the Π_a in terms of a smaller set of unconstrained P_n. (We will see an example in the next chapter, where the constraints on the Ψ^a are gauge fixing conditions that are used to eliminate first class constraints, and the constraints on the Π_a are secondary constraints arising from the consistency of the first class constraints with the field equations.) We assume that the unconstrained variables satisfy the usual canonical commutation relations $[Q^n, P_m] = i\delta^n_m$, $[Q^n, Q^m] = [P_n, P_m] = 0$. The constrained and unconstrained momenta are related by

$$P_n = \frac{\partial L}{\partial \dot{Q}^n} = \frac{\partial L}{\partial \dot{\Psi}^b}\frac{\partial \Psi^b}{\partial Q^n} = \Pi_b \frac{\partial \Psi^b}{\partial Q^n} . \tag{7.A.17}$$

It follows that

$$[\Psi^a, \Pi_b]\frac{\partial \Psi^b}{\partial Q^n} = [\Psi^a, P_n] = i\frac{\partial \Psi^a}{\partial Q^n}$$

or, in other words,

$$\{[\Psi^a, \Pi_b] - i\delta^a_b\}\frac{\partial \Psi^b}{\partial Q^n} = 0 . \tag{7.A.18}$$

Now, the constraint $\chi_{1r}(\Psi) = 0$ is satisfied for $\Psi^a = \Psi^a(Q)$ for all Q, so

$$\frac{\partial \chi_{1r}}{\partial \Psi^b}\frac{\partial \Psi^b}{\partial Q^n} = 0 . \tag{7.A.19}$$

Furthermore, the vectors $(V_r)_b \equiv \partial\chi_{1r}/\partial\Psi^b$ form a complete set perpendicular to all the vectors $(U_n)^b \equiv \partial\Psi^b/\partial Q^n$, because if there were some other vector V_b with $V_b(U_n)^b = 0$ for all n, then there would be additional constraints on the Ψ^a. Hence Eq. (7.A.18) implies that

$$[\Psi^a, \Pi_b] = i\delta^a_b + ic^a_r\frac{\partial \chi_{1r}}{\partial \Psi^b} \tag{7.A.20}$$

with some unknown coefficients c^a_r. To determine these coefficients, we make use of the other constraint, that $\chi_{2r}(\Pi) = 0$. It follows that

$$0 = [\Psi^a, \chi_{2r}(\Pi)] = i[\Psi^a, \Pi_b]\frac{\partial \chi_{2r}(\Pi)}{\partial \Pi_b} .$$

Using Eq. (7.A.20), we have then

$$\frac{\partial \chi_{2r}(\Pi)}{\partial \Pi_a} = -c^a_s\frac{\partial \chi_{1s}(\Psi)}{\partial \Psi^b}\frac{\partial \chi_{2r}(\Pi)}{\partial \Pi_b} . \tag{7.A.21}$$

We recognize the factor multiplying c^a_s as the Poisson bracket

$$\frac{\partial \chi_{1s}(\Psi)}{\partial \Psi^b}\frac{\partial \chi_{2r}(\Pi)}{\partial \Pi_b} = [\chi_{1s}, \chi_{2r}]_P \equiv C_{1s,2r} .$$

Also, since χ_{1s} depends only on the Ψ and χ_{2r} depends only on the Π, these are the only non-vanishing Poisson brackets of constraints, so

$$C_{1r,1s} = C_{2r,2s} = 0 .$$

Thus Eq. (7.A.21) may be written

$$\frac{\partial \chi_N}{\partial \Pi_a} = -c_s^a \, C_{1s,N} \tag{7.A.22}$$

with N running over all constraint functions. For second class constraints, this has the unique solution

$$c_s^a = -\frac{\partial \chi_N}{\partial \Pi_a}(C^{-1})^{N,1s} = -\frac{\partial \chi_{2r}}{\partial \Pi_a}(C^{-1})^{2r,1s} . \tag{7.A.23}$$

Using this in Eq. (7.A.20) shows that

$$[\Psi^a, \Pi_b] = i\left[\delta_b^a - \frac{\partial \chi_{2r}}{\partial \Pi_a}(C^{-1})^{2r,1s}\frac{\partial \chi_{1s}}{\partial \Psi^b}\right] . \tag{7.A.24}$$

The Poisson brackets of Ψ^a and Π_b with the constraint functions are

$$\begin{aligned} &[\Psi^a, \chi_{1r}]_{\rm P} = 0\,, \qquad && [\Psi^a, \chi_{2r}]_{\rm P} = \frac{\partial \chi_{2r}}{\partial \Pi_a}\,, \\ &[\chi_{1r}, \Pi_b]_{\rm P} = \frac{\partial \chi_{1r}}{\partial \Psi^b}\,, \qquad && [\chi_{2r}, \Pi_b]_{\rm P} = 0\,, \end{aligned} \tag{7.A.25}$$

so the quantity in brackets on the right-hand side of Eq. (7.A.24) is the Dirac bracket

$$[\Psi^a, \Pi_b] = i[\Psi^a, \Pi_b]_{\rm D} \tag{7.A.26}$$

as was to be shown. Also, we can easily see that because C^{-1} has no 11 or 22 components, the other Dirac brackets are

$$[\Psi^a, \Psi^b]_{\rm D} = [\Pi_a, \Pi_b]_{\rm D} = 0\,, \tag{7.A.27}$$

so trivially

$$[\Psi^a, \Psi^b] = i[\Psi^a, \Psi^b]_{\rm D}\,, \qquad [\Pi_a, \Pi_b] = i[\Pi_a, \Pi_b]_{\rm D}\,. \tag{7.A.28}$$

* * *

In addition to commutation rules, we also need an explicit formula for the Hamiltonian. The usual canonical formalism tells us to take

$$H = P_n \dot{Q}^n - L\,, \tag{7.A.29}$$

the sum running over independent canonical variables. In theories of both types considered in this Appendix, this Hamiltonian may be written in terms of the constrained variables as

$$H = \Pi_a \dot{\Psi}^a - L\,. \tag{7.A.30}$$

For theories of type A, this is trivial; the sum over a runs over values n, for which $\Psi^n = Q^n$ and $\Pi_n = P_n$ are the independent canonical variables, together with values r, for which $\Pi_r = \mathscr{P}_r = 0$. For theories of type B, we note that Eq. (7.A.17) gives

$$P_n \dot{Q}^n = \Pi_b \frac{\partial \Psi^b}{\partial Q^n} \dot{Q}^n = \Pi_b \dot{\Psi}^b$$

which again yields Eq. (7.A.30).

Problems

1. Consider the theory of a set of real scalar fields Φ^n, with Lagrangian density $\mathscr{L} = -\frac{1}{2} \sum_{mn} \partial_\mu \Phi^n \partial^\mu \Phi^m f_{nm}(\Phi)$, where $f_{nm}(\Phi)$ is an arbitrary real non-singular matrix function of the field. (This is called the *non-linear σ-model.*) Carry out the canonical quantization of this theory. Derive the interaction $V[\phi(t), \dot{\phi}(t)]$ in the interaction picture.

2. Consider a theory of real scalar fields Φ^n and Dirac fields Ψ^i, with Lagrangian density $\mathscr{L} = \mathscr{L}_0 + \mathscr{L}_1$, where $\mathscr{L}_0$ is the usual free-field Lagrangian density, and $\mathscr{L}_1$ is an interaction term involving Φ^n and Ψ^i, but not their derivatives. Derive an explicit expression for the symmetric energy–momentum tensor $\Theta^{\mu\nu}$.

3. In the theory described in Problem 2, suppose that the Lagrangian density is invariant under a global infinitesimal symmetry $\delta\Phi^n = i\epsilon \sum_m t^n{}_m \Phi^m$, $\delta\Psi^i = i\epsilon \sum_j \tau^i{}_j \Psi^j$. Derive an explicit expression for the conserved current associated with this symmetry.

4. Consider the theory of a complex scalar field Φ and a real vector field V^μ, with Lagrangian density

$$\mathscr{L} = -(D_\mu \Phi)^\dagger D^\mu \Phi - \tfrac{1}{4} F_{\mu\nu} F^{\mu\nu} - \tfrac{1}{2} m^2 V_\mu V^\mu - \mathscr{H}(\Phi^\dagger \Phi) ,$$

where $D_\mu \equiv \partial_\mu - igV_\mu$ and $F_{\mu\nu} \equiv \partial_\mu V_\nu - \partial_\nu V_\mu$, and $\mathscr{H}$ is an arbitrary function. Carry out the canonical quantization of this theory. Derive the interaction in the interaction picture.

5. In the theory of Problem 4, derive expressions for the symmetric energy–momentum tensor $\Theta^{\mu\nu}$ and for the conserved current associated with the symmetry under $\delta\Phi = i\epsilon\Phi$, $\delta V^\mu = 0$.

6. Prove that the Dirac bracket satisfies the Jacobi identity (7.6.23).

7. Prove that the Dirac bracket is independent of the choice of constraint functions χ_N used to describe a given submanifold of phase space.

References

1. H. B. G. Casimir, *Proc. K. Ned. Akad. Wet.* **51**, 635 (1948); M. J. Spaarnay, *Nature* **180**, 334 (1957).

2. F. Belinfante, Physica **6**, 887 (1939); also see L. Rosenfeld, *Mémoires de l'Academie Roy. Belgique* **6**, 30 (1930).

3. See, e.g., S. Weinberg, *Gravitation and Cosmology* (Wiley, New York, 1972): Chapter 12.

4. See, e.g., J. Wess and J. Bagger, *Supersymmetry and Supergravity* (Princeton University Press, Princeton, 1983), and original references quoted therein.

5. P. A. M. Dirac, *Lectures on Quantum Mechanics* (Yeshiva University, New York, 1964). Also see P. A. M. Dirac, *Can. J. Math.* **2**, 129 (1950); *Proc. Roy. Soc. London*, ser. A, **246**, 326 (1958); P. G. Bergmann, *Helv. Phys. Acta Suppl.* IV, 79 (1956).

6. T. Maskawa and H. Nakajima, *Prog. Theor. Phys.* **56**, 1295 (1976). I am grateful to J. Feinberg for bringing this reference to my attention.

8
Electrodynamics

The original approach to quantum electrodynamics was to take for granted Maxwell's classical theory of electromagnetism, and quantize it. It will probably not surprise the reader that this book will follow a different path. We shall first infer the need for a principle of gauge invariance from the peculiar difficulties that arise in formulating a quantum theory of massless particles with spin, and then deduce the main features of electrodynamics from the gauge invariance principle. After that we shall follow a more conventional modern approach, in which one takes gauge invariance as the starting point and uses it to deduce the existence of a vector potential describing massless particles of unit spin.

It is too soon to tell which of these two alternatives corresponds to the logical order of nature. Most theorists have tended to take gauge invariance as a starting point, but in modern string theories[1] the argument runs the other way; one first notices a state of mass zero and unit spin among the normal modes of a string, and then from that deduces the gauge invariance of the effective field theory that describes such particles. At any rate, as we shall see, using either approach one is led to the quantized version of Maxwell's theory, still the paradigmatic example of a successful quantum field theory.

8.1 Gauge Invariance

Let's start by recalling the problems encountered in constructing covariant free fields for a massless particle of helicity ± 1. We saw in Section 5.9 that there is no difficulty in constructing an antisymmetric tensor free field $f_{\mu\nu}(x)$ for such particles. This field can be expressed in terms of the four-potential $a_\mu(x)$, given by Eq. (5.9.23), through the familiar relation

$$f_{\mu\nu}(x) = \partial_\mu a_\nu(x) - \partial_\nu a_\mu(x) . \tag{8.1.1}$$

However, Eq. (5.9.31) shows that the $a_\mu(x)$ transforms as a four-vector only up to a gauge transformation

$$U_0(\Lambda)a_\mu(x)\, U_0^{-1}(\Lambda) = \Lambda_\mu{}^\nu a_\nu(\Lambda x) + \partial_\mu \Omega(x, \Lambda)\,. \tag{8.1.2}$$

There is, in fact, no way to construct a true four-vector as a linear combination of the creation and annihilation operators for helicity ± 1. This is one way of understanding the presence of singularities at $m = 0$ in the propagator of a massive vector field

$$\Delta_{\mu\nu}(x, y) = (2\pi)^{-4} \int d^4 q\; e^{iq\cdot(x-y)}\, \frac{\eta_{\mu\nu} + q_\mu q_\nu / m^2}{q^2 + m^2 - i\epsilon}\,,$$

which prevent us from dealing with massless particles of helicity ± 1 by simply passing to the limit $m \to 0$ of the theory of a massive particle of spin one.

We could avoid these problems by demanding that all interactions involve only* $F_{\mu\nu}(x) \equiv \partial_\mu A_\nu(x) - \partial_\nu A_\mu(x)$ and its derivatives, not $A_\mu(x)$, but this is not the most general possibility, and not the one realized in nature. Instead of banishing $A_\mu(x)$ from the action, we shall require instead that the part of the action I_M for matter and its interaction with radiation be invariant under the general gauge transformation

$$A_\mu(x) \to A_\mu(x) + \partial_\mu \epsilon(x) \tag{8.1.3}$$

(at least when the matter fields satisfy the field equations) so that the extra term in Eq. (8.1.2) should have no effect. The change in the matter action under the transformations (8.1.3) may be written

$$\delta I_M = \int d^4x\, \frac{\delta I_M}{\delta A_\mu(x)} \partial_\mu \epsilon(x)\,. \tag{8.1.4}$$

Hence the Lorentz invariance of I_M requires that

$$\partial_\mu \frac{\delta I_M}{\delta A_\mu(x)} = 0\,. \tag{8.1.5}$$

This is trivially true if I_M involves only $F_{\mu\nu}(x)$ and its derivatives, along with matter fields. In this case

$$\frac{\delta I_M}{\delta A_\mu(x)} = 2\partial_\nu\, \frac{\delta I_M}{\delta F_{\mu\nu}(x)}\,.$$

But if I_M involves $A_\mu(x)$ itself then Eq. (8.1.5) is a non-trivial constraint on the theory.

Now, what sort of theory will provide conserved currents to which we can couple the field $A^\mu(x)$? We saw in Section 7.3 that infinitesimal internal

* We now use A_μ and $F_{\mu\nu}$ for the electromagnetic potential vector and the field strength tensor because these are interacting fields.

symmetries of the action imply the existence of conserved currents. In particular, if the transformation**

$$\delta\Psi^\ell(x) = i\epsilon(x)\, q_\ell\, \Psi^\ell(x) \tag{8.1.6}$$

leaves the matter action invariant for a constant ϵ, then for general infinitesimal functions $\epsilon(x)$ the change in the matter action must take the form

$$\delta I_M = -\int d^4x\, J^\mu(x)\partial_\mu\epsilon(x)\,. \tag{8.1.7}$$

When the matter fields satisfy their field equations, the matter action is stationary with respect to *any* variation of the Ψ_ℓ, so in this case (8.1.7) must vanish, and hence

$$\partial_\mu J^\mu = 0\,. \tag{8.1.8}$$

In particular, we saw in Section 7.3 that if I_M is the integral of a function $\mathscr{L}_M$ of Ψ^ℓ and $\partial_\mu\Psi^\ell$, then the conserved current is given by†

$$J^\mu = -i\sum_\ell \frac{\partial\mathscr{L}_M}{\partial(\partial_\mu\Psi^\ell)}\, q_\ell\, \Psi^\ell\,,$$

and this generates the transformations (8.1.6) in the sense that

$$[Q, \Psi^\ell(x)] = -q_\ell\, \Psi^\ell(x)\,, \tag{8.1.9}$$

where Q is the time-independent charge operator

$$Q = \int d^3x\, J^0\,. \tag{8.1.10}$$

We can therefore construct a Lorentz-invariant theory by coupling the vector field A_μ to the conserved current J^μ, in the sense that $\delta I_M/\delta A_\mu(x)$ is taken to be proportional to $J^\mu(x)$. Any constant of proportionality may be absorbed into the definition of the overall scale of the charges q_ℓ, so we may simply set these quantities equal:

$$\frac{\delta I_M}{\delta A_\mu(x)} = J^\mu(x)\,. \tag{8.1.11}$$

The conservation of electric charge only allows us to fix the values of all charges in terms of the value of any one of them, conventionally taken to

** Because the field transformation matrix is taken now to be diagonal it is not convenient here to use the summation convention for sums over field indices, so there is no sum over ℓ in Eq. (8.1.6).

† Here Ψ^ℓ is understood to run over all independent fields other than A^μ. We use a capital psi to indicate that these are Heisenberg picture fields, whose time-dependence includes the effects of interactions. Of course, this Ψ^ℓ is not to be confused with a state-vector or wave function.

be the electron charge, denoted $-e$. It is Eq. (8.1.11) that gives a definite meaning‡ to the value of e.

The requirement (8.1.11) may be restated as a principle of invariance:[1a] the matter action is invariant under the joint transformations

$$\delta A_\mu(x) = \partial_\mu \epsilon(x)\,, \tag{8.1.12}$$

$$\delta \Psi_\ell(x) = i\epsilon(x) q_\ell \Psi_\ell(x)\,. \tag{8.1.13}$$

A symmetry of this type with an arbitrary function $\epsilon(x)$ is called a *local symmetry*, or a gauge invariance of the second kind. A symmetry under a transformation with ϵ constant is called a *global* symmetry, or a gauge invariance of the first kind. Several exact local symmetries are now known, but the only purely global symmetries appear to be accidents enforced by other principles. (See Section 12.5.)

We have not yet said anything about the action for photons themselves. As a guess, we can take this to be the same as for massive vector fields, but with $m = 0$:

$$I_\gamma = -\tfrac{1}{4}\int d^4x\, F_{\mu\nu}F^{\mu\nu}\,. \tag{8.1.14}$$

This is the same as the action used in classical electrodynamics, but its real justification is that it is (up to a constant) the unique gauge-invariant functional that is quadratic in $F_{\mu\nu}$, without higher derivatives. Also, as we will see in the next section, it leads to a consistent quantum theory. If there are any terms in the action with higher derivatives and/or of higher order in $F_{\mu\nu}$ they can be lumped into what we have called the matter action. Using Eqs. (8.1.11) and (8.1.14), the field equation for electromagnetism now reads

$$0 = \frac{\delta}{\delta A_\nu}\,[I_\gamma + I_M] = \partial_\mu F^{\mu\nu} + J^\nu\,. \tag{8.1.15}$$

We recognize these as the usual inhomogeneous Maxwell equations, with current J^ν. There are also other, homogeneous, Maxwell equations

$$0 = \partial_\mu F_{\nu\epsilon} + \partial_\epsilon F_{\mu\nu} + \partial_\nu F_{\epsilon\mu}\,, \tag{8.1.16}$$

which follow directly from the definition $F_{\mu\nu} \equiv \partial_\mu A_\nu - \partial_\nu A_\mu$.

In the above discussion, we have started with the existence of massless spin one particles, and have been led to infer the invariance of the matter action under a local gauge transformation (8.1.12), (8.1.13). As usually presented, the derivation runs in the opposite direction. That is, one starts

---

‡ Of course, Eq. (8.1.11) fixes the definition of e only after we have defined how we are normalizing $A_\mu(x)$. The question of electromagnetic field normalization is taken up in Section 10.4.

with a global internal symmetry

$$\delta\Psi^\ell(x) = i\epsilon q^\ell \Psi^\ell(x) \tag{8.1.17}$$

and asks what must be done to promote this to a local symmetry

$$\delta\Psi^\ell(x) = i\epsilon(x) q_\ell \Psi^\ell(x) . \tag{8.1.18}$$

If the Lagrange density $\mathscr{L}$ depended only on fields $\Psi^\ell(x)$ and not on their derivatives then it would make no difference whether ϵ is constant or not; invariance with ϵ constant would imply invariance with ϵ a function of spacetime position. But all realistic Lagrangians do involve field derivatives, and here we have the problem that derivatives of fields transform differently from fields themselves:

$$\delta\, \partial_\mu \Psi^\ell(x) = i\epsilon(x) q_\ell \partial_\mu \Psi^\ell(x) + i q_\ell \Psi^\ell(x) \partial_\mu \epsilon(x) . \tag{8.1.19}$$

In order to cancel the second term here, we 'invent' a vector field $A_\mu(x)$ with transformation rule

$$\delta A_\mu(x) = \partial_\mu \epsilon(x) \tag{8.1.20}$$

and require that the Lagrangian density depend on $\partial_\mu \Psi^\ell$ and A_μ only in the combination

$$D_\mu \Psi^\ell \equiv \partial_\mu \Psi^\ell - i q_\ell A_\mu \Psi^\ell , \tag{8.1.21}$$

which transforms just like Ψ^ℓ

$$\delta\, D_\mu \Psi^\ell(x) = i\epsilon(x) q_\ell D_\mu \Psi^\ell(x) . \tag{8.1.22}$$

A matter Lagrangian density $\mathscr{L}_M(\Psi, D\Psi)$ that is formed only out of Ψ^ℓ and $D_\mu \Psi^\ell$ will be invariant under the transformations (8.1.18), (8.1.20), with $\epsilon(x)$ an arbitrary function, if it is invariant with ϵ a constant. With the Lagrangian of this form, we have

$$\frac{\delta I_M}{\delta A_\mu} = \sum_\ell \frac{\partial \mathscr{L}_M}{\partial D_\mu \Psi^\ell} (-i q_\ell \Psi^\ell) = -i \sum_\ell \frac{\partial \mathscr{L}_M}{\partial\, \partial_\mu \Psi^\ell} q_\ell \Psi^\ell ,$$

which is the same as Eq. (8.1.11). (We could also include $F_{\mu\nu}$ and its derivatives in $\mathscr{L}_M$.) From this point of view, the masslessness of the particles described by A_μ is a consequence of gauge invariance rather than an assumption: a term $-\frac{1}{2}m^2 A_\mu A^\mu$ in the Lagrangian density would violate gauge invariance.

8.2 Constraints and Gauge Conditions

There are aspects of electrodynamics that stand in the way of quantizing the theory as we did for various theories of massive particles in the previous chapter. As usual, we may define the canonical conjugates to the electromagnetic vector potential by

$$\Pi^\mu \equiv \frac{\partial \mathscr{L}}{\partial(\partial_0 A_\mu)} \,. \tag{8.2.1}$$

Quantization by the usual rules would give

$$[A_\mu(\mathbf{x},t), \Pi^\nu(\mathbf{y},t)] = i\delta_\mu^\nu \delta^3(\mathbf{x}-\mathbf{y}) \,.$$

But this is not possible here, because A_μ and Π^ν are subject to several constraints.

The first constraint arises from the fact that the Lagrangian density is independent* of the time-derivative of A_0, and therefore

$$\Pi^0(x) = 0 \,. \tag{8.2.2}$$

This is called a *primary constraint*, because it follows directly from the structure of the Lagrangian. There is also a *secondary constraint* here, which follows from the field equation for the quantity fixed by the primary constraint:**

$$\partial_i \Pi^i = -\partial_i \frac{\partial \mathscr{L}}{\partial F_{i0}} = -\frac{\partial \mathscr{L}}{\partial A_0} = -J^0 \,, \tag{8.2.3}$$

the time-derivative term dropping out because $F_{00} = 0$. Even though the matter Lagrangian may generally depend on A^0, the charge density depends only on the canonical matter fields† Q^n and their canonical conjugates P_n:

$$J^0 = -i \sum_\ell \frac{\partial \mathscr{L}}{\partial(\partial_0 \Psi^\ell)} q_\ell \Psi^\ell = -i \sum_n P_n q_n Q^n \,. \tag{8.2.4}$$

Hence Eq. (8.2.3) is a functional relation among canonical variables. Both Eq. (8.2.2) and Eq. (8.2.3) are inconsistent with the usual assumptions that $[A_\mu(\mathbf{x},t), \Pi^\nu(\mathbf{y},t)] = i\delta_\mu^\nu \delta^3(\mathbf{x}-\mathbf{y})$ and $[Q^n(\mathbf{x},t), \Pi^\nu(\mathbf{y},t)] = [P_n(\mathbf{x},t), \Pi^\nu(\mathbf{y},t)] = 0$.

We encountered a similar problem in the theory of the massive vector field. In that case we found two equivalent ways of dealing with it: either by the method of Dirac brackets or, more directly, by treating only

* For $\mathscr{L}_\gamma = -F_{\mu\nu}F^{\mu\nu}/4$, we have $\partial \mathscr{L}_\gamma / \partial(\partial_0 A_\mu) = -F^{0\mu}$, which vanishes for $\mu = 0$ because $F^{\mu\nu}$ is antisymmetric. For matter Lagrangians $\mathscr{L}_M$ that involve only Ψ^ℓ and $D_\mu \Psi^\ell$, the prescription (8.1.21) tells us that $\mathscr{L}_M$ does not depend on any derivatives of any A^ν. Even if the matter Lagrangian depends also on $F_{\mu\nu}$, $\partial \mathscr{L}_M / \partial(\partial_\nu A_\mu)$ will be again antisymmetric in μ and ν, and therefore will vanish for $\mu = \nu = 0$.

** As usual, i, j, etc. run over the values 1, 2, 3.

† Due to exhaustion of alphabetic resources, I have had to adopt a notation here that is different from that of the previous chapter. The symbols Q^n and P_n are now reserved for the canonical matter fields and their canonical conjugates, respectively, while the canonical electromagnetic fields and canonical conjugates are A_i and Π_i.

A_i and Π^i as canonical variables, solving the analog of Eq. (8.2.3) to calculate A^0 in terms of these variables. It is clear that here we cannot use Dirac brackets; the constraint functions χ here are Π^0 and $\partial_i \Pi_i + J^0$ (as compared with $\partial_i \Pi_i - m^2 A^0 + J^0$) and these obviously have vanishing Poisson brackets. In Dirac's terminology, the constraints (8.2.2) and (8.2.3) are *first class*. Nor can we eliminate A^0 as a dynamical variable by solving for it in terms of the other variables. Instead of giving A^0 for all time, Eq. (8.2.3) is a mere initial condition; if Eq. (8.2.3) is satisfied at one time, then it is satisfied for all times, because (using the field equations for the other fields A^i), we have

$$\begin{aligned} \partial_0 \left[\partial_i \frac{\partial \mathscr{L}}{\partial F_{i0}} - J^0 \right] &= -\partial_i \partial_0 \frac{\partial \mathscr{L}}{\partial F_{0i}} - \partial_0 J^0 \\ &= + \partial_i \partial_j \frac{\partial \mathscr{L}}{\partial F_{ji}} - \partial_i J_i - \partial_0 J^0 \end{aligned}$$

and the current conservation condition then gives

$$\partial_0 \left[\partial_i \frac{\partial \mathscr{L}}{\partial F_{i0}} - J^0 \right] = 0 \,. \tag{8.2.5}$$

It should not be surprising that we still have four components of A^μ with only three field equations, because this theory has a local gauge symmetry that makes it, in principle, impossible to infer the values of the fields at arbitrary times from their values and rates of change at any one time. Given any solution $A_\mu(\mathbf{x}, t)$ of the field equations, we can always find another solution $A_\mu(\mathbf{x}, t) + \partial_\mu \epsilon(\mathbf{x}, t)$ with the same value and time-derivative at $t = 0$ (by choosing ϵ so that its first and second derivatives vanish there) but which differs from $A_\mu(\mathbf{x}, t)$ at later times.

Because of this partial arbitrariness of $A_\mu(\mathbf{x}, t)$, it is not possible to apply the canonical quantization procedure directly to A_μ (or, as for finite mass, to $\mathbf{A}$). Of the various approaches to this difficulty, two are particularly useful. One is the modern Lorentz-invariant method of BRST-quantization, to be discussed in Volume II. The other, which will be followed here, is to exploit the gauge invariance of the theory, to 'choose a gauge'. That is, we make a finite gauge transformation

$$A_\mu(x) \to A_\mu(x) + \partial_\mu \lambda(x) \,, \qquad \Psi_\ell(x) \to \exp\Big(i q_\ell \lambda(x)\Big) \Psi_\ell(x)$$

to impose a condition on $A_\mu(x)$ that will allow us to apply the methods of canonical quantization. There are various gauges that have been found useful in various applications:‡

‡ Here Φ is any complex scalar field with $q \neq 0$; this gauge condition is used when the gauge symmetry is spontaneously broken by a non-vanishing vacuum expectation value of Φ.

Lorentz (or Landau) gauge:	$\partial_\mu A^\mu = 0$
Coulomb gauge:	$\nabla \cdot \mathbf{A} = 0$
Temporal gauge:	$A^0 = 0$
Axial gauge:	$A^3 = 0$
Unitarity gauge:	Φ real

The canonical quantization procedure works most easily in the axial or Coulomb gauge, but of course Coulomb gauge keeps manifest rotation invariance in a way that axial gauge does not, so we will adopt Coulomb gauge here.[2]

To check that this is possible, note that if A^μ does not satisfy the Coulomb gauge condition, then the gauge-transformed field $A^\mu + \partial^\mu \lambda$ will, provided we choose λ so that $\nabla^2 \lambda = -\nabla \cdot \mathbf{A}$. From now on, we assume that this transformation has been made, so that

$$\nabla \cdot \mathbf{A} = 0 \,. \tag{8.2.6}$$

It will be convenient henceforth to limit ourselves to theories in which the matter Lagrangian $\mathscr{L}_M$ may depend on matter fields and their time-derivatives and also on A^μ but not on derivatives of A^μ. (The standard theories of the electrodynamics of scalar and Dirac fields have Lagrangians of this type.) Then the only term in the Lagrangian that depends on $F_{\mu\nu}$ is the kinematic term $-\frac{1}{4}F_{\mu\nu}F^{\mu\nu}$, and the constraint equation (8.2.3) reads

$$-\,\partial_i F^{i0} = J^0 \,. \tag{8.2.7}$$

Together with the Coulomb gauge condition (8.2.6), this yields

$$-\,\nabla^2 A^0 = J^0 \,, \tag{8.2.8}$$

which can be solved to give

$$A^0(\mathbf{x}, t) = \int d^3y \, \frac{J^0(\mathbf{y}, t)}{4\pi|\mathbf{x} - \mathbf{y}|} \,. \tag{8.2.9}$$

The remaining degrees of freedom are A^i, with $i = 1, 2, 3$, subject to the gauge condition $\nabla \cdot \mathbf{A} = 0$.

As mentioned earlier, the charge density depends only on the canonical matter fields Q^n and their canonical conjugates P_n, so Eq. (8.2.9) represents an explicit solution for the auxiliary field A^0.

8.3 Quantization in Coulomb Gauge

There is still an impediment to the canonical quantization of electrodynamics in the Coulomb gauge. Even after we use Eq. (8.2.9) to eliminate A^0 (and Π_0) from the list of canonical variables, we cannot apply the usual canonical commutation relations to A^i and Π_i, because there are two remaining constraints on these variables.* One of them is the Coulomb gauge condition

$$\chi_{1\mathbf{x}} \equiv \partial_i A^i(\mathbf{x}) = 0 \,. \tag{8.3.1}$$

The other is the secondary constraint Eq. (8.2.3), which requires that

$$\chi_{2\mathbf{x}} \equiv \partial_i \Pi^i(\mathbf{x}) + J^0(\mathbf{x}) = 0 \,. \tag{8.3.2}$$

Neither constraint is consistent with the usual commutation relations $[A_i(\mathbf{x}), \Pi_j(\mathbf{y})] = i\delta_{ij}\delta^3(\mathbf{x}-\mathbf{y})$, because operating on the right-hand side with either $\partial/\partial x^i$ or $\partial/\partial y^j$ does not give zero.

These constraints are of a type known as *second class*, for which there is a universal prescription for the commutation relations, discussed in Section 7.6. Note that the constraint functions have the Poisson brackets

$$\begin{aligned} C_{1\mathbf{x},2\mathbf{y}} &= -C_{2\mathbf{y},1\mathbf{x}} \equiv [\chi_{1\mathbf{x}}, \chi_{2\mathbf{y}}]_{\mathrm{P}} = -\nabla^2\delta^3(\mathbf{x}-\mathbf{y}) \,, \\ C_{1\mathbf{x},1\mathbf{y}} &\equiv [\chi_{1\mathbf{x}}, \chi_{1\mathbf{y}}]_{\mathrm{P}} = 0 \,, \\ C_{2\mathbf{x},2\mathbf{y}} &\equiv [\chi_{2\mathbf{x}}, \chi_{2\mathbf{y}}]_{\mathrm{P}} = 0 \,, \end{aligned} \tag{8.3.3}$$

where here, for any functionals U and V,

$$[U,V]_{\mathrm{P}} \equiv \int d^3x \left[\frac{\delta U}{\delta A^i(\mathbf{x})}\frac{\delta V}{\delta \Pi_i(\mathbf{x})} - \frac{\delta V}{\delta A^i(\mathbf{x})}\frac{\delta U}{\delta \Pi_i(\mathbf{x})}\right] .$$

The 'matrix' C_{NM} is non-singular, which identifies these as second class constraints. Also, the field variables A^i may be expressed in terms of independent canonical variables, which may, for instance, be taken as $Q_{1\mathbf{x}} = A^1(\mathbf{x})$, $Q_{2\mathbf{x}} = A^2(\mathbf{x})$, with A^3 given by the solution of Eq. (8.3.1):

$$A^3(\mathbf{x}) = -\int^{x^3} ds\, [\partial_1 A^1(x^1, x^2, s) + \partial_2 A^2(x^1, x^2, s)] \,.$$

Using Eq. (8.3.2), the canonical conjugates Π_i to A^i may likewise be expressed in terms of the canonical conjugates $P_{1\mathbf{x}}$ and $P_{2\mathbf{x}}$ to $Q_{1\mathbf{x}}$ and $Q_{2\mathbf{x}}$. In such cases, Part B of the Appendix to the previous chapter tells us that if the independent variables $Q_{1\mathbf{x}}$, $Q_{2\mathbf{x}}$, $P_{1\mathbf{x}}$, and $P_{2\mathbf{x}}$ satisfy the usual canonical commutation relations, then the commutators of the constrained variables and their canonical conjugates are given (aside from a factor i) by the corresponding Dirac brackets (7.6.20). This prescription has the great advantage that we do not have to use explicit expressions for the dependent variables in terms of the independent ones.

* In this section i, j, etc. run over the values 1,2,3. We continue the practice of taking all operators at the same time, and omitting the time argument.

To calculate the Dirac brackets, we note that the matrix C has the inverse

$$(C^{-1})_{1\mathbf{x},2\mathbf{y}} = -(C^{-1})_{2\mathbf{y},1\mathbf{x}} = -\int \frac{d^3k}{(2\pi)^3} \frac{e^{i\mathbf{k}\cdot(\mathbf{x}-\mathbf{y})}}{\mathbf{k}^2} = -\frac{1}{4\pi|\mathbf{x}-\mathbf{y}|}\,,$$
$$(C^{-1})_{1\mathbf{x},1\mathbf{y}} = (C^{-1})_{2\mathbf{x},2\mathbf{y}} = 0\,. \tag{8.3.4}$$

Also, the non-vanishing Poisson brackets of the A^i and Π_i with the constraint functions are

$$[A^i(\mathbf{x}), \chi_{2\mathbf{y}}]_{\rm P} = -\frac{\partial}{\partial x^i}\delta^3(\mathbf{x}-\mathbf{y})$$

and

$$[\Pi_i(\mathbf{x}), \chi_{1\mathbf{y}}]_{\rm P} = +\frac{\partial}{\partial x^i}\delta^3(\mathbf{x}-\mathbf{y})\,.$$

Hence according to Eqs. (7.6.19) and (7.6.20), the equal-time commutators are

$$\left[A^i(\mathbf{x}), \Pi_j(\mathbf{y})\right] = i\delta^i_j\delta^3(\mathbf{x}-\mathbf{y}) + i\frac{\partial^2}{\partial x^j\partial x^i}\left(\frac{1}{4\pi|\mathbf{x}-\mathbf{y}|}\right)\,,$$
$$\left[A^i(\mathbf{x}),\, A^j(\mathbf{y})\right] = \left[\Pi_i(\mathbf{x}),\, \Pi_j(\mathbf{y})\right] = 0\,. \tag{8.3.5}$$

Note that these are consistent with the Coulomb gauge conditions (8.3.1) and (8.3.2), as is guaranteed by the general properties of the Dirac bracket.

Now, what is $\mathbf{\Pi}$ in electrodynamics? For the class of theories discussed in the previous section where only the kinematic term $-\frac{1}{4}\int d^3x F_{\mu\nu}F^{\mu\nu}$ in the Lagrangian depends on $\dot{\mathbf{A}}$, varying the Lagrangian with respect to $\dot{\mathbf{A}}$ without worrying about the constraint $\nabla\cdot\mathbf{A}=0$ gives

$$\Pi_j = \frac{\delta L}{\delta \dot{A}^j(\mathbf{x})} = \dot{A}^j(\mathbf{x}) + \frac{\partial}{\partial x^j}A^0(\mathbf{x})\,. \tag{8.3.6}$$

But with $\mathbf{A}$ constrained by the condition $\nabla\cdot\mathbf{A}=0$, variational derivatives with respect to $\dot{\mathbf{A}}$ are not really well defined. If the variation of L under a change $\delta\dot{\mathbf{A}}$ in $\dot{\mathbf{A}}$ is $\delta L = \int d^3x\, \mathscr{P}\cdot\delta\dot{\mathbf{A}}$, then since $\nabla\cdot\delta\dot{\mathbf{A}}=0$, we also have $\delta L = \int d^3x\, [\mathscr{P}+\nabla\mathscr{F}]\cdot\delta\dot{\mathbf{A}}$ for any scalar function $\mathscr{F}(\mathbf{x})$. Thus all we can conclude from inspection of the Lagrangian is that $\mathbf{\Pi}$ equals $\dot{\mathbf{A}}(\mathbf{x})+\nabla A^0(\mathbf{x})$ plus the gradient of some scalar. This ambiguity is removed by condition (8.3.2), which requires that $\nabla\cdot\mathbf{\Pi} = -J^0 = \nabla^2 A^0$. Because $\nabla\cdot\dot{\mathbf{A}}=0$, we conclude that Eq. (8.3.6) does indeed give the correct formula for Π^i.

Although the commutation relations (8.3.5) are reasonably simple, we must face the complication that $\mathbf{\Pi}$ does not commute with matter fields and their canonical conjugates. If F is any functional of these matter degrees of freedom, then its Dirac bracket with $\mathbf{A}$ vanishes, but its Dirac bracket with $\mathbf{\Pi}$ is

$$
\begin{aligned}
[F,\mathbf{\Pi}(\mathbf{z})]_{\rm D} &= -\int d^3x\, d^3y\, [F,\chi_{2\mathbf{x}}]_{\rm P}\frac{1}{4\pi|\mathbf{x}-\mathbf{y}|}[\chi_{1\mathbf{y}},\mathbf{\Pi}(\mathbf{z})]_{\rm P} \\
&= -\int d^3x\, d^3y\, [F,J^0(\mathbf{x})]_{\rm P}\frac{1}{4\pi|\mathbf{x}-\mathbf{y}|}\nabla\delta^3(\mathbf{y}-\mathbf{z}) \\
&= -\int d^3y\, [F,A^0(\mathbf{y})]_{\rm P}\nabla\delta^3(\mathbf{y}-\mathbf{z}) \\
&= [F,\nabla A^0(\mathbf{z})]_{\rm P} = [F,\nabla A^0(\mathbf{z})]_{\rm D}\,.
\end{aligned}
$$

In order to facilitate the transition to the interaction picture, instead of expressing the Hamiltonian in terms of $\mathbf{A}$ and $\mathbf{\Pi}$, we shall write it in terms of $\mathbf{A}$ and $\mathbf{\Pi}_\perp$, where $\mathbf{\Pi}_\perp$ is the solenoidal part of $\mathbf{\Pi}$:

$$\mathbf{\Pi}_\perp \equiv \mathbf{\Pi} - \nabla A^0 = \dot{\mathbf{A}}\,, \tag{8.3.7}$$

for which $[F,\mathbf{\Pi}_\perp(\mathbf{z})]$ vanishes. By using the facts that $\mathbf{\Pi}_\perp(\mathbf{x})$ commutes with $\mathbf{\Pi}(\mathbf{y}) - \mathbf{\Pi}_\perp(\mathbf{y}) = \nabla A^0(\mathbf{y})$ and that $\partial_i A^0(\mathbf{x})$ commutes with $\partial_j A^0(\mathbf{y})$, it is easy to see that $\mathbf{\Pi}_\perp(\mathbf{x})$ satisfies the same commutation relations (8.3.5) as $\mathbf{\Pi}(\mathbf{x})$, and also the simple constraint

$$\nabla\cdot\mathbf{\Pi}_\perp = 0\,. \tag{8.3.8}$$

Now we need to construct a Hamiltonian. According to the general results of the Appendix to Chapter 7, we can apply the usual relation between the Hamiltonian and Lagrangian using the constrained variables $\mathbf{A}$ and $\mathbf{\Pi}_\perp$, without first having explicitly to write the Hamiltonian in terms of the unconstrained Qs and Ps. In electrodynamics, this gives

$$H = \int d^3x\,\left[\Pi_{\perp i}\dot{A}^i + P_n\dot{Q}^n - \mathscr{L}\right], \tag{8.3.9}$$

where, as mentioned earlier, Q^n and P_n are to be understood as the matter canonical fields and their canonical conjugates. (We can use $\mathbf{\Pi}_\perp$ in place of $\mathbf{\Pi}$ in Eq. (8.3.9) because $\nabla\cdot\mathbf{A} = 0$.)

To be specific, consider a theory with a Lagrangian density of the form

$$\mathscr{L} = -\tfrac{1}{4}F_{\mu\nu}F^{\mu\nu} + J_\mu A^\mu + \mathscr{L}_{\rm matter}\,, \tag{8.3.10}$$

where J_μ is a current that does not involve A^μ, and $\mathscr{L}_{\rm matter}$ is the Lagrangian for whatever other fields do appear in J^μ, aside from their electromagnetic interactions, which are given explicitly by the term $J_\mu A^\mu$ in Eq. (8.3.10). (The electrodynamics of spin $\frac{1}{2}$ particles has a Lagrangian of this form, but the electrodynamics of spinless particles is more complicated.) Replacing $\dot{\mathbf{A}}$ everywhere with $\mathbf{\Pi}_\perp$, this gives a Hamiltonian (8.3.9) of the form

$$H = \int d^3x\,\left[\mathbf{\Pi}_\perp^2 + \tfrac{1}{2}(\nabla\times\mathbf{A})^2 - \tfrac{1}{2}(\mathbf{\Pi}_\perp + \nabla A^0)^2 - \mathbf{J}\cdot\mathbf{A} + J^0A^0\right] + H_{\rm M}\,,$$

where H_M is the Hamiltonian for matter fields, excluding their electromagnetic interactions

$$H_M \equiv \int d^3x \, (P_n \dot{Q}^n - \mathscr{L}_{\text{matter}}) \, .$$

Using the solution (8.2.9) for A^0, this is

$$H = \int d^3x \left[\tfrac{1}{2}\mathbf{\Pi}_\perp^2 + \tfrac{1}{2}(\nabla \times \mathbf{A})^2 - \mathbf{J} \cdot \mathbf{A} + \tfrac{1}{2}J^0 A^0 \right] + H_M \, . \tag{8.3.11}$$

The term $\frac{1}{2}J^0A^0$ may look peculiar, but this is nothing but the familiar Coulomb energy

$$\begin{aligned} V_{\text{Coul}} &= \tfrac{1}{2} \int d^3x \; J^0 \; A^0 \\ &= \tfrac{1}{2} \int d^3x \int d^3y \frac{J^0(\mathbf{x})J^0(\mathbf{y})}{4\pi|\mathbf{x} - \mathbf{y}|} \, . \end{aligned} \tag{8.3.12}$$

The reader can verify, using the commutation relations (8.3.5), that the rate of change of any operator function F of $\mathbf{A}$ and $\mathbf{\Pi}$ is given by $i\dot{F} = [F, H]$, as it should be.

8.4 Electrodynamics in the Interaction Picture

We now break up the Hamiltonian (8.3.11) into a free-particle term H_0 and an interaction V

$$H = H_0 + V \, , \tag{8.4.1}$$

$$H_0 = \int d^3x \left[\frac{1}{2}\mathbf{\Pi}_\perp^2 + \frac{1}{2}(\nabla \times \mathbf{A})^2 \right] + H_{\text{matter},0} \, , \tag{8.4.2}$$

$$V = -\int d^3x \; \mathbf{J} \cdot \mathbf{A} + V_{\text{Coul}} + V_{\text{matter}} \, , \tag{8.4.3}$$

where $H_{\text{matter},0}$ and V_{matter} are the free-particle and interaction terms in H_{matter}, and V_{Coul} is the Coulomb interaction (8.3.12). The total Hamiltonian (8.4.1) is time-independent, so Eqs. (8.4.2) and (8.4.3) can be evaluated at any time we like (as long as both are evaluated at the same time), in particular at $t = 0$. As in Chapter 7, the transition to the interaction picture is made by applying the similarity transformation

$$\begin{aligned} V(t) &= \exp(iH_0 t) \; V[\mathbf{A}, \mathbf{\Pi}_\perp, Q, P]_{t=0} \; \exp(-iH_0 t) \\ &= V \left[\mathbf{a}(t), \; \boldsymbol{\pi}(t), \; q(t), \; p(t) \right] \, , \end{aligned} \tag{8.4.4}$$

where P here denotes the canonical conjugates to the matter fields Q, and any operator $o(\mathbf{x}, t)$ in the interaction picture is related to its value $O(\mathbf{x}, 0)$

in the Heisenberg picture at $t = 0$ by

$$o(\mathbf{x}, t) = \exp(iH_0 t)\; O(\mathbf{x}, 0)\; \exp(-iH_0 t), \tag{8.4.5}$$

so that

$$i\, \dot{o}(\mathbf{x}, t) = [o(\mathbf{x}, t), H_0]\;. \tag{8.4.6}$$

(We are dropping the subscript $\perp$ on $\pi(x)$.) Since Eq. (8.4.5) is a similarity transformation, the equal-time commutation relations are the same as in the Heisenberg picture:

$$\left[a^i(\mathbf{x}, t), \pi^j(\mathbf{y}, t)\right] = i\left[\delta_{ij}\delta^3(\mathbf{x}-\mathbf{y}) + \frac{\partial^2}{\partial x^i \partial x^j}\frac{1}{4\pi|\mathbf{x}-\mathbf{y}|}\right], \tag{8.4.7}$$

$$\left[a^i(\mathbf{x}, t), a^j(\mathbf{y}, t)\right] = 0, \tag{8.4.8}$$

$$\left[\pi^i(\mathbf{x}, t), \pi^j(\mathbf{y}, t)\right] = 0, \tag{8.4.9}$$

and likewise for the matter fields and their conjugates. For the same reason, the constraints (8.2.6) and (8.3.8) still apply

$$\nabla \cdot \mathbf{a} = 0\,, \tag{8.4.10}$$

$$\nabla \cdot \boldsymbol{\pi} = 0\,. \tag{8.4.11}$$

To establish the relation between $\boldsymbol{\pi}$ and $\dot{\mathbf{a}}$, we must use Eq. (8.4.6) to evaluate $\dot{\mathbf{a}}$:

$$\begin{aligned} i\dot{a}_i(\mathbf{x}, t) &= [a_i(\mathbf{x}, t), H_0] \\ &= i\int d^3y\,\left[\delta_{ij}\delta^3(\mathbf{x}-\mathbf{y}) + \frac{\partial^2}{\partial x^i \partial x^j}\frac{1}{4\pi|\mathbf{x}-\mathbf{y}|}\right]\,\pi_j(\mathbf{y}, t)\,. \end{aligned}$$

We can replace $\partial/\partial x^j$ in the second term with $-\partial/\partial y^j$, integrate by parts, and use Eq. (8.4.11), yielding

$$\dot{\mathbf{a}} = \boldsymbol{\pi} \tag{8.4.12}$$

just as in the Heisenberg picture. The field equation is likewise determined by

$$\begin{aligned} i\dot{\pi}_i(\mathbf{x}, t) &= [\pi_i(\mathbf{x}, t), H_0] \\ &= -i\int d^3y\,\left[\delta_{ij}\delta^3(\mathbf{x}-\mathbf{y}) + \frac{\partial^2}{\partial x^i \partial x^j}\frac{1}{4\pi|\mathbf{x}-\mathbf{y}|}\right] \\ &\qquad \times (\nabla\times\nabla\times\mathbf{a}(\mathbf{y}, t))_j\,, \end{aligned}$$

which (using Eqs. (8.4.10) and (8.4.12)) just yields the usual wave equation

$$\Box\, \mathbf{a} = 0\,. \tag{8.4.13}$$

Since A^0 is not an independent Heisenberg-picture field variable, but rather a functional (8.2.9) of the matter fields and their canonical conjugates that vanishes in the limit of zero charges, we do not introduce any corresponding operator a^0 in the interaction picture, but rather take

$$a^0 = 0\,. \tag{8.4.14}$$

The most general real solution of Eqs. (8.4.10), (8.4.13), and (8.4.14) may be written

$$a^\mu(x) = (2\pi)^{-3/2}\int \frac{d^3p}{\sqrt{2p^0}}\sum_\sigma \left[e^{ip\cdot x}e^\mu(\mathbf{p},\sigma)\; a(\mathbf{p},\sigma) + e^{-ip\cdot x}e^{\mu *}(\mathbf{p},\sigma)\; a^\dagger(\mathbf{p},\sigma)\right], \tag{8.4.15}$$

where $p^0 \equiv |\mathbf{p}|$; $e^\mu(\mathbf{p},\sigma)$ are any two independent 'polarization vectors' satisfying

$$\mathbf{p}\cdot\; \mathbf{e}(\mathbf{p},\sigma) = 0\,, \tag{8.4.16}$$

$$e^0(\mathbf{p},\sigma) = 0\,, \tag{8.4.17}$$

and $a(\mathbf{p},\sigma)$ are a pair of operator coefficients, with σ a two-valued index. By adjusting the normalization of $a(\mathbf{p},\sigma)$, we can normalize the $e^\mu(\mathbf{p},\sigma)$ so that the completeness relation reads

$$\sum_\sigma e^i(\mathbf{p},\sigma)e^j(\mathbf{p},\sigma)^* = \delta_{ij} - p_i\, p_j/|\mathbf{p}|^2\,. \tag{8.4.18}$$

For instance, we could take the $\mathbf{e}(\mathbf{p},\sigma)$ to be the same polarization vectors that we encountered in Section 5.9:

$$e^\mu(\mathbf{p},\pm 1) = R(\hat{\mathbf{p}})\begin{bmatrix} 1/\sqrt{2} \\ \pm i/\sqrt{2} \\ 0 \\ 0 \end{bmatrix}\,, \tag{8.4.19}$$

where $R(\hat{\mathbf{p}})$ is a standard rotation that carries the three-axis into the direction of $\mathbf{p}$. Using Eqs. (8.4.18) and (8.4.12), we can easily see that the commutation relations (8.4.7)–(8.4.9) are satisfied if (and in fact only if) the operator coefficients in Eq. (8.4.15) satisfy

$$\left[a(\mathbf{p},\sigma),\; a^\dagger(\mathbf{p}\,',\sigma')\right] = \delta^3(\mathbf{p}-\mathbf{p}\,')\;\delta_{\sigma\sigma'}, \tag{8.4.20}$$

$$\left[a(\mathbf{p},\sigma),\; a(\mathbf{p}\,',\sigma')\right] = 0\,. \tag{8.4.21}$$

As remarked before for massive particles, this result should be regarded not so much as an alternative derivation of Eqs. (8.4.20) and (8.4.21), but rather as a verification that Eq. (8.4.2) gives the correct Hamiltonian for free massless particles of helicity ± 1. In the same spirit one can also use Eqs. (8.4.12) and (8.4.15) in Eq. (8.4.2) to calculate the free-photon

Hamiltonian

$$\begin{aligned} H_0 &= \int d^3p \sum_\sigma \tfrac{1}{2} p^0 \left[a(\mathbf{p},\sigma)\,,\, a^\dagger(\mathbf{p},\sigma\right]_+ \\ &= \int d^3p \sum_\sigma p^0 \left(a^\dagger(\mathbf{p},\sigma) a(\mathbf{p},\sigma) + \tfrac{1}{2}\delta^3(\mathbf{p}-\mathbf{p})\right) \end{aligned} \quad (8.4.22)$$

which (aside from an inconsequential infinite c-number term) is just what we should expect.

Finally, we record that the interaction (8.4.4) in the interaction picture is

$$V(t) = -\int d^3x\, j_\mu(\mathbf{x},t)\, a^\mu(\mathbf{x},t) + V_{\text{Coul}}(t) + V_{\text{matter}}(t)\,, \quad (8.4.23)$$

where in terms of the current J in the Heisenberg picture

$$j_\mu(\mathbf{x},t) \equiv \exp(iH_0t)\; J_\mu(\mathbf{x},0) \exp(-iH_0t)\,, \quad (8.4.24)$$

while $V_{\text{Coul}}(t)$ is the Coulomb term

$$\begin{aligned} V_{\text{Coul}}(t) &= \exp(iH_0t) V_{\text{Coul}} \exp(-iH_0t) \\ &= \frac{1}{2}\int d^3x\, d^3y\, \frac{j^0(\mathbf{x},t) j^0(\mathbf{y},t)}{4\pi|\mathbf{x}-\mathbf{y}|} \end{aligned} \quad (8.4.25)$$

and $V_{\text{matter}}(t)$ is the non-electromagnetic part of the matter field interaction in the interaction picture:

$$V_{\text{matter}}(t) = \exp(iH_0t)\; V_{\text{matter}} \exp(-iH_0t)\,. \quad (8.4.26)$$

We have written $j_\mu a^\mu$ instead of $\mathbf{j}\cdot\mathbf{a}$ in Eq. (8.4.23), but these are equal because a^μ has been defined to have $a^0 = 0$.

8.5 The Photon Propagator

The general Feynman rules described in Chapter 6 dictate that an internal photon line in a Feynman diagram contributes a factor to the corresponding term in the S-matrix, given by the propagator:

$$-i\Delta_{\mu\nu}(x-y) \equiv \left(\Phi_{\text{VAC}},\, T\left\{a_\mu(x)\,,\, a_\nu(y)\right\} \Phi_{\text{VAC}}\right)\,, \quad (8.5.1)$$

where T as usual denotes a time-ordered product. Inserting our formula (8.4.15) for the electromagnetic potential then yields

$$-i\Delta_{\mu\nu}(x-y) = \int \frac{d^3p}{(2\pi)^3 2|\mathbf{p}|} P_{\mu\nu}(\mathbf{p}) \left[e^{ip\cdot(x-y)}\theta(x-y) + e^{ip\cdot(y-x)}\theta(y-x)\right]\,, \quad (8.5.2)$$

where

$$P_{\mu\nu}(\mathbf{p}) \equiv \sum_{\sigma=\pm 1} e_\mu(\mathbf{p},\sigma)\, e_\nu(\mathbf{p},\sigma)^* \tag{8.5.3}$$

and p^μ in the exponentials is taken with $p^0 = |\mathbf{p}|$. We recall from Eqs. (8.4.18) and (8.4.17) that

$$P_{ij}(\mathbf{p}) = \delta_{ij} - \frac{p^i p^j}{|\mathbf{p}|^2}\,,$$
$$P_{0i}(\mathbf{p}) = P_{i0}(\mathbf{p}) = P_{00}(\mathbf{p}) = 0\,. \tag{8.5.4}$$

As we saw in Chapter 6, the theta functions in Eq. (8.5.2) may be expressed as integrals over an independent time-component q^0 of an off-shell four-momentum q^μ, so that Eq. (8.5.2) may be rewritten

$$\Delta_{\mu\nu}(x-y) = (2\pi)^{-4} \int d^4q\, \frac{P_{\mu\nu}(\mathbf{q})}{q^2 - i\epsilon}\, e^{iq\cdot(x-y)}\,. \tag{8.5.5}$$

Thus in using the Feynman rules in momentum space, the contribution of an internal photon line carrying four-momentum q that runs between vertices where the photon is created and destroyed by fields a^μ and a^ν is

$$\frac{-i}{(2\pi)^4}\frac{P_{\mu\nu}(\mathbf{q})}{q^2 - i\epsilon}\,. \tag{8.5.6}$$

It will be very useful (though apparently perverse) to rewrite Eq. (8.5.4) as

$$P_{\mu\nu}(\mathbf{q}) = \eta_{\mu\nu} + \frac{q^0 q_\mu n_\nu + q^0 q_\nu n_\mu - q_\mu q_\nu + q^2 n_\mu n_\nu}{|\mathbf{q}|^2}\,, \tag{8.5.7}$$

where $n^\mu \equiv (0,0,0,1)$ is a fixed time-like vector, q^2 as usual is $\mathbf{q}^2 - (q^0)^2$, but q^0 is here entirely arbitrary. We shall choose q^0 in Eq. (8.5.7) to be given by four-momentum conservation: it is the difference of the matter p^0s flowing in and out of the vertex where the photon line is created. The terms proportional to q_μ and/or q_ν then do not contribute to the S-matrix, because the factors q_μ or q_ν act like derivatives ∂_μ and ∂_ν, and the photon fields a_μ and a_ν are coupled to currents j^μ and j^ν that satisfy the conservation condition $\partial_\mu j^\mu = 0$.* The term proportional to $n_\mu n_\nu$ contains a factor q^2 that cancels the q^2 in the denominator of the propagator, yielding a term that is the same as would be produced by a term in the action:

$$-i\frac{1}{2}\int d^4x \int d^4y\, [-ij^0(x)][-ij^0(y)] \frac{-i}{(2\pi)^4} \int \frac{d^4q}{|\mathbf{q}|^2}\, e^{iq\cdot(x-y)}\,.$$

* This argument as given here is little better than hand-waving. The result has been justified by a detailed analysis of Feynman diagrams,[3] but the easiest way to treat this problem is by path-integral methods, as discussed in Section 9.6.

The integral over q^0 here yields a delta function in time, so this is equivalent to a correction to the interaction Hamiltonian $V(t)$, of the form

$$-\frac{1}{2}\int d^3x \int d^3y \, \frac{j^0(\mathbf{x},t)\, j^0(\mathbf{y},t)}{4\pi|\mathbf{x}-\mathbf{y}|} \, .$$

This is just right to cancel the Coulomb interaction (8.4.25). Our result is that the photon propagator can be taken effectively as the covariant quantity

$$\Delta^{\text{eff}}_{\mu\nu}(x-y) = (2\pi)^{-4}\int d^4q \frac{\eta_{\mu\nu}}{q^2 - i\epsilon} e^{iq\cdot(x-y)} \tag{8.5.8}$$

with the Coulomb interaction dropped from now on. We see that the apparent violation of Lorentz invariance in the instantaneous Coulomb interaction is cancelled by another apparent violation of Lorentz invariance, that as noted in Section 5.9 the fields $a^\mu(x)$ are not four-vectors, and therefore have a non-covariant propagator. From a practical point of view, the important point is that in the momentum space Feynman rules, the contribution of an internal photon line is simply given by

$$\frac{-i}{(2\pi)^4} \frac{\eta_{\mu\nu}}{q^2 - i\epsilon} \tag{8.5.9}$$

and the Coulomb interaction is dropped.

8.6 Feynman Rules for Spinor Electrodynamics

We are now in a position to state the Feynman rules for calculating the S-matrix in quantum electrodynamics. For definiteness, we will consider the electrodynamics of a single species of spin $\frac{1}{2}$ particles of charge $q = -e$ and mass m. We will call these fermions electrons, but the same formalism applies to muons and other such particles. The simplest gauge- and Lorentz-invariant Lagrangian for this theory is*

$$\mathscr{L} = -\frac{1}{4} F_{\mu\nu}F^{\mu\nu} - \bar{\Psi}\left(\gamma^\mu[\partial_\mu + ie A_\mu] + m\right)\Psi \, . \tag{8.6.1}$$

The electric current four-vector is then simply

$$J^\mu = \frac{\partial \mathscr{L}}{\partial A_\mu} = -ie\,\bar{\Psi}\gamma^\mu\Psi \, . \tag{8.6.2}$$

* In Chapter 12 we will discuss reasons why more complicated terms are excluded from the Lagrangian density.

The interaction (8.4.23) in the interaction picture is here

$$V(t) = +ie \int d^3x \, (\bar{\psi}(\mathbf{x},t)\gamma^\mu \psi(\mathbf{x},t)) \; a_\mu(\mathbf{x},t) + V_{\text{Coul}}(t) \,. \tag{8.6.3}$$

(There is no V_{matter} here.) As we have seen, the Coulomb term $V_{\text{Coul}}(t)$ just serves to cancel a part of the photon propagator that is non-covariant and local in time.

Following the general results of Section 6.3, we can state the momentum space Feynman rules for the connected part of the S-matrix in this theory as follows:

(i) Draw all Feynman diagrams with up to some given number of vertices. The diagrams consist of electron lines carrying arrows and photon lines without arrows, with the lines joined at vertices, at each of which there is one incoming and one outgoing electron line and one photon line. There is one external line coming into the diagram from below or going upwards out of the diagram for each particle in the initial or final states, respectively; electrons are represented by external lines carrying arrows pointing upwards into or out of the diagram, while positrons are represented by lines carrying arrows pointing downwards into or out of the diagram. There are also as many internal lines as are needed to give each vertex the required number of attached lines. Each internal line is labelled with an off-mass-shell four-momentum flowing in a definite direction along the line (taken conventionally to flow along the direction of the arrow for electron lines.) Each external line is labelled with the momentum and spin z-component or helicity of the electron or photon in the initial and final states.

(ii) Associate factors with the components of the diagram as follows:

Vertices

Label each vertex with a four-component Dirac index α at the electron line with its arrow coming into the vertex, a Dirac index β at the electron line with its arrow going out of the vertex, and a spacetime index μ at the photon line. For each such vertex, include a factor

$$(2\pi)^4 \, e \, (\gamma^\mu)_{\beta\alpha} \delta^4(k - k' + q) \,, \tag{8.6.4}$$

where k and k' are the electron four-momenta entering and leaving the vertex, and q is the photon four-momentum entering the vertex (or minus the photon momentum leaving the vertex).

External lines:

Label each external line with the three-momentum **p** and spin z-component or helicity σ of the particle in the initial or final state. For each line for an electron in the final state running out of a vertex carrying a Dirac label β on this line, include a factor*

$$\frac{\bar{u}_\beta(\mathbf{p},\sigma)}{(2\pi)^{3/2}}\,. \tag{8.6.5}$$

For each line for a positron in the final state running into a vertex carrying a Dirac label α on this line, include a factor

$$\frac{v_\alpha(\mathbf{p},\sigma)}{(2\pi)^{3/2}}\,. \tag{8.6.6}$$

For each line for an electron in the initial state running into a vertex carrying a Dirac label α on this line, include a factor

$$\frac{u_\alpha(\mathbf{p},\sigma)}{(2\pi)^{3/2}}\,. \tag{8.6.7}$$

For each line for a positron in the initial state running out of a vertex carrying a Dirac label β on this line, include a factor

$$\frac{\bar{v}_\beta(\mathbf{p},\sigma)}{(2\pi)^{3/2}}\,. \tag{8.6.8}$$

The us and vs are the four-component spinors discussed in Section 5.5. For each line for a photon in the final state connected to a vertex carrying a spacetime label μ on this line, include a factor

$$\frac{e_\mu^*(\mathbf{p},\sigma)}{(2\pi)^{3/2}\sqrt{2p^0}}\,. \tag{8.6.9}$$

For each line for a photon in the initial state connected to a vertex carrying a spacetime label μ on this line, include a factor

$$\frac{e_\mu(\mathbf{p},\sigma)}{(2\pi)^{3/2}\sqrt{2p^0}}\,. \tag{8.6.10}$$

The e_μ are the photon polarization four-vectors described in the previous section.

Internal lines:

For each internal electron line carrying a four-momentum k and running from a vertex carrying a Dirac label β to another vertex carrying a Dirac

* A matrix β has been extracted from the interaction in (8.6.4), so that $\bar{u}$ and $\bar{v}$ appear instead of $u^\dagger$ and $v^\dagger$.

label α, include a factor

$$\frac{-i}{(2\pi)^4}\,\frac{[-i\,\not{k}+m]_{\alpha\beta}}{k^2+m^2-i\epsilon}\,. \tag{8.6.11}$$

(We are here using the very convenient 'Dirac slash' notation; for any four-vector v^μ, $\not{v}$ denotes $\gamma_\mu v^\mu$.) For each internal photon line carrying a four-momentum q that runs between two vertices carrying spacetime labels μ and ν include a factor

$$\frac{-i}{(2\pi)^4}\,\frac{\eta_{\mu\nu}}{q^2-i\epsilon}\,. \tag{8.6.12}$$

(iii) Integrate the product of all these factors over the four-momenta carried by the internal lines, and sum over all Dirac and spacetime indices.

(iv) Add up the results obtained in this way from each Feynman diagram.

Additional combinatoric factors and fermionic signs may need to be included, as described in parts (v) and (vi) of Section 6.1.

The difficulty of evaluating Feynman diagrams increases rapidly with the number of internal lines and vertices, so it is important to have some idea of what numerical factors tend to suppress the contributions of the more complicated diagrams. We shall estimate these numerical factors including not only the factors of the electronic charge e associated with vertices, but also the factors of 2 and π from vertices, propagators, and momentum space integrals.

Consider a connected Feynman diagram with V vertices, I internal lines, E external lines, and L loops. These quantities are not independent, but are subject to relations already used in Section 6.3:

$$L = I - V + 1\,, \qquad 2I + E = 3V\,.$$

There is a factor $e(2\pi)^4$ from each vertex, a factor $(2\pi)^{-4}$ from each internal line, and a four-dimensional momentum space integral for each loop. The volume element in four-dimensional Euclidean space in terms of a radius parameter κ is $\pi^2\kappa^2 d\kappa^2$, so each loop contributes a factor π^2. Thus the diagram will contain a factor

$$(2\pi)^{4V}e^V(2\pi)^{-4I}\pi^{2L} = (2\pi)^4 e^{E-2}\left(\frac{e^2}{16\pi^2}\right)^L\,.$$

The number E of external lines is fixed for a given process, so we see that the expansion parameter that governs the suppression of Feynman graphs for each additional loop is

$$\frac{e^2}{16\pi^2} = \frac{\alpha}{4\pi} = 5.81\times 10^{-4}\,.$$

Fortunately this is small enough that good accuracy can usually be obtained from Feynman diagrams with at most a few loops.

* * *

We must say a little more about the spin states of photons and electrons in realistic experiments, where not every particle in the initial and final states has a definite known helicity or spin z-component. This consideration is especially important for photons, which in practice are often characterized by a state of transverse or elliptical polarization rather than helicity. As we saw in the previous section, for photons of helicity ± 1, the polarization vectors are

$$e(\mathbf{p}, \pm 1) = R(\hat{\mathbf{p}}) \begin{bmatrix} 1/\sqrt{2} \\ \pm i/\sqrt{2} \\ 0 \\ 0 \end{bmatrix} ,$$

where $R(\hat{\mathbf{p}})$ is the standard rotation that takes the z-axis to the $\mathbf{p}$ direction. These are not the only possible photon states; in general, a photon state can be a linear combination of helicity states $\Psi_{\mathbf{p},\pm 1}$

$$\alpha_+ \Psi_{\mathbf{p},+1} + \alpha_- \Psi_{\mathbf{p},-1} \tag{8.6.13}$$

which is properly normalized if

$$|\alpha_+|^2 + |\alpha_-|^2 = 1 . \tag{8.6.14}$$

To calculate the S-matrix element for absorbing or emitting such a photon, we simply replace $e_\mu(\mathbf{p}, \pm 1)$ in the Feynman rules with

$$e_\mu(\mathbf{p}) = \alpha_+ \, e_\mu(\mathbf{p}, +1) + \alpha_- \, e_\mu(\mathbf{p}, -1) . \tag{8.6.15}$$

The polarization vectors for definite helicity satisfy the normalization condition

$$e^*_\mu(\mathbf{p}, \lambda') \, e^\mu(\mathbf{p}, \lambda) = \delta_{\lambda' \lambda} \tag{8.6.16}$$

and therefore in general

$$e^*_\mu(\mathbf{p}) \, e^\mu(\mathbf{p}) = 1 . \tag{8.6.17}$$

The two extreme cases are *circular polarization*, for which $\alpha_- = 0$ or $\alpha_+ = 0$, and *linear polarization*, for which $|\alpha_+| = |\alpha_-| = 1/\sqrt{2}$. For linear polarization, by an adjustment of the overall phase of the state (8.6.13), we can make α_+ and α_- complex conjugates, so that they can be expressed as

$$\alpha_\pm = \exp(\mp i\phi) \big/ \sqrt{2} . \tag{8.6.18}$$

Then in the Feynman rules we should use a polarization vector

$$e_\mu(\mathbf{p}) = R(\hat{\mathbf{p}}) \begin{bmatrix} \cos\phi \\ \sin\phi \\ 0 \\ 0 \end{bmatrix} . \tag{8.6.19}$$

That is, ϕ is the azimuthal angle of the photon polarization in the plane perpendicular to $\mathbf{p}$. Note that the photon polarization vector here is *real*, which is only possible for linear polarization. In between the extremes of circular and linear polarization are the states of *elliptic* polarization, for which $|\alpha_+|$ and $|\alpha_-|$ are non-zero and unequal.

More generally, an initial photon may be prepared in a statistical mixture of spin states. In the most general case, an initial photon may have any number of possible polarization vectors $e^{(r)}_\mu(\mathbf{p})$, each with probability P_r. The rate for absorbing such a photon in a given process will then be of the form

$$\Gamma = \sum_r P_r |e^{(r)}_\mu(\mathbf{p})\, M^\mu|^2 = M^{\mu *} M^\nu \rho_{\nu\mu} , \tag{8.6.20}$$

where ρ is the *density matrix*

$$\rho_{\nu\mu} \equiv \sum_r P_r\, e^{(r)}_\nu(\mathbf{p})\, e^{(r)*}_\mu(\mathbf{p}) . \tag{8.6.21}$$

Since ρ is obviously a Hermitian positive matrix of unit trace (because $\sum_r P_r = 1$) with $\rho_{\nu 0} = \rho_{0\mu} = 0$ and $\rho_{\nu\mu} p^\nu = \rho_{\nu\mu} p^\mu = 0$, it may be written as

$$\rho_{\nu\mu} = \sum_{s=1,2} \lambda_s\, e_\nu(\mathbf{p}; s)\, e^*_\mu(\mathbf{p}; s) , \tag{8.6.22}$$

where $e_\mu(\mathbf{p}; s)$ are the two orthonormal eigenvectors of ρ with

$$e_0(\mathbf{p}; s) = e_\mu(\mathbf{p}; s) p^\mu = 0 \tag{8.6.23}$$

and λ_s are the corresponding eigenvalues, with

$$\lambda_s \geq 0 , \qquad \sum_{s=1,2} \lambda_s = 1 .$$

We may then write the rate for the photon absorption process as

$$\Gamma = \sum_{s=1,2} \lambda_s |e_\nu(\mathbf{p}; s) M^\nu|^2 . \tag{8.6.24}$$

Thus any statistical mixture of initial photon states is always equivalent to having just two orthonormal polarizations $e_\nu(\mathbf{p}; s)$ with probabilities λ_s.

In particular, if we know nothing whatever about the initial photon polarization, then the two probabilities λ_s for the polarization vectors

$e_\nu(\mathbf{p};s)$ are equal, so that $\lambda_1 = \lambda_2 = \frac{1}{2}$, and the density matrix (and hence the absorption rate) is an average over initial polarizations

$$\rho_{ij} = \frac{1}{2} \sum_{s=1,2} e_i(\mathbf{p};s)\, e_j^*(\mathbf{p};s) = \frac{1}{2}\left(\delta_{ij} - \hat{p}_i\hat{p}_j\right) . \tag{8.6.25}$$

Fortunately, this result does not depend on the particular pair of polarization vectors $e_i(\mathbf{p};s)$ over which we average; for unpolarized photons we can average the absorption rate over any pair of orthonormal polarization vectors. Similarly, if we make no attempt to measure the polarization of a photon in the final state, then the rate may be calculated by summing over any pair of orthonormal final photon polarization vectors.

The same remarks apply to electrons and positrons; if (as is usually the case) we make no attempt to prepare an electron or positron so that some spin states are more likely than others, then the rate is to be calculated by *averaging* over any two orthonormal initial spin states, such as those with spin z-component $\sigma = \pm\frac{1}{2}$; if we make no attempt to measure a final electron's or positron's spin state, then we must *sum* the rate over any two orthonormal final spin states, such as those with spin z-component $\sigma = \pm\frac{1}{2}$. Such sums may be performed using the relations (5.5.37) and (5.5.38):

$$\sum_\sigma u_\alpha(\mathbf{p},\sigma)\bar{u}_\beta(\mathbf{p},\sigma) = \left(\frac{-i\, \not{p} + m}{2p^0}\right)_{\alpha\beta} , \tag{8.6.26}$$

$$\sum_\sigma v_\alpha(\mathbf{p},\sigma)\bar{v}_\beta(\mathbf{p},\sigma) = \left(\frac{-i\, \not{p} - m}{2p^0}\right)_{\alpha\beta} , \tag{8.6.27}$$

where $p^0 = \sqrt{\mathbf{p}^2 + m^2}$. For instance, if the initial state contains an electron with momentum $\mathbf{p}$ and spin z-component σ, and a positron with momentum $\mathbf{p}'$ and spin z-component σ', then the S-matrix element for the process will be of the form $(\bar{v}_\alpha(\mathbf{p}',\sigma')\,\mathscr{M}_{\alpha\beta}\, u_\beta(\mathbf{p},\sigma))$. Hence if neither electron nor positron spins are observed, the rate will be proportional to

$$\frac{1}{4}\sum_{\sigma',\sigma} \left|(\bar{v}_\alpha(\mathbf{p}',\sigma')\,\mathscr{M}_{\alpha\beta}\, u_\beta(\mathbf{p},\sigma))\right|^2$$

$$= \frac{1}{4}\mathrm{Tr}\left\{\beta\,\mathscr{M}^\dagger\,\beta\left(\frac{-i\,\not{p}' - m}{2p^{0\prime}}\right)\mathscr{M}\left(\frac{-i\,\not{p} + m}{2p^0}\right)\right\} .$$

Techniques for the calculation of such traces are described in the Appendix to this chapter.

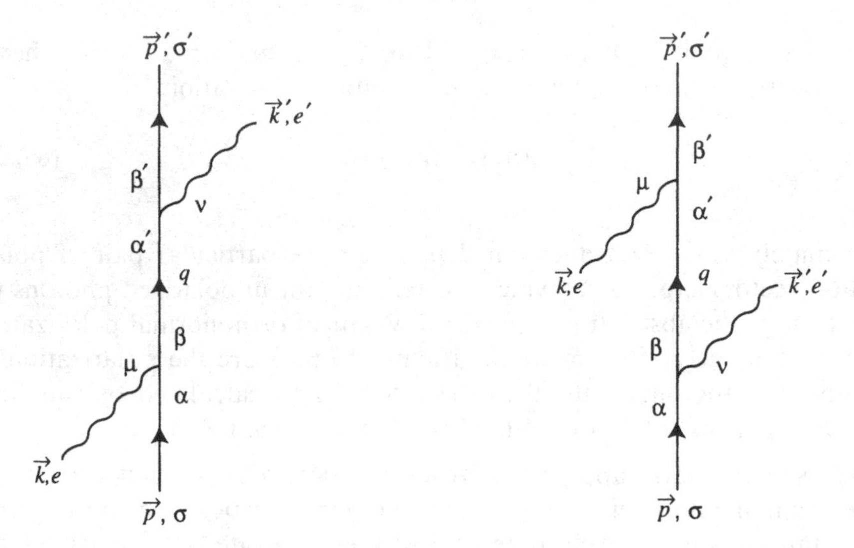

Figure 8.1. The two lowest-order Feynman diagrams for Compton scattering. Straight lines are electrons; wavy lines are photons.

8.7 Compton Scattering

As an example of the methods described in this chapter, we shall consider here the scattering of a photon by an electron (or other particle of spin $\frac{1}{2}$ and charge $-e$), to lowest order in e. We label the initial and final photon momenta and polarization vectors by k^μ, e^μ and k'^μ, e'^μ, where $k^0 = |\mathbf{k}|$ and $k^{0'} = |\mathbf{k}'|$. Also, the initial and final electron momenta and spin z-components are labelled p^μ, σ and p'^μ, σ', where $p^0 = \sqrt{\mathbf{p}^2 + m^2}$ and $p'^0 = \sqrt{\mathbf{p}'^2 + m^2}$, with m the electron mass. The lowest order Feynman diagrams for this process are shown in Figure 8.1. Using the rules outlined in the previous section, the corresponding S-matrix element is

$$S(\mathbf{p},\sigma + \mathbf{k},e \to \mathbf{p}',\sigma' + \mathbf{k}',e') =$$
$$\frac{\bar{u}(\mathbf{p}',\sigma')_{\beta'}}{(2\pi)^{3/2}} \frac{e^{*'}_\nu}{(2\pi)^{3/2}\sqrt{2k^{0'}}} \frac{u(\mathbf{p},\sigma)_\alpha}{(2\pi)^{3/2}} \frac{e_\mu}{(2\pi)^{3/2}\sqrt{2k^0}}$$
$$\times \int d^4q \left[\frac{-i}{(2\pi)^4}\right] \left[\frac{-i\,\not{q} + m}{q^2 + m^2 - i\epsilon}\right]_{\alpha'\beta}$$
$$\times \Bigg\{ \left[e(2\pi)^4 \gamma^\nu_{\beta'\alpha'} \delta^4(q - p' - k')\right] \left[e(2\pi)^4 \gamma^\mu_{\beta\alpha} \delta^4(q - p - k)\right]$$
$$+ \left[e(2\pi)^4 \gamma^\mu_{\beta'\alpha'} \delta^4(q + k - p')\right] \left[e(2\pi)^4 \gamma^\nu_{\beta\alpha} \delta^4(q + k' - p)\right] \Bigg\}. \quad (8.7.1)$$

Performing the (trivial) q-integral, collecting factors of i and 2π, and rewriting the result in matrix notation, we have more simply

$$S = \frac{-ie^2\delta^4(p'+k'-p-k)}{(2\pi)^2\sqrt{2k^{0'}\cdot 2k^0}}\bar{u}(\mathbf{p}',\sigma')\left[\not{e}^{*'}\left(\frac{-i(\not{p}+\not{k})+m}{(p+k)^2+m^2}\right)\not{e}\right.$$
$$\left. + \not{e}\left(\frac{-i(\not{p}-\not{k}')+m}{(p-k')^2+m^2}\right)\not{e}^{*'}\right]u(\mathbf{p},\sigma)\,. \tag{8.7.2}$$

(Here $\not{e}^*$ means $e^*_\mu\gamma^\mu$, not $(\not{e})^*$. Also, we drop the $-i\epsilon$, because the denominators here do not vanish.) Because $p^2 = -m^2$ and $k^2 = k'^2 = 0$, the denominators can be simplified

$$(p+k)^2 + m^2 = 2p\cdot k\,, \tag{8.7.3}$$
$$(p-k')^2 + m^2 = -2p\cdot k'\,. \tag{8.7.4}$$

Also, the 'Feynman amplitude' M is defined in general by Eq. (3.3.2), which (because some scattering is assumed to take place) here reads

$$S = -2\pi i\delta^4(p'+k'-p-k)M\,, \tag{8.7.5}$$

so

$$M = \frac{e^2}{4(2\pi)^3\sqrt{k^0k^{0'}}}\bar{u}(\mathbf{p}'\sigma')\left\{\not{e}^{*'}\left[-i(\not{p}+\not{k})+m\right]\not{e}\Big/p\cdot k\right.$$
$$\left. -\not{e}\left[-i(\not{p}-\not{k}')+m\right]\not{e}^{*'}\Big/p\cdot k'\right\}u(\mathbf{p},\sigma)\,. \tag{8.7.6}$$

The differential cross-section is given in terms of M by Eq. (3.4.15), which here reads

$$d\sigma = (2\pi)^4 u^{-1}|M|^2\delta^4(p'+k'-p-k)d^3p'd^3k'\,. \tag{8.7.7}$$

Since one of the particles here is massless, Eq. (3.4.17) for the initial velocity gives

$$u = |p\cdot k|\Big/p^0k^0\,. \tag{8.7.8}$$

To go further, it will be convenient to adopt a specific coordinate frame. Since electrons in atoms move non-relativistically, the laboratory frame for high-energy (X ray or gamma ray) photon–electron scattering experiments is usually (though not always) one in which the initial electron can be taken to be at rest. We will adopt this frame here, so that

$$\mathbf{p} = 0\,, \qquad p^0 = m\,. \tag{8.7.9}$$

The velocity (8.7.8) is then simply

$$u = 1\,. \tag{8.7.10}$$

To save writing, we denote the photon energies by

$$\omega = k^0 = |\mathbf{k}| = -p\cdot k\big/m\,, \tag{8.7.11}$$

$$\omega' = k^{0'} = |\mathbf{k}'| = -p\cdot k'\big/m\,. \tag{8.7.12}$$

The three-momentum delta function in Eq. (8.7.7) just serves to eliminate the differential d^3p', setting $\mathbf{p}' = \mathbf{k} - \mathbf{k}'$. This leaves the remaining energy delta function

$$\delta(p^{'0} + k^{'0} - p^0 - k^0) = \delta\left(\sqrt{(\mathbf{k}-\mathbf{k}')^2 + m^2} + \omega' - m - \omega\right)\,. \tag{8.7.13}$$

This fixes ω' to satisfy

$$\sqrt{\omega^2 - 2\omega\omega'\cos\theta + \omega'^2 + m^2} = \omega + m - \omega'\,,$$

where θ is the angle between $\mathbf{k}$ and $\mathbf{k}'$. Squaring both sides and cancelling $\omega^{'2}$ terms gives*

$$\omega' = \omega\,\frac{m}{m + \omega(1-\cos\theta)} \equiv \omega_c(\theta)\,. \tag{8.7.14}$$

The energy delta function (8.7.13) can be written

$$\begin{aligned}\delta(p^{'0} + k^{'0} - p^0 - k^0) &= \frac{\delta(\omega' - \omega_c(\theta))}{|\partial[\sqrt{\omega^2 - 2\omega\omega'\cos\theta + \omega^{'2} + m^2} + \omega']/\partial\omega'|} \\ &= \frac{\delta(\omega' - \omega_c(\theta))}{|(\omega' - \omega\,\cos\theta)/p'^0 + 1|} \\ &= \frac{p'^0\omega'}{m\omega}\delta(\omega' - \omega_c(\theta))\,.\end{aligned} \tag{8.7.15}$$

Also, the differential d^3k' can be written

$$d^3k' = \omega^{'2}d\omega' d\Omega\,, \tag{8.7.16}$$

where $d\Omega$ is the solid angle into which the final photon is scattered. The final delta function in Eq. (8.7.15) just serves to eliminate the differential $d\omega'$ in Eq. (8.7.16), leaving us with a differential cross-section

$$d\sigma = (2\pi)^4|M|^2\,\frac{p^{'0}\omega^{'3}}{m\omega}\,d\Omega \tag{8.7.17}$$

with $p^{'0} = m + \omega - \omega'$, and ω' given by Eq. (8.7.14).

---

* Equivalently, there is an increase in wavelength

$$\frac{1}{\omega'} - \frac{1}{\omega} = \frac{1-\cos\theta}{m}\,.$$

The verification of this formula in the scattering of X rays by electrons by A.H. Compton in 1922–3 played a key role in confirming Einstein's 1905 proposal of a quantum of light, which soon after Compton's experiments came to be known as the photon.

Usually we do not measure the spin z-component of the initial or final electron. In such cases, we must sum over σ' and average over σ, or in other words take half the sum over σ and σ':

$$d\bar{\sigma}(\mathbf{p} + \mathbf{k}, e \to \mathbf{p}' + \mathbf{k}', e') \equiv \tfrac{1}{2}\sum_{\sigma',\sigma} d\sigma(\mathbf{p},\sigma + \mathbf{k}, e \to \mathbf{p}',\sigma' + \mathbf{k}', e') . \tag{8.7.18}$$

To calculate this, we use the standard formula

$$\sum_{\sigma} u_\alpha(\mathbf{p},\sigma)\bar{u}_\beta(\mathbf{p},\sigma) = \frac{(-i\not{p} + m)_{\alpha\beta}}{2p^0} \tag{8.7.19}$$

and likewise for the sum over σ'. It follows that for an arbitrary 4×4 matrix A

$$\begin{aligned}\sum_{\sigma,\sigma'} \left|\bar{u}(\mathbf{p}',\sigma')Au(\mathbf{p},\sigma)\right|^2 &= \sum_{\sigma,\sigma'} (\bar{u}(\mathbf{p}',\sigma')Au(\mathbf{p},\sigma))(\bar{u}(\mathbf{p},\sigma)\beta A^\dagger \beta u(\mathbf{p}',\sigma')) \\ &= \sum_{\sigma,\sigma'} A_{\beta\alpha}u_\alpha(\mathbf{p},\sigma)\bar{u}_\gamma(\mathbf{p},\sigma)(\beta A^\dagger\beta)_{\gamma\delta}u_\delta(\mathbf{p}',\sigma')\bar{u}_\beta(\mathbf{p}',\sigma') \\ &= \mathrm{Tr}\left\{A\left(\frac{-i\not{p}+m}{2p^0}\right)\beta A^\dagger \beta\left(\frac{-i\not{p}'+m}{2p'^0}\right)\right\}. \end{aligned} \tag{8.7.20}$$

Recalling that $\beta\gamma_\mu^\dagger\beta = -\gamma_\mu$, Eq. (8.7.6) gives now

$$\begin{aligned}\sum_{\sigma,\sigma'}|M|^2 &= \frac{e^4}{64(2\pi)^6\omega\omega' p^0 p'^0} \\ &\times \mathrm{Tr}\left[\left\{\not{e}'^*\frac{[-i(\not{p}+\not{k})+m]}{p\cdot k}\not{e} - \not{e}\frac{[-i(\not{p}-\not{k}')+m]}{p\cdot k'}\not{e}'^*\right\}(-i\not{p}+m)\right. \\ &\times \left.\left\{\not{e}^*\frac{[-i(\not{p}+\not{k})+m]}{p\cdot k}\not{e}' - \not{e}'\frac{[-i(\not{p}-\not{k}')+m]}{p\cdot k'}\not{e}^*\right\}(-i\not{p}'+m)\right] .\end{aligned} \tag{8.7.21}$$

(Recall again that $\not{e}^*$ means $e_\mu^*\gamma^\mu$, not $(e_\mu\gamma^\mu)^*$, and likewise for $\not{e}'^*$.) We work in a 'gauge' in which

$$e\cdot p = e^*\cdot p = e'\cdot p = e'^*\cdot p = 0 \tag{8.7.22}$$

such as for instance Coulomb gauge in the laboratory frame, where $e^0 = e'^0 = 0$ and $\mathbf{p} = 0$. This implies that

$$\begin{aligned}[-i\not{p}+m]\,\not{e}[-i\not{p}+m] &= \not{e}[i\not{p}+m]\,[-i\not{p}+m] \\ &= \not{e}(\not{p}^2+m^2) = \not{e}(p_\mu p^\mu + m^2) = 0\end{aligned}$$

and likewise for $\not{e}'^*$, $\not{e}'$, and $\not{e}^*$. Eq. (8.7.21) can therefore be written in the greatly simplified form

$$\sum_{\sigma,\sigma'} |M|^2 = \frac{-e^4}{64(2\pi)^6 \omega\omega' p^0 p'^0} \mathrm{Tr}\left[\left\{\frac{\not{e}'^* \not{k} \not{e}}{p\cdot k} + \frac{\not{e} \not{k}' \not{e}'^*}{p\cdot k'}\right\}(-i\not{p}+m)\right.$$
$$\left.\times\left\{\frac{\not{e}^* \not{k} \not{e}'}{p\cdot k} + \frac{\not{e}' \not{k}' \not{e}^*}{p\cdot k'}\right\}(-i\not{p}'+m)\right] . \tag{8.7.23}$$

The trace of any product of an odd number of gamma matrices vanishes, so this breaks up into terms of zeroth and second order in m:

$$\sum_{\sigma,\sigma'} |M|^2 = \frac{e^4}{64(2\pi)^6 \omega\omega' p^0 p'^0}\left(\frac{T_1}{(p\cdot k)^2} + \frac{T_2}{(p\cdot k)(p\cdot k')} + \frac{T_3}{(p\cdot k)(p\cdot k')}\right.$$
$$\left. + \frac{T_4}{(p\cdot k')^2} - \frac{m^2 t_1}{(p\cdot k)^2} - \frac{m^2 t_2}{(p\cdot k)(p\cdot k')} - \frac{m^2 t_3}{(p\cdot k)(p\cdot k')} - \frac{m^2 t_4}{(p\cdot k')^2}\right) \tag{8.7.24}$$

where

$$T_1 = \mathrm{Tr}\left\{\not{e}'^* \not{k} \not{e} \not{p} \not{e}^* \not{k} \not{e}' \not{p}'\right\} , \tag{8.7.25}$$

$$T_2 = \mathrm{Tr}\left\{\not{e}'^* \not{k} \not{e} \not{p} \not{e}' \not{k}' \not{e}^* \not{p}'\right\} , \tag{8.7.26}$$

$$T_3 = \mathrm{Tr}\left\{\not{e} \not{k}' \not{e}'^* \not{p} \not{e}^* \not{k} \not{e}' \not{p}'\right\} , \tag{8.7.27}$$

$$T_4 = \mathrm{Tr}\left\{\not{e} \not{k}' \not{e}'^* \not{p} \not{e}' \not{k}' \not{e}^* \not{p}'\right\} , \tag{8.7.28}$$

$$t_1 = \mathrm{Tr}\left\{\not{e}'^* \not{k} \not{e} \not{e}^* \not{k} \not{e}'\right\} , \tag{8.7.29}$$

$$t_2 = \mathrm{Tr}\left\{\not{e}'^* \not{k} \not{e} \not{e}' \not{k}' \not{e}^*\right\} , \tag{8.7.30}$$

$$t_3 = \mathrm{Tr}\left\{\not{e} \not{k}' \not{e}'^* \not{e}^* \not{k} \not{e}'\right\} , \tag{8.7.31}$$

$$t_4 = \mathrm{Tr}\left\{\not{e} \not{k}' \not{e}'^* \not{e}' \not{k}' \not{e}^*\right\} . \tag{8.7.32}$$

The Appendix to this chapter shows how to calculate any trace $\mathrm{Tr}\{\not{a}\not{b}\not{c}\not{d}\ldots\}$ as a sum of products of scalar products of the four-vectors $a, b, c, d, \ldots$. In general, traces of products of 6 or 8 gamma matrices like the t_k or T_k would be given by a sum of 15 or 105 terms, respectively, but fortunately here most scalar products vanish; in addition to Eq. (8.7.22), we also have $k\cdot k = k'\cdot k' = 0$. (Furthermore, $e\cdot e^* = e'\cdot e'^* = 1$.) To simplify the calculation further, let us specialize to the case of *linear* polarization, where e^μ and e'^μ are real. Dropping the asterisks in Eqs. (8.7.25)–(8.7.32), we have then

$$T_1 = \mathrm{Tr}\left\{\not{e}' \not{k} \not{e} \not{p} \not{e} \not{k} \not{e}' \not{p}'\right\} .$$

Since $e^\mu p_\mu = 0$ and $e^\mu e_\mu = 1$, we have

$$\not{e}\,\not{p}\,\not{e} = -\not{p}\,\not{e}\,\not{e} = -\not{p}$$

so

$$T_1 = -\mathrm{Tr}\left\{\not{e}'\,\not{k}\,\not{p}\,\not{k}\,\not{e}'\,\not{p}'\right\}.$$

Also, $k^\mu k_\mu = 0$, so

$$\not{k}\,\not{p}\,\not{k} = -\not{k}\,\not{k}\,\not{p} + 2\,\not{k}p\cdot k = 2\,\not{k}p\cdot k$$

and hence

$$T_1 = -2p\cdot k\ \mathrm{Tr}\left\{\not{e}'\,\not{k}\,\not{e}'\,\not{p}'\right\}.$$

Using Eq. (8.A.6), this is

$$T_1 = -8p\cdot k\ [2e'\cdot k\ e'\cdot p' - k\cdot p'].$$

It is convenient to make the substitutions

$$e'\cdot p' = e'\cdot[p+k-k'] = e'\cdot k$$
$$k\cdot p' = -\tfrac{1}{2}(p'-k)^2 - \tfrac{1}{2}m^2 = -\tfrac{1}{2}(p-k')^2 - \tfrac{1}{2}m^2 = p\cdot k'$$

so

$$T_1 = -16\,p\cdot k(e'\cdot k)^2 + 8\,p\cdot k\,p\cdot k'. \tag{8.7.33}$$

A similar (though more lengthy) calculation gives

$$\begin{aligned}T_2 = T_3 = &-8(e\cdot k')^2(p\cdot k) + 16\,(e\cdot e')^2p\cdot k'p\cdot k + 8(e\cdot e')^2k\cdot k'm^2\\ &-8(e\cdot e')m^2(k\cdot e')(k'\cdot e) + 8(e'\cdot k)^2p\cdot k'\\ &-4(k\cdot p)^2 + 4(k\cdot k')(p\cdot p') - 4(k\cdot p')(p\cdot k'),\end{aligned} \tag{8.7.34}$$

$$T_4 = 16\,p\cdot k'(e\cdot k')^2 + 8(p\cdot k)(p\cdot k'), \tag{8.7.35}$$

$$t_1 = t_4 = 0, \tag{8.7.36}$$

$$t_2 = t_3 = -8\,e\cdot e'\,k\cdot e'\,k'\cdot e + 8(k\cdot k')(e\cdot e')^2 - 4(k\cdot k'). \tag{8.7.37}$$

Combining all these terms in Eq. (8.7.24) gives

$$\sum_{\sigma,\sigma'}|M|^2 = \frac{e^4}{64(2\pi)^6\omega\omega'p^0p'^0}\left[\frac{8(k\cdot k')^2}{(k\cdot p)(k'\cdot p)} + 32(e\cdot e')^2\right]. \tag{8.7.38}$$

All this applies in any Lorentz frame. In the *laboratory* frame, we have the special results

$$k\cdot k' = \omega\omega'(cos\theta - 1) = m\omega\omega'\left(\frac{1}{\omega} - \frac{1}{\omega'}\right),$$

$$p\cdot k = -m\omega \qquad p\cdot k' = -m\omega'.$$

Combining Eq. (8.7.38) with Eq. (8.7.17), the laboratory frame cross-section is

$$\tfrac{1}{2}\sum_{\sigma,\sigma'} d\sigma(\mathbf{p},\sigma + \mathbf{k},e \to \mathbf{p}',\sigma' + \mathbf{k}',e') = \frac{e^4\omega'^2 d\Omega}{64\pi^2 m^2\omega^2}$$

$$\times\left[\frac{\omega}{\omega'}+\frac{\omega'}{\omega} - 2 + 4(e\cdot e')^2\right] . \tag{8.7.39}$$

This is the celebrated formula derived (using old-fashioned perturbation theory) by O. Klein and Y. Nishina[4] in 1929.

As discussed in Section 8.6, if the incoming photon is (as usual) not prepared in a state with any particular polarization, then we must average over two orthonormal values of **e**. This average gives

$$\tfrac{1}{2}\sum_{e} e_i e_j = \tfrac{1}{2}(\delta_{ij} - \hat{\mathbf{k}}_i\hat{\mathbf{k}}_j)$$

and the differential cross-section is then

$$\tfrac{1}{4}\sum_{e,\sigma,\sigma'} d\sigma(\mathbf{p},\sigma + \mathbf{k},e \to \mathbf{p}',\sigma' + \mathbf{k},e') = \frac{e^4\omega'^2 d\Omega}{64\pi^2 m^2\omega^2}\left[\frac{\omega}{\omega'}+\frac{\omega'}{\omega} - 2(\hat{\mathbf{k}}\cdot e')^2\right] . \tag{8.7.40}$$

We see that the scattered photon is preferentially polarized in a direction perpendicular to the incident as well as the final photon direction, i.e., perpendicular to the plane in which the scattering takes place. This is a well-known result, responsible among other things for the polarization of light from eclipsing binary stars.**

To calculate the cross-section for experiments in which the final photon polarization is not measured, we must sum Eq. (8.7.40) over e', using

$$\sum_{e'} e'_i e'_j = \delta_{ij} - \hat{\mathbf{k}}'_i\hat{\mathbf{k}}'_j .$$

This gives

$$\tfrac{1}{4}\sum_{e,e',\sigma,\sigma'} d\sigma(\mathbf{p},\sigma + \mathbf{k},e \to \mathbf{p}',\sigma' + \mathbf{k}',e')$$

$$= \frac{e^4\omega'^2 d\Omega}{32\pi^2 m^2\omega^2}\left[\frac{\omega}{\omega'}+\frac{\omega'}{\omega} - 1 + \cos^2\theta\right] , \tag{8.7.41}$$

where θ is the angle between $\hat{\mathbf{k}}$ and $\hat{\mathbf{k}}'$. In the non-relativistic case, $\omega << m$,

** The light from one of the stars is polarized when it is scattered by free electrons in the outer atmosphere of the other, cooler, star when both are along the same line of sight. This polarization is normally undetectable because it cancels when the astronomer adds up light from all parts of the star's disk. The polarization has been observed in eclipsing binary stars at times when the cooler star blocks the light from just one side of the hotter star.

Eq. (8.7.41) gives

$$\tfrac{1}{4}\sum_{e,e',\sigma,\sigma'} d\sigma = \frac{e^4\, d\Omega}{32\pi^2 m^2}(1+\cos^2\theta)\,. \tag{8.7.42}$$

The solid angle integral is

$$\int [1+\cos^2\theta]d\Omega = \int_0^{2\pi} d\phi \int_0^{\pi} [1+\cos^2\theta]\ \sin\theta\ d\theta = \frac{16\pi}{3}\,,$$

giving a total cross-section for $\omega \ll m$:

$$\sigma_{\rm T} = \frac{e^4}{6\pi m^2}\,. \tag{8.7.43}$$

This is often written $\sigma_{\rm T} = 8\pi r_0^2/3$, where $r_0 = e^2/4\pi m = 2.818 \times 10^{-13}$ cm is known as the *classical electron radius.* Expression (8.7.43) is called the *Thomson cross-section*, after J. J. Thomson, the discoverer of the electron. Eqs. (8.7.42) and (8.7.43) were originally derived using classical mechanics and electrodynamics, by calculating the reradiation of light by a non-relativistic point charge in a plane wave electromagnetic field.

8.8 Generalization : *p*-form Gauge Fields*

The antisymmetric field strength tensor $F_{\mu\nu}$ of electromagnetism is a special case of a general class of tensors of special importance in physics and mathematics. A *p-form* is an antisymmetric covariant tensor of rank *p*. From a *p*-form $t_{\mu_1,\mu_2,\cdots\mu_p}$ one may construct a $(p+1)$-form called the *exterior derivative*** dt by taking the derivative and then antisymmetrizing with respect to all indices:

$$\begin{aligned}(dt)_{\mu_1\mu_2\cdots\mu_{p+1}} &\equiv \partial_{[\mu_1} t_{\mu_2\mu_3\cdots\mu_{p+1}]} \\ &\equiv \partial_{\mu_1} t_{\mu_2\mu_3\cdots\mu_{p+1}} - \partial_{\mu_2} t_{\mu_1\mu_3\cdots\mu_{p+1}} + \cdots + (-1)^p \partial_{\mu_{p+1}} t_{\mu_1\mu_2\cdots\mu_p}\end{aligned} \tag{8.8.1}$$

with square brackets indicating antisymmetrization with respect to the indices within the brackets. Because derivatives commute, repeated exterior derivatives vanish

$$d(dt) = 0\,. \tag{8.8.2}$$

A *p*-form whose exterior derivative vanishes is called *closed*, while a *p*-form that is itself an exterior derivative is called *exact*. From Eq. (8.8.2)

* This section lies somewhat out of the book's main line of development, and may be omitted in a first reading.

** Exterior derivatives and *p*-forms play a special role in general relativity. in part because the exterior derivative of a tensor transforms like a tensor even though it is calculated using ordinary rather than covariant derivatives.[5]

it follows that any exact p-form is closed; a famous theorem[6] of Poincaré states that in a region that is smoothly contractible to a point, any closed p-form is exact.† For instance, the homogeneous Maxwell equations (8.1.16) tell us that the electromagnetic field strength two-form $F_{\mu\nu}$ is closed; Poincaré's theorem then shows that it is also exact, so that it can be written as an exterior derivative, i.e., as $F_{\mu\nu} = \partial_\mu A_\nu - \partial_\nu A_\mu$. Again using Eq. (8.8.2), we see that the two-form $F_{\mu\nu}$ is invariant if A_μ is changed by an exterior derivative, i.e., by a gauge transformation $\delta A_\mu = \partial_\mu \Omega$.

The formalism of p-forms and exterior derivatives makes it natural to consider the possibility of massless particles described by p-form gauge fields‡ $A_{\mu_1 \cdots \mu_p}$, with an invariance under gauge transformations

$$\delta A = d\Omega \tag{8.8.3}$$

or in more detail

$$\delta A_{\mu_1 \cdots \mu_p} = \partial_{\,[\mu_1} \Omega_{\mu_2 \cdots \mu_p]} ,$$

where $\Omega_{\mu_1 \cdots \mu_{p-1}}$ is an arbitrary $(p-1)$-form. From such a p-form gauge field, we can construct a gauge-invariant field strength tensor given by

$$F = dA \tag{8.8.4}$$

or in detail

$$F_{\mu_1 \cdots \mu_{p+1}} = \partial_{\,[\mu_1} A_{\mu_2 \cdots \mu_{p+1}]} \,. \tag{8.8.5}$$

(Alternatively, we can start with a $(p+1)$-form F, and from an assumed condition $dF = 0$ infer the existence of a p-form A with $F = dA$.) By analogy with electrodynamics, we might expect the Lagrangian density for A to take the form

$$\mathscr{L} = -\frac{1}{2(p+1)} F_{\mu_1 \cdots \mu_{p+1}} F^{\mu_1 \cdots \mu_{p+1}} + J^{\mu_1 \cdots \mu_p} A_{\mu_1 \cdots \mu_p} \,, \tag{8.8.6}$$

where J is an antisymmetric tensor current (either a c-number, or a function of fields other than A) that in order to make the action gauge-invariant must satisfy the conservation condition

$$\partial_{\mu_1} J^{\mu_1 \cdots \mu_p} = 0 \,. \tag{8.8.7}$$

† In multiply connected spaces closed forms are not necessarily exact; although it is possible to write a closed p-form as an exterior derivative locally, this cannot in general be done smoothly throughout the space. The set of closed p-forms, modulo exact p-forms, makes up what is called the pth de Rham cohomology group of the space. There is a deep relation between the de Rham cohomology groups of a space and its topology,[6] which will be discussed further in Volume II.

‡ We are speaking loosely in calling $A_{\mu_1 \cdots \mu_p}$ a p-form, because for $F = dA$ to be a tensor it is only necessary for A to be a tensor up to a gauge transformation. In fact, we have already seen that in four spacetime dimensions, it is not possible to construct a four-vector field from the creation and annihilation operators of physical massless particles of helicity ± 1, so we have to deal with an $A^\mu(x)$ that according to Eq. (8.1.2) transforms as a four-vector only up to a gauge transformation.

The Euler–Lagrange equations are then

$$\partial_\mu F^{\mu\mu_1\cdots\mu_p} = -J^{\mu_1\cdots\mu_p} \, . \tag{8.8.8}$$

These p-form gauge fields play an important role in theories with more than four spacetime dimensions. For instance, in the simplest string theories in 26 spacetime dimensions, there is a normal mode of the string represented at low energies by a two-form gauge field $A_{\mu\nu}$. But in four spacetime dimensions, p-forms offer no new possibilities.

To see this, note first that in D spacetime dimensions there are no antisymmetric tensors with more than D indices, so in general we must take $p+1 \leq D$. Like any other $(p+1)$-form with $p+1 \leq D$, the field strength F may be expressed in terms of a dual $(D-p-1)$-form $\mathscr{F}$, as

$$F^{\mu_1\cdots\mu_{p+1}} = \epsilon^{\mu_1\cdots\mu_D}\mathscr{F}_{\mu_{p+2}\cdots\mu_D} \, . \tag{8.8.9}$$

Likewise, the p-form current J may be expressed in terms of a dual $(D-p)$-form current $\mathscr{J}$, as

$$J^{\mu_1\cdots\mu_p} = \epsilon^{\mu_1\cdots\mu_D}\mathscr{J}_{\mu_{p+1}\cdots\mu_D} \, . \tag{8.8.10}$$

The field equation (8.8.8) and conservation condition (8.8.7) then read simply

$$d\mathscr{F} = \mathscr{J} \, , \qquad d\mathscr{J} = 0 \, . \tag{8.8.11}$$

Because the dual current $\mathscr{J}$ is closed, it may be written in terms of a $(D-p-1)$-form $\mathscr{S}$ as

$$\mathscr{J} = d\mathscr{S} \, . \tag{8.8.12}$$

Eqs. (8.8.11) and (8.8.12) tell us that the difference of $\mathscr{F}$ and $\mathscr{S}$ is closed, and therefore according to Poincaré's theorem, may be written as

$$\mathscr{F} = \mathscr{S} + d\phi \, , \tag{8.8.13}$$

with ϕ a $(D-p-2)$-form. There is an exception for the case $p = D-1$, where $\mathscr{F}$ and $\mathscr{S}$ are zero-forms, i.e., scalars, and the condition $d\mathscr{F} = d\mathscr{S}$ simply tells us that $\mathscr{F}$ and $\mathscr{S}$ differ only by a constant. In this case the gauge field describes no degrees of freedom at all. We may therefore concern ourselves only with the cases $p \leq D-2$.

For $p \leq D-2$, the homogeneous 'Maxwell' equations $dF = 0$ read

$$\partial_{\mu_1}\mathscr{F}^{\mu_1\cdots\mu_{D-p-1}} = 0 \, , \tag{8.8.14}$$

which with Eq. (8.8.13) yields the field equation for ϕ:

$$\partial_{\mu_1}(d\phi)^{\mu_1\cdots\mu_{D-p-1}} = -\partial_{\mu_1}\mathscr{S}^{\mu_1\cdots\mu_{D-p-1}} \, . \tag{8.8.15}$$

This is invariant under a new set of gauge transformations $\phi \to \phi + d\omega$, except that where $D-p-2=0$ the gauge transformation that leaves F

invariant is $\phi \to \phi + c$, where c is an arbitrary constant. *We see that in D spacetime dimensions, the theory of a p-form gauge field A is equivalent to the theory of a* $(D-p-2)$*-form gauge field* ϕ.

We can now understand why p-form gauge fields offer no new possibilities in four spacetime dimensions. As we have seen, we need only consider the cases $p \leq D-2$, or $p = 0$, 1, or 2. A zero-form gauge field is a scalar S, for which Eq. (8.8.5) reads $F_\mu = \partial_\mu S$, and the field equations (8.8.8) read simply $\Box S = -J$. The gauge invariance here is invariance under a shift $S \to S + c$, with c a constant. This is just the theory of a massless scalar field with only derivative interactions. A one-form gauge field is a four-vector $A^\mu(x)$ coupled to a conserved four-vector current, just as in electrodynamics. Finally, according to the general result quoted above, a two-form gauge field in four spacetime dimensions is equivalent to a zero-form gauge field, which as we have seen is equivalent to a massless derivatively coupled scalar field.

Appendix Traces

In calculating S-matrix elements and transition rates for processes involving particles of spin $\frac{1}{2}$, we often encounter traces of products of Dirac gamma matrices. It will therefore be useful to give formulas for these traces that can be used in all such calculations.

For products of *even* numbers of gamma matrices, the trace is given by

$$\mathrm{Tr}\left\{\gamma_{\mu_1}\gamma_{\mu_2}\cdots\gamma_{\mu_{2N}}\right\} = 4 \sum_{\text{pairings}} \delta_P \prod_{\text{pairs}} \eta_{\text{paired }\mu\text{s}} \,. \tag{8.A.1}$$

Here the sum is over all different ways of pairing the indices $\mu_1, \cdots \mu_{2N}$. A pairing can be regarded as a permutation of the integers $1, 2, \cdots 2N$ into some order $P1$, $P2, \cdots P\cdot(2N)$, in which we pair μ_{P1} with μ_{P2}, μ_{P3} with μ_{P4}, and so on. Permuting pairs or permuting μs within a pair yields the same pairing, so the number of different pairings is

$$(2N)!/N!2^N = (2N-1)(2N-3)\cdots 1 \equiv (2N-1)!! \,. \tag{8.A.2}$$

We can avoid summing over equivalent pairings by requiring that

$$P1 < P2,\ P3 < P4, \cdots,\ P\cdot(2N-1) < P\cdot(2N) \tag{8.A.3}$$

and

$$P1 < P3 < P5 < \cdots \,. \tag{8.A.4}$$

With this convention, the factor δ_P is $+1$ or -1 according to whether the pairing involves an even or odd permutation of indices. The product in Eq.

(8.A.1) is over all N pairs, the nth pair contributing a factor $\eta_{\mu_{\mathcal{P}(2n-1)}\mu_{\mathcal{P}(2n)}}$. For instance (writing $\mu, \nu, \rho, \sigma, \cdots$ in place of $\mu_1, \mu_2, \mu_3, \mu_4, \cdots$), for $N = 1, 2$, and 3 we have*

$$\mathrm{Tr}\{\gamma_\mu\gamma_\nu\} = 4\,\eta_{\mu\nu}\,, \tag{8.A.5}$$

$$\mathrm{Tr}\{\gamma_\mu\gamma_\nu\gamma_\rho\gamma_\sigma\} = 4\Big[\eta_{\mu\nu}\eta_{\rho\sigma} - \eta_{\mu\rho}\eta_{\nu\sigma} + \eta_{\mu\sigma}\eta_{\nu\rho}\Big]\,, \tag{8.A.6}$$

$$\begin{aligned}\mathrm{Tr}\{\gamma_\mu\gamma_\nu\gamma_\rho\gamma_\sigma\gamma_\kappa\gamma_\eta\} = 4\Big[&\eta_{\mu\nu}\eta_{\rho\sigma}\eta_{\kappa\eta} - \eta_{\mu\nu}\eta_{\rho\kappa}\eta_{\sigma\eta} + \eta_{\mu\nu}\eta_{\rho\eta}\eta_{\sigma\kappa}\\ &-\eta_{\mu\rho}\eta_{\nu\sigma}\eta_{\kappa\eta} + \eta_{\mu\rho}\eta_{\nu\kappa}\eta_{\sigma\eta} - \eta_{\mu\rho}\eta_{\nu\eta}\eta_{\sigma\kappa} + \eta_{\mu\sigma}\eta_{\nu\rho}\eta_{\kappa\eta} - \eta_{\mu\sigma}\eta_{\nu\kappa}\eta_{\rho\eta}\\ &+\eta_{\mu\sigma}\eta_{\nu\eta}\eta_{\rho\kappa} - \eta_{\mu\kappa}\eta_{\nu\rho}\eta_{\sigma\eta} + \eta_{\mu\kappa}\eta_{\nu\sigma}\eta_{\rho\eta} - \eta_{\mu\kappa}\eta_{\nu\eta}\eta_{\rho\sigma} + \eta_{\mu\eta}\eta_{\nu\rho}\eta_{\sigma\kappa}\\ &-\eta_{\mu\eta}\eta_{\nu\sigma}\eta_{\rho\kappa} + \eta_{\mu\eta}\eta_{\nu\kappa}\eta_{\rho\sigma}\Big]\,.\end{aligned} \tag{8.A.7}$$

For an odd number of gamma matrices, the result is much simpler

$$\mathrm{Tr}\{\gamma_{\mu_1}\gamma_{\mu_2}\cdots\gamma_{\mu_{2N+1}}\} = 0\,. \tag{8.A.8}$$

The proof of Eq. (8.A.1) is by mathematical induction. First note that

$$\mathrm{Tr}\{\gamma_\mu\gamma_\nu\} = -\mathrm{Tr}\{\gamma_\nu\gamma_\mu\} + 2\mathrm{Tr}\{\eta_{\mu\nu}\,1\} = -\mathrm{Tr}\{\gamma_\mu\gamma_\nu\} + 8\eta_{\mu\nu}\,,$$

so $\mathrm{Tr}\{\gamma_\mu\gamma_\nu\} = 4\,\eta_{\mu\nu}$, in agreement with Eq. (8.A.1). Next, suppose that Eq. (8.A.1) is true for $N \leq M-1$. We then have

$$\begin{aligned}\mathrm{Tr}\{\gamma_{\mu_1}\gamma_{\mu_2}\cdots\gamma_{\mu_{2M}}\} &= 2\,\eta_{\mu_1\mu_2}\mathrm{Tr}\{\gamma_{\mu_3}\cdots\gamma_{\mu_{2M}}\} - \mathrm{Tr}\{\gamma_{\mu_2}\gamma_{\mu_1}\gamma_{\mu_3}\cdots\gamma_{\mu_{2M}}\}\\ &= 2\,\eta_{\mu_1\mu_2}\,\mathrm{Tr}\{\gamma_{\mu_3}\cdots\gamma_{\mu_{2M}}\} - 2\,\eta_{\mu_1\mu_3}\,\mathrm{Tr}\{\gamma_{\mu_2}\gamma_{\mu_4}\cdots\gamma_{\mu_{2N}}\}\\ &\quad + \mathrm{Tr}\{\gamma_{\mu_2}\gamma_{\mu_3}\gamma_{\mu_1}\gamma_{\mu_4}\cdots\gamma_{\mu_{2M}}\}\\ &= 2\,\eta_{\mu_1\mu_2}\,\mathrm{Tr}\{\gamma_{\mu_3}\cdots\gamma_{\mu_{2M}}\} - 2\,\eta_{\mu_1\mu_3}\,\mathrm{Tr}\{\gamma_{\mu_2}\gamma_{\mu_4}\cdots\gamma_{\mu_{2M}}\}\\ &\quad + 2\,\eta_{\mu_1\mu_4}\,\mathrm{Tr}\{\gamma_{\mu_2}\gamma_{\mu_3}\gamma_{\mu_5}\cdots\gamma_{\mu_{2M}}\} - \cdots\\ &\quad + 2\,\eta_{\mu_1\mu_{2M}}\,\mathrm{Tr}\{\gamma_{\mu_2}\cdots\gamma_{\mu_{2M-1}}\} - \mathrm{Tr}\{\gamma_{\mu_2}\cdots\gamma_{\mu_{2M}}\gamma_{\mu_1}\}\,.\end{aligned}$$

All commutators have zero trace, so the last term subtracted here is the same as the left-hand side, and so

$$\begin{aligned}\mathrm{Tr}\{\gamma_{\mu_1}\gamma_{\mu_2}\cdots\gamma_{\mu_{2M}}\} &= \eta_{\mu_1\mu_2}\,\mathrm{Tr}\{\gamma_{\mu_3}\cdots\gamma_{\mu_{2M}}\}\\ &- \eta_{\mu_1\mu_3}\,\mathrm{Tr}\{\gamma_{\mu_2}\gamma_{\mu_4}\cdots\gamma_{\mu_{2M}}\} + \eta_{\mu_1\mu_4}\,\mathrm{Tr}\{\gamma_{\mu_2}\gamma_{\mu_3}\gamma_{\mu_5}\cdots\gamma_{\mu_{2M}}\}\\ &- \cdots + \eta_{\mu_1\mu_{2M}}\,\mathrm{Tr}\{\gamma_{\mu_2}\cdots\gamma_{\mu_{2M-1}}\}\,.\end{aligned} \tag{8.A.9}$$

If we assume that Eq. (8.A.1) correctly gives the trace of any product of $2N-2$ Dirac matrices, then Eq. (8.A.9) shows that Eq. (8.A.1) also correctly gives the trace of any product of $2N$ Dirac matrices.

The easiest way to see that the trace of an odd number of Dirac matrices vanishes is to note that $-\gamma_\mu$ is related to γ_μ by a similarity transformation, $-\gamma_\mu = \gamma_5\gamma_\mu(\gamma_5)^{-1}$. Traces are unaffected by such similarity

* There are now computer programs[7] available for the calculation of traces of products of large numbers of Dirac matrices.

transformations, so the trace of an odd number of Dirac matrices is equal to minus itself, and hence vanishes.

One occasionally encounters another class of traces, of the form

$$\mathrm{Tr}\{\gamma_5\gamma_{\mu_1}\gamma_{\mu_2}\cdots\gamma_{\mu_n}\}\,.$$

This vanishes for odd n for the same reason as given above for traces without a γ_5. It also vanishes for $n=0$ and $n=2$:

$$\mathrm{Tr}\{\gamma_5\}=0 \tag{8.A.10}$$

$$\mathrm{Tr}\{\gamma_5\gamma_\mu\gamma_\nu\}=0\,. \tag{8.A.11}$$

(To see this just recall that $\gamma_5\equiv i\gamma_0\gamma_1\gamma_2\gamma_3$, and note that there is no way of pairing the indices in $\mathrm{Tr}\{\gamma_0\gamma_1\gamma_2\gamma_3\}$ or in $\mathrm{Tr}\{\gamma_0\gamma_1\gamma_2\gamma_3\gamma_\mu\gamma_\nu\}$ so that the spacetime indices in each pair are equal.) For $n=4$ it is possible to pair the indices in $\mathrm{Tr}\{\gamma_0\gamma_1\gamma_2\gamma_3\gamma_\mu\gamma_\nu\gamma_\rho\gamma_\sigma\}$ so that the spacetime indices in each pair are equal, but only if μ,ν,ρ,σ are some permutation of $0,1,2,3$. Furthermore this trace must be odd under permutations of μ,ν,ρ,σ since gamma matrices with different indices anticommute. Thus the trace $\mathrm{Tr}\{\gamma_5\gamma_\mu\gamma_\nu\gamma_\rho\gamma_\sigma\}$ must be proportional to the totally antisymmetric tensor $\epsilon_{\mu\nu\rho\sigma}$. The constant of proportionality may be worked out by letting μ,ν,ρ,σ take the values $0,1,2,3$, and recalling that $\epsilon_{0123}\equiv-1$. In this way we find

$$\mathrm{Tr}\{\gamma_5\gamma_\mu\gamma_\nu\gamma_\rho\gamma_\sigma\}=4i\epsilon_{\mu\nu\rho\sigma}\,. \tag{8.A.12}$$

The trace of products of γ_5 with six, eight, or more Dirac matrices may be calculated by the same methods used above to verify Eq. (8.A.1).

Problems

1. Calculate the differential and total cross-sections for the process $e^+e^-\to\mu^+\mu^-$ to lowest order in e. Assume that electron and muon spins are not observed. Use the simplest Lagrangian for the electrodynamics of electrons and muons.

2. Carry out the canonical quantization of the theory of a charged scalar field Φ and its interaction with electromagnetism, with Lagrangian density:

$$\mathscr{L}=-(D_\mu\Phi)^\dagger(D^\mu\Phi)-m^2\Phi^\dagger\Phi-\lambda(\Phi^\dagger\Phi)^2-\tfrac{1}{4}F_{\mu\nu}F^{\mu\nu}\,,$$

where

$$D_\mu\Phi\equiv\partial_\mu\Phi-iqA_\mu\Phi\,,\qquad F_{\mu\nu}\equiv\partial_\mu A_\nu-\partial_\nu A_\mu\,.$$

Use Coulomb gauge. Express the Hamiltonian in terms of the fields $\mathbf{A}$, Φ, and $\Phi^\dagger$ and their canonical conjugates. Evaluate the interaction

$V(t)$ in the interaction-picture in terms of the interaction picture fields and their derivatives.

3. Use the results of Problem 2 to calculate the differential and total cross-sections for photon scattering by a massive charged scalar particle to lowest order in e.

4. Write a gauge-invariant Lagrangian for a charged massive vector field interacting with the electromagnetic field.

5. Calculate the differential cross-section for electron–electron scattering to lowest order in e. Assume that final and initial spins are not measured.

References

1. See, e.g., M. B. Green, J. H. Schwarz, and E. Witten, *Superstring Theory* (Cambridge University Press, Cambridge, 1987): Section 2.2.

1*a*. V. Fock, *Z. f. Phys.* **39**, 226 (1927); H. Weyl, *Z. Phys.* **56**, 330 (1929). The term 'gauge invariance' derives from an analogy with earlier speculations about scale invariance by H. Weyl, in *Raum, Zeit, Materie*, 3rd ed. (Springer-Verlag, Berlin, 1920). Also see F. London, *Z. f. Phys.* **42**, 375 (1927). This history has been reviewed by C. N. Yang, talk at City College (unpublished).

2. The use of Coulomb gauge in electrodynamics was strongly advocated by Schwinger on pretty much the same grounds as here: that we ought not to introduce photons with helicities other than ± 1. See J. Schwinger, *Phys. Rev.* **78**, 1439 (1948); **127**, 324 (1962); *Nuovo Cimento* **30**, 278 (1963).

3. R. P. Feynman, *Phys. Rev.* **101**, 769 (1949): Section 8.

4. O. Klein and Y. Nishina, *Z. f. Phys.* **52**, 853 (1929); Y. Nishina, *ibid.*, 869 (1929); also see I. Tamm, *Z. f. Phys.* **62**, 545 (1930).

5. See, e.g., S. Weinberg, *Gravitation and Cosmology* (Wiley, New York, 1972): Section 4.11.

6. For a readable general introduction to the geometry and topology of p-forms, see H. Flanders, *Differential Forms* (Academic Press, New York, 1963).

7. T. West, *Comput. Phys. Commun.* **77**, 286 (1993)

9

Path-Integral Methods

In Chapters 7 and 8 we applied the canonical quantization operator formalism to derive the Feynman rules for a variety of theories. In many cases, such as the scalar field with derivative coupling or the vector field with zero or non-zero mass, the procedure though straightforward was rather awkward. The interaction Hamiltonian turned out to contain a covariant term, equal to the negative of the interaction term in the Lagrangian, plus a non-covariant term, which served to cancel non-covariant terms in the propagator. In the case of electrodynamics this non-covariant term (the Coulomb energy) turned out to be not even spatially local, though it is local in time. Yet the final results are quite simple: the Feynman rules are just those we should obtain with covariant propagators, and using the negative of the interaction term in the Lagrangian to calculate vertex contributions. The awkwardness in obtaining these simple results, which was bad enough for the theories considered in Chapters 7 and 8, becomes unbearable for more complicated theories, like the non-Abelian gauge theories to be discussed in Volume II, and also general relativity. One would very much prefer a method of calculation that goes directly from the Lagrangian to the Feynman rules in their final, Lorentz-covariant form.

Fortunately, such a method does exist. It is provided by the path-integral approach to quantum mechanics. This was first presented in the context of non-relativistic quantum mechanics in Feynman's Princeton Ph. D. thesis,[1] as a means of working directly with a Lagrangian rather than a Hamiltonian. In this respect, it was inspired by earlier work of Dirac.[2] The path-integral approach played a part (along with inspired guesswork) in Feynman's later derivation of his diagrammatic rules.[3] However, although Feynman diagrams became widely used in the 1950s, most physicists (including myself) tended to derive them using the operator methods of Schwinger and Tomonaga, which were shown by Dyson in 1949 to lead to the same diagrammatic rules that had been obtained by Feynman by his own methods.

The path-integral approach was revived in the late 1960s, when Faddeev and Popov[4] and De Witt[5] showed how to apply it to non-Abelian gauge theories and general relativity. For most theorists, the turning point came in 1971, when 't Hooft[6] used path-integral methods to derive the Feynman rules for spontaneously broken gauge theories (discussed in Volume II), including in particular the theory of weak and electromagnetic interactions, in a gauge that made the high energy behavior of these theories transparent. Soon after, as also discussed in Volume II, it was discovered that the path-integral method allows us to take account of contributions to the S-matrix that have an essential singularity at zero coupling constant and therefore cannot be discovered in any finite order of perturbation theory. Since then, the path-integral methods described here have become an indispensable part of the equipment of all physicists who make use of quantum field theory.

At this point the reader may be wondering why if the path-integral method is so convenient we bothered in Chapter 7 to introduce the canonical formalism. Indeed, Feynman seems at first to have thought of his path-integral approach as a substitute for the ordinary canonical formulation of quantum mechanics. There are two reasons for starting with the canonical formalism. The first is a point of principle: although the path-integral formalism provides us with manifestly Lorentz-invariant diagrammatic rules, it does not make clear why the S-matrix calculated in this way is unitary. As far as I know, the only way to show that the path-integral formalism yields a unitary S-matrix is to use it to reconstruct the canonical formalism, in which unitarity is obvious. There is a kind of conservation of trouble here; we can use the canonical approach, in which unitarity is obvious and Lorentz invariance obscure, or the path-integral approach, which is manifestly Lorentz-invariant but far from manifestly unitary. Since the path-integral approach is here derived from the canonical approach, we know that the two approaches yield the same S-matrix, so that the S-matrix must indeed be both Lorentz-invariant and unitary.

The second reason for introducing the canonical formalism first is more practical: there are important theories in which the simplest version of the Feynman path-integral method, in which propagators and interaction vertices are taken directly from the Lagrangian, is simply wrong. One example is the non-linear σ-model, with Lagrangian density $\mathscr{L} = -\frac{1}{2}g_{k\ell}(\phi)\partial_\mu\phi^k\partial^\mu\phi^\ell$. In such theories, using the naive Feynman rules derived directly from the Lagrangian density would yield an S-matrix that is not only wrong but even non-unitary, and that also depends on the way in which we define the scalar field.[7] In this chapter we shall derive the path-integral formalism from the canonical formalism, and in this way we will see what additional sorts of vertices are needed to supplement the simplest version of the Feynman path-integral method.

9.1 The General Path-Integral Formula

We start with a general quantum mechanical system, with Hermitian operator 'coordinates' Q_a and conjugate 'momenta' P_b, satisfying the canonical commutation relations*:

$$[Q_a,\, P_b] = i\,\delta_{ab}\,, \tag{9.1.1}$$

$$[Q_a,\, Q_b] = [P_a,\, P_b] = 0\,. \tag{9.1.2}$$

(We shall restrict ourselves in this and the next three sections to bosonic operators, which satisfy commutation rather than anticommutation relations. Our results will be generalized to include fermionic operators in Section 9.5.) In a field theory, the index a consists of a position $\mathbf{x}$ and a discrete Lorentz and species index m, and we conventionally write

$$Q_{\mathbf{x},m} \equiv Q_m(\mathbf{x}), \tag{9.1.3}$$

$$P_{\mathbf{x},m} \equiv P_m(\mathbf{x})\,. \tag{9.1.4}$$

Also, the Kronecker delta in Eq. (9.1.1) is interpreted in a field theory as

$$\delta_{\mathbf{x},m\,;\,\mathbf{y},n} \equiv \delta^3(\mathbf{x}-\mathbf{y})\delta_{mn}\,. \tag{9.1.5}$$

However, for the present it will be convenient to use the more compact notation of Eqs. (9.1.1) and (9.1.2). These are 'Schrödinger-picture' operators, taken at a fixed time (say, $t = 0$). The time-dependent operators in the Heisenberg picture will be considered a little later.

Since the Q_a all commute, we can find a simultaneous eigenstate $|q\rangle$, with eigenvalues q_a:

$$Q_a|q\rangle = q_a\,|q\rangle\,. \tag{9.1.6}$$

(We are using lower case qs and ps here to denote eigenvalues rather than to denote operators in the interaction picture as in Chapter 7, but since we will not be using the interaction picture in this chapter no confusion should arise.) The eigenvectors can be taken to be orthonormal,

$$\langle q'|q\rangle = \prod_a \delta(q'_a - q_a) \equiv \delta(q'-q)\,, \tag{9.1.7}$$

so that the completeness relation reads

$$1 = \int \prod_a dq_a\,|q\rangle\,\langle q|\,. \tag{9.1.8}$$

---

* We are tacitly assuming here that any first class constraints are eliminated by choosing a gauge, and any remaining second class constraints are 'solved' by writing the constrained degrees of freedom in terms of the unconstrained Q_a and P_a, as in Section 7.6. The direct application of path-integral methods to constrained systems is described by Faddeev.[8]

Similarly, we can find a complete orthonormal set of eigenstates of the P_a:

$$P_a|p\rangle = p_a|p\rangle \,, \tag{9.1.9}$$

$$\langle p'|p\rangle = \prod_a \delta(p'_a - p_a) \equiv \delta(p' - p) \,, \tag{9.1.10}$$

$$1 = \int \prod_a dp_a \, |p\rangle \, \langle p| \,. \tag{9.1.11}$$

As usual, it follows from Eq. (9.1.1) that these two complete sets of eigenstates have the scalar product**

$$\langle q\,|p\rangle = \prod_a \frac{1}{\sqrt{2\pi}} \exp(iq_a p_a) \,. \tag{9.1.12}$$

In the Heisenberg picture, the Q and P operators are given a time-dependence

$$Q_a(t) \equiv \exp(iHt)\, Q_a \exp(-iHt) \,, \tag{9.1.13}$$

$$P_a(t) \equiv \exp(iHt)\, P_a \exp(-iHt) \,, \tag{9.1.14}$$

where H is the total Hamiltonian. These have eigenstates $|q;t\rangle$ and $|p;t\rangle$

$$Q_a(t)|q;t\rangle = q_a|q;t\rangle \,, \tag{9.1.15}$$

$$P_a(t)|p;t\rangle = p_a|p;t\rangle \,, \tag{9.1.16}$$

given by

$$|q;t\rangle = \exp(iHt)|q\rangle \,, \tag{9.1.17}$$

$$|p;t\rangle = \exp(iHt)|p\rangle \,. \tag{9.1.18}$$

(Note that $|q;t\rangle$ is the eigenstate of $Q_a(t)$ with eigenvalue q_a, and *not* the result of letting the state $|q\rangle$ evolve for a time t. This is why its time-dependence is given by a factor $\exp(iHt)$ rather than $\exp(-iHt)$.) These states obviously satisfy the completeness and orthonormality conditions

$$\langle q';t|q;t\rangle = \delta(q' - q) \,, \tag{9.1.19}$$

$$\langle p';t|p;t\rangle = \delta(p' - p) \,, \tag{9.1.20}$$

$$\int \prod_a dq_a \, |q;t\rangle \, \langle q;t| = 1 \,, \tag{9.1.21}$$

$$\int \prod_a dp_a \, |p;t\rangle \, \langle p;t| = 1 \,, \tag{9.1.22}$$

** The proof follows the same lines as in the quantum mechanics of point particles. From Eq. (9.1.1), we see that P_b acts as $-i\partial/\partial q_b$ on wave functions in a q-basis. The right-hand side of Eq. (9.1.12) is then seen to be the wave function in this basis of an eigenstate of P. The factor $\prod 1/\sqrt{2\pi}$ is fixed by the normalization requirement, Eq. (9.1.10).

and also

$$\langle q;t|p;t\rangle = \prod_a \frac{1}{\sqrt{2\pi}} \exp(iq_a p_a)\,. \tag{9.1.23}$$

If, by measurements at time t, we find that our system is in a definite state $|q;t\rangle$, then the probability amplitude for measurements at time t' to give a state $|q';t'\rangle$ is the scalar product $\langle q';t'|q;t\rangle$. Our central dynamical problem is to calculate this scalar product.

This is easy when t' and t are infinitesimally close, say $t' = \tau + d\tau$ and $t = \tau$. Using Eq. (9.1.17), we have

$$\langle q';\tau + d\tau|q;\tau\rangle = \langle q';\tau|\exp(-iH d\tau)|q;\tau\rangle\,. \tag{9.1.24}$$

The Hamiltonian H is given as a function $H(Q,P)$, but since (9.1.13) and (9.1.14) are similarity transformations, and H commutes with itself, it can equally well be written as the *same* function of $Q(t)$ and $P(t)$

$$H \equiv H(Q,P) = e^{iHt}H(Q,P)e^{-iHt} = H\Big(Q(t),\ P(t)\Big)\,. \tag{9.1.25}$$

This function can be written in various different forms, with different constant coefficients, by using the commutation relations (9.1.1) and (9.1.2) to move the Qs and Ps past each other. It will be convenient to adopt a standard form, in which all Qs appear to the *left* of all Ps. For instance, given a term in the Hamiltonian of form $P_a Q_b P_c$, we would rewrite it as $P_a Q_b P_c = Q_b P_a P_c - i\,\delta_{ab} P_c$. With this convention, the $Q_a(t)$s in the Hamiltonian in Eq. (9.1.24) may be replaced[†] with their eigenvalues q'_a. To deal with the $P(t)$, we use Eq. (9.1.23) to expand $|q;\tau\rangle$ in P-eigenstates $|p;\tau\rangle$, and find

$$\begin{aligned}\langle q';\tau + d\tau \mid q;\tau\rangle &= \int \prod_a dp_a\, \langle q';\tau|\exp\Big(-iH(Q(\tau),P(\tau))d\tau\Big)|p;\tau\rangle \\ &\qquad \times\langle p;\tau|q;\tau\rangle \\ &= \int \prod_a \frac{dp_a}{2\pi} \exp\left[-iH(q',p)d\tau + i\sum_a (q'_a - q_a)p_a\right]\,,\end{aligned} \tag{9.1.26}$$

with each p_a integrated from $-\infty$ to $+\infty$.

Now let's return to the more general case of a finite time-interval. To calculate $\langle q';t'|q;t\rangle$, with $t < t'$, we break up the time-interval from t to t' into steps $t, \tau_1, \tau_2, \cdots \tau_N, t'$, with

$$\tau_{k+1} - \tau_k = d\tau = (t' - t)/(N+1) \tag{9.1.27}$$

[†] This is only possible because with $d\tau$ infinitesimal, $\exp(-iH\,d\tau)$ is linear in H.

and sum over a complete set of states $|q;\tau_k\rangle$ at each time τ_k:

$$\langle q';t'|q;t\rangle = \int dq_1 \cdots dq_N \langle q';t'|q_N;\tau_N\rangle\langle q_N;\tau_N|q_{N-1};\tau_{N-1}\rangle \cdots \langle q_1;\tau_1|q;t\rangle. \tag{9.1.28}$$

Inserting Eq. (9.1.26), this becomes

$$\langle q';t'|q;t\rangle = \int \left[\prod_{k=1}^{N}\prod_a dq_{k,a}\right]\left[\prod_{k=0}^{N}\prod_a dp_{k,a}/2\pi\right]$$
$$\times \exp\left[i\sum_{k=1}^{N+1}\left\{\sum_a (q_{k,a}-q_{k-1,a})p_{k-1,a} - H(q_k,p_{k-1})d\tau\right\}\right], \tag{9.1.29}$$

where

$$q_0 \equiv q\,, \qquad q_{N+1} \equiv q'\,. \tag{9.1.30}$$

Our result, Eq. (9.1.29), can be put in a much more elegant form. Define smooth interpolating functions, $q(\tau)$ and $p(\tau)$, such that

$$q_a(\tau_k) \equiv q_{k,a}\,, \qquad p_a(\tau_k) \equiv p_{k,a}\,. \tag{9.1.31}$$

In the limit $d\tau \to 0$ (i.e., $N \to \infty$), the argument of the exponential in Eq. (9.1.29) becomes just an integral over τ

$$\sum_{k=1}^{N+1}\left\{\sum_a (q_{k,a}-q_{k-1,a})p_{k-1,a} - H(q_k,p_{k-1})d\,\tau\right\}$$
$$= \sum_{k=1}^{N+1}\left\{\sum_a \dot{q}_a(\tau_k)p_a(\tau_k) - H\Big(q(\tau_k),p(\tau_k)\Big)\right\}d\,\tau + O(d\tau^2)$$
$$\to \int_t^{t'}\left\{\sum_a \dot{q}_a(\tau)p_a(\tau) - H\Big(q(\tau),p(\tau)\Big)\right\}d\,\tau\,. \tag{9.1.32}$$

Further, we may define integrals *over the functions* $q(\tau), p(\tau)$ by

$$\int \prod_{\tau,a} dq_a(\tau)\prod_{\tau,b}\frac{dp_b(\tau)}{2\pi}\cdots \equiv \lim_{d\tau\to 0}\int \prod_{k,a} dq_{k,a}\prod_{k,b}\frac{dp_{k,b}}{2\pi}\cdots\,. \tag{9.1.33}$$

Eq. (9.1.29) then becomes a constrained path integral

$$\langle q';t'|q;t\rangle = \int_{\substack{q_a(t)=q_a \\ q_a(t')=q'_a}} \prod_{\tau,a} dq_a(\tau)\prod_{\tau,b}\frac{dp_b(\tau)}{2\pi}$$
$$\times \exp\left[i\int_t^{t'} d\tau\left\{\sum_a \dot{q}_a(\tau)p_a(\tau) - H\Big(q(\tau),p(\tau)\Big)\right\}\right]. \tag{9.1.34}$$

This is called a *path integral*, because we integrate over all paths that take $q(\tau)$ from q at $\tau = t$ to q' at $\tau = t'$, as well as over all $p(\tau)$. The great advantage of writing matrix elements in this way is that, as shown in Section 9.3, the path integrals are easy to calculate when expanded in powers of the coupling constants in H.

The path-integral formalism allows us to calculate not only transition probability amplitudes like $\langle q';t'|q;t\rangle$, but also the matrix elements between states $\langle q';t'|$ and $|q,t\rangle$ of time-ordered products of general operators $\mathcal{O}(P(t),Q(t))$. It will be convenient to define these operators with (unlike H) all Ps moved to the *left* and all Qs to the *right*. Then by inserting any such operator $\mathcal{O}\Big(P(\tau),Q(\tau)\Big)$ in Eq. (9.1.26), we have

$$\langle q';\tau+d\tau|\mathcal{O}\Big(P(\tau),Q(\tau)\Big)|q;\tau\rangle = \int\prod_d dp_a$$

$$\times\ \langle q';\tau|\exp\Big(-iH\Big(Q(\tau),P(\tau)\Big)d\tau\Big)\ |p;\tau\rangle\langle p;\tau|\mathcal{O}\Big(P(\tau),Q(\tau)\Big)|q;\tau\rangle$$

$$=\int\prod_a\frac{dp_a}{2\pi}\exp\left[-iH(q',p)d\tau+i\sum_a(q'_a-q_a)p_a\right]\mathcal{O}(p,q)\,. \qquad (9.1.35)$$

In order to calculate the matrix element of a product $\mathcal{O}_A\Big(P(t_A),Q(t_A)\Big)\,\mathcal{O}_B\Big(P(t_B),Q(t_B)\Big)\cdots$ of operators with $t_A > t_B > \cdots$, we can insert the $\mathcal{O}$-operators between the appropriate states in Eq. (9.1.28), and use Eq. (9.1.35). For instance, if the time t_A falls between τ_k and τ_{k+1}, then insert $\mathcal{O}_A\Big(P(t_A),Q(t_A)\Big)$ between $\langle q_{k+1};\tau_{k+1}|$ and $|q_k;\tau_k\rangle$. Note that in Eq. (9.1.28) each successive sum over states is at a later time, so this is only possible because of our assumption that $t_A > t_B > \cdots$. Following the same steps as before, we now find the general path-integral formula

$$\langle q',t'|\mathcal{O}_A\Big(P(t_A),Q(t_A)\Big)\mathcal{O}_B\Big(P(t_B),Q(t_B)\Big)\cdots|q,t\rangle$$

$$=\int_{\substack{q_a(t)=q_a\\ q_a(t')=q'_a}}\prod_{\tau,a}dq_a(\tau)\prod_{\tau,b}\frac{dp_b(\tau)}{2\pi}\ \mathcal{O}_A\Big(p(t_A),q(t_A)\Big)\mathcal{O}_B\Big(p(t_B),q(t_B)\Big)\cdots$$

$$\times\ \exp\left[i\int_t^{t'}d\tau\left\{\sum_a\dot{q}_a(\tau)p_a(\tau)-H\Big(q(\tau),p(\tau)\Big)\right\}\right]\,. \qquad (9.1.36)$$

This result is only valid if the times are ordered, with

$$t' > t_A > t_B > \cdots > t\,. \qquad (9.1.37)$$

However, nothing on the right-hand side of Eq. (9.1.36) refers to the order of time-arguments. Hence if we are presented with a path integral like the right-hand side of Eq. (9.1.36), with $t_A, t_B, \cdots$ in arbitrary order (all between t and t', with $t < t'$), then this path integral will equal a matrix element like the left-hand side of Eq. (9.1.36), but with the operators arranged in order (from left to right) of decreasing time. That is, for

$t_A, t_B, \cdots$ in arbitrary order, we have

$$\langle q', t'|T\Big\{\mathcal{O}_A\Big(P(t_A), Q(t_A)\Big), \mathcal{O}_B\Big(P(t_B), Q(t_B)\Big), \cdots\Big\}|q, t\rangle$$

$$= \int_{\substack{q_a(t)=q_a \\ q_a(t')=q'_a}} \prod_{\tau,a} dq_a(\tau) \prod_{\tau,b} \frac{dp_b(\tau)}{2\pi}\, \mathcal{O}_A\Big(p(t_A), q(t_A)\Big)\mathcal{O}_B\Big(p(t_B), q(t_B)\Big) \cdots$$

$$\times \exp\left[i\int_t^{t'} d\tau \left\{\sum_a \dot{q}_a(\tau)p_a(\tau) - H\Big(q(\tau), p(\tau)\Big)\right\}\right] , \quad (9.1.38)$$

where T denotes the usual time-ordered product.

It should perhaps be stressed that the c-number functions $q_a(\tau)$, $p_a(\tau)$ in Eq. (9.1.38) are mere variables of integration, and in particular are *not* constrained to obey the equations of motion of classical Hamiltonian dynamics

$$\dot{q}_a(\tau) - \frac{\partial H(q(\tau), p(\tau))}{\partial p_a(\tau)} = 0\,, \quad (9.1.39)$$

$$\dot{p}_a(\tau) + \frac{\partial H(q(\tau), p(\tau))}{\partial q_a(\tau)} = 0\,. \quad (9.1.40)$$

(For this reason, the Hamiltonian $H(q(\tau), p(\tau))$ in Eq. (9.1.38) is *not* constant in τ.) Nevertheless, there is a limited sense in which path integrals do respect these equations of motion. Suppose that one of the functions in Eq. (9.1.38), say $\mathcal{O}_A\big(p(t_A),\ q(t_A)\big)$, happens to be the left-hand side of either Eq. (9.1.39) or Eq. (9.1.40). We note that (for $t < t_A < t'$)

$$\left(\dot{q}_a(t_A) - \frac{\partial H(q(t_A), p(t_A))}{\partial p_a(t_A)}\right) \exp\Big(iI[q,p]\Big) = -i\frac{\delta}{\delta p_a(t_A)} \exp\Big(iI[q,p]\Big),$$

$$\left(\dot{p}_a(t_A) + \frac{\partial H(q(t_A), p(t_A))}{\partial q_a(t_A)}\right) \exp\Big(iI[q,p]\Big) = i\frac{\delta}{\delta q_a(t_A)} \exp\Big(iI[q,p]\Big),$$

where iI is the argument of the exponential in Eq. (9.1.38):

$$I[q,p] \equiv \int_t^{t'} d\tau \left\{\sum_a \dot{q}_a(\tau)p_a(\tau) - H\Big(q(\tau), p(\tau)\Big)\right\} .$$

As long as t_A does not approach t or t', the integrations over $q_a(t_A)$ and $p_a(t_A)$ are unconstrained, and so with reasonable assumptions about the convergence of these integrals, the integral of such variational derivatives must vanish. Hence the path integral (9.1.38) vanishes if $\mathcal{O}_A(p,q)$ is taken to be the left-hand side of either of the equations of motion (9.1.39) or (9.1.40).

This simple rule applies only if the integration variables $q_a(t_A)$, $p_a(t_A)$ are independent of any of the variables $q_a(t_B)$, $p_a(t_B)$, etc. appearing in any of the other functions $\mathcal{O}_B$, $\mathcal{O}_C$, etc. in Eq. (9.1.38), and hence only if we prohibit t_A from approaching t_B, t_C, etc. as well as t or t'. When t_A approaches, say, t_B, the path integral will be found to involve a non-zero term proportional to $\delta(t_A - t_B)$ or its derivatives. These delta functions are the same as would be found in the operator formalism from time-derivatives of the step functions implicit in the definition of the time-ordered product.

In evaluating the path integrals (9.1.34) and (9.1.38), we only need to know the classical Hamiltonian, the c-number function $H(q,p)$. If we were to define a theory by the path integrals, the question would naturally arise, which of many possible quantum mechanical Hamiltonians $H(Q,P)$ (differing in the order of Qs and Ps) governed the quantum theory that corresponds to these path integrals. Our derivation has provided an answer: the quantum Hamiltonian is to be taken with all Qs on the left, all Ps on the right. But it would be a mistake to give this prescription too much significance. There are a great many ways of interpreting the measure $\Pi\, dq_a(\tau)\; \Pi\, dp_b(\tau)$ appearing in path integrals like (9.1.34) or (9.1.38). Our prescription, of putting all Qs to the left of all Ps, is appropriate only if the measure is interpreted according to Eqs. (9.1.31)–(9.1.33). Other measures would lead to other prescriptions for operator ordering. The question is not an urgent one, because different prescriptions for ordering operators in the Hamiltonian just correspond to different choices of the constants that appear as coefficients of the various terms in the Hamiltonian, and we generally formulate theories with these constants left as arbitrary parameters anyway.

It is difficult to use the general path integral in Eq. (9.1.38) for numerical calculations or as a source of rigorous theorems. For these purposes it is better to use the path-integral method to calculate amplitudes in Euclidean space, where t is replaced with an imaginary quantity $-ix_4$, and the argument of the exponential in Eq. (9.1.38) is a negative real quantity. In this way, instead of jagged paths producing rapid oscillations of the integrand from one path to another, all jagged paths are exponentially suppressed. Though we shall not go into it here, quantum field theory may be formulated from the beginning in terms of Feynman amplitudes in Euclidean spacetime.[8a] Under certain plausible assumptions, it is possible to reconstruct the Feynman amplitudes in Minkowskian spacetime from their Euclidean counterparts.[8b] But we may as well stick to the Minkowski space formulation of the path integral if we are only going to use it to calculate Feynman amplitudes in perturbation theory.

9.2 Transition to the S-Matrix

As already mentioned, we can easily convert the general quantum mechanical results of Section 9.1 to a notation appropriate to quantum field theory, by letting the index a run over points $\mathbf{x}$ in space and over a spin-and-species index m, and replacing $Q_a(t)$ and $P_a(t)$ with $Q_m(\mathbf{x},t)$ and $P_m(\mathbf{x},t)$, respectively. Eq. (9.1.38) then reads*

$$\langle q',t'|T\{\mathcal{O}_A[P(t_A),Q(t_A)],\ \mathcal{O}_B[P(t_B),Q(t_B)],\cdots\}|q,t\rangle$$
$$= \int_{\substack{q_m(\mathbf{x},t)=q_m(\mathbf{x})\\ q_m(\mathbf{x},t')=q'_m(\mathbf{x})}} \prod_{\tau,\mathbf{x},m} dq_m(\mathbf{x},\tau) \prod_{\tau,\mathbf{x},m} \frac{dp_m(\mathbf{x},\tau)}{2\pi}$$
$$\times\, \mathcal{O}_A\Big[p(t_A),q(t_A)\Big]\ \mathcal{O}_B\Big[p(t_B),q(t_B)\Big]\cdots$$
$$\times \exp\left[i\int_t^{t'} d\tau\left\{\int d^3x\ \sum_m \dot{q}_m(\mathbf{x},\tau)p_m(\mathbf{x},t) - H\Big[q(\tau),p(\tau)\Big]\right\}\right]. \quad (9.2.1)$$

However, in field theory Eq. (9.2.1) is not exactly what we want. Experimentalists do not measure probability amplitudes for transitions between eigenstates $\langle q',t'|$ and $|q,t\rangle$ of the quantum field Q, but rather S-matrix elements, the probability amplitudes for transitions between states that at $t\to-\infty$ or $t\to+\infty$ contain definite numbers of particles of various types. These are called 'in' and 'out' states, $|\alpha,\text{ in}\rangle$ and $|\beta,\text{ out}\rangle$, where α and β denote sets of particles characterized by the various particles' momenta, spin z-component (or helicity), and species. To calculate a matrix element of a time-ordered product (perhaps empty) between such states, we need to multiply Eq. (9.2.1) by the 'wave functions' $\langle\beta,\text{ out}|q',t'\rangle$ and $\langle q,t|\alpha,\text{in}\rangle$ at any fixed times t and t', taken for convenience here to be $-\infty$ and $+\infty$, respectively, and then perform an integral over the 'arguments' $q_m(\mathbf{x})$ and $q'_m(\mathbf{x})$ of these wave functions. But instead of constraining the path integral over $q_m(\mathbf{x},\tau)$ by the conditions

$$q_m(\mathbf{x},+\infty)=q'_m(\mathbf{x})\,, \qquad q_m(\mathbf{x},-\infty)=q_m(\mathbf{x})\,, \qquad (9.2.2)$$

and then integrating over $q'_m(\mathbf{x})$ and $q_m(\mathbf{x})$, we can just as well do an unconstrained integral over $q_m(\mathbf{x},\tau)$ (and also over $p_m(\mathbf{x},\tau)$), and set the arguments of the wave functions equal to the values given by Eq. (9.2.2):

* We are now writing H and the $\mathcal{O}$s with square brackets, to remind us that $H[q(t),p(t)]$ and $\mathcal{O}[p(t),q(t)]$ are *functionals* of $q_m(\mathbf{x},t)$ and $p_m(\mathbf{x},t)$ at a fixed time t.

$$\langle\beta,\text{out}|T\left\{\mathcal{O}_A\Big[P(t_A),Q(t_A)\Big],\ \mathcal{O}_B\Big[P(t_B),Q(t_B)\Big],\cdots\right\}|\alpha,\ \text{in}\rangle$$
$$=\int\prod_{\tau,\mathbf{x},m}dq_m(\mathbf{x},\tau)\prod_{\tau,\mathbf{x},m}(dp_m(\mathbf{x},\tau)/2\pi)$$
$$\times\,\mathcal{O}_A\Big[p(t_A),q(t_A)\Big]\ \mathcal{O}_B\Big[p(t_B),q(t_B)\Big]\cdots$$
$$\times\,\exp\left[i\int_{-\infty}^{\infty}d\tau\left\{\int d^3x\,\sum_m\dot{q}_m(\mathbf{x},\tau)p_m(\mathbf{x},\tau)-H\Big[q(\tau),p(\tau)\Big]\right\}\right]$$
$$\times\,\langle\beta,\ \text{out}|q(+\infty);+\infty\rangle\ \langle q(-\infty);-\infty|\alpha,\ \text{in}\rangle\,. \tag{9.2.3}$$

Incidentally, this result leads immediately** to Eq. (6.4.3), a theorem that we use repeatedly to relate sums of off-shell Feynman graphs to matrix elements of Heisenberg-picture operators between exact energy eigenstates.

It is necessary now to consider how to calculate the wave functions appearing as the final pair of factors in Eq. (9.2.3). Let's first consider the simplest and most important case, the vacuum. (We saw in Section 6.4 that S-matrix elements may be easily calculated from the vacuum expectation values of time-ordered products.) We assume as usual that for $t\to\pm\infty$, matrix elements may be calculated as if there were no interactions. The 'in' and 'out' vacua may thus be defined by the conditions

$$\begin{aligned}a_{\text{in}}(\mathbf{p},\sigma,n)|\text{VAC},\text{in}\rangle&=0\,,\\ a_{\text{out}}(\mathbf{p},\sigma,n)|\text{VAC},\text{out}\rangle&=0\,,\end{aligned} \tag{9.2.4}$$

where a_{in} and a_{out} are the operators appearing in the coefficients of $\exp(i\mathbf{p}\cdot\mathbf{x}-iEt)$ in the plane-wave expansion of the operator $Q_m(\mathbf{x},t)$ at $t\to-\infty$ and $t\to+\infty$, respectively. For instance, for the real scalar field of a neutral spinless particle, we have in effect

$$\Phi(\mathbf{x},t)\overset{t\to\mp\infty}{\longrightarrow}(2\pi)^{-3/2}\int d^3p\,(2E)^{-1/2}\Big[a_{\substack{\text{in}\\ \text{out}}}(\mathbf{p})e^{ip\cdot x}+\text{H.c.}\Big]\,, \tag{9.2.5}$$

** It is only necessary to note that, for a Hamiltonian $H[P(t),Q(t)]+\sum_A\int d^3x\,\epsilon_A(\mathbf{x},t)\mathcal{O}_A(\mathbf{x},t)$, the S-matrix is given by Eq. (9.2.3) as

$$\langle\beta,\text{out}|\alpha,\ \text{in}\rangle_\epsilon=\int\prod_{\tau,\mathbf{x},m}dq_m(\mathbf{x},\tau)\prod_{\tau,\mathbf{x},m}(dp_m(\mathbf{x},\tau)/2\pi)$$
$$\times\,\exp\left[i\int_{-\infty}^{\infty}d\tau\left\{\int d^3x\,\dot{q}_m(\mathbf{x},\tau)p_m(\mathbf{x},\tau)-H\Big[q(\tau),p(\tau)\Big]\right.\right.$$
$$\left.\left.-\sum_A\int d^3x\,\epsilon_A(\mathbf{x},\tau)O_A(\mathbf{x},\tau)\right\}\right]$$
$$\times\,\langle\beta,\ \text{out}|q(+\infty);+\infty\rangle\ \langle q(-\infty);-\infty|\alpha,\ \text{in}\rangle\,.$$

The left-hand side of Eq. (6.4.3) is the derivative of this expression with respect to ϵ_a, ϵ_b, etc., at $\epsilon=0$, which yields the right-hand side of Eq. (9.2.3), and using Eq. (9.2.3) again then immediately gives the right-hand side of Eq. (6.4.3).

$$\Pi(\mathbf{x},t) \xrightarrow{t\to\mp\infty} \dot{\Phi}(\mathbf{x},t)$$
$$\xrightarrow{t\to\mp\infty} -i(2\pi)^{-3/2}\int d^3p\,(E/2)^{1/2}\Big[a_{\substack{\text{in}\\ \text{out}}}(\mathbf{p})e^{ip\cdot x} - \text{H.c.}\Big] \tag{9.2.6}$$

where $p^0 \equiv E \equiv \sqrt{\mathbf{p}^2+m^2}$, and we here use the conventional Φ and Π rather than Q and P for a scalar field, and drop the unnecessary labels m, σ, n. Inverting the Fourier transforms and taking a linear combination of the resulting expressions, we have

$$a_{\substack{\text{in}\\ \text{out}}}(\mathbf{p}) = \lim_{t\to\mp\infty}\frac{e^{iEt}}{(2\pi)^{3/2}}\int d^3x\, e^{-i\mathbf{p}\cdot\mathbf{x}}$$
$$\times\left[\sqrt{\frac{E}{2}}\Phi(\mathbf{x},t) + i\sqrt{\frac{1}{2E}}\Pi(\mathbf{x},t)\right] . \tag{9.2.7}$$

As mentioned in Section 9.1, the 'momentum' $\Pi(\mathbf{x},t)$ acts on wave-functions in a ϕ-basis as the variational derivative $-i\delta/\delta\phi(\mathbf{x},t)$, so in this basis the conditions (9.2.4) read

$$0 = \int d^3x\, e^{-i\mathbf{p}\cdot\mathbf{x}}\left[\frac{\delta}{\delta\phi(\mathbf{x})} + E(\mathbf{p})\phi(\mathbf{x})\right]\Big\langle\phi(t\to\mp\infty);\mp\infty\Big|\text{VAC}, {}^{\text{in}}_{\text{out}}\Big\rangle . \tag{9.2.8}$$

The analogous ordinary differential equation has a well-known Gaussian solution, so let's try a Gaussian ansatz here:

$$\Big\langle\phi(t\to\mp\infty);\mp\infty\Big|\text{VAC}, {}^{\text{in}}_{\text{out}}\Big\rangle = \mathcal{N}\exp\left(-\tfrac{1}{2}\int d^3x\, d^3y\, \mathcal{E}(\mathbf{x},\mathbf{y})\phi(\mathbf{x})\phi(\mathbf{y})\right) , \tag{9.2.9}$$

with kernel $\mathcal{E}$ and constant $\mathcal{N}$ to be determined. Substituting this in Eq. (9.2.8), we see that the functional differential equation for the vacuum wave functional is satisfied if for all ϕ

$$0 = \int d^3x\, e^{-i\mathbf{p}\cdot\mathbf{x}}\left[\int d^3y\, \mathcal{E}(\mathbf{x},\mathbf{y})\phi(\mathbf{y}) - E(\mathbf{p})\phi(\mathbf{x})\right] \tag{9.2.10}$$

or, in other words, if

$$\int d^3x\, e^{-i\mathbf{p}\cdot\mathbf{x}}\mathcal{E}(\mathbf{x},\mathbf{y}) = E(\mathbf{p})\, e^{-i\mathbf{p}\cdot\mathbf{y}} . \tag{9.2.11}$$

The solution is easily found by inverting the Fourier transform

$$\mathcal{E}(\mathbf{x},\mathbf{y}) = (2\pi)^{-3}\int d^3p\, e^{i\mathbf{p}\cdot(\mathbf{x}-\mathbf{y})}E(\mathbf{p}) . \tag{9.2.12}$$

(Recall that $E(\mathbf{p}) \equiv \sqrt{\mathbf{p}^2+m^2}$). This is actually the most useful representation for the kernel $\mathcal{E}$, but we may note in passing that for $\mathbf{x}\neq\mathbf{y}$, $\mathcal{E}$ may also be written in terms of a Hankel function of negative order

$$\mathcal{E}(\mathbf{x},\mathbf{y}) = \frac{m}{2\pi^2 r}\frac{d}{dr}\left(\frac{1}{r}K_{-1}(mr)\right) , \tag{9.2.13}$$

where $r \equiv |\mathbf{x}-\mathbf{y}|$. The constant $\mathcal{N}$ in Eq. (9.2.9) may be formally obtained from the normalization condition for the vacuum state, but we will not need this result.

According to Eq. (9.2.9), in calculating vacuum expectation values in the theory of a scalar field, the product of the last two factors in Eq. (9.2.3) is

$$\begin{aligned}\langle \text{VAC, out}|\phi(\infty);+\infty\rangle\,\langle\phi(-\infty);-\infty|\text{VAC, in}\rangle \\ = |\mathcal{N}|^2 \exp\Big(-\tfrac{1}{2}\int d^3x\,d^3y\,\mathscr{E}(\mathbf{x},\mathbf{y})\Big[\phi(\mathbf{x},+\infty)\phi(\mathbf{y},+\infty) \\ +\phi(\mathbf{x},-\infty)\phi(\mathbf{y},-\infty)\Big]\Big) \qquad (9.2.14) \\ = |\mathcal{N}|^2 \exp\Big(-\tfrac{1}{2}\epsilon\int d^3x\,d^3y\int_{-\infty}^{\infty} d\tau\,\mathscr{E}(\mathbf{x},\mathbf{y})\phi(\mathbf{x},\tau)\phi(\mathbf{y},\tau)e^{-\epsilon|\tau|}\Big),\end{aligned}$$

where ϵ is a positive infinitesimal. To obtain the final expression, we have used the fact that for any reasonably smooth function $f(\tau)$,

$$f(+\infty)+f(-\infty) = \lim_{\epsilon\to 0+}\epsilon\int_{-\infty}^{\infty} d\tau f(\tau)\,e^{-\epsilon|\tau|}\,. \qquad (9.2.15)$$

Inserting Eq. (9.2.14) in Eq. (9.2.3) now gives

$$\begin{aligned}\Big\langle \text{VAC, out}\Big|T\Big\{\mathcal{O}_A\Big[\Pi(t_A),\Phi(t_A)\Big],\ \mathcal{O}_B\Big[\Pi(t_B),\Phi(t_B)\Big],\cdots\Big\}\Big|\text{VAC, in}\Big\rangle \\ = |\mathcal{N}|^2\int\prod_{\tau,\mathbf{x}} d\phi(\mathbf{x},\tau)\prod_{\tau,\mathbf{x}}(d\pi(\mathbf{x},\tau)/2\pi)\ \mathcal{O}_A\Big[\pi(t_A),\phi(t_A)\Big] \\ \times\mathcal{O}_B\Big[\pi(t_B),\phi(t_B)\Big]\cdots\exp\Big[i\int_{-\infty}^{\infty} d\tau\Big\{\int d^3x\,\dot\phi(\mathbf{x},\tau)\,\pi(\mathbf{x},\tau) \\ -H\Big[\phi(\tau),\pi(\tau)\Big] + \tfrac{1}{2}i\epsilon\int d^3x\,d^3y\ \mathscr{E}(\mathbf{x},\mathbf{y})e^{-\epsilon|\tau|}\phi(\mathbf{x},\tau)\phi(\mathbf{y},\tau)\Big\}\Big]\,. \qquad (9.2.16)\end{aligned}$$

We shall see in Section 9.4 that the whole effect of the last term in the argument of the exponential in Eq. (9.2.16) is to provide the $-i\epsilon$ in the denominator of the scalar field propagator in momentum space, $[p^2+m^2-i\epsilon]^{-1}$. We will not go into the corresponding details for fields of general spin, but will simply state that in general

$$\begin{aligned}\Big\langle \text{VAC, out}|T\Big\{\mathcal{O}_A\Big[P_A(t_A),Q(t_A)\Big],\ \mathcal{O}_B\Big[P_B(t_B),Q(t_B)\Big],\cdots\Big\}\,|\text{VAC, in}\Big\rangle \\ = |\mathcal{N}|^2\int\Big[\prod_{\tau,\mathbf{x},m} dq_m(\mathbf{x},\tau)\Big]\Big[\prod_{\tau,\mathbf{x},m}\frac{dp_m(\mathbf{x},\tau)}{2\pi}\Big]\ \mathcal{O}_A\Big[p(t_A),q(t_A)\Big] \\ \times\mathcal{O}_B\Big[p(t_B),q(t_B)\Big]\cdots\exp\Big[i\int_{-\infty}^{\infty} d\tau\Big\{\int d^3x\,\sum_m \dot q_m(\mathbf{x},\tau)p_m(\mathbf{x},\tau) \\ -H\Big[q(\tau),p(\tau)\Big] + i\epsilon\ \text{terms}\Big\}\Big], \qquad (9.2.17)\end{aligned}$$

where the '$i\epsilon$ terms' just have the effect of putting the correct $-i\epsilon$ in the denominators of all propagators.

This is a good place to mention that field-independent factors in Eq. (9.2.17), like the constant $|\mathcal{N}|^2$, are not important. This is because such factors contribute also to the matrix element $\langle \text{VAC, out}|\text{VAC, in}\rangle$. In calculating the *connected* part of vacuum expectation values of time-ordered products (or the S-matrix) we eliminate the contribution of disconnected vacuum fluctuation subgraphs by dividing by $\langle \text{VAC, out}|\text{VAC, in}\rangle$, and any constant factors in the vacuum expectation values cancel in this ratio.

We could go on and calculate matrix elements between multi-particle states, by inserting the appropriate 'wave functionals' in Eq. (9.2.3). These can be calculated by applying the adjoints of annihilation operators such as (9.2.7) to the vacuum state; just as for the harmonic oscillator, these wave functionals turn out to be Hermite polynomials in the field times the vacuum Gaussian. We do not need to work all this out here, because as shown in Section 6.4, the vacuum expectation values (9.2.17) are all we need in order to be able to calculate S-matrix elements.

9.3 Lagrangian Version of the Path-Integral Formula

The integrand in the exponential in Eqs. (9.1.38) or (9.2.17) looks like the Lagrangian L associated with the Hamiltonian H. This appearance is somewhat misleading because here the 'momenta' $p_a(t)$ or $p_n(\mathbf{x},t)$ are independent variables, not yet related to $q_a(t)$ or $q_n(\mathbf{x},t)$ or their derivatives. However, there is a large and important class of theories in which the integral over the 'momenta' can be done by just replacing them with the values dictated by the canonical formalism, in which case the integrand in the exponential in the path integrals really is the Lagrangian.

These are theories with a Hamiltonian that is quadratic in the 'momenta' — in the language of field theory

$$H[Q,P] = \tfrac{1}{2}\sum_{nm}\int d^3x\, d^3y\, A_{\mathbf{x}n,\mathbf{y}m}[Q]\, P_n(\mathbf{x})\, P_m(\mathbf{y}) + \sum_n \int d^3x\, B_{\mathbf{x}n}[Q]\, P_n(\mathbf{x}) + C[Q] \tag{9.3.1}$$

with a 'matrix' A that is real, symmetric, positive, and non-singular. The argument of the exponential in Eq. (9.2.17) is then quadratic in the ps:

$$\int d\tau \left\{ \int d^3x \sum_n p_n(\mathbf{x},\tau)\dot{q}_n(\mathbf{x},\tau) - H\Big[q(\tau),p(\tau)\Big] \right\}$$
$$= -\tfrac{1}{2}\sum_{nm}\int d^3x\, d^3y\, d\tau\, d\tau'\, \mathscr{A}_{\tau\mathbf{x}n,\tau'\mathbf{y}m}[q]\; p_n(\mathbf{x},\tau)\; p_m(\mathbf{y},\tau')$$
$$-\sum_n \int d^3x \int d\tau\, \mathscr{B}_{\tau\mathbf{x}n}[q]\, p_n(\mathbf{x},\tau) \;-\; \mathscr{C}[q]\,, \tag{9.3.2}$$

where

$$\mathscr{A}_{\tau\mathbf{x}n,\tau'\mathbf{y}m}[q] \equiv A_{\mathbf{x}n,\mathbf{y}m}[q(\tau)]\delta(\tau-\tau')\,, \tag{9.3.3}$$
$$\mathscr{B}_{\tau\mathbf{x}n}[q] \equiv B_{\mathbf{x}n}[q(\tau)] - \dot{q}_n(\mathbf{x},\tau)\,, \tag{9.3.4}$$
$$\mathscr{C}[q] \equiv \textstyle\int d\tau\, C[q(\tau)]\,. \tag{9.3.5}$$

Now, in general the integral of the exponential of a quadratic expression like (9.3.2) will be proportional to the exponential evaluated at the stationary point of its argument. For a finite number of real variables ξ_s, this formula reads

$$\int_{-\infty}^{\infty}\left(\prod_s d\xi_s\right)\,\exp\left\{-\tfrac{1}{2}i\sum_{sr}\mathscr{A}_{sr}\xi_s\xi_r - i\sum_s \mathscr{B}_s\xi_s - i\mathscr{C}\right\}$$
$$= \left(\text{Det}\,\left[i\mathscr{A}/2\pi\right]\right)^{-1/2}\,\exp\left\{-i\,\tfrac{1}{2}\sum_{sr}\mathscr{A}_{sr}\bar{\xi}_s\bar{\xi}_r - i\sum_s\mathscr{B}_s\bar{\xi}_s - i\mathscr{C}\right\}\,, \tag{9.3.6}$$

where $\bar{\xi}$ is the stationary point

$$\bar{\xi}_s = -\sum_r (\mathscr{A}^{-1})_{sr}\mathscr{B}_r\,. \tag{9.3.7}$$

(For a proof of this formula, see the Appendix to this chapter.) Hence, as long as the $\mathcal{O}_A$, $\mathcal{O}_B$, etc. in Eq. (9.2.17) are independent of the ps, for such a Hamiltonian we can evaluate the path integral over the ps in Eq. (9.2.17) by setting these variables at the stationary point of the quadratic expression in the argument of the exponential. But the variational derivative of this quadratic is

$$\frac{\delta}{\delta p_n(\mathbf{x},\tau)}\int_{-\infty}^{\infty} d\tau \left\{ \int d^3x\; \dot{q}_n(\mathbf{x},\tau)p_n(\mathbf{x},\tau) - H\Big[q(\tau),p(\tau)\Big] + i\epsilon\text{ terms} \right\}$$
$$= \dot{q}_n(\mathbf{x},\tau) - \frac{\delta}{\delta p_n(\mathbf{x},\tau)}\,H\Big[q(\tau),p(\tau)\Big]\,.$$

(The $i\epsilon$ terms depend only on the qs.) Thus the stationary 'point' $\bar{p}_n(\mathbf{x},t)$ where this vanishes is just the value of $p_n(\mathbf{x},t)$ dictated by the canonical formula

$$\dot{q}_n(\mathbf{x},\tau) = \left[\frac{\delta H\Big[q(\tau),p(\tau)\Big]}{\delta p_n(\mathbf{x},\tau)}\right]_{p=\bar{p}} . \tag{9.3.8}$$

With $p_n(\mathbf{x},t)$ set equal to this value, the argument of the exponential in Eq. (9.2.17) *is* the ordinary Lagrangian

$$L\Big[q(\tau),\dot{q}(\tau)\Big] \equiv \int d^3x \left(\sum_n \dot{q}_n(\mathbf{x},\tau)\bar{p}_n(\mathbf{x},t) - H\Big[q(\tau),\bar{p}(\tau)\Big]\right) \tag{9.3.9}$$

and we can write Eq. (9.2.17) as

$$\begin{aligned}&\Big\langle \text{VAC, out}\Big|T\Big\{\mathscr{O}_A\Big[Q(t_A)\Big],\mathscr{O}_B\Big[Q(t_B)\Big],\cdots\Big\}\Big|\text{VAC, in}\Big\rangle\\ &= |\mathscr{N}|^2\int\prod_{\tau,\mathbf{x},n} dq_n(\mathbf{x},\tau)\Big(\text{Det}\Big[2i\pi\mathscr{A}[q]\Big]\Big)^{-1/2}\\ &\quad\times\mathscr{O}_A\Big[q(t_A)\Big]\,\mathscr{O}_B\Big[q(t_B)\Big]\cdots\\ &\quad\times\exp\left[i\int_{-\infty}^{\infty}d\tau\Big\{L\Big[q(\tau),\dot{q}(\tau)\Big]+i\epsilon\text{ terms}\Big\}\right] .\end{aligned} \tag{9.3.10}$$

(We have combined the $1/2\pi$ factors in the integrals over the p_n with the determinant coming from Eq. (9.3.6).) This is the desired Lagrangian form of the path-integral formula.

In deriving Eq. (9.3.10), it was necessary to assume that the operators $\mathscr{O}_A, \mathscr{O}_B, \cdots$ were independent of the canonical 'momenta'. This is not as restrictive as it may seem. For instance, in a scalar field theory for which the canonical conjugate to Φ is $\Pi = \dot{\Phi}$, it is possible to calculate the matrix element of a time-ordered product of operators, one of which is $\dot{\Phi}(t)$, by taking the difference of matrix elements in which this operator is replaced with $\Phi(t + d\tau)$ and $\Phi(t)$, and then dividing by $d\tau$, with $d\tau \to 0$. Equivalently, as long as t is not equal to any of the other time-arguments of the operator in Eq. (9.3.10), we can simply differentiate Eq. (9.3.10) with respect to t.

The one serious remaining complication in Eq. (9.3.10) is the determinant of $\mathscr{A}[q]$. If $\mathscr{A}[q]$ is field-independent, then this is no problem; we have already noted that overall constants make no contribution to the connected parts of vacuum expectation values, in which we divide by a vacuum-vacuum amplitude proportional to the same constant factor. This is the case for instance for the theory of a set of scalar fields Φ_n with non-derivative coupling to each other and/or derivative coupling to external currents J_n. The Lagrangian density here is

$$\mathscr{L} = -\sum_n\Big[\tfrac{1}{2}\partial_\lambda\Phi_n\partial^\lambda\Phi_n + J_n{}^\lambda\partial_\lambda\Phi_n\Big] - V(\Phi) .$$

An obvious extension of the results of Section 7.5 from one to several derivatively coupled scalars shows that this Lagrangian implies the Hamiltonian

$$H = \int d^3x \sum_n \Big[\tfrac{1}{2}\Pi_n^2 + \tfrac{1}{2}(\nabla\Phi_n)^2 + \mathbf{J}_n \cdot \nabla\Phi_n + J_n{}^0\Pi_n + \tfrac{1}{2}(J_n{}^0)^2 \Big] + \int d^3x \, V(\Phi) \, .$$

(The Φ_n are taken to be real scalars, but complex scalars can be accommodated by separating them into real and imaginary parts.) In general there is a non-trivial term that is linear in the Π_n, but the coefficient of the quadratic term is a constant, just the unit 'matrix':

$$\mathscr{A}_{xn,x'n'} = \delta^4(x-x')\,\delta_{nn'} \, .$$

The factor $\Big(\text{Det}\big[2i\pi\mathscr{A}[q]\big]\Big)^{-1/2}$ in Eq. (9.3.10) is here a field-independent constant, and therefore is without effect.

However, matters are not always so simple. As a second example, let us consider the so-called non-linear σ-model, with Lagrangian density

$$\mathscr{L} = -\tfrac{1}{2}\sum_{nm} \partial_\lambda\Phi_n\partial^\lambda\Phi_m \Big[\delta_{nm} + U_{nm}(\Phi)\Big] - V(\Phi) \, .$$

A straightforward calculation gives the Hamiltonian as

$$H = \int d^3x \left[\tfrac{1}{2}\Pi_n\Big(1+U(\Phi)\Big)^{-1}_{nm}\Pi_m + \tfrac{1}{2}\nabla\Phi_n \cdot \nabla\Phi_m \Big(1+U(\Phi)\Big)_{nm} + V(\Phi) \right] .$$

Here $\mathscr{A}$ is the field-dependent quantity

$$\mathscr{A}_{nx,my} = \Big[1 + U(\Phi(x))\Big]^{-1}_{nm} \delta^4(x-y) \, .$$

In cases of this sort, the determinant may be reexpressed as a contribution to the effective Lagrangian, using the relation $\text{Det}\,\mathscr{A} = \exp\,\text{Tr}\,\ln\mathscr{A}$. By replacing the continuum of spacetime positions with a discrete lattice of points surrounded by separate regions of very small spacetime volume Ω, we may interpret the delta function in $\mathscr{A}_{nx,my}$ as $\delta^4(x-y) = \Omega^{-1}\delta_{x,y}$, so that

$$(\ln\mathscr{A})_{nx,my} = \delta_{x,y}\Big[-\ln\big(1+U(\Phi(x))\big) - 1\cdot\ln\Omega\Big]_{nm}$$

with the logarithm of a matrix defined now by its power series expansion

$$\ln(1+U) = U - \frac{U^2}{2} + \frac{U^3}{3} - \cdots \, .$$

To evaluate the trace, we note that $\sum_x \cdots = \Omega^{-1}\int d^4x \cdots$. The determinant factor here is then

$$\operatorname{Det}\mathscr{A} \propto \exp\left[-\Omega^{-1}\int d^4x \, \mathrm{tr}\ln\left[1+U(\Phi(x))\right]\right] ,$$

where ‘tr’ is to be understood as the trace in an ordinary matrix sense. The constant of proportionality (which arises from the $-\ln\Omega$ term) is field-independent and therefore of no present interest. We can regard this determinant as providing a correction to the effective Lagrangian density

$$\Delta\mathscr{L} = -\tfrac{1}{2}i\Omega^{-1}\mathrm{tr}\ln\left[1+U(\Phi(x))\right] .$$

The factor Ω^{-1} may be written as an ultraviolet divergent integral

$$\Omega^{-1} = \delta^4(x-x) = (2\pi)^{-4}\int d^4p \cdot 1 .$$

We shall not show this here, but the extra terms in the Feynman diagrams for this theory contributed by $\Delta\mathscr{L}$ could also have been derived in the canonical formalism by taking account of the equal-time-commutator terms in the propagator of time-derivatives of the scalar field.[7] Ignoring this correction would lead to a spurious dependence of the S-matrix on the way that the scalar field is defined, and would also be inconsistent with any symmetries of the Lagrangian under transformations of the scalar fields.

Even where the factor $(\operatorname{Det}\mathscr{A})^{-1/2}$ in the path-integral formula (9.3.10) is field-independent, the Lagrangian in this formula may not be the same as the one with which we started. As an example, let's consider the theory of a set of real vector fields, with Lagrangian density

$$\mathscr{L} = -\sum_n \Big[\tfrac{1}{4}\left(\partial_\mu A_{n\lambda}-\partial_\lambda A_{n\mu}\right)\left(\partial^\mu A_n{}^\lambda-\partial^\lambda A_n{}^\mu\right) + \tfrac{1}{2}m_n^2\, A_{n\lambda}A_n{}^\lambda + J_n{}^\lambda A_{n\lambda}\Big] ,$$

where the currents $J_n{}^\mu$ are either externally produced c-number quantities or depend on other fields (in which case terms describing these other fields are to be added to the Lagrangian). By a simple extension of the results of Section 7.5, we see that the Hamiltonian is

$$H = \int d^3x \sum_n \Big[\frac{1}{2}\boldsymbol{\Pi}_n^2 + \frac{1}{2}\left(\nabla\times\mathbf{A}_n\right)^2 + \frac{1}{2}m_n^2\mathbf{A}_n{}^2 + \frac{1}{2m_n^2}\left(\nabla\cdot\boldsymbol{\Pi}_n\right)^2 + \mathbf{J}_n\cdot\mathbf{A}_n - \frac{1}{m_n^2}J_n^0\nabla\cdot\boldsymbol{\Pi}_n + \frac{1}{2m_n^2}(J_n^0)^2\Big]$$

again with the understanding that other terms must be added involving any fields that appear in $J_n{}^\mu$. Here the coefficient of the quadratic term is somewhat more complicated than in our first example:

$$\mathscr{A}_{nix,mjy} = \delta_{nm}\left[\delta_{ij}\delta^4(x-y) - \frac{1}{2m_n^2}\nabla_i\nabla_j\delta^4(x-y)\right] ,$$

but it is field-independent, so that the factor $(\mathrm{Det}\,\mathscr{A})^{-1/2}$ has no effect. On the other hand, the Lagrangian (9.3.9) is here not the one with which we started; it is expressed entirely in terms of $\mathbf{A}$ and its spacetime derivatives, with no dependence on any time-component A^0. For this reason, the Lorentz invariance of Eq. (9.3.10) is far from obvious.

To remedy this, we may reintroduce the auxiliary field. Suppose we add to the Hamiltonian a term

$$\Delta H = -\tfrac{1}{2}\sum_n m_n^2 \int d^3x \left[A_n^0 - m_n^{-2}\nabla\cdot\mathbf{\Pi}_n + m_n^{-2}J_n^0\right]^2$$

and integrate over the A_n^0 as well as the $\mathbf{A}_n$ and $\mathbf{\Pi}_n$. This can only introduce a field-independent overall factor, since ΔH is a quadratic in A^0 (with a field-independent coefficient in the term of second order in A^0) whose stationary value vanishes. However, suppose that we now integrate over the $\mathbf{\Pi}_n$ *before* integrating over the A_n^0. The Hamiltonian in the path integral (9.2.17) is here replaced with

$$H + \Delta H = \int d^3x \sum_n \Big[\, \tfrac{1}{2}\mathbf{\Pi}_n^2 + \tfrac{1}{2}(\nabla\times\mathbf{A}_n)^2 + \tfrac{1}{2}m_n^2\mathbf{A}_n{}^2$$
$$- \tfrac{1}{2}m_n^2(A_n^0)^2 + \mathbf{J}_n\cdot\mathbf{A}_n - J_n^0\,A_n^0 + A_n^0\nabla\cdot\mathbf{\Pi}_n\Big] .$$

This is still quadratic in $\mathbf{\Pi}_n$ with a field-independent (and somewhat simpler) coefficient of the quadratic term, so the integral over the $\mathbf{\Pi}_n$s can be done by just replacing $\mathbf{\Pi}_n$ with its value at the stationary point of the functional $\sum_n \int d^3x\, \mathbf{\Pi}_n\cdot\dot{\mathbf{A}}_n - H - \Delta H$:

$$\mathbf{\Pi}_n = \dot{\mathbf{A}}_n + \nabla A_n^0 .$$

With $\mathbf{\Pi}_n$ eliminated in this way, $\sum_n \int d^3x\, \mathbf{\Pi}_n\cdot\dot{\mathbf{A}}_n - H - \Delta H$ is just the Lorentz-invariant Lagrangian with which we started.

In order to take account of the possible need to introduce auxiliary fields like A_n^0, from now on we shall write the path-integral formula after elimination of the canonical conjugates in terms of fields ψ_ℓ that include both canonical fields q_n and auxiliary fields c_r:

$$\Big\langle \mathrm{VAC,out}\Big|T\Big\{\mathscr{O}_A\Big[\Psi_A(t_A)\Big],\mathscr{O}_B\Big[\Psi_B(t_B)\Big],\cdots\Big\}\Big|\,\mathrm{VAC,in}\Big\rangle$$
$$\propto \int \prod_{\tau,\mathbf{x},n} d\psi_n(\mathbf{x},\tau)\;\mathscr{O}_A\Big[\psi(t_A)\Big]\;\mathscr{O}_B\Big[\psi(t_B)\Big]\cdots$$
$$\times\exp\left[i\int_{-\infty}^{\infty} d\tau\,\Big\{L\Big[\psi(\tau),\dot\psi(\tau)\Big] + i\epsilon\ \text{terms}\Big\}\right] , \qquad (9.3.11)$$

it now being understood that L includes any terms arising from a possible field-dependent factor $(\text{Det}\,\mathscr{A})^{-1/2}$.

9.4 Path-Integral Derivation of Feynman Rules

We are now ready to use the path-integral formalism to derive the Feynman rules in a wide class of theories. We will concentrate here on the vacuum expectation values of time-ordered products of field operators (and their adjoints),

$$M_{\ell_A \ell_B \cdots}(x_A x_B \cdots) = \frac{\langle \text{VAC, out} \,|\, T\left\{\Psi_{\ell_A}(x_A), \Psi_{\ell_B}(x_B) \cdots\right\} |\, \text{VAC, in}\rangle}{\langle \text{VAC, out} | \text{VAC, in}\rangle} \tag{9.4.1}$$

from which S-matrix elements may be obtained (as shown in Section 6.4) by stripping off the final propagators associated with each field, replacing them with the coefficient functions that multiply creation or annihilation operators in the corresponding free fields, and summing over the indices on these coefficient functions.

For the simpler theories whose Hamiltonian is quadratic in the Πs, Eq. (9.3.11) gives

$$M_{\ell_A \ell_B \cdots}(x_A x_B \cdots) = \frac{\int \left[\prod_{x,\ell} d\psi_\ell(x)\right] \psi_{\ell_A}(x_A)\, \psi_{\ell_B}(x_B) \cdots e^{iI[\psi]}}{\int \prod_{x,\ell} d\psi_\ell(x)\, e^{iI[\psi]}}\,, \tag{9.4.2}$$

where $I[\psi]$ is the action

$$I[\psi] = \int_{-\infty}^{\infty} d\tau \left\{ L\Big[\psi(\tau), \dot{\psi}(\tau)\Big] + i\epsilon \text{ terms}\right\} \tag{9.4.3}$$

with L now including any terms that may arise from a field-dependent determinant in Eq. (9.3.10).

Let us now suppose that the Lagrangian is the integral of a Lagrangian density, consisting of a quadratic term $\mathscr{L}_0$ which would be present in the absence of interactions, plus a Lagrangian interaction density $\mathscr{L}_1$:

$$\begin{aligned} L\Big[\psi(\tau), \dot{\psi}(\tau)\Big] = \int d^3x \,\Big[\mathscr{L}_0\Big(\psi(\vec{x},\tau),\ \partial_\mu\psi(\vec{x},\tau)\Big) \\ + \mathscr{L}_1\Big(\psi(\vec{x},\tau),\ \partial_\mu\psi(\vec{x},\tau)\Big)\Big]\,. \end{aligned} \tag{9.4.4}$$

That is, the action (9.4.3) is

$$I[\psi] = I_0[\psi] + I_1[\psi], \tag{9.4.5}$$

$$I_0[\psi] = \int d^4x\, \mathscr{L}_0\Big(\psi(x), \partial_\mu\psi(x)\Big) + i\epsilon \text{ terms}, \tag{9.4.6}$$

$$I_1[\psi] = \int d^4x\, \mathscr{L}_1\Big(\psi(x), \partial_\mu\psi(x)\Big)\,. \tag{9.4.7}$$

Since $\mathscr{L}_0$ and the '$i\epsilon$ terms' are quadratic in the fields, we may always write I_0 in the generalized quadratic form

$$I_0[\psi] = -\tfrac{1}{2}\int d^4x\, d^4x' \sum_{\ell,\ell'} \mathscr{D}_{\ell x,\ell' x'}\; \psi_\ell(x)\; \psi_{\ell'}(x')\,. \tag{9.4.8}$$

For instance, for a real scalar field of mass m, the unperturbed Lagrangian is

$$\mathscr{L}_0 = -\tfrac{1}{2}\partial_\mu\phi\partial^\mu\phi - \tfrac{1}{2}m^2\phi^2 \tag{9.4.9}$$

and the $i\epsilon$ terms in I_0 are given by Eq. (9.2.16), as

$$\tfrac{1}{2}i\epsilon\int dt\int d^3x\, d^3x'\; \mathscr{E}(\mathbf{x},\mathbf{x}')\; \phi(\mathbf{x},t)\phi(\mathbf{x}',t) \tag{9.4.10}$$

so here

$$\mathscr{D}_{x,x'} = \frac{\partial}{\partial x^\mu}\frac{\partial}{\partial x'_\mu}\,\delta^4(x-x') + m^2\,\delta^4(x-x') - i\epsilon\mathscr{E}(\mathbf{x},\mathbf{x}')\delta(t-t')\,. \tag{9.4.11}$$

(We are now dropping the factor $e^{-\epsilon|\tau|}$ in the $i\epsilon$ term, since it produces a correction of higher order in ϵ.) To deal with interactions, we will expand the exponential in powers of I_1,

$$\exp(iI[\psi]) = \exp(iI_0[\psi])\sum_{N=0}^{\infty}\frac{i^N}{N!}\Big(I_1[\psi]\Big)^N \tag{9.4.12}$$

and then expand I_1 in powers of the fields. The general integrals that we encounter in the numerator and the denominator of Eq. (9.4.2) are of the form

$$\mathscr{I}_{\ell_1\ell_2\cdots}(x_1x_2\cdots) \equiv \int\left(\prod_{\ell,x} d\psi_\ell(x)\right)\, e^{iI_0[\psi]}\; \psi_{\ell_1}(x_1)\;\psi_{\ell_2}(x_2)\cdots\,, \tag{9.4.13}$$

where the field factors $\psi_{\ell_1}(x_1), \psi_{\ell_2}(x_2)$, etc. arise from $I_1[\psi]$ and/or from the field factors $\psi_{\ell_A}(x_A)$ etc. originally present in the numerator of Eq. (9.4.2). With $I_0[\psi]$ of the form (9.4.8), the integral (9.4.13) is of the same form as the integral evaluated in the Appendix to this chapter, with the discrete index s replaced with the pair of labels ℓ, x. We can therefore use Eqs. (9.A.12) and (9.A.15), which give here

$$\mathscr{I}_{\ell_1\ell_2\cdots}(x_1x_2\cdots) = \left[\text{Det}\left(\frac{i\mathscr{D}}{2\pi}\right)\right]^{-1/2} \sum_{\substack{\text{pairings of}\\ \text{fields}}} \prod_{\text{pairs}} \left[-i\mathscr{D}^{-1}\right]_{\text{paired fields}} . \tag{9.4.14}$$

This just amounts to the coordinate-space Feynman rules for calculating the numerator of Eq. (9.4.2) in their covariant form: we expand in the interaction I_1, and then sum over the ways of pairing the fields in the I_1s with each other and with the fields $\psi_{\ell_A}(x_A)$, etc., with the contribution of each pairing being given by the spacetime integral of the product of the coefficients of the fields in $I_1[\psi]$ and the product of the 'propagators' $-i\Delta$, where

$$\Delta_{\ell_1\ell_2}(x_1, x_2) = (\mathscr{D}^{-1})_{\ell_1x_1,\ell_2x_2} . \tag{9.4.15}$$

(The factor $[\text{Det}(i\mathscr{D}/2\pi)]^{-1/2}$ in Eq. (9.4.14) actually represents the contribution of graphs with unlimited numbers of single loops unattached to any other lines, but in any case this factor cancels in the ratio (9.4.2).)

It remains to calculate the propagators (9.4.15). We interpret Eq. (9.4.15) as an integral equation

$$\sum_{\ell_2} \int d^4x_2\, \mathscr{D}_{\ell_1x_1,\ell_2x_2} \Delta_{\ell_2\ell_3}(x_2, x_3) = \delta^4(x_1 - x_3)\delta_{\ell_1\ell_3} . \tag{9.4.16}$$

In the absence of external fields, translation invariance will make $\mathscr{D}$ necessarily a function only of $x_1 - x_2$, which can be written as a Fourier integral

$$\mathscr{D}_{\ell_1x_1,\ell_2x_2} \equiv (2\pi)^{-4} \int d^4p\, e^{ip\cdot(x_1-x_2)}\, \mathscr{D}_{\ell_1\ell_2}(p) . \tag{9.4.17}$$

The solution of Eq. (9.4.16) is then

$$\Delta_{\ell_1\ell_2}(x_1, x_2) = (2\pi)^{-4} \int d^4p\, e^{ip\cdot(x_1-x_2)}\, \mathscr{D}^{-1}_{\ell_1\ell_2}(p) , \tag{9.4.18}$$

where $\mathscr{D}^{-1}$ is the ordinary inverse of the matrix $\mathscr{D}$. As we will see, the $i\epsilon$ terms have the effect of making the inverse well-defined for all real values of p. We have thus reduced the problem of calculating the propagator to that of taking the inverse of a finite matrix.

First consider a massive scalar field, for which the kernel $\mathscr{D}$ takes the form (9.4.11). We can write this as a Fourier integral

$$\mathscr{D}_{x,y} = (2\pi)^{-4} \int d^4p\, e^{ip\cdot(x-y)} \Big(p^2 + m^2 - i\epsilon\, E(\mathbf{p})\Big) ,$$

so the propagator is

$$\Delta(x, y) = (2\pi)^{-4} \int d^4p\, e^{ip\cdot(x-y)} \Big(p^2 + m^2 - i\epsilon E(\mathbf{p})\Big)^{-1} .$$

We recognize this as the same scalar propagator previously obtained by operator methods. (The difference between ϵ and $\epsilon E(\mathbf{p})$ is immaterial, since both are just positive infinitesimals.)

For a second example, consider a real massive vector field. The unperturbed Lagrangian is

$$\mathscr{L}_0 = -\tfrac{1}{4}(\partial_\mu A_\nu - \partial_\nu A_\mu)(\partial^\mu A^\nu - \partial^\nu A^\mu) - \tfrac{1}{2}m^2 A_\mu A^\mu \, .$$

We can again write $I_0[\psi]$ in the form (9.4.8), with kernel

$$\begin{aligned}\mathscr{D}_{\rho x,\sigma y} &= \left[\eta_{\rho\sigma}\frac{\partial^2}{\partial x^\mu \partial y_\mu} - \frac{\partial^2}{\partial x^\sigma \partial y^\rho} + m^2\eta_{\rho\sigma}\right]\delta^4(x-y) + i\epsilon \text{ terms} \\ &= (2\pi)^{-4}\int d^4p\, e^{ip\cdot(x-y)}\Big[\eta_{\rho\sigma}p^2 - p_\rho p_\sigma + m^2\eta_{\rho\sigma} + i\epsilon \text{ terms}\Big] \, .\end{aligned}$$

We will not bother to show it here, but the '$+\, i\epsilon$ terms' here take the simple form $-i\epsilon E(\mathbf{p})\eta_{\rho\sigma}$. The vector field propagator is then given by simply inverting the 4×4 matrix in the integrand

$$\Delta_{\rho\sigma}(x,y) = (2\pi)^{-4}\int \frac{d^4p\, e^{ip\cdot(x-y)}}{p^2 + m^2 - i\epsilon E(\mathbf{p})}\left[\eta_{\rho\sigma} + \frac{p_\rho p_\sigma}{m^2}\right] \, .$$

(Terms proportional to ϵ are dropped in the numerator. They are important in the denominator in defining how the integrand is to be treated near the mass shell, $p^2 = -m^2$.) This is the same as the propagator derived by operator methods, except that the non-covariant terms proportional to $\delta(x^0 - y^0)$ are now absent. These non-covariant terms were previously needed to cancel non-covariant terms in the interaction Hamiltonian, but the vertex contributions in the Feynman rules are now obtained directly by inspection of the covariant Lagrangian, and no such cancellation is needed.

Theories with derivative coupling are equally simple. The factor arising from the pairing of a field derivative $\partial_\mu \psi_\ell(x)$ with any other field $\psi_m(y)$ (perhaps itself a field derivative) is

$$\begin{aligned}\langle \partial_\mu \psi_\ell(x)\, \psi_m(y)\rangle &= \frac{\int \Big[\prod_{x,\ell} d\psi_\ell(x)\Big]\partial_\mu \psi_\ell(x)\, \psi_m(y)\, e^{iI[\psi]}}{\int \Big[\prod_{x,\ell} d\psi_\ell(x)\Big]\, e^{iI[\psi]}} \\ &= \frac{\partial}{\partial x^\mu}\langle \psi_\ell(x)\, \psi_m(y)\rangle \, . \end{aligned} \tag{9.4.19}$$

Such propagators have no non-covariant pieces. For instance, for a real scalar field, the pairing of $\partial_\mu \phi$ with $\partial_\nu \phi$ gives a momentum space propagator $k_\mu k_\nu/(k^2 + m^2 - i\epsilon)$. Also, as we saw in the previous section, vertices in the theory of a scalar field with derivative couplings to other fields may be read off from the Lagrangian, and are separately covariant.

9.5 Path Integrals for Fermions

We now turn to the problem of extending the path-integral formalism to cover theories containing fermions as well as bosons. It would be easy to proceed in a purely formal way, by analogy with the bosonic case, with the justification that this gives the 'right' Feynman rules. Instead, we will here derive the path-integral formalism for fermions directly from the principles of quantum mechanics, as we did for bosons.[9]

As before, we will start with a general quantum mechanical system, with 'coordinates' Q_a and canonical conjugate 'momenta' P_a, but now satisfying anticommutation rather than commutation relations:

$$\left\{Q_a, P_b\right\} = i\,\delta_{ab}\,, \tag{9.5.1}$$

$$\left\{Q_a, Q_b\right\} = \left\{P_a, P_b\right\} = 0\,. \tag{9.5.2}$$

(These are Schrödinger-picture operators, or in other words Heisenberg-picture operators at time $t = 0$.) Later we will replace the discrete index a with a spatial position $\mathbf{x}$ and a field index m.

We wish first to construct a complete basis for the states on which the Qs and Ps act. Note that for any given a, we have

$$Q_a^2 = P_a^2 = 0\,. \tag{9.5.3}$$

It follows that there will always be a 'ket' state $|0\rangle$ annihilated by all Q_a:

$$Q_a|0\rangle = 0\,, \tag{9.5.4}$$

and a 'bra' state $\langle 0|$ annihilated (from the right) by all P_a:

$$\langle 0|\,P_a = 0\,. \tag{9.5.5}$$

For instance, we can take

$$|0\rangle \propto \left(\prod_a Q_a\right)|f\rangle\,, \qquad \langle 0| \propto \langle g|\left(\prod_a P_a\right)\,,$$

where $|f\rangle$ and $\langle g|$ are any kets and bras for which these expressions do not vanish. (They cannot vanish for all $|f\rangle$ and $\langle g|$ unless the operators $\prod_a Q_a$ and $\prod_a P_a$ vanish, which we assume not to be the case.) These states satisfy Eqs. (9.5.4) and (9.5.5) by virtue of Eq. (9.5.3). They are not in general unique, because there may be other bosonic degrees of freedom that distinguish the various possible $|0\rangle$ and $\langle 0|$, but for simplicity we will limit ourselves here to the case where the only degrees of freedom are those described by the fermionic operators Q_a and P_a, and will assume that the states satisfying Eqs. (9.5.4) and (9.5.5) are unique up to constant factors, which we choose so that

$$\langle 0|0\rangle = 1\,. \tag{9.5.6}$$

(Note that this normalization convention could not be imposed if we had defined $\langle 0|$ as the left-eigenstate of the Q_a with eigenvalue zero, because in this case $\langle 0|\{Q_a, P_b\}|0\rangle$ would vanish, which with Eq. (9.5.1) would imply that $\langle 0|0\rangle = 0$.)

As we saw in Section 7.5, in the Dirac theory Q_a is not Hermitian, but instead has an adjoint $-iP_a$, in which case $\langle 0|$ *can* be regarded as simply the adjoint of $|0\rangle$. However, there are fermionic operators (such as the 'ghost' fields to be introduced in Volume II) for which P_a is unrelated to the adjoint of Q_a. In what follows we will not need to assume anything about the adjoints of Q_a or P_a, or about any relation between $|0\rangle$ and $\langle 0|$.

A complete basis for the states of this system is provided by $|0\rangle$ and the states (antisymmetric in indices $a, b, \cdots$)

$$| a, b, \cdots \rangle \equiv P_a P_b \cdots |0\rangle \tag{9.5.7}$$

with any number of *different* Ps acting on $|0\rangle$. That is, the result of acting on these states with any operator function of the Ps and Qs can be written as a linear combination of the same set of states. In particular, if an index a is unequal to any of the indices appearing in $| b, c, \cdots \rangle$, then

$$Q_a | b, c, \cdots \rangle = 0 \,, \tag{9.5.8}$$

$$P_a | b, c, \cdots \rangle = | a, b, c \cdots \rangle \,. \tag{9.5.9}$$

On the other hand, if a *is* equal to one of the indices in the sequence, $b, c, \cdots$, we can always rewrite the state (possibly changing its sign) so that a is the first of these indices, in which case we have

$$Q_a | a, b, c, \cdots \rangle = i | b, c \cdots \rangle \,, \tag{9.5.10}$$

$$P_a | a, b, c, \cdots \rangle = 0 \,. \tag{9.5.11}$$

Similarly, we may define a complete dual basis, consisting of $\langle 0|$ and the states (also antisymmetric in the indices)

$$\langle a, b, \cdots | \equiv \langle 0| \cdots (-iQ_b)(-iQ_a) \,. \tag{9.5.12}$$

Using Eqs. (9.5.4)–(9.5.6) and the anticommutation relation (9.5.1), we see that the scalar products of these states take the values

$$\langle c, d, \cdots | a, b, \cdots \rangle = \langle 0| \cdots (-iQ_d)(-iQ_c)\, P_a P_b \cdots |0\rangle$$
$$= \begin{cases} 0 & \text{if} \;\; \{c, d, \cdots\} \neq \{a, b, \cdots\} \\ 1 & \text{if} \;\; c = a, \; d = b, \text{ etc.} \end{cases} \tag{9.5.13}$$

where $\{\cdots\}$ here denotes the set of indices within the brackets, irrespective of order.

In deriving the Feynman rules, we would like to be able to rewrite sums over intermediate states like (9.5.7) as integrals over eigenstates of the Q_a or the P_a. However, it is not possible for these operators to have eigenvalues (other than zero) in the usual sense. Suppose we try to find a state $|q\rangle$ that satisfies (for all a)

$$Q_a|q\rangle = q_a|q\rangle \,. \tag{9.5.14}$$

From Eq. (9.5.2) we see that

$$q_a q_b + q_b q_a = 0 \tag{9.5.15}$$

which is impossible for ordinary numbers. However, nothing can stop us from introducing an algebra of 'variables' (known as *Grassmann variables*) q_a, which act like c-numbers as far as the physical Hilbert space is concerned, but which still satisfy the anticommutation relations (9.5.15). We will require further that

$$\{q_a, q'_b\} = \{q_a, Q_b\} = \{q_a, P_b\} = 0, \tag{9.5.16}$$

where q and q' denote any two 'values' of these variables. We can now construct eigenstates $|q\rangle$ satisfying Eq. (9.5.14):

$$|q\rangle = \exp\left(-i\sum_a P_a q_a\right)|0\rangle \tag{9.5.17}$$

with the exponential defined as usual by its power series expansion. (To verify Eq. (9.5.14), use the fact that all $P_a q_a$ commute with one another and have zero square, so that

$$\begin{aligned}
[Q_a - q_a]\,|q\rangle &= [Q_a - q_a]\,\exp(-iP_a q_a)\exp\left(-i\sum_{b\neq a} P_b q_b\right)|0\rangle \\
&= [Q_a - q_a]\,[1 - iP_a q_a]\,\exp\left(-i\sum_{b\neq a} P_b q_b\right)|0\rangle \\
&= [-i\{Q_a, P_a\}q_a - q_a]\,\exp\left(-i\sum_{b\neq a} P_b q_b\right)|0\rangle = 0
\end{aligned}$$

as required by Eq. (9.5.14).) We can also define left-eigenstates $\langle q|$ (*not* the adjoints of $|q\rangle$), as

$$\langle q| \equiv \langle 0|\left(\prod_a Q_a\right)\exp\left(-i\sum_a q_a P_a\right) = \langle 0|\left(\prod_a Q_a\right)\exp\left(+i\sum_a P_a q_a\right)\,, \tag{9.5.18}$$

where $\prod_a$ is the product in whatever order we take as standard. By the same argument as for Eq. (9.5.14), we see that

$$\langle q|\,Q_a \;=\; \langle q|q_a \,. \tag{9.5.19}$$

These eigenstates have the scalar product

$$\begin{aligned}\langle q'|q\rangle &= \langle 0|\left(\prod_a Q_a\right)\exp\left(i\sum_b P_b(q'_b - q_b)\right)|0\rangle \\ &= \langle 0|\left(\prod_a Q_a\right)\left(\prod_b(1 + iP_b(q'_b - q_b))\right)|0\rangle\,.\end{aligned}$$

Moving each Q_a to the right (starting with the rightmost) yields factors $i^2(q'_a - q_a)$, which we move to the right out of the scalar product, so

$$\langle q'|q\rangle = \prod_a (q_a - q'_a)\,. \tag{9.5.20}$$

We shall see that Eq. (9.5.20) plays the role of a delta function in integrals over the qs.

In the same way, we can construct right- and left-eigenstates of the P_a:

$$P_a|p\rangle = p_a|p\rangle\,, \tag{9.5.21}$$

$$\langle p|\,P_a = \langle p|p_a\,, \tag{9.5.22}$$

where the p_a are like q_a anticommuting c-numbers (taken for convenience to anticommute with the q_a and all fermionic operators as well as each other), and

$$|p\rangle = \exp\left(-i\sum_a Q_a p_a\right)\left(\prod_b P_b\right)|0\rangle\,, \tag{9.5.23}$$

$$\langle p| = \langle 0|\,\exp\left(-i\sum_a p_a Q_a\right) \tag{9.5.24}$$

with scalar product (now derived by moving the Ps to the left)

$$\langle p'|p\rangle = \prod_a (p'_a - p_a)\,. \tag{9.5.25}$$

The scalar products of these two sorts of eigenstate with each other are

$$\begin{aligned}\langle q|p\rangle &= \langle q|\exp\left(-i\sum_a Q_a p_a\right)\left(\prod_a P_a\right)|0\rangle \\ &= \left(\prod_a \exp(-iq_a p_a)\right)\langle q|\left(\prod_a P_a\right)|0\rangle \\ &= \left(\prod_a \exp(-iq_a p_a)\right)\langle 0|\left(\prod_a Q_a\right)\left(\prod_a P_a\right)|0\rangle\end{aligned}$$

and so

$$\langle q|p\rangle = \chi_N \exp\left(-i\sum_a q_a p_a\right) = \chi_N \exp\left(i\sum_a p_a q_a\right)\,, \tag{9.5.26}$$

where χ_N is a phase that depends only on the number N of Q_a operators:

$$\chi_N \equiv \langle 0| \left(\prod_a Q_a\right) \left(\prod_a P_a\right) |0\rangle = i^N(-1)^{N(N-1)/2} .$$

Somewhat more simply, we also find

$$\langle p|q\rangle = \prod_a \exp(-ip_a q_a) . \tag{9.5.27}$$

It is easy to see that the states $|q\rangle$ are in a sense a complete set (and so also are the $|p\rangle$.) From the definitions (9.5.17), we see that the state $|a, b, \cdots\rangle$ in the general basis is (up to a phase) just the coefficient of the product $q_a q_b \cdots$ in an expansion of $|q\rangle$ in a sum of products of qs. Therefore we can write any state $|f\rangle$ in the form

$$|f\rangle = f_0|q\rangle_0 + \sum_a f_a|q\rangle_a + \sum_{a\neq b} f_{ab}|q\rangle_{ab} + \cdots ,$$

where the fs are numerical coefficients, and a subscript $a, b, \cdots$ on $|q\rangle$ denotes the coefficient of $q_a q_b \cdots$ in $|q\rangle$.

In summing over states, it will be very convenient to introduce a sort of integration over fermionic variables, known as *Berezin integration*,[10] that is designed to pick out the coefficients of such products of anticommuting c-numbers. For any set of such variables ξ_n (either ps or qs or both together), the most general function $f(\xi)$ (either a c-number or a state-vector like $|q\rangle$) can be put in the form

$$f(\xi) = \left(\prod_n \xi_n\right) c \; + \; \text{terms with fewer } \xi \text{ factors} \tag{9.5.28}$$

and the integral over the ξs is defined simply by

$$\int \left(\widetilde{\prod_n} d\xi_n\right) f(\xi) \equiv c \tag{9.5.29}$$

with the tilde in Eq. (9.5.29) indicating that we use the convenient convention that the differentials are written in an order *opposite* to that of the product of integration variables in Eq. (9.5.28). Since this product is antisymmetric under the interchange of any two ξs, the integral is likewise antisymmetric under the interchange of any two $d\xi$s, so these 'differentials' effectively anticommute

$$d\xi_n d\xi_m + d\xi_m d\xi_n = 0 . \tag{9.5.30}$$

Also, the coefficient c may itself depend on other unintegrated c-number variables that anticommute with the ξs over which we integrate, in which case it is important to standardize the definition of c by moving all ξs to the left of c before integrating over them, as we have done in Eq. (9.5.28).

For instance, the most general function of a pair of anticommuting c-numbers ξ_1 and ξ_2 takes the form

$$f(\xi_1,\xi_2) = \xi_1\xi_2 c_{12} + \xi_1 c_1 + \xi_2 c_2 + d$$

because the squares and all higher powers of ξ_1 and ξ_2 vanish. This function has the integrals

$$\int d\xi_1\, f(\xi_1,\xi_2) = \xi_2 c_{12} + c_1\,, \quad \int d\xi_2\, f(\xi_1,\xi_2) = -\xi_1 c_{12} + c_2\,,$$

$$\int d\xi_2\, d\xi_1\, f(\xi_1,\xi_2) = c_{12}\,.$$

Note that the multiple integral is the same as a repeated integral:

$$\int d\xi_2\, d\xi_1\, f(\xi_1,\xi_2) = \int d\xi_2 \Big[\int d\xi_1\, f(\xi_1,\xi_2) \Big]\,,$$

a result that can easily be extended to integrals over any number of fermionic variables. (It was in order to obtain this result without extra sign factors that we took the product of differentials in Eq. (9.5.29) to be in the opposite order to the product of variables in Eq. (9.5.28).) Indeed, we could have first defined the integral over a single anticommuting c-number ξ_1, and then defined multiple integrals in the usual way by iteration. The most general function of anticommuting c-numbers is linear in any one of them

$$f(\xi_1,\xi_2,\cdots) = b(\xi_2\cdots) + \xi_1 c(\xi_2\cdots)$$

(because $\xi_1^2 = 0$), and its integral over ξ_1 is defined as

$$\int d\xi_1 f(\xi_1,\xi_2,\cdots) = c(\xi_2,\cdots)\,.$$

Repeating this process leads to the same multiple integral as defined by Eqs. (9.5.28) and (9.5.29).

This definition of integration shares some other properties with multiple integrals (from $-\infty$ to $+\infty$) over ordinary real variables, but there are significant differences.

Obviously, Berezin integration is linear, in the sense that

$$\int \Big(\tilde{\prod_n} d\xi_n\Big) \big[f(\xi)+g(\xi)\big] = \int \Big(\tilde{\prod_n} d\xi_n\Big) f(\xi) + \int \Big(\tilde{\prod_n} d\xi_n\Big) g(\xi) \tag{9.5.31}$$

and also

$$\int \Big(\tilde{\prod_n} d\xi_n\Big) \big[f(\xi)a(\xi')\big] = \Big[\int \Big(\tilde{\prod_n} d\xi_n\Big) f(\xi)\Big]\, a(\xi')\,, \tag{9.5.32}$$

where $a(\xi')$ is any function (including a constant) of any anticommuting c-numbers ξ'_m over which we are *not* integrating. However, linearity with respect to left-multiplication is not so obvious. If we are integrating over ν variables, then since ξ'_m is assumed to anticommute with all ξ_n, we have

$$a\Big((-)^\nu \xi'\Big)\left(\prod_n \xi_n\right) = \left(\prod_n \xi_n\right) a(\xi')$$

and so

$$\int \left(\prod_n d\xi_n\right)\left[a\Big((-)^\nu \xi'\Big)f(\xi)\right] = a(\xi')\int \left(\prod_n d\xi_n\right) f(\xi) \,. \tag{9.5.33}$$

It is therefore very convenient (though not strictly necessary) to take the differentials $d\xi_n$ to anticommute with all anticommuting variables (including the ξ_n):

$$(d\xi_n)\xi'_m + \xi'_m(d\xi_n) = 0 \tag{9.5.34}$$

in which case Eq. (9.5.33) reads more simply

$$\int a(\xi')\left(\prod_n d\xi_n\right) f(\xi) = a(\xi')\int \left(\prod_n d\xi_n\right) f(\xi) \,. \tag{9.5.35}$$

Another similarity with ordinary integration is that, for an arbitrary anticommuting c-number ξ' independent of ξ,

$$\int \left(\prod_n d\xi_n\right) f(\xi + \xi') = \int \left(\prod_n d\xi_n\right) f(\xi) \tag{9.5.36}$$

since shifting ξ by a constant only affects the terms in f with fewer than the total number of ξ-variables.

On the other hand, consider a change of variables

$$\xi_n \to \xi'_n = \sum_m \mathscr{S}_{nm}\xi_m \,, \tag{9.5.37}$$

where $\mathscr{S}$ is an arbitrary non-singular matrix of ordinary numbers. The product of the new variables is

$$\prod_n \xi'_n = \sum_{m_1 m_2 \cdots} \left(\prod_n \mathscr{S}_{nm_n}\xi_{m_n}\right) .$$

But $\prod_n \xi_{m_n}$ here is just the same as the product (in the original order) $\prod_n \xi_n$, except for a sign $\epsilon[m]$ which is $+1$ or -1 according to whether the permutation $n \to m_n$ is an even or odd permutation of the original order:

$$\prod_n \xi'_n = \left[\sum_{m_1 m_2 \cdots} \left(\prod_n \mathscr{S}_{nm_n}\right)\epsilon[m]\right]\prod_n \xi_n = (\text{Det}\, \mathscr{S})\prod_n \xi_n \,.$$

This applies whatever order we take for the ξ_n, as long as we take the ξ'_n in the same order. It follows that the coefficient of $\prod_n \xi'_n$ in any function

$f(\xi)$ is just $(\text{Det}\,\mathscr{S})^{-1}$ times the coefficient of $\prod_n \xi_n$, a statement we write as

$$\int \left(\tilde{\prod_n} d\xi'_n\right) f = (\text{Det}\,\mathscr{S})^{-1} \int \left(\tilde{\prod_n} d\xi_n\right) f \,. \tag{9.5.38}$$

This is the usual rule for changing variables of integration, except that $(\text{Det}\,\mathscr{S})$ appears to the power -1 instead of $+1$. We shall use Eq. (9.5.38) and the linearity properties (9.5.31), (9.5.32), and (9.5.35) later to evaluate the integrals encountered in deriving the Feynman rules for theories with fermions.

We can now use this definition of integration to write the completeness condition as a formula for an integral over eigenvalues. As already mentioned, any state $|f\rangle$ can be expanded in a series of the states $|0\rangle$, $|a\rangle$, $|a,b\rangle$, etc. and these states are (up to a phase) the coefficients of the products 1, q_a, $q_a q_b$, etc. in the Q-eigenstate $|q\rangle$. According to the definition of integration here, we can pick out the coefficient of any product $q_b q_c q_d \cdots$ in the state $|q\rangle$ by integrating the product of $|q\rangle$ with all q_a with a *not* equal to $b, c, d, \cdots$. Thus, by choosing a function $f(q)$ as a suitable sum of such products of qs, we can write any state $|f\rangle$ as an integral:

$$|f\rangle = \int \left(\tilde{\prod_a} dq_a\right) |q\rangle f(q) = \int |q\rangle \left(\tilde{\prod_a} dq_a\right) f(q) \,. \tag{9.5.39}$$

(We can move $|q\rangle$ to the left of the differentials without any sign changes because the exponential in Eq. (9.5.17) used to define $|q\rangle$ involves only even numbers of fermionic quantities.) To find the function $f(q)$ for a given state-vector $|f\rangle$, take the scalar product of Eq. (9.5.39) with some bra $\langle q'|$ (with q' any fixed Q-eigenvalue). According to Eqs. (9.5.35) and (9.5.20), this is

$$\langle q'|f\rangle = \int \left(\prod_a (q_a - q'_a)\right) \left(\tilde{\prod_b} dq_b\right) f(q) \,.$$

Moving every factor $(q_a - q'_a)$ to the right past every differential dq_b yields a sign factor $(-)^{N^2} = (-)^N$, where N is now the total number of q_a variables, so

$$\langle q'|f\rangle = (-)^N \int \left(\tilde{\prod_b} dq_b\right) \left(\prod_a (q_a - q'_a)\right) f(q) \,.$$

We can rewrite $f(q)$ as $f(q' + (q - q'))$ and expand in powers of $q - q'$. All terms beyond the lowest order vanish when multiplied with the product

$\prod(q_a - q'_a)$, so

$$\left(\prod_a (q_a - q'_a)\right) f(q) = \left(\prod_a (q_a - q'_a)\right) f(q') \,, \tag{9.5.40}$$

which partly justifies our earlier remark that Eq. (9.5.20) plays the role of a delta function for integrals over the qs. Using Eq. (9.5.32), we now have

$$\langle q'|f\rangle = (-)^N \left[\int \left(\widetilde{\prod_b} dq_b\right) \left(\prod_a (q_a - q'_a)\right) f(q')\right] \,.$$

The term in the integrand proportional to $\prod q_a$ has coefficient $f(q')$, so according to our definition of integration $\langle q'|f\rangle = (-)^N f(q')$. Inserting this back in Eq. (9.5.39) gives our completeness relation

$$|f\rangle = (-)^N \int |q\rangle \left(\widetilde{\prod_b} dq_b\right) \langle q|f\rangle \,,$$

or as an operator equation

$$1 = \int |q\rangle \left(\widetilde{\prod_a} -dq_a\right) \langle q| \,. \tag{9.5.41}$$

In exactly the same way, we can also show that

$$1 = \int |p\rangle \left(\widetilde{\prod_a} dp_a\right) \langle p| \,. \tag{9.5.42}$$

We are now in a position to calculate transition matrix elements. As before, we define time-dependent operators

$$Q_a(t) \equiv \exp(iHt)\, Q_a \, \exp(-iHt) \tag{9.5.43}$$

$$P_a(t) \equiv \exp(iHt)\, P_a \, \exp(-iHt) \tag{9.5.44}$$

and their right- and left-eigenstates

$$|q;t\rangle \equiv \exp(iHt)|q\rangle \,, \qquad |p;t\rangle \equiv \exp(iHt)|p\rangle \,, \tag{9.5.45}$$

$$\langle q;t| \equiv \langle q|\exp(-iHt) \,, \qquad \langle p;t| \equiv \langle p|\exp(-iHt) \,. \tag{9.5.46}$$

The scalar product between q-eigenstates defined at infinitesimally close times is then

$$\langle q';\tau + d\tau|q;\tau\rangle = \langle q'|\exp(-iHd\tau)|q\rangle \,.$$

Now insert Eq. (9.5.42) to the left of the operator $\exp(-iHd\tau)$. It is convenient here to define the Hamiltonian operator $H(P,Q)$ with all Ps to the left of all Qs, so that (for $d\tau$ infinitesimal)

$$\langle p|\exp\Big(-iH(P,Q)d\tau\Big)|q\rangle = \langle p|q\rangle \exp\Big(-iH(p,q)d\tau\Big) \,.$$

(We could move the c-number $H(p,q)$ to either side of the matrix element without any sign changes because each term in the Hamiltonian is assumed to contain an even number of fermionic operators.) This gives

$$\langle q';\tau+d\tau|q;\tau\rangle = \int \langle q'|p\rangle \left(\widetilde{\prod_a} dp_a\right) \langle p|\exp(-iHd\tau)|q\rangle$$
$$= \int \langle q'|p\rangle \left(\widetilde{\prod_a} dp_a\right) \langle p|q\rangle \exp\Big(-iH(p,q)d\tau\Big) .$$

Using Eqs. (9.5.26) and (9.5.27), and noting that the products $p_a q_a$ and $p_a q'_a$ commute with all anticommuting c-numbers, we find

$$\langle q';\tau+d\tau|q;\tau\rangle = \int \left(\widetilde{\prod_a} i\, dp_a\right) \exp\left[i\sum_a p_a(q'_a - q_a) - iH(p,q)d\tau\right] . \tag{9.5.47}$$

The rest of the derivation follows the same lines as in Section 9.1. To calculate the matrix element $\langle q';t'|O_A(P(t_A),Q(t_A))\,O_B(P(t_B),Q(t_B))\cdots|q;t\rangle$ of a product of operators (with $t' > t_A > t_B > \cdots > t$), divide the time-interval from t to t' into a large number of very close time steps; at each time step insert the completeness relation (9.5.41); use Eq. (9.5.47) to evaluate the resulting matrix elements (with $\mathcal{O}_A, \mathcal{O}_B$, etc. inserted where appropriate); move all differentials to the left (this introduces no sign changes, because at each step we have an equal number of dps and dqs); and then introduce functions $q_a(t)$ and $p_a(t)$ that interpolate between the values of q_a and p_a at each step. We then find

$$\langle q';t'|T\left\{\mathcal{O}_A\Big(P(t_A),Q(t_A)\Big),\ \mathcal{O}_B\Big(P(t_B),\ Q(t_B)\Big),\cdots\right\}|q;t\rangle$$
$$= (-i)^N \chi_N \int_{q_a(t)=q_a, q_a(t')=q'_a} \left(\widetilde{\prod_{a\tau}} dq_a(\tau)dp_a(\tau)\right)$$
$$\times\ \mathcal{O}_A\Big(p(t_A),q(t_A)\Big)\,\mathcal{O}_B\Big(p(t_B),q(t_B)\Big)\cdots$$
$$\times\ \exp\left[i\int_t^{t'} d\tau\left\{\sum_a p_a(\tau)\dot{q}_a(\tau) - H\Big(p(\tau),q(\tau)\Big)\right\}\right] . \tag{9.5.48}$$

The symbol T here denotes the ordinary product if the times are in the order originally assumed, $t_A > t_B > \cdots$. However, the right-hand side is totally symmetric in the $\mathcal{O}_A, \mathcal{O}_B, \cdots$ (except for minus signs where anticommuting c-numbers are interchanged) so this formula holds for general times (between t and t'), provided T is interpreted as the time-ordered product, with an overall minus sign if time-ordering the operators involves an odd number of permutations of fermionic operators.

Up to this point we have kept track of the overall phase factor $(-i)^N\chi_N$.

But in fact these phases contribute only to the vacuum–vacuum transition amplitude, and hence will not be of importance to us.

The transition to quantum field theory follows along the same lines as described for bosonic fields in Section 9.2. The vacuum expectation value of a time-ordered product of operators is given by a formula just like Eq. (9.2.17):

$$\Big\langle \text{VAC, out}\Big|T\Big\{\mathscr{O}_A\Big[P(t_A),Q(t_A)\Big],\ \mathscr{O}_B\Big[P(t_B),Q(t_B)\Big],\cdots\Big\}\Big|\text{VAC, in}\Big\rangle$$
$$\propto \int\Big[\prod_{\tau,\mathbf{x},m} dq_m(\mathbf{x},\tau)\Big]\Big[\prod_{\tau,\mathbf{x},m} dp_m(\mathbf{x},\tau)\Big]\ \mathscr{O}_A\Big[p(t_A),q(t_A)\Big]$$
$$\times\ \mathscr{O}_B\Big[p(t_B),q(t_B)\Big]\cdots\exp\Big[i\int_{-\infty}^{\infty}d\tau\Big\{\int d^3x\sum_m p_m(\mathbf{x},\tau)\,\dot{q}_m(\mathbf{x},\tau)$$
$$-H\Big[q(\tau),p(\tau)\Big] + i\epsilon\ \text{terms}\Big\}\Big] \tag{9.5.49}$$

where the proportionality constant is the same for all operators $\mathscr{O}_A$, $\mathscr{O}_B$, etc., and the '$i\epsilon$ terms' again arise from the wave function of the vacuum. As before, we have replaced each discrete index like a with a space position $\mathbf{x}$ and a field index m. We are also dropping the tilde on the product of differentials, since it only affects the constant phase in the path integral.

A major difference between the fermionic and bosonic cases is that here we will not want to integrate out the ps before the qs. Indeed, in the standard model of electroweak interactions (and in other theories, such as the older Fermi theory of beta decay) the canonical conjugates p_m are auxiliary fields unrelated to the $\dot{q}_m$, and the Lagrangian is linear in the $\dot{q}_m$, so that the quantity $\int d^3x \sum_m p_m \dot{q}_m - H$ in Eq. (9.5.49) as it stands *is* the Lagrangian L . Each term in the Hamiltonian for a fermionic field that carries a non-vanishing quantum number (like the electron field in quantum electrodynamics) generally contains an equal number of ps (proportional to $q^\dagger$) and qs. In particular, the free-particle term H_0 in the Hamiltonian is bilinear in p and q, so that

$$\int_{-\infty}^{\infty}d\tau\Big\{\int d^3x\sum_m p_m(\mathbf{x},\tau)\dot{q}_m(\mathbf{x},\tau) - H_0\Big[q(\tau),p(\tau)\Big] + i\epsilon\ \text{terms}\Big\}$$
$$= -\sum_{mn}\int d^4x\, d^4y\ \mathscr{D}_{mx,ny}\, p_m(x)\, q_n(y) \tag{9.5.50}$$

with $\mathscr{D}$ some numerical 'matrix'. The interaction Hamiltonian $V \equiv H - H_0$ is a sum of products of equal numbers of fermionic qs and ps (with coefficients that may depend on bosonic fields) so when we expand Eq. (9.5.49) in powers of the V we encounter a sum of fermionic integrals of the form

$$\mathscr{I}_{n_1 m_1 n_2 m_2 \cdots n_N m_N}(x_1, y_1, x_2, y_2 \cdots, x_N, y_N) \equiv \int \Big[\prod_{\tau,\mathbf{x},m} dq_m(\mathbf{x},\tau)\Big]$$

$$\times \Big[\prod_{\tau,\mathbf{x},m} dp_m(\mathbf{x},\tau)\Big] q_{m_1}(x_1)\, p_{n_1}(y_1)\, q_{m_2}(x_2)\, p_{n_2}(y_2) \cdots q_{m_N}(x_N)\, p_{n_N}(y_N)$$

$$\times \exp\Big(-i\sum_{mn}\int d^4x\, d^4y\, \mathscr{D}_{mx,ny}\, p_m(x)\, q_n(y)\Big)\,, \tag{9.5.51}$$

one such term for each possible set of vertices in the Feynman diagram, with coefficients contributed by each vertex given by i times the coefficient of the product of fields in the corresponding term in the interaction.

To calculate this sort of integral, first consider a generating function for all these integrals:

$$\mathscr{I}(f,g) \equiv \int \Big[\prod_{\mathbf{x},\tau,m} dq_m(\mathbf{x},\tau)\, dp_m(\mathbf{x},\tau)\Big]$$

$$\times \exp\Big(-i\sum_{mn}\int d^4x\, d^4y\, \mathscr{D}_{mx,ny}\, p_m(x)\, q_n(y)$$

$$-i\sum_{m}\int d^4x\, p_m(x)\, f_m(x) - i\sum_{n}\int d^4y\, g_n(y)\, q_n(y)\Big)\,, \tag{9.5.52}$$

where $f_m(x)$ and $g_n(y)$ are arbitrary anticommuting c-number functions. We shift to new variables of integration

$$p'_m(x) = p_m(x) + \sum_n \int d^4y\, g_n(y)\, (\mathscr{D}^{-1})_{ny,mx}\,,$$

$$q'_n(y) = q_n(y) + \sum_m \int d^4x\, (\mathscr{D}^{-1})_{ny,mx}\, f_m(x)\,.$$

Using the translation invariance condition (9.5.36), we then find

$$\mathscr{I}(f,g) = \exp\Big(i\sum_{mn}\int d^4x\, d^4y\, (\mathscr{D}^{-1})_{ny,mx}\, g_n(y)\, f_m(x)\Big)$$

$$\times \int \Big[\prod_{\mathbf{x},\tau,m} dq'_m(\mathbf{x},\tau)\, dp'_m(\mathbf{x},\tau)\Big]$$

$$\times \exp\Big(-i\sum_{mn}\int d^4x\, d^4y\, \mathscr{D}_{mx,ny}\, p'_m(x)\, q'_n(y)\Big)\,. \tag{9.5.53}$$

The integral is a constant (i.e., independent of the functions f and g) which can be shown using Eq. (9.5.38) to be proportional to Det $\mathscr{D}$. Of more importance to us is the first factor. Expanding this factor in powers of gf and comparing with the direct expansion of Eq. (9.5.52), we see that

$$\mathscr{I}_{n_1 m_1 n_2 m_2 \cdots n_N m_N}(x_1, y_1, x_2, y_2 \cdots, x_N, y_N)$$
$$\propto \sum_{\text{pairings}} \delta_{\text{pairing}} \prod_{\text{pairs}} \left(-i\mathscr{D}^{-1}\right)_{\text{paired } mx,ny} \tag{9.5.54}$$

with a proportionality constant that is independent of the x, y, m, or n, and also independent of the number of these variables. The sum is over all different ways of pairing ps with qs, not counting as different pairings that only differ in the order of the pairs. In other words, we sum over the $N!$ permutations either of the ps or the qs. The sign factor δ_{pairing} is $+1$ if this permutation is even; -1 if it is odd.

This sign factor and sum over pairings are just the same as we encountered in our earlier derivation of the Feynman rules, with the sum over pairings corresponding to the sum over ways of connecting the lines associated with vertices in the Feynman diagrams, and the factors $(\mathscr{D}^{-1})_{mx,ny}$ playing the role of the propagator for the pairing of $q_m(x)$ with $p_n(y)$. In the Dirac formalism for spin $\frac{1}{2}$, the free-particle action is

$$\int_{-\infty}^{\infty} d\tau \left\{ \int d^3x \sum_m p_m(\mathbf{x}, \tau)\dot{q}_m(\mathbf{x}, \tau) - H_0\Big[q(\tau), p(\tau)\Big] \right\}$$
$$= -\int d^4x \, \bar{\psi}(x) \left[\gamma^\mu \partial_\mu + m\right] \psi(x) \,, \tag{9.5.55}$$

where in the usual notation the canonical variables here are

$$q_m(x) = \psi_m(x) \,, \qquad p_m(x) = -[\bar{\psi}(x)\gamma^0]_m = i\psi_m^\dagger(x) \tag{9.5.56}$$

with m a four-valued Dirac index. Comparing this with Eq. (9.5.50) , we find here

$$\mathscr{D}_{mx,ny} = \left[\gamma^0 \left(\gamma^\mu \frac{\partial}{\partial x^\mu} + m - i\epsilon\right)\right]_{mn} \delta^4(x-y)$$
$$= \int \frac{d^4k}{(2\pi)^4} \left(\gamma^0[i\gamma^\mu k_\mu + m - i\epsilon]\right)_{mn} e^{ik\cdot(x-y)} \,. \tag{9.5.57}$$

(Though we shall not work it out in detail, the $i\epsilon$ term here arises in much the same way as for the scalar field in Section 9.2.) The propagator is then

$$(\mathscr{D}^{-1})_{mx,ny} = \int \frac{d^4k}{(2\pi)^4} \left([i\gamma^\mu k_\mu + m - i\epsilon]^{-1}[-\gamma^0]\right)_{mn} e^{ik\cdot(x-y)} \,, \tag{9.5.58}$$

just as we found in the operator formalism. The extra factor $-\gamma^0$ arises because this propagator is the vacuum expectation value of $T\{\psi_m(x), -[\bar{\psi}(y)\gamma^0]_n\}$, not $T\{\psi_m(x), \bar{\psi}_n(y)\}$.

As one example of a problem that is easier to solve by path-integral than by operator methods, let us calculate the field dependence of the vacuum→vacuum amplitude for a Dirac field that interacts only with an external field. Take the Lagrangian as

$$\mathscr{L} = -\bar{\psi}[\gamma^\mu \partial_\mu + m + \Gamma]\psi\,, \tag{9.5.59}$$

where $\Gamma(x)$ is an x-dependent matrix representing the interaction of the fermion with the external field. According to Eq. (9.5.49), the vacuum persistence amplitude in the presence of this external field is

$$\langle \text{VAC, out}|\text{VAC, in}\rangle_\Gamma \propto \int \Big[\prod_{\tau,\mathbf{x},m} dq_m(\mathbf{x},\tau)\Big]\Big[\prod_{\tau,\mathbf{x},m} dp_m(\mathbf{x},\tau)\Big]$$
$$\times \exp\Big\{-i\int d^4x\, p^T \gamma^0[\gamma^\mu\partial_\mu + m + \Gamma - i\epsilon]\,q\Big\} \tag{9.5.60}$$

with a proportionality constant that is independent of $\Gamma(x)$. We write this as

$$\langle \text{VAC, out}|\text{VAC, in}\rangle_\Gamma \propto \int \Big[\prod_{\tau,\mathbf{x},m} dq_m(\mathbf{x},\tau)\Big]\Big[\prod_{\tau,\mathbf{x},m} dp_m(\mathbf{x},\tau)\Big]$$
$$\times \exp\Big\{-i\sum_{mn}\int d^4x\, d^4y\; p_m(x)\, q_n(y)\, \mathscr{K}[\Gamma]_{mx,ny}\Big\}\,, \tag{9.5.61}$$

where

$$\mathscr{K}[\Gamma]_{mx,ny} = \left(\gamma^0\left[\gamma^\mu\frac{\partial}{\partial x^\mu} + m + \Gamma(x) - i\epsilon\right]\right)_{mn}\delta^4(x-y)\,. \tag{9.5.62}$$

To evaluate this, we change the variables of integration $q_n(x)$ to

$$q'_m(x) \equiv \sum_n \int d^4y\; \mathscr{K}[\Gamma]_{mx,ny}\, q_n(y)\,. \tag{9.5.63}$$

The remaining integral is now Γ-independent, so the whole dependence of the vacuum persistence amplitude is contained in the determinant arising according to Eq. (9.5.38) from the change of variables:

$$\langle \text{VAC, out}|\text{VAC, in}\rangle_\Gamma \propto \text{Det}\,\mathscr{K}[\Gamma]\,. \tag{9.5.64}$$

To recover the results of perturbation theory, let us write

$$\mathscr{K}[\Gamma] \equiv \mathscr{D} + \mathscr{G}[\Gamma]\,, \tag{9.5.65}$$

$$\mathscr{G}[\Gamma]_{mx,ny} = \Big(\gamma^0\,\Gamma(x)\Big)_{mn}\delta^4(x-y)\,, \tag{9.5.66}$$

and expand in powers of $\mathscr{G}[\Gamma]$. Eq. (9.5.64) gives then

$$\langle \text{VAC, out}|\text{VAC, in}\rangle_\Gamma \propto \text{Det}\,\Big(\mathscr{D}[1 + \mathscr{D}^{-1}\mathscr{G}[\Gamma]]\Big)$$
$$= [\text{Det}\,\mathscr{D}]\,\exp\left(\sum_{n=1}^\infty \frac{(-1)^{n+1}}{n}\text{Tr}\,(\mathscr{D}^{-1}\mathscr{G}[\Gamma])^n\right)\,. \tag{9.5.67}$$

This is just what we should expect from the Feynman rules: the contributions from internal lines and vertices in this theory are $-i\mathscr{D}^{-1}$ and $-i\mathscr{G}[\Gamma]$; the trace of the product of n factors of $-\mathscr{D}^{-1}\mathscr{G}[\Gamma]$ thus corresponds to a loop with n vertices connected by n internal lines; $1/n$ is the usual combinatoric factor associated with such loops (see Section 6.1); the sign factor is $(-1)^{n+1}$ rather than $(-1)^n$ because an extra minus sign is associated with fermion loops; and the sum over n appears as the argument of an exponential because the vacuum persistence amplitude receives contributions from graphs with any number of disconnected loops. The Γ-independent factor Det $\mathscr{D}$ is less easy to derive from the Feynman rules; it represents the contribution of any number of fermion loops that carry no vertices.

More to the point, a formula like Eq. (9.5.64) allows us to derive non-perturbative results by using topological theorems to derive information about the eigenvalues of kernels like $\mathscr{K}[\Gamma]$. This will be pursued further in Volume II.

9.6 Path-Integral Formulation of Quantum Electrodynamics

The path-integral approach to quantum field theory really comes into its own when applied to gauge theories of massless spin one particles, such as quantum electrodynamics. The derivation of the Feynman rules for quantum electrodynamics in the previous chapter involved a fair amount of hand-waving, in arguing that the terms in the photon propagator $\Delta^{\mu\nu}(q)$ proportional to q^μ or q^ν could be dropped, and that the purely time-like terms would just cancel the Coulomb term in the Hamiltonian, so that the effective photon propagator could be taken as $\eta^{\mu\nu}/q^2$. To give a real justification of this result using the methods of Chapter 8 would involve us in a complicated analysis of Feynman diagrams. But as we shall now see, the path-integral approach yields the desired form of the photon propagator, without ever having to think about the details of Feynman diagrams.

In Chapter 8 we found that in Coulomb gauge, the Hamiltonian for the interaction of photons with charged particles takes the form

$$H[\mathbf{A},\mathbf{\Pi}_\perp,\cdots] = H_{\rm M} + \int d^3x \left[\tfrac{1}{2}\mathbf{\Pi}_\perp{}^2 + \tfrac{1}{2}(\nabla\times\mathbf{A})^2 - \mathbf{A}\cdot\mathbf{J}\right] + V_{\rm Coul}\,. \tag{9.6.1}$$

Here $\mathbf{A}$ is the vector potential, subject to the Coulomb gauge condition

$$\nabla\cdot\mathbf{A} = 0\,, \tag{9.6.2}$$

while $\mathbf{\Pi}_\perp$ is the solenoidal part of its canonical conjugate, satisfying the same constraint

$$\nabla \cdot \mathbf{\Pi}_\perp = 0 . \tag{9.6.3}$$

Also, H_M is the matter Hamiltonian and $V_{\rm Coul}$ is the Coulomb energy

$$V_{\rm Coul}(t) = \tfrac{1}{2}\int d^3x\, d^3y\, J^0(\mathbf{x},t)J^0(\mathbf{y},t)\Big/4\pi|\mathbf{x}-\mathbf{y}| . \tag{9.6.4}$$

Just as for any other Hamiltonian system, we can calculate vacuum expectation values of time-ordered products as path integrals*

$$\langle T\{\mathcal{O}_A\mathcal{O}_B\cdots\}\rangle_{\rm VAC} = \int \left[\prod_{x,i} da_i(x) \prod_{x,i} d\pi_i(x) \prod_{x,\ell} d\psi_\ell(x)\right] \mathcal{O}_A\mathcal{O}_B\cdots$$
$$\times \exp\left\{i\int d^4x \left[\boldsymbol{\pi}\cdot\dot{\mathbf{a}} - \tfrac{1}{2}\boldsymbol{\pi}^2 - \tfrac{1}{2}(\nabla\times\mathbf{a})^2 + \mathbf{a}\cdot\mathbf{J} + \mathscr{L}_{\rm M}\right] - i\int dt\, V_{\rm Coul}\right\}$$
$$\times \left[\prod_x \delta\Big(\nabla\cdot\mathbf{a}(x)\Big)\right]\left[\prod_x \delta\Big(\nabla\cdot\boldsymbol{\pi}(x)\Big)\right] , \tag{9.6.5}$$

where $\psi_\ell(x)$ are generic matter fields. In writing Eq. (9.6.5) in terms of a matter Lagrangian density, we are assuming that H_M is local and either linear in the matter πs (as in spinor electrodynamics) or quadratic with field-independent coefficients (as in scalar electrodynamics). We have inserted delta functions** in Eq. (9.6.5) to enforce the constraints (9.6.2) and (9.6.3).

The argument of the exponential in Eq. (9.6.5) is evidently quadratic in the independent components of $\boldsymbol{\pi}$ (say, π_1 and π_2), with field-independent coefficients in the term of second order in $\boldsymbol{\pi}$. Thus, according to Eq. (9.A.9), the integral over $\boldsymbol{\pi}$ can be done (up to a constant factor) by setting $\boldsymbol{\pi}$ equal to the stationary point of the argument of the exponential,

* Note that $\pi(x)$ is the interpolating c-number field for the quantum operator $\mathbf{\Pi}_\perp$, whose commutation relations with each other and with $\mathbf{A}$ are the same as those of $\mathbf{\Pi}$, but which unlike $\mathbf{\Pi}$ commutes with all canonical matter variables.

** This is not strictly accurate. If we take the canonical variables to be, say, a_1, a_2 and π_1, π_2, with a_3 and π_3 regarded as functionals of these variables given by Eqs. (9.6.2) and (9.6.3), then we should insert the delta functions

$$\prod_x \delta\left(a_3(x) + \partial_3^{-1}\Big(\partial_1 a_1(x) + \partial_2 a_2(x)\Big)\right)\delta\left(\pi_3(x) + \partial_3^{-1}\Big(\partial_1\pi_1(x) + \partial_2\pi_2(x)\Big)\right) .$$

However, this differs from the product of delta functions in Eq. (9.6.5) only by a factor $\mathrm{Det}\,\partial_3^2$, which although infinite is field-independent and hence cancels in ratios like (9.4.1).

$\boldsymbol{\pi} = \dot{\mathbf{a}}$:

$$\langle T\{\mathcal{O}_A \mathcal{O}_B \cdots\}\rangle_{\rm VAC} = \int \left[\prod_{x,i} da_i(x) \prod_{x,\ell} d\psi_\ell(x)\right]$$
$$\times\ \mathcal{O}_A \mathcal{O}_B \cdots\ \exp\left\{ i \int d^4x \left[\frac{1}{2}\dot{\mathbf{a}}^2 - \frac{1}{2}(\nabla \times \mathbf{a})^2 + \mathbf{a}\cdot\mathbf{j} + \mathscr{L}_{\rm M}\right]\right.$$
$$\left. -i \int dt\, V_{\rm Coul} \ +\ i\epsilon \text{ terms}\right\} \left[\prod_x \delta\Big(\nabla \cdot \mathbf{a}(x)\Big)\right] . \tag{9.6.6}$$

To bring out the essential covariance of this result, we use a trick. Introduce a new variable of integration $a^0(x)$, and replace the Coulomb term $-\int dt\, V_{Coul}$ in the action with

$$\int d^4x \left[-a^0(x) j^0(x) + \tfrac{1}{2}\Big(\nabla a^0(x)\Big)^2\right] . \tag{9.6.7}$$

Since (9.6.7) is quadratic in a^0, the integral over a^0 can be done (up to a constant factor) by setting $a^0(x)$ equal to the stationary point of (9.6.7), i.e., to the solution of

$$-j^0(x) - \nabla^2 a^0(x) = 0$$

or in other words, to

$$a^0(\mathbf{x}, t) = \int d^3y\, \frac{j^0(\mathbf{y}, t)}{4\pi|\mathbf{x} - \mathbf{y}|} . \tag{9.6.8}$$

Using this in Eq. (9.6.7) just gives the Coulomb action $-\int dt\, V_{\rm Coul}$. Hence we can rewrite the argument of the exponential in Eq. (9.6.6) as

$$\tfrac{1}{2}\dot{\mathbf{a}}^2 - \tfrac{1}{2}(\nabla \times \mathbf{a})^2 + \mathbf{a}\cdot\mathbf{j} + \mathscr{L}_{\rm M} - a^0 j^0 + \tfrac{1}{2}(\nabla a^0)^2$$
$$= -\tfrac{1}{4} f_{\mu\nu} f^{\mu\nu} + a_\mu j^\mu + \mathscr{L}_{\rm M} + \text{total derivatives}$$

with $f_{\mu\nu} = \partial_\mu a_\nu - \partial_\nu a_\mu$, and integrate over a^0 as well as over $\mathbf{a}$ and matter fields. That is, the path integral (9.6.6) is now

$$\langle T\{\mathcal{O}_A \mathcal{O}_B \cdots\}\rangle_{\rm VAC} \propto \int \left[\prod_{x,\mu} da_\mu(x)\right] \left[\prod_{x,\ell} d\psi_\ell(x)\right]$$
$$\times\ \mathcal{O}_A \mathcal{O}_B \cdots\ \exp\Big(iI[a, \psi]\Big) \prod_x \delta\Big(\nabla \cdot \mathbf{a}(x)\Big) , \tag{9.6.9}$$

where I is the original action

$$I[a, \psi] = \int d^4x\, [-\tfrac{1}{4} f_{\mu\nu} f^{\mu\nu} + a_\mu j^\mu + \mathscr{L}_{\rm M}] \ +\ i\epsilon \text{ terms} . \tag{9.6.10}$$

Now everything is manifestly Lorentz- and gauge-invariant, except for the final product of delta functions which enforce the Coulomb gauge condition.** To make further progress, we shall use a simple version of a trick[4,5] that in Volume II will be used to treat the more difficult case of non-Abelian gauge theories. For simplicity, we shall deal here with the case where the operators $\mathcal{O}_A[A,\Psi]$, $\mathcal{O}_B[A,\Psi], \cdots$ as well as the action $I[a,\psi]$ and measure $[\prod da][\prod d\psi]$ are gauge-invariant.

First, replace the field variables of integration $a_\mu(x)$ and $\psi(x)$ everywhere in Eq. (9.6.9) with the new variables

$$a_{\mu\Lambda}(x) \equiv a_\mu(x) + \partial_\mu \Lambda(x) \,, \tag{9.6.11}$$

$$\psi_{\ell\Lambda}(x) \equiv \exp\Big(i q_\ell \Lambda(x)\Big)\psi_\ell(x) \tag{9.6.12}$$

with arbitrary finite $\Lambda(x)$. This step is a mathematical triviality, like changing an integral $\int_{-\infty}^{\infty} f(x)dx$ to read $\int_{-\infty}^{\infty} f(y)dy$, and does not require use of the postulated gauge invariance of the theory. Next, use gauge invariance to replace $a_{\mu\Lambda}(x)$ and $\psi_{\ell\Lambda}(x)$ in the action, measure, and $\mathcal{O}$-functions with the original fields $a_\mu(x)$ and $\psi_\ell(x)$, respectively. Eq. (9.6.9) then becomes

$$\Big\langle T\Big\{\mathcal{O}_A[A,\Psi],\, \mathcal{O}_B[A,\Psi], \cdots\Big\}\Big\rangle_{\text{VAC}}$$
$$\propto \int \Big[\prod_{x,\mu} da_\mu(x)\Big]\Big[\prod_{x,\ell} d\psi_\ell(x)\Big]\; \mathcal{O}_A[a,\psi]\mathcal{O}_B[a,\psi]\cdots$$
$$\times \exp\Big(iI[a,\psi]\Big)\prod_x \delta\Big(\nabla\cdot\mathbf{a}(x) + \nabla^2\Lambda(x)\Big) \,. \tag{9.6.13}$$

Now, the function $\Lambda(x)$ was chosen at random, so despite appearances the right-hand side of Eq. (9.6.13) cannot depend on this function. We shall exploit this fact to put the path integral in a much more convenient form. Multiply Eq. (9.6.13) by the functional

$$B[\Lambda,a] = \exp\Big(-\tfrac{1}{2}i\alpha\int d^4x\,\big(\partial_0 a^0 - \nabla^2\Lambda\big)^2\Big) \tag{9.6.14}$$

(where α is an arbitrary constant), and integrate over $\Lambda(x)$. By shifting the integration variable $\Lambda(x)$, and noting the actual Λ-independence of (9.6.13), we see that the effect is simply to multiply Eq. (9.6.13) with the field-independent constant

$$\int \Big[\prod_x d\Lambda(x)\Big]\; \exp\Big(-\tfrac{1}{2}i\alpha\int d^4x\,\big(\nabla^2\Lambda\big)^2\Big) \,. \tag{9.6.15}$$

** Note that now $a^0(x)$ is *not* equal to the value (9.6.8), but is an independent variable of integration. We will not integrate over $a^0(x)$ first, which would lead back to Eq. (9.6.6), but instead will treat it in tandem with $\mathbf{a}(x)$.

This factor cancels out in the connected part of the vacuum expectation value, and thus has no physical effect. But (9.6.13) is only Λ-independent *after* we integrate over $a^\mu(x)$ and $\psi(x)$. We can just as well integrate over $\Lambda(x)$ *before* we integrate over $a^\mu(x)$ and $\psi(x)$, in which case the factor $\prod_x \delta\Big(\nabla\cdot\mathbf{a}(x)+\nabla^2\Lambda\Big)$ in Eq. (9.6.13) is replaced with

$$\int\left[\prod_x d\Lambda(x)\right]\,\exp\left(-\tfrac{1}{2}i\alpha\int d^4x\,\Big(\partial_0 a^0-\nabla^2\Lambda\Big)^2\right)\prod_x\delta\Big(\nabla\cdot\mathbf{a}(x)+\nabla^2\Lambda\Big)$$
$$\propto \exp\left(-\tfrac{1}{2}i\alpha\int d^4x\,(\partial_\mu a^\mu)^2\right), \qquad (9.6.16)$$

where "$\propto$" again means proportional with a field-independent factor. Dropping constant factors, Eq. (9.6.9) now becomes

$$\langle T\{\mathcal{O}_A\mathcal{O}_B\cdots\}\rangle_{\text{VAC}}\propto\int\left[\prod_{x,\mu}da_\mu(x)\right]\left[\prod_{x,\ell}d\psi_\ell(x)\right]\mathcal{O}_A\mathcal{O}_B\cdots\exp\left(iI_{\text{eff}}[a,\psi]\right), \qquad (9.6.17)$$

where

$$I_{\text{eff}}[a,\psi]=I[a,\psi]-\tfrac{1}{2}\alpha\int(\partial_\mu a^\mu)^2d^4x\,. \qquad (9.6.18)$$

This is now manifestly Lorentz-invariant.

We consider the new term in (9.6.18) as a contribution to the unperturbed part of the action, whose photonic part now reads

$$I_0[a]=\int d^4x\left[-\tfrac{1}{4}(\partial_\mu a_\nu-\partial_\nu a_\mu)(\partial^\mu a^\nu-\partial^\nu a^\mu)-\tfrac{1}{2}\alpha(\partial_\mu a^\mu)^2+i\epsilon\text{ terms}\right]$$
$$=-\tfrac{1}{2}\int d^4x\,d^4y\;a^\mu(x)\,a^\nu(y)\,D_{\mu x,\nu y}, \qquad (9.6.19)$$

where

$$\mathscr{D}_{\mu x,\nu y}=\left[\eta_{\mu\nu}\frac{\partial^2}{\partial x^\rho\partial y_\rho}-(1-\alpha)\frac{\partial^2}{\partial x^\mu\partial y^\nu}\right]\delta^4(x-y)+i\epsilon\text{ terms}$$
$$=(2\pi)^{-4}\int d^4q\left[\eta_{\mu\nu}\,q^2-(1-\alpha)q_\mu q_\nu-i\epsilon\eta_{\mu\nu}\right]e^{iq\cdot(x-y)}\,. \qquad (9.6.20)$$

The photon propagator is then found immediately by inverting the 4×4 matrix in the integrand of Eq. (9.6.20)

$$\Delta_{\mu x,\nu y}=(2\pi)^{-4}\int d^4q\left[\frac{\eta^{\mu\nu}}{q^2-i\epsilon}+\frac{(1-\alpha)}{\alpha}\,\frac{q^\mu q^\nu}{(q^2-i\epsilon)^2}\right]e^{iq\cdot(x-y)}\,. \qquad (9.6.21)$$

We are free to choose α as seems most convenient. Two common choices are $\alpha=1$, which yields the propagator in *Feynman gauge*:

$$\Delta^{\text{Feynman}}_{\mu x,\nu y}=(2\pi)^{-4}\int d^4q\left[\frac{\eta^{\mu\nu}}{q^2-i\epsilon}\right]e^{iq\cdot(x-y)} \qquad (9.6.22)$$

or $\alpha = \infty$, in which case the factor (9.6.14) acts as a delta function, and we obtain the propagator in *Landau gauge* (often also called *Lorentz gauge*):

$$\Delta^{\text{Landau}}_{\mu x,\nu y} = (2\pi)^{-4} \int d^4q \left[\frac{\eta^{\mu\nu}}{q^2 - i\epsilon} - \frac{q^\mu q^\nu}{(q^2 - i\epsilon)^2} \right] e^{iq\cdot(x-y)} \,. \tag{9.6.23}$$

Practical calculations are made far more convenient by working with such manifestly Lorentz-invariant interactions and propagators.

9.7 Varieties of Statistics*

We can now take up a question raised in Chapter 4: what are the possibilities for the change of state-vectors when we interchange identical particles?

For this purpose, we will consider the preparation of the initial or final states in a scattering process. Suppose that a set of indistinguishable particles in either of these states is brought to a particular configuration with momenta $\mathbf{p}_1$, $\mathbf{p}_2$, etc. from a standard configuration with momenta $\mathbf{k}_1$, $\mathbf{k}_2$, etc., by some sort of slowly varying external fields, keeping the particles far enough apart in the process to justify the use of non-relativistic quantum mechanics. (Spin indices are not shown explicitly here; they should be understood to accompany momentum labels.) To calculate the amplitude for this process we can use the path-integral method,** taking the qs and ps of Section 9.1 as particle positions and momenta, rather than fields and their canonical conjugates. These always satisfy canonical commutation rather than anticommutation relations, whether or not the particles are bosons or fermions or something else, so at this point we are not committing ourselves to any particular statistics. The path-integral formula (9.1.34) gives an amplitude $\langle \mathbf{p}_1, \mathbf{p}_2, \cdots | \mathbf{k}_1, \mathbf{k}_2, \cdots \rangle_D$ as an integral over paths in which one particle is brought continuously from momentum $\mathbf{k}_1$ to momentum $\mathbf{p}_1$, another identical particle is brought continuously from momentum $\mathbf{k}_2$ to momentum $\mathbf{p}_2$, and so on. The subscript 'D' indicates that this is the amplitude we would calculate for distinguishable particles. In particular, this amplitude is symmetric under permutations of the $\mathbf{p}$s *and* simultaneous permutations of the $\mathbf{k}$s, but has no particular symmetry under separate permutations of the $\mathbf{p}$s *or* $\mathbf{k}$s. But

* This section lies somewhat out of the book's main line of development, and may be omitted in a first reading.

** Here I am following the discussion of Laidlaw and C. DeWitt,[11] except that they apply the path-integral method to the whole scattering process, rather than just the preparation of initial or final states. In a relativistic theory the possibility of particle creation and annihilation makes it necessary to apply the path-integral method to fields rather than particle orbits. For us, this is not a problem, because we limit such calculations to sufficiently early or late times, when the particles participating in a scattering process are all far apart.

if the particles are really indistinguishable, then there are other paths that are topologically distinct, but that yield the same final configuration. For *space* dimensionality $d \geq 3$, the only such paths† are those that take $\mathbf{k}_1, \mathbf{k}_2, \cdots$ into some non-trivial permutation $\mathscr{P}$ of $\mathbf{p}_1, \mathbf{p}_2, \cdots$. Hence the true amplitude should be written

$$\langle \mathbf{p}_1, \mathbf{p}_2, \cdots | \mathbf{k}_1, \mathbf{k}_2, \cdots \rangle = \sum_{\mathscr{P}} C_{\mathscr{P}} \langle \mathbf{p}_{\mathscr{P}1}, \mathbf{p}_{\mathscr{P}2}, \cdots | \mathbf{k}_1, \mathbf{k}_2, \cdots \rangle_{\mathrm{D}} , \tag{9.7.1}$$

the sum running over all $N!$ permutations of the N indistinguishable particles in the state, and $C_{\mathscr{P}}$ a set of complex constants. These amplitudes must satisfy a composition rule appropriate for indistinguishable particles:

$$\langle \mathbf{p}_1, \mathbf{p}_2, \cdots | \mathbf{k}_1, \mathbf{k}_2, \cdots \rangle = \frac{1}{N!} \int d^3q_1\, d^3q_2 \cdots \langle \mathbf{p}_1, \mathbf{p}_2, \cdots | \mathbf{q}_1, \mathbf{q}_2, \cdots \rangle \times \langle \mathbf{q}_1, \mathbf{q}_2, \cdots | \mathbf{k}_1, \mathbf{k}_2, \cdots \rangle . \tag{9.7.2}$$

Using Eq. (9.7.1), this is the requirement

$$\sum_{\mathscr{P}} C_{\mathscr{P}} \langle \mathbf{p}_{\mathscr{P}1}, \mathbf{p}_{\mathscr{P}2}, \cdots | \mathbf{k}_1, \mathbf{k}_2, \cdots \rangle_{\mathrm{D}} = \frac{1}{N!} \sum_{\mathscr{P}', \mathscr{P}''} C_{\mathscr{P}'} C_{\mathscr{P}''} \int d^3q_1\, d^3q_2 \cdots \times \langle \mathbf{p}_{\mathscr{P}'1}, \mathbf{p}_{\mathscr{P}'2}, \cdots | \mathbf{q}_1, \mathbf{q}_2, \cdots \rangle_{\mathrm{D}} \langle \mathbf{q}_{\mathscr{P}''1}, \mathbf{q}_{\mathscr{P}''2}, \cdots | \mathbf{k}_1, \mathbf{k}_2, \cdots \rangle_{\mathrm{D}}.$$

Applying a permutation $\mathscr{P}''$ to both the initial and final states in the first amplitude on the right, this is

$$\sum_{\mathscr{P}} C_{\mathscr{P}} \langle \mathbf{p}_{\mathscr{P}1}, \mathbf{p}_{\mathscr{P}2}, \cdots | \mathbf{k}_1, \mathbf{k}_2, \cdots \rangle_{\mathrm{D}} = \frac{1}{N!} \sum_{\mathscr{P}', \mathscr{P}''} C_{\mathscr{P}'} C_{\mathscr{P}''} \int d^3q_1\, d^3q_2 \cdots \times \langle \mathbf{p}_{\mathscr{P}''\mathscr{P}'1}, \mathbf{p}_{\mathscr{P}''\mathscr{P}'2}, \cdots | \mathbf{q}_{\mathscr{P}''1}, \mathbf{q}_{\mathscr{P}''2}, \cdots \rangle_{\mathrm{D}} \langle \mathbf{q}_{\mathscr{P}''1}, \mathbf{q}_{\mathscr{P}''2}, \cdots | \mathbf{k}_1, \mathbf{k}_2, \cdots \rangle_{\mathrm{D}} .$$

But the amplitudes $\langle \mathbf{p}_1, \mathbf{p}_2, \cdots | \mathbf{k}_1, \mathbf{k}_2, \cdots \rangle_{\mathrm{D}}$ satisfy the composition rule for *distinguishable* particles

$$\langle \mathbf{p}_1, \mathbf{p}_2, \cdots | \mathbf{k}_1, \mathbf{k}_2, \cdots \rangle_{\mathrm{D}} = \int d^3q_1\, d^3q_2 \cdots \langle \mathbf{p}_1, \mathbf{p}_2, \cdots | \mathbf{q}_1, \mathbf{q}_2, \cdots \rangle_{\mathrm{D}} \times \langle \mathbf{q}_1, \mathbf{q}_2, \cdots | \mathbf{k}_1, \mathbf{k}_2, \cdots \rangle_{\mathrm{D}} , \tag{9.7.3}$$

so the composition rule for the physical amplitudes may be written

$$\sum_{\mathscr{P}} C_{\mathscr{P}} \langle \mathbf{p}_{\mathscr{P}1}, \mathbf{p}_{\mathscr{P}2}, \cdots | \mathbf{k}_1, \mathbf{k}_2, \cdots \rangle_{\mathrm{D}} = \frac{1}{N!} \sum_{\mathscr{P}', \mathscr{P}''} C_{\mathscr{P}'} C_{\mathscr{P}''} \times \langle \mathbf{p}_{\mathscr{P}''\mathscr{P}'1}, \mathbf{p}_{\mathscr{P}''\mathscr{P}'2}, \cdots | \mathbf{k}_1, \mathbf{k}_2, \cdots \rangle_{\mathrm{D}} ,$$

† This is expressed formally in the statement that the first homotopy group of configuration space in $d \geq 3$ is the permutation group.[12] By 'configuration space' for N indistinguishable particles is meant the space of N d-vectors, excluding d-vectors that coincide with (or are within an arbitrary limiting distance of) each other, and identifying configurations that differ only by a permutation of the vectors.

which will be satisfied if and only if

$$C_{\mathscr{P}'\mathscr{P}''} = C_{\mathscr{P}'}\, C_{\mathscr{P}''}\,. \tag{9.7.4}$$

That is, the coefficients $C_{\mathscr{P}}$ must furnish a one-dimensional representation of the permutation group. But the permutation group has only two such representations: one is the identity, with $C_{\mathscr{P}} = +1$ for all permutations, and the other is the alternating representation, with $C_{\mathscr{P}} = +1$ or $C_{\mathscr{P}} = -1$ according to whether $\mathscr{P}$ is an even or odd permutation. These two possibilities correspond to Bose or Fermi statistics, respectively.‡

The nice feature of this argument is that it makes it clear why the case of two space dimensions is an exception. In this case there is a much richer variety of topologically distinct paths.¶ For instance, a path in which one particle circles another a definite number of times cannot be deformed into a path where it does not. In consequence, in two space dimensions it is possible to have *anyons*,[15] particles with more general permutation properties than just Fermi or Bose statistics.[8a]

Appendix Gaussian Multiple Integrals

We wish first to calculate the multiple integral, over a finite number of real variables ξ_r, of the exponential of a general quadratic function of ξ:

$$\mathscr{I} \equiv \int_{-\infty}^{\infty} \prod_r d\xi_r \, \exp\Big\{-Q(\xi)\Big\}\,, \tag{9.A.1}$$

$$Q(\xi) = \tfrac{1}{2}\sum_{rs} K_{rs}\xi_r\xi_s + \sum_r L_r\xi_r + M\,, \tag{9.A.2}$$

where K_{rs}, L_r, and M are arbitrary constants, except that the matrix K is required to be symmetric and non-singular. For this purpose, we begin by considering the case where K_{rs}, L_r, and M are all real, with K_{rs} also positive. The result in the general case can then be obtained by analytic continuation.

Any real symmetric matrix can be diagonalized by an orthogonal matrix. Therefore, there is a matrix $\mathscr{S}$ with transpose $\mathscr{S}^{\rm T} = \mathscr{S}^{-1}$ such that

$$\Big(\mathscr{S}^{\rm T} K \mathscr{S}\Big)_{rs} = \delta_{rs}\kappa_r\,. \tag{9.A.3}$$

‡ There has been much discussion in the literature of possibilities other than Bose or Fermi statistics, often under the label *parastatistics*. It has been shown[13] that parastatistics theories in $d \geq 3$ space dimensions are equivalent to theories in which all particles are ordinary fermions or bosons, but carrying an extra quantum number, so that wave functions could have unusual properties under permutations of momenta and spins.

¶ This is expressed in the statement that the first homotopy group of configuration space in two space dimensions is not the permutation group, but a larger group known as the *braid group*.[14]

Because K is assumed positive and non-singular, the eigenvalues κ_r are positive-definite. We can use the matrix $\mathscr{S}$ to perform a change of variables:

$$\xi_r = \sum_s \mathscr{S}_{rs}\xi'_s \, . \tag{9.A.4}$$

The Jacobian $|\mathrm{Det}\,\mathscr{S}|$ of this transformation is unity, so the multiple integral (9.A.1) is now given by a product of ordinary integrals:

$$\begin{aligned} \mathscr{I} &= e^{-M} \prod_r \int_{-\infty}^{\infty} d\xi' \exp\left\{ -\frac{\kappa_r}{2}\xi'^2 - (\mathscr{S}^{\mathrm{T}}L)_r \xi' \right\} \\ &= e^{-M} \prod_r \sqrt{\frac{2\pi}{\kappa_r}} \exp\left\{ \frac{1}{2\kappa_r} (\mathscr{S}^{\mathrm{T}}L)_r^2 \right\} \, . \end{aligned} \tag{9.A.5}$$

But the determinant and the reciprocal of Eq. (9.A.3) give

$$\mathrm{Det}\, K = \prod_r \kappa_r \, , \qquad K^{-1}_{rs} = \sum_\ell \mathscr{S}_{r\ell}\mathscr{S}_{s\ell}\kappa_\ell^{-1} \, ,$$

so Eq. (9.A.5) may be written

$$\mathscr{I} = \left(\mathrm{Det}\left(\frac{K}{2\pi} \right) \right)^{-1/2} \exp\left\{ \tfrac{1}{2} \sum_{rs} L_r L_s K^{-1}_{rs} - M \right\} . \tag{9.A.6}$$

Eq. (9.A.1) defines a function of K_{rs}, L_r, and M that is analytic in K_{rs} in a finite region around the surface where K_{rs} is real and positive, where the integral converges, and for such K_{rs} is analytic everywhere in L_r and M. Since (9.A.6) equals (9.A.1) for K_{rs}, L_r, and M all real, with K_{rs} also positive, Eq. (9.A.6) provides an analytic continuation of Eq. (9.A.1) to the whole complex plane, with a cut required by the square root. The sign of the square root is fixed by this analytic continuation. In field theory K_{rs} is actually imaginary, except for a small real part due to the '$i\epsilon$ term'.

It is useful to express Eq. (9.A.6) in terms of the stationary point of the function (9.A.2):

$$\bar{\xi}_r = -\sum_s K^{-1}_{rs} L_s \, , \tag{9.A.7}$$

$$\partial Q(\xi)/\partial \xi_r = 0 \quad \text{at} \quad \xi = \bar{\xi} \, , \tag{9.A.8}$$

as

$$\mathscr{I} = \left(\mathrm{Det}\left(\frac{K}{2\pi} \right) \right)^{-1/2} \exp\left\{ -Q(\bar{\xi}) \right\} . \tag{9.A.9}$$

This is the result to remember: *Gaussian integrals can be evaluated up to a determinant factor by setting the integration variable equal to the point where the argument of the exponential is stationary.*

We next wish to use this result to calculate the integrals

$$I_{r_1\cdots r_{2N}} \equiv \int \left(\prod_r d\xi_r\right) \xi_{r_1}\xi_{r_2}\cdots\xi_{r_{2N}} \exp\left\{-\tfrac{1}{2}\sum_{rs} K_{rs}\xi_r\xi_s\right\} . \qquad (9.A.10)$$

(Integrals of this sort with an odd number of ξ-factors in the integrand obviously vanish.) From the power-series expression of $\exp\left(-\sum_r L_r\xi_r\right)$ in Eq. (9.A.1), we have the sum rule

$$\begin{aligned}
\sum_{N=0}^{\infty}\sum_{r_1 r_2\cdots r_{2N}} \frac{1}{(2N)!} I_{r_1 r_2\cdots r_{2N}} L_{r_1}L_{r_2}\cdots L_{r_{2N}} \\
&= \int \left(\prod_r d\xi_r\right) \exp\left\{-\sum_r L_r\xi_r - \frac{1}{2}\sum_{rs} K_{rs}\xi_r\xi_s\right\} \\
&= \left[\mathrm{Det}\left(\frac{K}{2\pi}\right)\right]^{-1/2} \exp\left\{\frac{1}{2}\sum_{rs} L_r L_s K^{-1}_{rs}\right\} \\
&= \left[\mathrm{Det}\left(\frac{K}{2\pi}\right)\right]^{-1/2} \sum_{N=0}^{\infty} \frac{1}{N!2^N}\left(\sum_{rs} L_r L_s K^{-1}_{rs}\right)^N . \qquad (9.A.11)
\end{aligned}$$

Comparing the coefficients of $L_{r_1}L_{r_2}\cdots L_{r_{2N}}$ on both sides, we see that $I_{r_1 r_2\cdots r_{2N}}$ must be proportional to a sum of products of elements of K^{-1}, which symmetry requires to take the form

$$I_{r_1 r_2\cdots r_{2N}} = c_N \sum_{\substack{\text{pairings}\\ \text{of } r_1\cdots r_{2N}}} \prod_{\text{pairs}} (K^{-1})_{\text{paired indices}} . \qquad (9.A.12)$$

Here the sum is over all ways of pairing the indices $r_1\cdots r_{2N}$, with two pairings being considered the same if they differ only by the order of the pairs, or by the order of indices within a pair. To calculate the constant factor c_N, we note that the number ν_N of terms in the sum over pairings in Eq. (9.A.12) is equal to the number $(2N)!$ of permutations of the indices, divided by the number of ways $N!$ of permuting index pairs and by the number 2^N of permutations within index pairs

$$\nu_N = \frac{(2N)!}{N!2^N} . \qquad (9.A.13)$$

Therefore, Eq. (9.A.12) gives

$$\sum_{r_1 r_2\cdots r_{2N}} L_{r_1}L_{r_2}\cdots L_{r_{2N}} I_{r_1 r_2\cdots r_{2N}} = \nu_N c_N \left(\sum_{rs} L_r L_s K^{-1}_{rs}\right)^N . \qquad (9.A.14)$$

Comparing this with Eq. (9.A.11) shows that the factors $(2N)!$ and $N!2^N$ are cancelled by ν_N, leaving us with

$$c_N = \left[\mathrm{Det}\left(\frac{K}{2\pi}\right)\right]^{-1/2} . \qquad (9.A.15)$$

For instance,

$$I_{r_1r_2} = I_0(K^{-1})_{r_1r_2}\,, \tag{9.A.16}$$

$$I_{r_1r_2r_3r_4} = I_0\Big[(K^{-1})_{r_1r_2}(K^{-1})_{r_3r_4} + (K^{-1})_{r_1r_3}(K^{-1})_{r_2r_4} + (K^{-1})_{r_1r_4}(K^{-1})_{r_2r_3}\Big]\,, \tag{9.A.17}$$

and so on, where I_0 is the integral with no indices

$$I_0 \equiv \int\left(\prod_r d\xi_r\right)\exp\left\{-\tfrac{1}{2}\sum_{rs} K_{rs}\xi_r\xi_s\right\} = \left[\mathrm{Det}\left(\frac{K}{2\pi}\right)\right]^{-1/2}\,. \tag{9.A.18}$$

Problems

1. Consider a non-relativistic particle of mass m, moving along the x-axis in a potential $V(x) = m\omega^2x^2/2$. Use path-integral methods to find the probability that if the particle is at x_1 at time t_1, then it is between x and $x + dx$ at time t.

2. Find the wave function in field space of a state consisting of a single spinless particle of mass $m \neq 0$. Use the result to derive the Feynman rules for emission or absorption of such a particle.

3. Find the wave function in field space of the vacuum in the theory of a neutral vector field of mass $m \neq 0$. Use the result to derive the form of the $i\epsilon$ terms in the propagator of this field.

4. The Lagrangian density of the free spin 3/2 Rarita–Schwinger field ψ^μ is

$$\mathscr{L} = -\bar{\psi}^\mu(\gamma^\nu\partial_\nu + m)\psi_\mu - \tfrac{1}{3}\bar{\psi}^\mu(\gamma_\mu\partial_\nu + \gamma_\nu\partial_\mu)\psi^\nu + \tfrac{1}{3}\bar{\psi}^\mu\gamma_\mu(\gamma^\sigma\partial_\sigma - m)\gamma^\nu\psi_\nu.$$

Use path-integral methods to find the propagator of this field.

References

1. R. P. Feynman, *The Principle of Least Action in Quantum Mechanics* (Princeton University, 1942; University Microfilms Publication No. 2948, Ann Arbor). Also see R. P. Feynman and A. R. Hibbs, *Quantum Mechanics and Path Integrals* (McGraw-Hill, New York, 1965). For a general reference, see J. Glimm and A. Jaffe, *Quantum Physics–A Functional Integral Point of View*, 2nd edn (Springer-Verlag, New York, 1987).

2. P. A. M. Dirac, *Phys. Zeits. Sowjetunion* **3**, 62 (1933).

3. R. P. Feynman, *Rev. Mod. Phys.* **20**, 367 (1948); *Phys. Rev.*, **74**, 939, 1430 (1948); **76**, 749, 769 (1949); **80**, 440 (1950).

4. L. D. Faddeev and V. N. Popov, *Phys. Lett.* **B25**, 29 (1967). Also see R. P. Feynman, *Acta Phys. Pol.* **24**, 697 (1963); S. Mandelstam, *Phys. Rev.* **175**, 1580, 1604 (1968).

5. B. De Witt, *Phys. Rev. Lett.* **12**, 742 (1964).

6. G. 't Hooft, *Nucl. Phys.* **B35**, 167 (1971).

7. I. S. Gerstein, R. Jackiw, B. W. Lee, and S. Weinberg, *Phys. Rev.* **D3**, 2486 (1971).

8. L. D. Faddeev, *Teor. Mat. Fizika*, **1**, 3 (1969); translation in *Theor. Math. Phys.* **1**, 1 (1970).

8*a*. J. Schwinger, *Proc. Nat. Acad. Sci.* **44**, 956 (1958).

8*b*. K. Osterwalder and R. Schrader, *Phys. Rev. Lett.* **29**, 1423 (1972); *Commun. Math. Phys.* **31**, 83 (1973); *Commun. Math. Phys.* **42**, 281 (1975). The Osterwalder-Schrader axioms require smoothness, Euclidean covariance, 'reflection positivity,' permutation symmetry, and cluster decomposition.

9. This section was largely inspired by discussions with J. Polchinski.

10. F. A. Berezin, *The Method of Second Quantization* (Academic Press, New York, 1966).

11. M. G. G. Laidlaw and C. M. De Witt, *Phys. Rev.* D **3**, 1375 (1970).

12. J. M. Leinaas and J. Myrheim, *Nuovo Cimento* **37 B**, 1 (1977).

13. Y. Ohnuki and S. Kamefuchi, *Phys. Rev.* **170**, 1279 (1968); *Ann. Phys.* **51**, 337 (1969); K. Drühl, R. Haag, and J. E. Roberts, *Commun. Math. Phys.* **18**, 204 (1970).

14. The braid group was introduced by E. Artin. See *The Collected Papers of E. Artin*, ed. by S. Lang and J. E. Tate (Addison-Wesley, Reading, MA, 1965).

15. F. Wilczek, *Phys. Rev. Lett.* **49**, 957 (1982); K. Fredenhagen, M. R. Gaberdiel, and S. M. Rüger, Cambridge preprint DAMTP-94-90 (1994). Also see J. M. Leinaas and J. Myrheim, Ref. 12.

10
Non-Perturbative Methods

We are now going to begin our study of higher-order contributions to physical processes, corresponding to Feynman diagrams involving one or more loops. It will be very useful in this work to have available a method of deriving results valid to all orders in perturbation theory (and in some cases beyond perturbation theory). In this chapter we will exploit the field equations and commutation relations of the interacting fields in the Heisenberg picture for this purpose. The essential bridge between the Heisenberg picture and the Feynman diagrams of perturbation theory is provided by the theorem proved in Section 6.4: the sum of all diagrams for a process $\alpha \to \beta$ with extra vertices inserted corresponding to operators $o_a(x)$, $o_b(y)$, etc. is given by the matrix element of the time-ordered product of the corresponding Heisenberg-picture operators

$$\left(\Psi_\beta{}^-, T\left\{ - iO_a(x),\ -iO_b(y)\cdots\right\}\Psi_\alpha{}^+\right) .$$

As a special case, where the operators $O_a(x)$, $O_b(x)$, etc. are elementary particle fields, this matrix element equals the sum of all Feynman diagrams with incoming lines on the mass shell corresponding to the state α, outgoing lines on the mass shell corresponding to the state β, and lines off the mass shell (including propagators) corresponding to the operators $O_a(x)$, $O_b(x)$, etc. After exploring some of the non-perturbative results that can be obtained in this way we will be in a good position to take up the perturbative calculation of radiative corrections.

10.1 Symmetries

One obvious but important use of the theorem quoted above is to extend the application of symmetry principles from S-matrix elements, where all external lines have four-momenta on the mass shell, to parts of Feynman diagrams, with some or all external lines off the mass shell.

For instance, consider the symmetry of spacetime translational invariance. This symmetry has as a consequence the existence of a Hermitian four-vector operator P^μ, with the property that, for any local function $O(x)$ of field operators and their canonical conjugates,

$$\Big[P_\mu, O(x)\Big] = i\frac{\partial}{\partial x^\mu}O(x)\,. \tag{10.1.1}$$

(See Eqs. (7.3.28) and (7.3.29).) Also, the states α and β are usually chosen to be eigenstates of the four-momentum:

$$P^\mu\Psi_\alpha{}^+ = p_\alpha^\mu\Psi_\alpha{}^+\,, \qquad P^\mu\Psi_\beta{}^- = p_\beta^\mu\Psi_\beta{}^-\,. \tag{10.1.2}$$

It follows that for any set of local functions $O_a(x)$, $O_b(x)$, etc. of fields and/or field derivatives

$$\begin{aligned}(p_{\beta\mu} - p_{\alpha\mu})&\Big(\Psi_\beta{}^-, T\Big\{O_a(x_1), O_b(x_2)\cdots\Big\}\Psi_\alpha{}^+\Big)\\ &= \Big(\Psi_\beta{}^-, \Big[P_\mu, T\Big\{O_a(x_1), O_b(x_2)\cdots\Big\}\Big]\Psi_\alpha{}^+\Big)\\ &= i\left(\frac{\partial}{\partial x_1^\mu} + \frac{\partial}{\partial x_2^\mu} + \cdots\right)\Big(\Psi_\beta{}^- T\Big\{O_a(x_1), O_b(x_2), \cdots\Big\}\Psi_\alpha{}^+\Big)\,. \end{aligned}\tag{10.1.3}$$

This has the solution

$$\begin{aligned}&\Big(\Psi_\beta{}^-, T\Big\{O_a(x_1), O_b(x_2), \cdots\Big\}\Psi_\alpha{}^+\Big)\\ &= \exp\Big(i(p_\alpha - p_\beta)\cdot x\Big)\,F_{ab\cdots}(x_1 - x_2, \cdots)\,,\end{aligned}\tag{10.1.4}$$

where x is any sort of average spacetime coordinate

$$x^\mu = c_1 x_1^\mu + c_2 x_2^\mu + \cdots\,, \qquad c_1 + c_2 + \cdots = 1 \tag{10.1.5}$$

and F depends only on differences among the xs. (In particular, a vacuum expectation value can depend only on the coordinate differences.) We can Fourier transform Eq. (10.1.4) by integrating separately over x^μ and the coordinate differences, with the result that

$$\begin{aligned}&\int d^4x_1\, d^4x_2\cdots\Big(\Psi_\beta{}^-, T\Big\{O_a(x_1), O_b(x_2), \cdots\Big\}\Psi_\alpha{}^+\Big)\\ &\times\exp(-ik_1\cdot x_1 - ik_2\cdot x_2 - \cdots) \propto \delta^4(p_\alpha - p_\beta - k_1 - k_2 - \cdots)\,.\end{aligned}\tag{10.1.6}$$

We saw in Section 6.4 that the matrix element of the time-ordered product is given by applying the usual coordinate-space Feynman rules to the sum of all graphs with incoming particles corresponding to particles in α, outgoing particles in β, and external lines that simply terminate in vertices at $x_1, x_2, \cdots$. The Fourier transform (10.1.6) is correspondingly given by applying the momentum-space Feynman rules to the same sum of Feynman diagrams, with off-shell external lines carrying four-momenta $k_1, k_2, \cdots$ into the diagrams. Eq. (10.1.6) is then just the statement that this sum of Feynman graphs conserves four-momentum. The result is obvious in perturbation theory, because four-momentum is conserved at every vertex, so it is not surprising to see the same result emerging without having to rely on perturbation theory.

With somewhat more effort, one can use the Lorentz transformation properties of the Heisenberg-picture fields and the 'in' and 'out' states to show that the sum of all graphs with a given set of on- and off-shell lines satisfies the same Lorentz transformation conditions as the lowest-order terms.

Similar arguments apply to the conservation of internal quantum numbers, like electric charge. As shown in Section 7.3, a field or other operator $O_a(x)$ that destroys a charge q_a (or creates a charge $-q_a$) will satisfy

$$[Q, O_a(x)] = -q_a O_a(x)$$

in the Heisenberg and interaction pictures alike. Also, if the free-particle states α and β have charges q_α and q_β, then so do the corresponding 'in' and 'out' states. We then have

$$\begin{aligned}(q_\beta - q_\alpha) &\left(\Psi_\beta{}^-,\ T\left\{O_a(x), O_b(y), \cdots\right\}\Psi_\alpha{}^+\right) \\ &= \left(\Psi_\beta{}^-, \left[Q, T\left\{O_a(x), O_b(y), \cdots\right\}\right]\Psi_\alpha{}^+\right) \\ &= -(q_a + q_b + \cdots)\left(\Psi_\beta{}^-,\ T\left\{O_a(x), O_b(y), \cdots\right\}\Psi_\alpha{}^+\right) .\end{aligned}$$

Thus the amplitude $\left(\Psi_\beta{}^-,\ T\left\{O_a(x), O_b(y), \cdots\right\}\Psi_\alpha{}^+\right)$ vanishes unless charge is conserved

$$q_\beta = q_\alpha - q_a - q_b - \cdots \ . \tag{10.1.7}$$

A somewhat less trivial example is provided by the symmetry of charge-conjugation invariance. As we saw in Chapter 5, there is an operator C that interchanges electron and positron operators

$$\begin{aligned}\mathsf{C}\, a(\mathbf{p}, \sigma, e^-)\mathsf{C}^{-1} &= \xi^*\, a(\mathbf{p}, \sigma, e^+) , \\ \mathsf{C}\, a(\mathbf{p}, \sigma, e^+)\mathsf{C}^{-1} &= \xi\, a(\mathbf{p}, \sigma, e^-) ,\end{aligned}$$

with ξ a phase factor. For the free electron field $\psi(x)$, this gives

$$\mathsf{C}\psi(x)\,\mathsf{C}^{-1} = -\xi^*\beta\mathscr{C}\psi(x)^* ,$$

where βC is a 4×4 matrix, which (for the Dirac matrix representation we have been using, with γ_5 diagonal) takes the form

$$\beta C = \begin{bmatrix} 0 & 0 & 0 & 1 \\ 0 & 0 & -1 & 0 \\ 0 & -1 & 0 & 0 \\ 1 & 0 & 0 & 0 \end{bmatrix} .$$

Applied to the free-particle electric current in spinor electrodynamics, this gives

$$\mathsf{C}(\bar{\psi}\gamma^\mu\psi)\mathsf{C}^{-1} = -\bar{\psi}C\gamma^{\mu T}C\psi = -\bar{\psi}\gamma^\mu\psi\,.$$

If C is to be conserved in electrodynamics, it must then also be defined to anticommute with the free photon field

$$\mathsf{C}(a^\mu)\mathsf{C}^{-1} = -a^\mu\,.$$

In theories like electrodynamics for which C commutes with the interaction as well as H_0, it also commutes with the similarity transformation $\Omega(t)$ between the Heisenberg and interaction pictures, and so it anticommutes with the electric current of the interacting fields

$$\mathsf{C}(\bar{\Psi}\gamma^\mu\Psi)\mathsf{C}^{-1} = -\bar{\Psi}\gamma^\mu\Psi \tag{10.1.8}$$

and the electromagnetic field in the Heisenberg-picture

$$\mathsf{C}(A^\mu)\mathsf{C}^{-1} = -A^\mu\,. \tag{10.1.9}$$

It follows then that the vacuum expectation value of the time-ordered product of any odd number of electromagnetic currents and/or fields vanishes. Therefore the sum of all Feynman graphs with an odd number of external photon lines (off or on the photon mass shell) and no other external lines vanishes.

This result is known as *Furry's theorem.*[1] It can be proved perturbatively by noting that a graph consisting of electron loops ℓ, to each of which are attached n_ℓ photon lines, must have numbers I and E of internal and external photon lines related by an analog of Eq. (6.3.11):

$$2I + E = \sum_\ell n_\ell\,.$$

Hence if E is odd at least one of the loops must have attached an odd number of photon lines. For any such loop there is a cancellation between the two diagrams in which the electron arrows circulate around the loop in opposite directions. Hence Furry's theorem is a somewhat less trivial consequence of a symmetry principle than translation or Lorentz invariance; it is not true of individual diagrams, but rather of certain sums of diagrams. Figure 10.1 illustrates the application of Furry's theorem that was historically most important, its use to show that the scattering of a photon by an external electromagnetic field receives no contributions of first order (or any odd order) in the external field.

10.2 Polology

One of the most important uses of the non-perturbative methods described in this chapter is to clarify the pole structure of Feynman amplitudes as

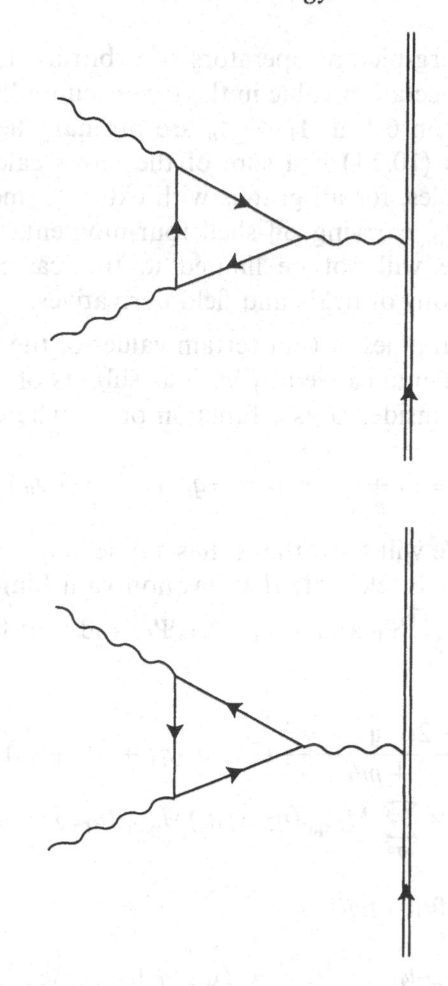

Figure 10.1. The lowest-order diagrams for the scattering of a photon by an electromagnetic field. Here straight lines represent virtual electrons; wavy lines represent real and virtual photons; and the double line represents a heavy particle like an atomic nucleus that serves as a source of an electromagnetic field. The contributions of these two diagrams cancel, as required by charge-conjugation invariance.

functions of the momenta carried by external lines. Often the S-matrix for a physical process can be well approximated by the contribution of a single pole. Also, an understanding of this pole structure will help us later in dealing with radiative corrections to particle propagators.

Consider the momentum-space amplitude

$$\int d^4x_1 \cdots d^4x_n \, e^{-iq_1 \cdot x_1} \cdots e^{-iq_n \cdot x_n} \left\langle T\left\{A_1(x_1) \cdots A_n(x_n)\right\}\right\rangle_0$$
$$\equiv G(q_1 \cdots q_n) \,. \tag{10.2.1}$$

The As are Heisenberg-picture operators of arbitrary Lorentz type, and $\langle\cdots\rangle_0$ denotes the expectation value in the true vacuum $\Psi_0{}^+ = \Psi_0{}^- \equiv \Psi_0$. As discussed in Section 6.4, if $A_1,\cdots A_n$ are ordinary fields appearing in the Lagrangian, then (10.2.1) is a sum of the terms calculated using the ordinary Feynman rules, for all graphs with external lines corresponding to the fields $A_1,\cdots A_n$, carrying off-shell four-momenta $q_1\cdots q_n$ into the graph. However, we will not be limited to this case; the A_i may be arbitrary local functions of fields and field derivatives.

We are interested in poles of G at certain values of the invariant squares of the total four-momenta carried by various subsets of the external lines. To be definite, let's consider G as a function of q^2, where

$$q \equiv q_1 + \cdots + q_r = -q_{r+1} - \cdots - q_n \tag{10.2.2}$$

with $1 \leq r \leq n-1$. We will show that G has a pole at $q^2 = -m^2$, where m is the mass of any one-particle state that has non-vanishing matrix elements with the states $A_1^\dagger \cdots A_r^\dagger \Psi_0$ and $A_{r+1}\cdots A_n\Psi_0$, and that the residue at this pole is given by

$$\begin{aligned} G \to & \frac{-2i\sqrt{\mathbf{q}^2+m^2}}{q^2+m^2-i\epsilon}\,(2\pi)^7\delta^4(q_1+\cdots+q_n) \\ & \times \sum_\sigma M_{0|\mathbf{q},\sigma}(q_2\cdots q_r)M_{\mathbf{q},\sigma|0}(q_{r+2}\cdots q_n) \end{aligned} \tag{10.2.3}$$

where the Ms are defined by*

$$\begin{aligned} &\int d^4x_1\cdots d^4x_r\, e^{-iq_1\cdot x_1}\cdots e^{-iq_r\cdot x_r}\left(\Psi_0, T\left\{A_1(x_1)\cdots A_r(x_r)\right\}\Psi_{\mathbf{p},\sigma}\right) \\ &\quad = (2\pi)^4\,\delta^4(q_1+\cdots+q_r-p)M_{0|\mathbf{p},\sigma}(q_2\cdots q_r)\,, \end{aligned} \tag{10.2.4}$$

$$\begin{aligned} &\int d^4x_{r+1}\cdots d^4x_n\, e^{-iq_{r+1}\cdot x_{r+1}}\cdots e^{-iq_n\cdot x_n} \\ &\quad \times \left(\Psi_{\mathbf{p},\sigma}, T\left\{A_{r+1}(x_{r+1})\cdots A_n(x_n)\right\}\Psi_0\right) \\ &\quad = (2\pi)^4\,\delta^4(q_{r+1}+\cdots+q_n+p)M_{\mathbf{p},\sigma|0}(q_{r+2}\cdots q_n) \end{aligned} \tag{10.2.5}$$

(with $p^0 \equiv \sqrt{\mathbf{p}^2+m^2}$), and the sum is over all spin (or other) states of the particle of mass m.

Before proceeding to the proof, it will help to clarify the significance of

* Recall that in the absence of time-varying external fields, there is no distinction between 'in' and 'out' one-particle states, so that $\Psi_{\mathbf{p},\sigma}{}^+ = \Psi_{\mathbf{p},\sigma}{}^- = \Psi_{\mathbf{p},\sigma}$.

(10.2.3) if we write it in the somewhat long-winded form

$$G(q_1 \cdots q_n) \to \sum_\sigma \int d^4k$$
$$\times \left[(2\pi)^4 \delta^4(q_1 + \cdots + q_r - k)(2\pi)^{3/2} \left(2\sqrt{\mathbf{k}^2 + m^2}\right)^{1/2} M_{0|\mathbf{k},\sigma}(q_2 \cdots q_r)\right]$$
$$\times \left[\frac{-i}{(2\pi)^4} \frac{1}{k^2 + m^2 - i\epsilon}\right]$$
$$\times \left[(2\pi)^4 \delta^4(k + q_{r+1} + \cdots + q_n)(2\pi)^{3/2}\right.$$
$$\left. \times \left(2\sqrt{\mathbf{k}^2 + m^2}\right)^{1/2} M_{\mathbf{k},\sigma|0}(q_{r+2} \cdots q_n)\right] . \tag{10.2.6}$$

This is just what we should expect from a Feynman diagram with a single internal line for a particle of mass m connecting the first r and the last $n-r$ external lines.** However, it is *not* necessary that the particle of mass m correspond to a field that appears in the Lagrangian of the theory. Eqs. (10.2.3) and (10.2.6) apply even if this particle is a bound state of the so-called elementary particles whose fields do appear in the Lagrangian. In this case, the pole arises not from single Feynman diagrams, like Figure 10.2, but rather from infinite sums of diagrams, such as the one shown in Figure 10.3. This is the first place where the methods of this chapter take us beyond results that could be derived as properties of each order of perturbation theory.

Now to the proof. Among the $n!$ possible orderings of the times $x_1^0 \cdots x_n^0$ in Eq. (10.2.1), there are $n!/r!(n-r)!$ for which the first r of the x_i^0 are *all* larger than the last $n-r$. Isolating the contribution of this part of the volume of integration in Eq. (10.2.1), we have

$$G(q_1 \cdots q_n) = \int d^4x_1 \cdots d^4x_n \, e^{-iq_1 \cdot x_1} \cdots e^{-iq_n \cdot x_n}$$
$$\times \theta\left(\min\,[x_1^0 \cdots x_r^0] - \max\,[x_{r+1}^0 \cdots x_n^0]\right)$$
$$\times \left(\Psi_0,\, T\Big\{A_1(x_1) \cdots A_r(x_r)\Big\}\, T\left\{A_{r+1}(x_{r+1}) \cdots A_n(x_n)\right\} \Psi_0\right)$$
$$+ \text{OT}, \tag{10.2.7}$$

where 'OT' denotes the other terms arising from different time-orderings. We can evaluate the matrix element here by inserting a complete set of

---

** See Figure 10.2. The factors $(2\pi)^{3/2}\left[2\sqrt{\mathbf{k}^2 + m^2}\right]^{1/2}$ just serve to remove kinematic factors associated with the mass m external line in $M_{0|\mathbf{k},\sigma}$ and $M_{\mathbf{k},\sigma|0}$. Also, the sum over σ of the product of coefficient-function factors from these two matrix elements yields the numerator of the propagator associated with the internal line in Figure 10.2.

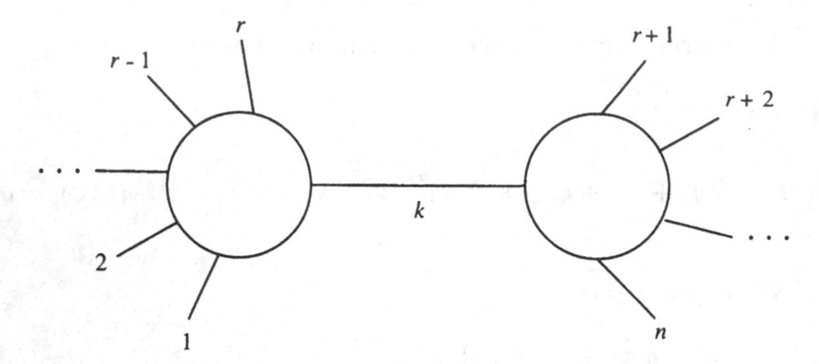

Figure 10.2. A Feynman diagram with the pole structure (10.2.6). Here the line carrying a momentum k represents an elementary particle, one whose field appears in the Lagrangian.

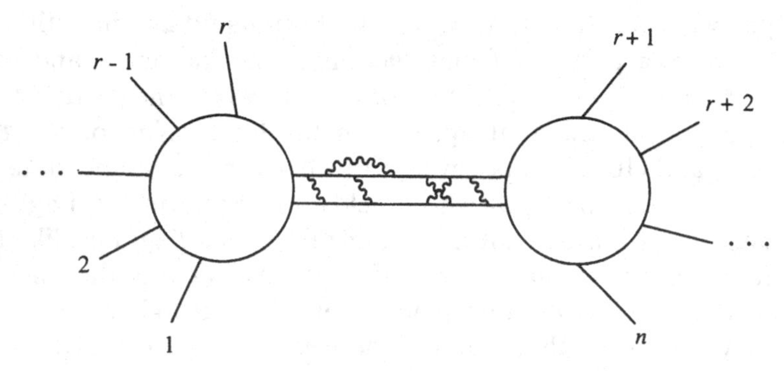

Figure 10.3. Figure 10.3. A Feynman diagram of the class whose sum has the pole structure (10.2.6). Here the pole is due to a composite particle, a bound state of two elementary particles. The elementary particles are represented by straight lines, and interact by the exchange of particles represented by wavy lines.

intermediate states between time-ordered products. Among these may be the single-particle state $\Psi_{\mathbf{p},\sigma}$ of a definite species of mass m. Further isolating the contribution of these one-particle intermediate states, we have

$$G(q_1 \cdots q_n) = \int d^4x_1 \cdots d^4x_n \, e^{-iq_1\cdot x_1} \cdots e^{-iq_n\cdot x_n}$$

$$\theta\left(\min[x_1^0 \cdots x_r^0] - \max[x_{r+1}^0 \cdots x_n^0]\right) \sum_\sigma \int d^3p$$

$$\left(\Psi_0, T\left\{A_1(x_1)\cdots A_r(x_r)\right\}\Psi_{\mathbf{p},\sigma}\right) \left(\Psi_{\mathbf{p},\sigma}, T\left\{A_{r+1}(x_{r+1})\cdots A_n(x_n)\right\}\Psi_0\right)$$
$$+ \text{OT}, \tag{10.2.8}$$

where 'OT' now denotes other terms, here arising not only from other time-orderings, but also from other intermediate states. It will be convenient to shift variables of integration, so that

$$x_i = x_1 + y_i\,, \qquad i = 2, 3, \cdots r\,,$$
$$x_i = x_{r+1} + y_i\,, \qquad i = r+2, \cdots n\,,$$

and use the results of the previous section to write

$$\Big(\Psi_0,\, T\Big\{A_1(x_1)\cdots A_r(x_r)\Big\}\Psi_{\mathbf{p},\sigma}\Big)$$
$$= e^{ip\cdot x_1}\Big(\Psi_0,\, T\Big\{A_1(0)A_2(y_2)\cdots A_r(y_r)\Big\}\Psi_{\mathbf{p},\sigma}\Big)\,, \tag{10.2.9}$$

$$\Big(\Psi_{\mathbf{p},\sigma},\, T\Big\{A_{r+1}(x_{r+1})\cdots A_n(x_n)\Big\}\Psi_0\Big)$$
$$= e^{-ip\cdot x_{r+1}}\Big(\Psi_{\mathbf{p},\sigma},\, T\Big\{A_{r+1}(0)\cdots A_n(y_n)\Big\}\Psi_0\Big)\,. \tag{10.2.10}$$

Also, the argument of the theta function becomes

$$\min\,[x_1^0\cdots x_r^0] - \max\,[x_{r+1}^0\cdots x_n^0]$$
$$= x_1^0 - x_{r+1}^0 + \min\,[0\; y_2^0\cdots y_r^0] - \max\,[0\; y_{r+1}^0\cdots y_n^0]\,.$$

We also insert the Fourier representation (6.2.15) of the step function

$$\theta(\tau) = -\frac{1}{2\pi i}\int_{-\infty}^{\infty}\frac{d\omega\, e^{-i\omega\tau}}{\omega + i\epsilon}\,.$$

The integrals over x_1 and x_{r+1} now just yield delta functions:

$$G(q_1\cdots q_n) = \int d^4y_2\cdots d^4y_r\, d^4y_{r+2}\cdots d^4y_n$$
$$\times\, e^{-iq_2\cdot y_2}\ldots e^{-iq_r\cdot y_r}e^{-iq_{r+2}\cdot y_{r+2}}\ldots e^{-iq_n\cdot y_n}$$
$$\times -\frac{1}{2\pi i}\int_{-\infty}^{\infty}\frac{d\omega}{\omega+i\epsilon}\exp\Big(-i\omega\Big[\min[0\; y_2^0\cdots y_r^0] - \max[0\; y_{r+1}^0\cdots y_n^0]\Big]\Big)$$
$$\times \sum_\sigma\int d^3p\Big(\Psi_0,\, T\Big\{A_1(0)\cdots A_r(y_r)\Big\}\Psi_{\mathbf{p},\sigma}\Big)$$
$$\times\Big(\Psi_{\mathbf{p},\sigma},T\Big\{A_{r+1}(0)\cdots A_n(y_n)\Big\}\Psi_0\Big)$$
$$\times\,(2\pi)^4\delta^3(\mathbf{p}-\mathbf{q}_1-\cdots-\mathbf{q}_r)\,\delta(\sqrt{\mathbf{p}^2+m^2}+\omega-q_1^0-\cdots-q_r^0)$$
$$\times\,(2\pi)^4\,\delta^3(\mathbf{q}_{r+1}+\cdots+\mathbf{q}_n+\mathbf{p})\,\delta\Big(q_{r+1}^0+\cdots+q_n^0+\sqrt{\mathbf{p}^2+m^2}+\omega\Big)$$
$$+\,\mathrm{OT}\,. \tag{10.2.11}$$

We are interested here only in the pole that arises from the vanishing of the denominator $\omega + i\epsilon$, so for our present purposes we can set the factor $\exp(-i\omega[\min - \max])$ equal to unity. The integrals over both $\mathbf{p}$ and ω are now trivial, and yield the pole

$$G(q_1 \cdots q_n) \to i(2\pi)^7\, \delta^4(q_1 + \cdots + q_n) \left[q^0 - \sqrt{\mathbf{q}^2 + m^2} + i\epsilon\right]^{-1}$$
$$\times \sum_\sigma M_{0|\mathbf{q},\sigma}(q_2 \cdots q_n)\, M_{\mathbf{q},\sigma|0}(q_{r+2} \cdots q_n) + \cdots \tag{10.2.12}$$

where now

$$q \equiv q_1 + \cdots + q_r = -q_{r+1} - \cdots - q_n\,,$$

$$M_{0|\mathbf{q},\sigma}(q_2 \cdots q_r) \equiv \int d^4y_2 \cdots d^4y_r\, e^{-iq_2\cdot y_2} \cdots e^{-iq_r\cdot y_r}$$
$$\times \left(\Psi_0,\ T\left\{A_1(0)A_2(y_2)\cdots A_r(y_r)\right\}\Psi_{\mathbf{q},\sigma}\right)\,, \tag{10.2.13}$$

$$M_{\mathbf{q},\sigma|0}(q_{r+2} \cdots q_n) \equiv \int d^4y_{r+2} \cdots d^4y_n\, e^{-iq_{r+2}\cdot y_{r+2}} \cdots e^{-iq_n\cdot y_n}$$
$$\times \left(\Psi_{\mathbf{q},\sigma},\ T\left\{A_{r+1}(0)A_{r+2}(y_{r+2})\cdots A_n(y_n)\right\}\Psi_0\right)\,, \tag{10.2.14}$$

and the final '$\cdots$' in Eq. (10.2.12) denotes terms that do not exhibit this particular pole. (The 'other terms' arising from other single-particle states produce poles in q at different positions, while those arising from multi-particle states produce branch points in q, and those arising from other time-orderings produce poles and branch cuts in other variables.) Using Eqs. (10.2.9) and (10.2.10), it is easy to see that these Ms are the same as defined by Eqs. (10.2.4) and (10.2.5). Also, near the pole we can write

$$\frac{1}{q^0 - \sqrt{\mathbf{q}^2 + m^2} + i\epsilon} = \frac{-q^0 - \sqrt{\mathbf{q}^2 + m^2} + i\epsilon}{-(q^0)^2 + (\sqrt{\mathbf{q}^2 + m^2} - i\epsilon)^2} \to \frac{-2\sqrt{\mathbf{q}^2 + m^2}}{q^2 + m^2 - i\epsilon}\,.$$

(We again redefine ϵ by a positive factor $2\sqrt{\mathbf{q}^2 + m^2}$, which is permissible since ϵ stands for any positive infinitesimal.) Eq. (10.2.12) is thus the same as the desired result (10.2.3).

This result has a classic application to the theory of nuclear forces. Let $\Phi_a(x)$ be any real field or combination of fields (for instance, proportional to a quark–antiquark bilinear $\bar{q}\gamma_5\tau_a q$) that has a non-vanishing matrix element between a one-pion state of isospin a and the vacuum, normalized so that

$$\langle \text{VAC}|\, \Phi_a(0)\, |\pi_b, \mathbf{p}\rangle = (2\pi)^{-3/2}(2p^0)^{-1/2}\delta_{ab}\,. \tag{10.2.15}$$

The matrix element of Φ_a between one-nucleon states with four-momenta p,p' then has a pole at $(p - p')^2 \to -m_\pi^2$ which isospin and Lorentz invariance (including space inversion invariance) dictate must take the form**

$$\langle N', \sigma', \mathbf{p}'|\,\Phi_a(0)\,|N, \sigma, \mathbf{p}\rangle \to i(2\pi)^{-3} G_\pi \times \frac{\left(\bar{u}'\,\gamma_5 \tau_a\, u\right)}{(p-p')^2 + m_\pi^2}\,, \tag{10.2.16}$$

where u and u' are the initial and final nucleon spinor coefficient functions, including the nucleon wave functions in isospin space, and τ_a with $a = 1, 2, 3$ are the 2×2 Pauli isospin matrices. The constant G_π is known as the *pion–nucleon coupling constant*. This pole is not actually in the physical region for the matrix element (10.2.16), for which $(p - p')^2 \geq 0$, but it can be reached by analytic extension of this matrix element, for instance by considering the off-shell matrix element

$$\int d^4x\, d^4x'\, e^{-ip\cdot x} e^{ip'\cdot x'} \langle\, T\{\Phi_a(0)\,\bar{N}(x)\,N'(x')\}\rangle_{\rm VAC}\,,$$

where N and N' are appropriate components of a field operator or product of field operators with non-vanishing matrix elements between one-nucleon states and the vacuum. The theorem proved above in this section shows then that exchange of a pion in the scattering of two nucleons with initial four-momenta p_1,p_2, and final four-momenta p_1', p_2' yields a pole at $(p_1 - p_1')^2 = (p_2 - p_2')^2 \to -m_\pi^2$:

$$S_{N_1'N_2',N_1N_2} \to -i(2\pi)^4 \delta^4(p_1' + p_2' - p_1 - p_2) \frac{G_\pi^2}{(p_1 - p_1')^2 + m_\pi^2}$$
$$\times\, (2\pi)^{-3}\left(\bar{u}_{1'}\,\gamma_5 \tau_a\, u_1\right) \times\, (2\pi)^{-3}\left(\bar{u}_2'\,\gamma_5 \tau_a\, u_2\right)\,. \tag{10.2.17}$$

(The easiest way to get the phases and numerical factors right in such formulas is to use Feynman diagrams; our theorem just says that the pole structure is the same as would be found in a field theory in which the Lagrangian involved an elementary pion field.) Again, this pion pole is not actually in the physical region for scattering of nucleons on the mass shell, for which $(p_1 - p_1')^2 \geq 0$, but it can be reached by analytic extension of the S-matrix element, for instance by considering the off-shell matrix

** Lorentz and isospin invariance requires this matrix to take the form $(\bar{u}'\,\Gamma\,\tau_a\,u)$, where Γ is a 4×4 matrix for which the bilinear $(\bar{\psi}'\,\Gamma\,\psi)$ transforms as a pseudoscalar. Like any 4×4 matrix, Γ can be expanded as a sum of terms proportional to the Dirac matrices 1, γ_μ, $[\gamma_\mu, \gamma_\nu]$, $\gamma_5\gamma_\mu$, and γ_5. The coefficients must be respectively pseudoscalar, pseudovector, pseudotensor, vector, and scalar. Out of the two momenta p and p' it is possible to construct no pseudoscalars or pseudovectors; just one pseudotensor, proportional to $\epsilon^{\mu\nu\rho\sigma} p_\rho\, p'_\sigma$; two independent vectors, proportional to p_μ or p'_μ; and a scalar proportional to unity, in each case with a proportionality factor depending on the only independent scalar variable, $(p - p')^2$. By using the momentum-space Dirac equations for u and u', it is easy to see that the tensor and pseudovector matrices in Γ give contributions proportional to γ_5.

element

$$\int d^4x_1\, d^4x_2\, d^4x_1'\, d^4x_2'\, e^{-ip_1\cdot x_1} e^{-ip_2\cdot x_2} e^{ip_1'\cdot x_1'} e^{ip_2'\cdot x_2'}$$
$$\times \langle T\{\bar{N}_1(x_1), \bar{N}_2(x_2), N_1'(x_1'), N_2'(x_2')\}\rangle_{\rm VAC}\,.$$

Although this pole is not in the physical region for nucleon–nucleon scattering, the pion mass is small enough so that the pole is quite near the physical region, and under some circumstances may dominate the scattering amplitude, as for instance for large ℓ in the partial wave expansion.

Interpreted in coordinate space, a pole like this at $(p_1 - p_1')^2 = (p_2 - p_2')^2 \to -m_\pi^2$ implies a force of range $1/m_\pi$. For instance, in Yukawa's original theory[2] of nuclear force the exchange of mesons (then assumed scalar rather than pseudoscalar) produced a local potential of the form $\exp(-m_\pi r)/4\pi r$, which in the first Born approximation yields an S-matrix for non-relativistic nucleon scattering proportional to the Fourier transform:

$$\int d^3x_1\, d^3x_2\, d^3x_1'\, d^3x_2'\, e^{-i\mathbf{x}_1\cdot\mathbf{p}_1} e^{-i\mathbf{x}_2\cdot\mathbf{p}_2} e^{i\mathbf{x}_1{}'\cdot\mathbf{p}_1{}'} e^{i\mathbf{x}_2{}'\cdot\mathbf{p}_2{}'}$$
$$\times \frac{\exp\Big(-m_\pi|\mathbf{x}_1 - \mathbf{x}_2|\Big)}{4\pi|\mathbf{x}_1 - \mathbf{x}_2|}\delta^3(\mathbf{x}_1 - \mathbf{x}_1{}')\delta^3(\mathbf{x}_2 - \mathbf{x}_2{}')$$
$$= -(2\pi)^3\delta^3(\mathbf{p}_1 + \mathbf{p}_2 - \mathbf{p}_1{}' - \mathbf{p}_2{}')\frac{1}{(\mathbf{p}_1 - \mathbf{p}_1{}')^2 + m_\pi^2}\,.$$

The factor $1/[(\mathbf{p}_1 - \mathbf{p}_1{}')^2 + m_\pi^2]$ is just the non-relativistic limit of the propagator $1/[(p_1 - p_1')^2 + m_\pi^2]$ in (10.2.17). (In (10.2.17) the energy transfer $p_1^0 - p_1'^0$ for $|\mathbf{p}_1| \ll m_N$ and $|\mathbf{p}_1'| \ll m_N$ equals $[\mathbf{p}_1{}^2 - \mathbf{p}_1'{}^2]/2m_N$, which is negligible compared with the magnitude $|\mathbf{p}_1 - \mathbf{p}_1'|$ of the momentum transfer.) When Yukawa's theory was first proposed, it was generally supposed that this sort of momentum-dependence arises from the appearance of a meson field in the theory. It was not until the 1950s that it became generally understood that the existence of a pole at $(p_1 - p_1')^2 \to -m_\pi^2$ follows from the existence of a pion *particle* and has nothing to do with whether this is an elementary particle with its own field in the Lagrangian.

10.3 Field and Mass Renormalization

We will now use a special case of the result of the previous section to clarify the treatment of radiative corrections in the internal and external line of general processes.

The special case that concerns us here is the one in which the four-momentum of a single external line approaches the mass shell. (In the notation of the previous section, this corresponds to taking $r = 1$.) We will consider a function

$$G_\ell(q_1 q_2 \cdots) = \int d^4x_1\, d^4x_2 \cdots e^{-iq_1\cdot x_1}\, e^{-iq_2\cdot x_2} \cdots \times \left(\Psi_0,\, T\left\{\mathcal{O}_\ell(x_1),\, A_2(x_2), \cdots\right\}\Psi_0\right), \qquad (10.3.1)$$

where $\mathcal{O}_\ell(x)$ is a Heisenberg-picture operator, with the Lorentz transformation properties of some sort of free field ψ_ℓ belonging to an irreducible representation of the homogeneous Lorentz group (or the Lorentz group including space inversion for theories that conserve parity), as labelled by the subscript ℓ, and A_2, A_3, etc. are arbitrary Heisenberg-picture operators. Suppose there is a one-particle state $\Psi_{\mathbf{q}_1,\sigma}$ that has non-vanishing matrix elements with the states $\mathcal{O}_\ell^\dagger \Psi_0$ and with $A_2 A_3 \cdots \Psi_0$. Then according to the theorem proved in the previous section, G_ℓ has a pole at $q_1^2 = -m^2$, with

$$G_\ell(q_1 q_2 \cdots) \to \frac{-2i\sqrt{\mathbf{q}_1{}^2 + m^2}}{q_1^2 + m^2 - i\epsilon}\,(2\pi)^3 \sum_\sigma \left(\Psi_0, \mathcal{O}_\ell(0)\Psi_{\mathbf{q}_1,\sigma}\right) \times \int d^4x_2 \cdots e^{-iq_2\cdot x_2} \cdots \left(\Psi_{\mathbf{q}_1,\sigma}\, T\left\{A_2(x_2)\cdots\right\}\Psi_0\right), \qquad (10.3.2)$$

We use Lorentz invariance to write

$$\left(\Psi_0, \mathcal{O}_\ell(0)\Psi_{\mathbf{q}_1,\sigma}\right) = (2\pi)^{-3/2}\, N\, u_\ell(\mathbf{q}_1, \sigma)\,, \qquad (10.3.3)$$

where $u_\ell(\mathbf{q}, \sigma)$ is (aside from the factor $(2\pi)^{-3/2}$) the coefficient function* appearing in the free field ψ_ℓ with the same Lorentz transformation properties as $\mathcal{O}_\ell$, and N is a constant. (It was in order to obtain Eq. (10.3.3) with a single free constant N that we had to assume that $\mathcal{O}_\ell$ transforms irreducibly.) We also define a 'truncated' matrix element M_ℓ by

$$\int d^4x_2 \cdots e^{-iq_2\cdot x_2} \cdots \left(\Psi_{\mathbf{q}_1,\sigma}\, T\left\{A_2(x_2)\cdots\right\}\Psi_0\right) \equiv N^{-1}(2\pi)^{-3/2} \sum_\ell u_\ell^*(\mathbf{q}_1, \sigma)\, M_\ell(q_2 \cdots)\,. \qquad (10.3.4)$$

Eq. (10.3.2) then reads, for $q^2 \to -m_1^2$

$$G_\ell \to \frac{-2i\sqrt{\mathbf{q}_1{}^2 + m^2}}{q_1^2 + m^2 - i\epsilon} \sum_{\sigma,\ell'} u_\ell(\mathbf{q}_1, \sigma) u_{\ell'}^*(\mathbf{q}_1, \sigma) M_{\ell'}\,. \qquad (10.3.5)$$

According to Eqs. (6.2.2) and (6.2.18), the quantity multiplying $M_{\ell'}$ in (10.3.5) is the momentum space matrix propagator $-i\Delta_{\ell\ell'}(q_1)$ for the free

* For instance, for a conventionally normalized free scalar field, $u_\ell(\mathbf{q}_1, \sigma) = [2\sqrt{\mathbf{q}_1^2 + m^2}]^{-1/2}$.

field with the Lorentz transformation properties of $\mathcal{O}_\ell$ (or at least its limiting behavior for $q_1^2 \to -m^2$), so (10.3.5) allows us to identify M_ℓ as the sum of all graphs with external lines carrying momenta $q_1, q_2 \cdots$ corresponding to the operators $\mathcal{O}_\ell, A_2, \cdots$, but with the final propagator for the $\mathcal{O}_\ell$ line stripped away. Eq. (10.3.4) is then just the usual prescription for how to calculate the matrix element for emission of a particle from the sum of Feynman diagrams: strip away the particle propagator, and contract with the usual external line factor $(2\pi)^{-3/2}u_\ell^*$. The only discrepancy with the usual Feynman rules is the factor N.

The above theorem is a famous result due to Lehmann, Symanzik, and Zimmerman,[3] known as the *reduction formula*, which we have proved here by a somewhat different method that has allowed us easily to generalize this result to the case of arbitrary spin. One important aspect of this result is that it applies to any sort of operator; $\mathcal{O}_\ell$ need not be some field that actually appears in the Lagrangian, and the particle it creates may be a bound state composed of those particles whose fields do occur in the Lagrangian. It provides an important lesson even where $\mathcal{O}_\ell$ is some field Ψ_ℓ in the Lagrangian: if we are to use the usual Feynman rules to calculate S-matrix elements, then we should first redefine the normalization of the fields by a factor $1/N$, so that (with apologies for the multiple use of the symbol Ψ):

$$\left(\Psi_0, \Psi_\ell(0)\Psi_{\mathbf{q},\sigma}\right) = (2\pi)^{-3/2}\, u_\ell(\mathbf{q},\sigma)\,. \tag{10.3.6}$$

A field normalized as in Eq. (10.3.6) is called a *renormalized field.*

The field renormalization constant N shows up in another place. Suppose that there is just one of the operators $A_2, A_3, \cdots$ in Eq. (10.3.1), and take it to be the adjoint of a member of the same field multiplet as $\mathcal{O}_\ell$. Then Eq. (10.3.2) reads

$$\begin{aligned}
&\int d^4x_1 \int d^4x_2\, e^{-iq_1\cdot x_1}\, e^{-iq_2\cdot x_2} \left(\Psi_0,\, T\left\{\mathcal{O}_\ell(x_1)\mathcal{O}^\dagger_{\ell'}(x_2)\right\}\Psi_0\right) \\
&\overset{q_1^2\to -m^2}{\longrightarrow} \frac{-2i\sqrt{\mathbf{q}_1{}^2+m^2}\,(2\pi)^3}{q_1^2+m^2-i\epsilon} \sum_\sigma \left(\Psi_0, \mathcal{O}_\ell(0)\Psi_{\mathbf{q}_1,\sigma}\right) \\
&\qquad\times \int d^4x_2\, e^{-iq_2\cdot x_2} e^{-iq_1\cdot x_2}\left(\Psi_{\mathbf{q}_1,\sigma}, \mathcal{O}^\dagger_{\ell'}(0)\Psi_0\right) \\
&= \frac{-2i|N|^2\sqrt{\mathbf{q}_1{}^2+m^2}}{q_1^2+m^2-i\epsilon} \sum_\sigma u_\ell(\mathbf{q}_1,\sigma)\, u^*_{\ell'}(\mathbf{q}_1,\sigma)\,(2\pi)^4\,\delta^4(q_1+q_2)\,.
\end{aligned}$$

This is just the usual behavior of a propagator (the sum of all graphs with two external lines) near its pole, except for the factor $|N|^2$. According to Eq. (10.3.6), this factor is absent in the propagator of the renormalized field Ψ_ℓ. Thus *a renormalized field is one whose propagator has the same behavior near its pole as for a free field, and the renormalized mass is defined by the position of the pole.*

To see how this works in practice, consider the theory of a real self-interacting scalar field Φ_B, the subscript B being added here to remind us that so far this is a 'bare' (i.e., unrenormalized) field. The Lagrangian density is taken as usual as

$$\mathscr{L} = -\tfrac{1}{2}\partial_\mu\Phi_B\partial^\mu\Phi_B - \tfrac{1}{2}m_B^2\Phi_B^2 - V_B(\Phi_B)\,. \tag{10.3.7}$$

In general there would be no reason to expect that the field Φ_B would satisfy condition (10.3.6), nor that the pole in q^2 would be at $-m_B^2$, so let us introduce a renormalized field and mass

$$\Phi \equiv Z^{-1/2}\Phi_B\,, \tag{10.3.8}$$

$$m^2 \equiv m_B^2 + \delta m^2\,, \tag{10.3.9}$$

with Z to be chosen so that Φ does satisfy Eq. (10.3.6), and δm^2 chosen so that the pole of the propagator is at $q^2 = -m^2$. (The use of the symbol Z in this context has become conventional; there is a different Z for each field in the Lagrangian.) The Lagrangian density (10.3.7) may then be rewritten

$$\mathscr{L} = \mathscr{L}_0 + \mathscr{L}_1, \tag{10.3.10}$$

$$\mathscr{L}_0 = -\tfrac{1}{2}\partial_\mu\Phi\partial^\mu\Phi - \tfrac{1}{2}m^2\Phi^2, \tag{10.3.11}$$

$$\mathscr{L}_1 = -\tfrac{1}{2}(Z-1)[\partial_\mu\Phi\partial^\mu\Phi + m^2\Phi^2] + \tfrac{1}{2}Z\delta m^2\Phi^2 - V(\Phi)\,, \tag{10.3.12}$$

where

$$V(\Phi) \equiv V_B(\sqrt{Z}\Phi)\,.$$

In calculating the corrections to the complete momentum space propagator of the renormalized scalar field, conventionally called $\Delta'(q)$, it is convenient to consider separately the *one-particle-irreducible* graphs: those connected graphs (excluding a graph consisting of a single scalar line) that cannot be disconnected by cutting through any one internal scalar line. An example is shown in Figure 10.4. It is conventional to write the sum of all such graphs, with the two external line propagator factors $-i(2\pi)^{-4}(q^2+m^2-i\epsilon)^{-1}$ omitted, as $i(2\pi)^4\Pi^*(q^2)$, with the asterisk to remind us that these are one-particle-irreducible graphs. Then the corrections to the complete propagator are given by a sum of chains of one, two, or more of these one-particle-irreducible subgraphs connected with the usual uncorrected

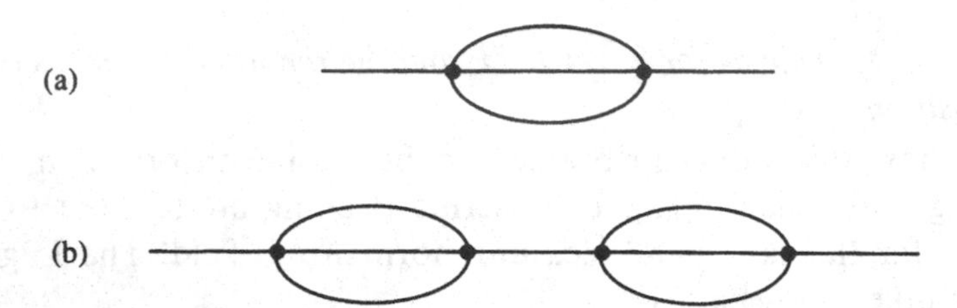

Figure 10.4. Diagrams that (a) are, or (b) are not, one-particle irreducible. These diagrams are drawn for a theory with some sort of quadrilinear interaction, like the theory of a scalar field ϕ with interaction proportional to ϕ^4.

propagator factors:

$$\begin{aligned}\frac{-i}{(2\pi)^4}\Delta'(q) &= \frac{-i}{(2\pi)^4}\frac{1}{q^2+m^2-i\epsilon}\\ &+\left[\frac{-i}{(2\pi)^4}\frac{1}{q^2+m^2-i\epsilon}\right]\left[i(2\pi)^4\Pi^*(q^2)\right]\left[\frac{-i}{(2\pi)^4}\frac{1}{q^2+m^2-i\epsilon}\right]\\ &+\left[\frac{-i}{(2\pi)^4}\frac{1}{q^2+m^2-i\epsilon}\right]\left[i(2\pi)^4\Pi^*(q^2)\right]\left[\frac{-i}{(2\pi)^4}\frac{1}{q^2+m^2-i\epsilon}\right]\\ &\times\left[i(2\pi)^4\Pi^*(q^2)\right]\left[\frac{-i}{(2\pi)^4}\frac{1}{q^2+m^2-i\epsilon}\right]+\cdots\end{aligned} \tag{10.3.13}$$

or more simply

$$\begin{aligned}\Delta'(q) &= [q^2+m^2-i\epsilon]^{-1}+[q^2+m^2-i\epsilon]^{-1}\Pi^*(q^2)[q^2+m^2-i\epsilon]^{-1}\\ &+[q^2+m^2-i\epsilon]^{-1}\Pi^*(q^2)[q^2+m^2-i\epsilon]^{-1}\Pi^*(q^2)[q^2+m^2-i\epsilon]^{-1}+\cdots\,.\end{aligned} \tag{10.3.14}$$

Summing the geometric series, this gives

$$\Delta'(q) = \left[q^2+m^2-\Pi^*(q^2)-i\epsilon\right]^{-1}\,. \tag{10.3.15}$$

In calculating Π^*, we encounter a tree graph arising from a single insertion of vertices corresponding to the terms in Eq. (10.3.12) proportional to $\partial_\mu\Phi\partial^\mu\Phi$ and Φ^2, plus a term $\Pi^*_{\rm LOOP}$ arising from loop graphs like that in Figure 10.4(a):

$$\Pi^*(q^2) = -(Z-1)[q^2+m^2]+Z\,\delta m^2+\Pi^*_{\rm LOOP}(q^2)\,. \tag{10.3.16}$$

The condition that m^2 is the true mass of the particle is that the pole of the propagator should be at $q^2=-m^2$, so that

$$\Pi^*(-m^2)=0\,. \tag{10.3.17}$$

Also, the condition that the pole of the propagator at $q^2=-m^2$ should have a unit residue (like the uncorrected propagator) is that

$$\left[\frac{d}{dq^2}\Pi^*(q^2)\right]_{q^2=-m^2} = 0\,. \tag{10.3.18}$$

These conditions allow us to evaluate Z and δm^2:

$$Z\,\delta m^2 = -\Pi^*_{\text{LOOP}}(-m^2)\,, \tag{10.3.19}$$

$$Z = 1 + \left[\frac{d}{dq^2}\Pi^*_{\text{LOOP}}(q^2)\right]_{q^2=-m^2}\,. \tag{10.3.20}$$

This incidentally shows that $Z\,\delta m^2$ and $Z-1$ are given by a series of terms containing one or more coupling constant factors, justifying the treatment of the first two terms in Eq. (10.3.12) as part of the interaction $\mathscr{L}_1$.

In actual calculations it is simplest just to say that from the loop terms $\Pi^*_{\text{LOOP}}(q^2)$ we must subtract a first-order polynomial in q^2 with coefficients chosen so that the difference satisfies Eqs. (10.3.17) and (10.3.18). As we shall see, this subtraction procedure incidentally cancels the infinities that arise from the momentum space integrals in Π^*_{LOOP}. However, as this discussion should make clear, *the renormalization of masses and fields has nothing directly to do with the presence of infinities, and would be necessary even in a theory in which all momentum space integrals were convergent.*

An important consequence of the conditions (10.3.17) and (10.3.18) is that it is not necessary to include radiative corrections in external lines on the mass shell. That is,

$$\Big[\Pi^*(q^2)[q^2+m^2-i\epsilon]^{-1} + \Pi^*(q^2)[q^2+m^2-i\epsilon]^{-1}\Pi^*(q^2)[q^2+m^2-i\epsilon]^{-1} + \cdots\Big]_{q^2\to -m^2} = 0\,. \tag{10.3.21}$$

Similar remarks apply to particles of arbitrary spin. For instance, for the 'bare' Dirac field the Lagrangian is

$$\mathscr{L} = -\bar{\Psi}_{\text{B}}[\not\partial + m_{\text{B}}]\Psi_{\text{B}} - V_{\text{B}}(\bar{\Psi}_{\text{B}}\Psi_{\text{B}})\,. \tag{10.3.22}$$

We introduce renormalized fields and masses

$$\Psi \equiv Z_2^{-1/2}\Psi_{\text{B}}\,, \tag{10.3.23}$$

$$m = m_{\text{B}} + \delta m\,. \tag{10.3.24}$$

(The subscript 2 on Z_2 is conventionally used to distinguish the renormalization constant of a fermion field.) The Lagrangian density is then rewritten

$$\mathscr{L} = \mathscr{L}_0 + \mathscr{L}_1\,, \tag{10.3.25}$$

$$\mathscr{L}_0 = -\bar{\Psi}[\not\partial + m]\Psi\,, \tag{10.3.26}$$

$$\mathscr{L}_1 = -(Z_2 - 1)[\bar{\Psi}[\partial\!\!\!/ + m]\Psi] + Z_2\delta m\bar{\Psi}\Psi - V_B(Z_2\bar{\Psi}\Psi)\,. \qquad (10.3.27)$$

Let $i(2\pi)^4\Sigma^*(k\!\!\!/)$ be the sum of all connected graphs, with one fermion line coming in with four-momentum k and one going out with the same four-momentum, that cannot be disconnected by cutting through any single internal fermion line, and with external line propagator factors $-i(2\pi)^{-4}$ and $[i\,k\!\!\!/ + m - i\epsilon]^{-1}$ omitted. (Lorentz invariance is being used to justify writing Σ^* as an ordinary function of the Lorentz scalar matrix $k\!\!\!/ \equiv k_\mu\gamma^\mu$.) Then the complete fermion propagator is

$$\begin{aligned}S'(k) &= [i\,k\!\!\!/ + m - i\epsilon]^{-1} + [i\,k\!\!\!/ + m - i\epsilon]^{-1}\Sigma^*(k\!\!\!/)[i\,k\!\!\!/ + m - i\epsilon]^{-1} \\ &\quad + [i\,k\!\!\!/ + m - i\epsilon]^{-1}\Sigma^*(k\!\!\!/)[i\,k\!\!\!/ + m - i\epsilon]^{-1}\Sigma^*(k\!\!\!/)[i\,k\!\!\!/ + m - i\epsilon]^{-1} + \cdots \\ &= [i\,k\!\!\!/ + m - \Sigma^*(k\!\!\!/) - i\epsilon]^{-1}\,. \end{aligned} \qquad (10.3.28)$$

In calculating $\Sigma^*(k\!\!\!/)$ we take into account the tree graphs from the terms in Eq. (10.3.27) proportional to $\bar{\Psi}\,\partial\!\!\!/\Psi$ and $\bar{\Psi}\Psi$ as well as loop contributions:

$$\Sigma^*(k\!\!\!/) = -(Z_2 - 1)[i\,k\!\!\!/ + m] + Z_2\delta m + \Sigma^*_{\rm LOOP}(k\!\!\!/)\,. \qquad (10.3.29)$$

The condition that the complete propagator has a pole at $k^2 = -m^2$ with the same residue as the uncorrected propagator is then that

$$\Sigma^*(im) = 0\,, \qquad (10.3.30)$$

$$\left.\frac{\partial\Sigma^*(k\!\!\!/)}{\partial\,k\!\!\!/}\right|_{k\!\!\!/=im} = 0\,, \qquad (10.3.31)$$

and hence

$$Z_2\delta m = -\Sigma^*_{\rm LOOP}(im)\,, \qquad (10.3.32)$$

$$Z_2 = 1 - i\left.\frac{\partial\Sigma^*_{\rm LOOP}(k\!\!\!/)}{\partial\,k\!\!\!/}\right|_{k\!\!\!/=im}\,. \qquad (10.3.33)$$

Just as for scalars, the vanishing of $[i\,k\!\!\!/ + m]^{-1}\Sigma^*(k\!\!\!/)$ in the limit $k\!\!\!/ \to im$ tells us that radiative corrections may be ignored in external fermion lines. Corresponding results for the photon propagator will be derived in Section 10.5.

10.4 Renormalized Charge and Ward Identities

The use of the commutation and conservation relations of Heisenberg-picture operators allows us to make a connection between the charges (or other similar quantities) in the Lagrangian density and the properties of physical states. Recall that the invariance of the Lagrangian density with respect to global gauge transformations $\Psi_\ell \to \exp(iq_\ell\alpha)\Psi_\ell$ (with α an arbitrary constant phase) implies the existence of a current

$$J^\mu = -i \sum_\ell \frac{\partial \mathscr{L}}{\partial(\partial_\mu \Psi_\ell)} q_\ell \Psi_\ell , \tag{10.4.1}$$

satisfying the conservation condition

$$\partial_\mu J^\mu = 0 . \tag{10.4.2}$$

This implies that the space-integral of the time component of J^μ is time-independent:

$$i \frac{d}{dt} Q = [Q, H] = 0 , \tag{10.4.3}$$

where

$$Q \equiv \int d^3x \, J^0 . \tag{10.4.4}$$

(There is a very important possible exception here, that the integral (10.4.4) may not exist if there are long-range forces due to massless scalars in the system. We will return to this point when we consider broken symmetries in Volume II.) Also, since it is a space-integral, Q is manifestly translation-invariant

$$[\mathbf{P}, Q] = 0 \tag{10.4.5}$$

and since J^μ is a four-vector, Q is invariant with respect to homogeneous Lorentz transformations

$$[J^{\mu\nu}, Q] = 0 . \tag{10.4.6}$$

It follows that Q acting on the true vacuum Ψ_0 must be another Lorentz-invariant state of zero energy and momentum, and hence (assuming no vacuum degeneracy) must be proportional to Ψ_0 itself. But the proportionality constant must vanish, because Lorentz invariance requires that $(\Psi_0, J_\mu \Psi_0)$ vanish. Hence

$$Q \Psi_0 = 0 . \tag{10.4.7}$$

Also, Q acting on any one-particle state $\Psi_{\mathbf{p},\sigma,n}$ must be another state with the same energy, momentum, and Lorentz transformation properties, and thus (assuming no degeneracy of one-particle states) must be proportional to the same one-particle state

$$Q \Psi_{\mathbf{p},\sigma,n} = q_{(n)} \Psi_{\mathbf{p},\sigma,n} . \tag{10.4.8}$$

The Lorentz invariance of Q ensures that the eigenvalue $q_{(n)}$ is independent of $\mathbf{p}$ and σ, depending only on the species of the particle. This eigenvalue is what is known as the electric charge (or whatever other quantum number of which J^μ may be the current) of the one-particle state. To relate this to the q_ℓ parameters in the Lagrangian, we note that the canonical commutation relations give

$$\left[J^0(\mathbf{x},t),\ \Psi_\ell(\mathbf{y},t)\right] = -q_\ell \Psi_\ell(\mathbf{y},t)\delta^3(\mathbf{x}-\mathbf{y})\,, \tag{10.4.9}$$

or integrating over $\mathbf{x}$:

$$\left[Q, \Psi_\ell(y)\right] = -q_\ell \Psi_\ell(y)\,. \tag{10.4.10}$$

The same is true of any local function $F(y)$ of the fields and field derivatives and their adjoints, containing definite numbers of each:

$$\left[Q, F(y)\right] = -q_F F(y)\,, \tag{10.4.11}$$

where q_F is the sum of the q_ℓ for all fields and field derivatives in $F(y)$, minus the sum of the q_ℓ for all field adjoints and their derivatives. Taking the matrix element of this equation between a one-particle state and the vacuum, and using Eqs. (10.4.7) and (10.4.8), we have

$$\left(\Psi_0,\ F(y)\Psi_{\mathbf{p},\sigma,n}\right)\left(q_F - q_{(n)}\right) = 0\,. \tag{10.4.12}$$

Hence we must have

$$q_{(n)} = q_F \tag{10.4.13}$$

as long as

$$\left(\Psi_0,\ F(y)\,\Psi_{\mathbf{p},\sigma,n}\right) \neq 0\,. \tag{10.4.14}$$

As we saw in the previous section, Eq. (10.4.14) is the condition that assures that momentum space Green's functions involving F have poles corresponding to the one-particle state $\Psi_{\mathbf{p},\sigma,n}$. For a one-particle state corresponding to one of the fields in the Lagrangian we could take $F = \Psi_\ell$, in which case $q_F = q_\ell$, but our results here apply to general one-particle states, whether or not their fields appear in the Lagrangian.

This almost, but not quite, tells us that despite all the possible high-order graphs that affect the emission and absorption of photons by charged particles, the physical electric charge is just equal to a parameter q_ℓ appearing in the Lagrangian (or to a sum of such parameters, like q_F.) The qualification that has to be added here is that the requirement, that the Lagrangian be invariant under the transformations $\Psi_\ell \to \exp(iq_\ell\alpha)\Psi_\ell$, does nothing to fix the over-all scale of the quantities q_ℓ. The physical electric charges are those that determine the response of matter fields to a given *renormalized* electromagnetic field A^μ. That is, the scale of the q_ℓ is fixed by requiring that the renormalized electromagnetic field appears in the matter Lagrangian $\mathscr{L}_M$ in the linear combinations $[\partial_\mu - iq_\ell A_\mu]\Psi_\ell$, so that the current J^μ is

$$J^\mu = \frac{\delta \mathscr{L}_M}{\delta A_\mu} \,. \tag{10.4.15}$$

But A^μ and q_ℓ are not the same as the 'bare electromagnetic field' $A^\mu_{\rm B}$ and 'bare charges' $q_{{\rm B}\ell}$ that appear in the Lagrangian when we write it in its simplest form

$$\mathscr{L} = -\tfrac{1}{4}(\partial_\mu A_{B\nu} - \partial_\nu A_{B\mu})(\partial^\mu A^\nu_{\rm B} - \partial^\nu A^\mu_{\rm B}) + \mathscr{L}_M\left(\Psi_\ell, [\partial_\mu - iq_{B\ell}A_{B\mu}]\Psi_\ell\right) \,. \tag{10.4.16}$$

The renormalized electromagnetic field (defined to have a complete propagator whose pole at $p^2 = 0$ has unit residue) is conventionally written in terms of $A^\mu_{\rm B}$ as

$$A^\mu = Z_3^{-1/2} A^\mu_{\rm B} \,, \tag{10.4.17}$$

so in order for the charge q_ℓ to characterize the response of the charged particles to a given renormalized electromagnetic field, we should define the renormalized charges by

$$q_\ell = \sqrt{Z_3}\, q_{{\rm B}\ell} \,. \tag{10.4.18}$$

We see that the physical electric charge q of any particle is just proportional to a parameter $q_{\rm B}$ related to those appearing in the Lagrangian, with a proportionality constant $\sqrt{Z_3}$ that is the same for all particles. This helps us to understand how a particle like the proton, that is surrounded by a cloud of virtual mesons and other strongly interacting particles, can have the same charge as the positron, whose interactions are all much weaker. It is only necessary to assume that for some reason the charges $q_{{\rm B}\ell}$ in the Lagrangian are equal and opposite for the electron and for those particles (two u quarks and one d quark) that make up the proton; the effect of higher-order corrections then appears solely in the *common* factor $\sqrt{Z_3}$.

In order for charge renormalization to arise only from radiative corrections to the photon propagator, there must be cancellations among the great variety of other radiative corrections to the propagators and electromagnetic vertices of the charged particles. We can see a little more deeply into the nature of these cancellations by making use of the celebrated relations between these charged particle propagators and vertices known as the *Ward identities.*

For instance, consider the Green's function for an electric current $J^\mu(x)$ together with a Heisenberg-picture Dirac field $\Psi_n(y)$ of charge q and its covariant adjoint $\bar{\Psi}_m(z)$. We define the electromagnetic vertex function Γ^μ

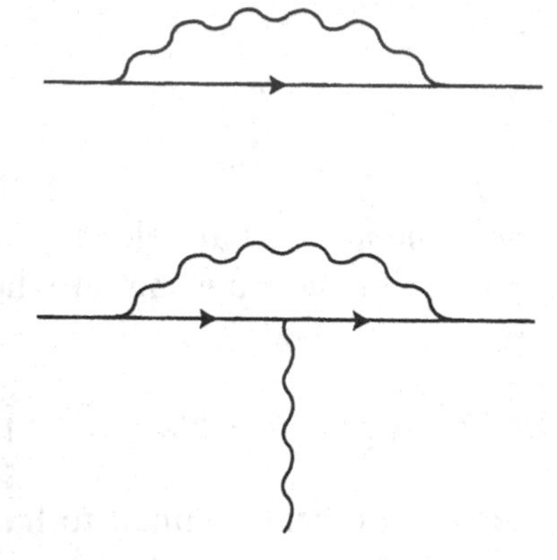

Figure 10.5. Diagrams for the first corrections to the electron propagator and vertex function in quantum electrodynamics. Here straight lines are electrons; wavy lines are photons.

of the charged particle by

$$\int d^4x\, d^4y\, d^4z\, e^{-ip\cdot x} e^{-ik\cdot y} e^{+i\ell\cdot z} \left(\Psi_0,\; T\left\{J^\mu(x)\, \Psi_n(y)\, \bar{\Psi}_m(z)\right\} \Psi_0\right)$$
$$\equiv -\, i(2\pi)^4 q S'_{nn'}(k) \Gamma^\mu_{n'm'}(k,\ell) S'_{m'm}(\ell)\, \delta^4(p+k-\ell)\,, \qquad (10.4.19)$$

where

$$-\, i(2\pi)^4 S'_{nm}(k)\, \delta^4(k-\ell) \equiv \int d^4y\, d^4z\, \left(\Psi_0,\; T\left\{\Psi_n(y)\bar{\Psi}_m(z)\right\}\Psi_0\right) e^{-ik\cdot y} e^{+i\ell\cdot z}\,. \qquad (10.4.20)$$

According to the theorem of Section 6.4, Eq. (10.4.20) gives the sum of all Feynman graphs with one incoming and one outgoing fermion line, i.e., the complete Dirac propagator. Also, Eq. (10.4.19) gives the sum of all such graphs with an extra photon line attached, so Γ^μ is the sum of "vertex" graphs with one incoming Dirac line, one outgoing Dirac line, and one photon line, but with the complete Dirac external line propagators and the bare photon external line propagator stripped away. To make the normalization of S' and Γ^μ perfectly clear, we mention that in the limit of no interactions, these functions take the values

$$S'(k) \to [i\gamma_\lambda k^\lambda + m - i\epsilon]^{-1}\,, \qquad \Gamma^\mu(k,\ell) \to \gamma^\mu\,.$$

The one-loop diagrams that provide corrections to these limiting values are shown in Figure 10.5.

We can derive a relation between Γ^μ and S' by use of the identity

$$\frac{\partial}{\partial x^\mu} T\Big\{J^\mu(x)\Psi_n(y)\bar{\Psi}_m(z)\Big\} = T\Big\{\partial_\mu J^\mu(x)\Psi_n(y)\bar{\Psi}_m(z)\Big\}$$
$$+\delta(x^0-y^0)\,T\Big\{\Big[J^0(x),\Psi_n(y)\Big]\,\bar{\Psi}_m(z)\Big\}$$
$$+\delta(x^0-z^0)\,T\Big\{\Psi_n(y)\,\Big[J^0(x),\bar{\Psi}_m(z)\Big]\Big\}\,, \tag{10.4.21}$$

where the delta functions arise from time-derivatives of step functions. The conservation condition (10.4.2) tells us that the first term vanishes, while the second and third terms can be calculated using the commutation relations (10.4.9), which here give

$$\Big[J^0(\mathbf{x},t),\ \Psi_n(\mathbf{y},t)\Big] = -q\Psi_n(\mathbf{y},t)\delta^3(\mathbf{x}-\mathbf{y}) \tag{10.4.22}$$

and its adjoint

$$\Big[J^0(\mathbf{x},t),\ \bar{\Psi}_n(\mathbf{y},t)\Big] = q\bar{\Psi}_n(\mathbf{y},t)\delta^3(\mathbf{x}-\mathbf{y})\,. \tag{10.4.23}$$

Eq. (10.4.21) then reads

$$\frac{\partial}{\partial x^\mu}\ T\Big\{J^\mu(x)\Psi_n(y)\bar{\Psi}_m(z)\Big\} = -q\,\delta^4(x-y)\,T\Big\{\Psi_n(y)\bar{\Psi}_m(z)\Big\}$$
$$+\,q\,\delta^4(x-z)\,T\Big\{\Psi_n(y)\bar{\Psi}_m(z)\Big\}\,. \tag{10.4.24}$$

Inserting this in the Fourier transform (10.4.19) gives

$$(\ell-k)_\mu\, S'(k)\,\Gamma^\mu(k,\ell)\,S'(\ell) = i\,S'(\ell)\ -\ i\,S'(k)$$

or in other words

$$(\ell-k)_\mu\Gamma^\mu(k,\ell) = i\,S'^{-1}(k) - iS'^{-1}(\ell)\,. \tag{10.4.25}$$

This is known as the *generalized Ward identity*, first derived (by these methods) by Takahashi.[4] The original Ward identity, derived earlier by Ward[5] from a study of perturbation theory, can be obtained from Eq. (10.4.25) by letting ℓ approach k. In this limit, Eq. (10.4.25) gives

$$\Gamma^\mu(k,k) = -i\,\frac{\partial}{\partial k_\mu}\,S'^{-1}(k)\,. \tag{10.4.26}$$

The fermion propagator is related to the self-energy insertion $\Sigma^*(\not{k})$ by Eq. (10.3.28)

$$S'^{-1}(k) = i\,\not{k} + m - \Sigma^*(\not{k})\,,$$

so Eq. (10.4.26) may be written

$$\Gamma^\mu(k,k) = \gamma^\mu + i\,\frac{\partial}{\partial k_\mu}\,\Sigma^*(\not{k})\,. \tag{10.4.27}$$

For a *renormalized* Dirac field, Eqs. (10.3.31) and (10.4.27) tell us that on the mass shell

$$\bar{u}'_k \,\Gamma^\mu(k,k)\, u_k = \bar{u}'_k \,\gamma^\mu\, u_k \,, \tag{10.4.28}$$

where $[i\gamma_\mu k^\mu + m]u_k = [i\gamma_\mu k^\mu + m]u'_k = 0$. Thus the renormalization of the fermion field ensures that the radiative corrections to the vertex function Γ_μ cancel when a fermion on the mass shell interacts with an electromagnetic field with zero momentum transfer, as is the case when we set out to measure the fermion's electric charge. If we had not used a renormalized fermion field then the corrections to the vertex function would have just cancelled the corrections due to radiative corrections to the external fermion lines, leaving the electric charge again unchanged.

10.5 Gauge Invariance

The conservation of electric charge may be used to prove a useful result for the quantities

$$M^{\mu\mu'\cdots}_{\beta\alpha}(q,q',\cdots) \equiv \int d^4x \int d^4x' \cdots e^{-iq\cdot x} e^{-iq'\cdot x'} \cdots \times \Big(\Psi^-_\beta,\, T\Big\{J^\mu(x), J^{\mu'}(x')\cdots\Big\}\Psi^+_\alpha\Big)\,. \tag{10.5.1}$$

In theories like spinor electrodynamics in which the electromagnetic interaction is linear in the field A^μ, this is the matrix element for emission (and/or absorption) of on- or off-shell photons having four-momenta q, q', etc. (and/or $-q$, $-q'$, etc.), with external line photon coefficient functions or propagators omitted, in an arbitrary transition $\alpha \to \beta$. Our result is that Eq. (10.5.1) vanishes when contracted with any one of the photon four-momenta:

$$q_\mu M^{\mu\mu'\cdots}_{\beta\alpha}(q,q',\cdots) = q'_{\mu'} M^{\mu\mu'\cdots}_{\beta\alpha}(q,q',\cdots) = \cdots = 0\,. \tag{10.5.2}$$

Since M is defined symmetrically with respect to the photon lines, it will be sufficient to show the vanishing of the first of these quantities.

For this purpose, note that by an integration by parts

$$q_\mu M^{\mu\mu'\cdots}_{\beta\alpha}(q,q',\cdots) = -i \int d^4x \int d^4x' \cdots \times e^{-iq\cdot x} e^{-iq'\cdot x'} \cdots \Big(\Psi^-_\beta,\, \frac{\partial}{\partial x^\mu} T\Big\{J^\mu(x), J^{\mu'}(x')\cdots\Big\}\Psi^+_\alpha\Big)\,. \tag{10.5.3}$$

The electric current $J^\mu(x)$ is conserved, but this does not immediately imply that Eq. (10.5.3) vanishes, because we still have to take account of the x^0-dependence contained in the theta functions that appear in the definition of the time-ordered product. For instance, for just two currents

$$T\Big\{J^\mu(x)J^\nu(y)\Big\} = \theta(x^0 - y^0)J^\mu(x)J^\nu(y) + \theta(y^0 - x^0)J^\nu(y)J^\mu(x)$$

so, taking account of the conservation of $J^\mu(x)$:

$$\frac{\partial}{\partial x^\mu} T\Big\{J^\mu(x)\,J^\nu(y)\Big\} = \delta(x^0 - y^0)\,J^0(x)\,J^\nu(y) - \delta(y^0 - x^0)\,J^\nu(y)\,J^0(x)$$
$$= \delta(x^0 - y^0)\Big[J^0(x),\, J^\nu(y)\Big]\,. \tag{10.5.4}$$

With more than two currents, we get an equal-time commutator like this (inside the time-ordered product) for each current aside from $J^\mu(x)$ itself. To evaluate this commutator, we recall that (as shown in the previous section) for any product F of field operators and their adjoints and/or derivatives

$$\Big[J^0(\vec{x},t),\, F(\vec{y},t)\Big] = -q_F\, F(\vec{x},t)\,\delta^3(\vec{x} - \vec{y})\,,$$

where q_F is the sum of the q_ℓs for the fields and field derivatives in F, minus the sum of the q_ℓs for the field adjoints and their derivatives. For the electric current, q_J is zero; $J^\nu(y)$ is itself an electrically neutral operator. It follows that

$$\Big[J^0(\vec{x},t),\, J^\nu(\vec{y},t)\Big] = 0 \tag{10.5.5}$$

and therefore Eq. (10.5.4) vanishes, so that Eq. (10.5.3) gives

$$q_\mu\, M^{\mu\mu'\cdots}_{\beta\alpha}(q, q', \cdots) = 0 \tag{10.5.6}$$

as was to be proved.

There is an important qualification here. In deriving Eq. (10.5.5) we should take into account the fact that a product of fields at the same spacetime point y like the current operator $J^\nu(y)$ can only be properly defined through some regularization procedure that deals with the infinities in such products. In many cases it turns out that there are non-vanishing contributions to the commutator of $J^0(\vec{x},t)$ with the regulated current $J^i(\vec{y},t)$, known as *Schwinger terms.*[6] Where the current includes terms arising from a charged scalar field Φ, there are additional regulator-independent Schwinger terms involving $\Phi^\dagger\Phi$. However, all these Schwinger terms are cancelled in multi-photon amplitudes by the contribution of additional interactions that are quadratic in the electromagnetic field, either arising from the regulator procedure (if gauge-invariant) or, as for charged scalars, directly from terms in the Lagrangian. We will be dealing mostly with charged spinor fields, and will use a regularization procedure (dimensional regularization) that does not lead to Schwinger terms, so in what follows we will ignore this issue and continue to use the naive commutation relation (10.5.5).

The same argument yields a result like Eq. (10.5.2) even if other particles besides photons are off the mass shell, provided that all *charged* particles are taken on the mass shell, i.e., kept in the states $\Psi_\beta{}^-$ and $\Psi_\alpha{}^+$. Otherwise, the left-hand side of Eq. (10.5.2) receives contributions from non-vanishing equal-time commutators, such as those we encountered in the derivation of the Ward identity in the previous section.

One consequence of Eq. (10.5.2) is that S-matrix elements are unaffected if we change any photon propagator $\Delta_{\mu\nu}(q)$ by

$$\Delta_{\mu\nu}(q) \to \Delta_{\mu\nu}(q) + \alpha_\mu q_\nu + q_\mu \beta_\nu \tag{10.5.7}$$

or if we change any photon polarization vector by

$$e_\rho(\mathbf{k}, \lambda) \to e_\rho(\mathbf{k}, \lambda) + c k_\rho \,, \tag{10.5.8}$$

where $k^0 \equiv |\mathbf{k}|$, and α_μ, β_ν, and c are entirely arbitrary (not necessarily constants, and not necessarily the same for all propagators or polarization vectors.) This is (somewhat loosely) called the gauge invariance of the S-matrix.

To prove this result it is only necessary to display the explicit dependence of the S-matrix on photon polarization vectors and propagators

$$\begin{aligned} S_{\beta\alpha} \propto \int & d^4q_1\, d^4q_2 \cdots \Delta_{\mu_1\nu_1}(q_1)\Delta_{\mu_2\nu_2}(q_2)\cdots \\ & \times e^*_{\rho_1}(\mathbf{k}'_1\lambda'_1)e^*_{\rho_2}(\mathbf{k}'_2\lambda'_2)\cdots e_{\sigma_1}(\mathbf{k}_1\lambda_1)e_{\sigma_2}(\mathbf{k}_2\lambda_2)\cdots \\ \times M_{ba}^{\mu_1\mu_2\cdots\nu_1\nu_2\cdots\rho_1\rho_2\cdots\sigma_1\sigma_2\cdots} & (-q_1, -q_2, \cdots, q_1, q_2, \cdots, -k'_1, -k'_2 \cdots, k_1, k_2 \cdots) \end{aligned} \tag{10.5.9}$$

where $M^{\rho\sigma\cdots}$ is the matrix element (10.5.1) calculated in the absence of electromagnetic interactions.* The invariance of Eq. (10.5.9) under the 'gauge transformations' (10.5.7) and (10.5.8) follows immediately from the conservation conditions (10.5.2). (In Section 9.6 we used the path-integral formalism to prove a special case of this theorem, that vacuum expectation values of time-ordered products of gauge-invariant operators are independent of the constant α in the propagator (9.6.21).) This result is not as elementary as it looks, as it applies not to individual diagrams, but only to sums of diagrams in which the current vertices are inserted in all possible places in the diagrams.

There is a particularly important application of Eq. (10.5.2) to the calculation of the photon propagator. The complete photon propagator,

* The states a and b are the same as α and β, but with photons deleted. Note that the arguments of M are all taken to be *incoming* four-momenta, which is why we have to insert various signs for some of the arguments of M in Eq. (10.5.9).

conventionally called $\Delta'_{\mu\nu}(q)$, takes the form

$$\Delta'_{\mu\nu}(q) = \Delta_{\mu\nu}(q) + \Delta_{\mu\rho}(q) M^{\rho\sigma}(q) \Delta_{\sigma\nu}(q) , \tag{10.5.10}$$

where $M^{\rho\sigma}$ is proportional to the matrix element (10.5.1) with two currents and α and β both the vacuum state, and $\Delta_{\mu\nu}$ is the bare photon propagator, written here in a general Lorentz-invariant gauge as

$$\Delta_{\mu\nu}(q) \equiv \frac{\eta_{\mu\nu} - \xi(q^2) q_\mu q_\nu / q^2}{q^2 - i\epsilon} . \tag{10.5.11}$$

From Eq. (10.5.2) we have here $q^\mu M_{\mu\nu}(q) = 0$, so that

$$q^\mu \Delta'_{\mu\nu}(q) = q^\mu \Delta_{\mu\nu}(q) = \frac{q_\nu (1 - \xi(q^2))}{q^2 - i\epsilon} . \tag{10.5.12}$$

On the other hand, just as we did for scalar and spinor fields in Section 10.3, we may express the complete photon propagator in terms of a sum $\Pi^*(q)$ of graphs with two external photon lines that (unlike M) are one-photon-irreducible:

$$\begin{aligned} \Delta'(q) &= \Delta(q) + \Delta(q)\Pi^*(q)\Delta(q) + \Delta(q)\Pi^*(q)\Delta(q)\Pi^*(q)\Delta(q) + \cdots \\ &= [\Delta(q)^{-1} - \Pi^*(q)]^{-1} \end{aligned} \tag{10.5.13}$$

or in other words

$$\Delta'_{\mu\nu}(q) = \Delta_{\mu\nu}(q) + \Delta_{\mu\rho}(q)\Pi^{*\rho\sigma}(q)\Delta'_{\sigma\nu}(q) . \tag{10.5.14}$$

Then in order to satisfy Eq. (10.5.12), we must have

$$q_\rho \Pi^{*\rho\sigma}(q) = 0 . \tag{10.5.15}$$

This together with Lorentz invariance tells us that $\Pi^*(q)$ must take the form

$$\Pi^{*\rho\sigma}(q) = (q^2 \eta^{\rho\sigma} - q^\rho q^\sigma)\pi(q^2) . \tag{10.5.16}$$

Then Eq. (10.5.13) yields a complete propagator of the form

$$\Delta'_{\mu\nu}(q) = \frac{\eta_{\mu\nu} - \tilde{\xi}(q^2) q_\mu q_\nu / q^2}{[q^2 - i\epsilon][1 - \pi(q^2)]} , \tag{10.5.17}$$

where

$$\tilde{\xi}(q^2) = \xi(q^2)[1 - \pi(q^2)] + \pi(q^2) . \tag{10.5.18}$$

Now, because $\Pi^*_{\mu\nu}(q)$ receives contributions only from one-photon-irreducible graphs, it is expected not to have any pole at $q^2 = 0$. (There is an important exception in the case of broken gauge symmetry, to be discussed in Volume II.) In particular, the absence of poles at $q^2 = 0$ in the $q_\mu q_\nu$ term in $\Pi^*_{\mu\nu}(q)$ tells us that the function $\pi(q^2)$ in Eq. (10.5.16) also has no such pole, and so the pole in the complete propagator (10.5.17) is still at $q^2 = 0$, indicating that *radiative corrections do not give the photon a mass.*

For a renormalized electromagnetic field, radiative corrections should also not alter the gauge-invariant part of the residue of the photon pole in Eq. (10.5.17), so

$$\pi(0) = 0 \,. \tag{10.5.19}$$

This condition leads to a determination of the electromagnetic field renormalization constant Z_3. Recall that when expressed in terms of the renormalized field (10.4.17), the electrodynamic Lagrangian takes the form

$$\mathscr{L} = -\tfrac{1}{4}Z_3(\partial_\mu A_\nu - \partial_\nu A_\mu)(\partial^\mu A^\nu - \partial^\nu A^\mu) + \mathscr{L}_M\left(\Psi_\ell, [\partial_\mu - iq_\ell A_\mu]\Psi_\ell\right) \,.$$

The function $\pi(q^2)$ in the one-photon-irreducible amplitude is then

$$\pi(q^2) = 1 - Z_3 + \pi_{\rm LOOP}(q^2) \,, \tag{10.5.20}$$

where $\pi_{\rm LOOP}$ is the contribution of loop diagrams. It follows that

$$Z_3 = 1 + \pi_{\rm LOOP}(0) \,. \tag{10.5.21}$$

In practice, we just calculate the loop contributions and subtract a constant in order to make $\pi(0)$ vanish.

Incidentally, Eq. (10.5.18) shows that for $q^2 \neq 0$ the gauge term in the photon propagator *is* altered by radiative corrections. The one exception is the case of Landau gauge, for which $\tilde{\xi} = \xi = 1$ for all q^2.

10.6 Electromagnetic Form Factors and Magnetic Moment

Suppose that we want to calculate the scattering of a particle by an external electromagnetic field (or by the electromagnetic field of another particle), to first order in this electromagnetic field, but to all orders in all other interactions (including electromagnetic) of our particle. For this purpose, we need to know the sum of the contributions of all Feynman diagrams with one incoming and one outgoing particle line, both on the mass shell, plus a photon line, which may be on or off the mass shell. According to the theorem of Section 6.4, this sum is given by the one-particle matrix element of the electromagnetic current $J^\mu(x)$. Let us see what governs the general form of this matrix element.

According to spacetime translation invariance, the one-particle matrix element of the electromagnetic current takes the form

$$\left(\Psi_{\mathbf{p}',\sigma'},\, J^\mu(x)\Psi_{\mathbf{p},\sigma}\right) = \exp(i(p-p')\cdot x)\left(\Psi_{\mathbf{p}',\sigma'}, J^\mu(0)\Psi_{\mathbf{p},\sigma}\right) \,. \tag{10.6.1}$$

The current conservation condition $\partial_\mu J^\mu = 0$ then requires

$$(p'-p)_\mu\left(\Psi_{\mathbf{p}',\sigma'},\, J^\mu(0)\Psi_{\mathbf{p},\sigma}\right) = 0 \,. \tag{10.6.2}$$

Also, setting $\mu = 0$ and integrating over all $\mathbf{x}$ gives

$$\left(\Psi_{\mathbf{p}',\sigma'},\, Q\Psi_{\mathbf{p},\sigma}\right) = (2\pi)^3\delta^3(\mathbf{p}-\mathbf{p}')\left(\Psi_{\mathbf{p}',\sigma'},\, J^0(0)\Psi_{\mathbf{p},\sigma}\right) .$$

Using Eq. (10.4.8), this gives

$$\left(\Psi_{\mathbf{p},\sigma'},\, J^0(0)\Psi_{\mathbf{p},\sigma}\right) = (2\pi)^{-3}q\,\delta_{\sigma'\sigma}\,, \tag{10.6.3}$$

where q is the particle charge.

We also have at our disposal the constraints on the current matrix elements imposed by Lorentz invariance. To explore these, we will limit ourselves to the simplest cases: spin zero and spin $\frac{1}{2}$. The analysis presented here provides an example of techniques that are useful for other currents, such as those of the semi-leptonic weak interactions.

Spin Zero

For spin zero, Lorentz invariance requires the one-particle matrix element of the current to take the general form

$$\left(\Psi_{\mathbf{p}'},\, J^\mu(0)\Psi_{\mathbf{p}}\right) = q(2\pi)^{-3}(2p'^0)^{-1/2}(2p^0)^{-1/2}\mathscr{J}^\mu(p',p), \tag{10.6.4}$$

where p^0 and p'^0 are the mass-shell energies ($p^0 = \sqrt{\mathbf{p}^2+m^2}$), and $\mathscr{J}^\mu(p',p)$ is a four-vector function of the two four-vectors p'^μ and p^μ. (We have extracted a factor of the charge q of the particle from $\mathscr{J}$ for future convenience.) Obviously, the most general such four-vector function takes the form of a linear combination of p'^μ and p^μ, or equivalently of $p'^\mu + p^\mu$ and $p'^\mu - p^\mu$, with scalar coefficients. But the scalars p^2 and p'^2 are fixed at the values $p^2 = p'^2 = -m^2$, so the scalar variables that can be formed from p^μ and p'^μ can be taken as functions only of $p \cdot p'$, or equivalently of

$$k^2 \equiv (p-p')^2 = -2m^2 - 2p\cdot p' . \tag{10.6.5}$$

Thus the function $\mathscr{J}^\mu(p',p)$ must take the form

$$\mathscr{J}^\mu(p',p) = (p'+p)^\mu F(k^2) + i\,(p'-p)^\mu H(k^2) . \tag{10.6.6}$$

The fact that J^μ is Hermitian implies that $\mathscr{J}^\mu(p',p)^* = \mathscr{J}^\mu(p,p')$, so that both $F(k^2)$ and $H(k^2)$ are real.

Now $(p'-p)\cdot(p'+p)$ vanishes, while $(p'-p)^2 = k^2$ is not generally zero, so the condition of current conservation is simply

$$H(k^2) = 0 . \tag{10.6.7}$$

Also, setting $\mathbf{p}' = \mathbf{p}$ and $\mu = 0$ in Eq. (10.6.4), and comparing with Eq. (10.6.3), we find that

$$F(0) = 1 . \tag{10.6.8}$$

The function $F(k^2)$ is called the *electromagnetic form factor* of the particle.

Spin $\frac{1}{2}$

For spin $\frac{1}{2}$, Lorentz invariance requires the one-particle matrix element of the current to take the general form

$$\left(\Psi_{\mathbf{p}',\sigma'}, J^\mu(0)\Psi_{\mathbf{p},\sigma}\right) = iq\,(2\pi)^{-3}\,\bar{u}(\mathbf{p}',\sigma')\Gamma^\mu(p',p)u(\mathbf{p},\sigma) \qquad (10.6.9)$$

where Γ^μ is a four-vector 4×4 matrix function of p^ν, p'^ν, and γ^ν, and u is the usual Dirac coefficient function. We have extracted a factor iq to make the normalization of Γ^μ the same as in the previous section.

Just as for any 4×4 matrix, we may expand Γ^μ in the 16 covariant matrices 1, γ_ρ, $[\gamma_\rho,\gamma_\sigma]$, $\gamma_5\gamma_\rho$, and γ_5. The most general four-vector Γ^μ can therefore be written as a linear combination of

$$\begin{array}{rl} 1: & p^\mu\,,\ p'^\mu \\ \gamma_\rho: & \gamma^\mu\,,\ p^\mu\,\not{p}\,,\ p'^\mu\,\not{p}\,,\ p^\mu\,\not{p}',\ p'^\mu\,\not{p}' \\ [\gamma_\rho,\gamma_\sigma]: & [\gamma^\mu,\,\not{p}],\ [\gamma^\mu,\,\not{p}'],\ [\,\not{p},\,\not{p}']p^\mu,\ [\,\not{p},\,\not{p}']p'^\mu \\ \gamma_5\gamma_\rho: & \gamma_5\gamma_\rho\,\epsilon^{\rho\mu\nu\sigma}p_\nu p'_\sigma \\ \gamma_5: & \text{NONE} \end{array}$$

with the coefficient of each term a function of the only scalar variable in the problem, the quantity (10.6.5). This can be greatly simplified by using the Dirac equations satisfied by u and $\bar{u}$:

$$\bar{u}(\mathbf{p}',\sigma')\,(i\,\not{p}'+m) = 0\,, \qquad (i\,\not{p}+m)\,u(\mathbf{p},\sigma) = 0\,.$$

In consequence, we can drop* all but the first three entries: p^μ, p'^μ, and γ^μ. We conclude that, on the fermion mass shell, Γ^μ may be expressed as a linear combination of γ^μ, p^μ, and p'^μ, which we choose to write as

$$\bar{u}(\mathbf{p}',\sigma')\Gamma^\mu(p',p)u(\mathbf{p},\sigma) = \bar{u}(\mathbf{p}',\sigma')\Big[\gamma^\mu F(k^2)$$
$$-\frac{i}{2m}\,(p+p')^\mu G(k^2) + \frac{(p-p')^\mu}{2m}H(k^2)\Big]\,u(\mathbf{p},\sigma). \qquad (10.6.10)$$

---

* This is obvious for the terms $p^\mu\,\not{p}$, $p'^\mu\,\not{p}$, $p^\mu\,\not{p}'$, and $p'^\mu\,\not{p}'$, which may be replaced respectively with imp^μ, imp'^μ, imp^μ, and imp'^μ, which are the same as terms already on our list. Also, we can write

$$[\gamma^\mu,\,\not{p}] = 2\gamma^\mu\,\not{p} - \{\gamma^\mu,\,\not{p}\} = 2\gamma^\mu\,\not{p} - 2p^\mu\,,$$

which may be replaced with $2im\gamma^\mu - 2p^\mu$, a linear combination of terms already on our list. The same applies to $[\gamma^\mu,\,\not{p}']$. Also,

$$[\,\not{p},\,\not{p}'] = -2\,\not{p}'\,\not{p} + \{\,\not{p},\,\not{p}'\} = -2\,\not{p}'\,\not{p} + 2p\cdot p'\,,$$

which may be replaced with $2m^2 + 2p\cdot p' = -k^2$. Hence the terms $[\,\not{p},\,\not{p}']p^\mu$ and $[\,\not{p},\,\not{p}']p'^\mu$ give nothing new. Finally, to deal with the last term we may use the relation

$$\gamma_5\gamma_\rho\,\epsilon^{\rho\mu\nu\sigma} = \tfrac{1}{6}i\Big(\gamma^\mu\gamma^\nu\gamma^\sigma + \gamma^\sigma\gamma^\mu\gamma^\nu + \gamma^\nu\gamma^\sigma\gamma^\mu - \gamma^\nu\gamma^\mu\gamma^\sigma - \gamma^\mu\gamma^\sigma\gamma^\nu - \gamma^\sigma\gamma^\nu\gamma^\mu\Big)\,.$$

Contracting this with p_ν and p'_σ and then moving all $\not{p}$ factors to the right and $\not{p}'$ factors to the left again gives a linear combination of p^μ, p'^μ, and γ^μ.

The hermiticity of $J^\mu(0)$ implies that

$$\beta\,\Gamma^{\mu\dagger}(p',p)\beta = -\Gamma^\mu(p,p')\,, \tag{10.6.11}$$

so that $F(k^2)$, $G(k^2)$, and $H(k^2)$ must all be *real* functions of k^2.

The conservation condition (10.6.2) is automatically satisfied by the first two terms in Eq. (10.6.10), because

$$(p'-p)_\mu\gamma^\mu = -i\Big[(i\,\not{p}'+m)-(i\,\not{p}+m)\Big]$$

and

$$(p'-p)\cdot(p'+p) = p'^2 - p^2\,.$$

On the other hand, $(p'-p)^2$ does not in general vanish, so current conservation requires the third term to vanish

$$H(k^2) = 0\,. \tag{10.6.12}$$

Also, letting $\mathbf{p}'\to\mathbf{p}$ in Eqs. (10.6.9) and (10.6.10), we find

$$\Big(\Psi_{\mathbf{p},\sigma'}\,J^\mu(0)\Psi_{\mathbf{p},\sigma}\Big) = i\,q(2\pi)^{-3}\bar{u}(\mathbf{p},\sigma')\Big[\gamma^\mu F(0) - \frac{i}{m}\,p^\mu G(0)\Big]\,u(\mathbf{p},\sigma)\,.$$

Using the identity $\{\gamma^\mu, i\,\not{p}+m\} = 2m\gamma^\mu + 2ip^\mu$, we also have

$$\bar{u}(\mathbf{p},\sigma')\gamma^\mu u(\mathbf{p},\sigma) = -\frac{ip^\mu}{m}\,\bar{u}(\mathbf{p},\sigma')u(\mathbf{p},\sigma)\,.$$

Recall also that

$$\bar{u}(\mathbf{p},\sigma')u(\mathbf{p},\sigma) = \delta_{\sigma'\sigma}m/p^0$$

and therefore

$$\Big(\Psi_{\mathbf{p},\sigma'},\,J^\mu(0)\Psi_{\mathbf{p},\sigma}\Big) = q(2\pi)^{-3}(p^\mu/p^0)\delta_{\sigma'\sigma}\Big[F(0)+G(0)\Big]\,. \tag{10.6.13}$$

Comparing this with Eq. (10.6.3) yields the normalization condition

$$F(0)+G(0) = 1\,. \tag{10.6.14}$$

It may be useful to note that the electromagnetic vertex matrix Γ^μ is commonly written in terms of two other matrices, as

$$\begin{aligned}\bar{u}(\mathbf{p}',\sigma')\Gamma^\mu(p',p)u(\mathbf{p},\sigma) = \bar{u}(\mathbf{p}',\sigma')\,\Big[\gamma^\mu F_1(k^2)\\ +\ \tfrac{1}{2}i\,[\gamma^\mu,\gamma^\nu]\,(p'-p)_\nu F_2(k^2)\Big]\,u(\mathbf{p},\sigma)\,.\end{aligned} \tag{10.6.15}$$

As already mentioned, we may rewrite the matrix appearing in the second term in terms of those used in defining $F(k^2)$ and $G(k^2)$:

$$\begin{aligned}&\bar{u}(\mathbf{p}',\sigma')\ \tfrac{1}{2}i\,[\gamma^\mu,\gamma^\nu]\,(p'-p)_\nu\,u(\mathbf{p},\sigma)\\ &= \bar{u}(\mathbf{p}',\sigma')\,\Big[-i\,\not{p}'\gamma^\mu + \tfrac{1}{2}i\{\gamma^\mu,\,\not{p}'\} - i\gamma^\mu\,\not{p} + \tfrac{1}{2}i\{\gamma^\mu,\,\not{p}\}\Big]\,u(\mathbf{p},\sigma)\\ &= \bar{u}(\mathbf{p}',\sigma')\,\Big[i(p'^\mu+p^\mu)+2m\gamma^\mu\Big]u(\mathbf{p},\sigma)\,.\end{aligned} \tag{10.6.16}$$

Comparing Eq. (10.6.15) with Eq. (10.6.10), we find

$$F(k^2) = F_1(k^2) + 2m\,F_2(k^2) \tag{10.6.17}$$

$$G(k^2) = -2m\,F_2(k^2)\,. \tag{10.6.18}$$

The normalization condition (10.6.14) now reads

$$F_1(0) = 1\,.$$

In order to evaluate the magnetic moment of our particle in terms of its form factors, let us consider the spatial part of the vertex function in the case of small momenta, $|\mathbf{p}|, |\mathbf{p}'| \ll m$. For this purpose, it is useful to use Eq. (10.6.16) to rewrite Eq. (10.6.10) (with $H = 0$) in a third form:

$$\bar{u}(\mathbf{p}',\sigma')\Gamma^\mu(p',p)u(\mathbf{p},\sigma) = \frac{-i}{2m}\bar{u}(\mathbf{p}',\sigma')\Big[(p+p')^\mu\{F(k^2)+G(k^2)\} - \frac{1}{2}[\gamma^\mu,\gamma^\nu]\,(p'-p)_\nu F(k^2)\Big]\,u(\mathbf{p},\sigma)\,. \tag{10.6.19}$$

For zero momenta the matrix elements of the commutators of Dirac matrices are given by (5.4.19) and (5.4.20) as

$$\bar{u}(0,\sigma')[\gamma^i,\gamma^j]u(0,\sigma) = 4i\epsilon_{ijk}\left(J_k^{(\frac{1}{2})}\right)_{\sigma',\sigma}\,, \qquad \bar{u}(0,\sigma')[\gamma^i,\gamma^0]u(0,\sigma) = 0\,,$$

where $\mathbf{J}^{(\frac{1}{2})} = \frac{1}{2}\boldsymbol{\sigma}$ is the angular momentum matrix for spin $\frac{1}{2}$. Hence to first order in the small momenta,

$$\bar{u}(\mathbf{p}',\sigma')\boldsymbol{\Gamma}(p',p)u(\mathbf{p},\sigma) \to \frac{-i}{2m}(\mathbf{p}+\mathbf{p}')\delta_{\sigma',\sigma} + \frac{1}{m}[(\mathbf{p}-\mathbf{p}')\times\mathbf{J}^{(\frac{1}{2})}]_{\sigma'\sigma}\,F(0)\,. \tag{10.6.20}$$

In a very weak time-independent external vector potential $A(\mathbf{x})$ the matrix element of the interaction Hamiltonian $H' = -\int d^3x\,\mathbf{J}(\mathbf{x})\cdot A(\mathbf{x})$ between one-particle states of small momentum is therefore

$$\begin{aligned}(\Psi_{\mathbf{p}',\sigma'}, H'\Psi_{\mathbf{p},\sigma}) &= \frac{-iqF(0)}{m(2\pi)^3}\int d^3x\,e^{i(\mathbf{p}-\mathbf{p}')\cdot\mathbf{x}}A(\mathbf{x})\cdot[(\mathbf{p}-\mathbf{p}')\times\mathbf{J}^{(\frac{1}{2})}]_{\sigma'\sigma} \\ &= -\frac{qF(0)}{m(2\pi)^3}\int d^3x\,e^{i(\mathbf{p}-\mathbf{p}')\cdot\mathbf{x}}(\mathbf{J}^{(\frac{1}{2})})_{\sigma'\sigma}\cdot\mathbf{B}(\mathbf{x})\,,\end{aligned} \tag{10.6.21}$$

where $\mathbf{B} = \nabla\times A$ is the magnetic field. Hence in the limit of a slowly varying weak magnetic field, the matrix element of the interaction Hamiltonian is

$$(\Psi_{\mathbf{p}',\sigma'}, H'\Psi_{\mathbf{p},\sigma}) = -\frac{qF(0)}{m}(\mathbf{J}^{(\frac{1}{2})})_{\sigma'\sigma}\cdot\mathbf{B}\,\delta^3(\mathbf{p}-\mathbf{p}')\,. \tag{10.6.22}$$

The magnetic moment μ for an arbitrary particle of general spin j is defined by the statement that the matrix element of the interaction of the particle with a weak static slowly varying magnetic field is

$$(\Psi_{\mathbf{p}',\sigma'}, H'\Psi_{\mathbf{p},\sigma}) = -\frac{\mu}{j}(\mathbf{J}^{(j)})_{\sigma'\sigma}\cdot\mathbf{B}\,\delta^3(\mathbf{p}-\mathbf{p}')\,. \tag{10.6.23}$$

Hence Eq. (10.6.22) gives the magnetic moment of a particle of charge q, mass m, and spin $\frac{1}{2}$ as:

$$\mu = \frac{qF(0)}{2m} \,. \tag{10.6.24}$$

This contains as a special case the celebrated Dirac result[7] $\mu = q/2m$ for a spin $\frac{1}{2}$ particle without radiative corrections.

We mention without proof that the form factors $F(k^2)$ and $G(k^2)$ of the proton may be measured for $k^2 > 0$ by comparison of experimental data for electron–proton scattering with the Rosenbluth formula[8] for the laboratory frame differential cross-section:

$$\frac{d\sigma}{d\Omega} = \frac{e^4}{4(4\pi)^2 E_0^2} \frac{\cos^2(\theta/2)}{\sin^4(\theta/2)} \left[1 + \frac{2E_0}{m}\sin^2(\theta/2)\right]^{-1} \times \left\{\left(F(k^2) + G(k^2)\right)^2 + \frac{k^2}{4m^2}\left(2F^2(k^2)\tan^2(\theta/2) + G^2(k^2)\right)\right\} ,$$

where E_0 is the energy of the incident electron (taken here with $E_0 \gg m_e$); θ is the scattering angle; and

$$k^2 = \frac{4\,E_0^2 \sin^2(\theta/2)}{1 + (2E_0/m)\sin^2(\theta/2)} \,.$$

10.7 The Källen–Lehmann Representation*

We saw in Section 10.2 that the presence of one-particle intermediate states leads to poles in Fourier transforms of matrix elements of time-ordered products, like (10.2.1). Multi-particle intermediate states lead to more complicated singularities, which are difficult to describe in general. But in the special case of a vacuum expectation value involving just two operators, we have a convenient representation that explicitly displays the analytic structure of the Fourier transform. In particular, this representation may be used for propagators, where the two operators are the fields of elementary particles. When combined with the positivity requirements of quantum mechanics, this representation yields interesting bounds on the asymptotic behavior of propagators and the magnitude of renormalization constants.

Consider a complex scalar Heisenberg-picture operator $\Phi(x)$, which may or may not be an elementary particle field. The vacuum expectation

* This section lies somewhat out of the book's main line of development, and may be omitted in a first reading.

value of a product $\Phi(x)\Phi^\dagger(y)$ may be expressed as

$$\langle\Phi(x)\Phi^\dagger(y)\rangle_0 = \sum_n \langle 0|\Phi(x)|n\rangle\, \langle n|\Phi^\dagger(y)|0\rangle\,, \tag{10.7.1}$$

where the sum runs over any complete set of states. (Here the sum over n includes integrals over continuous labels as well as sums over discrete labels.) Choosing these states as eigenstates of the momentum four-vector P^μ, translational invariance tells us that

$$\begin{aligned}\langle 0|\Phi(x)|n\rangle &= \exp(ip_n\cdot x)\langle 0|\Phi(0)|n\rangle\,,\\ \langle n|\Phi^\dagger(y)|0\rangle &= \exp(-ip_n\cdot y)\langle n|\Phi^\dagger(0)|0\rangle\end{aligned} \tag{10.7.2}$$

and so

$$\langle\Phi(x)\Phi^\dagger(y)\rangle_0 = \sum_n \exp(ip_n\cdot(x-y))\,|\langle 0|\Phi(0)|n\rangle|^2\,. \tag{10.7.3}$$

It is convenient to rewrite this in terms of a *spectral function.* Note that the sum $\sum_n \delta^4(p-p_n)\,|\langle 0|\Phi(0)|n\rangle|^2$ is a scalar function of the four-vector p^μ, and therefore may depend only on p^2 and (for $p^2\leq 0$) on the step function $\theta(p^0)$. In fact, the intermediate states in Eq. (10.7.3) all have $p^2\leq 0$ and $p^0>0$, so this sum takes the form

$$\sum_n \delta^4(p-p_n)\,|\langle 0|\Phi(0)|n\rangle|^2 = (2\pi)^{-3}\,\theta(p^0)\,\rho(-p^2) \tag{10.7.4}$$

with $\rho(-p^2) = 0$ for $p^2>0$. (The factor $(2\pi)^{-3}$ is extracted from ρ for future convenience.) The spectral function $\rho(-p^2)$ is clearly real and positive. With this definition, we can rewrite Eq. (10.7.3) as

$$\begin{aligned}\langle\Phi(x)\Phi^\dagger(y)\rangle_0 &= (2\pi)^{-3}\int d^4p\ \exp[ip\cdot(x-y)]\ \theta(p^0)\,\rho(-p^2)\\ &= (2\pi)^{-3}\int d^4p\int_0^\infty d\mu^2\ \exp[ip\cdot(x-y)]\,\theta(p^0)\\ &\qquad\times\rho(\mu^2)\,\delta(p^2+\mu^2)\,.\end{aligned} \tag{10.7.5}$$

Interchanging the order of integration over p^μ and μ^2, this may be expressed as

$$\langle\Phi(x)\Phi^\dagger(y)\rangle_0 = \int_0^\infty d\mu^2\ \rho(\mu^2)\,\Delta_+(x-y;\mu^2)\,, \tag{10.7.6}$$

where Δ_+ is the familiar function

$$\Delta_+(x-y;\mu^2) \equiv (2\pi)^{-3}\int d^4p\ \exp[ip\cdot(x-y)]\,\theta(p^0)\,\delta(p^2+\mu^2)\,. \tag{10.7.7}$$

In just the same way, we can show that

$$\langle\Phi^\dagger(y)\Phi(x)\rangle_0 = \int_0^\infty d\mu^2\ \bar\rho(\mu^2)\,\Delta_+(y-x;\mu^2) \tag{10.7.8}$$

with a second spectral function $\bar{\rho}(\mu^2)$ defined by

$$\sum_n \delta^4(p - p_n)\, |\langle n|\Phi(0)|0\rangle|^2 = (2\pi)^{-3}\,\theta(p^0)\,\bar{\rho}(-p^2)\,. \tag{10.7.9}$$

We now make use of the causality requirement, that the commutator $[\Phi(x), \Phi^\dagger(y)]$ must vanish for space-like separations $x - y$. The vacuum expectation value of the commutator is

$$\langle[\Phi(x), \Phi^\dagger(y)]\rangle_0 = \int_0^\infty d\mu^2\, \Big(\rho(\mu^2)\,\Delta_+(x - y; \mu^2) - \bar{\rho}(\mu^2)\,\Delta_+(y - x; \mu^2)\Big)\,. \tag{10.7.10}$$

As noted in Section 5.2, for $x - y$ space-like the function $\Delta_+(x - y)$ does not vanish, but it does become *even.* In order for (10.7.10) to vanish for arbitrary space-like separations, it is thus necessary that

$$\rho(\mu^2) = \bar{\rho}(\mu^2)\,. \tag{10.7.11}$$

This is a special case of the CPT theorem, proved here without the use of perturbation theory; for whatever states with $p^2 = -\mu^2$ have the quantum numbers of the operator Φ, there must be corresponding states with $p^2 = -\mu^2$ that have the quantum numbers of the operator $\Phi^\dagger$.

Using Eq. (10.7.11), the vacuum expectation of the time-ordered product is

$$\langle T\left\{\Phi(x)\Phi^\dagger(y)\right\}\rangle_0 = -i\int_0^\infty d\mu^2\, \rho(\mu^2)\,\Delta_F(x - y; \mu^2)\,, \tag{10.7.12}$$

where $\Delta_F(x - y; \mu^2)$ is the Feynman propagator for a spinless particle of mass μ:

$$-\,i\Delta_F(x - y; \mu^2) \equiv \theta(x^0 - y^0)\Delta_+(x - y; \mu^2) - \theta(y^0 - x^0)\Delta_+(y - x; \mu^2)\,. \tag{10.7.13}$$

Borrowing the notation introduced in Section 10.3 for complete propagators, we introduce the momentum space function

$$-\,i\Delta'(p) \equiv \int d^4x\, \exp[-ip\cdot(x - y)]\,\langle T\left\{\Phi(x)\Phi^\dagger(y)\right\}\rangle_0\,. \tag{10.7.14}$$

Recall that

$$\int d^4x\, \exp[-ip\cdot(x - y)]\,\Delta_F(x - y; \mu^2) = \frac{1}{p^2 + \mu^2 - i\epsilon}\,. \tag{10.7.15}$$

This yields our spectral representation:[9]

$$\Delta'(p) = \int_0^\infty \rho(\mu^2)\,\frac{d\mu^2}{p^2 + \mu^2 - i\epsilon}\,. \tag{10.7.16}$$

One immediate consequence of this result and the positivity of $\rho(\mu^2)$ is that $\Delta'(p)$ cannot vanish for $|p^2| \to \infty$ faster** than the bare propagator $1/(p^2 + m^2 - i\epsilon)$. From time to time the suggestion is made to include higher derivative terms in the unperturbed Lagrangian, which would make the propagator vanish faster than $1/p^2$ for $|p^2| \to \infty$, but the spectral representation shows that this would necessarily entail a departure from the positivity postulates of quantum mechanics.

We can use the spectral representation together with equal-time commutation relations to derive an interesting sum rule for the spectral function. If $\Phi(x)$ is a conventionally normalized (not renormalized) canonical field operator, then

$$\left[\frac{\partial\Phi(\mathbf{x},t)}{\partial t}, \Phi^\dagger(\mathbf{y},t)\right] = -i\delta^3(\mathbf{x}-\mathbf{y})\,. \tag{10.7.17}$$

We note that

$$\frac{\partial}{\partial x^0}\Delta_+(x-y)\Big|_{x^0=y^0} = -i\delta^3(\mathbf{x}-\mathbf{y})$$

so the spectral representation (10.7.10) and the commutation relations (10.7.17) together tell us that

$$\int_0^\infty \rho(\mu^2)\, d\mu^2 = 1\,. \tag{10.7.18}$$

This implies that for $|p^2| \to \infty$, the momentum space propagator (10.7.16) of the unrenormalized fields has the free-field asymptotic behavior

$$\Delta'(p) \to \frac{1}{p^2}\,.$$

This result is only meaningful within a suitable scheme for regulating ultraviolet divergences; in perturbation theory the unrenormalized fields have infinite matrix elements, and their propagator is ill-defined.

Now consider the possibility that there is a one-particle state $|\mathbf{k}\rangle$ of mass m with a non-vanishing matrix element with the state $\langle 0|\Phi(0)$. Lorentz

** In fact, it is not even certain that $\Delta'(p)$ vanishes for $|p^2| \to \infty$ at all, even though this would seem to follow from the spectral representation. The problem arises from the interchange of the integrals over p^μ and μ^2. What is certain is that $\Delta'(p)$ is an analytic function of $-p^2$ with a discontinuity across the positive real axis $-p^2 = \mu^2$ given by $\pi\rho(\mu^2)$, as can be shown by the methods of the next section. From this, it follows that $\Delta'(p)$ is given by a dispersion relation with spectral function $\rho(\mu^2)$ and possible subtractions:

$$\Delta'(p) = P(p^2) + (-p^2 + \mu_0^2)^n \int_0^\infty \frac{\rho(\mu^2)}{(\mu^2 + \mu_0^2)^n}\frac{d\mu^2}{p^2 + \mu^2 - i\epsilon}\,,$$

where n is a positive integer, μ_0^2 is an arbitrary positive constant, and $P(p^2)$ is a μ_0^2-dependent polynomial in p^2 of order $n-1$ that is absent for $n = 0$.

invariance requires this matrix element to take the form

$$\langle 0|\Phi(0)|\mathbf{k}\rangle = (2\pi)^{-3/2}\left(2\sqrt{\mathbf{k}^2+m^2}\right)^{-1/2} N\,, \tag{10.7.19}$$

where N is a constant. According to the general results of Section 10.3, the propagator $\Delta'(p)$ of the unrenormalized fields should have a pole at $p^2 \to -m^2$ with residue $Z \equiv |N|^2 > 0$. That is,

$$\rho(\mu^2) = Z\delta(\mu^2 - m^2) + \sigma(\mu^2)\,, \tag{10.7.20}$$

where $\sigma(\mu^2) \geq 0$ is the contribution of multi-particle states. Together with Eq. (10.7.18), this has the consequence that

$$1 = Z + \int_0^\infty \sigma(\mu^2)\, d\mu^2 \tag{10.7.21}$$

and so

$$Z \leq 1 \tag{10.7.22}$$

with the equality reached only for a free particle, for which $\langle 0|\Phi(x)$ has no matrix elements with multi-particle states.

Because Z is positive, Eq. (10.7.21) can also be regarded as providing an upper bound on the coupling of the field Φ to multi-particle states:

$$\int_0^\infty \sigma(\mu^2)\, d\mu^2 \leq 1 \tag{10.7.23}$$

with the equality reached for $Z = 0$. The limit $Z = 0$ has an interesting interpretation as a condition for a particle to be composite rather than elementary.[10] In this context, a 'composite' particle may be understood to be one whose field does not appear in the Lagrangian. Consider such a particle, say a neutral particle of spin zero, and suppose that its quantum numbers allow it to be destroyed by an operator $F(\Psi)$ constructed out of other fields. We can freely introduce a field Φ for this particle by adding a term to the Lagrangian density of the form† $\Delta\mathscr{L} = (\Phi - F(\Psi))^2$, because the path integral over Φ can be done by setting it equal to the stationary point $\Phi = F(\Psi)$, at which $\Delta\mathscr{L} = 0$. But suppose instead we write $\Delta\mathscr{L} = \Delta\mathscr{L}_0 + \Delta\mathscr{L}_1$, where $\Delta\mathscr{L}_0 \equiv -\frac{1}{2}\partial_\mu\Phi\partial^\mu\Phi - \frac{1}{2}m^2\Phi^2$ is the usual free-field Lagrangian, and treat $\Delta\mathscr{L}_1 \equiv \Delta\mathscr{L} - \Delta\mathscr{L}_0$ as an interaction. A term $\frac{1}{2}\partial_\mu\Phi\partial^\mu\Phi$ in the interaction is nothing new. We encountered such a term in Eq. (10.3.12), multiplied by a factor $(1-Z)$; the only new thing is that here $Z = 0$. Instead of adjusting Z to satisfy the field renormalization condition $\Pi^{*\prime}(0) = 0$, here we must regard this as a condition on the

† This is known in condensed matter physics as a 'Hubbard–Stratonovich transformation'.[11] It will be used to introduce fields for pairs of electrons in our discussion of superconductivity in Volume II.

coupling constants of the composite particle. Unfortunately, it has not been possible to implement this procedure in quantum field theories, because as we have seen $Z = 0$ means that the particle couples as strongly as possible to its constituents, and this rules out the use of perturbation theory. The condition $Z = 0$ does prove useful in non-relativistic quantum mechanics; for instance, it fixes the coupling of the deuteron to the neutron and proton.[12]

Although the spectral representation has been derived here only for a spinless field, it is easy to generalize these results to other fields. Indeed, in the next chapter we shall show that to order e^2, the Z-factor for the electromagnetic field (conventionally called Z_3) is given by

$$Z_3 = 1 - \frac{e^2}{12\pi^2} \ln\left(\frac{\Lambda^2}{m_e^2}\right)$$

(where $\Lambda \gg m_e$ is an ultraviolet cutoff), in agreement with the bound (10.7.22).

10.8 Dispersion Relations*

The failure of early attempts to apply perturbative quantum field theory to the strong and weak nuclear forces had led theorists by the late 1950s to attempt the use of the analyticity and unitarity of scattering amplitudes as a way of deriving general non-perturbative results that would not depend on any particular field theory. This started with a revival of interest in dispersion relations. In its original form,[13] a dispersion relation was a formula giving the real part of the index of refraction in terms of an integral over its imaginary part. It was derived from an analyticity property of the index of refraction as a function of frequency, which followed from the condition that electromagnetic signals in a medium cannot travel faster than light in a vacuum. By expressing the index of refraction in terms of the forward photon scattering amplitude, the dispersion relation could be rewritten as a formula for the real part of the forward scattering amplitude as an integral of its imaginary part, and hence via unitarity in terms of the total cross-section. One of the exciting things about this relation was that it provided an alternative to conventional perturbation theory; given the scattering amplitude to order e^2, one could calculate the cross-section and the imaginary part of the scattering amplitude to order e^4, and then use the dispersion relation to

* This section lies somewhat out of the book's main line of development, and may be omitted in a first reading.

calculate the real part of the forward scattering amplitude to this order, without having ever to calculate a loop graph.

The modern approach to dispersion relations began in 1954 with the work of Gell-Mann, Goldberger, and Thirring.[14] Instead of considering the propagation of light in a medium, they derived the analyticity of the scattering amplitude directly from the condition of microscopic causality, which states that commutators of field operators vanish when the points at which the operators are evaluated are separated by a space-like interval. This approach allowed Goldberger[15] soon thereafter to derive a very useful dispersion relation for the forward pion–nucleon scattering amplitude.

To see how to use the principle of microscopic causality, consider the forward scattering in the laboratory frame of a massless boson of any spin on an arbitrary target α of mass $m_\alpha > 0$ and $\mathbf{p}_\alpha = 0$. (This has important applications to the scattering not only of photons but also pions in the limit $m_\pi = 0$, to be discussed in Volume II.) By a repeated use of Eq. (10.3.4) or the Lehmann–Symanzik–Zimmerman theorem,[3] the S-matrix element here is

$$S = \frac{1}{(2\pi)^3\sqrt{4\omega\omega'}|N|^2} \lim_{k^2\to 0} \lim_{k'^2\to 0} \times \int d^4x \int d^4y \, e^{-ik'\cdot y} e^{ik\cdot x} (i\Box_y)(i\Box_x) \langle\alpha| T\{A^\dagger(y), A(x)\}|\alpha\rangle \,. \quad (10.8.1)$$

Here k and k' are the initial and final boson four-momenta, with $\omega = k^0$, $\omega' = k'^0$; $A(x)$ is any Heisenberg-picture operator with a non-vanishing matrix element $\langle \mathrm{VAC}|A(x)|k\rangle = (2\pi)^{-3/2}(2\omega)^{-1/2}Ne^{ik\cdot x}$ between the one-boson state $|k\rangle$ and the vacuum; and N is the constant in this matrix element. In photon scattering $A(x)$ would be one of the transverse components of the electromagnetic field, while for massless pion scattering it would be a pseudoscalar function of hadron fields. The differential operators $-i\Box_x$ and $-i\Box_y$ are inserted to supply factors of ik'^2 and ik^2 that are needed to cancel the external line boson propagators. Letting these operators act on $A^\dagger(y)$ and $A(x)$, we have

$$S = \frac{-1}{(2\pi)^3\sqrt{4\omega\omega'}|N|^2} \lim_{k^2\to 0} \lim_{k'^2\to 0} \times \int d^4x \int d^4y \, e^{-ik'\cdot y} e^{ik\cdot x} \langle\alpha| T\{J^\dagger(y), J(x)\}|\alpha\rangle \, + \mathrm{ETC} \,, \quad (10.8.2)$$

where $J(x) \equiv \Box_x A(x)$, and 'ETC' denotes the Fourier transform of equal time commutator terms arising from the derivative acting on the step functions in the time-ordered product. The commutators of operators like $A(x)$ and $A^\dagger(y)$ (or their derivatives) vanish for $x^0 = y^0$ unless $\mathbf{x} = \mathbf{y}$, so the 'ETC' term is the Fourier transform of a differential operator acting on $\delta^4(x - y)$, and is hence a polynomial function of the boson four-momenta. We are concerned here with the analytic properties of the S-matrix element, so the details of this polynomial will be irrelevant.

Using translation invariance, Eq. (10.8.2) gives the S-matrix element as $S = -2\pi i\delta^4(k'-k)M(\omega)$, where

$$M(\omega) = \frac{-i}{2\omega|N|^2}\,F(\omega)\,, \tag{10.8.3}$$

$$F(\omega) \equiv \int d^4x\, e^{i\omega \ell\cdot x}\langle\alpha|T\{J^\dagger(0),\,J(x)\}|\alpha\rangle \;+\; \text{ETC}\,, \tag{10.8.4}$$

it now being understood that $k^\mu = \omega\ell^\mu$, where ℓ is a fixed four-vector with $\ell^\mu\ell_\mu = 0$ and $\ell^0 = 1$.

The time-ordered product can be rewritten in terms of commutators in two different ways:

$$\begin{aligned} T\{J^\dagger(0),J(x)\} &= \theta(-x^0)[J^\dagger(0),J(x)] + J(x)J^\dagger(0) \\ &= -\theta(x^0)[J^\dagger(0),J(x)] + J^\dagger(0)J(x)\,. \end{aligned} \tag{10.8.5}$$

Correspondingly, we can write

$$F(\omega) = F_A(\omega) + F_+(\omega) = F_R(\omega) + F_-(\omega)\,, \tag{10.8.6}$$

where

$$F_A(\omega) \equiv \int d^4x\;\theta(-x^0)\;\langle\alpha|[J^\dagger(0),\,J(x)]|\alpha\rangle\,e^{i\omega\ell\cdot x} \;+\; \text{ETC}\,, \tag{10.8.7}$$

$$F_R(\omega) \equiv -\int d^4x\;\theta(x^0)\;\langle\alpha|[J^\dagger(0),\,J(x)]|\alpha\rangle\,e^{i\omega\ell\cdot x} \;+\; \text{ETC}\,, \tag{10.8.8}$$

$$F_+(\omega) \equiv \int d^4x\;\langle\alpha|J(x)\,J^\dagger(0)|\alpha\rangle\,e^{i\omega\ell\cdot x}\,, \tag{10.8.9}$$

$$F_-(\omega) \equiv \int d^4x\;\langle\alpha|J^\dagger(0)\,J(x)|\alpha\rangle\,e^{i\omega\ell\cdot x}\,. \tag{10.8.10}$$

Microscopic causality tells us that the integrands in (10.8.7) and (10.8.8) vanish unless x^μ is within the light cone, and the step functions then require that x^μ is in the backward light cone in (10.8.7), so that $x\cdot\ell > 0$, and in the forward light cone in Eq. (10.8.8), so that $x\cdot\ell < 0$. We conclude that $F_A(\omega)$ is analytic for $\text{Im}\,\omega > 0$ and $F_R(\omega)$ is analytic for $\text{Im}\,\omega < 0$, because in both cases the factor $e^{i\omega\ell\cdot x}$ provides a cutoff for the integral over x^μ. (Recall that the 'ETC' term is a polynomial, and hence analytic at all finite points.) We may then define a function

$$\mathscr{F}(\omega) \equiv \begin{cases} F_A(\omega) & \text{Im}\,\omega > 0 \\ F_R(\omega) & \text{Im}\,\omega < 0 \end{cases} \tag{10.8.11}$$

which is analytic in the whole complex ω plane, except for a cut on the real axis.

We can now derive the dispersion relation. According to Eq. (10.8.6), the discontinuity of $\mathscr{F}(\omega)$ across the cut at any real E is

$$\mathscr{F}(E+i\epsilon)-\mathscr{F}(E-i\epsilon)=F_A(E)-F_R(E)=F_-(E)-F_+(E)\,. \tag{10.8.12}$$

If $\mathscr{F}(\omega)/\omega^n$ vanishes as $|\omega|\to\infty$ in the upper or lower half-plane, then by dividing by any polynomial $P(\omega)$ of order n we obtain a function that vanishes for $|\omega|\to\infty$ and is analytic except for the cut on the real axis and poles at the zeroes ω_ν of $P(\omega)$. (Where $\mathscr{F}(\omega)$ itself vanishes as $|\omega|\to\infty$, we can take $P(\omega)=1$.) According to the method of residues, we then have

$$\frac{\mathscr{F}(\omega)}{P(\omega)}+\sum_\nu\frac{\mathscr{F}(\omega_\nu)}{(\omega_\nu-\omega)\,P'(\omega_\nu)}=\frac{1}{2\pi i}\oint_C\frac{\mathscr{F}(z)\,dz}{(z-\omega)\,P(z)}\,, \tag{10.8.13}$$

where ω is any point off the real axis, and C is a contour consisting of two segments: one running just above the real axis from $-\infty+i\epsilon$ to $+\infty+i\epsilon$ and then around a large semi-circle back to $-\infty+i\epsilon$, and the other just below the real axis from $+\infty-i\epsilon$ to $-\infty-i\epsilon$ and then around a large semi-circle back to $+\infty-i\epsilon$. Because the function $\mathscr{F}(z)/P(z)$ vanishes for $|z|\to\infty$, we can neglect the contribution from the large semi-circles. Using Eq. (10.8.12), Eq. (10.8.13) becomes

$$\mathscr{F}(\omega)=Q(\omega)+\frac{P(\omega)}{2\pi i}\int_{-\infty}^{+\infty}\frac{F_-(E)-F_+(E)}{(E-\omega)\,P(E)}\,dE\,, \tag{10.8.14}$$

where $Q(\omega)$ is the $(n-1)$th-order polynomial

$$Q(\omega)\equiv-P(\omega)\sum_\nu\frac{\mathscr{F}(\omega_\nu)}{(\omega_\nu-\omega)\,P'(\omega_\nu)}\,.$$

A dispersion relation of this form, with $P(\omega)$ and $Q(\omega)$ of order n and $n-1$ respectively, is said to have n *subtractions*. If we can take $P=1$ then $Q=0$, and the dispersion relation is said to be unsubtracted.

If we now let ω approach the real axis from above, Eq. (10.8.14) gives

$$F_A(\omega)=Q(\omega)+\frac{P(\omega)}{2\pi i}\int_{-\infty}^{+\infty}\frac{F_-(E)-F_+(E)}{(E-\omega-i\epsilon)\,P(E)}\,dE\,. \tag{10.8.15}$$

Recalling Eqs. (10.8.6) and (3.1.25), this is

$$F(\omega)=Q(\omega)+\frac{1}{2}F_-(\omega)+\frac{1}{2}F_+(\omega)+\frac{P(\omega)}{2\pi i}\int_{-\infty}^{+\infty}\frac{F_-(E)-F_+(E)}{(E-\omega)\,P(E)}\,dE \tag{10.8.16}$$

with $1/(E-\omega)$ now interpreted as the principal value function $\mathscr{P}/(E-\omega)$.

This result is useful because the functions $F_\pm(E)$ may be expressed in terms of measurable cross-sections. Summing over a complete set of multi-particle intermediate states β in Eqs. (10.8.9) and (10.8.10) (including integrations over the momenta of the particles in β) and using translation invariance again, we have

$$F_+(E) = (2\pi)^4 \sum_\beta \left|\langle\beta|J(0)^\dagger|\alpha\rangle\right|^2 \delta^4(-p_\alpha + E\ell + p_\beta)\,, \tag{10.8.17}$$

$$F_-(E) = (2\pi)^4 \sum_\beta |\langle\beta|J(0)|\alpha\rangle|^2 \delta^4(p_\alpha + E\ell - p_\beta)\,. \tag{10.8.18}$$

But the matrix elements for the absorption of the massless scalar boson B in $B + \alpha \to \beta$ or its antiparticle B^c in $B^c + \alpha \to \beta$ are

$$-2i\pi M_{B^c+\alpha\to\beta} = \frac{(2\pi)^4}{(2\pi)^{3/2}\sqrt{2E_{B^c}}N}\langle\beta|J^\dagger(0)|\alpha\rangle\,, \tag{10.8.19}$$

$$-2i\pi M_{B+\alpha\to\beta} = \frac{(2\pi)^4}{(2\pi)^{3/2}\sqrt{2E_B}N}\langle\beta|J(0)|\alpha\rangle\,. \tag{10.8.20}$$

Comparing with Eq. (3.4.15), we see that $F_\pm(E)$ may be expressed in terms of total cross-sections** at energies $\mp E$:

$$F_+(E) = \theta(-E)\frac{2|E||N|^2}{(2\pi)^3}\sigma_{\alpha+B^c}(|E|)\,, \tag{10.8.21}$$

$$F_-(E) = \theta(E)\frac{2E|N|^2}{(2\pi)^3}\sigma_{\alpha+B}(E)\,. \tag{10.8.22}$$

The scattering amplitude (10.8.3) is now, for real $\omega > 0$,

$$\begin{aligned} M(\omega) = &\frac{-iQ(\omega)}{2\omega|N|^2} - \frac{i}{2(2\pi)^3}\sigma_{\alpha+B}(\omega) \\ &- \frac{P(\omega)}{\omega(2\pi)^4}\int_0^\infty \left[\frac{\sigma_{\alpha+B}(E)}{(E-\omega)P(E)} + \frac{\sigma_{\alpha+B^c}(E)}{(E+\omega)P(-E)}\right] E\,dE\,. \end{aligned} \tag{10.8.23}$$

It is more usual to express this dispersion relation in terms of the amplitude $f(\omega)$ for forward scattering in the laboratory frame, defined so that the laboratory frame differential cross-section in the forward direction is $|f(\omega)|^2$. This amplitude is given in terms of $M(\omega)$ by

** In some cases where selection rules allow the transition $\alpha \to \alpha + B$ and $\alpha \to \alpha + B^c$, the functions $F_\pm(E)$ also contain terms proportional to $\delta(E)$ arising from the contribution of the one-particle state α in the sum over intermediate states β. This does not occur for transversely polarized photons, or for pseudoscalar pions in the limit $m_\pi \to 0$.

$f(\omega) = -4\pi^2\omega\, M(\omega) = 2\pi^2 iF(\omega)/|N|^2$, so Eq. (10.8.23) now reads

$$f(\omega) = R(\omega) + \frac{i\omega}{4\pi}\sigma_{\alpha+B}(\omega) + \frac{P(\omega)}{4\pi^2}\int_0^\infty \left[\frac{\sigma_{\alpha+B}(E)}{(E-\omega)P(E)} + \frac{\sigma_{\alpha+B^c}(E)}{(E+\omega)P(-E)}\right] E\, dE\,,$$

where $R(\omega) \equiv 2i\pi^2 Q(\omega)/|N|^2$. The optical theorem (3.6.4) tells us that the second term on the right-hand side equals $i\text{Im}\, f(\omega)$, so this can just as well be written in the more conventional form

$$\text{Re}\, f(\omega) = R(\omega) + \frac{P(\omega)}{4\pi^2}\int_0^\infty \left[\frac{\sigma_{\alpha+B}(E)}{(E-\omega)P(E)} + \frac{\sigma_{\alpha+B^c}(E)}{(E+\omega)P(-E)}\right] E\, dE\,, \quad (10.8.24)$$

In particular, we see that $R(\omega)$ is real if we choose $P(\omega)$ real.

The forward scattering amplitude also satisfies an important symmetry condition. By changing the integration variable in Eqs. (10.8.7) and (10.8.8) from x to $-x$ and then using the translation-invariance property

$$\langle\alpha|[J^\dagger(0), J(-x)]|\alpha\rangle = \langle\alpha|[J^\dagger(x), J(0)]|\alpha\rangle$$

we see that for $\text{Im}\,\omega \leq 0$, $F_A(-\omega)$ is the same as $F_R(\omega)$, except for an interchange of J with $J^\dagger$. That is,

$$F_A(-\omega) = F_R^c(\omega) \quad \text{for} \quad \text{Im}\,\omega \leq 0\,,$$

where a superscript c indicates that the amplitude refers to the scattering of the antiparticle B^c on α. (We leave it to the reader to show that this relation is not upset by the equal-time commutator terms in Eqs. (10.8.7) and (10.8.8).) In the same way, we find

$$F_R(-\omega) = F_A^c(\omega) \quad \text{for} \quad \text{Im}\,\omega \geq 0\,,$$

and for real ω

$$F_\pm(-\omega) = F_\mp(\omega)\,.$$

Using these relations in (10.8.6), and recalling that $f(\omega)$ is proportional to $F(\omega)$, we find the *crossing symmetry* relation, that for real ω

$$f(-\omega) = f^c(\omega). \quad (10.8.25)$$

We are free to take $P(\omega)$ as any polynomial of sufficiently high order, but $R(\omega)$ then depends not only on $P(\omega)$ but also on the values of $\mathscr{F}(\omega)$ at the zeroes of $P(\omega)$. For $P(\omega)$ real and of nth order, the only free parameters in Eq. (10.8.16) are the n real coefficients in the real $(n-1)$th order polynomial $R(\omega)$. Hence Eq. (10.8.16) contains just n unknown real independent constants, the coefficients in the polynomial $R(\omega)$ for a given $P(\omega)$. We therefore wish to take the order n of the otherwise arbitrary polyomial $P(\omega)$ to be as small as possible.

We might try taking $P(\omega) = 1$, but this doesn't work. The analysis of Section 3.7 suggests that the forward scattering amplitude should grow like ω or perhaps as fast as $\omega \ln^2 \omega$. In this case for $f(\omega)/P(\omega)$ to vanish as $\omega \to \infty$, it is sufficient to take $P(\omega)$ as a second order polynomial, so that $R(\omega)$ is linear in ω. Choosing $P(E) = E^2$ for convenience, Eq. (10.8.24) then becomes

$$\mathrm{Re}\, f(\omega) = a + b\omega + \frac{\omega^2}{4\pi^2} \int_0^\infty \left[\frac{\sigma_{\alpha+B}(E)}{(E-\omega)} + \frac{\sigma_{\alpha+B^c}(E)}{(E+\omega)} \right] \frac{dE}{E}\,, \tag{10.8.26}$$

with a and b unknown real constants. The crossing symmetry condition (10.8.25) tells us that the corresponding constants in the dispersion relation for the antiparticle scattering amplitude $f^c(\omega)$ are

$$a^c = a\,, \qquad b^c = -b\,. \tag{10.8.27}$$

If we assume for instance that the cross-sections $\sigma_{\alpha+B}(E)$ and $\sigma_{\alpha+B^c}(E)$ behave for $E \to \infty$ as different constants times $(\ln E)^r$, then (10.8.26) would give

$$\mathrm{Re}\, f(\omega) \sim [\sigma_{\alpha+B}(\omega) - \sigma_{\alpha+B^c}(\omega)]\omega \ln \omega \sim \omega(\ln \omega)^{r+1} \tag{10.8.28}$$

so the real part of the scattering amplitude would grow faster than the imaginary part by a factor $\ln \omega$. This is implausible; we saw in Section 3.7 that the real part of the forward scattering amplitude is expected to become much smaller than the imaginary part for $\omega \to \infty$, as confirmed by experiment. We conclude that if $\sigma_{\alpha+B}(E)$ and $\sigma_{\alpha+B^c}(E)$ do behave for $E \to \infty$ as constants times $(\ln E)^r$ then the constants must be the same. Because we are concerned here with the high-energy limit, this result does not depend on the assumption that B is a massless boson, so in the same sense, *the ratio of the cross-sections of any particle and its antiparticle on a fixed target should approach unity at high energy.* This result is a somewhat generalized version of what is known as Pomeranchuk's theorem.[16] (Pomeranchuk considered only the case $r = 0$, while Section 3.7 and the observed behavior of cross-sections both suggest that $r = 2$ is more likely.)

Although Pomeranchuk took his estimates of the asymptotic behavior of scattering amplitudes from arguments like those of Section 3.7, today high energy behavior is usually inferred from Regge pole theory.[17] It would take us too far from our subject to go into details about this; suffice it to say that for hadronic processes the asymptotic behavior of $f(\omega)$ as ω goes to infinity is a sum over terms proportional to $\omega^{\alpha_n(0)}$, where $\alpha_n(t)$ are a set of 'Regge trajectories', each representing the exchange of an infinite family of different one-hadron states in the collision process. The leading trajectory (actually, a complex of many trajectories) in hadron–

hadron scattering is the 'Pomeron,' for which $\alpha(0)$ is close to unity. It is this trajectory that gives cross-sections that are approximately constant for $E \to \infty$. According to Pomeranchuk's theorem, the Pomeron couples equally to any hadron and its antiparticle. We can estimate $\alpha_n(0)$ for the lower Regge trajectories from the spectrum of hadronic states. A necessary though not sufficient condition[18] for a mesonic resonance of spin j to occur at a mass m is that m^2 equals the value of t where one of trajectories $\alpha_n(t)$ equals j. Apart from the Pomeron, the leading trajectory in pion–nucleon scattering is that on which we find the $j = 1$ ρ meson at $m = 770$ MeV, the $j = 3$ g meson at $m = 1690$ MeV, and a $j = 5$ meson at $m = 2350$ MeV. Extrapolating these values of $\alpha(t)$ down to $t = 0$, we can estimate that this trajectory has $\alpha(0) \approx 0.5$. This trajectory couples with opposite sign to π^+ and π^-, so for pion–nucleon scattering we expect $f(\omega) - f^c(\omega)$ to behave roughly like $\sqrt{\omega}$.

For photon scattering there is no distinction between B and B^c, so here Eq. (10.8.27) gives $b = 0$, and Eq. (10.8.26) reads

$$f(\omega) = a + \frac{\omega^2}{2\pi^2} \int_0^\infty \frac{\sigma(E)}{E^2 - \omega^2} dE \,. \tag{10.8.29}$$

This is essentially the original Kramers–Kronig[13] relation. As we shall see in Section 13.5, for a target of charge e and mass m the constant a has the known value $\mathrm{Re}\, f(0) = -e^2/m$.

Problems

1. Consider a neutral vector field $v_\mu(x)$. What conditions must be imposed on the sum $\Pi^*_{\mu\nu}(k)$ of one-particle-irreducible graphs with two external vector field lines in order that the field should be properly renormalized and describe a particle of renormalized mass m? How do we split the free-field and interacting terms in the Lagrangian to achieve this?

2. Derive the generalized Ward identity that governs the electromagnetic vertex function of a charged scalar field.

3. What is the most general form of the matrix element $\langle \mathbf{p}_2\sigma_2 | J^\mu(x) | \mathbf{p}_1\sigma_1 \rangle$ of the electromagnetic current $J^\mu(x)$ between two spin $\frac{1}{2}$ one-particle states of *different* masses m_1 and m_2 and equal parity? What if the parities were opposite? (Assume parity conservation throughout.)

4. Derive the spectral (Källen–Lehmann) representation for the vacuum expectation value $\langle T\{J^\mu(x)\, J^\nu(y)^\dagger\}\rangle_0$, where $J^\mu(x)$ is a complex conserved current.

5. Derive the spectral (Källen–Lehmann) representation for the vacuum expectation value $\langle T\{\psi_n(x)\,\bar{\psi}_m(y)\}\rangle_0$, where $\psi(x)$ is a Dirac field.

6. Without using any assumptions about the asymptotic behavior of the scattering amplitude or cross-sections, show that it is impossible for forward photon scattering amplitudes to satisfy unsubtracted dispersion relations.

7. Derive the spectral (Källen–Lehmann) representation for a complex scalar field by using the methods of dispersion theory.

8. Use dispersion theory and the results of Section 8.7 to calculate the amplitude for forward photon–electron scattering in the electron rest frame to order e^4.

References

1. W. H. Furry, *Phys. Rev.* **51**, 125 (1937).

2. H. Yukawa, *Proc. Phys.-Math. Soc. Japan* **17**, 48 (1935).

3. H. Lehmann, K. Symanzik, and W. Zimmerman, *Nuovo Cimento* **1**, 205 (1955).

4. Y. Takahashi, *Nuovo Cimento*, Ser. 10, **6**, 370 (1957).

5. J. C. Ward, *Phys. Rev.* **78**, 182 (1950).

6. J. Schwinger, *Phys. Rev. Lett.* **3**, 296 (1959).

7. P. A. M. Dirac, *Proc. Roy. Soc. (London)* **A117**, 610 (1928).

8. M. N. Rosenbluth, *Phys. Rev.* **79**, 615 (1950).

9. G. Källen, *Helv. Phys. Acta* **25**, 417 (1952); *Quantum Electrodynamics* (Springer-Verlag, Berlin, 1972); H. Lehmann, *Nuovo Cimento* **11**, 342 (1954).

10. J. C. Howard and B. Jouvet, *Nuovo Cimento* **18**, 466 (1960); M. J. Vaughan, R. Aaron, and R. D. Amado, *Phys. Rev.* **1254**, 1258 (1961); S. Weinberg, in *Proceedings of the 1962 High-Energy Conference at CERN* (CERN, Geneva, 1962): p. 683.

11. R. L. Stratonovich, *Sov. Phys. Dokl.* **2**, 416 (1957); J. Hubbard, *Phys. Rev. Lett.* **3**, 77 (1959).

12. S. Weinberg, *Phys. Rev.* **137**, B672 (1965).

13. H. A. Kramers, *Atti Congr. Intern. Fisici, Como* (Nicolo Zanichelli, Bologna, 1927); reprinted in H. A. Kramers, *Collected Scientific Papers* (North-Holland, Amsterdam, 1956); R. de Kronig, *Ned. Tyd. Nat. Kunde* **9**, 402 (1942); *Physica* **12**, 543 (1946); J. S. Toll, The Dispersion Relation for Light and its Application to Problems Involving Electron Pairs (Princeton University Ph. D. Thesis 1952). For historical reviews, see J. D. Jackson in *Dispersion Relations*, ed. by G. R. Screaton (Oliver and Boyd, Edinburgh, 1961); M. L. Goldberger in *Dispersion Relations and Elementary Particles*, ed. by C. De Witt and R. Omnes (Hermann, Paris, 1960).

14. M. Gell-Mann, M. L. Goldberger, and W. Thirring, *Phys. Rev.* **95**, 1612 (1954). The non-perturbative nature of this result was shown by M. L. Goldberger, *Phys. Rev.* **97**, 508 (1955).

15. M. L. Goldberger, *Phys. Rev.* **99**, 979 (1955).

16. I. Ia. Pomeranchuk, *J. Expt. Theor. Phys. (USSR.)* **34**, 725 (1958). English version: *Soviet Physics - JETP* **34**(7), 499 (1958). For a generalization, see S. Weinberg, *Phys. Rev.* **124**, 2049 (1961).

17. See, e.g., P. D. B. Collins, *An Introduction to Regge Theory and High Energy Physics* (Cambridge University Press, Cambridge, 1977). The original references are T. Regge, *Nuovo Cimento* **14**, 951 (1959): **18**, 947 (1960).

18. The graph of spin versus squared mass is known as a Chew–Frautschi plot; see G. F. Chew and S. C. Frautschi, *Phys. Rev. Lett.* **8**, 41 (1962).

11

One-Loop Radiative Corrections in Quantum Electrodynamics

In this chapter we shall proceed to carry out some of the classic one-loop calculations in the theory of charged leptons — massive spin $\frac{1}{2}$ particles that interact only with the electromagnetic field. There are three known species or 'flavors' of leptons: the electron and muon, and the heavier, more recently discovered tauon. For definiteness we shall refer to the charged particles in our calculations here as 'electrons,' though most of our calculations will apply equally to muons and tauons. After some generalities in Section 11.1, we will move on to the calculation of the vacuum polarization in Section 11.2, the anomalous magnetic moment of the electron in Section 11.3, and the electron self-energy in Section 11.4. Along the way, we will introduce a number of the mathematical techniques that prove useful in such calculations, including the use of Feynman parameters, Wick rotation, and both the dimensional regularization of 't Hooft and Veltman and the older regularization method of Pauli and Villars. Although we shall encounter infinities, it will be seen that the final results are finite if expressed in terms of the renormalized charge and mass. In the next chapter we shall extend what we have learned here about renormalization to general theories in arbitrary orders of perturbation theory.

11.1 Counterterms

The Lagrangian density for electrons and photons is taken in the form*

$$\mathscr{L} = -\tfrac{1}{4}F_{\rm B}^{\mu\nu}F_{{\rm B}\,\mu\nu} - \bar{\psi}_{\rm B}\left[\gamma_\mu\left[\partial^\mu + ie_{\rm B}A_{\rm B}^\mu\right] + m_{\rm B}\right]\psi_{\rm B} \tag{11.1.1}$$

where $F_{\rm B}^{\mu\nu} \equiv \partial^\mu A_{\rm B}^\nu - \partial^\nu A_{\rm B}^\mu$; $A_{\rm B}^\mu$ and $\psi_{\rm B}$ are the bare (i.e., unrenormalized) fields of the photon and electron, and $-e_{\rm B}$ and $m_{\rm B}$ are the bare charge and

* In this chapter we will not be making transformations between Heisenberg- and interaction-picture operators, so we shall return to a conventional notation, in which an upper case A and a lower case ψ are used to denote the photon and charged particle fields, respectively.

mass of the electron. As described in the previous chapter, we introduce renormalized fields and charge and mass:

$$\psi \equiv Z_2^{-1/2}\psi_{\rm B}\,, \tag{11.1.2}$$

$$A^\mu \equiv Z_3^{-1/2}A^\mu_{\rm B}\,, \tag{11.1.3}$$

$$e \equiv Z_3^{+1/2}e_{\rm B}\,, \tag{11.1.4}$$

$$m \equiv m_{\rm B} + \delta m\,, \tag{11.1.5}$$

with the constants Z_2, Z_3, and δm adjusted so that the propagators of the renormalized fields have poles in the same position and with the same residues as the propagators of the free fields in the absence of interactions. The Lagrangian may then be written in terms of renormalized quantities, as

$$\mathscr{L} = \mathscr{L}_0 + \mathscr{L}_1 + \mathscr{L}_2\,, \tag{11.1.6}$$

where

$$\mathscr{L}_0 = -\tfrac{1}{4}F^{\mu\nu}F_{\mu\nu} - \bar\psi\left[\gamma_\mu\partial^\mu + m\right]\psi\,, \tag{11.1.7}$$

$$\mathscr{L}_1 = -ieA_\mu\bar\psi\gamma^\mu\psi\,, \tag{11.1.8}$$

and $\mathscr{L}_2$ is a sum of 'counterterms'

$$\mathscr{L}_2 = -\tfrac{1}{4}(Z_3-1)F^{\mu\nu}F_{\mu\nu} - (Z_2-1)\bar\psi\left[\gamma_\mu\partial^\mu + m\right]\psi$$
$$+Z_2\delta m\bar\psi\,\psi - ie(Z_2-1)A_\mu\bar\psi\gamma^\mu\psi\,. \tag{11.1.9}$$

It will turn out that all of the terms in $\mathscr{L}_2$ are of second order and higher order in e, and that these terms just suffice to cancel the ultraviolet divergences that arise from loop graphs.

11.2 Vacuum Polarization

We now begin our first calculation of a radiative correction involving loop graphs, the so-called vacuum polarization effect, consisting of the corrections to the propagator associated with an internal photon line. Vacuum polarization produces measurable shifts in the energy levels of hydrogen, and makes an important correction to the energies of muons bound in atomic orbits around heavy nuclei. Also, as we shall see in Volume II, the calculation of the vacuum polarization provides a key

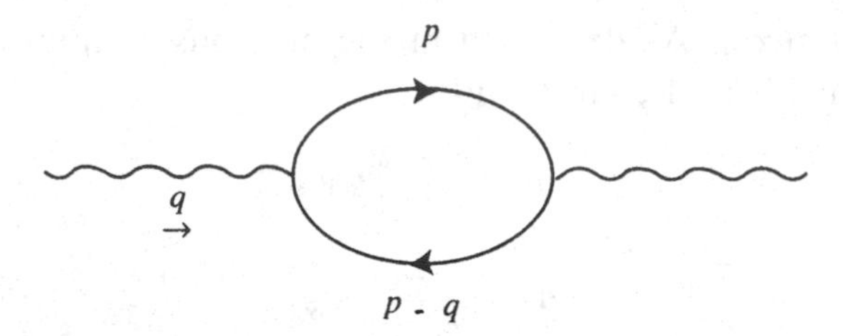

Figure 11.1. The one-loop diagram for the vacuum polarization in quantum electrodynamics. Here wavy lines represent photons; lines carrying arrows represent electrons.

element in the calculation of the high energy behavior of electrodynamics and other gauge theories.

As in Section 10.5, we define $i(2\pi)^4\Pi^{*\mu\nu}(q)$ as the sum of all connected graphs with two external photon lines with polarization indices μ and ν and carrying four-momentum q into and out of the diagram, not including photon propagators for the two external lines, and with the asterisk indicating that we exclude diagrams that can be disconnected by cutting through some internal photon line. The complete photon propagator $\Delta'^{\mu\nu}(q)$ is given by Eq. (10.5.13):

$$\Delta' = \Delta\,[1 - \Pi^*\Delta]^{-1}\,, \tag{11.2.1}$$

where $\Delta^{\mu\nu}(q)$ is the photon propagator without radiative corrections. Our task here is to calculate the leading contributions to $\Pi^{*\rho\sigma}(q)$.

In lowest order there is a one-loop contribution to Π^*, corresponding to the diagram in Figure 11.1:

$$\begin{aligned} i(2\pi)^4\Pi^{*\rho\sigma}_{1\text{ loop}}(q) = &-\int d^4p\,\mathrm{Tr}\left\{\left[\frac{-i}{(2\pi)^4}\,\frac{-i\not{p}+m}{p^2+m^2-i\epsilon}\right]\right.\\ &\times\left[(2\pi)^4 e\gamma^\rho\right]\left[\frac{-i}{(2\pi)^4}\,\frac{-i(\not{p}-\not{q})+m}{(p-q)^2+m^2-i\epsilon}\right]\left.\left[(2\pi)^4 e\gamma^\sigma\right]\right\} \end{aligned} \tag{11.2.2}$$

with the first minus sign on the right required by the presence of a fermion loop. More simply, this is

$$\Pi^{*\rho\sigma}_{1\text{ loop}}(q) = \frac{-ie^2}{(2\pi)^4}\int d^4p\,\frac{\mathrm{Tr}\{[-i\not{p}+m]\gamma^\rho\,[-i(\not{p}-\not{q})+m]\,\gamma^\sigma\}}{(p^2+m^2-i\epsilon)\,((p-q)^2+m^2-i\epsilon)}\,. \tag{11.2.3}$$

The first step in doing this integral is to use a trick introduced by Feynman.[1] We use the elementary formula

$$\frac{1}{AB} = \int_0^1 \frac{dx}{[(1-x)A+xB]^2} \tag{11.2.4}$$

to write the product of scalar propagators in Eq. (11.2.3) as

$$\frac{1}{(p^2+m^2-i\epsilon)((p-q)^2+m^2-i\epsilon)} = \int_0^1 \Big[(p^2+m^2-i\epsilon)(1-x) + \Big((p-q)^2+m^2-i\epsilon\Big)x\Big]^{-2} dx$$
$$= \int_0^1 \Big[p^2+m^2-i\epsilon-2p\cdot qx+q^2x\Big]^{-2} dx$$
$$= \int_0^1 \Big[(p-qx)^2+m^2-i\epsilon+q^2x(1-x)\Big]^{-2} dx\,.$$

(This is a special case of a class of integrals given in the Appendix to this chapter.) We can now shift the variable of integration in momentum space*

$$p \to p+qx\,,$$

so that Eq. (11.2.3) becomes

$$\Pi^{*\rho\sigma}_{1\,\text{Loop}}(q) = \frac{-ie^2}{(2\pi)^4}\int_0^1 dx \int d^4p \Big[p^2+m^2-i\epsilon+q^2x(1-x)\Big]^{-2}$$
$$\times \operatorname{Tr}\{[-i(\not p+\not q x)+m]\,\gamma^\rho\,[-i(\not p-\not q(1-x))+m]\,\gamma^\sigma\}\,. \quad (11.2.5)$$

Using the results of the Appendix to Chapter 8, the trace here can easily be calculated as

$$\operatorname{Tr}\{[-i(\not p+\not q x)+m]\,\gamma^\rho\,[-i(\not p-\not q(1-x))+m]\,\gamma^\sigma\}$$
$$=4\Big[-(p+qx)^\rho\,(p-q(1-x))^\sigma+(p+qx)\cdot(p-q(1-x))\eta^{\rho\sigma}$$
$$-(p+qx)^\sigma\,(p-q(1-x))^\rho+m^2\eta^{\rho\sigma}\Big]\,. \quad (11.2.6)$$

Our next step is called a *Wick rotation.*[2] As long as $-q^2 < 4m^2$, the quantity $m^2+q^2x(1-x)$ is positive for all x between 0 and 1, so the poles in the integrand of Eq. (11.2.5) are at $p^0 = \pm\sqrt{\mathbf{p}^2+m^2+q^2x(1-x)-i\epsilon}$, i.e., just above the negative real axis and just below the positive real axis. (See Figure 11.2.) We can rotate the contour of integrations of p^0 counterclockwise without crossing either of these poles, so that instead of integrating p^0 on the real axis from $-\infty$ to $+\infty$, we integrate it on the imaginary axis from $-i\infty$ to $+i\infty$. That is, we can write $p^0 = ip^4$, with p^4 integrated over real values from $-\infty$ to $+\infty$. (If an $i\epsilon$ instead of $-i\epsilon$ had appeared in the denominator of the propagator, then we would have been setting $p^0 = -ip^4$, with p^4 again integrated over real values from $-\infty$ to

---

* Strictly speaking, this step is only valid in convergent integrals. In principle, in order to justify the shift of variables, we should introduce some regulator scheme to make all integrals converge, such as the dimensional regularization scheme discussed below.

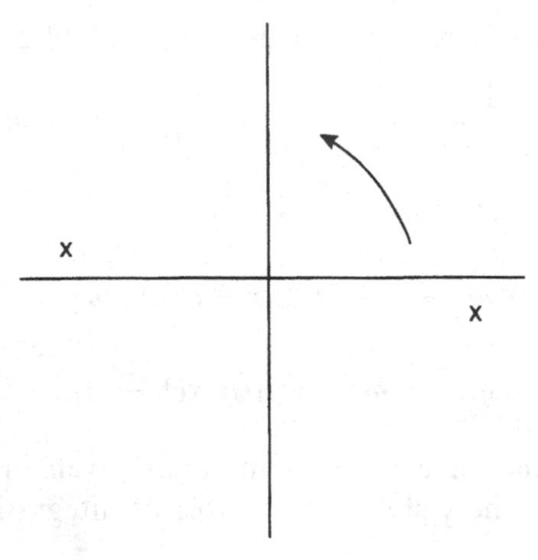

Figure 11.2. Wick rotation of the p^0 contour of integration. Small x's mark the poles in the p^0 complex plane; the arrow indicates the direction of rotation of the contour of integration, from the real to the imaginary p^0-axes.

$+\infty$. The effect would be a change of sign of $\Pi^{*\rho\sigma}_{1\,\text{loop}}(q)$.) Eq. (11.2.5) now becomes

$$\Pi^{*\rho\sigma}_{1\,\text{loop}}(q) = \frac{4e^2}{(2\pi)^4}\int_0^1 dx \int (d^4p)_E\,\left[p^2+m^2+q^2x(1-x)\right]^{-2}$$
$$\times\Big[-(p+qx)^\rho(p-q(1-x))^\sigma + (p+qx)\cdot(p-q(1-x))\,\eta^{\rho\sigma}$$
$$-(p+qx)^\sigma(p-q(1-x))^\rho + m^2\eta^{\rho\sigma}\Big]\,, \qquad (11.2.7)$$

where

$$(d^4p)_E = dp^1dp^2dp^3dp^4$$

and all scalar products are evaluated using the Euclidean norm

$$a\cdot b = a^1b^1 + a^2b^2 + a^3b^3 + a^4b^4$$

with the understanding that $q^4 \equiv -iq^0$. Also, $\eta^{\rho\sigma}$ can be taken as either the Kronecker delta, with the indices running over 1, 2, 3, 4, or as the usual Minkowski tensor, with the indices running over 1, 2, 3, 0.

The integral (11.2.7) is badly divergent. Eventually all infinities will cancel, but to see this it is necessary at intermediate stages of the calculation to use some sort of regularization technique that makes the integrals finite. It would not do simply to cut off the integrals at some maximum momentum Λ, integrating only over p^μ with $p^2 < \Lambda^2$, because this would amount to introducing a step function $\theta(\Lambda^2 - p^2)$ into the electron propagator, and the Ward identity (10.4.25) shows that in order to maintain gauge invariance, any modification of the electron propagator must be accompanied with a modification of the electron–photon vertex. In fact, with an ordinary cutoff Λ, radiative corrections would induce a photon mass, a clear violation of the requirements of gauge invariance.

Experience has shown that the most convenient method for regulating divergent integrals without impairing gauge invariance is the dimensional regularization technique introduced by 't Hooft and Veltman[3] in 1972, based on a continuation from four to an arbitrary number d of spacetime dimensions. This amounts to carrying out angular averages in integrals like (11.2.7) by dropping all terms that are odd in p, and replacing the terms that have even numbers of p-factors with**

$$p^\mu p^\nu \to p^2\eta^{\mu\nu}\Big/d\,, \tag{11.2.8}$$

$$p^\mu p^\nu p^\rho p^\sigma \to (p^2)^2[\eta^{\mu\nu}\eta^{\rho\sigma}+\eta^{\mu\rho}\eta^{\nu\sigma}+\eta^{\mu\sigma}\eta^{\nu\rho}]\Big/d(d+2)\,, \tag{11.2.9}$$

Also, after writing the integrand in this way as a function only of p^2, the volume element d^4p_E is to be replaced with $\Omega_d\kappa^{d-1}d\kappa$, where $\kappa \equiv \sqrt{p^2}$, and Ω_d is the area of a unit sphere in d dimensions

$$\Omega_d = 2\pi^{d/2}\Big/\Gamma(d/2)\,. \tag{11.2.10}$$

The integral (11.2.7) now converges for complex spacetime dimensionality d. We can continue the integral through complex d-values to $d = 4$, the infinities then reappearing as factors $(d-4)^{-1}$.

For the integral (11.2.7), dimensional regularization gives

$$\Pi^{*\rho\sigma}_{1\text{ loop}}(q) = \frac{4e^2\Omega_d}{(2\pi)^4}\int_0^1 dx\int_0^\infty \kappa^{d-1}d\kappa\left[\kappa^2+m^2+q^2x(1-x)\right]^{-2}$$
$$\times\left[\frac{-2\kappa^2}{d}\eta^{\rho\sigma}+2q^\rho q^\sigma x(1-x)+\Big(\kappa^2-q^2x(1-x)\Big)\eta^{\rho\sigma}+m^2\eta^{\rho\sigma}\right].$$

The integrals over κ can be carried out for any complex d (or for any real d, aside from the even integers). We use the well-known formulas (given in greater generality in the Appendix to this chapter):

$$\int_0^\infty \kappa^{d-1}[\kappa^2+\nu^2]^{-2}d\kappa = \tfrac{1}{2}(\nu^2)^{\frac{d}{2}-2}\,\Gamma(d/2)\;\Gamma(2-d/2)\,, \tag{11.2.11}$$

$$\int_0^\infty \kappa^{d+1}[\kappa^2+\nu^2]^{-2}d\kappa = \tfrac{1}{2}(\nu^2)^{\frac{d}{2}-1}\Gamma(1+d/2)\,\Gamma(1-d/2)\,, \tag{11.2.12}$$

** These expressions may most easily be derived by noting that their form is dictated by Lorentz invariance and the symmetry among the indices μ, ν, ρ, etc., while the factors may be found by requiring that both sides give the same result when contracted with ηs.

and find

$$\Pi^{*\rho\sigma}_{1\,\text{loop}}(q) = \frac{2e^2\Omega_d}{(2\pi)^4}$$
$$\times \int_0^1 dx \Big[(1-2/d)\,\eta^{\rho\sigma}\left(m^2+q^2x(1-x)\right)^{\frac{d}{2}-1}\Gamma(1+d/2)\,\Gamma(1-d/2)$$
$$+ \left(2q^\rho q^\sigma x(1-x) - q^2\eta^{\rho\sigma}x(1-x) + m^2\eta^{\rho\sigma}\right)\left(m^2+q^2x(1-x)\right)^{\frac{d}{2}-2}$$
$$\times\,\Gamma(d/2)\,\Gamma(2-d/2)\Big].$$

The two terms in the integrand can be combined, using

$$(1-2/d)\,\Gamma(1+d/2)\,\Gamma(1-d/2) = -\Gamma(d/2)\,\Gamma(2-d/2)\,.$$

We find

$$\Pi^{*\rho\sigma}_{1\,\text{loop}}(q) = \frac{4e^2\Omega_d}{(2\pi)^4}\,\Gamma(d/2)\,\Gamma(2-d/2)\,(q^\rho q^\sigma - q^2\eta^{\rho\sigma})$$
$$\times \int_0^1 dx\; x(1-x)(m^2+q^2x(1-x))^{\frac{d}{2}-2}\,. \tag{11.2.13}$$

We note the very important result that this contribution to $\Pi^{*\rho\sigma}$ satisfies the relation

$$q_\rho \Pi^{*\rho\sigma}_{1\,\text{loop}}(q) = 0 \tag{11.2.14}$$

that was derived in Section 10.5 on the basis of the conservation and neutrality of the electric current. It was precisely to achieve this result that we adopted the dimensional regularization scheme. The reason that dimensional regularization gives this result is that the conservation of current does not depend on the dimensionality of spacetime.

The gamma function $\Gamma(2-d/2)$ in Eq. (11.2.13) blows up for $d \to 4$. Fortunately, as we saw in Section 11.1, there is another term that must be added to $\Pi^{*\rho\sigma}(q)$, arising from the term $-\frac{1}{4}(Z_3-1)F_{\mu\nu}F^{\mu\nu}$ in the interaction Lagrangian. This term has a structure like Eq. (11.2.13)

$$\Pi^{*\rho\sigma}_{\mathscr{L}_2}(q) = -(Z_3-1)(q^2\eta^{\rho\sigma} - q^\rho q^\sigma)\,, \tag{11.2.15}$$

so to order e^2, the complete Π^* has the form

$$\Pi^{*\rho\sigma}(q) = (q^2\eta^{\rho\sigma} - q^\rho q^\sigma)\pi(q^2)\,, \tag{11.2.16}$$

with

$$\pi(q^2) = -\frac{4e^2\Omega_d}{(2\pi)^4}\,\Gamma(\tfrac{d}{2})\,\Gamma(2-\tfrac{d}{2})\int_0^1 dx\; x(1-x)(m^2+q^2x(1-x))^{\frac{d}{2}-2}$$
$$-(Z_3-1)\,. \tag{11.2.17}$$

As we saw in Section 10.5, the definition of the renormalized electromagnetic field requires that $\pi(0) = 0$ (in order that the residue of the pole in the complete photon propagator at $q^2 = 0$ should be the same as for the bare propagator, aside from gauge-dependent terms). Therefore, to order e^2,

$$Z_3 = 1 - \frac{4e^2\Omega_d}{(2\pi)^4}\Gamma\left(\tfrac{d}{2}\right)\Gamma\left(2-\tfrac{d}{2}\right)(m^2)^{\frac{d}{2}-2}\int_0^1 x(1-x)dx\,, \tag{11.2.18}$$

so that, to order e^2,

$$\pi(q^2) = -\frac{4e^2\Omega_d}{(2\pi)^4}\Gamma\left(\tfrac{d}{2}\right)\Gamma\left(2-\tfrac{d}{2}\right)\int_0^1 dx\, x(1-x)$$
$$\times\left[\left(m^2+q^2x(1-x)\right)^{\frac{d}{2}-2} - (m^2)^{\frac{d}{2}-2}\right]. \tag{11.2.19}$$

Now we can remove the regularization, allowing d to approach its physical value $d = 4$. As mentioned before, there is an infinity in the one-loop contribution, arising from the limiting behavior of the Gamma function

$$\Gamma\left(2-\tfrac{d}{2}\right) \to \frac{1}{(2-d/2)} - \gamma\,,$$

where γ is the Euler constant, $\gamma = 0.5772157$. The infinite part of $Z_3 - 1$ is given by using $1/(2-d/2)$ for $\Gamma(2-d/2)$, and replacing d everywhere else by 4:

$$(Z_3 - 1)_\infty = -\frac{4e^2\cdot 2\pi^2}{6(2\pi)^4}\,\frac{1}{2-d/2} = \frac{e^2}{6\pi^2}\,\frac{1}{d-4}\,. \tag{11.2.20}$$

We shall see in Volume II that this result may be used to derive the leading term in the renormalization group equation for the electric charge.

The poles at $d = 4$ obviously cancel in $\pi(q^2)$, because for $d = 4$ both $(m^2+q^2x(1-x))^{\frac{d}{2}-2}$ and $(m^2)^{\frac{d}{2}-2}$ have the same limit, unity. For the same reason, the term $-\gamma$ in $\Gamma(2-d/2)$ cancels in the total $\pi(q^2)$, though it does make a finite contribution to Z_3-1. There are other finite contributions to Z_3-1, that arise from the product of the pole in $\Gamma(2-d/2)$ with the linear terms in the expansion of $\Omega_d\Gamma(d/2)$ around $d = 4$, but these also cancel in the total $\pi(q^2)$. Indeed, in carrying out our dimensional regularization, we might have replaced $(2\pi)^{-4}$ with $(2\pi)^{-d}$, and the factor $\mathrm{Tr}\,1 = 4$ might have been replaced with the dimensionality $2^{d/2}$ of gamma matrices in arbitrary even spacetime dimensionalities d, and these too would have contributed to the finite part of $Z_3 - 1$, but not of $\pi(q^2)$. Moreover, e^2 cannot be supposed to be d-independent, because as shown by inspection of Eq. (11.2.13), it has the d-dependent dimensionality $[\text{mass}]^{4-d}$. If we take $e^2 \propto \mu^{4-d}$, where μ is some quantity with the units of mass, then there are additional finite terms in $Z_3 - 1$, arising from the product of the pole in $\Gamma(2-d/2)$ with the term $(4-d)\ln\mu$ in the expansion of μ^{4-d} in powers of $4-d$, but again, these cancel between $Z_3 - 1$ and the one-loop contributions to $\pi(q^2)$.

The only terms that *do* contribute to $\pi(q^2)$ in the limit $d \to 4$ are those arising from the product of the pole in $\Gamma(2-d/2)$ with the linear terms in the expansion of $(m^2+q^2x(1-x))^{\frac{d}{2}-2}$ and $(m^2)^{\frac{d}{2}-2}$ in powers of $d-4$:

$$(m^2+q^2x(1-x))^{\frac{d}{2}-2}-(m^2)^{\frac{d}{2}-2} \to \left(\tfrac{d}{2}-2\right)\ln\left(1+\frac{q^2x(1-x)}{m^2}\right). \quad (11.2.21)$$

This gives at last

$$\pi(q^2) = \frac{e^2}{2\pi^2}\int_0^1 x(1-x)\ln\left(1+\frac{q^2x(1-x)}{m^2}\right)dx. \quad (11.2.22)$$

The physical significance of the vacuum polarization can be explored by considering its effect on the scattering of two charged particles of spin $\frac{1}{2}$. The Feynman diagrams of Figure 11.3 make contributions to the scattering S-matrix element of the form

$$S_a(1,2\to 1',2') = (2\pi)^{-12/2}\delta^4(p_{1'}+p_{2'}-p_1-p_2)\left[e_1(2\pi)^4\bar{u}_{1'}\gamma^\mu u_1\right] \times\left[-i(2\pi)^{-4}\frac{1}{q^2}\right]\left[e_2(2\pi)^4\bar{u}_{2'}\gamma_\mu u_2\right],$$

$$S_b(1,2\to 1',2') = (2\pi)^{-12/2}\delta^4(p_{1'}+p_{2'}-p_1-p_2)\left[e_1(2\pi)^4\bar{u}_{1'}\gamma^\mu u_1\right] \times\left[-i(2\pi)^{-4}\frac{1}{q^2}\right]^2\left[i(2\pi)^4(q^2\eta_{\mu\nu}-q_\mu q_\nu)\pi(q^2)\right]\left[e_2(2\pi)^4\bar{u}_{2'}\gamma^\nu u_2\right],$$

where e_1 and e_2 are the charges of the two particles being scattered; $\pi(q^2)$ is calculated using for e in Eq. (11.2.22) the magnitude of the charge of the particle circulating in the loop in Figure 11.3; and q^μ is the momentum transfer $q \equiv p_1 - p_{1'} = p_{2'} - p_2$. Using the conservation property $q_\mu\bar{u}_{1'}\gamma^\mu u_1 = 0$ the two diagrams together yield an S-matrix element:

$$S_{a+b}(1,2\to 1',2') = \frac{-ie_1e_2}{4\pi^2q^2}\left[1+\pi(q^2)\right]\delta^4(p_{1'}+p_{2'}-p_1-p_2) \times\left[\bar{u}_{1'}\gamma^\mu u_1\right]\left[\bar{u}_{2'}\gamma_\mu u_2\right]. \quad (11.2.23)$$

In the non-relativistic limit, $\bar{u}_{1'}\gamma^0 u_1 \simeq -i\delta_{\sigma_1'\sigma_1}$ while $\bar{u}_{1'}\gamma^i u_1 \simeq 0$, and likewise for particle 2. Also, in this limit q^0 is negligible compared with $|\mathbf{q}|$. Eq. (11.2.23) in this limit becomes

$$S_{a+b}(1,2\to 1',2') = \frac{-ie_1e_2}{4\pi^2\mathbf{q}^2}\left[1+\pi(\mathbf{q}^2)\right]\delta^4(p_{1'}+p_{2'}-p_1-p_2)\delta_{\sigma_1'\sigma_1}\delta_{\sigma_2'\sigma_2}. \quad (11.2.24)$$

This may be compared with the S-matrix in the Born approximation due to a local spin-independent central potential $V(r)$:

$$S_{\text{Born}}(1,2\to 1',2') = -2\pi i\delta(E_{1'}+E_{2'}-E_1-E_2)T_{\text{Born}}(1,2\to 1',2') \,. \tag{11.2.25}$$

$$T_{\text{Born}}(1,2\to 1',2') = \delta_{\sigma_1'\sigma_1}\delta_{\sigma_2'\sigma_2}\int d^3x_1 \int d^3x_2\, V\Big(|\mathbf{x}_1-\mathbf{x}_2|\Big) \times (2\pi)^{-12/2}e^{-i\mathbf{p}_{1'}\cdot\mathbf{x}_1}e^{-i\mathbf{p}_{2'}\cdot\mathbf{x}_2}e^{i\mathbf{p}_1\cdot\mathbf{x}_1}e^{i\mathbf{p}_2\cdot\mathbf{x}_2} \,. \tag{11.2.26}$$

Setting $\mathbf{x}_1 = \mathbf{x}_2 + \mathbf{r}$, this gives

$$S_{\text{Born}} = \frac{-i}{4\pi^2}\delta^4(p_{1'}+p_{2'}-p_1-p_2)\delta_{\sigma_1'\sigma_1}\delta_{\sigma_2'\sigma_2} \times \int d^3r\, V(r)\, e^{-i\mathbf{q}\cdot\mathbf{r}} \,. \tag{11.2.27}$$

Comparing this with Eq. (11.2.23) shows that in the non-relativistic limit the diagrams of Figure 11.3 yield the same S-matrix element as a potential $V(r)$ such that

$$\int d^3r\, V(r)\, e^{-i\mathbf{q}\cdot\mathbf{r}} = e_1e_2\frac{1+\pi(\mathbf{q}^2)}{\mathbf{q}^2}$$

or, inverting the Fourier transform,

$$V(r) = \frac{e_1e_2}{(2\pi)^3}\int d^3q\; e^{i\mathbf{q}\cdot\mathbf{r}}\left[\frac{1+\pi(\mathbf{q}^2)}{\mathbf{q}^2}\right] \,. \tag{11.2.28}$$

Eq. (11.2.28) is to first order in the radiative correction the same potential energy that would be produced by the electrostatic interaction of two extended charge distributions $e_1\eta(\mathbf{x})$ and $e_2\eta(\mathbf{y})$ at a distance r:

$$V(|\mathbf{r}|) = e_1e_2\int d^3x \int d^3y\; \frac{\eta(\mathbf{x})\eta(\mathbf{y})}{4\pi|\mathbf{x}-\mathbf{y}+\mathbf{r}|} \,, \tag{11.2.29}$$

where

$$\eta(\mathbf{r}) = \delta^3(\mathbf{r}) + \frac{1}{2(2\pi)^3}\int d^3q\; \pi(\mathbf{q}^2)\, e^{i\mathbf{q}\cdot\mathbf{r}} \,. \tag{11.2.30}$$

Note that

$$\int d^3r\; \eta(\mathbf{r}) = 1 + \tfrac{1}{2}\pi(0) = 1 \,, \tag{11.2.31}$$

so the total charges of particles 1 and 2, as determined from the long-range part of the Coulomb potential, are the same constants e_1 and e_2 that govern the interactions of the renormalized electromagnetic field.

For $|\mathbf{r}| \neq 0$ the integral (11.2.30) can be carried out by a straightforward

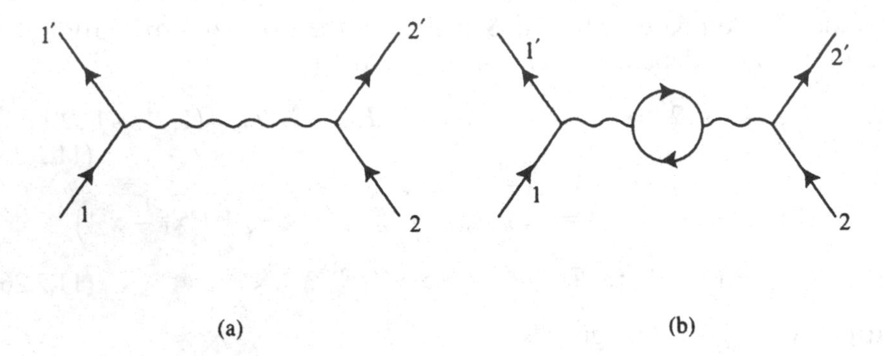

Figure 11.3. Two diagrams for the scattering of charged particles. Here lines carrying arrows are charged particles; wavy lines are photons. Diagram (b) represents the lowest-order vacuum polarization correction to the tree approximation graph (a).

contour integration:

$$\eta(\mathbf{r}) = -\frac{e^2}{8\pi^3 r^3}\int_0^1 x(1-x)\,dx\left[1+\frac{mr}{\sqrt{x(1-x)}}\right]\exp\left(\frac{-mr}{\sqrt{x(1-x)}}\right).$$

This expression is negative everywhere. However, we have seen that the integral of $\eta(\mathbf{r})$ over all $\mathbf{r}$ equals $+1$. Therefore, $\eta(\mathbf{r})$ must contain a term $(1+L)\delta^3(\mathbf{r})$ that is singular at $\mathbf{r}=0$, with L chosen to satisfy Eq. (11.2.31):

$$L = \frac{e^2}{8\pi^3}\int\frac{d^3r}{r^3}\int_0^1 x(1-x)\,dx\left[1+\frac{mr}{\sqrt{x(1-x)}}\right]\exp\left(\frac{-mr}{\sqrt{x(1-x)}}\right). \tag{11.2.32}$$

The complete expression for the charge distribution function is then

$$\eta(\mathbf{r}) = (1+L)\delta^3(\mathbf{r}) - \frac{e^2}{8\pi^3 r^3}\int_0^1 x(1-x)\,dx \times\left[1+\frac{mr}{\sqrt{x(1-x)}}\right]\exp\left(\frac{-mr}{\sqrt{x(1-x)}}\right). \tag{11.2.33}$$

The physical interpretation of this result is that a bare point charge attracts particles of charge of opposite sign out of the vacuum, repelling their antiparticles to infinity, so that the bare charge is partially shielded, yielding a renormalized charge smaller by a factor $1/(1+L)$. As a check, we may note that if we cut off the divergent integral (11.2.32) by taking the integral to extend only over $r \geq a$, we find that the part that is divergent for $a \to 0$ is

$$L_\infty = \frac{e^2}{12\pi^2}\ln a^{-1}. \tag{11.2.34}$$

Hence if we identify the momentum space cutoff Λ with a^{-1}, the divergent part of L is related to the divergent part of $Z_3 - 1$ by

$$(Z_3 - 1)_\infty = -2L_\infty \,, \tag{11.2.35}$$

because to order e^2 the renormalized charge (10.4.18) is given by

$$e_\ell = Z_3^{1/2} e_{B\ell} \simeq \left(1 + \tfrac{1}{2}(Z_3 - 1)\right) e_{B\ell} \simeq (1 + L)^{-1} e_{B\ell} \,. \tag{11.2.36}$$

Eq. (11.2.35) is confirmed below.

Vacuum polarization has a measurable effect on muonic atomic energy levels. As we shall see in Chapter 14, the effect of Feynman graph (b) in Figure 11.3 is to shift the energy of an atomic state with wave function $\psi(\mathbf{r})$ by

$$\Delta E = \int d^3r \, \Delta V(\mathbf{r}) \, |\psi(\mathbf{r})|^2 \,, \tag{11.2.37}$$

where $\Delta V(\mathbf{r})$ is the perturbation in the potential (11.2.28):

$$\Delta V(r) = \frac{e_1 e_2}{(2\pi)^3} \int d^3q \, e^{i\mathbf{q}\cdot\mathbf{r}} \left[\frac{\pi(\mathbf{q}^2)}{\mathbf{q}^2}\right] . \tag{11.2.38}$$

This perturbation falls off exponentially for $r \gg m^{-1}$. On the other hand, the wave function of electrons in ordinary atoms will generally be confined within a much larger radius $a \gg m^{-1}$; for instance, for hydrogenic orbits of electrons around a nucleus of charge Ze we have $a = 137/Zm$ (where here $m = m_e$). The energy shift will then depend only on the behavior of the wave function for $r \ll a$. For orbital angular momentum ℓ, the wave function behaves like r^ℓ for $r \ll a$, so Eq. (11.2.37) gives ΔE proportional to a factor $(ma)^{-(2\ell+1)}$. The effect of vacuum polarization is therefore very much larger for $\ell = 0$ than for higher orbital angular momenta. For $\ell = 0$ the wave function is approximately equal to the constant $\psi(0)$ for r less than or of the order of m^{-1}, so Eq. (11.2.37) becomes

$$\Delta E = |\psi(0)|^2 \int d^3r \, \Delta V(\mathbf{r}) \,. \tag{11.2.39}$$

Using Eqs. (11.2.38) and (11.2.22), the integral of the shift in the potential (for $e_1 e_2 = -Ze^2$) is

$$\int d^3r \, \Delta V(r) = -Ze^2 \pi'(0) = -\frac{4Z\alpha^2}{15m^2} \,. \tag{11.2.40}$$

Also, in states of hydrogenic atoms with $\ell = 0$ and principal quantum number n the wave function at the origin is

$$\psi(0) = \frac{2}{\sqrt{4\pi}} \left(\frac{Z\alpha m}{n}\right)^{3/2} , \tag{11.2.41}$$

so the energy shift (11.2.39) is

$$\Delta E = -\frac{4Z^4\alpha^5 m}{15\pi n^3} . \tag{11.2.42}$$

For instance, in the 2s state of hydrogen this energy shift is -1.122×10^{-7} eV, corresponding to a frequency shift $\Delta E/2\pi\hbar$ of $-$ 27.13 MHz. This is sometimes called the *Uehling effect*.[4] As discussed in Chapter 1, such tiny energy shifts became measurable because in the absence of various radiative corrections the pure Dirac theory would predict exact degeneracy of the 2s and 2p states of hydrogen. As we shall see in Chapter 14, most of the +1058 MHz 'Lamb shift' between the 2s and the 2p states comes from other radiative corrections, but the agreement between theory and experiment is good enough to verify the presence of the $-$37.13 MHz shift due to vacuum polarization.

Although vacuum polarization contributes only a small part of the radiative corrections in ordinary atoms, it dominates the radiative corrections in muonic atoms, in which a muon takes the place of the orbiting electron. This is because most radiative corrections give energy shifts in muonic atoms that on dimensional grounds are proportional to m_μ, while the integrated vacuum polarization energy $\int d^3r\, \Delta V$ due to an *electron* loop is still proportional to m_e^{-2} as in Eq. (11.2.40), giving an energy shift proportional to $m_\mu^3 m_e^{-2} = (210)^2 m_\mu$. However, in this case the muonic atomic radius is not much larger than the electron Compton wavelength, so the approximate result (11.2.39) only gives the order of magnitude of the energy shift due to vacuum polarization.

* * *

For the purposes of comparison with later calculations, note that if we had cut off the integral (11.2.7) at $\kappa = \Lambda$, then in place of Eq. (11.2.20) we would have encountered an integral of the form

$$(Z_3 - 1)_\infty = -\frac{e^2}{6\pi^2}\int_\mu^\Lambda \kappa^{d-5}\, d\kappa = \frac{e^2}{6\pi^2}\frac{\mu^{d-4} - \Lambda^{d-4}}{d-4} ,$$

where μ is an infrared effective cutoff of the order of the mass of the charged particle circulating in the loop of Figure 11.1. (The easiest way to find the constant factor here is to require that the limit of this expression for $d < 3$ and $\Lambda \to \infty$ matches Eq. (11.2.20).) With such an ultraviolet cutoff in place, we can pass to the limit $d \to 4$, and obtain

$$(Z_3 - 1)_\infty = -\frac{e^2}{6\pi^2}\ln(\Lambda/\mu) . \tag{11.2.43}$$

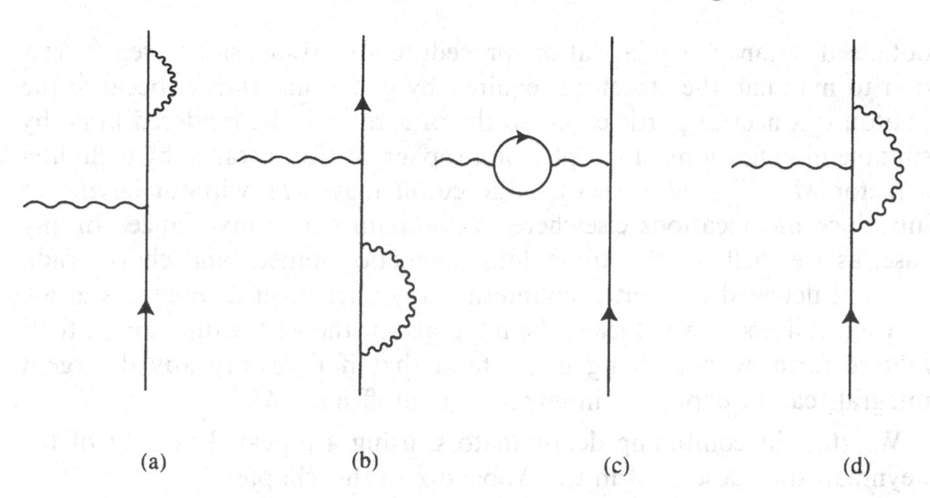

Figure 11.4. One-loop diagrams for the photon–lepton vertex function Γ^μ. Here wavy lines represent photons; other lines represent electrons or muons. Diagrams (a) and (b) are cancelled by lepton field renormalization terms; diagram (c) arises from the vacuum polarization calculated in Section 11.2; and (d) is the term calculated in Section 11.3.

11.3 Anomalous Magnetic Moments and Charge Radii

For our next example, we shall calculate the shift in the magnetic moment and the charge radius of an electron or muon due to lowest-order radiative corrections. The one-loop graphs and renormalization corrections for the photon–lepton vertex are shown in Figure 11.4. Of these graphs, those involving insertions in incoming or outgoing lepton lines vanish because the lepton is on the mass shell, as discussed in Section 10.3. The graph involving an insertion in the external photon line is the vacuum polarization effect, discussed in the previous section. This leaves one one-loop graph (the last in Figure 11.4) that needs to be calculated here:

$$\Gamma^\mu_{1\,\text{loop}}(p',p) = \int d^4k \left[e\gamma^\rho(2\pi)^4\right]\left[\frac{-i}{(2\pi)^4}\frac{-i(\not{p}'-\not{k})+m}{(p'-k)^2+m^2-i\epsilon}\right][\gamma^\mu]$$
$$\times\left[\frac{-i}{(2\pi)^4}\frac{-i(\not{p}-\not{k})+m}{(p-k)^2+m^2-i\epsilon}\right]\left[e\gamma_\rho(2\pi)^4\right]\left[\frac{-i}{(2\pi)^4}\frac{1}{k^2-i\epsilon}\right], \tag{11.3.1}$$

where p' and p are the final and initial lepton four-momenta, respectively. (The contribution of the vertex connecting the external photon line and the internal lepton line is taken as γ^μ, because a factor $e(2\pi)^4$ was extracted in defining Γ^μ.)

This integral has an obvious ultraviolet divergence, roughly like $\int d^4k/(k^2)^2$. Unlike the case of the vacuum polarization, here we do not need a fancy regularization procedure like dimensional regularization to maintain the structure required by gauge invariance, because the photon is a neutral particle and so the integral may be rendered finite by suitable modifications of the photon propagator (for instance by including a factor $M^2/(k^2+M^2)$ with a large cutoff mass M), without having to introduce modifications elsewhere to maintain gauge invariance. In any case, as we shall see the anomalous magnetic moment and charge radii can be calculated without encountering any ultraviolet divergences at all. In what follows we shall leave the integrals for the vertex function in their infinite form, with it being understood that if necessary any divergent integrals can be expressed in terms of a cutoff mass M.

We start by combining denominators, using a repeated version of the Feynman trick described in the Appendix to this chapter

$$\frac{1}{ABC} = 2\int_0^1 dx \int_0^x dy \left[Ay + B(x-y) + C(1-x)\right]^{-3} . \tag{11.3.2}$$

Applied to the denominators in Eq. (11.3.1), this gives

$$\begin{aligned}
&\frac{1}{(p'-k)^2+m^2-i\epsilon}\,\frac{1}{(p-k)^2+m^2-i\epsilon}\,\frac{1}{k^2-i\epsilon} \\
&= 2\int_0^1 dx \int_0^x dy \left[\left((p'-k)^2+m^2-i\epsilon\right)y + \left((p-k)^2+m^2-i\epsilon\right)(x-y)\right. \\
&\qquad \left. +\left(k^2-i\epsilon\right)(1-x)\right]^{-3} \\
&= 2\int_0^1 dx \int_0^x dy \left[\left(k-p'y-p(x-y)\right)^2 + m^2x^2 + q^2y(x-y) - i\epsilon\right]^{-3} ,
\end{aligned} \tag{11.3.3}$$

where $q \equiv p - p'$ is the momentum transferred to the photon. Shifting the variable of integration

$$k \to k + p'y + p(x-y)$$

the integral (11.3.1) becomes

$$\begin{aligned}
\Gamma^{\mu}_{1\,\text{loop}}(p',p) = {}& \frac{2ie^2}{(2\pi)^4}\int_0^1 dx \int_0^x dy \int \frac{d^4k}{\left[k^2+m^2x^2+q^2y(x-y)-i\epsilon\right]^3} \\
&\times \gamma^{\rho}\left[-i\left(\not{p}'(1-y) - \not{k} - \not{p}(x-y)\right) + m\right]\gamma^{\mu} \\
&\times \left[-i\left(\not{p}(1-x+y) - \not{k} - \not{p}'y\right) + m\right]\gamma_{\rho} .
\end{aligned} \tag{11.3.4}$$

Our next step is a Wick rotation. As explained in the previous section, the $-i\epsilon$ in the denominator dictates that when we rotate the k^0 contour of integration to the imaginary axis we must rotate counterclockwise, so that the integral over k^0 from $-\infty$ to $+\infty$ is replaced with an integral over imaginary values from $-i\infty$ to $+i\infty$, or equivalently over real values of $k^4 \equiv -ik^0$ from $-\infty$ to $+\infty$. We also exploit the rotational symmetry of the denominator in Eq. (11.3.4); we drop terms in the numerator of odd order in k, replace $k^\lambda k^\sigma$ with $\eta^{\lambda\sigma}k^2/4$, and replace the volume element $d^4k = idk^1dk^2dk^3dk^4$ with $2i\pi^2\kappa^3 d\kappa$, where κ is the Euclidean length of the four-vector k. Putting this all together, Eq. (11.3.4) now becomes

$$\Gamma^\mu_{1\,\text{loop}}(p',p) = \frac{-4\pi^2e^2}{(2\pi)^4}\int_0^1 dx \int_0^x dy \int_0^\infty \kappa^3 d\kappa \Big\{ -\kappa^2\gamma^\rho\gamma^\sigma\gamma^\mu\gamma_\sigma\gamma_\rho/4$$
$$+\gamma^\rho\Big[-i\Big(\not{p}'(1-y) - \not{p}(x-y)\Big) + m\Big]\gamma^\mu$$
$$\times\Big[-i\Big(\not{p}(1-x+y) - \not{p}'y\Big) + m\Big]\gamma_\rho\Big\}$$
$$\times\Big[\kappa^2 + m^2x^2 + q^2y(x-y)\Big]^{-3}. \tag{11.3.5}$$

We are interested here only in the matrix element $\bar{u}'\Gamma^\mu u$ of the vertex function between Dirac spinors that satisfy the relations

$$\bar{u}'[i\not{p}' + m] = 0\,, \qquad [i\not{p} + m]u = 0\,.$$

We can therefore simplify this expression by using the anticommutation relations of the Dirac matrices to move all factors $\not{p}'$ to the left and all factors $\not{p}$ to the right, replacing them when they arrive on the right or left with im. After a straightforward but tedious calculation, Eq. (11.3.5) then becomes

$$\bar{u}'\Gamma^\mu_{\text{one loop}}(p',p)u = \frac{-4\pi^2e^2}{(2\pi)^4}\int_0^1 dx \int_0^x dy \int_0^\infty \kappa^3 d\kappa$$
$$\bar{u}'\Big\{\gamma^\mu\Big[-\kappa^2 + 2m^2(x^2-4x+2) + 2q^2(y(x-y)+1-x)\Big]$$
$$+4im\,p'^\mu(y-x+xy) + 4im\,p^\mu(x^2-xy-y)\Big\}u$$
$$\times\Big[\kappa^2 + m^2x^2 + q^2y(x-y)\Big]^{-3}. \tag{11.3.6}$$

We next exploit the symmetry of the final factor under the reflection $y \to x - y$. Under this reflection, the functions $y - x + xy$ and $x^2 - xy - y$ that multiply p'^μ and p^μ are interchanged, so both may be replaced with their average:

$$\tfrac{1}{2}(y - x + xy) + \tfrac{1}{2}(x^2 - xy - y) = -\tfrac{1}{2}x(1-x)\,.$$

This gives finally

$$\bar{u}'\Gamma^{\mu}_{\text{one loop}}(p',p)u = \frac{-4\pi^2 e^2}{(2\pi)^4}\int_0^1 dx \int_0^x dy \int_0^\infty \kappa^3 d\kappa$$
$$\times\, \bar{u}'\Big\{\gamma^\mu\Big[-\kappa^2 + 2m^2(x^2-4x+2) + 2q^2(y(x-y)+1-x)\Big]$$
$$-2im\,(p'^\mu + p^\mu)x(1-x)\Big\}u$$
$$\times\,\Big[\kappa^2 + m^2x^2 + q^2y(x-y)\Big]^{-3}. \tag{11.3.7}$$

Note that p^μ and p'^μ now enter only in the combination $p^\mu + p'^\mu$, as required by current conservation.

There are other diagrams that need to be taken into account. Of course, there is the zeroth-order term γ^μ in Γ^μ. The term proportional to $Z_2 - 1$ in the correction term (11.1.9) yields a term in Γ^μ

$$\Gamma^\mu_{\mathscr{L}_2} = (Z_2 - 1)\gamma^\mu\,. \tag{11.3.8}$$

Also, the effect of insertions of corrections to the external photon propagator is a term:

$$\Gamma^\mu_{\text{vac pol}}(p',p) = \frac{1}{(p'-p)^2 - i\epsilon}\Pi^{\mu\nu}(p'-p)\,\gamma_\nu\,. \tag{11.3.9}$$

The form of each of these terms is in agreement with the general result (10.6.10) (with $H(q^2) = 0$)

$$\bar{u}'\Gamma^\mu(p',p)u = \bar{u}'\left[\gamma^\mu F(q^2) - \frac{i}{2m}(p+p')^\mu G(q^2)\right]u\,. \tag{11.3.10}$$

To order e^2, the form factors are

$$F(q^2) = Z_2 + \pi(q^2) + \frac{4\pi^2e^2}{(2\pi)^4}\int_0^1 dx\int_0^x dy\int_0^\infty \kappa^3 d\kappa$$
$$\times\,\frac{\Big[\kappa^2 - 2m^2(x^2-4x+2) - 2q^2(y(x-y)+1-x)\Big]}{\Big[\kappa^2 + m^2x^2 + q^2y(x-y)\Big]^3}\,, \tag{11.3.11}$$

$$G(q^2) = \frac{-4\pi^2e^2}{(2\pi)^4}\int_0^1 dx\int_0^x dy\int_0^\infty \frac{4m^2x(1-x)\,\kappa^3\,d\kappa}{\Big[\kappa^2+m^2x^2+q^2y(x-y)\Big]^3}\,, \tag{11.3.12}$$

where $\pi(q^2)$ is the vacuum polarization function (11.2.22).

The integral for the form factor $G(q^2)$ is finite as it stands:

$$G(q^2) = \frac{-e^2m^2}{4\pi^2}\int_0^1 dx\int_0^x dy\,\frac{x(1-x)}{m^2x^2+q^2y(x-y)}\,. \tag{11.3.13}$$

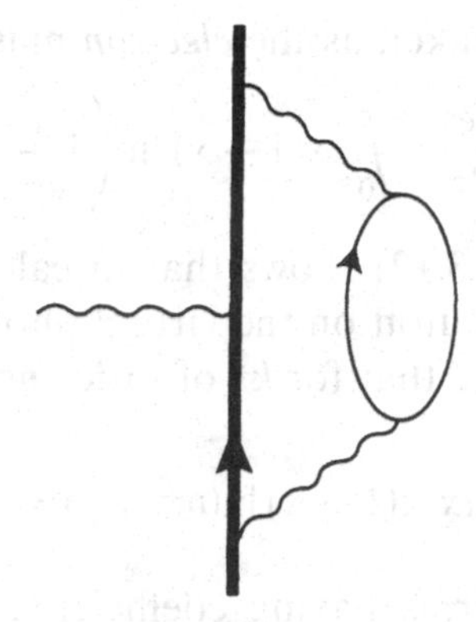

Figure 11.5. A two-loop diagram for the muon magnetic moment. Here the heavy straight line represents a muon; the light wavy lines are photons; and the other light lines are electrons. This diagram makes a relatively large contribution to the fourth-order muon gyromagnetic ratio, proportional to $\ln(m_\mu/m_e)$.

This makes it easy to calculate the anomalous magnetic moment. As noted in Section 10.6, it is only the γ^μ term that contributes to the magnetic moment, so the effect of radiative corrections is to multiply the Dirac value $e/2m$ of the magnetic moment by a factor $F(0)$. But the definition of e as the true lepton charge requires that

$$F(0) + G(0) = 1 \,, \tag{11.3.14}$$

so the magnetic moment may be expressed as

$$\mu = \frac{e}{2m}\Big(1 - G(0)\Big) \,. \tag{11.3.15}$$

From Eq. (11.3.13), we find

$$-\,G(0) = \frac{e^2}{8\pi^2} = 0.001161 \,. \tag{11.3.16}$$

This is the famous $\alpha/2\pi$ correction first calculated by Schwinger.[5]

Of course, this is only the first term in the radiative corrections to the magnetic moment. Even in just the next order, fourth order in e, there are so many terms that the calculations become quite complicated. However, because of the large muon–electron mass ratio, there is one fourth-order term in the magnetic moment of the *muon* that is somewhat larger than any of the others. It arises from the insertion of an *electron* loop in the virtual photon line of the second-order diagram, as shown in Figure 11.5. The effect of this electron loop is to change the photon propagator $1/k^2$ in Eq. (11.3.1) to $(1 + \pi_e(k^2))/k^2$, where $\pi_e(k^2)$ is given by Eq. (11.2.22), but with the mass m taken as the *electron* mass:

$$\pi_e(k^2) = \frac{e^2}{2\pi^2} \int_0^1 x(1-x) \ln\left(1 + \frac{k^2 x(1-x)}{m_e^2}\right) dx \, .$$

Inspection of Eq. (11.3.12) shows that in calculating the muon magnetic moment the effective cutoff on the virtual photon momentum k is m_μ. The ratio m_μ/m_e is so large that for k^2 of order m_μ^2 we may approximate

$$\pi_e(k^2) \simeq \frac{e^2}{2\pi^2} \int_0^1 dx \, x(1-x) \ln(m_\mu^2/m_e^2) = \frac{e^2}{12\pi^2} \ln(m_\mu^2/m_e^2) \qquad (11.3.17)$$

with the neglected terms having coefficients of order unity in place of $\ln(m_\mu^2/m_e^2)$. Since this is a constant, the change in $-G(0)$ produced by adding an electron loop in the virtual photon line is simply given by multiplying our previous result (11.3.16) for $-G(0)$ by Eq. (11.3.17), so that now

$$\mu_\mu = \frac{e}{2m_\mu}\left(1 + \frac{e^2}{8\pi^2} + \frac{e^4}{96\pi^4}\left[\ln\frac{m_\mu^2}{m_e^2} + O(1)\right]\right) . \qquad (11.3.18)$$

(As we shall see in Volume II, this argument is a primitive version of the method of the renormalization group.) The result (11.3.18) may be compared with the full fourth-order result:[6]

$$\mu_\mu = \frac{e}{2m_\mu}\left(1 + \frac{e^2}{8\pi^2} + \frac{e^4}{96\pi^4}\left[\ln\frac{m_\mu^2}{m_e^2} - \frac{25}{6} + \frac{197}{24} + \frac{\pi^2}{2} + \frac{9\zeta(3)}{2} - 3\pi^2 \ln 2 + O\left(\frac{m_e}{m_\mu}\right)\right]\right) . \qquad (11.3.19)$$

It turns out that the '$O(1)$' terms multiplying $e^4/96\pi^4$ add up to -6.137, which is not very much smaller than $\ln(m_\mu^2/m_e^2) = 10.663$, so the approximation (11.3.18) gives the fourth-order terms only to a factor of order 2. The correct fourth-order result (11.3.19) gives $\mu_\mu = 1.00116546 \, e/2m_\mu$, in comparison with the second-order result $\mu_\mu = 1.001161 \, e/2m_\mu$ and the current experimental value,[7] $\mu_\mu = 1.001165923(8) e/2m_\mu$.

Now let us turn to the other form factor. The integral in Eq. (11.3.11) for $F(q^2)$ has an ultraviolet divergence. However, in order to satisfy the charge-non-renormalization condition (11.3.14), it is necessary that Z_2 take the value

$$Z_2 = 1 + \frac{e^2}{8\pi^2} - \frac{4\pi^2 e^2}{(2\pi)^4} \int_0^1 dx \int_0^x dy \int_0^\infty \kappa^3 d\kappa \times \frac{\kappa^2 - 2m^2(x^2 - 4x + 2)}{\left[\kappa^2 + m^2 x^2\right]^3} . \qquad (11.3.20)$$

(Recall that $\pi(0) = 0$.) This is itself ultraviolet divergent, with an infinite part

$$(Z_2 - 1)_\infty = -\frac{e^2}{8\pi^2}\int^\infty \frac{d\kappa}{\kappa}. \tag{11.3.21}$$

Inserting Eq. (11.3.20) back into Eq. (11.3.11) gives

$$F(q^2) = 1 + \frac{e^2}{8\pi^2} + \pi(q^2) + \frac{4\pi^2 e^2}{(2\pi)^4}\int_0^1 dx\int_0^x dy\int_0^\infty \kappa^3 d\kappa$$
$$\times\left\{\frac{\left[\kappa^2 - 2m^2(x^2-4x+2) - 2q^2(y(x-y)+1-x)\right]}{\left[\kappa^2 + m^2x^2 + q^2y(x-y)\right]^3}\right.$$
$$\left. - \frac{\left[\kappa^2 - 2m^2(x^2-4x+2)\right]}{\left[\kappa^2+m^2x^2\right]^3}\right\}. \tag{11.3.22}$$

The integral over κ is now convergent:

$$F(q^2) = 1 + \frac{e^2}{8\pi^2} + \pi(q^2) + \frac{2\pi^2 e^2}{(2\pi)^4}\int_0^1 dx\int_0^x dy$$
$$\times\left\{\frac{-m^2[x^2-4x+2] - q^2[y(x-y)+1-x]}{m^2x^2+q^2y(x-y)} + \frac{x^2-4x+2}{x^2}\right.$$
$$\left. - \ln\left[\frac{m^2x^2+q^2y(x-y)}{m^2x^2}\right]\right\}. \tag{11.3.23}$$

However, we see that the integral over x and y now diverges logarithmically at $x = 0$ and $y = 0$, because there are two powers of x and/or y in the denominators, and just two differentials $dx\,dy$ in the numerator. This divergence can be traced to the vanishing of the denominator $[\kappa^2 + m^2x^2 + q^2y(x-y)]^3$ in Eq. (11.3.11) at $x = 0$, $y = 0$, and $\kappa = 0$. Because this infinity comes from the region of small rather than large κ, it is termed an *infrared divergence* rather than an ultraviolet divergence.

We shall give a comprehensive treatment of the infrared divergences in Chapter 13. It will be shown there that infrared divergences in the cross-section for processes like electron–electron scattering, such as those that are introduced by the infrared divergence in the electron form factor $F(q^2)$, are cancelled when we include the emission of low-energy photons as well as elastic scattering. Also, as we shall see in Chapter 14, when we calculate radiative corrections to atomic energy levels the infrared divergence in $F(q^2)$ is cut off because the bound electron is not exactly on the free-particle mass shell. For the present we shall continue our calculation by simply introducing a fictitious photon mass μ to cut off the infrared divergence in $F(q^2)$, leaving it for Chapter 14 to see how to use this result.

With a photon mass μ, the denominator $k^2 - i\epsilon$ in Eq. (11.3.1) would be replaced with $k^2 + \mu^2 - i\epsilon$. The effect would then be to add a term $\mu^2(1-x)$ to the cubed quantity in the denominators of Eqs. (11.3.3)–(11.3.7), (11.3.11), (11.3.20), and (11.3.22). Eq. (11.3.23) then is replaced with

$$F(q^2) = 1 + \frac{e^2}{8\pi^2} + \pi(q^2) + \frac{2\pi^2 e^2}{(2\pi)^4}\int_0^1 dx \int_0^x dy \times \left\{ \frac{-m^2[x^2-4x+2] - q^2[y(x-y)+1-x]}{m^2x^2 + q^2y(x-y) + \mu^2(1-x)} + \frac{m^2[x^2-4x+2]}{m^2x^2+\mu^2(1-x)} - \ln\left[\frac{m^2x^2+q^2y(x-y)+\mu^2(1-x)}{m^2x^2+\mu^2(1-x)}\right]\right\} . \tag{11.3.24}$$

This integral is now completely convergent. It can be expressed in terms of Spence functions, but the result is not particularly illuminating. For our purposes in Chapter 14, it will be sufficient to calculate the behavior of $F(q^2)$ for small q^2. We already know from the Ward identity that $F(0) = 1 - G(0) = 1 + e^2/8\pi^2$, so let us consider the first derivative $F'(q^2)$ at $q^2 = 0$. According to Eq. (11.3.24), this is

$$F'(0) = \pi'(0) + \frac{2\pi^2 e^2}{(2\pi)^4}\int_0^1 dx \int_0^x dy \times \left\{ -\frac{2y(x-y)+1-x}{m^2x^2+\mu^2(1-x)} + \frac{m^2[x^2-4x+2]y(x-y)}{[m^2x^2+\mu^2(1-x)]^2}\right\} . \tag{11.3.25}$$

The vacuum polarization contribution is given by Eq. (11.2.22) as

$$\pi'(0) = \frac{e^2}{60\pi^2 m^2} . \tag{11.3.26}$$

Dropping all terms proportional to powers of μ/m in Eq. (11.3.25), we then have*

$$F'(0) = \frac{e^2}{24\pi^2 m^2}\left[\ln\left(\frac{\mu^2}{m^2}\right) + \frac{2}{5} + \frac{1}{4}\right] \tag{11.3.27}$$

with the term $\frac{2}{5}$ the contribution of vacuum polarization. On the other hand, Eq. (11.3.13) shows that $G(q^2)$ has a finite derivative at $q^2 = 0$,

$$G'(0) = \frac{e^2}{48\pi^2 m^2} . \tag{11.3.28}$$

* The y-integral is trivial. The x-integral is most easily calculated in the limit $\mu \ll m$ by dividing the range of integration into two parts, one from 0 to s, where $\mu/m \ll s \ll 1$, and the second from s to 1.

These results are most conveniently expressed in terms of the charge form factor $F_1(q^2)$, defined by the alternative representation (10.6.15) of the vertex function

$$\bar{u}(\mathbf{p}',\sigma')\Gamma^\mu(p',p)u(\mathbf{p},\sigma)$$
$$= \bar{u}(\mathbf{p}',\sigma')\left[\gamma^\mu F_1(q^2) + \tfrac{1}{2}i[\gamma^\mu,\gamma^\nu](p'-p)_\nu F_2(q^2)\right]u(\mathbf{p},\sigma)\,. \quad (11.3.29)$$

According to Eqs. (10.6.17) and (10.6.18),

$$F_1(q^2) = F(q^2) + G(q^2)\,. \quad (11.3.30)$$

For $|q^2| \ll m^2$, this form factor is approximately

$$F_1(q^2) \simeq 1 + \frac{e^2}{24\pi^2}\left(\frac{q^2}{m^2}\right)\left[\ln\left(\frac{\mu^2}{m^2}\right) + \frac{2}{5} + \frac{3}{4}\right]\,. \quad (11.3.31)$$

This may be expressed in terms of a *charge radius a*, defined by the limiting behavior of the charge form factor for $q^2 \to 0$:

$$F_1(q^2) \to 1 - q^2a^2/6\,. \quad (11.3.32)$$

(This definition is motivated by the fact that the average of $\exp(i\mathbf{q}\cdot\mathbf{x})$ over a spherical shell of radius a goes as $1 - \mathbf{q}^2a^2/6$ for $\mathbf{q}^2a^2 \ll 1$.) We see that the charge radius of the electron is given by

$$a^2 = -\frac{e^2}{4\pi^2m^2}\left[\ln\left(\frac{\mu^2}{m^2}\right) + \frac{2}{5} + \frac{3}{4}\right]\,. \quad (11.3.33)$$

We will see in Chapter 14 that for electrons in atoms the role of the photon mass is played by an effective infrared cutoff that is much less than m, so the logarithm here is large and negative, yielding a positive value for a^2.

11.4 Electron Self-Energy

We conclude this chapter with a calculation of the electron self-energy function. This by itself does not have any direct experimental implications, but some of the results here will be useful in Chapter 14 and Volume II.

As in Section 10.3, we define $i(2\pi)^4[\Sigma^*(p)]_{\beta,\alpha}$ as the sum of all graphs with one incoming and one outgoing electron line carrying momenta p and Dirac indices α and β respectively, with the asterisk indicating that we exclude diagrams that can be disconnected by cutting through some internal electron line, and with propagators omitted for the two external lines. The complete electron propagator is then given by the sum

$$[-i(2\pi)^{-4}S'(p)] = [-i(2\pi)^{-4}S(p)]$$
$$+ [-i(2\pi)^{-4}S(p)][i(2\pi)^4\Sigma^*(p)][-i(2\pi)^{-4}S(p)] + \dots\,, \quad (11.4.1)$$

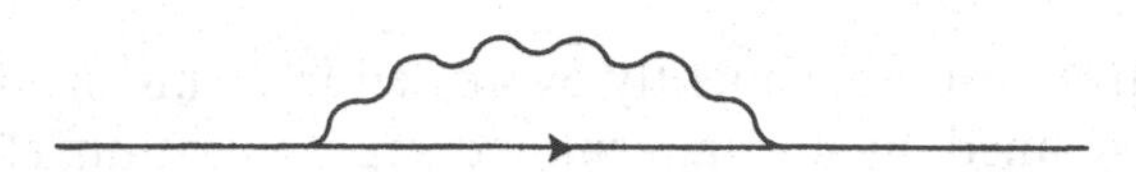

Figure 11.6. The one-loop diagram for the electron self-energy function. As usual, the straight line represents an electron; the wavy line is a photon.

where

$$S(p) \equiv \frac{-i \not{p} + m_e}{p^2 + m_e^2 - i\epsilon} . \tag{11.4.2}$$

The sum is trivial, and gives

$$S'(p) = [i \not{p} + m_e - \Sigma^*(p) - i\epsilon]^{-1} . \tag{11.4.3}$$

In lowest order there is a one-loop contribution to Σ^*, given by Figure 11.6:

$$i(2\pi)^4 \Sigma^*_{1\,\text{loop}}(p) = \int d^4k \, [\frac{-i}{(2\pi)^4} \frac{\eta_{\rho\sigma}}{k^2 - i\epsilon}]$$
$$\times \, [(2\pi)^4 e \gamma^\rho] \left[\frac{-i}{(2\pi)^4} \frac{-i \not{p} + i \not{k} + m_e}{(p-k)^2 + m_e^2 - i\epsilon}\right] [(2\pi)^4 e \gamma^\sigma]$$

or more simply

$$\Sigma^*_{1\,\text{loop}}(p) = \frac{ie^2}{(2\pi)^4} \int d^4k \, [\frac{1}{k^2 - i\epsilon}]$$
$$\times \left[\frac{\gamma^\rho(-i \not{p} + i \not{k} + m_e)\gamma_\rho}{(p-k)^2 + m_e^2 - i\epsilon}\right] . \tag{11.4.4}$$

(This is in Feynman gauge; amplitudes with charged particles off the mass shell are not gauge-invariant.) For use in our calculation of the Lamb shift, it will be convenient to use a method of regularization introduced by Pauli and Villars.[8] We replace the photon propagator $(k^2 - i\epsilon)^{-1}$ with

$$\frac{1}{k^2 - i\epsilon} - \frac{1}{k^2 + \mu^2 - i\epsilon} ,$$

so that the electron self-energy function is

$$\Sigma^*_{1\,\text{loop}}(p) = \frac{ie^2}{(2\pi)^4} \int d^4k \, \left[\frac{1}{k^2 - i\epsilon} - \frac{1}{k^2 + \mu^2 - i\epsilon}\right]$$
$$\times \left[\frac{\gamma^\rho(-i \not{p} + i \not{k} + m_e)\gamma_\rho}{(p-k)^2 + m_e^2 - i\epsilon}\right] . \tag{11.4.5}$$

Later we can drop the regulator by letting the regulator mass μ go to infinity. In Chapter 14 we will also be interested in the case where $\mu \ll m_e$.

We again use the Feynman trick to combine denominators, and recall that $\gamma^\rho\gamma^\kappa\gamma_\rho = -2\gamma^\kappa$ and $\gamma^\rho\gamma_\rho = 4$. This gives

$$\Sigma^*_{1\,\text{loop}}(p) = \frac{ie^2}{(2\pi)^4}\int d^4k\,[2i(\not p - \not k) + 4m_e]$$
$$\times \int_0^1 dx \left[\frac{1}{((k-px)^2 + p^2x(1-x) + m_e^2x - i\epsilon)^2} - \frac{1}{((k-px)^2 + p^2x(1-x) + m_e^2x + \mu^2(1-x) - i\epsilon)^2}\right] \quad (11.4.6)$$

Shifting the variable of integration $k \to k + px$ and rotating the contour of integration gives

$$\Sigma^*_{1\,\text{loop}}(p) = \frac{-2\pi^2e^2}{(2\pi)^4}\int_0^1 dx\,[2i(1-x)\not p + 4m_e]\int_0^\infty d\kappa\,\kappa^3$$
$$\times \left[\frac{1}{(\kappa^2 + p^2x(1-x) + m_e^2x)^2} - \frac{1}{(\kappa^2 + p^2x(1-x) + m_e^2x + \mu^2(1-x))^2}\right]. \quad (11.4.7)$$

The κ-integral is trivial:

$$\Sigma^*_{1\,\text{loop}}(p) = \frac{-\pi^2e^2}{(2\pi)^4}\int_0^1 dx\,[2i(1-x)\not p + 4m_e]$$
$$\times \ln\left(\frac{p^2x(1-x) + m_e^2x + \mu^2(1-x)}{p^2x(1-x) + m_e^2x}\right). \quad (11.4.8)$$

The interaction (11.1.9) also contributes a renormalization counterterm $-(Z_2 - 1)(i\not p + m_e) + Z_2\delta m_e$ in $\Sigma^*(p)$, with Z_2 and δm_e determined by the condition that the complete propagator $S'(p)$ regarded as a function of $i\not p$ should have a pole at $i\not p = -m_e$ with residue unity. (As we shall see in the next chapter, this makes Σ^* finite as $\mu \to \infty$ to all orders in e.) In lowest order, this gives

$$\delta m_e = -\left.\Sigma^*_{1\,\text{loop}}\right|_{i\not p = -m_e}$$
$$= \frac{2m_e\pi^2e^2}{(2\pi)^4}\int_0^1 dx\,[1+x]\,\ln\left(\frac{m_e^2x^2 + \mu^2(1-x)}{m_e^2x^2}\right), \quad (11.4.9)$$

$$Z_2 - 1 = -i\left.\frac{\partial\Sigma^*_{1\,\text{loop}}}{\partial\not p}\right|_{i\not p = -m_e}$$
$$= -\frac{2\pi^2e^2}{(2\pi)^4}\int_0^1 dx\left\{(1-x)\ln\left(\frac{m_e^2x^2 + \mu^2(1-x)}{m_e^2x^2}\right) - \frac{2\mu^2(1-x)^2(1+x)}{x(m_e^2x^2 + \mu^2(1-x))}\right\}. \quad (11.4.10)$$

(To this order, we do not distinguish between δm_e and $Z_2\delta m_e$.) Dropping terms that vanish for $\mu^2 \to \infty$, Eqs. (11.4.8)–(11.4.10) yield

$$\Sigma^*_{1\,\text{loop}}(p) = \frac{-\pi^2 e^2}{(2\pi)^4}\int_0^1 dx\,[2i(1-x)\,\not{p} + 4m_e]\,\ln\left(\frac{\mu^2(1-x)}{p^2x(1-x)+m_e^2x}\right), \tag{11.4.11}$$

$$\delta m_e = \frac{2m_e\pi^2e^2}{(2\pi)^4}\int_0^1 dx\,[1+x]\,\ln\left(\frac{\mu^2(1-x)}{m_e^2x^2}\right), \tag{11.4.12}$$

$$Z_2 - 1 = \frac{-2\pi^2e^2}{(2\pi)^4}\int_0^1 dx\,\left\{(1-x)\ln\left(\frac{\mu^2(1-x)}{m_e^2x^2}\right) - \frac{2(1-x^2)}{x}\right\}. \tag{11.4.13}$$

Inspection then shows that in the complete self-energy function the $\ln\mu^2$ terms cancel, leaving us with

$$\begin{aligned}\Sigma^*_{\text{order }e^2}(p) &= \Sigma^*_{1\,\text{loop}}(p) - (Z_2-1)(i\,\not{p}+m_e) + Z_2\delta m_e \\ &= \frac{-2\pi^2e^2}{(2\pi)^4}\int_0^1 dx\left\{[i(1-x)\,\not{p}+2m_e]\ln\left(\frac{m_e^2(1-x)}{p^2x(1-x)+m_e^2x}\right)\right. \\ &\quad -m_e[1+x]\ln\left(\frac{1-x}{x^2}\right) \\ &\quad \left.-(i\,\not{p}+m_e)\left[(1-x)\ln\left(\frac{1-x}{x^2}\right) - \frac{2(1-x^2)}{x}\right]\right\}.\end{aligned} \tag{11.4.14}$$

There is still a divergence from the behavior of the last term as $x \to 0$, which can be traced to the singular behavior of the integral over the photon momentum k in Eq. (11.4.5) at $k^2 = 0$, when we take p^2 at the point $p^2 = -m_e^2$ where we evaluated $Z_2 - 1$. Such infrared divergences will be discussed in detail in Chapter 13. For the present, the point that concerns us is that the ultraviolet divergence has cancelled.

* * *

The result (11.4.9) for δm_e is of some interest in itself. Note that $\delta m_e/m_e > 0$, as we would expect for the electromagnetic self energy due to the interaction of a charge with its own field. But unlike the classical estimates of electromagnetic self-energy by Poincaré, Abraham, and others,[9] Eq. (11.4.9) is only logarithmically divergent in the limit $\mu \to \infty$ where the cutoff is removed. In this limit:

$$\delta m_e \to \frac{6m_e\pi^2e^2}{(2\pi)^4}\ln\left(\frac{\mu}{m_e}\right). \tag{11.4.15}$$

In our calculation of the Lamb shift in Section 14.3 we will be interested in the opposite limit, $\mu \ll m_e$. Here Eq. (11.4.9) gives

$$\delta m_e \to \frac{e^2\mu}{8\pi}\left[1 - \frac{3\mu}{2\pi m_e} + \ldots\right] . \tag{11.4.16}$$

Appendix Assorted Integrals

In order to combine the denominators of N propagators, we need to replace a product like $D_1^{-1}D_2^{-1}\ldots D_N^{-1}$ with an integral of a function that involves a linear combination of $D_1, D_2, \ldots D_N$. For this purpose it is often convenient to make use of the formula

$$\frac{1}{D_1 D_2 \ldots D_N} = (N-1)! \int_0^1 dx_1 \int_0^{x_1} dx_2 \cdots \int_0^{x_{N-2}} dx_{N-1}$$
$$\times\ [D_1 x_{N-1} + D_2(x_{N-2} - x_{N-1}) + \cdots + D_N(1 - x_1)]^{-N} . \tag{11.A.1}$$

In this chapter we have used special cases of this formula for $N = 2$ and $N = 3$.

After combining denominators, shifting the four-momentum variable of integration, Wick rotating, and using four-dimensional rotational invariance, we commonly encounter integrals of the form

$$\int d^4k \, \frac{(k^2)^n}{(k^2 + v^2)^m}$$

with $(k^2 + v^2)^m$ coming from the combined propagator denominators, and $(k^2)^n$ coming from the propagator numerators and vertex momentum factors. This is divergent for $2n + 4 \geq 2m$, but the integral can be given a finite value by analytically continuing the spacetime dimensionality from 4 to a complex value d. To evaluate the resulting integral, we use the well-known formula

$$\int_0^\infty d\kappa \frac{\kappa^{\ell-1}}{(\kappa^2 + v^2)^m} = v^{\ell - 2m} \frac{\Gamma(\ell/2)\,\Gamma(m - \ell/2)}{2\,\Gamma(m)} , \tag{11.A.2}$$

where $\ell = d + 2n$. We used this formula in the special cases $n = 0, m = 2$ and $n = 1, m = 2$ in Section 11.2.

Ultraviolet divergences manifest themselves in Eq. (11.A.2) as poles in the factor $\Gamma(m - \ell/2) = \Gamma(m - n - d/2)$ as $d \to 4$ with fixed integer n. For $2 + n = m$, this factor goes as

$$\Gamma\left(\frac{4-d}{2}\right) \to \frac{2}{d-4} + \gamma , \tag{11.A.3}$$

where $\gamma = 0.5772157\cdots$ is the Euler constant. The limiting behavior for $2 + n > m$ can be obtained from (11.A.3) and the recursion relation for Gamma functions.

Problems

1. Calculate the contributions to the vacuum polarization function $\pi(q^2)$ and to Z_3 of one-loop graphs involving a charged spinless particle of mass m_s. What effect does this have on the energy shift of the 2*s* state of hydrogen, if $m_s >> Z\alpha m_e$?

2. Suppose that a neutral scalar field ϕ of mass m_ϕ has an interaction $g\phi\bar{\psi}\psi$ with the electron field. To one-loop order, what effect does this have on the magnetic moment of the electron? On Z_2?

3. Consider a neutral scalar field ϕ with mass m_ϕ and self-interaction $g\phi^3/6$. To one-loop order, calculate the *S*-matrix element for scalar–scalar scattering.

4. To one-loop order, calculate the effect of the neutral scalar field of Problem 2 on the mass shift δm_e of the electron.

References

1. R. P. Feynman, *Phys. Rev.* **76**, 769 (1949).

2. G. C. Wick, *Phys. Rev.* **96**, 1124 (1954).

3. G. 't Hooft and M. Veltman, *Nucl. Phys.* **B44**, 189 (1972).

4. E. A. Uehling, *Phys. Rev.* **48**, 55 (1935). The one-loop function $\pi(q^2)$ was first given for $q^2 \neq 0$ by J. Schwinger, *Phys. Rev.* **75**, 651 (1949).

5. J. Schwinger, *Phys. Rev.* **73**, 416 (1948).

6. This is calculated (including terms that vanish for $m_e \ll m_\mu$) by H. Suura and E. Wichmann, *Phys. Rev.* **105**, 1930 (1957); A. Petermann, *Phys. Rev.* **105**, 1931 (1957); H. H. Elend, *Phys. Lett.* **20**, 682 (1966); **21**, 720 (1966); G. W. Erickson and H. H. T. Liu, UCD-CNL-81 report (1968).

7. J. Bailey *et al.* (CERN–Mainz–Daresbury Collaboration), *Nucl. Phys.* **B150**, 1 (1979). These experiments are done by observing the precession of the muon spin in a storage ring.

8. W. Pauli and F. Villars, *Rev. Mod. Phys.* **21**, 434 (1949). Also see J. Rayski, *Phys. Rev.* **75**, 1961 (1949).

9. See, e.g., A. I. Miller, *Theory of Relativity — Emergence (1905) and Early Interpretation (1905–1911)* (Addison-Wesley, Reading, MA, 1981): Chapter 1.

12

General Renormalization Theory

We saw in the previous chapter that calculations in quantum electrodynamics involving one-loop graphs yield divergent integrals over momentum space, but that these infinities cancel when we express all parameters of the theory in terms of 'renormalized' quantities, such as the masses and charges that are actually measured. In 1949 Dyson[1] sketched a proof that this cancellation would take place to all orders in quantum electrodynamics. It was immediately apparent (and will be shown here in Sections 12.1 and 12.2) that Dyson's arguments apply to a larger class of theories with finite numbers of relatively simple interactions, the so-called *renormalizable* theories, of which quantum electrodynamics is just one simple example.

For some years it was widely thought that any sensible physical theory would have to take the form of a renormalizable quantum field theory. The requirement of renormalizability played a crucial role in the development of the modern 'standard model' of weak, electromagnetic, and strong interactions. But as we shall see here, the cancellation of ultraviolet divergences does not really depend on renormalizability; as long as we include every one of the infinite number of interactions allowed by symmetries, the so-called non-renormalizable theories are actually just as renormalizable as renormalizable theories.

It is generally believed today that the realistic theories that we use to describe physics at accessible energies are what are known as 'effective field theories.' As discussed in Section 12.3, these are low-energy approximations to a more fundamental theory that may not be a field theory at all. Any effective field theory necessarily includes an infinite number of non-renormalizable interactions. Nevertheless, as discussed in Sections 12.3 and 12.4, we expect that at sufficiently low energy all the non-renormalizable interactions in such effective field theories are highly suppressed. Renormalizable theories like quantum electrodynamics and the standard model thus retain their special status in physics, though for reasons that are somewhat different from those that originally motivated the assumption of renormalizability in these theories.

12.1 Degrees of Divergence

Let us consider a very general sort of theory, containing interactions of varying types labelled i. Each interaction may be characterized by the number n_{if} of fields of each type f, and by the number d_i of derivatives acting on these fields.

We will start by calculating the 'superficial degree of divergence' D of an arbitrary connected one-particle irreducible Feynman diagram in such a theory. This is the number of factors of momentum in the numerator minus the number in the denominator of the integrand, plus four for every independent four-momentum over which we integrate. The superficial divergence is the actual degree of divergence of the integration over the region of momentum space in which the momenta of all internal lines go to infinity together. That is, if $D > 0$, then the part of the amplitude where all internal momenta go to infinity with a common factor $\kappa \to \infty$ will diverge like

$$\int^{\infty} \kappa^{D-1} d\kappa \, . \tag{12.1.1}$$

In the same sense, an integral with degree of divergence $D = 0$ is logarithmically divergent, and an integral with $D < 0$ is convergent, at least as far as this region of momentum space is concerned. We will come back later to the problem posed by subintegrations that behave worse than the integral over this region.

To calculate D, we will need to know the following about the diagram:

$$\begin{aligned} I_f &\equiv \text{number of internal lines of field type } f \, , \\ E_f &\equiv \text{number of external lines of field type } f \, , \\ N_i &\equiv \text{number of vertices of interaction type } i \, . \end{aligned}$$

We will write the asymptotic behavior of the propagator $\Delta_f(k)$ of a field of type f in the form

$$\Delta_f(k) \sim k^{-2+2s_f} \, . \tag{12.1.2}$$

Looking back at Chapter 6, we see that $s_f = 0$ for scalar fields, $s_f = \frac{1}{2}$ for Dirac fields, and $s_f = 1$ for massive vector fields. More generally, it can be shown that for massive fields of Lorentz transformation type (A, B), we have $s_f = A + B$. Speaking loosely, we may call s_f the 'spin.' However, dropping terms that because of gauge invariance have no effect, the effective photon propagator $\eta_{\mu\nu}/k^2$ has $s_f = 0$. A similar result applies to a massive vector field coupled to a conserved current, provided the current does not depend on the vector field. It can also be shown that, in the same sense, the graviton field $g_{\mu\nu}$ has a propagator also with $s_f = 0$.

According to (12.1.2), the propagators make a total contribution to D equal to

$$\sum_f I_f(2s_f - 2)\,. \tag{12.1.3}$$

Also, the derivatives in each interaction of type i introduce d_i momentum factors into the integrand, yielding a total contribution to D equal to

$$\sum_i N_i\, d_i\,. \tag{12.1.4}$$

Finally, we need the total number of independent momentum variables of integration. Each internal line can be labelled with a four-momentum, but these are not all independent; the delta function associated with each vertex imposes a linear relation among these internal momenta, except that one delta function only serves to enforce conservation of the external momenta. Thus, the momentum space integration volume elements contribute to D a term

$$4\left[\sum_f I_f - \left(\sum_i N_i - 1\right)\right]\,, \tag{12.1.5}$$

which, of course, is just four times the number of independent loops in the diagram. Adding the contributions (12.1.3), (12.1.4), and (12.1.5), we find

$$D = \sum_f I_f(2s_f + 2) + \sum_i N_i(d_i - 4) + 4\,. \tag{12.1.6}$$

Eq. (12.1.6) is not very convenient as it stands, because it gives a value for D that seems to depend on the internal details of the Feynman diagram. Fortunately, it can be simplified by using the topological identities

$$2I_f + E_f = \sum_i N_i\, n_{if}\,. \tag{12.1.7}$$

(Each internal line contributes two of the lines attached to vertices, while each external line contributes only one.) Using Eq. (12.1.7) to eliminate I_f, we see that Eq. (12.1.6) becomes

$$D = 4 - \sum_f E_f(s_f + 1) - \sum_i N_i \Delta_i\,, \tag{12.1.8}$$

where Δ_i is a parameter characterizing interactions of type i:

$$\Delta_i \equiv 4 - d_i - \sum_f n_{if}(s_f + 1)\,. \tag{12.1.9}$$

This result could have been obtained by simple dimensional analysis, without considering the structure of Feynman diagrams. The propagator of a field is a four-dimensional Fourier transform of the vacuum expectation value of a time-ordered product of a pair of free fields, so a conventionally normalized field f whose dimensionality* in powers of momentum is $\mathscr{D}_f$ will have a propagator of dimensionality $-4+2\mathscr{D}_f$. Hence if the propagator behaves like k^{-2+2s_f} when k is much larger than the mass, then the field must have a dimensionality with $-4+2\mathscr{D}_f = -2+2s_f$, or $\mathscr{D}_f = 1+s_f$. An interaction i with n_{if} such fields and d_i derivatives will then have dimensionality $d_i+\sum_f n_{if}(1+s_f)$. But the action must be dimensionless, so each term in the Lagrangian density must have dimensionality $+4$ to cancel the dimensionality -4 of d^4x. Hence the interaction must have a coupling constant of dimensionality $4-d_i-\sum_f n_{if}(1+s_f)$, which is just the parameter Δ_i. The momentum space amplitude corresponding to a connected Feynman graph with E_f external lines of type f is the Fourier transform over $4\sum_f E_f$ coordinates of a vacuum expectation value of the time-ordered product of fields with a total dimensionality $\sum_f E_f(1+s_f)$, so it has dimensionality $\sum_f E_f(-3+s_f)$. Of this dimensionality, -4 comes from a momentum space delta function, and $\sum_f E_f(-2+2s_f)$ is the dimensionality of the propagators for the external lines, so the momentum space integral itself together with all coupling constant factors has dimensionality

$$\sum_f E_f(-3+s_f) - (-4) - \sum_f E_f(-2+2s_f) = 4 - \sum_f E_f(s_f+1)\,.$$

The coupling constants for a given Feynman graph have total dimensionality $\sum_i N_i\Delta_i$, leaving the momentum space integral with dimensionality $4-\sum_f E_f(s_f+1)-\sum_i N_i\Delta_i$. As long as we are interested in the region of integration where all momenta go to infinity together, the degree of divergence of the momentum space integral is its dimensionality, thus justifying Eq. (12.1.8).

If all interactions have $\Delta_i \geq 0$, then Eq. (12.1.8) provides an upper bound on D that depends only on the numbers of external lines of each type, i.e., on the physical process whose amplitude is being calculated

$$D \leq 4 - \sum_f E_f(s_f+1)\,. \tag{12.1.10}$$

For example, in the simple version of quantum electrodynamics studied in the previous chapter, the Lagrangian included terms of the types shown in Table 12.1. All interactions here have $\Delta_i \geq 0$, and hence a Feynman diagram with E_γ external photon lines and E_e external Dirac lines will

* In this chapter, 'dimensionality' will always refer to the dimensionality in powers of mass or momentum, in units with $\hbar = c = 1$. We are using fields that are conventionally normalized, in the sense that the term in the free-field Lagrangian with the largest number of derivatives (which determines the asymptotic behavior of the propagator) has a dimensionless coefficient.

Table 12.1. Terms in the Lagrangian density for quantum electrodynamics. Here d_i, $n_{i\gamma}$, and n_{ie} are the numbers of derivatives, photon fields, and electron fields in the interaction, and Δ_i is the dimensionality of the corresponding coefficient. (Recall that $s_\gamma = 0$, $s_e = \frac{1}{2}$.)

Interaction	d_i	$n_{i\gamma}$	n_{ie}	Δ_i
$-ie\bar{\psi}A\!\!\!/\psi$	0	1	2	$4-1-3=0$
$-\frac{1}{4}(Z_3-1)F_{\mu\nu}F^{\mu\nu}$	2	2	0	$4-2-2=0$
$-(Z_2-1)\bar{\psi}\partial\!\!\!/\psi$	1	0	2	$4-1-3=0$
$[-(Z_2-1)m+Z_2\delta m]\bar{\psi}\psi$	0	0	2	$4-3=1$

have superficial degree of divergence bounded by Eq. (12.1.10):

$$D \leq 4 - \tfrac{3}{2}E_e - E_\gamma \,. \tag{12.1.11}$$

Only a finite number of sets of external lines can yield superficially divergent integrals; these will be enumerated in Section 12.2. We are going to show that the limited number of divergences that appear in theories with $\Delta_i \geq 0$ for all interactions are automatically removed by a redefinition of a finite number of physical constants and a renormalization of fields. For this reason, such theories are called *renormalizable*. In Section 12.3 we will catalog all the renormalizable theories, and discuss the significance of renormalizability as a criterion for physical theories.

The term 'renormalizable' is also applied to individual interactions. Renormalizable interactions are those with $\Delta_i \geq 0$, whose coupling constants have positive or zero dimensionality. Sometimes one distinguishes between interactions with $\Delta_i = 0$, called simply renormalizable, and those with $\Delta_i > 0$, called *superrenormalizable*. Since adding additional fields or derivatives always lowers Δ_i, there can only be a finite number of renormalizable interactions involving fields of any given types. We have seen that all the interactions in the simplest version of quantum electrodynamics are renormalizable, with the $\bar{\psi}\psi$ terms superrenormalizable.

On the other hand, if any interaction has $\Delta_i < 0$, the degree of divergence (12.1.8) becomes larger and larger the more such vertices we include. No matter how large we take the various E_f, eventually with enough vertices of type i for which $\Delta_i < 0$, Eq. (12.1.8) will become positive (or zero), and the integral will diverge. Such interactions, whose couplings have negative-definite dimensionality, are called *non-renormalizable*;** the-

** In perturbative statistical mechanics, non-renormalizable interactions are called *irrelevant*, because they become less important in the limit of low energies. Renormalizable and super-renormalizable interactions are called *marginal* and *relevant*, respectively.

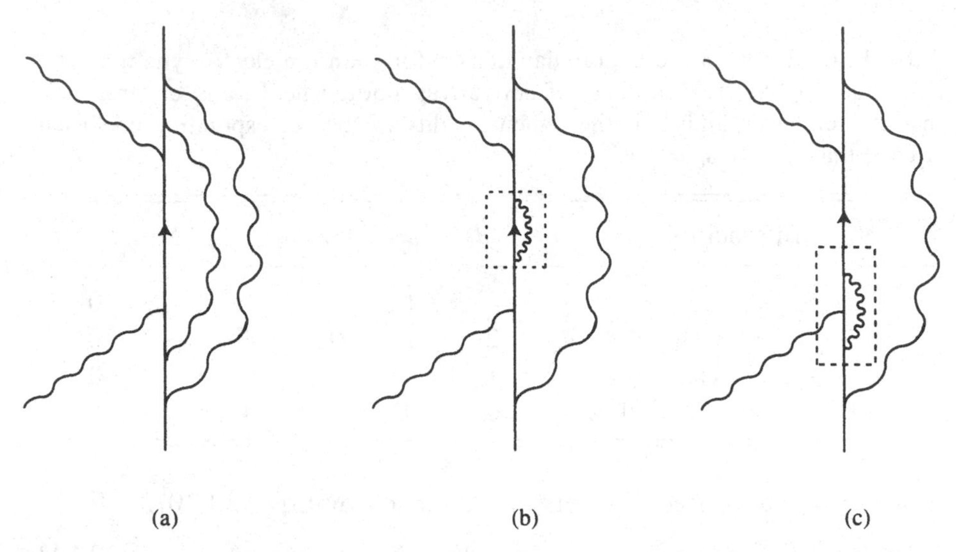

Figure 12.1. Some two-loop graphs for Compton scattering. Here straight lines are electrons; wavy lines are photons. The momentum space integral for diagram (a) is convergent, while for (b) and (c) it is divergent, due to the subintegration associated with the subgraphs surrounded by dotted lines.

ories with any non-renormalizable interactions are also known as non-renormalizable. But this does not mean that such theories are hopeless; we shall see that these divergences may also be absorbed into a redefinition of the parameters of the theory, but here we need an infinite number of couplings.

It should be kept in mind that we have here calculated the degree of divergence of Feynman diagrams arising only from regions of momentum space in which all internal four-momenta go to infinity together. Divergences can also arise from regions in which only the four-momenta of lines belonging to some subgraph go to infinity. For instance, in quantum electrodynamics Eq. (12.1.11) gives $D \leq -1$ for Compton scattering (where $E_e = 2$, $E_\gamma = 2$), and indeed graphs like Figure 12.1(a) are convergent, but a graph like Figure 12.1(b) or 12.1(c) is logarithmically divergent, because these graphs contain subgraphs (indicated by dotted boxes) with $D \geq 0$. We can think of the divergence of these graphs as being due to an anomalously bad asymptotic behavior that occurs when the eight components of the two independent internal four-momenta of these graphs go to infinity on a particular four-dimensional subspace, namely, that subspace in which the only four-momentum actually going to infinity is the one circulating in the loops that are inserted in the internal lines or at an electron–photon vertex.

It has been shown[2] that the requirement for the actual convergence of the amplitude corresponding to any graph is that power-counting should give $D < 0$ not only for the complete multiple integral for the whole amplitude, but also for any subintegration defined by holding any one or more linear combinations of the loop momenta fixed. (The graphs shown in Figures 12.1(b) and 12.1(c) fail this test because $D \geq 0$ for the subintegrations in which only the momenta for the loops within the dotted squares are integrated.) We will not repeat the rather long proof here, because it is well treated in earlier books,[3] and in any case the method of proof has little to do with how we actually do calculations. The next section will describe how this requirement is fulfilled.

12.2 Cancellation of Divergences

Consider a Feynman diagram, or part of a Feynman diagram, with positive superficial degree of divergence, $D \geq 0$. The part of the momentum space integral where all internal momenta go to infinity together will then diverge, like $\int^{\infty} k^{D-1}\, dk$. If we differentiate $D+1$ times with respect to any external momentum, we lower the net number of momentum factors in the integrand by $D+1$,* and hence render this part of the momentum space integral convergent. There may still be divergences arising from subgraphs, like those in Figures 12.1(b) and 12.1(c); for the moment we will ignore this possibility, returning to it later in this section. Since differentiation $D+1$ times renders the integral finite, it follows that the contribution of such a graph or subgraph can be written as a polynomial of order D in external momenta, with divergent coefficients, plus a finite remainder.

To see how this works without irrelevant complications, consider the logarithmically divergent one-dimensional integral

$$\mathscr{I}(q) \equiv \int_0^\infty \frac{dk}{k+q}$$

with $D = 1 - 1 = 0$. Differentiating once gives

$$\mathscr{I}'(q) \equiv -\int_0^\infty \frac{dk}{(k+q)^2} = -\frac{1}{q}\,,$$

so

$$\mathscr{I}(q) = -\ln q + c\,.$$

* For instance, if an internal scalar field line carries a momentum $k+p$, where p is a linear combination of external four-momenta and k is a four-momentum variable of integration, then the derivative of the propagator $[(k+p)^2+m^2]^{-1}$ with respect to p^μ gives $2(k_\mu+p_\mu)[(k+p)^2+m^2]^{-2}$, which goes as k^{-3} rather than k^{-2} for $k \to \infty$.

The constant c is obviously divergent, but the rest of the integral is perfectly finite. In exactly the same way, we can evaluate the $D = 1$ integral

$$\int_0^\infty \frac{k\,dk}{k+q} = a + bq + q \ln q$$

with divergent constants a and b.

Now, a polynomial term in external momenta is just what would be produced by adding suitable terms to the Lagrangian: if a graph with E_f external lines of type f has degree of divergence $D \geq 0$, then the ultraviolet divergent polynomial is the same as would be produced by adding various interactions i with $n_{if} = E_f$ fields of type f and $d_i \leq D$ derivatives. If there already are such interactions in the Lagrangian, then the ultraviolet divergences simply add corrections to the coupling constants of these interactions. Hence these infinities can be cancelled by including suitable infinite terms in these coupling constants. All that we ever measure is the sum of the bare coupling constant and the corresponding coefficient from one of the divergent polynomials, so if we demand that the sum equals the (presumably finite) measured value, then the bare coupling must automatically contain an infinity that cancels the infinity from the divergent integral over internal momenta. (One qualification: where the divergence occurs in a graph or subgraph with just two external lines, which appears as a radiative correction to a particle propagator, we must demand not that some effective coupling constant equals its measured value, but rather that the complete propagator has a pole at the same position and with the same residue as for free particles.) In this way, all infinities are absorbed into a redefinition of couplings constants, masses, and fields.

For this renormalization program to work, it is essential that the Lagrangian include *all* interactions that correspond to the ultraviolet divergent parts of Feynman amplitudes. The interactions in the Lagrangian are, of course, limited by various symmetry principles, such as Lorentz invariance, gauge invariance, etc., but these constrain the ultraviolet divergences in the same way. (It takes some work to prove that non-Abelian gauge symmetries constrain infinities in the same way that they constrain interactions. This will be shown in Volume II.) In the general case, there are no other limitations on the ultraviolet divergences, *so the Lagrangian must include every possible term consistent with symmetry principles.* (There are exceptions to this rule in supersymmetric theories.[4])

But there is an important class of theories with only a finite number of interactions, where the renormalization program also works. These are the so-called renormalizable theories, whose interactions all have $\Delta_i \geq 0$.

Eq. (12.1.8) then gives

$$D \leq 4 - \sum_f E_f(s_f + 1)\,,$$

so divergent polynomials arise in only a limited number of Feynman graphs or subgraphs: those with few enough external lines so that $D \geq 0$. The contribution of such divergent polynomials is just the same as would be produced by replacing the divergent graph or subgraph with a single vertex arising from a term in the Lagrangian with E_f fields of type f and $0, 1, \cdots D$ derivatives. But, comparing with Eq. (12.1.9), we see that *these are precisely the same as the interactions that satisfy the renormalizability requirement* $\Delta_i \geq 0$, or in other words,

$$0 \leq d_i \leq 4 - \sum_f n_{if}(s_f + 1)\,.$$

In order for all infinities to cancel in a renormalizable theory, it is usually necessary that *all* renormalizable interactions that are allowed by symmetries must actually appear in the Lagrangian.** For instance, if there is a scalar (or pseudoscalar) field ϕ and fermion field ψ with interactions $\bar{\psi}\psi\phi$ (or $\bar{\psi}\gamma_5\psi\phi$) then we cannot exclude an interaction ϕ^4; otherwise there would be no counterterm to cancel the logarithmic divergence arising from fermion loops with four attached scalar or pseudoscalar lines.

Let's see in more detail how the cancellation of infinities works in the simplest version of quantum electrodynamics. Eq. (12.1.11) shows that the only graphs or subgraphs that could possibly yield divergent integrals are the following:

$\mathbf{E_e = 2,\ E_\gamma = 1}$

This is the electron–photon vertex $\Gamma^{(\ell)}_\mu(p', p)$. (The superscript ℓ indicates that this includes only contributions from graphs with loops.) It has $D = 0$, so its divergent part is momentum-independent. Lorentz invariance then only allows this divergent constant to be proportional to γ_μ, so

$$\Gamma^{(\ell)}_\mu = L\gamma_\mu + \Gamma^{(f)}_\mu \tag{12.2.1}$$

with L a logarithmically divergent constant, and $\Gamma^{(f)}_\mu$ finite. This does not uniquely define the constant L, since we can always move a finite term

** In addition, interactions and mass terms that are not allowed by global symmetries may appear in the Lagrangian, as long as they are superrenormalizable, that is, with $\Delta_i > 0$. This is because the presence of a superrenormalizable coupling lowers the degree of divergence, so that the symmetry breaking does not affect those divergences that are cancelled by the strictly renormalizable couplings with $\Delta_i = 0$. Note that it is the *bare* strictly renormalizable couplings that would exhibit the symmetry; renormalized couplings that are defined in terms of mass-shell matrix elements generally show the effect of symmetry breaking.

$\delta L \gamma_\mu$ from $\Gamma^{(f)}_\mu$ to $L\gamma_\mu$. To complete the definition, we may note that as shown in Section 10.4, the mass-shell matrix element of $\Gamma_\mu(p,p)$ and hence of $\Gamma^{(f)}_\mu(p,p)$ between mass-shell Dirac spinors is proportional to the same matrix element of γ^μ, so we may define L by the prescription that

$$\bar{u}(\mathbf{p},\sigma')\Gamma^{(f)}_\mu(p,p)u(\mathbf{p},\sigma) = 0 \qquad (12.2.2)$$

for $p^2 + m_e^2 = 0$.

$\mathbf{E_e = 2,\ E_\gamma = 0}$

This is the electron self-energy insertion $\Sigma^*(p)$. It has $D = 1$, so its divergent part is linear in the momentum p^μ carried by the incoming and outgoing fermion. Lorentz invariance (including parity conservation) will only allow it to be a function of $\not{p}$, so we may write the loop contribution as

$$\Sigma^{(\ell)}(p) = A - (i\not{p} + m)B + \Sigma^{(f)}(\not{p})\,, \qquad (12.2.3)$$

where A and B are divergent constants, and $\Sigma^{(f)}$ is finite. Again, this does not uniquely define the constants A and B, because we can always shift $\Sigma^{(f)}$ by a finite first-order polynomial in $\not{p}$. We will define A and B by the prescription that

$$\Sigma^{(f)} = \frac{\partial \Sigma^{(f)}}{\partial \not{p}} = 0 \ \text{ for } \ i\not{p} = -m\,. \qquad (12.2.4)$$

Actually, B is not a new divergent constant. As long as we use a regularization procedure that respects current conservation, Γ_μ and Σ will be related by the Ward identity (10.4.27)

$$\Gamma^\mu(p,p) = \gamma^\mu + i\frac{\partial}{\partial p_\mu}\Sigma(p)$$

and therefore

$$L\gamma_\mu + \Gamma^{(f)}_\mu(p,p) = B\gamma_\mu + i\frac{\partial \Sigma^{(f)}(p)}{\partial p^\mu}\,. \qquad (12.2.5)$$

Taking the matrix element of this equation between $\bar{u}(\mathbf{p},\sigma')$ and $u(\mathbf{p},\sigma)$ and using Eqs. (12.2.2) and (12.2.4), we find

$$L = B\,. \qquad (12.2.6)$$

$\mathbf{E_\gamma = 2,\ E_e = 0}$

This is the photon self-energy insertion $\Pi^*_{\mu\nu}(q)$. It has $D = 2$, so its divergent part is a second-order polynomial in q. Lorentz invariance only allows $\Pi^*_{\mu\nu}$ to take the form of a linear combination of $\eta_{\mu\nu}$ and $q_\mu q_\nu$ with coefficients depending only on q^2, so the loop contributions take the form

$$\Pi^{(\ell)}_{\mu\nu}(q) = C_1\eta_{\mu\nu} + C_2\eta_{\mu\nu}q^2 + C_3 q_\mu q_\nu + \text{finite terms}\,,$$

where C_1, C_2, and C_3 are divergent constants. As long as we use a regularization technique that respects current conservation, we must have

$$q^\mu \Pi^{(\ell)}_{\mu\nu}(q) = 0\,.$$

The same must then be true for the divergent terms, so $C_1 q_\nu + (C_2 + C_3)q^2 q_\nu$ must be finite for all q. It follows that C_1 and $C_2 + C_3$ must be finite, and can therefore be lumped into the finite part of $\Pi^{(\ell)}_{\mu\nu}(q)$. Thus

$$\Pi^{(\ell)}_{\mu\nu}(q) = (\eta_{\mu\nu}q^2 - q_\mu q_\nu)\Big(C + \pi(q^2)\Big)\,, \qquad (12.2.7)$$

where $\pi(q^2)$ is finite and C is the sole remaining divergence in $\Pi^{(\ell)}_{\mu\nu}$. To pin down the definition of C, we may move any finite constant $\pi(0)$ into C, so that

$$\pi(0) = 0\,. \qquad (12.2.8)$$

$\mathbf{E_\gamma = 4,\ E_e = 0}$

This is the amplitude $M_{\mu\nu\rho\sigma}$ for scattering of light by light. It has $D = 0$, so using Lorentz invariance and Bose statistics, it may be written (there is no non-loop contribution)

$$M_{\mu\nu\rho\sigma} = K(\eta_{\mu\nu}\eta_{\rho\sigma} + \eta_{\mu\rho}\eta_{\nu\sigma} + \eta_{\mu\sigma}\eta_{\nu\rho}) \ + \ \text{finite terms}$$

with K a potentially divergent constant. However, current conservation gives

$$q^\mu M_{\mu\nu\rho\sigma} = 0$$

and so $K(q_\nu\eta_{\rho\sigma} + q_\rho\eta_{\nu\sigma} + q_\sigma\eta_{\nu\rho})$ is finite. In order for this to be true for $q \neq 0$, K must itself be finite. This is a nice example of the role of symmetry principles in the renormalization program; if K had turned out to be infinite it could not be removed by renormalization of the coupling constant for an interaction $(A_\mu A^\mu)^2$, because no such interaction is allowed by gauge invariance, but K *is* finite because of current conservation conditions that are imposed by gauge invariance.

$\mathbf{E_\gamma = 1,\ E_e = 0}$ **and** $\mathbf{E_e = 1,\ E_\gamma = 0, 1, 2}$

These have $D = 3$ and $D = \frac{5}{2}, \frac{3}{2}$ and $\frac{1}{2}$, respectively, but Lorentz invariance makes all such graphs vanish.

$\mathbf{E_\gamma = 3,\ E_e = 0}$

This has $D = 1$, but vanishes because of charge-conjugation invariance.

The reader will perhaps have noticed that the independent divergent constants A, B, C are in one-to-one correspondence with the independent parameters Z_2, Z_3, and δm in the counterterm part (11.1.9) of the Lagrangian for quantum electrodynamics. These counterterms make a direct contribution $Z_2\delta m - (Z_2 - 1)(i \not{p} + m)$ to $\Sigma^*(p)$. The requirement that the position and residue of the one-particle pole be the same as in the free-field propagator means we must choose Z_2 and δm so that the total $\Sigma^*(p)$ satisfies Eq. (12.2.4), i.e.,

$$Z_2\,\delta m = -A, \tag{12.2.9}$$

$$Z_2 - 1 = -B, \tag{12.2.10}$$

so that the complete electron self-energy insertion is just the finite function $\Sigma^{(f)}(p)$:

$$\Sigma(p) = \Sigma^{(f)}(p)\,. \tag{12.2.11}$$

Also, $\mathscr{L}_2$ makes a direct contribution to Γ_μ equal to $(Z_2 - 1)\gamma_\mu$. Using Eq. (12.2.6), we see that the full vertex is

$$\Gamma_\mu = \gamma_\mu + (Z_2 - 1)\gamma_\mu + \Gamma_\mu^{(f)} = \gamma_\mu + \Gamma_\mu^{(f)}\,. \tag{12.2.12}$$

This is not only finite, but satisfies the condition

$$\bar{u}(\mathbf{p}, \sigma')\Gamma_\mu(p, p)u(\mathbf{p}, \sigma) = \bar{u}(\mathbf{p}, \sigma')\gamma_\mu u(\mathbf{p}, \sigma')\,, \tag{12.2.13}$$

as can also be seen from Eqs. (10.6.13) and (10.6.14). Finally, $\mathscr{L}_2$ makes a contribution $-(Z_3 - 1)(q^2\eta_{\mu\nu} - q_\mu q_\nu)$ to $\Pi^*_{\mu\nu}(q)$. In order that the photon propagator should have a pole with the same residue as for free fields we need the coefficient of $q^2\eta_{\mu\nu} - q_\mu q_\nu$ in the total $\Pi_{\mu\nu}(q)$ to vanish, so

$$Z_3 = 1 + C \tag{12.2.14}$$

and the photon propagator is then finite:

$$\Pi_{\mu\nu}(q) = (\eta_{\mu\nu}q^2 - q_\mu q_\nu)\pi(q^2)\,. \tag{12.2.15}$$

So far, we have only checked that the divergences, arising from the region of momentum space in which all internal momenta are large (and with generic ratios), are polynomials in external momenta that are cancelled by suitable counterterms. Such graphs are called *superficially convergent*. Before we conclude that all ultraviolet divergences actually are removed by renormalization, we need to consider the ultraviolet divergences arising in higher-order graphs when some subset of the momentum space integration variables rather than all of them go to infinity. For instance, in quantum electrodynamics the superficial divergences in subintegrations come from subgraphs that are either photon self-energy parts Π^*, or electron self-energy parts Σ^*, or electron–electron–photon vertices Γ^μ. The

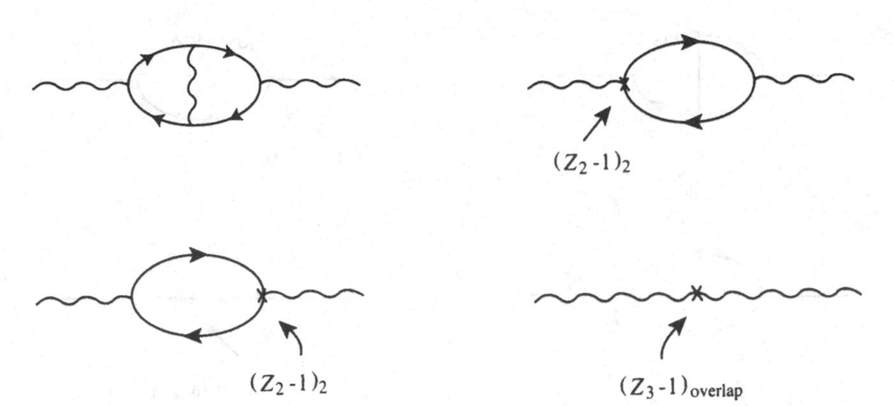

Figure 12.2. Some fourth-order graphs for the photon self-energy in quantum electrodynamics that involve overlapping divergences. Lines carrying arrows are electrons; wavy lines photons. The crosses mark the contribution of counterterms.

problem with such divergences is that they cannot be removed by differentiation with respect to external momenta; we are left with terms where the derivatives act only on internal lines in the parts of the graph which are *not* in the divergent subgraphs, and therefore do not reduce the degree of divergence of these subgraphs. As mentioned in the previous section, a graph or sum of graphs is actually convergent only if it and all its subintegrations are superficially convergent, in the sense of counting powers of momentum. But wherever such a divergent subgraph appears, it always comes accompanied with an infinite counterterm. In electrodynamics, these are the terms in Eq. (11.1.9): a term $-(Z_3-1)(q^2\eta_{\mu\nu}-q_\mu q_\nu)$ for each $\Pi^*_{\mu\nu}(q)$; a term $Z_2\delta m-(Z_2-1)(i\not{p}+m)$ for each $\Sigma^*(p)$; and a term $(Z_2-1)\gamma^\mu$ for each Γ^μ. Just as for the graph as a whole, these counterterms cancel the infinities from the divergent subgraphs.[1]

Unfortunately, there is a flaw in this simple argument — the possibility of overlapping divergences. That is, it is possible that two divergent subgraphs may share an internal line, so that we cannot regard them as independent divergent integrals. In quantum electrodynamics this happens only when two electron–electron–photon vertices overlap inside a photon or an electron self-energy insertion,[†] as shown in Figures 12.2 and 12.3.

A complete treatment of renormalization that takes account of overlap-

† The sharing of a line in two self-energy insertions or in a self-energy insertion and a vertex part would not leave enough external lines to attach such a subgraph to the rest of the diagram. Historically, the Ward identity (10.4.26) was used to by-pass the problem of overlapping divergences in the electron self-energy, by expressing the electron self-energy in terms of the vertex function, where overlapping divergences do not occur. This approach will not be followed here, as it is unnecessary, and in any case does not solve the problem for the self-energy of the photon or other neutral particles.

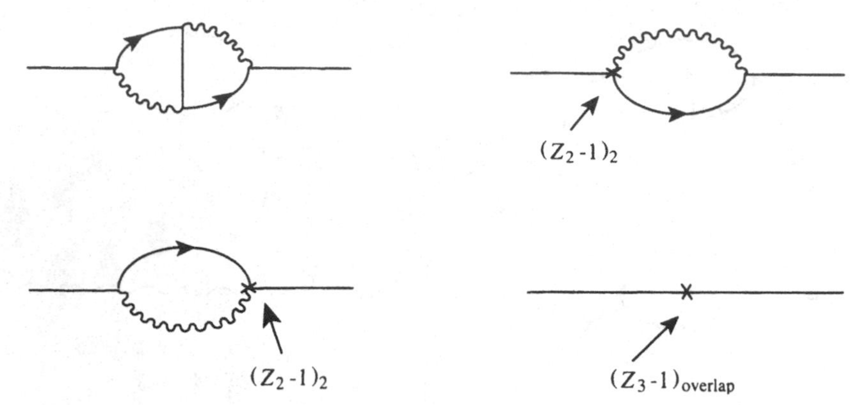

Figure 12.3. Some fourth-order graphs for the electron self-energy in quantum electrodynamics that involve overlapping divergences. Wavy lines are photons; other lines are electrons. The crosses mark the contribution of counterterms.

ping divergences should include a prescription for eliminating superficial ultraviolet divergences, not only in the overall integration but in all subintegrations as well, together with a proof that this prescription is (at least formally) implemented by renormalization of masses, fields, and coupling constants. The theorem of Ref. 2 then ensures that all Green's functions of renormalized fields are finite when expressed in terms of renormalized masses and couplings. The first proof that the renormalization of fields, couplings, and masses renders the whole integration and all its subintegrations superficially convergent was offered by Salam.[5] A more specific prescription for eliminating ultraviolet divergences was given by Bogoliubov and Parasiuk,[6] and corrected by Hepp,[7] and shown by them to be equivalent to a renormalization of fields, masses, and coupling constants. Finally, Zimmerman[8] proved that this prescription does eliminate all superficial divergences in the whole integration and all its subintegrations, and used the theorem of Ref. 2 to conclude from this that the renormalized Feynman momentum-space integrals are convergent.

Briefly, the 'BPHZ' prescription for eliminating superficial divergences requires that we consider all possible ways (called 'forests') of surrounding a whole graph and/or its subgraphs with boxes that may be nested within each other but do not overlap. (An example is given below.) For each forest we define a subtraction term by replacing the integrand for any subgraph of superficial divergence D within a box (starting with the innermost boxes and working outwards) with the first $D+1$ terms of its Taylor series expansion in the momenta flowing into or out of that box.‡

‡ As described here, this prescription applies to unrenormalizable as well as renormalizable theories. In renormalizable theories it implies that there is no subtraction unless the box contains one of the limited number of graphs corresponding to renormalizable terms in the Lagrangian.

The subtracted Feynman amplitude is given by the original graph, minus all these subtraction terms, including the subtraction term for a forest consisting of a single box surrounding the whole graph.

It is fairly easy to see that the subtracted Feynman amplitude that is calculated in this way is the same as would be obtained by replacing all fields, coupling constants, and masses in the original Lagrangian by their renormalized counterparts. The difference between this procedure and the sort of renormalization we carried out in Chapter 11 is that the renormalized fields, coupling constants, and masses are defined in terms of amplitudes at an unconventional renormalization point, where all four-momenta vanish. (In this respect, the one-dimensional divergent integrals discussed at the start of this section provide an elementary example of the BPHZ method of separating divergent terms.) But there is nothing special about this renormalization point; once a Feynman amplitude is made convergent by expressing it in terms of these unconventional renormalized quantities, it can be rewritten in terms of conventionally renormalized fields, couplings, and masses without introducing new infinities.

It is not necessary to use the BPHZ subtraction prescription in practice. Replacing fields, masses, and couplings with their renormalized counterparts (defined using any convenient renormalization points) automatically provides counterterms that cancel all infinities. Instead of proving that the BPHZ subtraction prescription really does make all integrals converge, we shall just look at one example that shows how renormalization works, even in the presence of overlapping divergences.

Consider the fourth-order contribution to the photon self-energy insertion $\Pi^*_{\mu\nu}(q)$ shown in Figure 12.2. (The forests here consist of the whole integral over p and p', the subintegration over p alone, and the subintegration over p' alone.) Including the corresponding counterterms for the vertex parts and photon field renormalization, this has the value

$$\begin{aligned}[\Pi^*_{\mu\nu}(q)]_{\text{overlap}} = & -\frac{e^4}{(2\pi)^8}\int d^4p\int d^4p'\,\frac{1}{(p-p')^2-i\epsilon} \\ & \times \text{Tr}\left\{S(p')\gamma_\nu\, S(p'+q)\gamma^\rho\, S(p+q)\gamma_\mu\, S(p)\gamma_\rho\right\} \\ & -2\,(Z_2-1)_2\frac{ie^2}{(2\pi)^4}\int d^4p\,\text{Tr}\left\{\gamma_\nu\, S(p+q)\gamma_\mu\, S(p)\right\} \\ & -(Z_3-1)_{\text{overlap}}\,(q^2\eta_{\mu\nu}-q_\mu q_\nu)\,, \end{aligned} \tag{12.2.16}$$

where $S(p) \equiv [-i\not{p}+m]/[p^2+m^2-i\epsilon]$; $(Z_2-1)_2$ is the term in Z_2-1 of second order in e; and $(Z_3-1)_{\text{overlap}}$ is a logarithmically divergent constant of fourth order in e that cancels the terms in $[\Pi^*_{\mu\nu}(q)]_{\text{overlap}}$ of second order in q^λ. The factor 2 in the second term arises because there is a renormalization counterterm Z_2-1 for each of the two vertices in the second-order photon self-energy. Note, however, that the first term here can be thought of *either* as the insertion of a vertex correction given by the p'-integral into a photon self-energy given by the p-integral, *or* as the insertion of a vertex correction given by the p-integral into a photon self-energy given by the p'-integral, but *not* as the insertion of two independent vertex corrections, because there is only one photon propagator.

To see how to handle the infinities in Eq. (12.2.16), note that

$$[(Z_2 - 1)_2 + R_2]\gamma_\mu = \frac{ie^2}{(2\pi)^4}\int \frac{d^4p'}{p'^2 - i\epsilon}\,\gamma_\rho\, S(p')\,\gamma_\mu\, S(p')\gamma^\rho\,, \qquad (12.2.17)$$

where R_2 is a finite remainder. (Lorentz invariance tells us that the integral on the right is proportional to γ_μ. The difference between this integral and $(Z_2 - 1)_2\gamma_\mu$ equals the complete renormalized electron–electron–photon vertex to second order in e at zero electron and photon momenta, and is therefore finite.) This allows us to rewrite Eq. (12.2.16) in the form

$$\begin{aligned}
[\Pi^*_{\mu\nu}(q)]_{\text{overlap}} = &-\frac{e^4}{(2\pi)^8}\int d^4p\int d^4p' \\
&\times\Bigg[\frac{1}{(p-p')^2 - i\epsilon}\mathrm{Tr}\left\{S(p')\,\gamma_\nu\, S(p'+q)\,\gamma^\rho\, S(p+q)\,\gamma_\mu\, S(p)\,\gamma_\rho\right\} \\
&-\frac{1}{p'^2 - i\epsilon}\mathrm{Tr}\left\{S(p')\,\gamma_\nu\, S(p')\,\gamma^\rho\, S(p+q)\,\gamma_\mu\, S(p)\,\gamma_\rho\right\} \\
&-\frac{1}{p^2 - i\epsilon}\mathrm{Tr}\left\{S(p')\,\gamma_\nu\, S(p'+q)\,\gamma^\rho\, S(p)\,\gamma_\mu\, S(p)\,\gamma_\rho\right\}\Bigg] \\
&-2\,R_2\frac{ie^2}{(2\pi)^4}\int d^4p\;\mathrm{Tr}\left\{\gamma_\nu\; S(p+q)\,\gamma_\mu\, S(p)\right\} \\
&-(Z_3 - 1)_{\text{overlap}}(q^2\eta_{\mu\nu} - q_\mu q_\nu)\,.
\end{aligned} \qquad (12.2.18)$$

First consider integration over p' alone. Each of the first two terms is logarithmically divergent, but their difference is finite. The third term is also logarithmically divergent (with a gauge-invariant regulator), but the divergence in this term (unlike the first two terms) takes the form of a second-order polynomial in q, with the remainder finite. This remaining divergence is cancelled by the term $-(Z_3 - 1)(q^2\eta_{\mu\nu} - q_\mu q_\nu)$ that cancels all second-order terms in the expansion of $\Pi^*_{\mu\nu}(q)$. So the p'-subintegration gives a finite result. The symmetry of Eq. (12.2.16) shows that in exactly the same way the p subintegration also gives a finite result. Generic subintegrations over p and p' with $ap + bp'$ held fixed (where a and b are arbitrary non-zero constants) are manifestly convergent, and the integration over p and p' together is made finite by the counterterm $-(Z_3 - 1)_{\text{overlap}}(q^2\eta_{\mu\nu} - q_\mu q_\nu)$. Thus Eq. (12.2.18) *and* any of its subintegrations satisfy the power-counting requirements for convergence, and therefore according to the theorem[2] quoted in the previous section, the whole expression actually converges.

* * *

In electrodynamics there is a natural definition of renormalized couplings as well as renormalized masses and fields. This is not always the case. For example, consider the theory of a single real scalar field $\phi(x)$ with Lagrangian density

$$\mathscr{L} = -\frac{1}{2}\partial_\lambda\phi\partial^\lambda\phi - \tfrac{1}{2}m^2\phi^2 - \tfrac{1}{24}g\,\phi^4. \tag{12.2.19}$$

To one-loop order, the S-matrix for scalar–scalar scattering is given by the Feynman rules as

$$S(q_1q_2 \to q_1'q_2') = \frac{-i(2\pi)^4\delta^4(q_1'+q_2'-q_1-q_2)}{(2\pi)^6(16E_1'E_2'E_1E_2)^{1/2}}F(q_1q_2 \to q_1'q_2')\,, \tag{12.2.20}$$

where

$$-i(2\pi)^4\,F(q_1q_2 \to q_1'q_2') = -i(2\pi)^4 g + \frac{1}{2}\Big[-i(2\pi)^4 g\Big]^2\left[\frac{-i}{(2\pi)^4}\right]^2$$
$$\times \int d^4k \left[\frac{1}{\Big[(q_1+k)^2+m^2-i\epsilon\Big]\Big[(q_2-k)^2+m^2-i\epsilon\Big]} + (q_2 \to -q_1') + (q_2 \to -q_2')\right], \tag{12.2.21}$$

and q_1, q_2 and q_1', q_2' are the incoming and outgoing four-momenta. Combining denominators and rotating the k^0-integration contour as usual, this is

$$F = g - \frac{g^2}{16\pi^2}\int_0^\infty k^3dk\int_0^1 dx\Big\{\Big[k^2+m^2-sx(1-x)\Big]^{-2}$$
$$+\Big[k^2+m^2-tx(1-x)\Big]^{-2} + \Big[k^2+m^2-ux(1-x)\Big]^{-2}\Big\}, \tag{12.2.22}$$

where s, t, and u are the *Mandelstam variables*

$$s = -(q_1+q_2)^2\,, \qquad t = -(q_1-q_1')^2\,, \qquad u = -(q_1-q_2')^2\,, \tag{12.2.23}$$

related by $s+t+u = 4m^2$; also, x is the Feynman parameter introduced in combining denominators. With an ultraviolet cutoff at $k = \Lambda$, this gives the result (for $\Lambda \gg m$)

$$F = g - \frac{g^2}{32\pi^2}\int_0^1 dx\left\{\ln\left(\frac{\Lambda^2}{m^2 - sx(1-x)}\right)\right.$$
$$\left. + \ln\left(\frac{\Lambda^2}{m^2 - tx(1-x)}\right) + \ln\left(\frac{\Lambda^2}{m^2 - ux(1-x)}\right) - 3\right\}. \quad (12.2.24)$$

We can define the renormalized coupling g_R as the value of F at any point s, t, u we like, provided we stay in the region where F is real. For instance, suppose that in order to maintain the symmetry among the scalars, we choose to renormalize at the off-mass-shell point¶ $q_1^2 = q_2^2 = q_1'^2 = q_2'^2 = \mu^2$, $s = t = u = -4\mu^2/3$. Defining the renormalized coupling g_R as the value of F at this point, we have

$$g = g_R + \frac{3g^2}{32\pi^2}\left[\ln\left(\frac{\Lambda^2}{\mu^2}\right) - 1 - \int_0^1 dx\ \ln\left(\frac{4x(1-x)}{3} + \frac{m^2}{\mu^2}\right)\right] + \cdots . \quad (12.2.25)$$

The cutoff dependence then cancels in Eq. (12.2.24) to order g_R^2, leaving a finite formula for F in terms of g_R:

$$F = g_R - \frac{g_R^2}{32\pi^2}\int_0^1 dx\left\{\ln\left(\frac{m^2 + 4x(1-x)\mu^2/3}{m^2 - sx(1-x)}\right)\right.$$
$$\left. + \ln\left(\frac{m^2 + 4x(1-x)\mu^2/3}{m^2 - tx(1-x)}\right) + \ln\left(\frac{m^2 + 4x(1-x)\mu^2/3}{m^2 - ux(1-x)}\right)\right\} + \cdots . \quad (12.2.26)$$

Here μ^2 may be taken to be any real quantity greater than $-3m^2$, in which range g_R is real. The explicit μ-dependence in Eq. (12.2.26) is, of course, cancelled by the μ-dependence of the renormalized coupling. This freedom to change the renormalization prescription (which of course exists also in electrodynamics and other realistic theories) will be of great importance to us when we come to the renormalization group method in Volume II.

12.3 Is Renormalizability Necessary?

In the previous section we found a special class of theories having only a finite number of terms in the Lagrangian, to which the renormalization program is nevertheless applicable. These are theories in which all

¶ Going back over the derivation of Eq. (12.2.25), one may check that in this derivation we have not used the conditions $q_1^2 = q_2^2 = q_1'^2 = q_2'^2 = -m^2$, so Eq. (12.2.24) is valid whatever we take for the external line masses.

interactions satisfy the renormalizability condition

$$\Delta_i \equiv 4 - d_i - \sum_f n_{if}(s_f + 1) \geq 0 ,$$

where d_i and n_{if} are the numbers of derivatives and fields of type f in interactions of type i, and s_f is (with some qualifications) the spin of fields of type f. For renormalization to work in such theories, it is also usually necessary that all renormalizable interactions that are allowed by symmetry principles should actually appear in the Lagrangian.

It is important that there are only a limited number of such interaction types. Δ_i becomes negative if we have too many fields or derivatives, or fields of too high spin. Barring special cancellations, there are no renormalizable interactions at all involving fields with $s_f \geq 1$, because the only possible term in the Lagrangian with $\Delta_i \geq 0$ that involves such a field along with two or more other fields would involve a single $s_f = 1$ field along with two scalars and no derivatives, which would not be Lorentz-invariant. We shall see in Volume II that general, massless, spin one gauge fields in a suitable gauge effectively have $s_f = 0$, like the photon. Also, in Volume II we shall see that even massive gauge fields may effectively have $s_f = 0$, depending on where their mass comes from. Leaving aside these special cases, Table 12.2 gives a list of *all* renormalizable terms in the Lagrangian density that are allowed by Lorentz invariance and gauge invariance involving scalars ($s = 0$), photons ($s = 0$), and spin $\frac{1}{2}$ fermions ($s = \frac{1}{2}$).

We see that the requirement of renormalizability puts severe restrictions on the variety of physical theories that we may consider. Such restrictions provide a valuable key to the structure of physical theories. For instance, Lorentz and gauge invariance by themselves would allow the introduction of a 'Pauli' term proportional to $\bar{\psi}[\gamma_\mu, \gamma_\nu]\psi F^{\mu\nu}$ in the Lagrangian of quantum electrodynamics, which would make the magnetic moment of the electron an adjustable parameter, but we exclude such terms because they are not renormalizable. The successful predictions of quantum electrodynamics, such as the calculation of the magnetic moment of the electron outlined in Section 11.3, may be regarded as validations of the principle of renormalizability. The same applies to the standard model of weak, electromagnetic, and strong interactions, to be discussed in Volume II; there are any number of terms that might be added to this theory, such as four-fermion interactions among quarks and leptons, that would invalidate all the predictions of the standard model, and are excluded only because they are non-renormalizable.

Must we believe that the Lagrangian is restricted to contain only renormalizable interactions? As we saw in the previous section, if we include in the Lagrangian *all* of the infinite number of interactions allowed

Table 12.2. Allowed renormalizable terms in a Lagrangian density involving scalars ϕ, Dirac fields ψ, and photon fields A^μ. Here n_{if} and d_i are the number of fields of type f and the number of derivatives in an interaction of type i, and Δ_i is the dimensionality of the associated coefficient.

	n_{if}		d_i	Δ_i	$\mathscr{H}_i$
Scalars	Photons	Spin $\frac{1}{2}$			
1	0	0	0	3	ϕ
2	0	0	0	2	ϕ^2
2	0	0	2	0	$\partial_\mu\phi\partial^\mu\phi$
3	0	0	0	1	ϕ^3
4	0	0	0	0	ϕ^4
2	1	0	1	0	$\phi\partial_\mu\phi A^\mu$
2	2	0	0	0	$\phi^2 A_\mu A^\mu$
1	0	2	0	0	$\phi\bar{\psi}\psi$
0	2	0	2	0	$F_{\mu\nu}F^{\mu\nu}$
0	0	2	0	1	$\bar{\psi}\psi$
0	0	2	1	0	$\bar{\psi}\gamma^\mu\partial_\mu\psi$
0	1	2	0	0	$\bar{\psi}\gamma^\mu A_\mu\psi$

by symmetries, then there will be a counterterm available to cancel every ultraviolet divergence. In this sense, as said earlier, non-renormalizable theories are just as renormalizable as renormalizable theories, as long as we include all possible terms in the Lagrangian.

In recent years it has become increasingly apparent that renormalizability is not a fundamental physical requirement, and that in fact any realistic quantum field theory will contain non-renormalizable as well as renormalizable terms. This change in point of view can be traced in part to the continued failure to find a renormalizable theory of gravitation. In the general class of metric theories of gravitation governed by Einstein's principle of equivalence there are no renormalizable interactions at all – generally covariant interactions must be constructed from the curvature tensor and its generally covariant derivatives, and hence, even in a 'gauge' where the graviton propagator goes as k^{-2}, these interactions involve too many derivatives of the metric for renormalizability. In particular, we can easily see that general relativity is non-renormalizable from the fact that its coupling constant $8\pi G_N = (2.43 \times 10^{18}\text{ GeV})^{-2}$ has negative dimensionality. Even if nothing else did, the cancellation of divergences due to virtual gravitons would require that the Lagrangian contain all interactions allowed by symmetries — not only interactions involving gravitons, but involving any particles.

But if renormalizability is not a fundamental physical principle, then how do we explain the success of renormalizable theories like quantum electrodynamics and the standard model? The answer can be seen by simple dimensional analysis. We have already noted that the coupling constant of an interaction of type i has dimensionality

$$[g_i] \sim [\text{mass}]^{\Delta_i}\,, \tag{12.3.1}$$

where Δ_i is the index (12.1.9). Non-renormalizable interactions are just those whose coupling constants have the dimensionality of *negative* powers of mass. Now, it is not unreasonable to guess from (12.3.1) that the coupling constants not only have dimensionalities governed by Δ_i, but are roughly of order

$$g_i \approx M^{\Delta_i}\,, \tag{12.3.2}$$

where M is some common mass. (This is found to be actually the case in the effective field theories discussed below and in more detail in Volume II.) In calculating physical processes at a characteristic momentum scale $k \ll M$, the inclusion of a non-renormalizable interaction of type i with $\Delta_i < 0$ will introduce a factor $g_i \approx M^{\Delta_i}$, which on dimensional grounds must be accompanied by a factor $k^{-\Delta_i}$, and so the effect of such an interaction is suppressed* for $k \ll M$ by a factor $(k/M)^{-\Delta_i} \ll 1$. (This argument will be made more carefully using the method of the renormalization group in Volume II.) The success of the renormalizable theories of electroweak and strong interactions shows only that M is very much larger than the energy scale at which these theories have been tested.

* It is essential at this point to assume that the ultraviolet divergences have been removed by renormalization, so that there are no factors of an ultraviolet cutoff Λ to mess up our dimensional analysis. Otherwise, dimensional analysis tells us that for $\Lambda \to \infty$, each additional non-renormalizable coupling constant factor g_i with $\Delta_i < 0$ would be accompanied with a growing factor $\Lambda^{-\Delta_i}$. This dimensional argument led Heisenberg[9] very early to classify interactions according to the dimensionality of their coupling constants, and to suggest[10] that new effects might arise at energies of order g_i^{1/Δ_i}, as for instance at the energy $G_F^{-1/2} \approx 300$ GeV, where G_F is the four-fermion coupling constant of the Fermi beta decay theory. After the development of renormalization theory it was noted by Sakata *et al.*[11] that the non-renormalizable theories are those whose coupling constants have negative dimensionality.

For instance, the leading non-renormalizable corrections to the conventional electrodynamics of electrons or muons would be those interactions of dimension 5, which are suppressed by only one factor of $1/M$. There is just one such interaction allowed by Lorentz, gauge, and CP invariance, a Pauli term of order $(ie/2M)\,\bar{\psi}[\gamma_\mu, \gamma_\nu]\psi\, F^{\mu\nu}$. According to Eqs. (10.6.24), (10.6.17), and (10.6.19), such a term would contribute an amount of order $4e/M$ to the magnetic moment of the electron or muon. The calculated value of the magnetic moment of the electron agrees with experiment to within terms of order $10^{-10}e/2m_e$, so M must be greater than about $8 \times 10^{10} m_e = 4 \times 10^7$ GeV.

This limit may be weakened if other symmetries restrict the form of the non-renormalizable interactions. For instance, the conventional Lagrangian of quantum electrodynamics is invariant under a chiral transformation $\psi \to \gamma_5\psi$, except for a change of sign of the fermion mass term $-m\bar{\psi}\psi$. If we assume that the full Lagrangian is invariant under a formal symmetry $\psi \to \gamma_5\psi$, $m \to -m$, then a Pauli term in the Lagrangian would have to appear with an extra factor m/M, so that its contribution to the magnetic moment would be only of order $4em/M^2$. Because of the extra factor of m, here it is the muon rather than the electron that provides the most useful limit on M. The calculated value of the magnetic moment of the muon agrees with experiment to within terms of order $10^{-8}e/2m_\mu$, so M must be greater than about $\sqrt{8 \times 10^8}m_\mu = 3 \times 10^3$ GeV. In any case, if M is anywhere near as large as 10^{18} GeV, then we are certainly justified in neglecting any non-renormalizable interactions that might appear in quantum electrodynamics.

These considerations help us to cope with some of the puzzles associated with higher-derivative terms in the Lagrangian. For instance, in the general theory of a real scalar field ϕ, we would expect to find terms in the Lagrangian density of the form $\phi\Box^n\phi$. Any one such term would make a direct contribution to the scalar self-energy function $\Pi^*(q^2)$ proportional to $(q^2)^n$. If we were to include this contribution to all orders, but ignore all other effects of non-renormalizable interactions, then the propagator $\Delta'(q^2) = 1/(q^2 + m^2 - \Pi^*(q^2))$ would not have the simple pole in q^2 at negative q^2 expected from the general arguments of Section 10.7, but n such poles (some of which may coincide), generally at complex values of q^2. But if the non-renormalizable term $\phi\Box^n\phi$ has a coefficient of order $M^{-2(n-1)}$, where $M \gg m$, then the extra poles are at q^2 of order M^2, where it is illegitimate to ignore the infinite number of other non-renormalizable interactions that must also appear in the Lagrangian. Thus the appearance of higher-derivative terms in a general non-renormalizable Lagrangian is not in conflict with the general principles underlying quantum field theory that were used in Section 10.7. But by the same token, we also cannot use higher-derivative terms to avoid ultraviolet divergences altogether, as has been repeatedly proposed. A term $M^{-2(n-1)}\phi\Box^n\phi$ in the Lagrangian density provides a cutoff at momenta $q^2 \approx M^2$, but at these momenta we cannot ignore all the other non-renormalizable interactions that must be present.

Although highly suppressed, non-renormalizable interactions may be detectable if they have effects that would otherwise be forbidden. For instance, we will see in Section 12.5 that the symmetries of charge-conjugation and space-inversion invariance are an automatic consequence of the structure of the electromagnetic interactions that is imposed by gauge invariance, Lorentz invariance, and renormalizability, but we can easily imagine non-renormalizable terms that would violate these symmetries, such as an electron electric dipole moment term $\bar{\psi}\gamma_5\,[\gamma_\mu, \gamma_\nu]\,\psi F^{\mu\nu}$, or the Fermi interaction $\bar{\psi}\gamma_5\gamma_\mu\psi\bar{\psi}\gamma^\mu\psi$. It is widely believed today that the conservation of baryon and lepton number is violated by very small effects of highly suppressed non-renormalizable interactions. Another example of a detectable non-renormalizable interaction is provided by gravitation. As mentioned before, gravitons have no renormalizable interactions at all. But, of course, we detect gravitation, because it has the special property that the gravitational fields of all the particles in a macroscopic body add up coherently.

Although non-renormalizable theories involve an infinite number of free parameters, they retain considerable predictive power:[12] they allow us to calculate the non-analytic parts of Feynman amplitudes, like the $\ln q$ and $q \ln q$ terms in the one-dimensional examples at the beginning of the previous section. Such calculations just reproduce the results required by the axiom of S-matrix theory, that the S-matrix has only those singularities required by unitarity.

Paradoxically, it is just in the case where symmetry principles forbid renormalizable interactions that non-renormalizable quantum field theories prove the most useful. In such cases we can derive a useful perturbation theory by expanding in powers of k/M. This has been worked out in detail for the theory of low-energy pions,[12,13] to be discussed in detail in Volume II, and the theory of low-energy gravitons.[14] For a simpler example, consider the theory of a real scalar field, satisfying the principle of invariance under the field translation

$$\phi(x) \to \phi(x) + \epsilon$$

with ϵ an arbitrary constant. This symmetry forbids any renormalizable interactions or scalar mass, but it allows an infinite number of non-renormalizable derivative interactions

$$\mathscr{L} = -\frac{1}{2}\partial_\mu\phi\partial^\mu\phi - \frac{g}{4}(\partial_\mu\phi\partial^\mu\phi)^2 - \cdots ,$$

where $g \approx M^{-4}$, and '$\cdots$' denotes terms with more derivatives or fields. (For simplicity, it is assumed here that the theory also has a symmetry under the reflection $\phi \to -\phi$.) According to the above dimensional analysis, the graph for a general reaction in which all energies and momenta are of order $k \ll M$ is suppressed by a factor $(k/M)^\nu$, where

$$\nu = -\sum_i V_i \Delta_i = \sum_i V_i(d_i + n_i - 4) ,$$

with n_i and d_i the numbers of scalar fields and derivatives in an interaction of type i, and V_i the number of vertices for these interactions in our graph. For $k \ll M$, the dominant contributions to any process are those with the smallest value of ν. The formula for ν can be put in a more useful form by using the familiar topological identities for a connected graph:

$$\sum_i V_i = I - L + 1 , \qquad \sum_i V_i n_i = 2I + E ,$$

where I, E, and L are the numbers of internal lines, external lines, and loops in our graph. Combining these relations gives

$$\nu = 2E - 4 + 4L + \sum_i V_i(d_i - n_i) ,$$

Now, the field translation symmetry requires that every field must be accompanied with at least one derivative, so the quantity $d_i - n_i$ as well as L and V_i are non-negative for all interactions. Thus for a given process (that is, a fixed number E of external lines) the dominant terms will be those constructed solely from *tree* graphs (i.e., $L = 0$), and interactions with the minimum number $d_i = n_i$ of derivatives. That is, in leading order we can take the Lagrangian density to depend only on *first* derivatives of the field. Higher-order corrections may involve loops and/or interactions with more derivatives on some fields. But to any given order ν in k/M, we need only consider a finite number of graphs, those with $L \leq (4-2E+\nu)/4$, and only a finite number of interaction types.

For instance, scalar–scalar scattering is given in leading order by the one-vertex tree graph calculated using the interaction $-g(\partial_\mu\phi\partial^\mu\phi)^2$ in first order. According to our formula for ν, the leading correction, suppressed at low energy by a factor $(k/M)^2$, arises from another single-vertex tree graph, produced by an interaction with two additional derivatives of the form** $\partial_\mu\partial_\nu\phi\, \partial^\mu\partial^\nu\phi\, \partial_\lambda\phi\partial^\lambda\phi$. The next corrections, suppressed at low energy by two further factors of k/M, arise both from the one-loop diagram of Figure 12.4 (including permutations of external lines), calculated using only the interaction $-g(\partial_\mu\phi\partial^\mu\phi)^2$, and also from tree graphs with

** In accordance with the remarks of Section 7.7, we are excluding interactions involving $\Box\phi$, because the field equation for ϕ can be used to express such interactions in terms of the others.

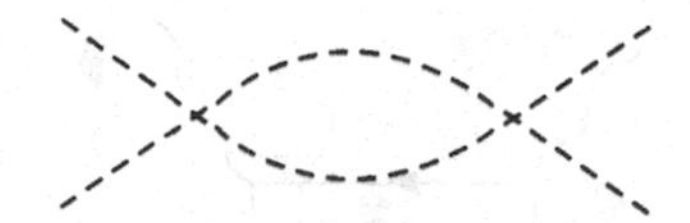

Figure 12.4. One-loop diagram for scalar–scalar scattering in the theory with derivative quadrilinear interactions.

a single vertex arising from a quartic interaction with eight derivatives, whose couplings contain infinite parts that cancel the ultraviolet divergence from the loop graph.† The loop graph also yields finite terms in the scattering amplitude proportional to terms like $s^4 \ln s + t^4 \ln t + u^4 \ln u$, $s^2t^2 \ln u + t^2u^2 \ln s + u^2s^2 \ln t$, etc., with calculable coefficients proportional to g^2. These finite terms simply represent the correction to the lowest-order scattering amplitude needed to ensure the unitarity of the S-matrix, but perturbative quantum field theory is by far the easiest way of calculating them.

Although non-renormalizable theories can provide useful expansions in powers of energy, they inevitably lose all predictive power at energies of the order of the common mass scale M that characterizes the various couplings. If we were to take these expansions literally, the results for S-matrix elements would violate unitarity bounds for $E \gg M$. There seem to be just two possibilities about what happens at such energies. One is that the growing strength of the effects of the non-renormalizable couplings somehow saturates, avoiding any conflict with unitarity.[15] The other is that new physics of some sort enters at the scale M. In this case, the non-renormalizable theories that describe nature at energies $E \ll M$ are just *effective field theories* rather than truly fundamental theories.

Probably the earliest example of an effective field theory was derived in the 1930s by Euler *et al.*,[16] as a theory of low-energy photon–photon interactions. (See Section 1.3.) In effect, they calculated the contribution to photon-photon scattering of Feynman diagrams such as Figure 12.5, and found that at energies much less than m_e the scattering of light by light was the same as would be calculated with an effective Lagrangian

$$\mathscr{L}_{\text{eff}} = \frac{2\alpha^2}{45m_e^4}\left[(\mathbf{E}^2 - \mathbf{B}^2)^2 + 7(\mathbf{E}\cdot\mathbf{B})^2\right]$$
$$+ \text{ higher orders in } \frac{eE}{m_e^2} \;\&\; \frac{eB}{m_e^2}$$

† These are the only ultraviolet divergences encountered in one-loop graphs if we use dimensional regularization. For other methods of regularization there are also quartic and quadratic divergences, which are cancelled by counterterms in the four-scalar interactions with four or six derivatives.

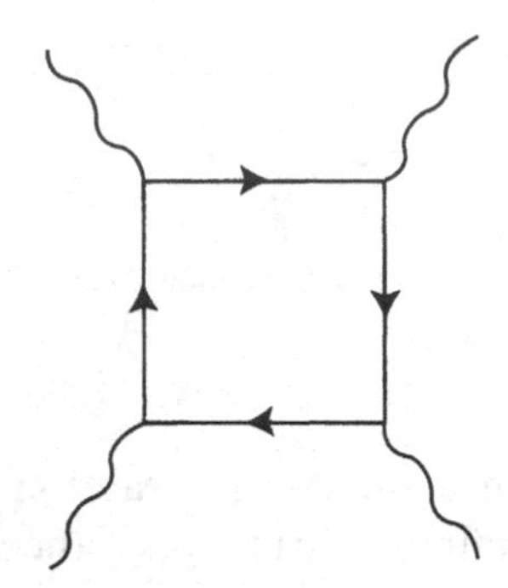

Figure 12.5. Diagram for photon–photon scattering, whose effect at low energy can be calculated from the effective Lagrangian of Euler *et al.*[16] Straight lines are electrons; wavy lines are photons.

Euler *et al.* used this effective Lagrangian only in the tree approximation, to calculate the leading terms in photon interaction matrix elements. It was not until much later that such Lagrangians, though non-renormalizable, were used beyond the tree approximation.[12,17]

In modern jargon, we say that in deriving this Lagrangian the electron is 'integrated out', because in the one-loop approximation we have

$$\exp\left(i\int \mathscr{L}_{\text{eff}}(\mathbf{E},\mathbf{B})d^4x\right) = \int\left[\prod_x d\psi_e(x)\right]\exp\left(i\int \mathscr{L}_{\text{QED}}(\psi_e,\mathbf{A})d^4x\right)\,.$$

A more general procedure is simply to write down the most general non-renormalizable effective Lagrangian, use it to calculate various amplitudes as an expansion in energies and momenta, and then choose the constants in the effective Lagrangian by matching the results it gives for these amplitudes to those derived from the underlying theory.

We will encounter effective field theories again, especially in considering broken symmetries in Volume II. As we shall see, effective field theories are useful even where they cannot be derived from an underlying theory, either because the theory is unknown, or because its interactions are too strong to allow the use of perturbation theory. Indeed, even if we knew nothing about the properties of charged particles, the scattering of photons at sufficiently low energy would have to be described by an effective Lagrangian consisting of the terms $(\mathbf{E}^2-\mathbf{B}^2)^2$ and $(\mathbf{E}\cdot\mathbf{B})^2$, because these are the unique quartic Lorentz- and gauge-invariant terms with no derivatives acting on $\mathbf{E}$ and $\mathbf{B}$. Terms with such derivatives would be suppressed at low photon energies E by additional factors of E/M, where M is some typical mass of the charged particles that are being integrated out. We can go further: we shall see that effective field theories are useful even where the light particles they describe are not present in the underlying theory at all, but composites of the heavy particles that are integrated out. The underlying theory might not even be a field theory at all — the problem of incorporating gravitation has led many theorists to believe that, in fact, it is a string theory. But wherever an effective field theory comes from, it is inevitably a non-renormalizable theory.

12.4 The Floating Cutoff*

Before closing this chapter, it is worth commenting on the relation between conventional renormalization theory and an approach pioneered by Wilson.[18] In Wilson's method one imposes a 'floating' finite ultraviolet cut-off (either sharp or smooth) at momenta with components of order Λ, and instead of taking $\Lambda \to \infty$, one requires that the bare constants of the theory (those appearing in the Lagrangian) depend on Λ in such a way that all observable quantities are Λ-independent.

It is convenient to work with dimensionless parameters. If a bare coupling or mass parameter $g_i(\Lambda)$ has dimensionality $[\text{mass}]^{\Delta_i}$, we define the corresponding dimensionless parameter $\mathscr{G}_i$ by

$$\mathscr{G}_i(\Lambda) \equiv \Lambda^{-\Delta_i} g_i(\Lambda) \,. \tag{12.4.1}$$

Ordinary dimensional analysis tells us that the value of $\mathscr{G}_i$ at one value Λ' of the cutoff can be expressed as a function of the values of the $\mathscr{G}_j$ at another value Λ of the cutoff, and the dimensionless ratio Λ'/Λ:

$$\mathscr{G}_i(\Lambda') = F_i\Big(\mathscr{G}(\Lambda), \Lambda'/\Lambda\Big) \,. \tag{12.4.2}$$

No dimensional parameters other than Λ' and Λ can appear in F, because no ultraviolet or infrared divergences can enter here; the difference between the constants at Λ and at Λ' arises from diagrams whose internal lines are restricted to have momenta between Λ and Λ'. Differentiating Eq. (12.4.2) with respect to Λ' and then setting Λ' equal to Λ yields a differential equation for $\mathscr{G}_i$:

$$\Lambda \frac{d}{d\Lambda} \mathscr{G}_i(\Lambda) = \beta_i(\mathscr{G}(\Lambda)) \,, \tag{12.4.3}$$

where $\beta_i(\mathscr{G}) \equiv [\partial/\partial z \, F_i(\mathscr{G}, z)]_{z=1}$. The functions $\beta_i(\mathscr{G})$ may be calculated for small couplings in perturbation theory. This is Wilson's version of the 'renormalization group' equation, which will be discussed in somewhat different terms in Volume II.

The Lagrangian for any finite value of the cutoff defines an effective field theory, in which instead of (or in addition to) integrating out 'heavy'

* This section lies somewhat out of the book's main line of development, and may be omitted in a first reading.

particles like the electron in the work of Euler *et al.*, one integrates out *all* particles with momenta greater than Λ. Even if one starts with a theory with a finite number of coupling parameters $\mathscr{G}_i^0$ at some cutoff Λ_0, at any other value of the cutoff the differential equation (12.4.3) will generally yield non-zero values for all couplings allowed by symmetry principles.**

We now distinguish between the renormalizable and unrenormalizable couplings, labelled $\mathscr{G}_a$ and $\mathscr{G}_n$, respectively, with a running over the finite number N of couplings (including masses) for which $\Delta_a \geq 0$, and n running over the infinite number of couplings with dimensionalities $\Delta_n < 0$. We want to show that if the couplings $\mathscr{G}_a(\Lambda_0)$ and $\mathscr{G}_n(\Lambda_0)$ at some initial cutoff value Λ_0 lie on a generic N-dimensional initial surface $\mathscr{S}_0$, then (with some qualifications) for $\Lambda \ll \Lambda_0$ they will approach a fixed surface $\mathscr{S}$ that is independent of both Λ_0 and the initial surface.† This fixed surface is stable, in the sense that from any point on the surface, the trajectory generated by Eq. (12.4.3) stays on the surface. Such a stable surface defines a finite-parameter set of theories whose physical content is cutoff-independent, which as argued in the previous section, is the essential property of renormalizable theories. Furthermore, this construction shows that a generic theory defined with cutoff Λ_0 will look for $\Lambda \ll \Lambda_0$ like a renormalizable theory.‡

To prove these results, consider any small perturbation $\delta\mathscr{G}_i(\Lambda)$ in the values of the $\mathscr{G}_i(\Lambda)$ satisfying Eq. (12.4.3). It will satisfy the differential equation

$$\Lambda\frac{d}{d\Lambda}\delta\mathscr{G}_i(\Lambda) = \sum_j M_{ij}(\mathscr{G}(\Lambda))\,\delta\mathscr{G}_j(\Lambda)\,, \tag{12.4.4}$$

where

$$M_{ij}(\mathscr{G}) \equiv \frac{\partial}{\partial\mathscr{G}_j}\beta_i(\mathscr{G})\,. \tag{12.4.5}$$

This equation couples the renormalizable and unrenormalizable couplings, making it difficult to see the difference in their behavior. To decouple them, we introduce the linear combinations

$$\xi_n \equiv \delta\mathscr{G}_n - \sum_{ab}\frac{\partial\mathscr{G}_n}{\partial\mathscr{G}_a^0}\left(\frac{\partial\mathscr{G}}{\partial\mathscr{G}^0}\right)^{-1}_{ab}\delta\mathscr{G}_b\,, \tag{12.4.6}$$

** The only known exceptions to this rule are in theories based on supersymmetry.[4]

† This theorem is due to Polchinski.[19] What follows here is a shortened and less rigorous version. (In Polchinski's proof, the initial surface is taken to be that with all non-renormalizable couplings vanishing. As we shall see here, the couplings approach the same fixed surface for generic initial surfaces.)

‡ Of course some theories have symmetries and a field content that do not allow any renormalizable interactions. This is the case for theories containing only fermion fields, or only the gravitational field. Such theories for $\Lambda \ll \Lambda_0$ look like free-field theories.

where $\mathscr{G}_a^0$ are the values of the renormalizable couplings at cutoff Λ_0, which we shall use as coordinates for the initial surface, and $\mathscr{G}_n$ are the values of the non-renormalizable couplings at cutoff Λ derived from the differential equation (12.4.3), with initial value for Λ_0 at the point on the initial surface with coordinates $\mathscr{G}_a^0$. To calculate the derivative of ξ_n with respect to Λ, we note that the derivatives $\partial\mathscr{G}_i/\partial\mathscr{G}_a^0$ satisfy the same differential equation (12.4.4) as the $\delta\mathscr{G}_i$. It is an elementary exercise then to show that

$$\Lambda\frac{d}{d\Lambda}\xi_n = \sum_m N_{nm}\,\xi_m\,, \tag{12.4.7}$$

where

$$N_{nm} \equiv M_{nm} - \sum_{ab}\frac{\partial\mathscr{G}_n}{\partial\mathscr{G}_a^0}\left(\frac{\partial\mathscr{G}}{\partial\mathscr{G}^0}\right)^{-1}_{ab} M_{bm}\,. \tag{12.4.8}$$

Now we must estimate the elements of N_{nm}. For a free-field theory no cutoff is needed, so for very small coupling all bare parameters $g_i(\Lambda)$ become Λ-independent. Hence for small couplings the dimensionless parameters $\mathscr{G}_i$ simply scale as $\Lambda^{-\Delta_i}$, and the matrix M_{ij} is given by

$$M_{ij} \approx -\Delta_i\delta_{ij}\,. \tag{12.4.9}$$

It follows that the matrix N_{nm} is given approximately by $-\Delta_n\delta_{nm}$. The defining characteristic of the non-renormalizable couplings is that $\Delta_n < 0$, so Eq. (12.4.7) tells us that, at least for couplings in some finite range, where N_{nm} is positive-definite, the ξ_n decay for $\Lambda \ll \Lambda_0$ like positive powers of Λ/Λ_0. In this limit, then, the perturbations are related by

$$\delta\mathscr{G}_n = \sum_{ab}\frac{\partial\mathscr{G}_n}{\partial\mathscr{G}_a^0}\left(\frac{\partial\mathscr{G}}{\partial\mathscr{G}^0}\right)^{-1}_{ab}\delta\mathscr{G}_b\,. \tag{12.4.10}$$

In particular, if we make a small change in the initial surface $\mathscr{S}_0$ and/or the starting point on that surface and/or the initial cutoff Λ_0, such that the perturbations $\delta\mathscr{G}_a$ in the renormalizable couplings vanish at some cutoff $\Lambda \ll \Lambda_0$, then the perturbations $\delta\mathscr{G}_n$ in all the other couplings at cutoff Λ also vanish. Thus the non-renormalizable couplings $\mathscr{G}_n(\Lambda)$ for $\Lambda \ll \Lambda_0$ can depend only on the renormalizable couplings $\mathscr{G}_a(\Lambda)$, not separately on the initial surface or the starting point on that surface or the initial cutoff Λ_0. At cutoff $\Lambda \ll \Lambda_0$ all the couplings therefore approach an N-dimensional surface $\mathscr{S}$, with coordinates $\mathscr{G}_a(\Lambda)$, which is independent of both the initial surface and of Λ_0. Note that the non-renormalizable couplings $\mathscr{G}_n$ are not generally small on $\mathscr{S}$; the important point is that they become functions of the renormalizable couplings. Changes in Λ with Λ remaining much less than Λ_0 will change the couplings, but the couplings will remain close to $\mathscr{S}$ (at least as long as the couplings do not become so large that N_{nm} is no longer a positive-definite matrix). Hence $\mathscr{S}$ is a stable surface, as was to be proved.

We have seen that all physical quantities may be expressed in terms of Λ and the $\mathscr{G}_n(\Lambda)$, and are Λ-independent. This is true in particular of the N conventional renormalized couplings and masses, like e and m_e in quantum electrodynamics. But we can then invert this relation, and express the $\mathscr{G}_n(\Lambda)$ in terms of the conventional parameters and Λ. In this way we can justify the usual renormalization program: all physical quantities are expressed in a cutoff-independent way in terms of the conventional renormalized couplings and masses.

The Wilson approach has some advantages in practice. One does not have to worry about subintegrations and overlapping divergences; the momentum cutoff applies to all internal lines. Also, some of the non-renormalization theorems of supersymmetry theory, which tell us that certain couplings are not affected by radiative corrections, work only for the cutoff-dependent bare couplings.[20]

On the other hand, there are disadvantages to the Wilson approach. One must give up the special simplicities of working with renormalizable theories like quantum electrodynamics; once one starts integrating out particles with momenta above some scale Λ, the resulting effective field theory will contain all Lorentz- and gauge-invariant interactions, with Λ-dependent couplings. (Nevertheless, in physical processes at energies $E \ll \Lambda$, the dominant couplings will still be the renormalizable ones.) Also, the cutoff generally destroys *manifest* gauge invariance, and either manifest Lorentz invariance or unitarity. None of this is a problem in condensed matter physics, the original context of Wilson's approach, because no one would expect a realistic condensed matter theory to be strictly renormalizable, and there are no fundamental physical principles that are necessarily violated by a cutoff. In fact, in crystals there *is* a cutoff on phonon momenta, provided by the inverse lattice spacing.

At bottom, the difference between the conventional and the Wilson approach is one of mathematical convenience rather than of physical interpretation. Indeed, conventional renormalization already provides a sort of adjustable cutoff; when we express our answer in terms of coupling constants that are defined as the values of physical amplitudes at some momenta of order μ (as for the scalar field theory discussed in the previous section), the cancellations that make integrals converge begin to operate at virtual momenta of order μ. Conversely, the Λ-dependent coupling constants of the Wilson approach must ultimately be expressed in terms of observable masses and charges, and when this is done the results are, of course, the same as those obtained by conventional means.

12.5 Accidental Symmetries*

In Section 12.3 we saw that there are good reasons to adopt renormalizable field theories as approximate descriptions of nature at sufficiently low energy. It often happens that the condition of renormalizability is so stringent that the effective Lagrangian automatically obeys one or more symmetries, which are not symmetries of the underlying theory, and may therefore be violated by the suppressed non-renormalizable terms in the effective Lagrangian. Indeed, most of the experimentally discovered symmetries of elementary particle physics are 'accidental symmetries' of this sort.

A classic example is provided by the inversions and flavor conservation in the electrodynamics of charged leptons. The most general renormalizable and gauge- and Lorentz-invariant Lagrangian density for photons and fields ψ_i of spin $\frac{1}{2}$ and charge $-e$ takes the form

$$\begin{aligned}\mathscr{L} = & -\tfrac{1}{4}Z_3 F_{\mu\nu}F^{\mu\nu} \\ & -\sum_{ij} Z_{Lij}\bar{\psi}_{Li}[\partial\!\!\!/ + ie A\!\!\!/]\psi_{Lj} - \sum_{ij} Z_{Rij}\bar{\psi}_{Ri}[\partial\!\!\!/ + ie A\!\!\!/]\psi_{Rj} \\ & -\sum_{ij} M_{ij}\bar{\psi}_{Li}\psi_{Rj} - \sum_{ij} M^{\dagger}_{ij}\bar{\psi}_{Ri}\psi_{Lj}\,, \end{aligned} \tag{12.5.1}$$

where i, j are summed over the three lepton flavors (e, μ, and τ), ψ_{iL} and ψ_{iR} are the left- and right-handed parts of the field ψ_i, defined by

$$\psi_{Li} = \tfrac{1}{2}(1+\gamma_5)\psi_i\,, \qquad \psi_{Ri} = \tfrac{1}{2}(1-\gamma_5)\psi_i\,, \tag{12.5.2}$$

and Z_L, Z_R, and M are numerical matrices. We are not assuming anything about lepton flavor conservation, so the matrices Z_{Lij}, Z_{Rij} and M_{ij} need not be diagonal. Also we are not assuming anything about invariance under P, C, or T invariance, so there is no necessary relation between Z_L and Z_R, or between M and $M^{\dagger}$. The only constraints on these matrices come from the reality of the Lagrangian density, which requires that Z_{Lij} and Z_{Rij} are Hermitian, and from the canonical anticommutation relations, which require that Z_{Lij} and Z_{Rij} are positive-definite.

Now suppose we replace the lepton fields ψ_L, ψ_R with new fields ψ'_L, ψ'_R defined by

$$\psi_L = S_L\psi'_L\,, \qquad \psi_R = S_R\psi'_R\,, \tag{12.5.3}$$

where $S_{L,R}$ are non-singular matrices that can be chosen as we like. The Lagrangian density when expressed in terms of these new fields then takes

* This section lies somewhat out of the book's main line of development, and may be omitted in a first reading.

the same form as in Eq. (12.5.1), but with new matrices

$$Z'_L = S_L^\dagger Z_L S_L\,, \qquad Z'_R = S_R^\dagger Z_R S_R\,, \qquad M' = S_L^\dagger M S_R\,. \tag{12.5.4}$$

We can choose S_L and S_R so that $Z'_L = Z'_R = 1$. (Take $S_{L,R} = U_{L,R} D_{L,R}$, where $U_{L,R}$ are the unitary matrices that diagonalize the positive-definite Hermitian matrices $Z_{L,R}$, and the $D_{L,R}$ are the diagonal matrices whose elements are the inverse square roots of the eigenvalues of $Z_{L,R}$.)

Now make another transformation, to lepton fields ψ''_i defined by

$$\psi'_L = S'_L \psi''_L\,, \qquad \psi'_R = S'_R \psi''_R\,. \tag{12.5.5}$$

The Lagrangian density again takes the same form when expressed in terms of these new fields, with new matrices

$$Z''_L = S_L'^\dagger S'_L\,, \qquad Z''_R = S_R'^\dagger S'_R\,, \qquad M'' = S_L'^\dagger M' S'_R\,. \tag{12.5.6}$$

This time we take $S'_{L,R}$ unitary, so that again $Z''_L = Z''_R = 1$. We choose these unitary matrices so that M'' is real and diagonal. (By the polar decomposition principle, M' like any square matrix may be put in the form $M' = VH$, where V is unitary and H is Hermitian. Take $S'_L = S_R'^\dagger\, V^\dagger$ and choose S'_R as the unitary matrix that diagonalizes H.) Dropping primes, the Lagrangian density now takes the form

$$\mathscr{L} = -\tfrac{1}{4} Z_3 F_{\mu\nu} F^{\mu\nu} - \sum_i \bar{\psi}_{Li} [\not\partial + ie \not A] \psi_{Li} - \sum_i \bar{\psi}_{Ri} [\not\partial + ie \not A] \psi_{Ri}$$
$$- \sum_i m_i \bar{\psi}_{Li} \psi_{Ri} - \sum_i m_i \bar{\psi}_{Ri} \psi_{Li}\,, \tag{12.5.7}$$

where m_i are real numbers, the eigenvalues of the Hermitian matrix H. Finally, this can be put in the more familiar form

$$\mathscr{L} = -\tfrac{1}{4} Z_3 F_{\mu\nu} F^{\mu\nu} - \sum_i \bar{\psi}_i [\not\partial + ie \not A] \psi_i - \sum_i m_i \bar{\psi}_i \psi_i\,. \tag{12.5.8}$$

With the Lagrangian taking this form, it is now apparent that any renormalizable Lagrangian for lepton electrodynamics automatically conserves P, C, and T, as well as the numbers of leptons (minus the numbers of antileptons) of each flavor: electron, muon, and tauon.** In particular, despite the appearance of Eq. (12.5.1), this theory does not allow such processes as $\mu \to e+\gamma$. The reader may perhaps worry whether it is correct to identify the lepton fields as the ψ_i (previously called ψ''_i) appearing in Eq. (12.5.8), which obviously conserves lepton flavor, rather than the ψ_i appearing in Eq. (12.5.1), which seems to allow processes like $\mu \to e + \gamma$.

** This was first shown by Feinberg, Kabir, and myself.[21] Feinberg[22] had earlier noted that weak interaction effects in a theory with only one neutrino species would give rise to an observable rate for the process $\mu \to e + \gamma$, a difficulty that was only resolved by the discovery of a second neutrino species.

Such worries may be put aside; as stressed in Section 10.3, there is no one field that can be identified as *the* field of the electron or muon. In fact, although Eq. (12.5.1) yields a non-vanishing matrix element for radiative decay of lepton 1 into lepton 2 *off* the lepton mass shell, by taking the lepton momenta on the mass shell we find a vanishing S-matrix for all such processes even when calculated using Eq. (12.5.1).

It was essential in deriving these results that the same electric charge appeared in Eq. (12.5.1) for both the left- and right-handed parts of the lepton fields or, in other words, that both left- and right-handed parts of the lepton fields transform in the same way under electromagnetic gauge transformations. As we shall see in Volume II, for similar reasons the modern renormalizable theory of strong interactions known as quantum chromodynamics automatically conserves C, and (aside from certain non-perturbative effects) P and T, as well as the numbers of quarks (minus the numbers of antiquarks) of each quark flavor. We shall also see in Volume II that the simplest version of the renormalizable standard model of weak and electromagnetic interactions automatically conserves lepton flavor (though not C and P) for reasons similar to those described here for electrodynamics. It remains an open possibility that non-renormalizable interactions arising from higher mass scales may violate any of these conservation laws.

Problems

1. List all the renormalizable (or superrenormalizable) Lorentz-invariant terms in the Lagrangian of a single scalar field for space-time dimensionalities 2, 3, and 6.

2. Show how the overlapping divergence in the electron self-energy is cancelled in quantum electrodynamics.

3. Consider the theory of a scalar field ϕ and spinor field ψ, with interaction Hamiltonian $g\phi\bar{\psi}\psi$. Write the one-loop part of the scalar self-energy function $\Pi^*(q)$ as a divergent polynomial in p^μ, plus an explicit convergent integral.

4. Suppose that the quantum electrodynamics of electrons and photons is actually an effective field theory, derived by integrating out unknown particles of mass $M \gg m_e$. Assume gauge invariance and Lorentz invariance, but not invariance under C, P, or T. What are the non-renormalizable terms in the Lagrangian of leading order in $1/M$? Of next to leading order?

References

1. F. J. Dyson, *Phys. Rev.* **75**, 486, 1736 (1949). A historical perspective is provided by *Renormalization*, ed. by L. M. Brown (Springer-Verlag, New York, 1993). For a comprehensive modern treatment, see J. Collins, *Renormalization* (Cambridge University Press, Cambridge, 1984).

2. S. Weinberg, *Phys. Rev.* **118**, 838 (1959). This proof relied only on general asymptotic properties of the integrands of Feynman graphs in Euclidean momentum space, obtained by Wick rotation of all integration contours. The proof was simplified through the use of more detailed properties of the integrand, by Y. Hahn and W. Zimmerman, *Commun. Math. Phys.* **10**, 330 (1968), and then extended to Minkowskian momentum space by W. Zimmerman, *Commun. Math. Phys.* **11**, 1 (1968).

3. J. D. Bjorken and S. D. Drell, *Relativistic Quantum Fields* (McGraw-Hill, New York, 1965): Sections 19.10 and 19.11.

4. See, e.g., J. Wess and J. Bagger, *Supersymmetry and Supergravity* (Princeton University Press, Princeton, 1983), and original references quoted therein.

5. A. Salam, *Phys. Rev.* **82**, 217 (1951); *Phys. Rev.* **84**, 426 (1951).; P. T. Matthews and A. Salam, *Phys. Rev.* **94**, 185 (1954).

6. N. N. Bogoliubov and O. Parasiuk, *Acta Math.* **97**, 227 (1957)

7. K. Hepp, *Comm. Math. Phys.* **2**, 301 (1966). Hepp remarks that 'it is difficult to find two theorists whose understanding of the essential steps of the proof [of Bogoliubov and Parasiuk] is isomorphic,' but Hepp's paper is itself not easy to read.

8. W. Zimmerman, *Comm. Math. Phys.* **15**, 208 (1969). Also see W. Zimmerman, in *Lectures on Elementary Particles and Quantum Field Theory — Brandeis University Summer Institute in Theoretical Physics* (M.I.T. Press, Cambridge, 1970).

9. W. Heisenberg, *Z. Physik* **110**, 251 (1938).

10. W. Heisenberg, Ref. 6 and *Z. Physik* **113**, 61 (1939).

11. S. Sakata, H. Umezawa, and S. Kamefuchi, *Prog. Theor. Phys.* **7**, 327 (1952).

12. S. Weinberg, *Physica* **96A**, 327 (1979)

13. J. Gasser and H. Leutwyler, *Ann. Phys.* (NY) **158**, 142 (1984); *Nucl. Phys.* **B250**, 465 (1985).

14. J. F. Donoghue, *Phys. Rev.* **D 50**, 3874 (1994).

15. One possible way that this can happen is through the phenomenon of 'asymptotic safety'; see S. Weinberg, in *General Relativity — An Einstein Centenary Survey*, ed. by S. W. Hawking and W. Israel (Cambridge University Press, Cambridge, 1979): Section 16.3.

16. H. Euler and B. Kockel, *Naturwiss.* **23**, 246 (1935); W. Heisenberg and H. Euler, *Z. Physik* **98**, 714 (1936).

17. The Euler *et al.* effective Lagrangian has been used in one-loop calculations by J. Halter, *Phys. Lett.* **B 316**, 155 (1993).

18. K. G. Wilson, *Phys. Rev.* **B4**, 3174, 3184 (1971); *Rev. Mod. Phys.* **47**, 773 (1975).

19. J. Polchinski, *Nucl. Phys.* **B231**, 269 (1984); lecture in *Recent Directions in Particle Theory — Proceedings of the 1992 TASI Conference*, ed. by J. Harvey and J. Polchinski (World Scientific, Singapore, 1993): p. 235.

20. V. Novikov, M. A. Shifman, A. I. Vainshtein, and V. I. Zakharov, *Nucl. Phys.* **B229**, 381 (1983); M. A. Shifman and A. I. Vainshtein, *Nucl. Phys.* **B277**, 456 (1986); and references quoted therein. See also M. A. Shifman and A. I. Vainshtein, *Nucl. Phys.* **B359**, 571 (1991).

21. G. Feinberg, P. Kabir, and S. Weinberg, *Phys. Rev. Lett.* **3**, 527 (1959).

22. G. Feinberg, *Phys. Rev.* **110**, 1482 (1958).

13
Infrared Effects

In the study of radiative corrections a special role is played by those corrections due to 'soft' photons: photons whose energy and momentum are much less than the masses and energies characteristic of the process in question. Not only are these corrections often so large that they must be summed to all orders of perturbation theory; they are so simple that this summation is not difficult. The contribution of photons of infinitely long wavelength takes the form of divergent integrals, but as we shall see these 'infrared divergences' all cancel.[1]

In most of this chapter we will deal with photons interacting with charged particles of arbitrary type and spin, including particles like atomic nuclei that have strong as well as electromagnetic interactions. But it is not difficult to adapt the calculations presented here to the infrared effects of other massless particles, such as the gluons of quantum chromodynamics. In Section 13.4 we shall explicitly consider very general theories of massless particles, and will show the cancellation of infrared divergences on general grounds.

After these generalities, we shall return to photons, and take up two topics of practical importance: the scattering of soft photons by charged particles with arbitrary non-electromagnetic interactions and arbitrary spin, and the treatment of heavy charged particles like atomic nuclei as a source of an external electromagnetic field.

13.1 Soft Photon Amplitudes

In this section we shall derive a universal formula that gives the amplitude for emission of any number of very-low-energy photons in a process $\alpha \to \beta$ involving any number of higher-energy charged particles of any types.

Let us start with the amplitude for emission of just one soft photon. If we attach the soft photon line with outgoing momentum q and polarization index μ to an outgoing charged-particle line that leaves some connected Feynman diagram for the process $\alpha \to \beta$, as in Figure 13.1(a), then we must multiply the S-matrix element for $\alpha \to \beta$ with an additional charged-particle propagator carrying the momentum $p+q$ that the charged particle had before emitting the photon, together with the contribution of the new charged-particle–photon vertex. For charged particles of spin zero, mass m, and charge $+e$, these factors are

$$\left[i(2\pi)^4 e(2p^\mu + q^\mu)\right] \left[\frac{-i}{(2\pi)^4} \frac{1}{(p+q)^2 + m^2 - i\epsilon}\right] ,$$

which in the limit $q \to 0$ becomes

$$\frac{e\, p^\mu}{p \cdot q - i\epsilon} . \tag{13.1.1}$$

(We are freely redefining the scale of the positive infinitesimal ϵ, being careful only to keep track of its sign.) This result is actually true for charged particles of any spin. For instance, for a particle of spin $\frac{1}{2}$ and charge $+e$, we must replace the coefficient function $\bar{u}(\mathbf{p}, \sigma)$ for the outgoing charged particle with

$$\bar{u}(\mathbf{p}, \sigma) \left[-(2\pi)^4 e\gamma^\mu\right] \left[\frac{-i}{(2\pi)^4} \frac{-i(\not{p} + \not{q}) + m}{(p+q)^2 + m^2 - i\epsilon}\right] .$$

In the limit $q \to 0$ the numerator of the propagator is given by a sum of dyads:

$$-i\,\not{p} + m = 2p^0 \sum_{\sigma'} u(\mathbf{p}, \sigma')\,\bar{u}(\mathbf{p}, \sigma') ,$$

so we have a sum of equal-momentum matrix elements of γ^μ, given by

$$\bar{u}(\mathbf{p}, \sigma)\gamma^\mu u(\mathbf{p}, \sigma') = -i\delta_{\sigma,\sigma'} p^\mu / p^0 ,$$

and the effect again is to multiply the matrix element for the process $\alpha \to \beta$ by the factor (13.1.1). More generally, for any spin in the limit $q \to 0$ the four-momentum $p+q$ of the new internal charged particle line approaches the mass shell, so the numerator of the propagator approaches a sum of dyads of coefficient functions which convert the new vertex matrix into a factor proportional to p^μ and a unit matrix in helicity indices, leading again to the factor (13.1.1). Furthermore, as we saw in Chapter 10, higher-order corrections do not affect either the residue of the mass-shell poles in the propagators or the matrix element of the electric current between states of the same particle at equal momentum, so (13.1.1) gives the correct factor associated with the emission of a soft photon from an outgoing charged-particle line to all orders of perturbation theory.

The same reasoning applies to a photon emitted from an incoming charged-particle line of the process $\alpha \to \beta$, except that after the incoming particle emits a photon of four-momentum q the charged-particle line has

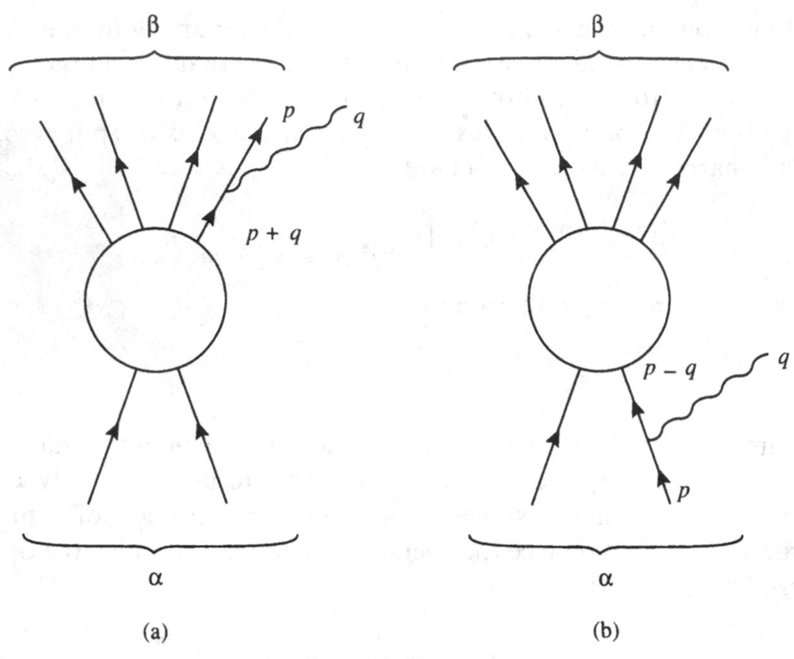

Figure 13.1. Dominant graphs for the emission of soft photons in an arbitrary process $\alpha \to \beta$. Straight lines are particles in the states α and β (including possible hard photons); wavy lines are soft photons.

four-momentum $p - q$, so in place of (13.1.1) we find a factor

$$\frac{e\,p^\mu}{-p \cdot q - i\epsilon} \, . \tag{13.1.2}$$

The photon can also, of course, be emitted from an internal line of the process $\alpha \to \beta$, but in this case there is no factor that goes as $(p \cdot q)^{-1}$ for $q \to 0$. The amplitude $M^\mu_{\beta\alpha}(q)$ (the S-matrix without the energy-momentum conservation delta function) for emitting a single soft photon with four-momentum q and polarization index μ in the process $\alpha \to \beta$ is therefore given in the limit $q \to 0$ by multiplying the matrix element $M_{\beta\alpha}$ for $\alpha \to \beta$ with a sum of terms like (13.1.1) and (13.1.2), one for each outgoing or incoming charged particle:

$$M^\mu_{\beta\alpha}(q) \to M_{\beta\alpha} \sum_n \frac{\eta_n e_n\, p^\mu_n}{p_n \cdot q - i\eta_n \epsilon} \, , \tag{13.1.3}$$

where p_n and $+e_n$ are the four-momentum and charge of the nth particle in the initial and final states, and η_n is a sign factor with the value $+1$ for particles in the final state β and -1 for particles in the initial state α.

Before going on to consider the emission of more than one soft photon, it is worth mentioning an important feature[2] of the formula (13.1.3). To calculate the amplitude for emission of a photon of definite helicity, we must contract this expression with the corresponding photon polarization vector $e_\mu(\mathbf{q}, \pm)$. But as we saw in Section 5.9, $e^\mu(\mathbf{q}, \pm)$ is not a four-vector; under a Lorentz transformation $\Lambda^\mu{}_\nu$, the polarization vector is transformed into $\Lambda^\mu{}_\nu e^\nu(\mathbf{q}, \pm)$ plus a term proportional to q^μ. In order for this last term not to spoil Lorentz invariance, it is therefore necessary that $M^\mu_{\beta\alpha}(q)$ should vanish when contracted with q_μ. But for $q \to 0$, (13.1.3) gives

$$q_\mu M^\mu_{\beta\alpha}(q) \to M_{\beta\alpha} \sum_n \eta_n e_n \,. \tag{13.1.4}$$

The coefficient of $M_{\beta\alpha}$ on the right-hand side is just the total charge in the final state minus the total charge in the initial state, so the condition that it vanishes is just the condition that charge is conserved. Thus without any independent assumptions about gauge invariance, we see that *for particles of spin one and mass zero, Lorentz invariance requires the conservation of whatever coupling constant like electric charge governs the interaction of these particles at low energies.*

Incidentally, the amplitude for emitting a soft *graviton* of four-momentum q and tensor indices μ, ν in a process $\alpha \to \beta$ is given by a formula[3] analogous to (13.1.3):

$$M^{\mu\nu}_{\beta\alpha}(q) \to M_{\beta\alpha} \sum_n \frac{\eta_n f_n p_n^\mu p_n^\nu}{p_n \cdot q - i\eta_n \epsilon} \,, \tag{13.1.5}$$

where f_n is the coupling constant of the soft graviton to particles of type n. Lorentz invariance here requires that this vanish when contracted with q_μ. But

$$q_\mu M^{\mu\nu}_{\beta\alpha}(q) \to M_{\beta\alpha} \sum_n \eta_n f_n p_n^\nu \,, \tag{13.1.6}$$

so the sum $\sum f_n p_n^\nu$ is conserved. However, the only linear combination of the four-momenta that can be conserved without forbidding all non-trivial scattering processes is the total four-momentum, so in order for (13.1.6) to vanish, the f_n must all be equal. (The common value of all f_n may be identified as $\sqrt{8\pi G_N}$, where G_N is Newton's constant of gravitation.) Thus Lorentz invariance requires the result that low-energy massless particles of spin two couple in the same way to all forms of energy and momentum. This goes a long way toward showing that Einstein's principle of equivalence is a necessary consequence of Lorentz invariance as applied to massless particles of spin two. Likewise, the amplitude for emitting a soft massless particle of four-momentum q and spin $j \geq 3$ in a process $\alpha \to \beta$ is of the form

$$M^{\mu\nu\rho\cdots}_{\beta\alpha}(q) \to M_{\beta\alpha} \sum_n \frac{\eta_n g_n p_n^\mu p_n^\nu p_n^\rho \cdots}{p_n \cdot q - i\eta_n \epsilon} \,.$$

Lorentz invariance here requires that the sum $\sum g_n p_n^\nu p_n^\rho \cdots$ must be conserved. But no such quantity can be conserved without prohibiting all non-trivial scattering processes, so the g_n must all vanish. Massless particles of spin $j \geq 3$ may exist, but they cannot have couplings that survive in the limit of low energy, and in particular they cannot mediate inverse square law forces.

Now let us consider the emission of two soft photons. The contribution to the matrix element from a graph in which the two photons are emitted from different external lines of the process $\alpha \to \beta$ is given by multiplying the matrix element for $\alpha \to \beta$ by a product of factors like (13.1.1) or (13.1.2). Perhaps surprisingly, the same is true even if the two photons are emitted from the *same* external line. For example, if photon 1 is emitted from an external line of charge $+e$ and energy-momentum four-vector p after photon 2 we get a factor

$$\left[\frac{\eta\, e\, p^{\mu_1}}{p\cdot q_1 - i\eta\epsilon}\right]\left[\frac{\eta\, e\, p^{\mu_2}}{p\cdot(q_2+q_1) - i\eta\epsilon}\right],$$

while if photon 2 is emitted after photon 1 the factor is

$$\left[\frac{\eta\, e\, p^{\mu_2}}{p\cdot q_2 - i\eta\epsilon}\right]\left[\frac{\eta\, e\, p^{\mu_1}}{p\cdot(q_1+q_2) - i\eta\epsilon}\right].$$

(See Figure 13.2. Again, η is $+1$ or -1 according to whether the charged-particle line is outgoing or incoming.) These two factors add up to

$$\left[\frac{\eta\, e\, p^{\mu_1}}{p\cdot q_1 - i\eta\epsilon}\right]\left[\frac{\eta\, e\, p^{\mu_2}}{p\cdot q_2 - i\eta\epsilon}\right],$$

which is just a product of the same factors encountered in the emission of a single photon.

More generally, in emitting an arbitrary number of photons from a single external line we encounter a sum of the form*

---

* This identity may be proved by mathematical induction. We have already seen that it is true for two photons. Suppose that it is true for $N-1$ photons. For N photons we may then write the sum over permutations as a sum over the choice of the first photon to be emitted together with a sum over permutations of the remaining photons:

$$[p\cdot q_1 - i\eta\epsilon]^{-1}[p\cdot(q_1+q_2) - i\eta\epsilon]^{-1}\cdots[p\cdot(q_1+q_2+\cdots+q_N) - i\eta\epsilon]^{-1} + \text{permutations}$$

$$= \sum_{r=1}^{N}\left[p\cdot\left(\sum_{s=1}^{N} q_s\right) - i\eta\epsilon\right]^{-1}\prod_{s\neq r}[p\cdot q_s - i\eta\epsilon]^{-1}$$

$$= \sum_{r=1}^{N}\left[p\cdot\left(\sum_{s=1}^{N} q_s\right) - i\eta\epsilon\right]^{-1}[p\cdot q_r - i\eta\epsilon]\prod_{s=1}^{N}[p\cdot q_s - i\eta\epsilon]^{-1} = \prod_{s=1}^{N}[p\cdot q_s - i\eta\epsilon]^{-1}$$

as was to be proved.

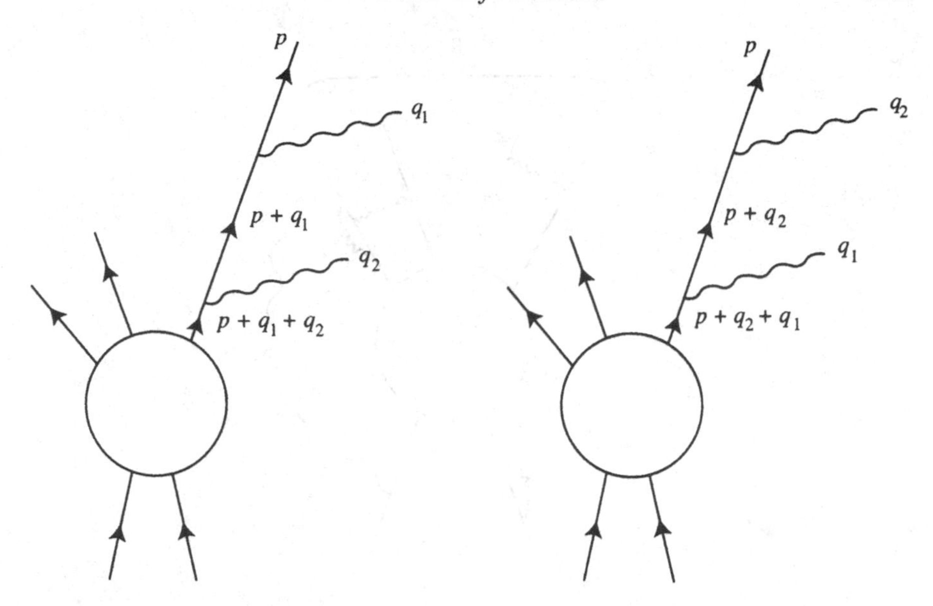

Figure 13.2. Graphs for the emission of two soft photons from the same outgoing charged particle. Straight lines are hard particles; wavy lines are soft photons.

$$[p \cdot q_1 - i\eta\epsilon]^{-1}[p \cdot (q_1 + q_2) - i\eta\epsilon]^{-1}[p \cdot (q_1 + q_2 + q_3) - i\eta\epsilon]^{-1} \cdots$$
$$+ \text{ permutations}$$
$$= [p \cdot q_1 - i\eta\epsilon]^{-1}[p \cdot q_2 - i\eta\epsilon]^{-1}[p \cdot q_3 - i\eta\epsilon]^{-1} \cdots \quad . \tag{13.1.7}$$

It follows then that the amplitude $M_{\beta\alpha}^{\mu_1 \cdots \mu_N}(q_1 \cdots q_N)$ for emitting N very soft photons with polarization indices $\mu_1, \cdots \mu_N$ and four-momenta $q_1, \cdots q_N$ in the process $\alpha \to \beta$ is given in the limit $q \to 0$ by multiplying the matrix element $M_{\beta\alpha}$ for $\alpha \to \beta$ by a product of factors like that in (13.1.3), one for each photon:

$$M_{\beta\alpha}^{\mu_1 \cdots \mu_N}(q_1 \cdots q_N) \to M_{\beta\alpha} \prod_{r=1}^{N} \left(\sum_n \frac{\eta_n e_n p_n^{\mu_r}}{p_n \cdot q_r - i\eta_n \epsilon} \right) . \tag{13.1.8}$$

13.2 Virtual Soft Photons

We shall now use the results of the previous section to calculate the effect to all orders of radiative corrections involving virtual soft photons exchanged among the charged particle lines of a process $\alpha \to \beta$, as in Figure 13.3. By a 'soft' photon we mean one that carries momentum less than Λ, where Λ is some convenient dividing point chosen low enough to justify the approximations made in the previous section. We shall find that these soft photons introduce infrared divergences, so as a stop-gap

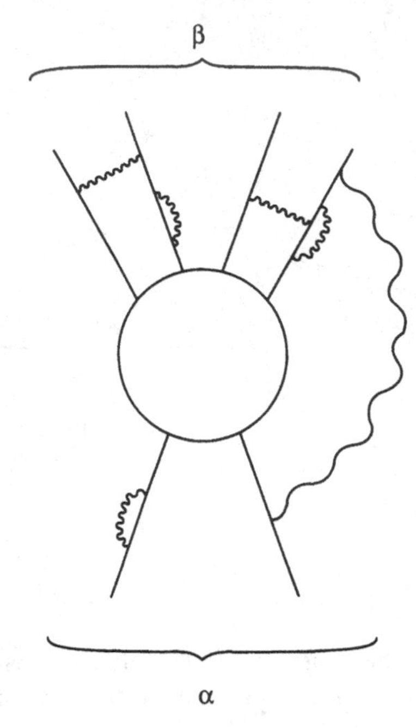

Figure 13.3. A typical dominant graph for the radiative corrections due to virtual soft photons to the S-matrix for the process $\alpha \to \beta$. Straight lines are particles in the states α and β (including possible hard photons); wavy lines are soft photons.

we will have to introduce also a lower bound λ on the photon momenta. It is important to recognize the difference between these two limits on the soft photon momenta. The upper cutoff Λ just serves to define what we mean by 'soft' photons; the Λ-dependence of the soft-photon radiative corrections is cancelled by the Λ-dependence of the rest of the amplitude, which is calculated including only virtual photons with momenta *greater* than Λ. On the other hand, the lower cutoff λ must eventually be removed by taking $\lambda \to 0$; as we shall see, the infrared divergences in this limit will be cancelled by the effects of real soft photon emission.

For each virtual soft photon we must supply a propagator factor

$$\frac{-i}{(2\pi)^4}\frac{\eta_{\mu\mu'}}{q^2 - i\epsilon}\,, \tag{13.2.1}$$

then multiply the amplitude (13.1.8) with the product of these propagators, contract photon polarization indices, and integrate over photon four-momenta. In addition for N virtual photons we must divide by a factor $2^N N!$, because the sum over all places to which we may attach the two ends of the soft photon lines includes spurious sums over the $N!$ permutations of the photon lines and over interchanges of the two ends of these lines. The effect of radiative corrections involving N soft photons is then to multiply the matrix element $M_{\beta\alpha}$ for the process without such radiative corrections by a factor

$$\frac{1}{N!2^N}\left[\frac{1}{(2\pi)^4}\sum_{nm} e_n e_m \eta_n \eta_m J_{nm}\right]^N , \tag{13.2.2}$$

where

$$J_{nm} \equiv -i(p_n \cdot p_m)\int_{\lambda\leq|\mathbf{q}|\leq\Lambda} \frac{d^4q}{[q^2-i\epsilon][p_n\cdot q - i\eta_n\epsilon][-p_m\cdot q - i\eta_m\epsilon]} . \tag{13.2.3}$$

Note that we have changed the sign of $p_m \cdot q$ in the denominator in (13.2.3), because if we define q as the momentum emitted by line n then $-q$ is the momentum emitted by line m.

Summing over N, we conclude that the matrix element for a process including radiative corrections due to any number of soft photons with momenta $|\mathbf{q}| \geq \lambda$ is given by

$$M^{\lambda}_{\beta\alpha} = M^{\Lambda}_{\beta\alpha}\exp\left[\frac{1}{2(2\pi)^4}\sum_{nm} e_n e_m \eta_n \eta_m J_{nm}\right] , \tag{13.2.4}$$

where $M^{\Lambda}_{\beta\alpha}$ is the amplitude including virtual photons only with momenta greater than Λ.

The integral over q^0 in (13.2.3) may be done by the method of residues. The integrand is analytic in q^0 except for four poles, at

$$q^0 = |\mathbf{q}| - i\epsilon, \qquad q^0 = -|\mathbf{q}| + i\epsilon ,$$
$$q^0 = \mathbf{v}_n \cdot \mathbf{q} - i\eta_n\epsilon, \qquad q^0 = \mathbf{v}_m \cdot \mathbf{q} + i\eta_m\epsilon ,$$

where $\mathbf{v}_n \equiv \mathbf{p}_n/p^0_n$, and likewise for $\mathbf{v}_m$. If particle n is outgoing and particle m is incoming, then $\eta_n = +1$, $\eta_m = -1$, so by closing the q^0 contour in the upper half-plane we avoid the contributions from the poles at $q^0 = \mathbf{v}_n \cdot \mathbf{q} - i\eta_n\epsilon$ or $q^0 = \mathbf{v}_m \cdot \mathbf{q} + i\eta_m\epsilon$. Similarly, if n is incoming and m is outgoing we can avoid these two poles by closing the contour in the lower half-plane. In these two cases it is only one of the poles at $q^0 = \pm(|\mathbf{q}| - i\epsilon)$ that contributes, and we find a purely real integral:

$$J_{nm} = -\pi(p_n \cdot p_m)\int_{\lambda\leq|\mathbf{q}|\leq\Lambda} \frac{d^3q}{|\mathbf{q}|^3(E_n - \hat{\mathbf{q}}\cdot\mathbf{p}_n)(E_m - \hat{\mathbf{q}}\cdot\mathbf{p}_m)}$$
$$(\text{for } \eta_n = -\eta_m = \pm 1) \ . \tag{13.2.5}$$

On the other hand, if particles n and m are both outgoing or both incoming, then the poles at $\mathbf{v}_n \cdot \mathbf{q} - i\eta_n\epsilon$ and $\mathbf{v}_m \cdot \mathbf{q} + i\eta_m\epsilon$ lie on opposite sides of the real q^0-axis, and we cannot avoid a contribution from one of them whichever way we close the contour:

$$J_{nm} = -\pi(p_n \cdot p_m) \int_{\lambda \le |\mathbf{q}| \le \Lambda} \frac{d^3q}{|\mathbf{q}|^3(E_n - \hat{\mathbf{q}} \cdot \mathbf{p}_n)(E_m - \hat{\mathbf{q}} \cdot \mathbf{p}_m)}$$
$$- \frac{4i\pi^3}{\beta_{nm}} \ln\left(\frac{\Lambda}{\lambda}\right) \quad (\text{for } \eta_n = \eta_m = \pm 1), \tag{13.2.6}$$

where β_{nm} is the relative velocity of particles n and m in the rest frame of either:

$$\beta_{nm} \equiv \sqrt{1 - \frac{m_n^2 m_m^2}{(p_n \cdot p_m)^2}} \,. \tag{13.2.7}$$

The imaginary term in Eq. (13.2.6) leads to an infrared-divergent phase factor[4] in Eq. (13.2.4), which drops out when we take the absolute value of the matrix element to calculate the rate for the process $\alpha \to \beta$. (This infinite phase factor is the relativistic counterpart of the well-known feature of non-relativistic Coulomb scattering, that the outgoing wave part of the Schrödinger wave function has a dependence on the radial coordinate r of the form $\exp(ipr - iv \ln r)/r$ instead of $\exp(ipr)/r$, where v is the product of charges divided by the relative velocity.[5]) The reaction rate *is* affected by the real part of J_{nm}, which for all η_n and η_m takes the value

$$\text{Re}\, J_{mn} = -\pi(p_n \cdot p_m) \int_{\lambda \le |\mathbf{q}| \le \Lambda} \frac{d^3q}{|\mathbf{q}|^3(E_n - \hat{\mathbf{q}} \cdot \mathbf{p}_n)(E_m - \hat{\mathbf{q}} \cdot \mathbf{p}_m)} \,. \tag{13.2.8}$$

An elementary calculation gives

$$\text{Re}\, J_{mn} = \frac{2\pi^2}{\beta_{nm}} \ln\left(\frac{1 + \beta_{nm}}{1 - \beta_{nm}}\right) \ln\left(\frac{\Lambda}{\lambda}\right) \,. \tag{13.2.9}$$

Using this in the absolute value squared of Eq. (13.2.4) gives the effect of soft virtual photons on the rates for the process $\Gamma_{\beta\alpha}$ as

$$\Gamma^{\lambda}_{\beta\alpha} = \left(\frac{\lambda}{\Lambda}\right)^{A(\alpha \to \beta)} \Gamma^{\Lambda}_{\beta\alpha} \,, \tag{13.2.10}$$

where $\Gamma^{\lambda}_{\beta\alpha}$ and $\Gamma^{\Lambda}_{\beta\alpha}$ are the rates for the process $\alpha \to \beta$ including radiative corrections of soft photons only with momenta greater than λ or Λ, respectively, and A is the exponent

$$A(\alpha \to \beta) = -\frac{1}{8\pi^2} \sum_{nm} \frac{e_n e_m \eta_n \eta_m}{\beta_{nm}} \ln\left(\frac{1 + \beta_{nm}}{1 - \beta_{nm}}\right) \,. \tag{13.2.11}$$

Note that this makes sense only because the correction factor $(\lambda/\Lambda)^A$ has turned out to be the ratio of a function of λ to the same function of Λ, since the two rates in Eq. (13.2.10) can only depend on λ and Λ, respectively.

The exponent A is always positive. For instance, in the scattering of a single charged particle by a neutral particle or an external potential, we must add terms in Eq. (13.2.11) where both n and m are the initial or the final charged particle (in which case $\eta_n\eta_m = +1$ and $\beta_{nm} = 0$), or n is the initial or final charged particle and m is the other one (in which case $\eta_n\eta_m = -1$ and $\beta_{nm} = \beta$, where $1 > \beta > 0$.) This gives

$$A = -\frac{e^2}{8\pi^2}\left[4 - \frac{2}{\beta}\ln\left(\frac{1+\beta}{1-\beta}\right)\right] ,$$

which is positive for all $1 > \beta > 0$. Because A is positive the effect of the infrared divergences introduced by soft virtual photons when summed to all orders is to make the rate for any given charged particle process $\alpha \to \beta$ *vanish* in the limit $\lambda \to 0$.

* * *

Before we go on to consider how soft real photon emission cancels these infrared divergences, we should pause to note a technicality in the above calculation which as far as I know has always been ignored in the literature. In calculating these radiative corrections we have included diagrams in which the virtual photon is absorbed and emitted at the *same* external charged-particle line, as well as those in which it is emitted and absorbed at different lines. But as we learned in Chapter 10, in calculating the S-matrix we are not supposed to include radiative corrections arising from insertions of self-energy subgraphs in external lines. This might suggest that we should drop the terms $n = m$ in Eq. (13.2.11), but then the cancellation of infrared divergences we will find in the next section would not be complete.

The resolution of this problem can be found in the observation that soft virtual photons produce infrared divergences not only directly, but also through their effect on the renormalization constants Z_n of the charged particle fields. (The renormalization constant Z_n is the one usually called Z_2 in theories like quantum electrodynamics with a single charged field of spin $\frac{1}{2}$.) It is the counterterms proportional to $Z_n - 1$ that cancel the effect of radiative corrections in external lines. To be specific, the renormalized field of a charged particle of type n is a factor $Z_n^{-1/2}$ times the unrenormalized field, so when we calculate the S-matrix using renormalized fields (corresponding to the omission of radiative corrections in external lines) we are introducing an extra factor $\prod_n Z_n^{-1/2}$, the product running over all charged particles in the initial and final states. (Of course there are also factors $Z_n^{-1/2}$ for neutral particles, but these are not infrared-divergent.) In a slightly different notation, this factor is

$$\prod_f Z_f^{-E_f/2},$$

where Z_f is the field renormalization constant for fields of type f, E_f is the number of external lines of type f, and the product now runs over *all* charged field types. However, these field renormalization constants also appear in the interiors of diagrams; expressing an interaction of type i that involves N_{if} charged-particle fields of type f in terms of renormalized fields introduces an infrared-divergent factor

$$\prod_f (Z_f)^{N_{if}/2}.$$

(For instance, the counterterm $-ie(Z_2-1)A_\mu\bar{\psi}\gamma^\mu\psi$ in Eq. (11.1.9) together with the ordinary electromagnetic interaction $-ieA_\mu\bar{\psi}\gamma^\mu\psi$ yields a total interaction $-iZ_2eA_\mu\bar{\psi}\gamma^\mu\psi$. It was the infrared divergence in this Z_2 factor that was responsible for the infrared divergence arising from the second term in the brackets in Eq. (11.3.23), and the last term in Eq. (11.4.14).) There is also an infrared divergence in the propagators of the renormalized fields; the propagator of a renormalized charged field of type f when expressed in terms of the propagator of the unrenormalized field introduces a factor Z_f^{-1}. Putting this all together, the total number of factors of Z_f for each charged field type f, introduced by the counterterms to interactions and to radiative corrections on internal as well as external lines, is

$$\tfrac{1}{2}\sum_i V_i N_{if} - I_f - \tfrac{1}{2}E_f,$$

where I_f and E_f are the numbers of internal and external lines of type f, and V_i is the number of vertices for interactions of type i. We have already noted in Section 6.3 that this quantity vanishes for each f. Thus the counterterms which cancelled the radiative corrections on external lines are themselves cancelled by the Z_f factors arising from internal lines and vertices. Eq. (13.2.11) is therefore correct as it stands, including the terms with $n = m$.

13.3 Real Soft Photons; Cancellation of Divergences

The resolution of the infrared divergence problem encountered in the previous section is found in the observation that it is not really possible to measure the rate $\Gamma_{\beta\alpha}$ for a reaction $\alpha \to \beta$ involving definite numbers of photons and charged particles, because photons of very low energy can always escape undetected. What can be measured is the rate $\Gamma_{\beta\alpha}(E, E_T)$

for such a reaction to take place with no unobserved photon having an energy greater than some small quantity E, and with not more than some small total energy E_T going into any number of unobserved photons. (Of course, $E \leq E_T$. In an experiment without soft photon detectors, one can rely on measurements of the energies of the 'hard' particles in α and β to put a limit E_T on the total energy going into soft photons, and in this case we just set $E = E_T$.) We now turn to a calculation of this rate.

The S-matrix for emitting N real soft photons in a process $\alpha \to \beta$ is obtained by contracting each of the N photon polarization indices $\mu_1, \mu_2, \cdots$ on the amplitude (13.1.8) with the appropriate coefficient function

$$\frac{\epsilon_\mu^*(\mathbf{q}, h)}{(2\pi)^{3/2}\sqrt{2|\mathbf{q}|)}} ,$$

where $\mathbf{q}$ is the photon momentum, $h = \pm 1$ is its helicity, and ϵ^μ is the corresponding photon polarization 'vector'.* This gives a photon emission matrix element (the S-matrix element with delta function omitted) as

$$M^\lambda_{\beta\alpha}(\mathbf{q}_1, h_1, \mathbf{q}_2, h_2, \cdots) = M^\lambda_{\beta\alpha} \times \prod_{r=1}^{N} (2\pi)^{-3/2}(2|\mathbf{q}_r|)^{-1/2} \sum_n \frac{\eta_n e_n [p_n \cdot \epsilon^*(\mathbf{q}_r, h_r)]}{p_n \cdot q_r} . \tag{13.3.1}$$

(The superscript λ is to remind us that these amplitudes are to be calculated with an infrared cutoff λ on the momenta of virtual photons. Eventually we shall take $\lambda \to 0$. The presence of soft virtual photons does not interfere with the result (13.3.1) because of the factorization discussed in Section 13.1.) The differential rate for emitting N soft photons into a volume $\prod_r d^3 q_r$ of momentum space is given by squaring this matrix element, summing over helicities, and multiplying with $\prod_r d^3 q_r$. We recall from Eq. (8.5.7) that for $q^2 = 0$, the helicity sums take the form

$$\sum_{h=\pm 1} \epsilon_\mu(\mathbf{q}, h)\epsilon_\nu^*(\mathbf{q}, h) = \eta_{\mu\nu} + q_\mu c_\nu + q_\nu c_\mu , \tag{13.3.2}$$

where $\mathbf{c} \equiv -\mathbf{q}/2|\mathbf{q}|^2$ and $c^0 \equiv 1/2|\mathbf{q}|$. The charge conservation condition $\sum_n \eta_n e_n = 0$ allows us to drop the terms in Eq. (13.3.2) involving q_μ or q_ν,

* We are using ϵ^μ instead of e^μ for photon polarization vectors, to avoid confusion with our use here of e_n for electric charges.

yielding a differential rate**

$$d\Gamma^{\lambda}_{\beta\alpha}(\mathbf{q}_1,\cdots \mathbf{q}_N) = \Gamma^{\lambda}_{\beta\alpha}\prod_{r=1}^{N}\frac{d^3q_r}{(2\pi)^3(2|\mathbf{q}_r|)}\sum_{nm}\frac{\eta_n\eta_m e_n e_m(p_n\cdot p_m)}{(p_n\cdot q_r)(p_m\cdot q_r)}\,. \tag{13.3.3}$$

To calculate the differential rate for the emission of N soft photons with definite energies $\omega_r \equiv |\mathbf{q}_r|$ we must integrate Eq. (13.3.3) over the directions of the photon momenta $\mathbf{q}_r$. These integrals are the same as those we encountered in the integrals (13.2.8),

$$-\pi(p_n\cdot p_m)\int\frac{d^2\hat{\mathbf{q}}}{(E_n-\hat{\mathbf{q}}\cdot\mathbf{p}_n)(E_m-\hat{\mathbf{q}}\cdot\mathbf{p}_m)} = \frac{2\pi^2}{\beta_{nm}}\ln\left(\frac{1+\beta_{nm}}{1-\beta_{nm}}\right)\,. \tag{13.3.4}$$

Integrating Eq. (13.3.3) over photon directions thus gives the differential rate for photons of energy $\omega_1,\cdots\omega_N$:

$$d\Gamma^{\lambda}_{\beta\alpha}(\omega_1\cdots\omega_N) = \Gamma^{\lambda}_{\beta\alpha}A(\alpha\to\beta)^N\frac{d\omega_1}{\omega_1}\cdots\frac{d\omega_N}{\omega_N}\,, \tag{13.3.5}$$

where $A(\alpha\to\beta)$ is the same constant encountered in the previous section:

$$A(\alpha\to\beta) = -\frac{1}{8\pi^2}\sum_{nm}\frac{e_n e_m\eta_n\eta_m}{\beta_{nm}}\ln\left(\frac{1+\beta_{nm}}{1-\beta_{nm}}\right)\,.$$

We see from Eq. (13.3.5) that an unrestricted integral over the energies of the emitted photons would introduce another infrared divergence. However, unitarity demands that if we use an infrared cutoff for the momenta of the virtual photons (as implied by the superscript λ) then we must use the same infrared cutoff for the real photons. To calculate the rate $\Gamma^{\lambda}_{\beta\alpha}(E,E_T)$ for the reaction $\alpha\to\beta$ with not more than energy E going into any one unobserved photon and not more than energy E_T going into any number of unobserved photons (with E and E_T chosen small enough to justify the approximations used in deriving Eq. (13.3.1)), we must integrate Eq. (13.3.5) over all photon energies, subject to the limits $E\geq\omega_r\geq\lambda$ and $\sum_r\omega_r\leq E_T$, then divide by $N!$ because this integral includes configurations that differ only by permutations of the N soft photons, and finally sum over N. This gives

$$\Gamma^{\lambda}_{\beta\alpha}(E,E_T) = \Gamma^{\lambda}_{\beta\alpha}\sum_{N=0}^{\infty}\frac{A(\alpha\to\beta)^N}{N!}\int_{E\geq\omega_r\geq\lambda,\sum_r\omega_r\leq E_T}\prod_{r=1}^{N}\frac{d\omega_r}{\omega_r}\,. \tag{13.3.6}$$

This integral would factor into the product of N integrals over the individual ω_r were it not for the restriction $\sum_r\omega_r\leq E$. This restriction may

** The result for $|\mathbf{q}|d\Gamma_{\beta\alpha}(\mathbf{q})/\Gamma_{\beta\alpha}$ in the case $N=1$ corresponds to the distribution of energy emitted classically by a discontinuously changing current density four-vector $J^{\mu}(x) = \sum_n^{(t)}\delta^3(\mathbf{x}-\mathbf{v}_n t)\,p_n^{\mu}e_n/E_n$, with the sum here running only over particles in the initial state for $t<0$ and in the final state for $t>0$.

be implemented by including as a factor in the integrand a step function

$$\theta(E_T - \sum_r \omega_r) = \frac{1}{\pi}\int_{-\infty}^{\infty} du \, \frac{\sin E_T u}{u} \exp\left(iu \sum_r \omega_r\right) . \tag{13.3.7}$$

Eq. (13.3.6) then becomes

$$\Gamma^{\lambda}_{\beta\alpha}(E, E_T) = \frac{1}{\pi}\int_{-\infty}^{\infty} du \, \frac{\sin E_T u}{u} \exp\left(A(\alpha \to \beta)\int_{\lambda}^{E} \frac{d\omega}{\omega} e^{i\omega u}\right) \Gamma^{\lambda}_{\beta\alpha} . \tag{13.3.8}$$

The integral in the exponent can be done in the limit $\lambda \ll E$ by writing it as the sum of the integral of $(e^{i\omega u} - 1)/\omega$, in which we can set $\lambda = 0$, and the integral of $1/\omega$, which is trivial. Rescaling the u and ω variables, this gives for $\lambda \ll E$:

$$\Gamma^{\lambda}_{\beta\alpha}(E, E_T) \to \mathscr{F}\left(E/E_T ; A(\alpha \to \beta)\right) \left(\frac{E}{\lambda}\right)^{A(\alpha\to\beta)} \Gamma^{\lambda}_{\beta\alpha} , \tag{13.3.9}$$

where

$$\begin{aligned}\mathscr{F}(x; A) &\equiv \frac{1}{\pi}\int_{-\infty}^{\infty} du \, \frac{\sin u}{u} \exp\left(A \int_0^x \frac{d\omega}{\omega} \left(e^{i\omega u} - 1\right)\right) \\ &= 1 - \frac{A^2\theta(x - \frac{1}{2})}{2}\int_{1-x}^{x} \frac{d\omega}{\omega} \ln\left(\frac{x}{1-\omega}\right) + \cdots .\end{aligned} \tag{13.3.10}$$

For E and E_T of the same order and $A \ll 1$ the factor $\mathscr{F}(E/E_T, A)$ in Eq. (13.3.9) is close to unity; for instance,

$$\mathscr{F}(1; A) \simeq 1 - \tfrac{1}{12}\pi^2 A^2 + \cdots .$$

Because $A(\alpha \to \beta) > 0$, the factor $(E/\lambda)^{A(\alpha\to\beta)}$ in Eq. (13.3.9) becomes infinite in the limit $\lambda \to 0$. However, Eq. (13.2.10) shows that the rate $\Gamma^{\lambda}_{\beta\alpha}$ vanishes in this limit:

$$\Gamma^{\lambda}_{\beta\alpha} = \left(\frac{\lambda}{\Lambda}\right)^{A(\alpha\to\beta)} \Gamma^{\Lambda}_{\beta\alpha} .$$

Using this in Eq. (13.3.9) shows that the infrared cutoff λ drops out in the limit $\lambda \ll E$:

$$\Gamma^{\lambda}_{\beta\alpha}(E, E_T) \to \mathscr{F}\left(E/E_T ; A(\alpha \to \beta)\right) \left(\frac{E}{\Lambda}\right)^{A(\alpha\to\beta)} \Gamma^{\Lambda}_{\beta\alpha} . \tag{13.3.11}$$

We remind the reader that the energy Λ is just a convenient dividing point between 'soft' photons which are taken into account explicitly in Eq. (13.3.11) and 'hard' photons whose effects are buried in $\Gamma^{\Lambda}_{\beta\alpha}$. The right-hand side of Eq. (13.3.11) is independent of Λ because $\Gamma^{\Lambda}_{\beta\alpha} \propto \Lambda^{A}$. However, in theories with a small coupling constant like quantum electrodynamics it is frequently a good strategy to take Λ to be sufficiently small compared with the typical energies W involved in the collision so that the approximations made here apply for photon energies less than Λ, but large enough so that $A(\alpha \to \beta)\ln(W/\Lambda) \ll 1$. Then it may be a good approximation to calculate $\Gamma^{\Lambda}_{\beta\alpha}$ in lowest-order perturbation theory, with the dominant radiative corrections for $E \ll \Lambda$ given by the factor $(E/\Lambda)^A$ in Eq. (13.3.11).

* * *

The same cancellation of infrared divergences occurs for soft gravitons.[3] The rate for any process $\alpha \to \beta$, with not more than an energy E going into soft gravitons, turns out to be proportional to E^B, where

$$B = \frac{G}{2\pi}\sum_{nm}\eta_n\eta_m m_n m_m \frac{1+\beta_{nm}^2}{\beta_{nm}\sqrt{1-\beta_{nm}^2}}\ln\left(\frac{1+\beta_{nm}}{1-\beta_{nm}}\right) . \tag{13.3.12}$$

13.4 General Infrared Divergences

The infrared divergence due to soft photons that we have been considering up to now in this chapter is just one example of a variety of infrared divergences that are encountered in various physical theories. Another example is provided by quantum electrodynamics with massless charged particles. Here even after the cancellation of infrared divergences due to soft photons, we find a logarithmic divergence in the exponent A in Eq. (13.3.11). According to Eqs. (13.2.11) and (13.2.7), for a process in which all charged particles are electrons, in the limit $m_e \to 0$ the exponent goes as

$$A \to -\frac{1}{4\pi^2}\sum_n e_n^2 - \frac{1}{4\pi^2}\sum_{n\neq m} e_n e_m \eta_n \eta_m \ln\left(\frac{2|p_n\cdot p_m|}{m_e^2}\right) \to -\frac{\ln m_e}{2\pi^2}\sum_n e_n^2 .$$

(In the last step we have used the charge conservation condition $\sum_n e_n\eta_n = 0$.) The infrared divergence in this formula arises from soft photons that are emitted in a direction *parallel* to the momentum of one of the 'hard' electrons in the initial or final state, but it occurs also even if the photon like the electron is not soft, because the propagator denominator $(p_n \pm q)^2$ vanishes for $p_n^2 = q^2 = 0$ if $\mathbf{p}_n$ is parallel to $\mathbf{q}$. To be a little more specific, for $p_n^2 = q^2 = 0$ the integral of this factor* over photon directions takes

* This factor is not squared, because the divergence occurs only in the interference between this term in the S-matrix element and terms in which the photon is emitted from some other charged particle line $m \neq n$. For $m = n$ the integral (13.2.8) is proportional to m_n^2.

the form

$$\int d^2\hat{q}\,(p \pm q)^{-2} = \mp \frac{\pi}{\sqrt{\mathbf{p}^2\mathbf{q}^2}} \int_0^\pi \frac{\sin\theta\, d\theta}{1-\cos\theta}\,,$$

where θ is the angle between the momenta of the photon and the charged particle. This integral diverges logarithmically at $\theta = 0$.

Of course, in the real world there are no massless electrically charged particles, but in reactions in which the typical value E^2 of the scalar products $|p_n \cdot p_m|$ is much larger than m_e^2, it is of interest to identify the places where large $\ln(m_e/E)$ factors appear. The dominant radiative correction in this case is often given by the term $-\ln(m_e/E)\sum_n e_n^2/2\pi^2$ in A. More importantly, in quantum chromodynamics there are massless particles, the gluons, that carry a conserved quantum number known as color that is analogous to electric charge, so that infrared divergences arise from the emission of parallel hard gluons from hard gluons or other hard colored particles in the initial or final states.

These infrared divergences are not, in general, eliminated by summing over suitable sets of final states. However, Lee and Nauenberg[6] have pointed out that the infrared divergences can be made to cancel if we not only sum over suitable final states, but also assume a certain probabilistic distribution of *initial* states. What follows is a modified version of their argument, which will immediately make clear why in the case of electrodynamics with massive electrons it was sufficient to sum over final states.

For these purposes it is convenient to return to 'old-fashioned' perturbation theory, in which the S-matrix is given by Eq. (3.2.7) and Eq. (3.5.3) as

$$S_{ba} = \delta(b-a) - 2i\pi\delta(E_a - E_b)T_{ba}\,, \tag{13.4.1}$$

where

$$T_{ba} = V_{ba} + \sum_{\nu=1}^{\infty} \int dc_1 \cdots dc_\nu \, \frac{V_{bc_1}V_{c_1c_2}\cdots V_{c_\nu a}}{(E_a - E_{c_1} + i\epsilon)\cdots(E_a - E_{c_\nu} + i\epsilon)}. \tag{13.4.2}$$

(The integrals over $c_1 \cdots c_\nu$ should be understood to include sums over the spins and types of particles in these states as well as integrals over the three-momenta of these particles.) Infrared divergences arise from (and only from) the vanishing of one or more of the energy denominators in this expression.

However, not all vanishing energy denominators give rise to infrared divergences. A general intermediate state c may have $E_c = E_a$, but usually this is just one point in the interior of the range of integration, and the integral over this range is rendered convergent by the prescription implied by the $i\epsilon$ in the denominator. In order for an intermediate state c to produce an infrared divergence, it is necessary that the energy $E_c = E_a$ be reached at the *endpoint* of the range of integration. This happens for instance if the first intermediate state c_1 in Eq. (13.4.2) consists of the particles in the initial state a, with any of the massless particles in this state replaced with *jets*, consisting of any number of nearly parallel massless particles with a total momentum equal to that of the particle the jet replaces. In this case, the endpoint at which $E_{c_1} = E_a$ is the point in momentum space at which all of the massless particles in each jet are parallel. More generally, we can have any number of the massless particles in a replaced with jets of nearly parallel massless particles, plus any number of additional soft massless particles. The set of all such states will be called $D(a)$. (To be precise, we need to introduce a small angle Θ as well as a small energy Λ to define what we mean by 'nearly parallel' and 'soft'. We will not bother to show the dependence of the set $D(a)$ on Θ and Λ.) The states in $D(a)$ are 'dangerous', in the sense that the vanishing of the energy denominator $E_a - E_{c_1}$ at the endpoint can introduce an infrared divergence; the endpoint at which $E_{c_1} = E_a$ is the point at which all massless particles in each jet are parallel, and all soft massless particles have zero energy.

Furthermore, if $c_1, \cdots c_n$ are each in the set $D(a)$, then an intermediate state c_{n+1} in $D(a)$ is also dangerous in the same sense. On the other hand, if some intermediate state c_m is *not* in $D(a)$, then a later state c_k with $k > m$ would not be dangerous even if it belonged to the set $D(a)$, because the configuration of hard particles or jets with three-momenta equal to those of particles in the state a would be just an ordinary point inside the range of integration. In exactly the same way, we may define a set of states $D(b)$ in which one or more of the massless particles in the state b are replaced with jets of nearly parallel massless particles, each having the same total three-momentum as the particle it replaces, and we add any number of soft massless particles. An intermediate state c_m is dangerous if it belongs to the set $D(b)$ and if the later states c_k with $k > m$ all belong to $D(b)$.

To isolate the effects of these dangerous states we rewrite Eq. (13.4.2) in the form

$$T_{ba} = V_{ba} + \sum_{\nu=1}^{\infty} \left(V \left[\frac{\mathscr{P}_a + \mathscr{P}_b + \mathscr{P}_{\notin a,b}}{E_a - H_0 + i\epsilon} V \right]^{\nu} \right)_{ba} , \tag{13.4.3}$$

where $\mathscr{P}_a$, $\mathscr{P}_b$, and $\mathscr{P}_{\notin a,b}$ are the projection operators respectively on $D(a)$, $D(b)$, and on all other states. (It is assumed here that none of the charged particles in b have momenta close to that of some charged particle in a, so that the sets $D(a)$ and $D(b)$ do not overlap.) Now, for $\Lambda \to 0$ and $\Theta \to 0$, the dangerous intermediate states occupy so little phase space that they may be neglected wherever they do not lead to infrared divergences. The power series (13.4.3) therefore becomes

$$T_{ba} = \sum_{r=0}^{\infty}\sum_{s=0}^{\infty}\sum_{\nu=0}^{\infty} \left(\left[V \frac{\mathscr{P}_b}{E_a - H_0 + i\epsilon} \right]^r V \left[\frac{\mathscr{P}_{\notin a,b}}{E_a - H_0 + i\epsilon} V \right]^\nu \times \left[\frac{\mathscr{P}_a}{E_a - H_0 + i\epsilon} V \right]^s \right)_{ba} . \quad (13.4.4)$$

This would be exact if all of the projection operators $\mathscr{P}_{\notin a,b}$ between the leftmost and the rightmost were replaced with $\mathscr{P}_a + \mathscr{P}_b + \mathscr{P}_{\notin a,b}$, and if $\mathscr{P}_b$ and $\mathscr{P}_a$ on the left and right were replaced with $\mathscr{P}_b + \mathscr{P}_a$, but as remarked above this would have a negligible effect on the final result when Λ and Θ are sufficiently small.

Eq. (13.4.4) may be written in a more compact form:

$$T_{ba} = \left(\Omega_b^{-\dagger}\, T_S\, \Omega_a^+ \right)_{ba} , \quad (13.4.5)$$

where, for future use, we define Ω_α^+ and Ω_β^- for general states α and β as:

$$(\Omega_\alpha^+)_{ca} \equiv \sum_{r=0}^{\infty} \left(\left[\frac{\mathscr{P}_\alpha}{E_a - H_0 + i\epsilon} V \right]^r \right)_{ca} , \quad (13.4.6)$$

$$(\Omega_\beta^-)_{db} \equiv \sum_{r=0}^{\infty} \left(\left[\frac{\mathscr{P}_\beta}{E_b - H_0 - i\epsilon} V \right]^r \right)_{db} , \quad (13.4.7)$$

and T_S is the 'safe' operator**

$$(T_S)_{dc} \equiv \sum_{\nu=0}^{\infty} \left(V \left[\frac{\mathscr{P}_{\notin c,d}}{E_c - H_0 + i\epsilon} V \right]^\nu \right)_{dc} . \quad (13.4.8)$$

All of the infrared divergences have now been isolated in the two operator factors Ω_b^- and Ω_a^+.

To eliminate these infrared divergences it is now only necessary to note that if it were not for the projection operators on the dangerous states, the operators Ω_b^- and Ω_a^+ would be just the unitary operators that according to Eq. (3.1.16) convert free-particle states into 'out' or 'in' states, respectively. These operators are therefore unitary if confined to the subspaces $D(\beta)$ and $D(\alpha)$ of states that would be dangerous for some given final state β and some given initial state α. That is, for general α and β

$$\Omega_\beta^- \mathscr{P}_\beta \Omega_\beta^{-\dagger} = \mathscr{P}_\beta , \quad (13.4.9)$$

$$\Omega_\alpha^+ \mathscr{P}_\alpha \Omega_\alpha^{+\dagger} = \mathscr{P}_\alpha . \quad (13.4.10)$$

** In $(\Omega_b^-)_{db}$ we are using the fact that T_{ba} is calculated with $E_b = E_a$, and in $(T_S)_{dc}$ we are using the fact that the projection operators $\mathscr{P}_a$ make $(\Omega_a^+)_{ca}$ vanish unless E_c is very close to E_a. Also, the factors $\Omega_b^{-\dagger}$ and Ω_a^+ in Eq. (13.4.5) make $\mathscr{P}_{\notin c,d} = \mathscr{P}_{\notin a,b}$.

The transition rate is therefore free of infrared divergences if summed over the subspaces of states that would be dangerous for any given final and initial states β and α:

$$\sum_{a\in D(\alpha)}\sum_{b\in D(\beta)} |T_{ba}|^2 = \text{Tr}\left\{\Omega_\beta^- \mathscr{P}_\beta \Omega_\beta^{-\dagger} T_S \Omega_\alpha^+ \mathscr{P}_\alpha \Omega_\alpha^{+\dagger} T_S^\dagger\right\}$$
$$= \text{Tr}\left\{\mathscr{P}_\beta T_S \mathscr{P}_\alpha T_S^\dagger\right\} = \sum_{a\in D(\alpha)}\sum_{b\in D(\beta)} |(T_S)_{ba}|^2 . \tag{13.4.11}$$

In order to be satisfied that this really does solve the general problem of infrared divergences, it is necessary to argue that it is only sums like that in Eq. (13.4.11) that are experimentally measurable. It is plausible that we should have to sum over dangerous final states in order to have a measurable transition rate, since it is not possible experimentally to distinguish an outgoing charged (or colored) massless particle from a jet of massless particles with nearly parallel momenta and the same total energy,[7] together with an arbitrary number of very soft quanta, all with the same total charge (or color). The sum over initial states is more problematic. Presumably one may argue that truly massless particles are always produced as jets accompanied by an ensemble of soft quanta that is uniform within some volume of momentum space. However, to the best of my knowledge no one has given a complete demonstration that the sums of transition rates that are free of infrared divergences are the only ones that are experimentally measurable.

This problem does not arise in quantum electrodynamics (with massive charged particles), where as we have seen it is only necessary to sum over final states in order to eliminate infrared divergences. The reason for this difference can be traced to the fact that in electrodynamics the states $a, b, c, \cdots$ are direct products of states (labelled with Greek letters) with fixed numbers of charged particles and hard photons, times states containing only soft photons having energy less than some small quantity Λ. Then for a reaction in which some set of soft photons f is produced in a reaction $\alpha \rightarrow \beta$ among charged particles and hard photons, Eq. (13.4.5) simplifies to

$$T_{\beta f,\alpha} = \left(\Omega^-(\beta)^\dagger \Omega^+(\alpha)\right)_{f0} (T_S)_{\beta\alpha} , \tag{13.4.12}$$

where 0 denotes the soft photon vacuum, and $\Omega^\pm$ are calculated as before, but in the reduced Hilbert space consisting only of soft photons, and with the interactions of these photons taken as the interaction Hamiltonian with all charged particles in the fixed states indicated by the arguments β or α. Just as before, these operators are unitary in the 'dangerous' Hilbert space $\mathscr{D}$ of soft photons, so[†]

$$\sum_{f\in\mathscr{D}} |T_{\beta f,\alpha 0}|^2 = |(T_S)_{\beta\alpha}|^2 \left(\Omega^+(\alpha)^\dagger\Omega^-(\beta)\Omega^-(\beta)^\dagger\Omega^+(\alpha)\right)_{00}$$
$$= |(T_S)_{\beta\alpha}|^2 \left(\Omega^+(\alpha)^\dagger\Omega^+(\alpha)\right)_{00} = |(T_S)_{\beta\alpha}|^2 \qquad (13.4.13)$$

without having to sum over initial states.

13.5 Soft Photon Scattering*

In our treatment in this chapter of soft photon interactions, we have up to now considered only processes in which the soft photons are emitted or absorbed in a process $\alpha \to \beta$ which was going on anyway. It is also possible to make useful general statements about processes in which the process $\alpha \to \beta$ is trivial, and the soft photons play an essential part in producing an interesting reaction. We will consider here the simplest and most important example of this sort, the scattering of a soft photon from a massive particle of arbitrary type and spin, where α and β are single-particle states. The complication here is that the leading term in the soft photon scattering amplitude does not come from the pole terms, but from non-pole terms that are related to the pole terms by the condition of current conservation.

The S-matrix for photon scattering may be put in the form

$$S(q,\lambda;p,\sigma \to q',\lambda';p',\sigma') = i(2\pi)^4\delta^4(q+p-q'-p') \times \frac{\epsilon^*_\nu(\mathbf{q}',\lambda')\,\epsilon_\mu(\mathbf{q},\lambda)\,M^{\nu\mu}_{\sigma',\sigma}(q;\mathbf{p}',\mathbf{p})}{(2\pi)^6\sqrt{4q^0q'^0}}\,, \qquad (13.5.1)$$

where q and q' are the initial and final photon four-momenta, p and p' are the initial and final target four-momenta, λ and λ' are the initial and final photon helicities, $\epsilon_\nu(\mathbf{q}',\lambda')$ and $\epsilon_\mu(\mathbf{q},\lambda)$ are the corresponding photon polarization vectors, and σ and σ' are the initial and final target spin z-components. According to the theorem of Section 6.4, the amplitude $M^{\nu\mu}$ may be expressed as

$$(2\pi)^{-3}M^{\nu\mu}_{\sigma',\sigma}(q;\mathbf{p}',\mathbf{p}) = \int d^4x\, e^{iq\cdot x}\left(\Psi_{\mathbf{p}',\sigma'}, T\{J^\nu(0), J^\mu(x)\}\Psi_{\mathbf{p},\sigma}\right) + \cdots \qquad (13.5.2)$$

† The reason that we are now not encountering any factor $(E/\Lambda)^A$ like that in Eq. (13.3.11) is that we are here identifying the maximum energy E of the real soft photon states over which we are summing with the maximum energy Λ of the 'dangerous' soft photon states over which we sum in calculating $\Omega^\pm$.

* This section lies somewhat out of the book's main line of development, and may be omitted in a first reading.

where $J^\mu(x)$ is the electromagnetic current, and the dots indicate possible 'seagull' terms such as those in the theory of charged scalar fields in which the two photons interact at a single vertex instead of with separate currents. We now repeat the standard polology arguments described in Chapter 10 and already used in Section 13.1. Inserting a complete set of intermediate states between the current operators in Eq. (13.5.2), integrating over x and isolating the one-particle intermediate states gives

$$M^{\nu\mu}(q;\mathbf{p}',\mathbf{p}) = \frac{G^\nu(\mathbf{p}',\mathbf{p}+\mathbf{q})G^\mu(\mathbf{p}+\mathbf{q},\mathbf{p})}{E(\mathbf{p}+\mathbf{q})-E(\mathbf{p})-q^0-i\epsilon} + \frac{G^\mu(\mathbf{p}',\mathbf{p}'-\mathbf{q})G^\nu(\mathbf{p}'-\mathbf{q},\mathbf{p})}{E(\mathbf{p}'-\mathbf{q})-E(\mathbf{p}')+q^0-i\epsilon} + N^{\nu\mu}(q;\mathbf{p}',\mathbf{p}) \,, \tag{13.5.3}$$

where G^μ is the one-particle matrix element of the current

$$(2\pi)^{-3}G^\mu_{\sigma',\sigma}(\mathbf{p}',\mathbf{p}) \equiv \left(\Psi_{\mathbf{p}',\sigma'}, J^\mu(0)\,\Psi_{\mathbf{p},\sigma}\right) \tag{13.5.4}$$

and $N^{\nu\mu}$ represents the contribution of states other than the one-particle state itself, plus any direct two-photon interaction terms. (Eq. (13.5.3) is to be understood in the sense of matrix multiplication, with spin indices not shown explicitly.) About $N^{\nu\mu}$ we know very little, except that it does not have the singularity at $q^\mu \to 0$ exhibited by the first two terms, and therefore may be expanded in powers of q^μ.

We now use the current conservation (or gauge invariance) conditions:

$$q_\mu M^{\nu\mu}(q;\mathbf{p}',\mathbf{p}) = 0 \,, \tag{13.5.5}$$

$$\mathbf{q}\cdot\mathbf{G}(\mathbf{p}+\mathbf{q},\mathbf{p}) = [E(\mathbf{p}+\mathbf{q})-E(\mathbf{p})]G^0(\mathbf{p}+\mathbf{q},\mathbf{p}) \,, \tag{13.5.6}$$

$$\mathbf{q}\cdot\mathbf{G}(\mathbf{p}',\mathbf{p}'-\mathbf{q}) = [E(\mathbf{p}')-E(\mathbf{p}'-\mathbf{q})]G^0(\mathbf{p}',\mathbf{p}'-\mathbf{q}) \,. \tag{13.5.7}$$

Applied to Eq. (13.5.3), these conditions yield the condition we need on $N^{\nu\mu}$:

$$q_\mu N^{\nu\mu}(q;\mathbf{p}',\mathbf{p}) = -G^\nu(\mathbf{p}',\mathbf{p}+\mathbf{q})\,G^0(\mathbf{p}+\mathbf{q},\mathbf{p}) + G^0(\mathbf{p}',\mathbf{p}'-\mathbf{q})\,G^\nu(\mathbf{p}'-\mathbf{q},\mathbf{p}) \,. \tag{13.5.8}$$

We also note that $M^{\nu\mu}$ satisfies the 'crossing symmetry' condition

$$M^{\nu\mu}(q;\mathbf{p}',\mathbf{p}) = M^{\mu\nu}(p'-p-q;\mathbf{p}',\mathbf{p}) \tag{13.5.9}$$

and since the pole terms in Eq. (13.5.3) evidently satisfy this condition, so also does $N^{\nu\mu}$:

$$N^{\nu\mu}(q;\mathbf{p}',\mathbf{p}) = N^{\mu\nu}(p'-p-q;\mathbf{p}',\mathbf{p}) \,. \tag{13.5.10}$$

We will use these conditions to determine the first terms in the expansion of $N^{\nu\mu}$ in powers of momenta.

First we need to say something about the expansion of the one-particle current matrix elements $G^\mu(\mathbf{p}',\mathbf{p})$ in powers of the momenta $\mathbf{p}'$ and $\mathbf{p}$.

Space inversion invariance (to the extent that it is applicable) tells us that the expansion of G^0 and G^i (with i =1,2,3) contain terms respectively of only even and odd order in the momenta. According to Eq. (10.6.3), the term in $G^0_{\sigma',\sigma}$ of zeroth order in momenta is $e\delta_{\sigma',\sigma}$, where e is the particle charge. The current conservation condition then tells us that to second order in momenta,

$$(\mathbf{p}' - \mathbf{p}) \cdot \mathbf{G}_{\sigma',\sigma}(\mathbf{p}', \mathbf{p}) = \left(\frac{\mathbf{p}'^2}{2m} - \frac{\mathbf{p}^2}{2m} \right) e\delta_{\sigma',\sigma} \, .$$

The terms in $\mathbf{G}$ of first order in momenta are thus given by $e(\mathbf{p}'+\mathbf{p})\delta_{\sigma',\sigma}/2m$, plus a possible first-order term orthogonal to $\mathbf{p}' - \mathbf{p}$, which rotational invariance tells us must be proportional to $(\mathbf{p}' - \mathbf{p}) \times \mathbf{J}_{\sigma',\sigma}$, where $\mathbf{J}$ is the familiar spin matrix of the charged particle. Summarizing these results, we have the expansions

$$G^0(\mathbf{p}', \mathbf{p}) = e1 + \text{quadratic} \, , \tag{13.5.11}$$

$$\mathbf{G}(\mathbf{p}', \mathbf{p}) = \frac{e1}{2m}(\mathbf{p}' + \mathbf{p}) + \frac{i\mu}{j}\mathbf{J} \times (\mathbf{p}' - \mathbf{p}) + \text{cubic} \, , \tag{13.5.12}$$

where '1' is the unit spin matrix, and 'quadratic' and 'cubic' refer to the order of the neglected terms in powers of the small momenta $\mathbf{p}$ and $\mathbf{p}'$. The coefficient μ/j in Eq. (13.5.12) is real because the current is Hermitian. With the coefficient written in this way (with j the spin of the charged particle), μ is the quantity known as the magnetic moment of the particle.

Now let us return to $N^{\nu\mu}$, and consider the expansion of Eq. (13.5.8) in powers of the small momenta q^μ, $\mathbf{p}$ and $\mathbf{p}'$. Taking $\nu = 0$ in Eq. (13.5.8) shows that $q_\mu N^{0\mu}$ is at least quadratic in these small quantities. There is no constant vector orthogonal to q^μ, so $N^{0\mu}$ must be at least of first order in small momenta. The crossing symmetry condition (13.5.10) then tells us that N^{i0} must also be at least of first order in small momenta. Taking $\nu = i$ in Eq. (13.5.8) and using Eq. (13.5.12) then tells us that

$$q_k N^{ik} = -\frac{e^2 q^i}{m} + \text{quadratic}$$

and hence

$$N^{ik} = -\frac{e^2}{m}\delta_{ik} + \text{linear} \, . \tag{13.5.13}$$

Since G^i is at least of first order in the small momenta, so are the pole terms in Eq. (13.5.3) for M^{ik}, leaving us in zero order with only the *non-pole* term N^{ik}:

$$M^{ik}(0;0,0) = N^{ik}(0;0,0) = -\frac{e^2}{m}\delta_{ik} \, . \tag{13.5.14}$$

From this we can calculate the soft photon scattering cross-section. But there is no need for this calculation; now that we know that the photon scattering amplitude in the limit of zero momentum depends only on the target particle mass and charge, and is of second order in the charge, we can immediately use the results of *any* second-order calculation of the photon scattering cross-section for target particles of any given spin, such as our result (8.7.42) for the differential photon scattering cross-section in quantum electrodynamics:

$$\frac{d\sigma}{d\Omega} = \frac{e^4}{32\pi^2 m^2}(1+\cos^2\theta)\,. \tag{13.5.15}$$

We now see that this is a universal formula, valid in the low-energy limit for target particles of mass m and charge e and of arbitrary type and spin, even if these particles are composite and strongly interacting, like atomic nuclei. Gell-Mann and Goldberger and Low[8] have shown that these results may be extended to give the next-to-leading term in the soft photon scattering amplitude in terms of the target particle's mass, charge, and magnetic moment.

13.6 The External Field Approximation*

It is intuitively obvious that a heavy charged particle like the nucleus of an atom acts approximately like the source of a classical external field. In this section we will see how to justify this approximation, and will gain some idea of its limitations.

Consider a Feynman diagram or a part of a Feynman diagram in which a heavy charged particle passing through the diagram from the initial to the final state emits N off-shell photons with four-momenta $q_1, q_2, \cdots q_N$ and polarization indices $\mu_1, \mu_2, \cdots \mu_N$. The sum of all such graphs or subgraphs (not including the N photon propagators) yields an amplitude

$$\begin{aligned}&\int d^4x_1\, d^4x_2 \cdots d^4x_N\, e^{-iq_1\cdot x_1} e^{-iq_2\cdot x_2} \cdots e^{-iq_N\cdot x_N} \\ &\qquad \times \langle \mathbf{p}', \sigma'| T\Big\{J^{\mu_1}(x_1),\, J^{\mu_2}(x_2),\, \cdots J^{\mu_N}(x_N)\Big\}|\mathbf{p}, \sigma\rangle \\ &\equiv \mathscr{G}^{\mu_1\mu_2\cdots\mu_N}_{\sigma',\sigma}(q_1, q_2, \cdots q_N; p)\end{aligned} \tag{13.6.1}$$

with the matrix element calculated including all interactions in which the heavy particle may participate, including strong nuclear forces. This amplitude has a multiple pole at $q_1, q_2, \cdots q_N \to 0$, arising from terms in

* This section lies somewhat out of the book's main line of development, and may be omitted in a first reading.

the matrix elements of the product of currents in which the intermediate states consist of just the same heavy particle as in the initial and final states. This multiple pole dominates (13.6.1) when all components of $q_1, q_2, \cdots q_N$ are small compared with all energies and momenta associated with the dynamics of the (perhaps composite) heavy particle. In this case the methods of Section 10.2 give**

$$\begin{aligned}
\mathscr{G}^{\mu_1\mu_2\cdots\mu_N}_{\sigma',\sigma}(q_1, q_2, \cdots q_N; p) &\to \frac{(-i)^{N-1}}{2p^0(2\pi)^3}(2\pi)^4 \\
&\times \delta^4(p' + q_1 + q_2 + \cdots + q_N - p) \sum_{\sigma_1,\sigma_2,\cdots\sigma_{N-1}} \\
&\times \frac{\mathscr{G}^{\mu_1}_{\sigma',\sigma_1}(p)\,\mathscr{G}^{\mu_2}_{\sigma_1,\sigma_2}(p)\cdots\mathscr{G}^{\mu_N}_{\sigma_{N-1},\sigma}(p)}{[2p\cdot q_1 - i\epsilon][2p\cdot(q_1+q_2) - i\epsilon]\cdots[2p\cdot(q_1+\cdots+q_{N-1}) - i\epsilon]} \\
&+ \text{permutations}\,, \qquad (13.6.2)
\end{aligned}$$

where

$$\frac{\mathscr{G}^{\mu}_{\sigma',\sigma}(p)}{2p^0(2\pi)^3} \equiv \langle \mathbf{p}, \sigma' | J^\mu(0) | \mathbf{p}, \sigma \rangle \qquad (13.6.3)$$

and '+ permutations' indicates that we are to sum over all permutations of the N photons. For applications to atomic systems it is important to recognize that (13.6.1) applies for particles of arbitrary spin that have strong as well as electromagnetic interactions, like atomic nuclei.

We also note that for particles of arbitrary spin and charge Ze, the matrix elements of the electric current between states of equal four-momenta are†

$$\langle p, \sigma' | J^\mu(0) | p, \sigma \rangle = \frac{Ze\, p^\mu \delta_{\sigma'\sigma}}{p^0(2\pi)^3}\,, \qquad (13.6.4)$$

so that

$$\mathscr{G}^{\mu}_{\sigma'\sigma}(p) = 2Ze\, p^\mu \delta_{\sigma'\sigma}\,. \qquad (13.6.5)$$

The important thing about Eq. (13.6.5) is that these matrices all commute,

** In perturbation theory, the denominators come from the denominators of the propagators:

$$(p' + q_1 + \cdots q_r)^2 + m^2 - i\epsilon \to 2p'\cdot(q_1 + \cdots q_r) - i\epsilon \to 2p\cdot(q_1 + \cdots q_r) - i\epsilon\,,$$

while the numerators of the propagators provide factors $\sum uu^\dagger$ that together with the photon emission vertex matrices yield the matrix elements (13.6.3). The matrix $\mathscr{G}^\mu$ differs from the matrix G^μ of the previous section by a factor $2p^0$.

† This is most easily proved by first noting that in the Lorentz frame in which the particle is at rest, rotational invariance requires that the matrix elements of the current have vanishing space components and a time component proportional to $\delta_{\sigma'\sigma}$, with no other dependence on σ or σ'. The constant of proportionality is supplied by Eq. (10.6.3), and a Lorentz transformation then gives Eq. (13.6.4).

so their product can be factored out of the sum over permutations:

$$\mathscr{G}^{\mu_1\mu_2\cdots\mu_N}_{\sigma',\sigma}(q_1,q_2,\cdots q_N;p) \quad \rightarrow$$
$$\frac{(-i)^{N-1}(Ze)^N p^{\mu_1}p^{\mu_2}\cdots p^{\mu_N}}{p^0(2\pi)^3}(2\pi)^4\delta^4(p'+q_1+q_2+\cdots+q_N-p)\delta_{\sigma',\sigma}$$
$$\times\left[\frac{1}{[p\cdot q_1-i\epsilon][p\cdot(q_1+q_2)-i\epsilon]\cdots[p\cdot(q_1+\cdots+q_{N-1})-i\epsilon]}\right.$$
$$\left.+\text{ permutations}\right]\,. \tag{13.6.6}$$

To leading order in the qs the delta function here may be written

$$\delta^4(p'+q_1+\cdots+q_N-p)=p^0\delta^3(\mathbf{p}'+\mathbf{q}_1+\cdots+\mathbf{q}_N-\mathbf{p})\delta(p\cdot(q_1+\cdots+q_N))\,. \tag{13.6.7}$$

Fortunately, it turns out that the result of summing over permutations here is much simpler than the individual terms. For $p\cdot(q_1+\cdots+q_N)=0$, we have

$$\left[\frac{1}{[p\cdot q_1-i\epsilon][p\cdot(q_1+q_2)-i\epsilon]\cdots[p\cdot(q_1+\cdots+q_{N-1})-i\epsilon]}\right.$$
$$\left.+\text{ permutations}\right]=(2i\pi)^{N-1}\delta(p\cdot q_1)\,\delta(p\cdot q_2)\cdots\delta(p\cdot q_{N-1})\,. \tag{13.6.8}$$

For instance, for $N=2$ this reads:

$$\frac{1}{[p\cdot q_1-i\epsilon]}+\frac{1}{[p\cdot q_2-i\epsilon]}=\frac{1}{[p\cdot q_1-i\epsilon]}+\frac{1}{[-p\cdot q_1-i\epsilon]}=2i\pi\delta(p\cdot q_1)\,.$$

The general result (13.6.8) can be obtained most easily as the Fourier transform of the identity

$$\theta(\tau_1-\tau_2)\,\theta(\tau_2-\tau_3)\cdots\theta(\tau_{N-1}-\tau_N)+\text{permutations}=1\,.$$

Inserting Eq. (13.6.8) in Eq. (13.6.6) gives our final result for the amplitude (13.6.1):

$$\mathscr{G}^{\mu_1\mu_2\cdots\mu_N}_{\sigma',\sigma}(q_1,q_2,\cdots q_N;p)\rightarrow(Ze)^N(2\pi)^N\;\delta_{\sigma',\sigma}\,p^{\mu_1}p^{\mu_2}\cdots p^{\mu_N}$$
$$\times\,\delta^3(\mathbf{p}'+\mathbf{q}_1+\mathbf{q}_2+\cdots+\mathbf{q}_N-\mathbf{p})\delta(p\cdot q_1)\,\delta(p\cdot q_2)\cdots\delta(p\cdot q_N)\,. \tag{13.6.9}$$

This result applies to relativistic as well as slowly moving heavy particles, and can be used to derive the 'Weizsäcker-Williams' approximation[9] for charged-particle scattering. In the special case of a non-relativistic heavy charged particle, with $|\mathbf{p}|\ll p^0$, Eq. (13.6.9) further simplifies to

$$\mathscr{G}^{\mu_1\mu_2\cdots\mu_N}_{\sigma',\sigma}(q_1,q_2,\cdots q_N;p)\rightarrow(Ze)^N(2\pi)^N n^{\mu_1}n^{\mu_2}\cdots n^{\mu_N}$$
$$\times\delta^3(\mathbf{p}'+\mathbf{q}_1+\mathbf{q}_2+\cdots+\mathbf{q}_N-\mathbf{p})\delta(q_1^0)\,\delta(q_2^0)\cdots\delta(q_N^0)\;\delta_{\sigma',\sigma}, \tag{13.6.10}$$

where n is a unit time-like vector

$$n^0 = 1\,, \qquad \mathbf{n} = 0\,.$$

Now suppose that a single heavy non-relativistic particle of charge Ze with normalized momentum space wave function $\chi_\sigma(\mathbf{p})$ appears in both the initial and final states. Using the Fourier representation of the delta function in Eq. (13.6.10), the matrix element of $\mathscr{G}$ in this state is

$$\int d^3p\, d^3p'\, \chi^*_{\sigma'}(\mathbf{p}')\,\chi_\sigma(\mathbf{p})\; \mathscr{G}^{\mu_1\mu_2\cdots\mu_N}_{\sigma',\sigma}(q_1, q_2, \cdots q_N; p) \to$$

$$\int d^3X \sum_\sigma |\psi_\sigma(\mathbf{X})|^2 \prod_{r=1}^{N} 2\pi\, Ze\, n^{\mu_r}\delta(q_r^0)e^{-i\mathbf{q}_r\cdot\mathbf{X}} \tag{13.6.11}$$

where $\psi(\mathbf{X})$ is the coordinate space wave function:

$$\psi_\sigma(\mathbf{X}) \equiv (2\pi)^{-3/2}\int d^3p\, \chi_\sigma(\mathbf{p})e^{i\mathbf{p}\cdot\mathbf{X}}\,. \tag{13.6.12}$$

Because of the factorization in Eq. (13.6.11), the effect of including a heavy charged particle in this state is then the same as that of adding any numbers of a new kind of vertex in the momentum space Feynman rules, in which light Dirac particles of charge $-e$ such as electrons interact with an external field, with each such vertex contributing to the overall amplitude a factor† (now including the photon propagator and electron–photon vertex)

$$i\int d^4q \left[\frac{-i}{(2\pi)^4}\frac{1}{q^2 - i\epsilon}\right]\left[2\pi\, Ze\, n_\mu \delta(q^0)e^{-i\mathbf{q}\cdot\mathbf{X}}\right]\left[(2\pi)^4\, e\,\gamma^\mu\, \delta^4(k - k' - q)\right] \tag{13.6.13}$$

where k and k' are the initial and final electron four-momenta. The complete scattering amplitude must then be averaged over the heavy particle position $\mathbf{X}$, with weight function $\sum_\sigma |\psi_\sigma(\mathbf{X})|^2$. The factor (13.6.13) is the same as would be produced by a new term in the interaction Lagrangian

$$\mathscr{L}_{\rm ext}(x) = \mathscr{A}_\mu(x)\, J_e^\mu(x)\,, \tag{13.6.14}$$

where $J_e^\mu \equiv -ie\bar{\Psi}\gamma^\mu\Psi$ is the electric current of the electrons, and $\mathscr{A}^\mu$ is an external vector potential

$$\mathscr{A}^\mu(x) = \frac{1}{(2\pi)^4}\int d^4q\; e^{iq\cdot x}\left[\frac{2\pi\, Ze\, n^\mu \delta(q^0)e^{-i\mathbf{q}\cdot\mathbf{X}}}{q^2 - i\epsilon}\right]\,. \tag{13.6.15}$$

† The first factor here is the usual factor of i accompanying the constants in the interaction Lagrangian of the heavy charged particle in the Feynman rules.

This, of course, is just the usual Coulomb potential:

$$\mathscr{A}^0(x) = \frac{Ze}{4\pi|\mathbf{x}-\mathbf{X}|}\,, \qquad \boldsymbol{\mathscr{A}}(x) = 0. \tag{13.6.16}$$

If there is more than one heavy charged particle (as in a molecule) we must express $\mathscr{A}^\mu(x)$ as a sum of terms like (13.6.16), each with its own charge Ze and position $\mathbf{X}$.

It is useful to keep in mind what diagrams we are summing in using the external field approximation. Consider the interaction of a single electron (relativistic or non-relativistic) with a single heavy charged particle such as a proton or deuteron. If we ignore all other interactions, then the Feynman diagrams for the scattering of the electron due to its interaction with the external field are just those with any number of insertions of the electron-external field vertex (13.6.14) in the electron line. (See Figure 13.4.) But as shown by the sum over permutations in Eq. (13.6.2), these diagrams in the external field approximation come from diagrams in the underlying theory in which the photons attached to the electron line are attached to the heavy charged particle line in all possible orders. (See Figure 13.5.) The 'uncrossed ladder' diagrams (labelled L) of Figure 13.5 do *not* dominate this sum unless the electron as well as the heavy charged particle is non-relativistic. (These diagrams include contributions from terms in old-fashioned perturbation theory whose intermediate states contain the same particles as the initial and final states, leading to small energy denominators when the electron and heavy charged particle are both non-relativistic, while all other diagrams of Figure 13.5 correspond to intermediate states with either extra photons, electron–positron pairs, or heavy particle–antiparticle pairs, which are suppressed by large energy denominators.) The uncrossed ladders can be summed by solving an integral equation, known as the *Bethe–Salpeter equation*,[10] but there is no rationale for selecting out this subset of diagrams unless both particles are non-relativistic, in which case the Bethe–Salpeter equation just reduces to the ordinary non-relativistic Schrödinger equation, plus relativistic corrections associated with the spin–orbit coupling that can be treated as small perturbations. It must be said that the theory of relativistic effects and radiative corrections in bound states is not yet in entirely satisfactory shape.

In the derivation of the external field (13.6.16) we evaluated the interaction of the heavy charged particle with the electromagnetic field only to leading order in the photon momentum. There are corrections of higher order in the photon momentum arising from the heavy particle's magnetic dipole moment, electric quadrupole moment, etc. Also, of course, there are radiative corrections arising from Feynman diagrams beyond those of Figure 13.4, such as diagrams in which photons are emitted and absorbed

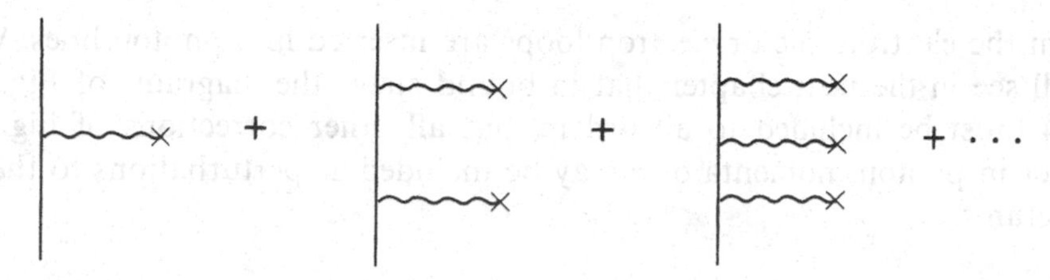

Figure 13.4. Diagrams for the scattering of an electron by an external electromagnetic field. Here straight lines represent the electron; wavy lines ending in crosses represent its interaction with an external field.

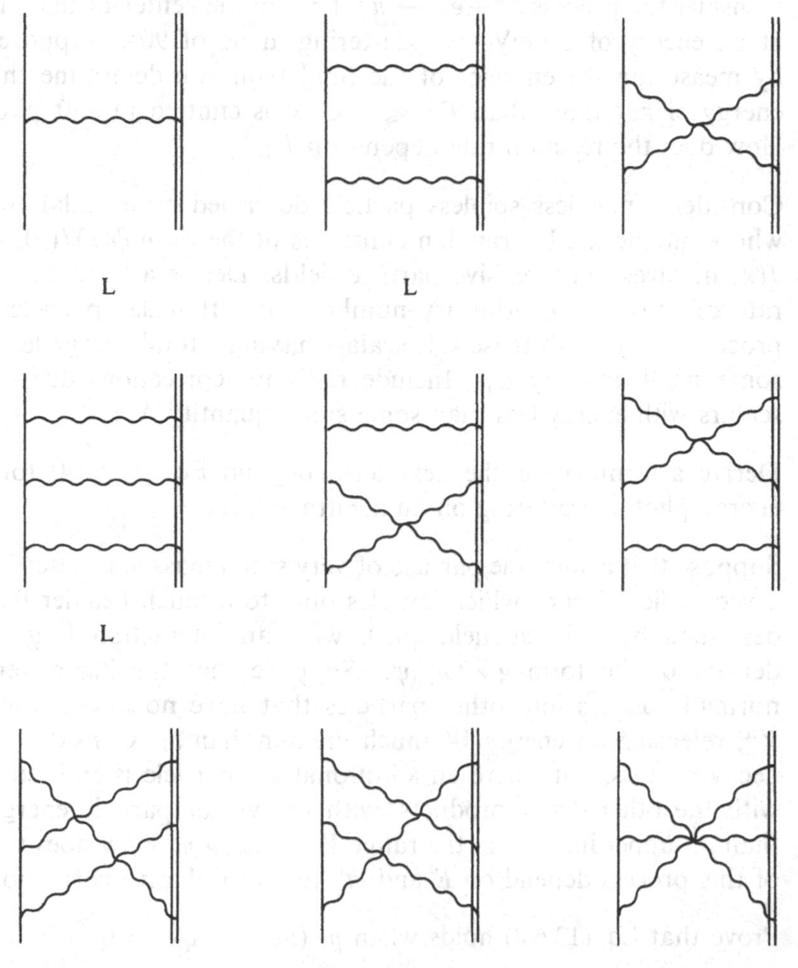

Figure 13.5. Diagrams for the scattering of an electron by a heavy, charged target particle, which in the limit of large target mass yield the same result as the diagrams of Figure 13.4. Here the single straight line is the electron; the double straight line is the heavy target particle; and wavy lines are virtual photons. Diagrams marked 'L' are called uncrossed ladder graphs; they dominate the sum when the electron as well as the target particle is non-relativistic.

from the electron line or electron loops are inserted into photon lines. We shall see in the next chapter that in bound states the diagrams of Figure 13.4 must be included to all orders, but all other corrections of higher order in photon momenta or e may be included as perturbations to these diagrams.

Problems

1. Consider the process $e^+ + e^- \to \pi^+ + \pi^-$ in the center-of-mass frame at an energy of 1 GeV and scattering angle of 90°. Suppose that by measuring the energies of the final pions we determine that an energy of not more than $E_T \ll 1$ GeV is emitted in soft photons. How does the reaction rate depend on E_T?

2. Consider a massless spinless particle described by a scalar field ϕ, whose interaction Lagrangian density is of the form $\phi(x)J(x)$, where $J(x)$ involves only massive particle fields. Derive a formula for the rate of emission of arbitrary numbers of soft scalar particles in a process $\alpha \to \beta$, with these soft scalars having a total energy less than some small quantity E_T. Include radiative corrections due to soft scalars with energy less than some small quantity Λ.

3. Derive a formula for the next term beyond Eq. (13.5.14) for low-energy photon scattering on an arbitrary target.

4. Suppose that a spin one particle of very small mass m is described by a vector field $V^\mu(x)$, which couples only to a much heavier fermion described by a Dirac field $\psi(x)$, with an interaction Lagrangian density of the form $g\, V^\mu \bar{\psi}\gamma_\mu\psi$. Suppose that the heavy fermion normally decays into other particles that have no interaction with V^μ, releasing an energy W much greater than m. Consider such a decay process, but where an additional V^μ particle is emitted along with the other decay products, with the vector particle energy less than an upper limit E, in the range $W \gg E \gg m$. How does the rate of this process depend on E and m? Ignore radiative corrections.

5. Prove that Eq. (13.6.8) holds when $p \cdot (q_1 + \cdots q_N) = 0$.

References

1. F. Bloch and A. Nordsieck, *Phys. Rev.* **37**, 54 (1937); D. R. Yennie, S. C. Frautschi, and H. Suura, *Ann. Phys. (NY)* **13**, 379 (1961). Also

see K. T. Mahantappa, Ph.D. Thesis at Harvard University (1961), unpublished.

2. S. Weinberg, *Phys. Lett.* **9**, 357 (1964); *Phys. Rev.* **135**, B1049 (1964).
3. S. Weinberg, *Phys. Rev.* **140**, B515 (1965).
4. This phase was encountered in the perturbation series for non-relativistic Coulomb scattering, by R. H. Dalitz, *Proc. Roy. Soc. London* **206**, 509 (1951).
5. See, e.g., L. I. Schiff, *Quantum Mechanics* (McGraw-Hill, New York, 1949): Section 20.
6. T. D. Lee and M. Nauenberg, *Phys. Rev.* **133**, B1549 (1964). Also see T. Kinoshita, *J. Math. Phys.* **3**, 650 (1962); G. Sterman and S. Weinberg, *Phys. Rev. Lett.* **39**, 1416 (1977).
7. G. Sterman and S. Weinberg, Ref. 6.
8. F. E. Low, *Phys. Rev.* **96**, 1428 (1954); M. Gell-Mann and M. L. Goldberger, *Phys. Rev.* **96**, 1433 (1954). Also see S. Weinberg, in *Lectures on Elementary Particles and Quantum Field Theory — 1970 Brandeis Summer Institute in Theoretical Physics*, ed. by S. Deser, M. Grisaru, and H. Pendleton (MIT Press, Cambridge, MA, 1970).
9. E. J. Williams, *Kgl. Dan. Vid. Sel. Mat.-fys. Medd.* **XIII**, No. 4 (1935).
10. H. A. Bethe and E. E. Salpeter, *Phys. Rev.* **82**, 309 (1951); **84**, 1232 (1951).

14
Bound States in External Fields

In our calculations of radiative corrections in Chapter 11 we went just one step beyond the lowest order in perturbation theory. However, there is a very important class of problems where even the simplest calculation requires that from the beginning we consider classes of Feynman diagrams of arbitrarily high order in coupling constants like e. These problems are those involving bound states – in electrodynamics, either ordinary atoms and molecules, or such exotic atoms as positronium or muonium.

It is easy to see that such problems necessarily involve a breakdown of ordinary perturbation theory. Consider for instance the amplitude for electron–proton scattering as a function of the center-of-mass energy E. As shown in Section 10.3, the existence of a bound state like the ground state of hydrogen implies the existence of a pole in this amplitude at $E = m_p + m_e - 13.6$ eV. However, no single term in the perturbation series for electron–proton scattering has such a pole. The pole therefore can only arise from a divergence of the sum over all diagrams at center-of-mass energies near $m_p + m_e$.

The reason for this divergence of the perturbation series is also easy to see, especially if for the moment we consider the time-ordered diagrams of old-fashioned perturbation theory instead of Feynman diagrams. Suppose that in the center-of-mass system the electron and proton both have momenta of magnitude $q \ll m_e$, and consider an intermediate state in which the electron and proton momenta are different but also of order q. The energy denominator factor contributed by this state will be of order $[q^2/m_e]^{-1}$. Each such state will also contribute a matrix element of the Coulomb interaction of order e^2/q^2 (the Fourier transform of e^2/r), and the corresponding momentum space integration will contribute a factor of order q^3. Putting this together, we see that each additional Coulomb interaction contributes an overall factor of order

$$[q^2/m_e]^{-1}\,[e^2/q^2]\,[q^3] = e^2 m_e/q\,.$$

Thus the perturbation theory should break down when q is less than or of the order of $e^2 m_e$, or in other words when the kinetic and potential energies, which are of order q^2/m_e, are no larger than about $e^4 m_e$, which, of course, is of the order of the binding energy of hydrogen.

Our problem here is to learn how to use perturbation theory to evaluate radiative corrections in bound-state problems, summing to all orders those diagrams that need to be summed to all orders, and keeping only a finite number of those that don't.

14.1 The Dirac Equation

We shall limit ourselves in this chapter to problems in which bound states arise because of the Coulomb interaction of electrons (or muons) with heavy charged particles such as atomic nuclei. As shown in Section 13.6, this interaction may be taken into account by adding to the interaction Lagrangian a term* representing the effects of a c-number external vector potential $\mathscr{A}^\mu(x)$:

$$\mathscr{L}_{\mathscr{A}} = -ie\bar{\Psi}\gamma^\mu\Psi\mathscr{A}_\mu - \tfrac{1}{2}(Z_3-1)(\partial^\mu\mathscr{A}^\nu - \partial^\nu\mathscr{A}^\mu)(\partial_\mu A_\nu - \partial_\nu A_\mu)$$
$$-ie(Z_2-1)\mathscr{A}_\mu\,\bar{\Psi}\gamma^\mu\Psi\,, \qquad (14.1.1)$$

which is obtained by replacing the quantum vector potential A^μ with $A^\mu + \mathscr{A}^\mu$ in the interaction part of Eq. (11.1.6). For instance, for a single heavy particle of charge Ze at the origin,

$$\mathscr{A}^0(x) = \frac{Ze}{4\pi|\mathbf{x}|}\,, \qquad \boldsymbol{\mathscr{A}}(x) = 0\,. \qquad (14.1.2)$$

It is the interaction (14.1.1) that must be taken into account to all orders. In this section we will consider the theory with only this interaction, leaving radiative corrections to subsequent sections.

Physicists learn in kindergarten to approach this sort of problem by solving the wave equation of Dirac in the presence of the external field. It might seem unnecessary to derive this equation here, but as emphasized in Chapter 1, Dirac's original motivation for this equation as a sort of relativistic Schrödinger equation does not stand up to inspection. Also, in the course of our derivation we will discover the normalization conditions that have to be imposed on the solutions of the Dirac equation, which seemed somewhat *ad hoc* in Dirac's approach. The solutions of the Dirac equation discussed here will be important ingredients in our treatment of radiative corrections in the next section.

* In this chapter we return to the use of an upper case Ψ to denote the electron field in the Heisenberg picture, reserving a lower case ψ for the Dirac field with time-dependence governed solely by the c-number external field $\mathscr{A}^\mu(x)$.

We will work here in a version of the Heisenberg picture, in which the time-dependence of operators is determined by a Hamiltonian including the external field interaction (14.1.1), but no other interactions. The electron field $\psi(x)$ in this picture satisfies the field equation

$$\left[\gamma^\lambda \frac{\partial}{\partial x^\lambda} + m + ie\gamma^\lambda \mathscr{A}_\lambda(x)\right] \psi(x) = 0 . \tag{14.1.3}$$

This is not the Dirac equation in the original sense of Dirac,[1] because $\psi(x)$ here is not a c-number wave function but a quantum operator. The c-number Dirac *wave functions* are defined by

$$u_N(x) \equiv (\Phi_0, \psi(x)\Phi_N) , \tag{14.1.4}$$

$$v_N(x) \equiv (\Phi_N, \psi(x)\Phi_0) , \tag{14.1.5}$$

where Φ_N are a complete orthonormal set of state-vectors, with Φ_0 the vacuum. It follows immediately from Eq. (14.1.3) that these functions satisfy the homogeneous Dirac equation

$$\begin{aligned}\left[\gamma^\lambda \frac{\partial}{\partial x^\lambda} + m + ie\gamma^\lambda \mathscr{A}_\lambda(x)\right] u_N(x) &= \left[\gamma^\lambda \frac{\partial}{\partial x^\lambda} + m + ie\gamma^\lambda \mathscr{A}_\lambda(x)\right] v_N(x) \\ &= 0 .\end{aligned} \tag{14.1.6}$$

We can also derive a normalization condition from the equal-time anti-commutation relations for the Dirac field. These are unaffected by the interaction (14.1.1), and therefore take the same form as for the free fields:

$$\{\psi(\mathbf{x},t), \bar{\psi}(\mathbf{y},t)\} = i\gamma^0 \delta^3(\mathbf{x}-\mathbf{y}) . \tag{14.1.7}$$

Taking the vacuum expectation value and inserting a sum over the states Φ_N, we find

$$\sum_N u_N(\mathbf{x},t) u_N^\dagger(\mathbf{y},t) + \sum_N v_N(\mathbf{x},t) v_N^\dagger(\mathbf{y},t) = \delta^3(\mathbf{x}-\mathbf{y}) , \tag{14.1.8}$$

it being understood that the sum over N includes an integral over continuum states as well as a sum over any discrete bound states.

We are chiefly interested in the case of a time-independent external field, like (14.1.2). In such cases, the states Φ_N may be taken as eigenstates of the Hamiltonian (including the interaction (14.1.1)) with energies E_N. Time-translation invariance then tells us that the $u_N(x)$ and $v_N(x)$ have the time-dependence:

$$u_N(\mathbf{x},t) = e^{-iE_N t} u_N(\mathbf{x}), \qquad v_N(\mathbf{x},t) = e^{+iE_N t} v_N(\mathbf{x}) . \tag{14.1.9}$$

The homogeneous Dirac equations (14.1.6) then become

$$i\gamma^0\left[\gamma\cdot\nabla+m+ie\gamma^\lambda\mathscr{A}_\lambda(\mathbf{x})\right]\,u_N(\mathbf{x})=E_N u_N(\mathbf{x})\,,\qquad(14.1.10)$$

$$i\gamma^0\left[\gamma\cdot\nabla+m+ie\gamma^\lambda\mathscr{A}_\lambda(\mathbf{x})\right]\,v_N(\mathbf{x})=-E_N v_N(\mathbf{x})\,.\qquad(14.1.11)$$

The minus sign on the right-hand side of Eq. (14.1.11) shows that the v_N are the famous 'negative-energy' solutions of Dirac. As shown by Eq. (14.1.8), these negative-energy solutions are needed to make up a complete set of wave functions. Of course, for moderate external fields there are no negative-energy *states* in the theory, so all E_N are positive, but there is still an important difference between the states with non-vanishing u_N or v_N: the definitions (14.1.4) and (14.1.5) show that a state can have $u_N \neq 0$ or $v_N \neq 0$ only if it has charge $-e$ or $+e$, respectively. It is in this sense that negative-energy solutions of the Dirac equation have something to do with the existence of antiparticles. However, this argument has nothing to do with the details of the Dirac equation, or even with the spin of the electron.

From the Dirac wave equations (14.1.10) and (14.1.11) we can easily see that wave functions of different energy are orthogonal. That is,

$$(E_M-E_N^*)\left(u_N^\dagger u_M\right)=\nabla\cdot\left(u_N^\dagger i\gamma^0\gamma u_M\right)\,,$$

so if $|\mathbf{x}|^2\left(u_N^\dagger i\gamma^0\gamma u_M\right)$ remains bounded as $|\mathbf{x}|\to 0$ and $|\mathbf{x}|\to\infty$, then

$$\int d^3x\,\left(u_N^\dagger(\mathbf{x})u_M(\mathbf{x})\right)=0\quad\text{if}\quad E_N\neq E_M^*\,.\qquad(14.1.12)$$

With similar boundary conditions for the v_N, we find in the same way that

$$\int d^3x\,\left(v_N^\dagger(\mathbf{x})v_M(\mathbf{x})\right)=0\quad\text{if}\quad E_N\neq E_M^*\,,\qquad(14.1.13)$$

$$\int d^3x\,\left(u_N^\dagger(\mathbf{x})v_M(\mathbf{x})\right)=0\quad\text{if}\quad E_N\neq -E_M^*\,.\qquad(14.1.14)$$

Taking $N=M$, Eqs. (14.1.12) and (14.1.13) tell us that the energies are all real. Dropping the complex conjugation of E_M in Eqs. (14.1.12)–(14.1.14), we see that us of different energy are orthogonal, vs of different energy are orthogonal, and (as long as the potential is not strong enough to produce negative energy *states*) all us are orthogonal to all vs. By a suitable choice of the discrete quantum numbers that characterize the states along with the energy, we can then always arrange that

$$\int d^3x\,\left(u_N^\dagger(\mathbf{x})\,u_M(\mathbf{x})\right)=0\quad\text{if}\quad N\neq M\,,\qquad(14.1.15)$$

$$\int d^3x\,\left(v_N^\dagger(\mathbf{x})\,v_M(\mathbf{x})\right)=0\quad\text{if}\quad N\neq M\,,\qquad(14.1.16)$$

$$\int d^3x\,\left(u_N^\dagger(\mathbf{x})\,v_M(\mathbf{x})\right)=0\,.\qquad(14.1.17)$$

Multiplying Eq. (14.1.8) on the right with $u_M(\mathbf{y})$ or $v_M(\mathbf{y})$, we find then that these wave functions must satisfy the normalization conditions

$$\int d^3y \left(u_N^\dagger(\mathbf{y})u_M(\mathbf{y})\right) = \int d^3y \left(v_N^\dagger(\mathbf{y})v_M(\mathbf{y})\right) = \delta_{NM} , \tag{14.1.18}$$

where δ_{NM} is a product of Kronecker deltas and momentum space delta functions, with normalization adapted to that used in defining $\sum_N$, in such a way that $\sum_N \delta_{NM} = 1$. These normalization conditions have nothing directly to do with any probabilistic interpretation of the Dirac wave functions, but arise instead from the anticommutation relations (14.1.7) for the fields.

Let us now specialize to the case of a pure electrostatic external field with $\mathscr{A} = 0$. In our standard representation of the Dirac matrices, we have

$$\boldsymbol{\gamma} = i\begin{pmatrix} 0 & -\boldsymbol{\sigma} \\ \boldsymbol{\sigma} & 0 \end{pmatrix}, \qquad i\gamma^0 = \beta = \begin{pmatrix} 0 & 1 \\ 1 & 0 \end{pmatrix},$$

where $\boldsymbol{\sigma}$ is the usual three-vector of 2×2 Pauli matrices, and '1' and '0' here are the 2×2 unit and zero matrices. We introduce two-component wave functions f_N and g_N by setting

$$u_N = \frac{1}{\sqrt{2}}\begin{pmatrix} f_N + ig_N \\ f_N - ig_N \end{pmatrix} . \tag{14.1.19}$$

The energy eigenvalue condition (14.1.10) then takes the form:

$$(\boldsymbol{\sigma} \cdot \nabla) f_N = (E_N + e\mathscr{A}^0 + m)\, g_N , \tag{14.1.20}$$

$$(\boldsymbol{\sigma} \cdot \nabla) g_N = -(E_N + e\mathscr{A}^0 - m)\, f_N . \tag{14.1.21}$$

In the non-relativistic case where $e\mathscr{A}^0 r \approx Z\alpha \ll 1$ the binding energy $m - E_N$ is of order $Z^2\alpha^2 m$, while the gradient operator is of the order of $Z\alpha m$, so g_N is smaller than f_N by a factor of order $Z\alpha$. (To find the positron wave functions v_N we replace E_N everywhere by $-E_N$, so in this case f_N is smaller than g_N by the same factor.) We shall return to this non-relativistic case at the end of this section.

Physical states may be classified as even or odd under space inversion:

$$\mathsf{P}\Phi_N = \eta_N \Phi_N , \tag{14.1.22}$$

where η_N is a sign factor, ± 1. Recall that with the intrinsic parity of the electron defined to be $+1$, the Dirac field has the space-inversion property

$$\mathsf{P}\psi(\mathbf{x}, t)\mathsf{P}^{-1} = \beta\, \psi(-\mathbf{x}, t)$$

so Eqs. (14.1.4) and (14.1.22) show that the Dirac wave functions satisfy the parity condition

$$u_N(\mathbf{x}) = \eta_N \beta\, u_N(-\mathbf{x}) \tag{14.1.23}$$

or equivalently

$$f_N(\mathbf{x}) = \eta_N f_N(-\mathbf{x}) , \qquad g_N(\mathbf{x}) = -\eta_N g_N(-\mathbf{x}) . \qquad (14.1.24)$$

Note that the parity of the state is the same as the parity of $f_N(\mathbf{x})$, not $g_N(\mathbf{x})$.

Where the potential $\mathscr{A}^0$ is rotationally invariant, the solutions of the wave equations here may be classified according to their total angular momentum j and parity η. For a given j, the components f and g may be expanded in spherical harmonics with orbital angular momentum $\ell = j+\frac{1}{2}$ and $\ell = j-\frac{1}{2}$, but for a definite parity $\eta = (-1)^{j\mp\frac{1}{2}}$, Eqs. (14.1.24) show that we can only have $\ell = j \mp \frac{1}{2}$ in f and $\ell = j \pm \frac{1}{2}$ in g. The usual rules of angular-momentum addition then show that for a state of total angular momentum j, total angular-momentum z-component μ, and parity $(-1)^{j\mp\frac{1}{2}}$, the 'large' two-component wave function f has the form

$$f(\mathbf{x}) = \begin{pmatrix} C_{j\mp\frac{1}{2},\frac{1}{2}}(j\,\mu\,;\mu-\frac{1}{2}\;\;\frac{1}{2})\,Y^{\mu-\frac{1}{2}}_{j\mp\frac{1}{2}}(\hat{\mathbf{x}}) \\ C_{j\mp\frac{1}{2},\frac{1}{2}}(j\,\mu\,;\mu+\frac{1}{2}\;-\frac{1}{2})\,Y^{\mu+\frac{1}{2}}_{j\mp\frac{1}{2}}(\hat{\mathbf{x}}) \end{pmatrix} F(|\mathbf{x}|) , \qquad (14.1.25)$$

where C and Y are the usual Clebsch–Gordan coefficients and spherical harmonics.[2] Also, given any wave function of definite total angular momentum and parity we can construct another wave function with the same j and μ but opposite parity by applying the operator $\boldsymbol{\sigma}\cdot\hat{\mathbf{x}}$, so the 'small' components may be put in the form

$$g(\mathbf{x}) = (\boldsymbol{\sigma}\cdot\hat{\mathbf{x}}) \begin{pmatrix} C_{j\mp\frac{1}{2},\frac{1}{2}}(j\,\mu\,;\mu-\frac{1}{2}\;\;\frac{1}{2})\,Y^{\mu-\frac{1}{2}}_{j\mp\frac{1}{2}}(\hat{\mathbf{x}}) \\ C_{j\mp\frac{1}{2},\frac{1}{2}}(j\,\mu\,;\mu+\frac{1}{2}\;-\frac{1}{2})\,Y^{\mu+\frac{1}{2}}_{j\mp\frac{1}{2}}(\hat{\mathbf{x}}) \end{pmatrix} G(|\mathbf{x}|) . \qquad (14.1.26)$$

It is conventional to define the orbital angular momentum ℓ of the state as the orbital angular momentum of the 'large' components $f(\mathbf{x})$,

$$\ell = j \mp \tfrac{1}{2} , \qquad (14.1.27)$$

so that the parity is always $(-1)^\ell$.

Inserting Eqs. (14.1.25) and (14.1.26) in Eqs. (14.1.20) and (14.1.21) yields the coupled differential equations

$$\frac{dG}{dr} + \frac{k+1}{r}G + (E + e\mathscr{A}^0 - m)F = 0 , \qquad (14.1.28)$$

$$\frac{dF}{dr} - \frac{k-1}{r}F - (E + e\mathscr{A}^0 + m)G = 0, \qquad (14.1.29)$$

where for parity $\eta = (-1)^{j\mp\frac{1}{2}}$,

$$k \equiv \pm(j + \tfrac{1}{2}) . \qquad (14.1.30)$$

Let us now concentrate on the simple Coulomb field (14.1.2), for which $e\mathscr{A}^0 = Z\alpha/r$. The treatment of the Dirac equation in this case[3] is familiar, so it will be summarized briefly here just for completeness. It is easy to see that the solutions near the origin go as r^{s-1}, with $s^2 = k^2 - Z^2\alpha^2$. (Note that $k^2 \geq 1$, so the exponent s is real for $Z\alpha \leq 1$.) We must reject the solutions with $s < 0$ as being inconsistent with the normalization condition (14.1.18). The condition that the wave functions do not blow up for $r \to \infty$ then fixes the allowed values of the energy eigenvalues:

$$E_{n,j} = m\left[1 + \left(\frac{Z\alpha}{n - j - \frac{1}{2} + \sqrt{(j+\frac{1}{2})^2 - Z^2\alpha^2}}\right)^2\right]^{-1/2}, \tag{14.1.31}$$

where n is a 'principal quantum number' with

$$j + \tfrac{1}{2} \leq n\,. \tag{14.1.32}$$

It is noteworthy that these energies do not depend on the parity or ℓ, but only on n and j. For each n and j there are two solutions, corresponding to the two signs of k or the two possible parities, except that for $n = j + \frac{1}{2}$ we only have $k > 0$ and parity $(-1)^{j-\frac{1}{2}}$, so that $\ell = j - \frac{1}{2}$. With Eq. (14.1.32), this is the same as the familiar non-relativistic restriction that $\ell \leq n - 1$.

For light atoms with $Z\alpha \ll 1$, Eq. (14.1.31) yields the power series

$$E = m\left[1 - \frac{Z^2\alpha^2}{2n^2} + \frac{Z^4\alpha^4}{n^4}\left(\frac{3}{8} - \frac{n}{2j+1}\right) + \cdots\right]. \tag{14.1.33}$$

The first two terms, of course, just represent the rest energy and the binding energy as given by the non-relativistic Schrödinger equation. The leading term that depends on j as well as n is the third term, the first relativistic correction. For $n = 1$ there is only one value of the total angular momentum, $j = \frac{1}{2}$, and since here $n = j + \frac{1}{2}$ there is also only one parity, $(-1)^{j-\frac{1}{2}} = +1$, corresponding to $\ell = 0$. It is therefore difficult to see the effects of the relativistic corrections in Eq. (14.1.33) in the $n = 1$ states of hydrogenic atoms, though as we shall see in Section 14.3, this has recently become possible. On the other hand, for $n = 2$ we have a $j = \frac{1}{2}$ state with both parities (i.e., $2s_{1/2}$ and $2p_{1/2}$) as well as a $2p_{3/2}$ state with $j = \frac{3}{2}$ and negative parity. Eq. (14.1.33) gives the splitting between the p states in hydrogen as

$$E(2p_{3/2}) - E(2p_{1/2}) = \frac{\alpha^4 m_e}{32} = 4.5283 \times 10^{-5}\,\text{eV}\,. \tag{14.1.34}$$

Such relativistic line splitting is known as the *fine structure* of the atomic state. From the beginning it was known that this prediction is in good agreement with the observed fine structure. On the other hand, the Dirac equation does not yield any energy difference in the $2s_{1/2}$ and $2p_{1/2}$ states, so this is a good place to look for the effects of further corrections, to be considered in Section 14.3.

Before closing this section we shall consider the approximate forms for the wave functions and matrix elements in the non-relativistic case for a general electrostatic potential $\mathscr{A}^0$. (For a Coulomb potential, this is the limit $Z\alpha \ll 1$.) Since here $E_N + m \simeq 2m \gg |e\mathscr{A}^0|$, the 'small' components of the electron wave function are given approximately in terms of the large components by

$$g_N \simeq (\boldsymbol{\sigma} \cdot \nabla) f_N / 2m \,. \tag{14.1.35}$$

Eq. (14.1.21) then becomes just the non-relativistic Schrödinger equation

$$\left[-\frac{\nabla^2}{2m} - e\mathscr{A}^0 \right] f_N \simeq (E_N - m) f_N \,. \tag{14.1.36}$$

Since there is no longer any coupling between spin and orbital degrees of freedom in the equation for f_N, we may find a complete set of solutions of this equation in the form

$$f_N = \chi_N \, \psi_N(\mathbf{x}) \,,$$

where χ_N is a two-component constant spinor, and $\psi_N(\mathbf{x})$ is an ordinary one-component solution of the Schrödinger equation. However, we often work with states that have definite values of the total angular momentum j, for which f_N is (for non-vanishing orbital angular momentum) a sum of such terms.

In the non-relativistic approximation, the four-component Dirac wave function takes the form

$$u_N \simeq \frac{1}{\sqrt{2}} \begin{bmatrix} (1 + i\boldsymbol{\sigma} \cdot \nabla / 2m) \, f_N \\ (1 - i\boldsymbol{\sigma} \cdot \nabla / 2m) \, f_N \end{bmatrix} \tag{14.1.37}$$

and Eq. (14.1.18) gives the normalization condition

$$\int d^3x (f_N^\dagger, f_M) \simeq \delta_{NM} - \tfrac{1}{4} (\mathbf{v}^2)_{NM} \,, \tag{14.1.38}$$

where

$$(\mathbf{v}^2)_{NM} \equiv -\frac{1}{m^2} \int d^3x f_N^\dagger(\mathbf{x}) \nabla^2 f_M(\mathbf{x}) \,.$$

In relating matrix elements in an external field to free-particle matrix elements, it is useful to note that the momentum space wave function in an energy eigenstate N may be written

$$u_N(\mathbf{p}) \equiv (2\pi)^{-3/2} \int d^3x \, e^{-i\mathbf{p}\cdot\mathbf{x}} u_N(\mathbf{x}) \simeq \sum_\sigma u(\mathbf{p}, \sigma) [f_N(\mathbf{p})]_\sigma \,, \tag{14.1.39}$$

where $u(\mathbf{p},\sigma)$ is the free-particle Dirac spinor

$$u(\mathbf{p},\sigma) \simeq \frac{1}{\sqrt{2}}\begin{bmatrix} (1-\mathbf{p}\cdot\boldsymbol{\sigma}/2m)\ \chi_\sigma \\ (1+\mathbf{p}\cdot\boldsymbol{\sigma}/2m)\ \chi_\sigma \end{bmatrix},$$

$$\chi_{+\frac{1}{2}} \equiv \begin{pmatrix} 1 \\ 0 \end{pmatrix}, \qquad \chi_{-\frac{1}{2}} \equiv \begin{pmatrix} 0 \\ 1 \end{pmatrix}$$

and $f_N(\mathbf{p})$ is the Fourier transform of the two-component Schrödinger wave function

$$f_N(\mathbf{p}) \equiv (2\pi)^{-3/2}\int d^3x\, e^{-i\mathbf{p}\cdot\mathbf{x}} f_N(\mathbf{x}),$$

* * *

In closing, as an aid in calculating the effects of various perturbations, we note that the leading terms in the electron matrix elements of the sixteen independent 4×4 matrices are

$$(\bar{u}_M\, u_N) \simeq (f_M^\dagger\, f_N) - \frac{1}{4m^2}(\nabla f_M^\dagger\cdot\boldsymbol{\sigma}\,\boldsymbol{\sigma}\cdot\nabla f_N)\,, \tag{14.1.40}$$

$$i(\bar{u}_M\gamma^0\, u_N) \simeq (f_M^\dagger\, f_N) + \frac{1}{4m^2}(\nabla f_M^\dagger\cdot\boldsymbol{\sigma}\,\boldsymbol{\sigma}\cdot\nabla f_N)\,, \tag{14.1.41}$$

$$(\bar{u}_M\, \boldsymbol{\gamma}\, u_N) \simeq \frac{1}{2m}[(\nabla f_M^\dagger\cdot\boldsymbol{\sigma}\boldsymbol{\sigma} f_N) - (f_M^\dagger\boldsymbol{\sigma}\boldsymbol{\sigma}\cdot\nabla f_N)]\,, \tag{14.1.42}$$

$$(\bar{u}_M\,[\gamma^0,\boldsymbol{\gamma}]\, u_N) \simeq \frac{i}{m}[(\nabla f_M^\dagger\cdot\boldsymbol{\sigma}\boldsymbol{\sigma} f_N) + (f_M^\dagger\boldsymbol{\sigma}\boldsymbol{\sigma}\cdot\nabla f_N)]\,, \tag{14.1.43}$$

$$(\bar{u}_M\,[\gamma^i,\gamma^j]\, u_N) \simeq 2i\epsilon_{ijk}\,(f_M^\dagger\sigma_k f_N)\,, \tag{14.1.44}$$

$$(\bar{u}_M\,\gamma_5\boldsymbol{\gamma}\, u_N) \simeq -i(f_M^\dagger\boldsymbol{\sigma} f_N)\,, \tag{14.1.45}$$

$$(\bar{u}_M\,\gamma_5\gamma^0\, u_N) \simeq \frac{1}{2m}[(\nabla f_M^\dagger\cdot\boldsymbol{\sigma}\, f_N) - (f_M^\dagger\boldsymbol{\sigma}\cdot\nabla f_N)]\,, \tag{14.1.46}$$

$$(\bar{u}_M\,\gamma_5\, u_N) \simeq \frac{i}{2m}[(\nabla f_M^\dagger\cdot\boldsymbol{\sigma}\, f_N) + (f_M^\dagger\boldsymbol{\sigma}\cdot\nabla f_N)]\,. \tag{14.1.47}$$

14.2 Radiative Corrections in External Fields

We now consider radiative corrections to the results of the previous section, due to the interaction of electrons with the quantum electromagnetic field as well as the external field of the heavy charged particles. These radiative corrections can be calculated using Feynman diagrams of the usual sort, with the whole effect of the external field being to modify the propagator of the electron field in the presence of the external electromagnetic field (and to supply the external-field-dependent renormalization counterterms shown in Eq. (14.1.1)). To be specific, the effect of inserting any number of vertices corresponding to the first term of the interaction (14.1.1) in an internal electron line of any graph is to replace the bare coordinate space propagator $-iS(x-y)$ with a corrected propagator

$$\begin{aligned}-iS_{\mathscr{A}}(x,y) &\equiv -iS(x-y)+(-i)^2\int d^4z_1\; S(x-z_1)e\gamma^\mu\mathscr{A}_\mu(z_1)S(z_1-y)\\ &+(-i)^3\int d^4z_1\int d^4z_2\; S(x-z_1)e\gamma^\mu\mathscr{A}_\mu(z_1)S(z_1-z_2)e\gamma^\nu\mathscr{A}_\nu(z_2)S(z_2-y)\\ &+\cdots\,, \end{aligned} \tag{14.2.1}$$

where as usual

$$S(x-y)\equiv\frac{1}{(2\pi)^4}\int d^4p\,\frac{-i\gamma_\lambda p^\lambda+m}{p^2+m^2-i\epsilon}e^{ip\cdot(x-y)}\,.$$

(We must write $S_{\mathscr{A}}$ as a function of x and y rather than of $x-y$, because the external field invalidates translation invariance.) The theorem proved in Section 6.4 tells us that Eq. (14.2.1) is the same as

$$-\,iS_{\mathscr{A}}(x,y)=\left(\Phi_0,\;T\{\psi(x),\;\bar\psi(y)\}\Phi_0\right)_{\mathscr{A}} \tag{14.2.2}$$

with the subscript $\mathscr{A}$ on the right indicating that the vacuum state Φ_0 and electron field $\psi(x)$ are to be defined in a Heisenberg picture in which the only interaction taken into account is the interaction (14.1.1) with the external field. Inserting a complete set of intermediate states Φ_N in Eq. (14.2.2) yields an expression for the propagator in terms of the Dirac wave functions u_N and v_N introduced in the previous section

$$-\,iS_{\mathscr{A}}(x,y)=\theta(x^0-y^0)\sum_N u_N(x)\bar u_N(y)-\theta(y^0-x^0)\sum_M v_M(x)\bar v_M(y)\,. \tag{14.2.3}$$

It is also possible to obtain the propagator (14.2.2) as the solution of the *inhomogeneous Dirac equation*:

$$\left[\gamma^\lambda\frac{\partial}{\partial x^\lambda}+m+ie\gamma^\lambda\mathscr{A}_\lambda(x)\right]S_{\mathscr{A}}(x,y)=\delta^4(x-y)\,, \tag{14.2.4}$$

which follows from the field equations (14.1.3) and anticommutation relations (14.1.7), or formally from the perturbation series (14.2.1). Also Eq. (14.2.3) tells us that the propagator satisfies boundary conditions: its Fourier decomposition contains only 'positive frequency terms' proportional to $\exp(-iE(x^0-y^0))$ with $E>0$ for $x^0-y^0\to\infty$, and only 'negative frequency terms' proportional to $\exp(+iE(x^0-y^0))$ with $E>0$ for $x^0-y^0\to-\infty$. The inhomogeneous Dirac equation with these boundary conditions may be used to obtain a numerical solution[4] for this propagator even in cases where the external field is too strong to allow the use of the perturbation series (14.2.1). Once the propagator $S_{\mathscr{A}}(x,y)$ has been calculated, the amplitudes for scattering in an external field can be calculated using ordinary Feynman diagrams, but with $S_{\mathscr{A}}(x,y)$ in place of $S(x-y)$ (and with $\mathscr{A}$-dependent renormalization counterterms inserted where appropriate).

Now let us see how to use the perturbation series with this corrected propagator to calculate the shifts of bound-state energy levels. Consider the full electron propagator $S'_{\mathscr{A}}(x,y)$, involving interactions of the electron with the quantum electromagnetic field as well as the external field:

$$-iS'_{\mathscr{A}}(x,y) \equiv \left(\Omega_0, T\left\{\Psi(x), \bar{\Psi}(y)\right\}\Omega_0\right)_{\mathscr{A}} \tag{14.2.5}$$

with $\Psi(x)$ the electron field in a Heisenberg picture including all interactions, and Ω_0 the vacuum eigenstate of the full Hamiltonian. For a time-independent external potential we can find a complete orthonormal set Ω_N of eigenstates of the full Hamiltonian with energies E'_N. Inserting a sum over these states in the operator product in Eq. (14.2.5), we find

$$\begin{aligned}-iS'_{\mathscr{A}}(x,y) = \theta(x^0-y^0)e^{-iE'_N(x^0-y^0)}\sum_N U_N(\mathbf{x})\bar{U}_N(\mathbf{y})\\ -\theta(y^0-x^0)e^{-iE'_N(y^0-x^0)}\sum_N V_N(\mathbf{x})\bar{V}_N(\mathbf{y})\,,\end{aligned} \tag{14.2.6}$$

where

$$(\Omega_0, \Psi(\mathbf{x},t)\Omega_N) \equiv e^{-iE'_N t}U_N(\mathbf{x})\,, \tag{14.2.7}$$

$$(\Omega_N, \Psi(\mathbf{x},t)\Omega_0) \equiv e^{+iE'_N t}V_N(\mathbf{x})\,. \tag{14.2.8}$$

(The sum includes an integral over continuum states as well as a sum over discrete bound states. As before, U_N and V_N are non-zero only if the state Ω_N has charge $-e$ or $+e$, respectively.) We can redefine the propagator as a function of energy rather than time

$$S'_{\mathscr{A}}(\mathbf{x},\mathbf{y};E) \equiv \int_{-\infty}^{\infty} dx^0\, e^{iE(x^0-y^0)}S'_{\mathscr{A}}(x,y)\,. \tag{14.2.9}$$

(Time-translation invariance dictates that $S'_{\mathscr{A}}(x,y)$ is a function of x^0-y^0 but not of x^0 and y^0 separately.) From Eq. (14.2.6) we see that

$$S'_{\mathscr{A}}(\mathbf{x},\mathbf{y};E) = \sum_N \frac{U_N(\mathbf{x})\bar{U}_N(\mathbf{y})}{E'_N-E-i\epsilon} - \sum_N \frac{V_N(\mathbf{x})\bar{V}_N(\mathbf{y})}{E'_N+E-i\epsilon}\,. \tag{14.2.10}$$

In particular, $S'_{\mathscr{A}}(\mathbf{x},\mathbf{y};E)$ has a pole at any electron bound-state energy, and also at the negative of any positron bound-state energy. (Of course, positrons do not have bound states in the Coulomb field of an ordinary positively charged nucleus.)

Let us now consider the lowest-order radiative corrections to the complete propagator. The Feynman rules here give the complete propagator to this order as $S'_{\mathscr{A}} = S_{\mathscr{A}} + \delta S_{\mathscr{A}}$, with a correction term

$$\delta S_{\mathscr{A}}(x,y) = \int d^4z \int d^4w \; S_{\mathscr{A}}(x,z)\,\Sigma^*_{\mathscr{A}}(z,w)\,S_{\mathscr{A}}(w,y), \tag{14.2.11}$$

where $i\Sigma^*_{\mathscr{A}}$ is the sum of all one-loop diagrams with one incoming and one outgoing electron line (excluding final electron propagators) calculated using $S_{\mathscr{A}}(x,y)$ in place of $S(x-y)$ for internal electron lines, plus second-order renormalization counterterms. Using energy variables in place of time variables, this is

$$\delta S_{\mathscr{A}}(\mathbf{x},\mathbf{y};E) = \int d^3z \int d^3w \; S_{\mathscr{A}}(\mathbf{x},\mathbf{z};E)\,\Sigma^*_{\mathscr{A}}(\mathbf{z},\mathbf{w};E)\,S_{\mathscr{A}}(\mathbf{w},\mathbf{y};E)\,, \tag{14.2.12}$$

where

$$\Sigma^*_{\mathscr{A}}(\mathbf{z},\mathbf{w};E) \equiv \int dz^0 \, e^{iE(z^0-w^0)}\Sigma^*_{\mathscr{A}}(z,w)\,. \tag{14.2.13}$$

The effect of these radiative corrections is to change the wave functions to $U_N = u_N + \delta u_N$ and $V_N = v_N + \delta v_N$ and the bound-state energies to $E'_N = E_N + \delta E_N$, so that the complete propagator is

$$\begin{aligned} S'_{\mathscr{A}}(\mathbf{x},\mathbf{y};E) \simeq\; & S_{\mathscr{A}}(\mathbf{x},\mathbf{y};E) \\ & + \sum_N \frac{\delta u_N(\mathbf{x})\bar{u}_N(\mathbf{y}) + u_N(\mathbf{x})\delta\bar{u}_N(\mathbf{y})}{E_N - E} \\ & - \sum_N \frac{\delta v_N(\mathbf{x})\bar{v}_N(\mathbf{y}) + v_N(\mathbf{x})\delta\bar{v}_N(\mathbf{y})}{E_N + E} \\ & - \sum_N \frac{u_N(\mathbf{x})\bar{u}_N(\mathbf{y})\delta E_N}{(E_N - E)^2} + \sum_N \frac{v_N(\mathbf{x})\bar{v}_N(\mathbf{y})\delta E_N}{(E_N + E)^2}\,. \end{aligned} \tag{14.2.14}$$

(We are dropping the $i\epsilon$ terms because we are now not taking E to lie in the continuum of scattering states.) We see that the shift δE_N of an electron bound-state energy is given by the coefficient of $-u_N(\mathbf{x})\bar{u}_N(\mathbf{y})/(E_N - E)^2$ in the complete propagator. To calculate this, we note that Eq. (14.2.3) gives

$$S_{\mathscr{A}}(\mathbf{x},\mathbf{y};E) = \sum_N \frac{u_N(\mathbf{x})\bar{u}_N(\mathbf{y})}{E_N - E - i\epsilon} - \sum_N \frac{v_N(\mathbf{x})\bar{v}_N(\mathbf{y})}{E_N + E - i\epsilon}\,. \tag{14.2.15}$$

Inserting this in Eq. (14.2.12) gives

$$\begin{aligned} \delta S_{\mathscr{A}}(\mathbf{x},\mathbf{y};E) = & \sum_{N,M} \frac{u_N(\mathbf{x})\bar{u}_M(\mathbf{y})}{(E_N - E)(E_M - E)} \\ & \times \int d^3z \int d^3w \, \bar{u}_N(\mathbf{z})\,\Sigma^*_{\mathscr{A}}(\mathbf{z},\mathbf{w};E)\,u_M(\mathbf{w}) + \end{aligned} \tag{14.2.16}$$

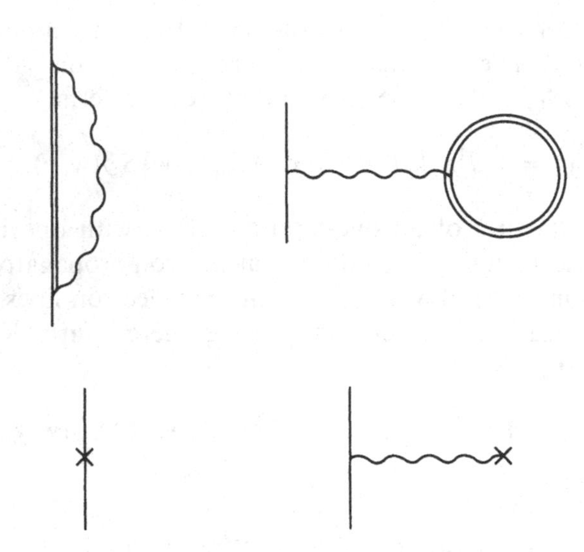

Figure 14.1. Lowest-order Feynman diagrams for the electron self-energy function $\Sigma^*_{\mathscr{A}}(x, y)$ in the presence of an external field. Here double straight lines represent electron propagators $S_{\mathscr{A}}$ that include effects of the external field; single straight lines are the incoming and outgoing electron lines; wavy lines are virtual photons; the cross represents renormalization counterterms.

where the dots denote additional terms involving at least one negative-energy pole. Comparing the coefficients of $(E_N - E)^{-2}$ here and in Eq. (14.2.14), we find

$$\delta E_N = -\int d^3x \int d^3y \; \bar{u}_N(\mathbf{x}) \, \Sigma^*_{\mathscr{A}}(\mathbf{x}, \mathbf{y}; E_N) \, u_N(\mathbf{y}) \, . \qquad (14.2.17)$$

The u_N are solutions of the homogeneous Dirac equation satisfying the normalization condition (14.1.8), so this is very much like ordinary first-order perturbation theory, but with $-\Sigma^*$ in place of a perturbation to the Hamiltonian.

Generally δE_N turns out to be complex. This is simply a consequence of the instability of atomic energy levels to radiative decay to lower levels; we saw in Chapter 3 that an unstable state of energy E and decay rate Γ produces poles in various amplitudes at the complex energy $E - i\Gamma/2$. The imaginary part of Eq. (14.2.17) therefore equals $-\Gamma/2$, while its real part gives the energy shift.

The Feynman diagrams for Σ^* are shown in Figure 14.1. (Note that there are new photon tadpole diagrams here because the external field breaks the Lorentz invariance and charge-conjugation invariance that forbids such diagrams in the ordinary Feynman rules.) Application of the position space Feynman rules to these diagrams gives

$$
\begin{aligned}
& i\Sigma^*_{\mathscr{A}}(x,y) = [e\gamma^\mu]\,[-iS_{\mathscr{A}}(x,y)]\,\left[e\gamma_\mu\right]\,[-iD(x-y)] \\
& -\left[e\gamma^\mu\delta^4(x-y)\right]\int d^4z\;[-iD(x-z)]\,\mathrm{Tr}\left\{[-iS_{\mathscr{A}}(z,z)]\,\left[e\gamma_\mu\right]\right\} \\
& -i(Z_2-1)(\gamma^\mu\partial_\mu+m)\delta^4(x-y)+i\delta m\,\delta^4(x-y) \\
& +e\gamma^\mu(Z_2-1)\delta^4(x-y)\mathscr{A}_\mu(x) \\
& +i(Z_3-1)[e\gamma_\mu]\delta^4(x-y)\int d^4z\;[-iD(x-z)]\,\partial_\nu(\partial^\nu\mathscr{A}^\mu(z)-\partial^\mu\mathscr{A}^\nu(z))
\end{aligned}
\tag{14.2.18}
$$

with the renormalization constants (Z_2-1), (Z_3-1), and δm calculated to second order in e. (The minus sign in the second term is the usual one that accompanies closed fermion loops.)

For strong external fields with $Z\alpha$ of order unity it is necessary to calculate the configuration space electron propagator $S_{\mathscr{A}}$ and the integrals in Eqs. (14.2.17) and (14.2.18) numerically.[4] However, for weak fields we can use the first few terms of the series (14.2.1) in Eq. (14.2.18) and calculate these integrals in closed form. For this purpose, it is more convenient to work in momentum space, defining:

$$S_{\mathscr{A}}(x,y)=(2\pi)^{-4}\int d^4p'\,d^4p\;e^{ip'\cdot x}e^{-ip\cdot y}\;S_{\mathscr{A}}(p',p)\,, \tag{14.2.19}$$

$$\Sigma^*_{\mathscr{A}}(x,y)=(2\pi)^{-4}\int d^4p'\,d^4p\;e^{ip'\cdot x}e^{-ip\cdot y}\,\Sigma^*_{\mathscr{A}}(p',p)\,, \tag{14.2.20}$$

$$u_N(\mathbf{x})=(2\pi)^{-3/2}\int d^3p\;e^{i\mathbf{p}\cdot\mathbf{x}}u_N(\mathbf{p})\,, \tag{14.2.21}$$

$$\mathscr{A}^\mu(x)=\int d^4q\;e^{iq\cdot x}\mathscr{A}^\mu(q)\,. \tag{14.2.22}$$

(We are here committing the impropriety of using the same symbol for a function and its Fourier transform, leaving it to the displayed arguments of the functions to indicate which is which.) Then Eqs. (14.2.1) and (14.2.18) become

$$
\begin{aligned}
S_{\mathscr{A}}(p',p) = {} & \frac{-i\not{p}+m}{p^2+m^2-i\epsilon}-ie\,\frac{-i\not{p}'+m}{p'^2+m^2-i\epsilon}\,\not{\mathscr{A}}(p'-p)\,\frac{-i\not{p}+m}{p^2+m^2-i\epsilon} \\
& +\cdots\,,
\end{aligned}
\tag{14.2.23}
$$

and

$$\Sigma^*_{\mathscr{A}}(p',p) = \frac{ie^2}{(2\pi)^4}\int \frac{d^4k}{k^2-i\epsilon}\gamma^\mu S_{\mathscr{A}}(p'-k,p-k)\gamma_\mu$$
$$+[-(Z_2-1)(i\not{p}+m)+Z_2\delta m]\,\delta^4(p'-p) - ie(Z_2-1)\not{\mathscr{A}}(p'-p)$$
$$-\frac{ie^2\gamma^\mu}{(2\pi)^4}\frac{1}{(p-p')^2-i\epsilon}\int d^4q\,\mathrm{Tr}\,\{S_{\mathscr{A}}(q,q+p'-p)\gamma_\mu\}$$
$$+\frac{ie(Z_3-1)}{(p-p')^2-i\epsilon}\left[(p-p')^2\not{\mathscr{A}}(p'-p)-(\not{p}-\not{p}')\,(p-p')\cdot\mathscr{A}(p'-p)\right]. \tag{14.2.24}$$

Because the external field is time-independent, $S_{\mathscr{A}}(x,y)$ and $\Sigma^*_{\mathscr{A}}(x,y)$ can depend on x^0 and y^0 only through the difference x^0-y^0, so $S_{\mathscr{A}}(p',p)$ and $\Sigma^*_{\mathscr{A}}(p',p)$ as well as $\mathscr{A}^\mu(p'-p)$ must be proportional to $\delta(p'^0-p^0)$:

$$\mathscr{A}^\mu(p'-p) = \delta(p'^0-p^0)\mathscr{A}^\mu(\mathbf{p}'-\mathbf{p}), \tag{14.2.25}$$

$$S_{\mathscr{A}}(p',p) = \delta(p'^0-p^0)S_{\mathscr{A}}(\mathbf{p}',\mathbf{p};p^0), \tag{14.2.26}$$

$$\Sigma^*_{\mathscr{A}}(p',p) = \delta(p'^0-p^0)\Sigma^*_{\mathscr{A}}(\mathbf{p}',\mathbf{p};p^0). \tag{14.2.27}$$

The energy shift is then given by Eqs. (14.2.17) and (14.2.13) as

$$\delta E_N = -\int d^3p'\int d^3p\,\bar{u}_N(\mathbf{p}')\,\Sigma^*_{\mathscr{A}}(\mathbf{p}',\mathbf{p};E_N)\,u_N(\mathbf{p}), \tag{14.2.28}$$

with $\Sigma^*_{\mathscr{A}}(\mathbf{p}',\mathbf{p};E_N)$ given by Eqs. (14.2.23), (14.2.24), and (14.2.27). This is the master formula we shall use in the next section to calculate energy shifts in weak external fields.

14.3 The Lamb Shift in Light Atoms

Let us now consider radiative corrections to the energy levels of a non-relativistic electron in a general electrostatic field, such as an electron in the Coulomb field of a light nucleus with $Z\alpha \ll 1$. It is natural in this limit to treat the Coulomb field as a weak perturbation, but we will see that this would lead to an infrared divergence, related to that found in Section 11.3. The infrared divergence is really fictitious, because the four-momenta $\mathbf{p}, E_N$ and $\mathbf{p}', E_N$ are not on the electron mass shell, but it does force us to proceed with some care.

Usually this problem is dealt with by dividing the integral over virtual photon energies into a low-energy range, within which we can treat the electrons non-relativistically but must include effects to all orders in the external field, and a high-energy range, within which we have to include relativistic effects but may include only effects of lowest order in the external field. Instead, we shall here introduce a fictitious photon mass μ, chosen to be much larger than typical electron kinetic energies, but much less than typical electron momenta. For a Coulomb field, this amounts to the requirement that

$$(Z\alpha)^2 m_e \ll \mu \ll Z\alpha m_e \,. \tag{14.3.1}$$

We write the photon propagator in the first two terms of Eq. (14.2.24) (including the formulas for the counterterms $Z_2 - 1$ and $Z_2\delta m$) in the form

$$\frac{1}{k^2 - i\epsilon} = \left[\frac{1}{k^2 + \mu^2 - i\epsilon}\right] + \left[\frac{1}{k^2 - i\epsilon} - \frac{1}{k^2 + \mu^2 - i\epsilon}\right] . \tag{14.3.2}$$

The energy shift is correspondingly a sum of two terms, a 'high-energy' and a 'low-energy' term: The high-energy term is calculated by using the first term in the photon propagator (14.3.2) in the first three terms in Eq. (14.2.24), and adding the result to the last two terms (the vacuum polarization terms) in Eq. (14.2.24) which are not infrared divergent anyway; the low-energy term is calculated using the second term in Eq. (14.3.2) in the first three terms of Eq. (14.2.24). One advantage of this procedure is that we shall be able to use the results of the relativistic calculations of Sections 11.3 and 11.4 directly, without the rather tricky conversion from a photon mass to an infrared energy cutoff. Of course, in the end we shall have to check that the dependence on the photon mass μ in the high-energy and low-energy contributions to the energy shift cancel, leaving the total energy shift μ-independent.

A High-Energy Term

Because μ is taken much larger than the atomic binding energies we can here keep only terms of lowest order in the external field. The one-loop radiative correction to atomic energy levels in a general time-independent external vector potential $\mathscr{A}^\mu(\mathbf{x})$ is given in momentum space by Eq. (14.2.28), with the self-energy insertion $\Sigma_{\mathscr{A}}(p', p)$ given by Eqs. (14.2.24) and (14.2.23). The terms of zeroth order in the external field simply cancel: the δm term cancels the first term with $\mathscr{A} = 0$; the $Z_3 - 1$ term cancels the third term with $\mathscr{A} = 0$; and the $Z_2 - 1$ terms vanish because $u(p)$ satisfies the Dirac equation. The term in $\Sigma_{\mathscr{A}}(p', p)$ that is of first order in $\mathscr{A}^\mu$ may be put in the form

$$\Sigma_{\mathscr{A}1}(p', p) = -ie\mathscr{A}_\mu(p' - p)\,\Gamma_1^\mu(p', p) \tag{14.3.3}$$

with

$$\Gamma_1^\mu(p',p) = \frac{ie^2}{(2\pi)^4}\int \frac{d^4k}{k^2+\mu^2-i\epsilon}$$
$$\times \gamma^\nu \left[\frac{-i(\not p'-\not k)+m_e}{(p'-k)^2+m_e^2-i\epsilon}\right]\gamma^\mu\left[\frac{-i(\not p-\not k)+m_e}{(p-k)^2+m_e^2-i\epsilon}\right]\gamma_\nu$$
$$+(Z_2-1)\gamma^\mu$$
$$-\frac{ie^2\gamma_\nu}{(2\pi)^4}\frac{1}{(p-p')^2-i\epsilon}$$
$$\times\int d^4l\,\mathrm{Tr}\left\{\left[\frac{-i\not l+m_e}{l^2+m_e^2}\right]\gamma^\mu\left[\frac{-i\not l-i\not p'+i\not p+m_e}{(l+p'-p)^2+m_e^2}\right]\gamma^\nu\right\}$$
$$-\frac{Z_3-1}{(p'-p)^2-i\epsilon}\left[(p-p')^2\eta^{\mu\nu}-(p'-p)^\mu(p'-p)^\nu\right]\gamma_\nu. \tag{14.3.4}$$

Comparison of the first two terms with Eqs. (11.3.1) and (11.3.8) and the second two terms with Eqs. (11.3.9), (11.2.3), and (11.2.15) reveals that $\Gamma_1^\mu(p',p)$ is the complete one-loop vertex function, including vacuum polarization and all counterterms, whose mass-shell matrix elements we have already calculated in Section 11.3. Using Eqs. (14.2.26) and (14.2.25), this contribution to the energy shift (14.2.28) is given by

$$[\delta E_N]_{\text{high energy}} = ie\int d^3p'\int d^3p\;(\bar u_N(\mathbf{p}')\,\Gamma_1^\mu(\mathbf{p}',E_N,\mathbf{p},E_N)\,u_N(\mathbf{p}))$$
$$\times\mathscr{A}_\mu(\mathbf{p}'-\mathbf{p})\,. \tag{14.3.5}$$

(This could have been guessed at, by simply replacing the γ^μ in the interaction of the electron with the external field with Γ_1^μ.) As discussed in Section 14.1, because $Z\alpha \ll 1$ we can approximate the Dirac wave function u_N in Eq. (14.3.5) as

$$[u_N(\mathbf{p})]_\alpha = \sum_\sigma u_\alpha(\mathbf{p},\sigma)\,[f_N(\mathbf{p})]_\sigma\,, \tag{14.3.6}$$

where f_N is the non-relativistic two-component wave function of an electron in the external Coulomb field, and $u(\mathbf{p},\sigma)$ is the four-component normalized solution of the momentum space Dirac equation

$$[i\gamma_\mu p^\mu + m_e]\,u(\mathbf{p},\sigma) = 0 \tag{14.3.7}$$

for spin z-component σ. Since $u_N(\mathbf{p})$ approximately satisfies the free-particle Dirac equation, Eq. (10.6.15) gives the general form of the matrix element of Γ_1^μ as

$$\bar u_M(\mathbf{p}')[\gamma^\mu + \Gamma_1^\mu(p',p)]u_N(\mathbf{p})$$
$$= \bar u_M(\mathbf{p}')\left[\gamma^\mu F_1(q^2) + \tfrac{1}{2}i\,[\gamma^\mu,\gamma^\nu]\,q_\nu F_2(q^2)\right]u_N(\mathbf{p})\,, \tag{14.3.8}$$

where $q \equiv p' - p$. The wave functions $u_N(\mathbf{p})$ fall off very rapidly for $|\mathbf{p}| \gg Z\alpha m_e$, so we only need $F_1(\mathbf{q}^2)$ and $F_2(\mathbf{q}^2)$ in the limit $|q^2| \ll m_e^2$. In this limit, Eqs. (11.3.31), (10.6.18), and (11.3.16) give

$$F_1(q^2) \simeq 1 + \frac{e^2}{24\pi^2}\left(\frac{q^2}{m_e^2}\right)\left[\ln\left(\frac{\mu^2}{m_e^2}\right) + \frac{2}{5} + \frac{3}{4}\right], \tag{14.3.9}$$

$$F_2(q^2) \simeq \frac{e^2}{16 m_e \pi^2}. \tag{14.3.10}$$

Let us first consider the contribution of the F_1 term in Eq. (14.3.8), which makes by far the largest contribution to the energy shift, and raises the most interesting problems in its calculation. For a pure electrostatic field with $\mathscr{A} = 0$, Eqs. (14.3.5), (14.3.8), and (14.3.9) give

$$[\delta E_N]_{F_1} = -\frac{e^2}{24\pi^2 m_e^2}\left[\ln\left(\frac{\mu^2}{m_e^2}\right) + \frac{2}{5} + \frac{3}{4}\right]$$
$$\times \int d^3p' \int d^3p\, \bar{u}_N(\mathbf{p}')\Big(-ie\mathscr{A}^0(\mathbf{p}'-\mathbf{p})\Big)\gamma^0(\mathbf{p}'-\mathbf{p})^2 u_N(\mathbf{p}). \tag{14.3.11}$$

To calculate this contribution, we may use the leading term in the non-relativistic matrix element (14.1.41), and find

$$[\delta E_N]_{F_1} = -\frac{e^2}{24\pi^2 m_e^2}\left[\ln\left(\frac{\mu^2}{m_e^2}\right) + \frac{2}{5} + \frac{3}{4}\right]$$
$$\times \int d^3p' \int d^3p\, f_N^\dagger(\mathbf{p}')\, e\mathscr{A}^0(\mathbf{p}'-\mathbf{p})[\mathbf{p}'-\mathbf{p}]^2 f_N(\mathbf{p}) \tag{14.3.12}$$

or in position space

$$[\delta E_N]_{F_1} = \frac{e^2}{24\pi^2 m_e^2}\left[\ln\left(\frac{\mu^2}{m_e^2}\right) + \frac{2}{5} + \frac{3}{4}\right]\int d^3x\, f_N^\dagger(\mathbf{x})\,[e\nabla^2\mathscr{A}^0(\mathbf{x})] f_N(\mathbf{x}). \tag{14.3.13}$$

In particular, for the Coulomb potential (14.1.2), we have $e\nabla^2\mathscr{A}^0(\mathbf{x}) = -Ze^2\delta^3(\mathbf{x})$, and the label N consists of a principal quantum number n and angular-momentum quantum numbers j, m, ℓ, while (11.2.41) gives $[f_{njm\ell}(0)]_\sigma = 2(Z\alpha m_e/n)^{3/2}\delta_{\ell,0}\delta_{\sigma,m}/\sqrt{4\pi}$. The energy shift (14.3.11) is then

$$[\delta E_{jn\ell}]_{F_1} = -\frac{2Z^4\alpha^5 m_e}{3\pi n^3}\left[\ln\left(\frac{\mu^2}{m_e^2}\right) + \frac{2}{5} + \frac{3}{4}\right]\delta_{\ell,0}. \tag{14.3.14}$$

(The lack of dependence of δE on the total angular momentum z-component m is guaranteed by rotational invariance.) The term $\frac{2}{5}$ in the brackets in Eqs. (14.3.12) and (14.3.13) arises from vacuum polarization, and yields just the energy shift calculated somewhat heuristically in Section 11.2.

Before going on to calculate the magnetic and low-energy contributions to the energy shift, it is worth noting that the result we have obtained so far yields a fair order-of- magnitude estimate of the Lamb shift without further work. We can anticipate that the low-energy terms will contain a term proportional to $\ln(\mu/B)$, with a coefficient such as to cancel the μ-dependence in Eq. (14.3.12). The constant B here is an energy which must be included to make the argument of the logarithm dimensionless; since it is the binding of the electron in the atom that will eventually provide our infrared cutoff, we may guess that B is a typical atomic binding energy, of order $B \simeq (Z\alpha)^2 m_e$. The total energy shift in a state N with principal quantum number n and orbital angular momentum ℓ is therefore of the form

$$\delta E_N = -\frac{2Z^4\alpha^5 m_e}{3\pi n^3}\left[\ln\left(Z^4\alpha^4\right)\delta_{\ell,0} + O(1)\right] . \tag{14.3.15}$$

For the $2s$ state of hydrogen, the logarithmic term alone gives

$$\delta E_{2s} \simeq -\frac{\alpha^5 m_e}{12\pi}\ln\left(\alpha^4\right) = 5.5\times 10^{-6}\ \text{eV} = 1300\ \text{MHz}\times 2\pi\hbar\, .$$

As we shall see, the 'O(1)' terms in Eq. (14.3.15) will lower the total energy shift by about 25%.

Now let us consider the contribution of the F_2 term in the matrix element of Γ_1^μ, which as we saw in Section 10.6 may be interpreted as a radiative correction to the magnetic moment of the electron. Using Eqs. (14.3.10), (14.3.8), and (14.3.6) in Eq. (14.3.5), we find that this term gives an energy shift

$$[\delta E_N]_{F_2} = -\frac{e^2}{32\pi^2 m_e}\int d^3p' \int d^3p\ \left(\bar{u}_N(\mathbf{p}')\,[\gamma^\mu,\gamma^\nu]u_N(\mathbf{p})\right) \times e\mathscr{A}_\mu(\mathbf{p}'-\mathbf{p})\,(p'-p)_\nu \tag{14.3.16}$$

or in position space

$$[\delta E_N]_{F_2} = \frac{ie^2}{64\pi^2 m_e}\int d^3x\ \left(\bar{u}_N(\mathbf{x})\,[\gamma^\mu,\gamma^\nu]u_N(\mathbf{x})\right) e\mathscr{F}_{\mu\nu}(\mathbf{x})\, , \tag{14.3.17}$$

where

$$\mathscr{F}_{\mu\nu}(\mathbf{x}) \equiv \partial_\mu \mathscr{A}_\nu(\mathbf{x}) - \partial_\nu \mathscr{A}_\mu(\mathbf{x})\, . \tag{14.3.18}$$

For a pure electrostatic field with $\mathscr{A} = 0$, this is

$$[\delta E_N]_{F_2} = \frac{-ie^2}{32\pi^2 m_e}\int d^3x\ \left(\bar{u}_N(\mathbf{x})\,[\gamma,\gamma^0]u_N(\mathbf{x})\right)\cdot \nabla[e\mathscr{A}^0(\mathbf{x})]\, . \tag{14.3.19}$$

In the non-relativistic limit $Z\alpha \ll 1$, we can use the approximate result

(14.1.43), which here reads

$$(\bar{u}_N[\gamma^0,\gamma]u_N) \simeq \frac{i}{m_e}[(\nabla f_N^\dagger \cdot \boldsymbol{\sigma}\boldsymbol{\sigma} f_N) + (f_N^\dagger \boldsymbol{\sigma}\boldsymbol{\sigma}\cdot\nabla f_N)]$$
$$= \frac{i}{m_e}[\nabla(f_N^\dagger f_N) - i(\nabla f_N^\dagger \times \boldsymbol{\sigma}) f_N - i f_N^\dagger(\boldsymbol{\sigma}\times\nabla f_N)] \,. \quad (14.3.20)$$

Using this in Eq. (14.3.8) and integrating by parts, this part of the energy shift is

$$[\delta E_N]_{F_2} = \frac{e^2}{32\pi^2 m_e^2}\int d^3x \Big[-|f_N(\mathbf{x})|^2\nabla^2(e\mathscr{A}^0(\mathbf{x}))$$
$$+2i f_N^\dagger(\mathbf{x})\boldsymbol{\sigma}\cdot(\nabla(e\mathscr{A}^0(\mathbf{x}))\times\nabla f_N(\mathbf{x}))\Big] \,. \quad (14.3.21)$$

Combining Eqs. (14.3.12) and (14.3.21) gives the total high-energy contribution to the energy shift in an arbitrary electrostatic potential $\mathscr{A}^0$:

$$[\delta E_N]_{\text{high energy}} =$$
$$\frac{e^2}{24\pi^2 m_e^2}\left[\ln\left(\frac{\mu^2}{m_e^2}\right)+\frac{2}{5}\right]\int d^3x\, f_N^\dagger(\mathbf{x})\,[e\nabla^2\mathscr{A}^0(\mathbf{x})]f_N(\mathbf{x})$$
$$+\frac{ie^2}{16\pi^2 m_e^2}\int d^3x\, f_N^\dagger(\mathbf{x})\boldsymbol{\sigma}\cdot(\nabla(e\mathscr{A}^0(\mathbf{x}))\times\nabla f_N(\mathbf{x}))\,.$$
$$(14.3.22)$$

B Low-Energy Term

The low-energy contribution to the energy shift is obtained from the first three terms in Eq. (14.2.24), making the replacement in the photon propagator

$$\frac{1}{k^2-i\epsilon} \quad\to\quad \frac{1}{k^2-i\epsilon}-\frac{1}{k^2+\mu^2-i\epsilon}\,. \quad (14.3.23)$$

This substitution will eventually serve to cut off the integral over components of the photon four-momentum k at values of order μ, but it is not possible to see this until we carefully take mass renormalization into account, so we shall defer making any non-relativistic approximations until then. Also, we are now including photon momenta as small or smaller than the binding energies of the atomic states, so we must treat the electrostatic forces responsible for this binding to all orders.

Instead of working with the momentum space formula (14.2.24), it will be convenient to return to the configuration space formula (14.2.18). This gives the low- energy contribution to the electron self-energy function as

$$[\Sigma^*_{\mathscr{A}}(x,y)]_{\text{low energy}} = ie^2\gamma^\rho S_{\mathscr{A}}(x,y)\gamma_\rho D(x-y;\mu)+\delta m_e(\mu)\,\delta^4(x-y)$$
$$-(Z_2(\mu)-1)(\gamma^\mu[\partial_\mu+ie\mathscr{A}_\mu]+m_e)\delta^4(x-y)\,, \quad (14.3.24)$$

where $D(x-y;\mu)$ is the modified photon propagator

$$D(x-y;\mu) = \frac{1}{(2\pi)^4}\int d^4k\, e^{ik\cdot(x-y)}\left[\frac{1}{k^2-i\epsilon} - \frac{1}{k^2+\mu^2-i\epsilon}\right], \quad (14.3.25)$$

and the counterterms $Z_2(\mu)-1$ and $\delta m(\mu)$ are calculated using this modified propagator. Converting from time to energy variables, the low-energy contribution to the function (14.2.13) is then

$$\begin{aligned}[\Sigma^*_{\mathscr{A}}(\mathbf{x},\mathbf{y};E)]_{\text{low energy}} = \frac{ie^2}{(2\pi)^4}\int d^4k\, \gamma^\rho\, S_{\mathscr{A}}(\mathbf{x},\mathbf{y};E-k^0)\,\gamma_\rho \\ \times\left[\frac{1}{k^2-i\epsilon} - \frac{1}{k^2+\mu^2-i\epsilon}\right] e^{i\mathbf{k}\cdot(\mathbf{x}-\mathbf{y})} \\ -(Z_2(\mu)-1)\left(\gamma\cdot\nabla + i\gamma^0 E + ie\gamma^\nu\mathscr{A}_\nu + m_e\right)\delta^3(\mathbf{x}-\mathbf{y}) \\ +\,\delta m_e(\mu)\,\delta^3(\mathbf{x}-\mathbf{y})\,. \quad (14.3.26)\end{aligned}$$

The low-energy contribution to the energy shift is then given by Eq. (14.2.17) as

$$\begin{aligned}[\delta E_N]_{\text{low energy}} = -\int d^3x\int d^3y\; \bar{u}_N(\mathbf{x})\,[\Sigma^*_{\mathscr{A}}(\mathbf{x},\mathbf{y};E_N)]_{\text{low energy}}\, u_N(\mathbf{y}) \\ = \frac{-ie^2}{(2\pi)^4}\int d^4k\int d^3x\int d^3y\; \bar{u}_N(\mathbf{x})\,\gamma^\rho\, S_{\mathscr{A}}(\mathbf{x},\mathbf{y};E_N-k^0)\,\gamma_\rho\, u_N(\mathbf{y}) \\ \times\left[\frac{1}{k^2-i\epsilon} - \frac{1}{k^2+\mu^2-i\epsilon}\right] e^{i\mathbf{k}\cdot(\mathbf{x}-\mathbf{y})} \\ -\delta m_e(\mu)\int d^3x\; \bar{u}_N(\mathbf{x})u_N(\mathbf{x}) \quad . \quad (14.3.27)\end{aligned}$$

Note that the terms proportional to $Z_2(\mu)-1$ have dropped out because the Dirac wave function $u_N(\mathbf{x})$ satisfies the Dirac equation Eq. (14.1.10). For the electron propagator in the presence of the Coulomb field, we use Eq. (14.2.15):

$$S_{\mathscr{A}}(\mathbf{x},\mathbf{y};E) = \sum_M \frac{u_M(\mathbf{x})\bar{u}_M(\mathbf{y})}{E_M - E - i\epsilon} - \sum_M \frac{v_M(\mathbf{x})\bar{v}_M(\mathbf{y})}{E_M + E - i\epsilon},$$

with the sums in the first and second terms running over all one-electron and one-positron states, respectively. The k^0 integrals can be done most easily by closing the contour of integration with a large semi-circle in the lower half-plane in the first term and in the upper half-plane in the second term:

$$\begin{aligned}\int dk^0 \left(\frac{1}{k^2+\mu^2-i\epsilon}\right)\left(\frac{1}{E_M \mp E_N \pm k^0 - i\epsilon}\right) = \\ = \frac{i\pi}{\sqrt{\mathbf{k}^2+\mu^2}}\left(\frac{1}{E_M \mp E_N + \sqrt{\mathbf{k}^2+\mu^2} - i\epsilon}\right)\end{aligned}$$

and likewise if μ is replaced with zero. The energy shift (14.3.27) now becomes

$$
\begin{aligned}
[\delta E_N]_{\text{low energy}} &= -\frac{e^2}{2(2\pi)^3}\int d^3k \sum_M \\
&\times \Bigg[\Gamma^\rho_{MN}(\mathbf{k})^*\Gamma_{\rho MN}(\mathbf{k})\left(\frac{1}{|\mathbf{k}|(E_M - E_N + |\mathbf{k}| - i\epsilon)}\right. \\
&\qquad \left. - \frac{1}{\sqrt{\mathbf{k}^2+\mu^2}(E_M - E_N + \sqrt{\mathbf{k}^2+\mu^2} - i\epsilon)}\right) \\
&\quad -\tilde{\Gamma}^\rho_{MN}(\mathbf{k})^*\tilde{\Gamma}_{\rho MN}(\mathbf{k})\left(\frac{1}{|\mathbf{k}|(E_M + E_N + |\mathbf{k}| - i\epsilon)}\right. \\
&\qquad \left. - \frac{1}{\sqrt{\mathbf{k}^2+\mu^2}(E_M + E_N + \sqrt{\mathbf{k}^2+\mu^2} - i\epsilon)}\right)\Bigg] \\
&\quad -\delta m_e(\mu)\int d^3x\, \bar{u}_N(\mathbf{x})u_N(\mathbf{x})\,, \qquad (14.3.28)
\end{aligned}
$$

where

$$\Gamma^\rho_{MN}(\mathbf{k}) \equiv \int d^3y\, e^{-i\mathbf{k}\cdot\mathbf{y}}\, \bar{u}_M(\mathbf{y})\gamma^\rho u_N(\mathbf{y})\,, \qquad (14.3.29)$$

$$\tilde{\Gamma}^\rho_{MN}(\mathbf{k}) \equiv \int d^3y\, e^{-i\mathbf{k}\cdot\mathbf{y}}\, \bar{v}_M(\mathbf{y})\gamma^\rho u_N(\mathbf{y})\,. \qquad (14.3.30)$$

(Of course, the 'sum' over M in Eq. (14.3.28) receives contributions only from electron states in the first term and positron states in the second term.) Eq. (14.3.28) could have been derived more directly from old-fashioned perturbation theory; the energy denominators $E_M - E_N + \omega$ and $E_M + E_N + \omega$ are the result of subtracting the energy E_N of the initial state from the energy of an intermediate state consisting of either an electron of energy E_M and a photon of energy ω, or else a positron of energy E_M, a photon of energy ω, and both the final and the initial electron. (See Figure 14.2.)

Before making any approximations to Eq. (14.3.28), it will be convenient to express the time components of the matrix elements Γ^ρ_{MN} and $\tilde{\Gamma}^\rho_{MN}$ in terms of the corresponding space components, using relations* derived

* To derive Eq. (14.3.31), note that

$$
\begin{aligned}
k_i\Gamma^i_{MN}(\mathbf{k}) &= -i\int d^3x\, e^{-i\mathbf{k}\cdot\mathbf{x}}\, \nabla\cdot(\bar{u}_M(\mathbf{x})\boldsymbol{\gamma} u_N(\mathbf{x})) \\
&= i\int d^3x\, e^{-i\mathbf{k}\cdot\mathbf{x}}\, \partial_0\left[\left(\bar{u}_M(\mathbf{x})\gamma^0 u_N(\mathbf{x})\right) e^{-i(E_N - E_M)x^0}\right]_{x^0=0} = (E_N - E_M)\Gamma^0_{MN}(\mathbf{k})\,.
\end{aligned}
$$

Eq. (14.3.32) is derived in the same way.

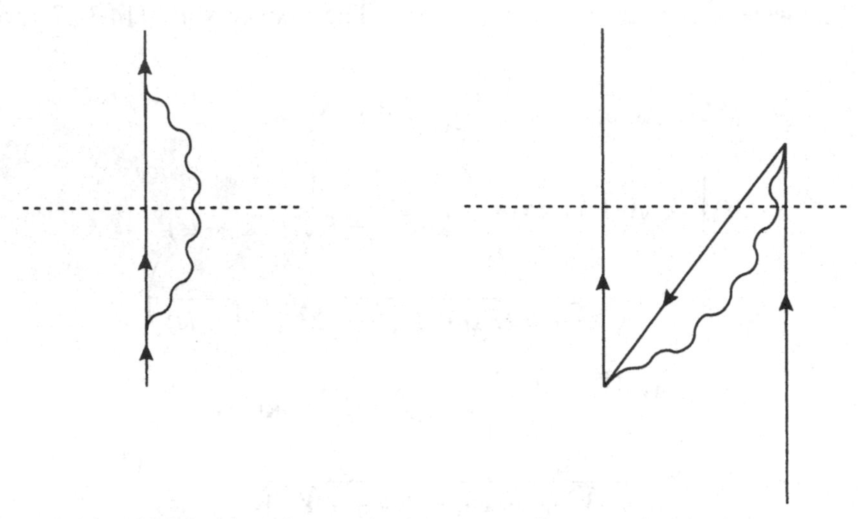

Figure 14.2. Old-fashioned perturbation theory diagrams for the low-energy part of the electron energy shift. Here solid lines are electrons; wavy lines are photons; the dashed line cuts through particle lines corresponding to the immediate states appearing in the first two terms in (14.3.28)

from the conservation of the electric current:

$$k_i\,\Gamma^i_{MN}(\mathbf{k}) = (E_N - E_M)\Gamma^0_{MN}(\mathbf{k})\,, \tag{14.3.31}$$

$$k_i\,\tilde{\Gamma}^i_{MN}(\mathbf{k}) = (E_N + E_M)\tilde{\Gamma}^0_{MN}(\mathbf{k})\,. \tag{14.3.32}$$

Furthermore, by using the completeness relation (14.1.8) it is straightforward to show that:

$$\sum_M \left[|\Gamma^0_{MN}(\mathbf{k})|^2 + |\tilde{\Gamma}^0_{MN}(\mathbf{k})|^2\right] = 1 \tag{14.3.33}$$

and

$$\begin{aligned}\sum_M &\left[|\Gamma^0_{MN}(\mathbf{k})|^2(E_M - E_N) - |\tilde{\Gamma}^0_{MN}(\mathbf{k})|^2(E_M + E_N)\right] \\ &= \sum_M \left[-\Gamma^{0*}_{MN}(\mathbf{k})\,\mathbf{k}\cdot\mathbf{\Gamma}_{MN}(\mathbf{k}) - \tilde{\Gamma}^{0*}_{MN}(\mathbf{k})\,\mathbf{k}\cdot\tilde{\mathbf{\Gamma}}_{MN}(\mathbf{k})\right] \\ &= -i\mathbf{k}\cdot\int d^3x\,\bar{u}_N(\mathbf{x})\,\boldsymbol{\gamma} u_N(\mathbf{x}) = 0\,,\end{aligned} \tag{14.3.34}$$

the last step following from the parity condition (14.1.23). In this way,

Eq. (14.3.28) can be rewritten

$$[\delta E_N]_{\text{low energy}} =$$

$$-\frac{e^2}{2(2\pi)^3}\int d^3k \sum_M \left[\frac{\left(|\mathbf{\Gamma}_{MN}(\mathbf{k})|^2 - |\mathbf{k}\cdot\mathbf{\Gamma}_{MN}(\mathbf{k})|^2/\mathbf{k}^2\right)}{|\mathbf{k}|\,(E_M - E_N + |\mathbf{k}| - i\epsilon)}\right.$$

$$\left. - \frac{\left(|\mathbf{\Gamma}_{MN}(\mathbf{k})|^2 - |\mathbf{k}\cdot\mathbf{\Gamma}_{MN}(\mathbf{k})|^2/(\mathbf{k}^2+\mu^2)\right)}{\sqrt{\mathbf{k}^2+\mu^2}\,(E_M - E_N + \sqrt{\mathbf{k}^2+\mu^2} - i\epsilon)}\right]$$

$$+\frac{e^2}{2(2\pi)^3}\int d^3k \sum_M \left[\frac{\left(|\tilde{\mathbf{\Gamma}}_{MN}(\mathbf{k})|^2 - |\mathbf{k}\cdot\tilde{\mathbf{\Gamma}}_{MN}(\mathbf{k})|^2/\mathbf{k}^2\right)}{|\mathbf{k}|\,(E_M + E_N + |\mathbf{k}|)}\right.$$

$$\left. - \frac{\left(|\tilde{\mathbf{\Gamma}}_{MN}(\mathbf{k})|^2 - |\mathbf{k}\cdot\tilde{\mathbf{\Gamma}}_{MN}(\mathbf{k})|^2/(\mathbf{k}^2+\mu^2)\right)}{\sqrt{\mathbf{k}^2+\mu^2}\,(E_M + E_N + \sqrt{\mathbf{k}^2+\mu^2})}\right]$$

$$-\frac{e^2}{(2\pi)^3}\int d^3k \sum_M |\tilde{\Gamma}^0_{MN}(\mathbf{k})|^2 \left(\frac{1}{\mathbf{k}^2} - \frac{1}{\mathbf{k}^2+\mu^2}\right)$$

$$+\tfrac{1}{2}\alpha\mu - \delta m_e(\mu)\int d^3x\, \bar{u}_N(\mathbf{x})u_N(\mathbf{x})\,, \tag{14.3.35}$$

In the next-to-last term, we have used the elementary integral

$$\int d^3k \left(\frac{1}{\mathbf{k}^2} - \frac{1}{\mathbf{k}^2+\mu^2}\right) = 2\mu\pi^2\,.$$

So far, this has been an exact rewriting of Eq. (14.3.28). We must now invoke several approximations. First, consider mass renormalization. We have already calculated $\delta m_e(\mu)$ to order α in Section 11.4; it is

$$\delta m_e(\mu) = \frac{2m_e\pi^2 e^2}{(2\pi)^4}\int_0^1 dx\,[1+x]\,\ln\left(\frac{m_e^2x^2 + \mu^2(1-x)}{m_e^2x^2}\right)\,. \tag{14.3.36}$$

Although in Section 11.4 we regarded μ as a regulator mass, to be taken much larger than m_e, we can just as well use Eq. (14.3.36) to provide a value for $\delta m_e(\mu)$ in the case that interests us here, $\mu \ll m_e$. In this limit Eq. (14.3.36) gives

$$\delta m_e(\mu) \to \frac{\alpha\mu}{2}\left[1 - \frac{3\mu}{2\pi m_e} + \ldots\right]\,. \tag{14.3.37}$$

We also recall that for $Z\alpha \ll 1$, $u_N(\mathbf{x})$ is given by Eq. (14.1.37) as

$$u_N(\mathbf{x}) = \frac{1}{\sqrt{2}}\begin{bmatrix}(1 - \boldsymbol{\sigma}\cdot\mathbf{v}/2 + \cdots)f_N(\mathbf{x})\\ (1 + \boldsymbol{\sigma}\cdot\mathbf{v}/2 + \cdots)f_N(\mathbf{x})\end{bmatrix}\,, \tag{14.3.38}$$

where dots indicate terms of higher order in $Z\alpha$; $\mathbf{v}$ is the non-relativistic velocity operator $-i\nabla/m_e$; and $f_N(\mathbf{x})$ is a two-component spinor solution of the non-relativistic Schrödinger equation, normalized according to Eq. (14.1.38) so that

$$\int |f_N(\mathbf{x})|^2 = 1 - \tfrac{1}{4}(\mathbf{v}^2)_{NN} + \cdots \quad . \tag{14.3.39}$$

This gives the coefficient of $\delta m_e(\mu)$ in Eq. (14.3.35) as

$$\int d^3x\, \bar{u}_N(\mathbf{x}) u_N(\mathbf{x}) = 1 - \tfrac{1}{2}(\mathbf{v}^2)_{NN} + \cdots \quad . \tag{14.3.40}$$

We note immediately that the leading term in $-\delta m_e(\mu) \int d^3x\, \bar{u}_N u_N$ cancels the term $\alpha\mu/2$ in Eq. (14.3.35). Indeed, we could have anticipated this cancellation, because the term $\alpha\mu/2$ in Eq. (14.3.35) survives in the limit $Z\alpha \to 0$, and the definition of $m_e(\mu)$ as the renormalized electron mass implies that there must be no energy shift in this limit. By the same argument we can anticipate that the term of order $\alpha\mu^2/m_e$ in $\delta m_e(\mu)$ (which is larger than of order $\alpha(Z\alpha)^4 m_e$, and therefore cannot simply be neglected) cancels the second and third terms in Eq. (14.3.35), which are also of this order.** On the other hand, the product of the $\alpha\mu^2/m_e$

** The cancellation can be shown as follows. We anticipate that the second and third terms in Eq. (14.3.35) are small enough so they can be evaluated using the extreme nonrelativistic approximation $\beta u_N(\mathbf{x}) = u_N(\mathbf{x})$ for the Dirac equation satisfied by $u_N(\mathbf{x})$. On the other hand, although the Coulomb force may be neglected in the positron wave functions $v_M(\mathbf{x})$, the sum over M in the third term receives important contributions from relativistic positrons, so we use the approximation $v_{\mathbf{p},\sigma}(\mathbf{x}) \simeq (2\pi)^{-3/2} v(\mathbf{p},\sigma) e^{i\mathbf{p}\cdot\mathbf{x}}$, where $v(\mathbf{p},\sigma)$ is the positron spinor introduced in Section 5.5, normalized so that $\bar{v}(\mathbf{p},\sigma')v(\mathbf{p},\sigma) = \delta_{\sigma',\sigma}$. Thus the sums over M in the second and third terms of Eq. (14.3.35) are approximately given by

$$\frac{1}{2}\left[\sum_M \tilde{\Gamma}^{i*}_{MN}(\mathbf{k})\tilde{\Gamma}^{j}_{MN}(\mathbf{k}) + (i \leftrightarrow j)\right] \simeq \delta_{ij}\left(\frac{\sqrt{\mathbf{k}^2+m_e^2}+m_e}{2\sqrt{\mathbf{k}^2+m_e^2}}\right),$$

$$\sum_M |\tilde{\Gamma}^0_{MN}(\mathbf{k})|^2 \simeq \left(\frac{\sqrt{\mathbf{k}^2+m_e^2}-m_e}{2\sqrt{\mathbf{k}^2+m_e^2}}\right).$$

To leading order in μ/m_e, the second and third terms in Eq. (14.3.35) are then, respectively,

$$\frac{e^2}{4m_e(2\pi)^3}\int d^3k \left[\frac{2}{k} - \frac{(3-k^2/(k^2+\mu^2))}{\sqrt{k^2+\mu^2}}\right]\left(\frac{\sqrt{\mathbf{k}^2+m_e^2}+m_e}{2\sqrt{\mathbf{k}^2+m_e^2}}\right) \simeq \frac{\alpha\mu^2}{4\pi m_e}$$

and

$$-\frac{e^2}{2(2\pi)^3}\int d^3k \left(\frac{1}{\mathbf{k}^2} - \frac{1}{\mathbf{k}^2+\mu^2}\right)\left(\frac{\sqrt{\mathbf{k}^2+m_e^2}-m_e}{2\sqrt{\mathbf{k}^2+m_e^2}}\right) \simeq -\frac{\alpha\mu^2}{\pi m_e}.$$

(Eq. (14.3.32) rules out the possibility that a relativistic correction to the latter expression might not be suppressed by the factor $\mathbf{k}^2/m_e^2$ that appears in this expression for $|\mathbf{k}|^2 \ll m_e^2$.) These two terms are cancelled by the term $+3\alpha\mu^2/4\pi m_e$ in $-\delta m_e(\mu)\int d^3x\bar{u}_N(\mathbf{x})u_N(\mathbf{x})$. Finally, we note that relativistic corrections to the above estimates for the sums over positron states would involve additional factors of $v^2/c^2 \approx (Z\alpha)^2$, yielding contributions of order $\alpha(Z\alpha)^2\mu^2/m_e \ll \alpha(Z\alpha)^4 m_e$, which justifies the non-relativistic approximations used here.

term in δm_e with the second term in the matrix element (14.3.40) is of order $(Z\alpha)^2\alpha\mu^2/m_e \ll \alpha(Z\alpha)^4 m_e$, and may therefore be neglected. To order $\alpha(Z\alpha)^4 m_e$, the only remaining effect of mass renormalization is to leave us with the product of the leading term in $\delta m_e(\mu)$ with the term of order $(Z\alpha)^2$ in $\int d^3x\, \bar{u}_N u_N$:

$$-\left[\frac{e^2\mu}{8\pi}\right]\left[-\frac{1}{2}(\mathbf{v}^2)_{NN}\right] = \frac{e^2\mu}{16\pi}(\mathbf{v}^2)_{NN} .$$

(This is the effect of mass renormalization on the electron kinetic energy, mentioned in Section 1.3.) It turns out that this is just the negative of what the first term in Eq. (14.3.35) would be if we neglected the difference between energy levels. To see this, note that the integral in this term is effectively cut off at $|\mathbf{k}| \sim \mu \ll Z\alpha m_e$, so we can evaluate the matrix element $\mathbf{\Gamma}_{MN}(\mathbf{k})$ in the limit $\mathbf{k} \to 0$. To lowest order in $Z\alpha$, Eq. (14.1.42) gives

$$\mathbf{\Gamma}_{MN}(0) = (\mathbf{v})_{MN} \tag{14.3.41}$$

and using the completeness of the solutions f_N of the non-relativistic Schrödinger equation, we then have

$$\sum_M \Gamma^{i*}_{MN}(\mathbf{k})\Gamma^{j}_{MN}(\mathbf{k}) \simeq (v^i v^j)_{NN} , \tag{14.3.42}$$

so to this order

$$-\frac{e^2}{2(2\pi)^3}\int d^3k \sum_M \left[\frac{\left(|\mathbf{\Gamma}_{MN}(\mathbf{k})|^2 - |\mathbf{k}\cdot\mathbf{\Gamma}_{MN}(\mathbf{k})|^2/\mathbf{k}^2\right)}{\mathbf{k}^2} - \frac{\left(|\mathbf{\Gamma}_{MN}(\mathbf{k})|^2 - |\mathbf{k}\cdot\mathbf{\Gamma}_{MN}(\mathbf{k})|^2/(\mathbf{k}^2+\mu^2)\right)}{\mathbf{k}^2+\mu^2}\right]$$

$$\simeq -\frac{e^2}{2(2\pi)^3}(\mathbf{v}^2)_{NN}\int d^3k\left[\frac{2}{3\mathbf{k}^2} - \frac{\left(1-\mathbf{k}^2/3(\mathbf{k}^2+\mu^2)\right)}{\mathbf{k}^2+\mu^2}\right]$$

$$= -\frac{e^2\mu}{16\pi}(\mathbf{v}^2)_{NN} .$$

Thus after mass renormalization we are left with just the first term in Eq. (14.3.35), less the same with energy differences $E_N - E_M$ dropped.

$$[\delta E_N]_{\text{low energy}} = \frac{e^2}{2(2\pi)^3} \int d^3k \sum_M (E_M - E_N)$$
$$\times \left[\frac{\left(|\boldsymbol{\Gamma}_{MN}(\mathbf{k})|^2 - |\mathbf{k} \cdot \boldsymbol{\Gamma}_{MN}(\mathbf{k})|^2/\mathbf{k}^2\right)}{\mathbf{k}^2(E_M - E_N + |\mathbf{k}| - i\epsilon)} \right.$$
$$\left. - \frac{\left(|\boldsymbol{\Gamma}_{MN}(\mathbf{k})|^2 - |\mathbf{k} \cdot \boldsymbol{\Gamma}_{MN}(\mathbf{k})|^2/(\mathbf{k}^2 + \mu^2)\right)}{(\mathbf{k}^2 + \mu^2)(E_M - E_N + \sqrt{\mathbf{k}^2 + \mu^2} - i\epsilon)} \right] . \tag{14.3.43}$$

Again using Eq. (14.3.41), this is

$$[\delta E_N]_{\text{low energy}} = \frac{e^2}{2(2\pi)^3} \sum_M (E_M - E_N)|\mathbf{v}_{MN}|^2$$
$$\times \int d^3k \left[\frac{2}{3\mathbf{k}^2(E_M - E_N + |\mathbf{k}| - i\epsilon)} \right.$$
$$\left. - \frac{1 - \mathbf{k}^2/3(\mathbf{k}^2 + \mu^2)}{(\mathbf{k}^2 + \mu^2)(E_M - E_N + \sqrt{\mathbf{k}^2 + \mu^2} - i\epsilon)} \right] . \tag{14.3.44}$$

Even though typical values of the electron momentum are much larger than typical atomic energy differences, this is not true of typical values of $|\mathbf{k}|$ in this integral, because the integral would be infrared divergent if we did not keep the $E_M - E_N$ terms in the denominators. The integral in Eq. (14.3.44) may be evaluated in the limit $\mu \gg |E_M - E_N| \sim (Z\alpha)^2 m_e$ by dividing the range of integration of $|\mathbf{k}|$ into two segments, from zero to λ and from λ to infinity, with λ chosen so that $|E_M - E_N| \ll \lambda \ll \mu$ but otherwise arbitrary. In this way we find

$$\int_0^\infty k^2 dk \left[\frac{2}{3k^2(E_M - E_N + k) - i\epsilon} \right.$$
$$\left. - \frac{1 - k^2/3(k^2 + \mu^2)}{(k^2 + \mu^2)(E_M - E_N + \sqrt{k^2 + \mu^2} - i\epsilon)} \right]$$
$$\simeq \frac{2}{3} \left[\ln\left(\frac{\mu}{2|E_M - E_N|} \right) + \frac{5}{6} + i\pi\theta(E_N - E_M) \right] .$$

The imaginary term here reflects the possibility of decay of the atom in state N to states M of lower energy. This term contributes to the decay rate, given by the imaginary part of the energy shift. We are interested here in the real part of the energy shift, and so will drop this imaginary term in what follows. Eq. (14.3.44) now gives

$$[\delta E_N]_{\text{low energy}} = \frac{e^2}{6\pi^2} \sum_M (E_M - E_N)|\mathbf{v}_{MN}|^2 \left[\ln\left(\frac{\mu}{2|E_N - E_M|} \right) + \frac{5}{6} \right] . \tag{14.3.45}$$

C Total Energy Shift

We need to make a connection between the sum in Eq. (14.3.45) and the matrix element in the high-energy term (14.3.22). For this purpose, let's first see what value the sum in Eq. (14.3.45) would have if we could ignore the logarithm. We note that $(E_M - E_N)\mathbf{v}_{NM} = [\mathbf{v}, H]_{NM}$, so

$$\sum_M (E_M - E_N)|\mathbf{v}_{MN}|^2 = \tfrac{1}{2}\sum_M \Big([v^i, H]_{NM}\, v^i_{MN} + v^i_{NM}\, [H, v^i]_{MN}\Big)$$
$$= -\frac{1}{2m_e^2}\Big([p^i,[p^i,H]]\Big)_{NN} .$$

The only term in the non-relativistic Hamiltonian H that does not commute with the momentum operator $\mathbf{p}$ is the potential term, $-e\mathscr{A}^0(\mathbf{x})$, so this gives

$$\sum_M (E_M - E_N)|\mathbf{v}_{MN}|^2 = -\frac{e}{2m_e^2}\Big(\nabla^2 \mathscr{A}^0(\mathbf{x})\Big)_{NN} . \qquad (14.3.46)$$

Inspection of Eqs. (14.3.45) and (14.3.22) now shows that the term proportional to $\ln\mu$ in the high-energy term is cancelled by the term proportional to $\ln\mu$ in the low-energy term:

$$\delta E_N = [\delta E_N]_{\text{high energy}} + [\delta E_N]_{\text{low energy}}$$
$$= \frac{e^2}{6\pi^2}\sum_M (E_M - E_N)|\mathbf{v}_{MN}|^2\left[\ln\left(\frac{m_e}{2|E_N - E_M|}\right) + \frac{5}{6} - \frac{1}{5}\right]$$
$$-\frac{e^2}{16\pi^2 m_e^2}\Big(\boldsymbol{\sigma}\cdot\nabla(e\mathscr{A}^0(\mathbf{x}))\times\mathbf{p}\Big)_{NN} . \qquad (14.3.47)$$

So far, this has been for a general electrostatic field $\mathscr{A}^0(\mathbf{x})$. Let us now specialize to a pure Coulomb field, with

$$\mathscr{A}^0(\mathbf{x}) = Ze/|\mathbf{x}| . \qquad (14.3.48)$$

In this case, Eq. (14.3.46) reads

$$\sum_M (E_M - E_N)|\mathbf{v}_{MN}|^2 = \frac{Ze^2}{2m_e^2}\Big(\delta^3(\mathbf{x})\Big)_{NN} = \frac{Ze^2}{2m_e^2}\Big(f_N^\dagger(0) f_N(0)\Big) . \qquad (14.3.49)$$

This is non-vanishing only for $\ell = 0$. Also, the matrix element in the last term in Eq. (14.3.47) has the value

$$\Big(\boldsymbol{\sigma}\cdot\nabla(e\mathscr{A}^0(\mathbf{x}))\times\mathbf{p}\Big)_{NN} = -Ze\Big(\frac{1}{r^3}\boldsymbol{\sigma}\cdot\mathbf{L}\Big)_{NN} , \qquad (14.3.50)$$

which is non-vanishing only for $\ell \neq 0$. It is therefore useful at this point to divide our consideration between the two cases, $\ell = 0$ and $\ell \neq 0$.

i. $\ell = 0$

It is convenient here to define a mean excitation energy ΔE_N:

$$\sum_M |\mathbf{v}_{MN}|^2 (E_M - E_N) \ln |E_N - E_M| \equiv \ln \Delta E_N \sum_M |\mathbf{v}_{MN}|^2 (E_M - E_N)$$
$$= \frac{Ze^2}{2m_e^2} \ln \Delta E_N \left(f_N^\dagger(0) f_N(0) \right) . \quad (14.3.51)$$

For s-wave hydrogenic states, the label N consists of a principal quantum number n and spin z-component m, and $[f_{nm}(0)]_\sigma = 2(Z\alpha m_e/n)^{3/2} \delta_{\sigma,m}/\sqrt{4\pi}$, so

$$\left(f_N^\dagger(0) f_N(0) \right) = \frac{1}{\pi} \left(\frac{Z\alpha m_e}{n} \right)^3 . \quad (14.3.52)$$

Using Eqs. (14.3.51) and (14.3.52) in Eq. (14.3.47) gives the energy shift in these states as

$$[\delta E]_{n,\ell=0} = \frac{4\alpha (Z\alpha)^4 m_e}{3\pi n^3} \left[\ln \left(\frac{m_e}{2\Delta E_{n,\ell=0}} \right) + \frac{19}{30} \right] . \quad (14.3.53)$$

ii. $\ell \neq 0$

For non-vanishing orbital angular momentum the sum (14.3.49) vanishes, so the definition (14.3.51) is inappropriate. Instead, it is conventional here to define a mean excitation energy ΔE_N by

$$\sum_M (E_M - E_N) |\mathbf{v}_{MN}|^2 \ln |E_N - E_M| \equiv \frac{2(Z\alpha)^4 m_e}{n^3} \ln \left(\frac{2\Delta E_N}{Z^2 \alpha^2 m_e} \right) . \quad (14.3.54)$$

(Because Eq. (14.3.49) vanishes, it makes no difference what units are used to measure $E_N - E_M$ in Eq. (14.3.54).) Also, in a state of total angular momentum j and orbital angular momentum ℓ, the scalar product $\boldsymbol{\sigma} \cdot \mathbf{L}$ has the familiar value $j(j+1) - \ell(\ell+1) - \frac{3}{4}$, and for principal quantum number n the operator $1/r^3$ has the expectation value

$$\int d^3 r |f_N|^2 / r^3 = \frac{2Z^3 \alpha^3 m_e^3}{n^3 \ell(\ell+1)(2\ell+1)} . \quad (14.3.55)$$

Putting this all together in Eq. (14.3.47), we have for $\ell \neq 0$:

$$[\delta E]_{jn\ell} = -\frac{4\alpha (Z\alpha)^4 m_e}{3\pi n^3} \ln \left(\frac{2\Delta E_{jn\ell}}{Z^2 \alpha^2 m_e} \right)$$
$$+ \frac{\alpha (Z\alpha)^4 m_e}{2\pi n^3} \left[\frac{j(j+1) - \ell(\ell+1) - \frac{3}{4}}{\ell(\ell+1)(2\ell+1)} \right] . \quad (14.3.56)$$

It only remains to use these results to give numerical value for the energy shifts. The mean excitation energies here must be calculated numerically; using non-relativistic hydrogen wave functions, they have the values[5]:

$$\Delta E_{1s} = 19.769266917(6)\ \text{Ry}\,,$$

$$\Delta E_{2s} = 16.63934203(1)\ \text{Ry}\,,$$

$$\Delta E_{2p} = 0.9704293186(3)\ \text{Ry}\,,$$

where 1 Ry $\equiv m_e\alpha^2/2 = 13.6057$ eV. Eq. (14.3.53) then gives

$$[\delta E]_{1s} = \frac{4\alpha^5 m_e}{3\pi}\left[\ln\left(\frac{m_e}{2\Delta E_{1s}}\right) + \frac{19}{30}\right] = 3.3612 \times 10^{-6}\ \text{eV}$$
$$= 2\pi\hbar \times 8127.4\ \text{MHz}\,, \tag{14.3.57}$$

$$[\delta E]_{2s} = \frac{\alpha^5 m_e}{6\pi}\left[\ln\left(\frac{m_e}{2\Delta E_{2s}}\right) + \frac{19}{30}\right] = 4.2982 \times 10^{-5}\ \text{eV}$$
$$= 2\pi\hbar \times 1039.31\ \text{MHz}\,, \tag{14.3.58}$$

while Eq. (14.3.56) gives

$$[\delta E]_{2p_{1/2}} = \frac{\alpha^5 m_e}{6\pi}\left[\ln\left(\frac{\alpha^2 m_e}{2\Delta E_{2p}}\right) - \frac{1}{8}\right] = -5.3267 \times 10^{-8}\ \text{eV}$$
$$= 2\pi\hbar \times -12.88\ \text{MHz}\,. \tag{14.3.59}$$

The classic Lamb shift is the energy difference between the $2s$ and $2p_{\frac{1}{2}}$ states of the hydrogen atom, states that would be degenerate in the absence of radiative corrections. Our calculation has given

$$[\delta E]_{2s} - [\delta E]_{2p_{1/2}} = 4.35152 \times 10^{-6}\ \text{eV} = 2\pi\hbar \times 1052.19\ \text{MHz}\,.$$

This is just the same as the old result of Kroll and Lamb[6] and French and Weisskopf,[7] which they obtained using the techniques of old-fashioned perturbation theory. Earlier in this section we made a crude estimate of 1300 MHz by considering only the high-energy contribution to the $2s$ energy shift, with an infrared cutoff guessed to be of order $\alpha^2 m_e = 2$ Ry. We can now see that this was an overestimate, arising mostly from the fact that the true value of the effective infrared cutoff $\Delta E_{2s} = 16.64$ Ry is considerably larger than we had guessed. On the other hand, as described in Section 1.3, in 1947 Hans Bethe[8] was able to make a rather good estimate of the Lamb shift, 1040 MHz, by considering only the *low-energy* contribution to the $2s$ energy shift, with an *ultraviolet* cutoff guessed to be m_e. (Bethe made the first estimate of the excitation energy, $\Delta E_{2s} \simeq 17.8$ Ry.)

The calculation of the Lamb shift described here has been improved by the inclusion of higher-order radiative corrections and nuclear size and recoil effects. At present the greatest uncertainty is due to a doubt about the correct value of the rms charge radius r_p of the proton. For $r_p = 0.862 \times 10^{-13}$ cm or $r_p = 0.805 \times 10^{-13}$ cm, one calculation[9] gives a Lamb shift of either 1057.87 MHz or 1057.85 MHz, while another[10] gives either 1057.883 MHz or 1057.865 MHz. Given the uncertainty in proton radius, the agreement is excellent with the present experimental value,[11] 1057.845(9) MHz. The accuracy of this experimental value is limited chiefly by the $\sim$ 100 MHz natural linewidth of the $2p$ state in hydrogen, so further improvements here will be very difficult.

In the last few years there has been an important improvement in measurements of the energy shift of the $1s$ ground state itself, by direct comparison of the frequency of the $1s$–$2s$ resonance with four times the frequencies of the $2s$–$4s$ and $2s$–$4d$ two-photon resonances. These s and d states are much narrower than the $2p$ state, so these frequency differences can be measured more accurately than the classic Lamb shift. For a brief time it seemed that there was a discrepancy here between theory and experiment. Calculations[12,13] showed that for a proton radius $r_p = 0.862(11) \times 10^{-13}$ cm or $0.805(11) \times 10^{-13}$ cm, the inclusion of proton size and other corrections increases the theoretical $1s$ energy shift from the above result of 8127.4 MHz to 8173.12(6) MHz or 8172.94(9) MHz, respectively. This result for a proton radius $r_p = 0.862(11) \times 10^{-13}$ cm, which is believed to be more reliable, was not quite in agreement with the measured value[13] of 8172.86(5) MHz. But a later calculation[14] that adopts this proton radius and includes terms of order $\alpha^2(Z\alpha)^5$ yields energy shifts for the $1s$, $2s$, and $4s$ states in agreement with experiment. So apparently quantum electrodynamics wins again.

Problems

1. Consider a charged scalar particle of mass $m \neq 0$, described by a field $\phi(x)$ whose only interaction is with an external time-independent electromagnetic field $\mathscr{A}^\mu(\mathbf{x})$. Let Φ_N be a complete set of normalized one-boson or one-antiboson states with energies E_N, and define $u_N(\mathbf{x})e^{-iE_Nt} \equiv (\Phi_0, \phi(\mathbf{x}, t)\Phi_N)$ and $v_N(\mathbf{x})e^{iE_Nt} \equiv (\Phi_N, \phi(\mathbf{x}, t)\Phi_0)$, where Φ_0 is the vacuum. Show that the u_N and v_N together form a complete set, and give formulas for the coefficients of u_N and v_N in the expansion of a general function $f(\mathbf{x})$.

2. Suppose we include radiative corrections in the theory of Problem 1. Let $i\Pi^*(x, y)$ be the sum of all diagrams with one incoming and one outgoing charged scalar line (excluding final scalar propagators) to order α. Derive a formula for the shift in the energies E_N of the

one-boson states due to these radiative corrections, in terms of $u_N(\mathbf{x})$ and $\Pi^*(x, y)$.

3. Use the results of Section 14.3 to calculate the radiative decay rate of the $2p$ states of hydrogen.

4. Suppose that the electron has an interaction with a light scalar field ϕ, of the form $g\phi\bar{\psi}_e\psi_e$. Suppose that the scalar mass m_ϕ is in the range $(Z\alpha)^2 m_e \ll m_\phi \ll Z\alpha m_e$. Calculate the change in the energy of the $1s$ state of hydrogenic atoms due to this interaction.

5. Carry out the calculation of Problem 4 for $m_\phi = 0$.

References

1. P. A. M. Dirac, *Proc. Roy. Soc. (London)* **A117**, 610 (1928).

2. See, e.g., A. R. Edmonds, *Angular Momentum in Quantum Mechanics*, (Princeton University Press, Princeton, 1957); M. E. Rose, *Elementary Theory of Angular Momentum* (John Wiley & Sons, New York, 1957).

3. See, e.g., L. I. Schiff, *Quantum Mechanics*, (McGraw-Hill, New York, 1949): Section 43. The original references are C. G. Darwin, *Proc. Roy. Soc. (London)* **A118**, 654 (1928); *ibid.*, **A120**, 621 (1928); W. Gordon, *Zeit. f. Phys.* **48**, 11 (1928).

4. G. E. Brown, J. S. Langer, and G. W. Schaefer, *Proc. Roy. Soc. (London)* **A 251**, 92 (1959); G. E. Brown and D. F. Mayers, *Proc. Roy. Soc. (London)* **A 251**, 105 (1959); A. M. Desiderio and W. R. Johnson, *Phys. Rev.* **A3**, 1267 (1971).

5. R. W. Huff, *Phys. Rev.* **186**, 1367 (1969).

6. N. M. Kroll and W. E. Lamb, *Phys. Rev.* **75**, 388 (1949)

7. J. B. French and V. F. Weisskopf, *Phys. Rev.* **75**, 1240 (1949)

8. H. A. Bethe, *Phys. Rev.* **72**, 339 (1947).

9. J. R. Sapirstein and D. R. Yennie, in *Quantum Electrodynamics*, ed. by T. Kinoshita (World Scientific, Singapore, 1990): p. 575, and references quoted therein.

10. H. Grotch, *Foundations of Physics* **24**, 249 (1994).

11. S. R. Lundeen and F. M. Pipkin, *Phys. Rev. Lett.* **46**, 232 (1981); S. R. Lundeen and F. M. Pipkin, *Metrologia* **22**, 9 (1986). For a review, see F. M. Pipkin, in *Quantum Electrodynamics*, ed. by T. Kinoshita (World Scientific, Singapore, 1990): p. 697.

12. M. Weitz, A. Huber, F. Schmidt-Kaler, D. Leibfried, and T. W. Hänsch, *Phys. Rev. Lett.* **72**, 328 (1994).

13. M. Weitz, F. Schmidt-Kaler, and T. W. Hänsch, *Phys. Rev. Lett.* **68**, 1120 (1992), and Ref. 12.

14. K. Pachucki, *Phys. Rev. Lett.* **72**, 3154 (1994).

Author Index

Where page numbers are given in italics, they refer to publications cited in bibliographies and lists of references.

Subject Index